Advances in Geroscience

Felipe Sierra • Ronald Kohanski
Editors

Advances in Geroscience

Springer

Editors
Felipe Sierra
Division of Aging Biology
National Institutes of Health/National Institute of Aging
Bethesda, MD
USA

Ronald Kohanski
Division of Aging Biology
National Institutes of Health/National Institute of Aging
Bethesda, MD
USA

ISBN 978-3-319-23245-4 ISBN 978-3-319-23246-1 (eBook)
DOI 10.1007/978-3-319-23246-1

Library of Congress Control Number: 2015955267

Springer Cham Heidelberg New York Dordrecht London

Printed on acid-free paper

Springer International Publishing AG Switzerland is part of Springer Science+Business Media (www.springer.com)

Foreword

In an introduction to a 2003 special issue of Science [1], my colleagues and I emphasized that "We have focused on physiological mechanisms underlying processes of aging, rather than on the large array of debilitating and costly disorders that so commonly emerge during the latter half of the life-spans of human beings." That was before the days of Geroscience, however! But what exactly *is* Geroscience? It is a term coined by my colleague and friend, Gordon Lithgow, a professor at the Buck Institute for Research on Aging who, true to his Scottish heritage, is thrifty with the use of all instruments of commerce, including terminology. Gordon used that nomenclature to summarize a new interdisciplinary research enterprise at the Buck Center; although it needs some grammatical editing, it is quite informative: "We consider that the relationship between aging and age-related disease (is) an important problem that can be tackled through an interdisciplinary approach" (http://www.geroscienceonline.org/index.php). This view emphasizes the concept that it is not only difficult to disassociate fundamental processes of aging from the numerous diseases of aging, but that joint investigations of these two domains of scholarship are essential if we are to unravel the pathogenesis of atherosclerosis, myocardial infarctions, strokes, non-ischemic heart failure, benign and malignant neoplasms, dementias of the Alzheimer type, frontal temporal dementias, Parkinson's disease and Lewy body dementias, peripheral neuropathies, cataracts and age-related macular degeneration, presbycusis, chronic obstructive pulmonary disease, type 2 diabetes mellitus, the metabolic syndrome, osteoporosis, osteoarthritis, sarcopenia, glomerulosclerosis, etc.

The editors of this volume, Felipe Sierra and Ronald Kohanski, have taken this concept to an exciting new level of implementation. As the Director of the Division of Aging Biology of the National Institute on Aging, Felipe approached his counterparts at more than twenty sister NIH institutes with narrow interests in specific diseases of aging with the following paraphrased sales pitch: "Let's have a discussion about how basic processes of aging are the major risk factor for your disease X. But first, I want to assure you that I do not want your money for my own institute! I want to start a 'Geroscience Interest Group' so that we can work together to accelerate progress towards the enhancement of healthy human aging." The response was overwhelmingly enthusiastic and resulted in a series of productive joint conferences, including a 2013 Summit meeting at the NIH [2]. Readers are urged to learn more about this trans-NIH initiative via Felipe's excellent video on the subject: http://www.nia.nih.gov/about/links/2013/07/video-dr-felipe-sierra-discusses-trans-nih-geroscience-interest-group-and-more.

This volume, despite its 19 chapters and its stellar list of authors, can best be viewed as just a beginning step for Geroscience. Given the extent of research funding that it richly deserves – far more than the current NIH pay lines – we can anticipate major basic and translational advances in our healthspans, at which point we can go to the Food and Drug Administration and point out that the side effect of our research – increased longevity – can really be a good thing for humanity. As the late Charlie Chaplin pointed out, "we are all amateurs; we don't live long enough to become anything else" [1].

Seattle, WA, USA George M. Martin, M.D.
June 5, 2015

References

1. Martin GM, LaMarco K, Strauss E, L Kelner K (2003) Research on aging: the end of the beginning. Science 28;299(5611):1339–1341
2. Burch JB, Augustine AD, Frieden LA, Hadley E, Howcroft TK, Johnson R, Khalsa PS, Kohanski RA, Li XL, Macchiarini F, Niederehe G, Oh YS, Pawlyk AC, Rodriguez H, Rowland JH, Shen GL, Sierra F, Wise BC (2014) Advances in geroscience: impact on healthspan and chronic disease. J Gerontol A Biol Sci Med Sci 69(Suppl 1):S1–S3

Preface

Geroscience is a new field that aims to bridge two communities: biologists focused on understanding the basic mechanisms that drive aging, and geriatricians attempting to improve the quality of life of elderly patients. Geroscience has been defined (Wikipedia) as "an interdisciplinary field that aims to understand the relationship between aging and age-related diseases." Because aging is the major risk factor for most chronic diseases, the "Geroscience Hypothesis" posits that common biological mechanisms of aging play important roles in the susceptibility of aged individuals to multiple chronic diseases.

The role of aging as a driver of chronic diseases is often downplayed under the assumption that aging is a non-modifiable risk factor. Yet we know that the rate of aging (however that is defined, as decay in function or susceptibility to disease) is modifiable by simple changes in the environment: a healthy diet, moderate exercise, and other elements of a generally healthy lifestyle will increase a person's chances of leading a longer and healthier life. In fact it is a dietary intervention, diet restriction, that provided the first handle to biologists' intent to understand the underpinnings of the aging process. That work, coupled with genetic experiments in short-lived simple organisms ranging from yeast to flies and worms, allowed scientists to unravel some of the major mechanisms involved. Furthermore, driven primarily by the Intervention Testing Program (created and supported by the National Institute on Aging), the field has moved recently to defining pharmacological interventions that expand lifespan in rodents (mainly mice). In addition to these impressive advances in terms of increasing lifespan, recently there has been a shift in focus, towards measuring healthspan as well as lifespan. Indeed, while some interventions (such as resveratrol) only increase health but not lifespan in mice, a handful of other interventions have been shown to increase both, although with the strong caveat that no manipulation has achieved that goal without having some secondary negative effects.

The enormous aging of the population worldwide, with the oldest-old being the age segment with the fastest growth, poses an urgent dilemma: if aging is indeed the largest risk factor for most chronic diseases, this increase in the proportion of elderly will necessarily pose an insurmountable challenge to the world's economic and health-care systems. Fortunately, at the same time that this demographic change is reaching a critical stage, our understanding of aging biology is allowing scientists to consider the possibility of intervening to delay aging, and hopefully, with it all major chronic diseases. This improved knowledge has allowed the organization of concepts into six to eight hallmarks or pillars, believed to be the main drivers of the process. While some details still need further clarification, there is broad agreement within the research community about these major drivers, and differences only reflect different biases and granularity.

These advances and the emerging opportunity to modify the process of aging by pharmacological means spawned the appearance of the new field of Geroscience. The initial concept came from a group at the Buck Institute for Research on Aging which, under the leadership of Dr. Gordon Lithgow, put forward a successful proposal to the Common Fund of the National Institute of Health. This project, Interdisciplinary Research Consortium on Geroscience, was funded for 5 years between 2007 and 2012. Subsequent to that effort, the editors of this book started a trans-NIH effort following the same line of thinking and resulting in the formation of the trans-NIH GeroScience Interest Group (GSIG). The effort attracted the attention of over 20 different institutes within the NIH, as well as wide support from both the scientific community and, importantly, non-federal advocacy and support groups. With the imprimatur of so many NIH institutes, the subsequent growth of the field was impressive and culminated in the organization of a large Summit, Advances in Geroscience: Impact on Healthspan and Chronic Disease, held in the NIH Campus on October/November 2013. In turn, that effort resulted in the publication of a White Paper in November 2014 in the Journal *Cell*.

For whom is the book written? Geroscience is an interdisciplinary field attempting to address the mechanisms by which aging biology is the main risk factor for chronic diseases. As such, this book examines those mechanisms and it provides an emerging overview of the new discipline of Geroscience. Each chapter aims at connecting the clinical manifestation of specific age-related chronic diseases with the major pillars of aging biology. These pillars are suspected (or in some cases, known) to play a role in the etiology of these diseases, not just singly but in multiple diseases because aging is the major risk factor for their appearance. As such, each chapter combines features of clinical science and basic biology, and it is hoped that this approach will be informative and enlightening to both these communities: physicians will hopefully learn about the basic underpinnings of aging that might affect the outcome of chronic diseases and/or treatments, while basic scientists might profit from learning about clinical aspects of their disease of interest in the context of aging. Altogether, the editors and authors anticipate that this book will raise further awareness of the molecular mechanisms which might become targets for further investigation and, ultimately, new targets to combat multiple comorbidities at once.

This book examines the biological mechanisms and clinical consequences of aging by providing specific coverage on a wide range of chronic diseases, from arthritis and cancer to dementia and stroke among others. This book begins with an introduction of the general principles, including both a description of current thinking in aging biology and a description of the Geroscience Hypothesis, with some background on the critical role that epidemiology has played in defining our basic understanding of age-related chronic diseases. Each of the following 16 chapters focuses on one particular disease, or group of related diseases, with an emphasis on how aging is a risk factor. This book finishes with a global discussion of pain in the elderly and a commentary on the important topic of translation into the clinic and the necessity of cross-fertilization between clinicians and basic scientists. Unfortunately, not all diseases afflicting the elderly have been covered, and this omission is simply the result of space limitations. It is hoped that this book will entice experts in those areas to think more deeply about the basis on which aging as a risk factor for their specialty disease, and perhaps publish their thoughts on this subject.

Contributors to this volume represent a large range of disciplines. An effort was made, whenever possible and appropriate, to engage in each chapter both basic scientists and clinicians; the editors are enormously thankful to all authors for the effort and sense of shared responsibility. Importantly, the editors thank all members of the GSIG executive committee for their support during the preparation of this book and the help provided on the selection of authors. Similarly, GSIG members are to be commended for the additional editorial work to complete this volume.

The editors believe that this book presents vital information and ideas that can help readers better understand how aging is a critical – but malleable – risk factor in chronic diseases of the elderly. The potential to alter the rate of aging is at the heart of Geroscience, and achieving that goal should improve lifespan and healthspan in the human population.

Bethesda, MD, USA

Felipe Sierra
Ronald A. Kohanski

Contents

Contributors

Jayakrishna Ambati Department of Ophthalmology and Visual Sciences, University of Kentucky, Lexington, KY, USA

Julie K. Andersen, Ph.D. Buck Institute for Research on Aging, Novato, CA, USA

Steven N. Austad, Ph.D. Department of Biology, University of Alabama at Birmingham, Birmingham, UK

Andrzej Bartke, Ph.D. Department of Medicine, Southern Illinois University School of Medicine, Springfield, IL, USA

Paula Busse, M.D. Department of Medicine, Icahn School of Medicine, Mount Sinai, NY, USA

Ying Ann Chiao, Ph.D. Department of Pathology, University of Washington, Seattle, WA, USA

Shankar Chinta, Ph.D. Buck Institute for Research on Aging, Novato, CA, USA

Dao-Fu Dai, M.D., Ph.D. Department of Pathology, University of Washington, Seattle, WA, USA

Luigi Ferrucci Intramural Research Program, National Institute on Aging, Bethesda, MD, USA

Roger B. Fillingim, Ph.D. Pain Research and Intervention Center of Excellence, University of Florida, Gainesville, FL, USA

Linda P. Fried Mailman School of Public Health, Columbia University Medical Center, New York, NY, USA

Julie Glowacki, Ph.D. Department of Orthopedic Surgery, Brigham and Women's Hospital, Boston, MA, USA

Nicola A. Hanania, M.D., M.S. Department of Medicine, Baylor College of Medicine, Houston, TX, USA

Shenghui He, Ph.D. Department of Genetics, The Lineberger Comprehensive Cancer Center, University of North Carolina School Medicine, CB #7295, Chapel Hill, NC, USA

Anna Hearps, Ph.D. Centre for Biomedical Research, Burnet Institute, Melbourne, VIC, Australia

John P. Higgins, M.D. Department Pathology, Stanford University Medical Center, Stanford, CA, USA

Kevin High, M.D. Department of Internal Medicine, Section on Infectious Diseases, Wake Forest School of Medicine, Winston-Salem, NC, USA

Geoffrey A. Kerchner Department of Neurology and Neurological Sciences, Stanford University School of Medicine, Stanford, CA, USA

Stuart K. Kim, Ph.D. Department Developmental Biology and Genetics, Stanford University Medical Center, Stanford, CA, USA

James L. Kirkland, M.D., Ph.D. Aging Research, Robert and Arlene Kogod Center on Aging, Mayo Clinic, Rochester, MN, USA

Edward Lakatta, M.D. Laboratory of Cardiovascular Science, Biomedical Research Center, National Institute of Aging, Baltimore, USA

Alan Landay, Ph.D. Department of Immunology/Microbiology, Rush University Medical Center, Chicago, IL, USA

Changhan Lee, Ph.D. Davis School of Gerontology, University of Southern California, Los Angeles, CA, USA

Richard F. Loeser, M.D. Division of Rheumatology, Allergy, and Immunology, Thurston Arthritis Research Center, University of North Carolina School of Medicine, Chapel Hill, NC, USA

Valter Longo, Ph.D. Davis School of Gerontology, University of Southern California, Los Angeles, CA, USA

IFOM, FIRC Institute of Molecular Oncology, Milan, Italy

Martin Lotz, M.D. Department of Molecular and Experimental Medicine, The Scripps Research Institute, La Jolla, CA, USA

Simon Melov, Ph.D. Buck Institute for Research on Aging, Novato, CA, USA

Nicolas Musi, M.D. Department of Medicine, Sam and Ann Barshop Institute for Longevity and Aging Studies, San Antonio Geriatric Research, Education and Clinical Center, University of Texas Health Science Center, San Antonio, TX, USA

Peter Rabinovitch, M.D., Ph.D. Department of Pathology, University of Washington, Seattle, WA, USA

Clifford J. Rosen, M.D. Department of Medicine, Musculoskeletal Center, Center for Clinical & Translational Research, Maine Medical Center Research Institute, Scarborough, ME, USA

Cecilia G. Sanchez, Ph.D. Section of Pulmonary Diseases, Critical Care and Environmental Medicine, Tulane University Medical School, New Orleans, LA, USA

Katherine Schafer, M.D. Department of Internal Medicine, Section on Infectious Diseases, Wake Forest School of Medicine, Winston-Salem, NC, USA

Norman E. Sharpless, M.D. Department of Genetics, The Lineberger Comprehensive Cancer Center, University of North Carolina School Medicine, Chapel Hill, NC, USA

Farida Sohrabji, Ph.D. Texas A&M Health Science Center, Bryan, TX, USA

Tamar Tchkonia, Ph.D. Department of Medicine, Robert and Arlene Kogod Center on Aging, Mayo Clinic, Rochester, MN, USA

Dennis C. Turk, Ph.D. Department of Anesthesiology and Pain Medicine, University of Washington, Seattle, WA, USA

Zoltan Ungvari, M.D., Ph.D. Department of Geriatric Medicine, Reynolds Oklahoma Center on Aging, University of Oklahoma Health Science Center, Oklahoma City, OK, USA

Tamara Vokes, M.D. Department of Medicine, Section of Endocrinology, University of Chicago, Chicago, IL, USA

Charles Wright Department of Ophthalmology and Visual Sciences, University of Kentucky, Lexington, KY, USA

Tony Wyss-Coray Department of Neurology and Neurological Sciences, Stanford University School of Medicine, Stanford, CA, USA

Center for Tissue Regeneration, Repair and Restoration, Veterans Affairs Palo Alto Health Care System, Palo Alto, CA, USA

Robert P. Yezierski, Ph.D. Pain Research and Intervention Center of Excellence, University of Florida, Gainesville, FL, USA

[illegible]

Abbreviations

AAV	Adeno-associated virus
Aβ	Amyloid beta
ACEi	Angiotensin converting enzyme inhibitor
ACL	Anterior cruciate ligament
ACOS	Asthma-COPD overlap syndrome
ACT	Asthma control test
AD	Alzheimer's disease
ADCC	Antibody-dependent cellular cytotoxicity
ADE	Adverse drug events
AF	Atrial fibrillation
AGE	Advanced glycation end product
AHR	Airway hyperresponsiveness
AIDS	Acquired immunodeficiency syndrome
AKT	Protein kinase B
AMD	Age-related macular degeneration
AMPK	Adenine monophosphate-activated protein kinase
ANI	Asymptomatic neurocognitive impairment
APP	Amyloid precursor protein
ARB	Angiotensin 2 receptor blocker
ART	Anti-retroviral therapy
ATP	Adenosine-5'-triphosphate
BALF	Bronchoalveolar lavage fluid
BBB	Blood brain barrier
BDNF	Brain-derived neurotrophic factor
BER	Base excision repair
BMD	Bone mineral density
BMI	Body mass index
BMP	Bone morphogenetic protein
BMSC/B-MSC	Bone marrow-derived stem cell
BMU	Basic multicellular unit
Breg	Regulatory B cells

BV/TV	Bone volume fraction
CAC	Coronary artery calcium
CBT	Cognitive behavior therapy
CCI	Chronic constriction injury
CD	Cluster of differentiation
CDC	Center for Disease Control and Prevention
CDK	Cyclin-dependent kinase
CEE	Conjugated equine estrogens
CFA	Complete Freund's adjuvant
CFB	Complement factor B
CGRP	Calcitonin gene-related peptide
CHF	Congestive heart failure
CLBP	Chronic low back pain
CM	Conditioned medium
CMV	Cytomegalovirus
CNS	Central nervous system
CNV	Choroidal neovascularization
COLD	Chronic obstructive lung disease
COMP	Cartilage oligomeric protein
COMT	Catechol-O-methyl transferase
COPD	Chronic obstructive pulmonary disease
COX2	Cyclooxygenase-2
CPM	Conditioned pain modulation
CR	Caloric restriction
CRP	C-reactive protein
CSE	Cigarette smoke extract
CT	Computed tomography
CTGF	Connective tissue growth factor
CVD	Cardiovascular disease
DA	Dopamine
DAergic	Dopaminergic
DBP	Diastolic blood pressure
DBS	Deep brain stimulation
DC	Dendritic cell
DDI	Drug-drug interactions
DG	Dentate gyrus
DHEA	Dehydroepiandrosterone
DLB	Dementia with Lewy bodies
DLCO	Carbon monoxide diffusing capacity
DMM	Destabilized medial meniscus
DPP	Diabetes prevention program
DR	Dietary restriction
DRG	Dorsal root ganglia
DSB	Double strand break
DSR	Differential stress resistance

dsRNA	Double-stranded RNA
DSS	Differential stress sensitization
DXA	Dual energy X-ray absorptiometry
E2	Estradiol
Ea/Aa	Ratio of early to late diastolic mitral annular velocity
EBV	Epstein-Barr virus
ECM	Extracellular matrix
EF	Ejection fraction
EF	EnhanceFitness
EMT	Epithelial-mesenchymal transition
EPC	Endothelial progenitor cell
eQTL	Expression quantitative trait loci
ER	Endoplasmic reticulum
ERK	Extracellular-regulated kinase
ETC	Electron transport chain
EWAS	Epigenome-wide association study
FDA	Food and Drug Administration
FFA	Free fatty acid
FGF	Fibroblast growth factor
FGFR2	Fibroblast growth factor receptor 2
FIRKO	Fat tissue-specific insulin receptor knockout
FOXO	Forkhead box protein O
FRP	Frailty-related phenotype
FS	Fractional shortening
FTIRM	Fourier transformed infrared microepectroscopy
GA	Geographic atrophy
GAD	Glutamic acid decarboxylase
GALT	Gut-associated lymphoid tissue
GDF-11	Growth and differentiation factor 11
GDNF	Glial-derived neurotrophic factor
GERD	Gastroesophageal reflux disease
GFAP	Glial fibrillary acidic protein
GFP	Green fluorescent protein
GH	Growth hormone
GHR	Growth hormone receptor
GTT	Glucose tolerance test
GWAS	Genome-wide association study
GzmB	Granzyme B
HAD	HIV-associated dementia
HANA	HIV-associated non-AIDS
HAND	HIV-associated neurocognitive disorder
HAT	Histone acetyl transferase
HCV	Hepatitis C virus
HDAC	Histone deacetylase
HERV	Human endogenous retroviruses

HFD	High-fat diet
HIV	Human immunodeficiency virus
HLA	Human leukocyte antigen
HPA	Hypothalamo-pituitary axis
HPC	Hematopoietic progenitor cell
HPLC	High pressure liquid chromatography
HPV	Human papillomavirus
HR	Heart rate
HSC	Hematopoietic stem cells
hsCRP	High-sensitivity C-reactive protein
HSP	Heat shock protein
HSV	Herpes simplex virus
HT	Hormone therapy
ICH	Intracerebral hemorrhage
ICS	Inhaled corticosteroids
I-FABP	Intestinal fatty acid binding protein
IFN	Interferon
IGF	Insulin-like growth factor
IGFBP	IGF binding protein
IGT	Insulin glucose tolerance test
IL	Interleukin
IND	Investigational new drug
iNOS	Inducible nitric oxide synthase
IOM	Institute of Medicine
IPF	Idiopathic pulmonary fibrosis
iPS/iPSC	Induced pluripotent stem cell
IR	Ionizing radiation
IRAP	IL-1 receptor antagonist protein
IRS1	Insulin receptor substrate 1
ISH	Isolated systolic hypertension
ITM	Intima-media thickness
ITP	Interventions Testing Program
KMT	Lysine methyl transferase
LBP	Low back pain
LBP	LPS binding protein
LDL	Low density lipoprotein
L-DOPA	Levo-DOPA
lncRNA	Long non-coding RNA
LOA	Late onset asthma
LPS	Lipopolysaccharide
LSA	Long-standing asthma
LSD	Lysine-specific demethylase
LV	Left ventricle
LVH	Left ventricular hypertrophy
MAC	Membrane attack complex

MAO-B	Monoamine oxidase B
MAP	Mean arterial pressure
MCAo	Middle cerebral artery occlusion
MCAT	Mitochondrial-directed catalase
MD2	Myeloid differentiation factor 2
MDP	Mitochondrial-derived peptides
miRNA/miR	MicroRNA
MMP	Matrix metalloproteinase
MMSE	Mini-mental status exam
MND	Mild neurocognitive disorder
mPTP	Mitochondrial permeability transition pore
MRI	Magnetic resonance imaging
MRS	Magnetic resonance spectroscopy
MSC	Marrow stromal cell
MSC	Mesenchymal stem cell
mtDNA	Mitochondrial DNA
mTOR	Mechanistic target of rapamycin
mTORC1	mTOR complex 1
mtROS	Mitochondrial ROS
NAD+	Nicotinamide adenine dinucleotide
NADC	Non-AIDS-defining cancers
ncRNA	Non-coding RNA
NER	Nucleotide excision repair
NET	Neutrophil extracellular trap
NHEJ	Non-homologous end joining
NIA	National Institute on Aging
NIH	National Institutes of Health
NK cell	Natural killer cell
NLRP3	Nucleotide-binding domain and leucine-rich repeat pyrin domain containing 3
NMDA	N-methyl-D-aspartate
NO	Nitric oxide
NOX	NAP(P)H oxidase
NPC	Neural precursor/progenitor cell
Nrf2	Nuclear factor erythroid-derived 2
NRTI	Nucleos(t)ide reverse transcriptase inhibitor
NSAID	Nonsteroidal anti-inflammatory drug
nvAMD	Neovascular AMD
OA	Osteoarthritis
OPG	Osteoprotegerin
OR	Odds ratio
ORF	Open reading frame
PAMP	Pathogen-associated molecular pattern
PAWH	People aging with HIV
PBMC	Peripheral blood mononuclear cell

PD	Parkinson's disease
PD1	Programmed cell death protein 1
PDGF	Platelet-derived growth factor
PEDF	Pigmented epithelium-derived factor
PET	Positron emission tomography
PGC-1α	PPAR-gamma coactivator 1-alpha
PGE2	Prostaglandin E2
PHB1	Prohibitin 1
PI	Protease inhibitor
PI3K	Phosphatidyl inositol 3 kinase
PKA	Protein kinase A
PKC	Protein kinase C
PMN	Polymorphonuclear cell
Polg	DNA polymerase gamma
PPARγ	Peroxisome proliferator-activated receptor gamma
pQCT	Peripheral quantitative computed tomography
PRR	Pattern recognition receptor
PSNL	Partial sciatic nerve ligation
PSVT	Paroxysmal supraventricular tachycardia
PTEN	Phosphatase and tensin homolog
PTH	Parathyroid hormone
RAAS	Renin-angiotensin-aldosterone system
RAGE	Receptor for advanced glycation end product
RANKL	Receptor activator of nuclear factor kappa B ligand
RAP	Retinal angiomatous proliferation
RAS	Renin-angiotensin system
RB	Retinoblastoma
RCT	Randomized control trial
RI	Recombinant inbred
RNS	Reactive nitrogen species
ROS	Reactive oxygen species
RPE	Retinal pigmented epithelium
RSV	Respiratory syncytial virus
RT	Reverse transcriptase
RTL	Recombinant T cell receptor ligand
SABA	Short-acting b adrenergic 2 receptor agonist
SA-βGal	Senescence-associated beta galactosidase
SAHF	Senescence-associated heterochromatic foci
SAM	Senescence-accelerated mouse
SASP	Senescence-associated secretory phenotype
SBP	Systolic blood pressure
sCD	Soluble cluster of differentiation
SD-OCT	Spectral domain optical coherence tomography
SERM	Selected estrogen receptor modulator
SGZ	Subgranular zone

SHR	Spontaneously hypertensive rat
SIRT1	Sirtuin1
SIV	Simian immunodeficiency virus
SNAE	Serious non-AIDS events
SNP	Single nucleotide polymorphism
SNpc	Substantia nigra pars compacta
SNS	Sympathetic nervous system
SOD	Superoxide dismutase
SPF	Specific pathogen free
SPT	Skin prick testing
SRM	Selected reaction monitoring
STAT	Signal transducer and activator of transcription
SVZ	Subventricular zone
T	Testosterone
T2DM	Type 2 diabetes mellitus
TB	Tuberculosis
TC	Tai Chi
TCR	T cell receptor
TERT	Telomerase reverse transcriptase
TGFβ	Transforming growth factor beta
TIA	Transient ischemic attack
TIMP	Tissue inhibitor of matrix metalloproteinase
TLR	Toll-like receptor
TNFα	Tumor necrosis factor alpha
TNFR1	Tumor necrosis factor receptor 1
tPA	Tissue plasminogen activator
Treg	Regulatory T cells
TSA	Trichostatin A
UPR	Unfolded protein response
UPS	Ubiquitin-proteasome system
UPS	Unfolded protein stress
VE	Ventricular ectopy
VEGF	Vascular endothelial growth factor
VEGFR	VEGF receptor
VFA	Vertebral fracture assessment
WHI	Women's Health Initiative
WRN	Werner's

The Geroscience Hypothesis: Is It Possible to Change the Rate of Aging?

Steven N. Austad

Contents

1 Introduction

In the 200,000 year history of anatomically modern humans, we have never lived remotely as long as we do today. The rate of change in our life expectancy has been breathtaking. For the past 175 years, the mean age-at-death has increased steadily by about 2.5 years per decade, or 6 h per day, among the longest-lived countries [1]. While in the early part of the twentieth century, the rise in life expectancy was

S.N. Austad, Ph.D. (✉)
Department of Biology, University of Alabama at Birmingham, Birmingham, Alabama, USA
e-mail: austad@uab.edu

F. Sierra, R. Kohanski (eds.), *Advances in Geroscience*,
DOI 10.1007/978-3-319-23246-1_1

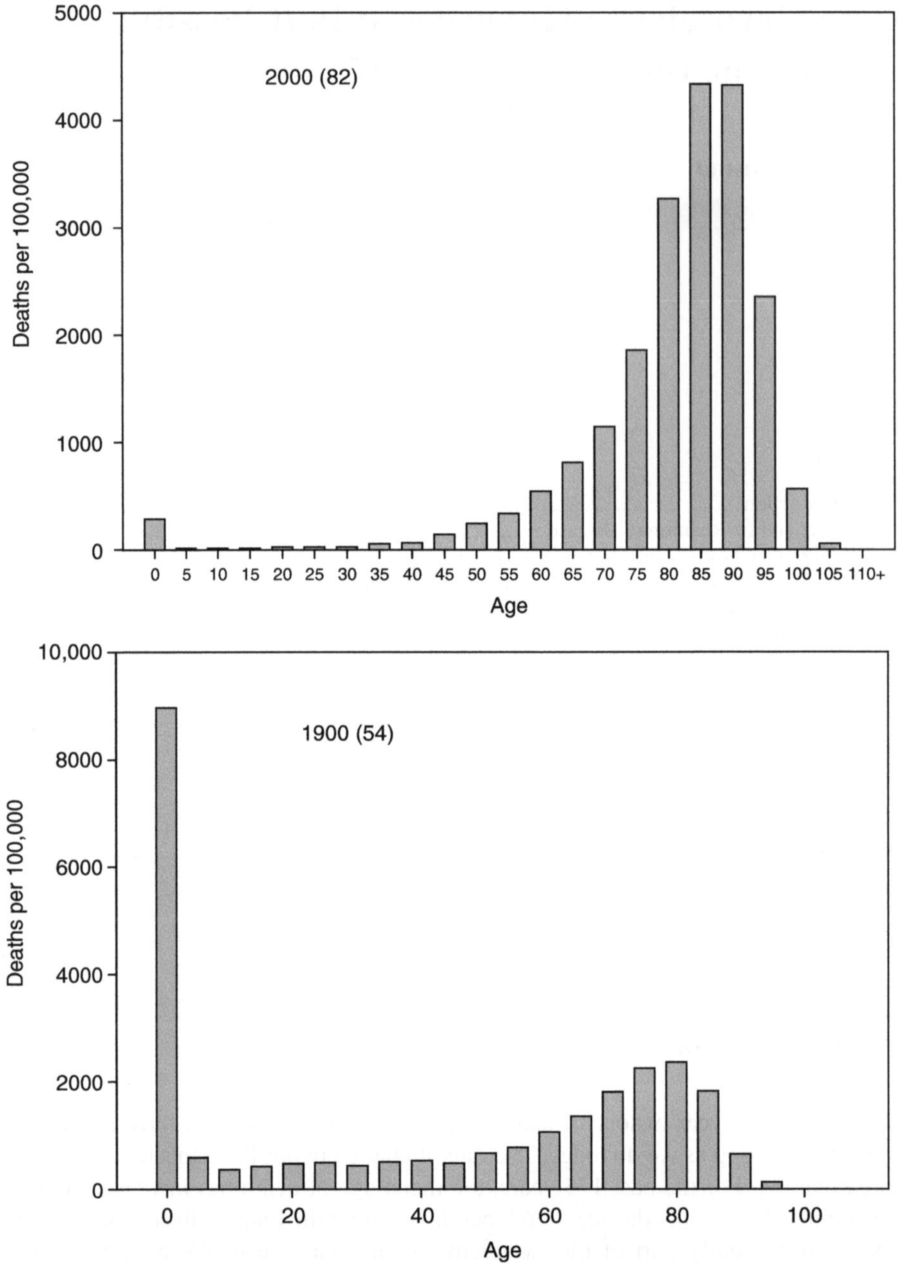

Fig. 1 Distribution of deaths as a function of age among Swedish women from the years 1900 and from 2000. Birth and death data from Swede are among the most reliable in the world. Note that more than 90 % of the population reached age 65 in the year 2000 compared with only 50 % reaching that age in 1900. Numbers in parentheses are life expectancies at birth (Data from the Human Mortality Database (downloaded November 2014))

driven mainly by reduced infant and young adult mortality, more recently bigger advances have been made in combating later life diseases (Fig. 1). As a consequence chronic health problems associated with aging, such as sarcopenia, osteoporosis, and Alzheimer's disease, which were once rare, have become common. As the global population continues to age over the coming decades, maladies of aging threaten to overwhelm our healthcare infrastructure, disrupt our national economies, and potentially poison relations among generations. Fortunately, understanding of the basic biology of aging has also progressed rapidly in the past several decades such that the promise of medical interventions that enhance and lengthen healthy life is no longer an empty promise promulgated only by avaricious quacks and charlatans. The economic impact of generalized health extension could be stunning. According to one analysis, slowing the rate of human aging by 20 % would be worth more than $7 trillion over the next 50 years in the United States alone [2].

The likelihood that we will ultimately be able to slow human aging depends on our understanding of underlying processes. I have claimed that such understanding has progressed rapidly in recent decades. What is the evidence for such a claim? How realistic is the promise of medically extended healthy life? Those are the topics of this chapter.

2 Aging and Its Relation to Disease

No 60 year old – even the healthiest, hardest-training, and most disease-free 60 year old – can sprint as fast or throw as far as she could as a healthy 25 year old. This is *prima facie* evidence that aging, the progressive decline in physical function that accompanies growing older, occurs even in the absence of disease. However aging is so intimately intertwined with numerous diseases and disabling conditions that almost any discussion that begins with aging ends on disease. Although aging occurs

Table 1 Death rates from selected diseases in the United States (2010) * indicates the disease is essentially nonexistent at this age.

	Age group					
	35–44	45–54	55–64	65–74	75–84	85+
Malignant neoplasms	29	112	300	666	1202	1730
Diabetes mellitus	4	13	32	68	144	286
Diseases of the heart	26	82	187	409	117	4285
Alzheimer's disease	*	0.3	2	20	185	987
Parkinson's disease	*	0.2	1	12	75	166
Influenza & Pneumonia	2	4	10	28	102	426
COPD	2	10	39	146	370	691
Stroke	5	13	30	82	288	994

Data from Murphy et al. [3]

Rates are per 100,000 population in the specific age group based on the 2010 U.S. Census even in the absence of disease, it clearly increases *vulnerability* to multiple diseases. In other words, aging is a risk factor for diseases. In fact, it is by far the biggest risk factor for virtually all of the chronic diseases that strain our health care system today and it increases the chances that a given disease will be fatal (Table 1). For instance in 2010, an American 75–84 years old had a 42-fold greater chance of dying of cancer, and a 45-fold greater chance of dying of heart disease, than a 35–44 year old. The chances of dying from Alzheimer's disease increase more than 600-fold between age 50 and 80 [3]. For comparison, smoking only increases overall mortality rate by threefold compared with nonsmokers (http://www.cdc.gov/tobacco/data_statistics/fact_sheets/health_effects/tobacco_related_mortality/) and having two copies of the ApoE4 allele, the most common genetic risk factor for Alzheimer's disease, increases an individual's chances of contracting that disease by only 12-fold relative to those with two copies of the ApoE3 allele [4]. Looked at from this perspective, aging is by far the biggest threat to human health in the developed world today. Geroscience, the topic of this book, is an interdisciplinary field seeking to understand the basis for the relationship between aging and disease vulnerability.

The Geroscience Hypothesis posits that, because aging underlies so many diseases and disabling conditions, interventions that would retard aging would also simultaneously prevent or delay the onset of multiple chronic diseases. In recent years there has been success at delaying cardiovascular diseases. Over the first decade of the twenty-first century, the age-adjusted death rate from heart diseases fell by more than 30 % and for stroke fell by more than 35 % [5, 3]. One contributing factor is the discovery of treatments that address underlying risk factors such as high blood pressure and high cholesterol. There has also been progress against a major behavioral risk factor – smoking. Importantly though, aging is a bigger risk to health than high blood pressure, cholesterol, and smoking combined. If we could similarly learn to treat the risk factor of aging, the health benefits would be enormous, not only for delaying fatal diseases but in delaying many nonfatal conditions such as hearing and vision loss, osteoporosis, and arthritis that degrade the quality of later life.

3 Experimental Organisms in Aging Research

3.1 Uses and Caveats in the Use of Model Organisms

For most of its history, basic aging research relied on standard laboratory animals such as fruit flies, mice, and rats. The chief advantage of these animals was that their laboratory husbandry was established and that they were short-lived. That is, rats and mice are short-lived among mammals, fruit flies are relatively short-lived among insects. Initially, basic aging research focused on describing physiological changes occurring during aging in the hope that the nature of these changes would reveal underlying aging mechanisms. Short-lived animals were useful because individuals could be monitored throughout their lives and the longevity of different groups could be compared and contrasted. Until recently, lengthening of life was assumed to be sufficient evidence that aging had been slowed. This view has recently been questioned as will be discussed later, but it has dominated the history of experimental aging research.

After dietary restriction (DR), simply reducing the amount of available food, was discovered to lengthen life in many laboratory rats and mice, attention shifted to searching for mechanism(s) by which DR had these effects and also searching for other methods of life extension. Again, the rate limiting step for such studies was the length of the animals' lives. But even the shortest-lived species commonly used in this research lived months (fruit flies) or years (mice and rats), and because the focus was on increasing lifespan, aging studies were particularly time-consuming compared with other areas of biomedical research.

It is important to understand why the focus so quickly fell on lengthening life rather than shortening it. In principle, understanding basic aging processes could be studied much more quickly by accelerating them rather than retarding them. The practical difficulty with this logical approach is that there are many ways to shorten animals' lives by inducing pathological processes that may have nothing to do with normal aging processes. The problem is how would we know the difference between those aberrant pathologies and normal aging processes? This doesn't mean that so-called accelerated aging models, which do exist, are not informative. It does mean that such models are difficult to evaluate with respect to normal aging and findings from them need to be interpreted with considerable care. For instance, the so-called Senescence Accelerated Mouse (SAM mouse) is a series of exceptionally short-lived mouse strains, created by accidental outbreeding of an AKR/J inbred strain with an unknown other strain. Despite their short lives – most live less than 1 year – they have had virtually no impact on the larger mouse aging research field, because like all so-called accelerated aging models, they replicate at best a few of the features of normal aging and the fidelity of that replication is not clear.

Experiments that lengthen life are much less problematic to interpret. Animals are unlikely to live longer if we haven't retarded at least *some* normal aging process, such as the increasing susceptibility to cancer. We may not have retarded them all (however many that may be), but we must have retarded some. To verify that one had identified a mechanism regulating aging, generally, the mantra for many years was that both mean (or median) and maximum longevity must be extended. Maximum longevity is generally defined as the mean longevity of the oldest $x\%$ of the starting population, where x often equals 10 %. The focus on maximum longevity implies that ameliorating a specific disease process may impact mean longevity, but only by affecting aging itself would both the mean and the length of life of the longest-lived animals be longer. For example, if group A displays longer mean or median survival, but no difference in maximum survival than group B, then group A must have experienced higher mortality rate than group B in the latter part of life. Higher mortality late in life is not a trait that one would associate with slower aging. For this reason exercise, which consistently increases mean longevity in both rats and people [6, 7] and has manifold beneficial health-preserving effects, is not generally considered to retard aging by researchers in the basic aging research community. By comparison, DR increases both mean and maximum longevity in many (but not all) species and genotypes, therefore has generally been considered the gold standard of aging retardation. As will be noted later, the over-reliance on longevity as the cannonical metric of aging is now being re-thought by many researchers.

3.2 *Worms* (Caenorhabditis elegans)

A major breakthrough in the field, which had previously relied on fruit flies and laboratory rodents, was the widespread adoption of the model nematode, *Caenorhabditis elegans*, for aging research in the 1990s [8–10]. A 1 mm long, free-living, soil nematode introduced to the biological research community in the 1960s by Sydney Brenner, *C. elegans* proved to be a wonderful model for the study of aging as well (Table 2). Not only did *C. elegans* develop rapidly and live only a couple of weeks rather than a couple of months like fruit flies, they could be maintained in much larger numbers in much smaller space than flies. They were also naturally inbred, which mitigated the problem of inbreeding depression and unpredictable genetic background effects, and they were more genetically tractable than flies, particularly after the discovery that expression of individual genes could be suppressed with ease by genetically altering their *E. coli* food to produce RNAi against the gene of interest [11]. A final advantage of worms is that because they are transparent, reporters such as green fluorescent protein (GFP) can be employed on living worms to assess gene expression and protein location [12].

A key feature of worm biology that turns out to be highly relevant to its aging biology is that under conditions of overcrowding, food shortage, or high temperature – conditions not conducive to successful reproduction – developing worms enters an alternative 3rd larval stage called dauer. Dauer is a nonfeeding, metabolically and transcriptionally quiescent, highly stress-resistant and long-lived stage of arrested development from which worms emerge only when crowding eases, food

Table 2 Relevant biological traits of traditional animal species used in basic aging research

	Caenorhabditis elegans	*Drosophila melanogaster*	*Mus musculus*
Body size	1 mm (length)	3 mm (length) 1–1.5 mg	30–45 g
Genome size (bp)	1.0×10^8	1.2×10^8	2.8×10^9
Stem cells	No	Yes, limited	Yes
Neuron number	302	135,000	75,000,000
Wild-type longevity[a]	2–3 weeks	2–3 months	2.0–2.5 years
Max. life extension	10×	2×	1.75×

[a]Wild-type longevity equals the longevity of laboratory-adapted animals

becomes abundant once again, or the temperature falls. Adult worm longevity, upon emerging from dauer and completing development, does not appear to be related to the length of time it spent in dauer [13]. In nature, worms are often found in dauer, which appears to be a specialized dispersal phase [14]. Thus dauer appears to be an important part of the worm's natural life cycle and the genetics of dauer entry and exit have been extensively investigated [15]. Longevity, as typically reported in the worm literature, is adult longevity. The 3 day larval period, whether or not worms went through dauer, is ignored.

The reason that dauer is a key life history feature for aging research is that many of the hundreds of known worm longevity genes are part of the dauer regulatory network. As dauer larvae are very long-lived, partial induction of the dauer regulatory network is likely to lengthen adult life. Possibly for this reason, an order of magnitude more longevity-enhancing genes have been found in worms than in any other species and the magnitude of genetically-induced life extension achieved in

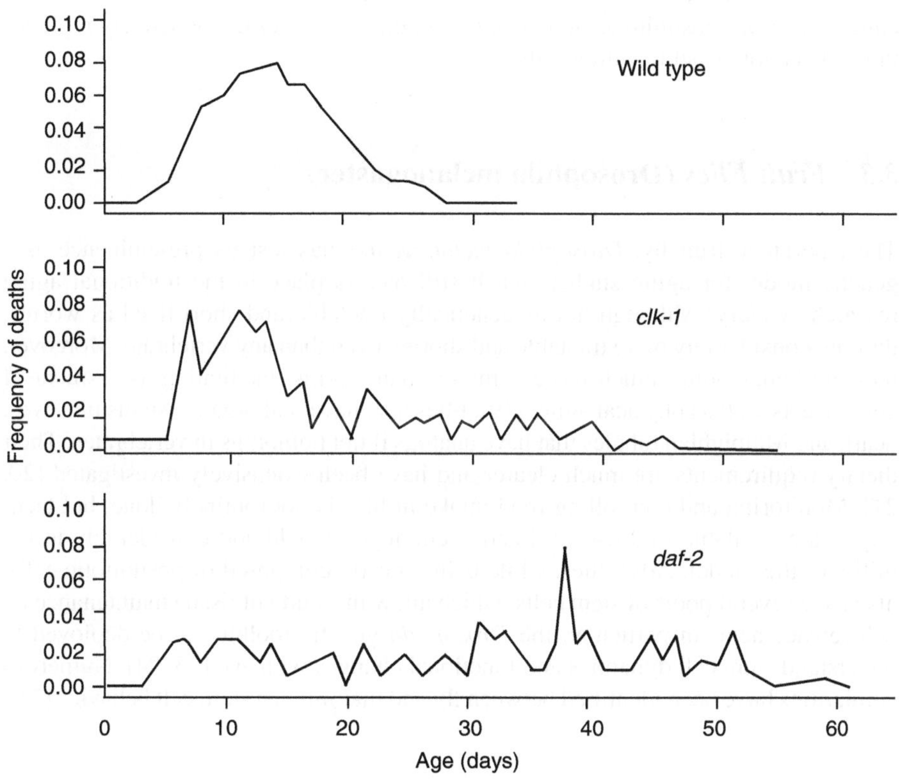

Fig. 2 Distribution of deaths in *wild-type*, long-lived *clk-1* and *daf-2 C. elegans* mutants. Note that only deaths in wild-type worms bear any resemblance to the distribution of deaths in modern humans, as shown in Fig. 1 (Adapted from Chen et al. [17] by permission of the Gerontological Society of America)

worms is proportionally much longer than any other species. For instance, one worm mutant has been reported to increase adult longevity by nearly tenfold [16].

In addition to changes in mean or median longevity among long-lived worm mutants, the *distribution* of deaths varies dramatically among the mutants (Fig. 2) [17]. Deaths in the wild-type strain were concentrated between 5 and 20 days. Considerably less clumped deaths were seen in the *clk-1* mutant, and in the longest-lived *daf-2* mutant, there is virtually no clumping of deaths but a slow steady trickle of them for 60 days. This is not a standard pattern of senescence-related mortality. Only the wild-type strain has a death distribution resembling to any degree that of senescent deaths of humans.

For all their many benefits, worms also have their limitations as aging models. They have a limited behavioral repertoire making assessment of their physical and cognitive health status difficult. The food they are fed in the laboratory, *E. coli*, is not part of their natural diet and in fact is slightly toxic to them. No one yet knows what their natural diet is, but when fed *Bacillus subtilis* rather than *E. coli*, wild-type worms lived about 40 % longer [18]. It is difficult to monitor food consumption in *C. elegans*, so controlled feeding trials are challenging. And all somatic cells in adult worms are postmitotic, so that studying the aging biology of actively replicating cells is not possible with worms.

3.3 *Fruit Flies (*Drosophila melanogaster*)*

The laboratory fruit fly, *Drosophila melanogaster* has lost its pre-eminence as a genetic model for aging studies, but it still has its place in the traditional aging research bestiary. Although not as genetically tractable and short-lived as worms, they are considerably more tractable and shorter-lived than any vertebrate. Moreover, they are behaviorally much more complex than worms, facilitating assessment of cognitive as well as physical aging [19]. Flies also have real organ systems like eyes, heart, and Malpighian tubules that have analogs if not homologs in vertebrates. Their dietary requirements are much clearer and have been extensively investigated [20, 21]. Monitoring and controlling food intake in flies is not routinely done, but techniques are available to do so [22] and if employed would add considerably to the utility of the model. Adult flies, while being mostly composed of postmitotic cells, also have several pools of stem cells, which allow the study of tissue maintenance by cell replacement. In particular, the *Drosophila* genetic toolkit can be deployed to understand stem cell dynamics and functional changes with age [23, 24]. Numerous similarities have been identified between fly and mammalian stem cell behavior [25].

3.4 *Laboratory Mice (*Mus musculus*)*

Since the 1980s when targeted genetic manipulation of laboratory mice became possible, they have eclipsed rats to assume the role of pre-eminent mammalian model for biomedical research of all types [26]. Basic aging research is no exception. Husbandry practices for mice are particularly well developed, including the maintenance of mice under specific-pathogen-free (SPF) conditions. Mouse physical, sensory, and cognitive phenotypes and the manner in which they change with age have been extensively explored as has the pathophysiology typical of various mouse strains (http://phenome.jax.org/). The longevity and disease profile of mouse strains – a strain is the product of at least 20 generations of brother-sister inbreeding – varies considerably [27]. Some strains should be strenuously avoided for aging research because they are particularly prone to die early of a single specific disease [28], meaning that studies of those strains are useful for investigating the disease process but not informative with respect to generalized aging processes. By far the most common strain used in aging research is C57BL/6 and by far the most common sex is male. In well-maintained SPF colonies, median longevity of C57BL/6 males is 26–30 months [29, 27].

The dominance of aging research by this single sex and strain is unfortunate. The presumed advantage of standardization, facilitating comparison of experimental results among laboratories and between studies, is purchased at the expense of generality. One never knows if a result is an idiosyncrasy of a particular genetic background or is a more general phenomenon. All inbred strains have their idiosyncrasies. For instance, C57BL/6 mice are particularly prone to idiopathic ulcerative dermatitis as they age [30], they are also particularly prone to lymphoma [31]. Like many laboratory strains, they exhibit defective melatonin synthesis due to mutations in both necessary biosynthetic enzymes [32]. Most recently it was discovered that the C57BL/6 J sub-strain – and only that sub-strain – has a single nucleotide change in a brain-specific tRNA that interferes with the protein translation machinery [33]. Although that phenotype was only revealed when a second mutation (in *gtpbp2*) led to a major neurodegenerative phenotype only in the C57BL/6 J sub-strain, it raises concern about whether "wild-type" animals of that commonly used sub-strain, might be affected in unknown ways by aberrant protein translation in the brain. So over-reliance on any single mouse strain or sub-strain limits our ability to spot cryptic aberrancies affecting what is classified as a healthy state.

One approach to the problem of cryptic strain idiosyncrasies that combines some generality with some genetic control is the use of genetically heterogeneous mice generated in a repeatable fashion. An example is the UM-HET3 mouse stock [34]. Originally created for gene mapping studies, this stock is created by interbreeding two F_1 hybrids of inbred strains. The resultant F_2 mice are each genetically unique full siblings representing a broad swathe of genetic diversity within the laboratory mouse. *Populations* with the same genetic diversity can be recreated at any time. One reason that inbred mouse strains became so popular was the belief that they would be phenotypically more uniform that outbred populations. However, at least for longevity, UM-HET3 mice are no more variable than C57BL/6 mice (Miller, R.A. 2015, personal communication). This is the mouse stock that is currently used in the National Institute on Aging's Intervention Testing Program (ITP) [35].

Over-reliance on a single genetic background is not a research phenomenon confined to mice or to aging research. Virtually all *C. elegans* research employs the same N2 genetic background. However both mice and fly researchers have discovered that genetic background makes a dramatic difference in the impact of longevity interventions. For instance, overexpressing human superoxide dismutase 1 (SOD1) in adult fly motorneurons significantly increased longevity in both males and females, by approximately 30 % and 40 %, respectively, in a particular laboratory strain. However the same mutation introduced into ten inbred, wild-caught strain found that females lived significantly longer in only 6 of the 10 strains and male lived longer in only 1 of the 10 strains compared to controls [36].

3.5 Other Species

Each of the three animal species described above is well-suited for research in the discovery of genetic interventions that modulate laboratory life. Together they represent more than 600 million years of evolutionary divergence from one another. Some phenomena such as reduced insulin/IGF signaling leading to lengthened life have been found in all three models. Does this mean that these species are sufficient for investigating *both* fundamental aging processes *and* age-related disease processes relevant to people? I would argue that these species are not enough for either and that we need to expand the traditional bestiary of aging models for the following reasons. First, our workhorse invertebrate models have undergone extensive gene loss since their divergence from our common ancestor. This can be seen by noting that more than 10 % of genes identified in a more distant human relative, the cnidarian *Acropora millepora*, have clear human orthologs that are missing from worm and fly genomes [37]. Thus there is a genomic universe of unknown size that may be relevant to aging processes not susceptible to investigation in worms or flies. Additionally, worms have no somatic cell division in adulthood and flies have limited cell division or regenerative capacity. Consequently, a key anti-senescence process – regenerative capacity – is difficult to study in these species. Finally with respect to these traditional invertebrate models, both worms and flies employ specialized nonaging life history stages during times of environmental stress (dauer in worms, reproductive diapause in flies) which have no human equivalent. Partial induction of these stages could retard aging via mechanisms not available to humans. In this sense, worm and fly findings could provide false clues to a deeper understanding of human aging biology. Thus new models, both vertebrate and invertebrate, with additional human orthologs, greater regenerative capacity, or lacking some type of diapause would be valuable to develop [38, 39].

Second, the use of the laboratory mouse as the sole representative of our own mammal clade warrants rethinking. Mice are the main species we rely on to model specific human disease processes and develop interventions to mitigate these processes. Yet despite their comparatively close relationship to us and the sophistication of genetic manipulation which we can deploy in mice, therapeutic successes in mice across a number of diseases such as stroke, cancer, and Alzheimer's disease (AD) have not translated well to patients. Alzheimer's disease is the most spectacular example of translational failure, with more than 300 drugs showing significant benefit in mouse models, but to date none have replicated that promise in humans [40]. A largely unexplored possibility that warrants attention is that mouse disease models may fail because researchers typically induce diseases of aging in young animals. For instance, various AD mouse models have been created using a wide variety of genetic strategies leading to cognitive deficits in the mice by 3–18 months of age [41]. Even the latest onset of these is barely in the comparable range of age as a fraction of lifespan of the most aggressive human familial mutation [42]. It may be that aging is a critical component of the disease etiology, say for instance, requiring some vascular injury as an initiating event. Therefore, aged mouse models might be more representative of the human disease. The problem may lie even deeper than the age of onset in the mouse models. It may be that the mouse brain just cannot be humanized with respect to AD sufficiently to make it a therapeutically useful tool. Provocatively, cage enrichment of current mouse models of AD significantly improve cognitive deficits and reduce neuropathological hallmarks of the disease [43–45]. In this case, other species may serve better. A transgenic rat model has been developed that replicates a fuller spectrum of AD pathology than any mouse model [46], for instance, and the emergence of CRISPR-Cas9 technology [47] may allow the development of a better AD model in a small, short-lived primate such as a mouse lemur or marmoset. However, some aspects of human biology may be unique to humans, and AD is a good candidate for a human-specific condition. Nothing representing the full spectrum of cognitive and pathological signs of AD has been found even in our closest relatives, chimpanzees and gorillas [48].

Third, all model organisms currently used in aging research are distinguished by their lack of success in resisting fundamental aging processes. That is, they deteriorate and die quickly. That is one of their advantages for the type of aging research that requires lifespan studies. However, there is another research paradigm available for basic aging research. Some species are already well-known for their exceptionally long and healthy lifespans in the natural world and for being able to resist aging processes much better than the longest-lived, most robust model species. So a complementary research paradigm is to investigate the cellular mechanisms underlying their resistance [49–51]. Given the phenomenal advances in sequencing power in recent years, insight into the genomes and transcriptomes of some of these exceptionally senescence-resistance organisms as well as tools for their further investigation could follow rapidly. An unresolved issue is whether the aging research community is more likely to gain novel insights about aging from the study of animals that are long-lived for their body size such as naked mole-rats or any of a large number of bat species, but not long-lived relative to humans, versus animals that are absolutely long-lived – substantially longer-lived than humans – as some whales, fishes, or diverse taxa of invertebrates such as sea urchins or bivalve mollusks [52].

4 Is It Really Possible to Change the Rate of Aging?

I stated rather blithely above that the promise of medical interventions that enhance and lengthen healthy life is no longer an empty promise. What evidence do I have for that claim? The most obvious evidence is that nature has achieved this feat many times over the course of evolution. Lifespan has approximately doubled, for instance, in the 6 million years since humans split from our chimpanzee ancestors. However, this evidence is not completely satisfactory as change over that time period could involve hundreds to thousands of variations in genes or gene expression. By far the most compelling evidence – presented below – is that we now are able to lengthen life and enhance a number of aspects of health in multiple ways by simple interventions in model organisms in the laboratory. Some of those ways are likely not to be relevant to humans, who are after all, many times more successful at resisting aging than any of our model organisms. The hope is that some of our successes with model animals will be relevant and translatable to humans and the more targets we identify in animals, the more likely this will be.

4.1 Dietary Restriction: The First Experimental Paradigm

4.1.1 Rodent Studies Past and Present

In the 1930s while investigating the impact of energy intake on growth rate, Clive McCay discovered that feeding rats substantially less than they would eat if given unlimited food increased their longevity [53]. Follow-up studies by McCay and a host of others confirmed this longevity-enhancing result of DR in both sexes of rats, mice, and a number of invertebrate species. Generally these studies found increases in both mean and maximum longevity (defined in the rest of this chapter as the longest-lived 10 % of the population) in both sexes and gradually over the next several decades information accumulated that many, although not all, maladies of aging laboratory rodents were also delayed [54]. By the 1990s a generalized – albeit premature – consensus emerged among aging researchers that DR universally extended life and health and that reduced caloric intake rather than restriction of specific macronutrients or the timing of food availability was the key to this "DR effect."

It is tempting to speculate on whether, if McCay's research had focused on obesity rather than growth rate, might DR might have been considered merely the absence of obesity or excess fat rather than a special ultra-lean state? This thought also illustrates how difficult it is to translate the circumstances of laboratory animals into human terms. The issue of whether DR is the absence of obesity or something else has not yet been determined, as the differing interpretations of the two existing macaque studies of DR illustrates. Although both studies report health benefits of DR, one study finds no effect on survival [55], the other does [56]. Control animals in the study that finds a survival effect weigh roughly 6–11 % more than the national average for captive monkeys of the same species whereas control animals in the study that finds no effect are roughly 8–16 % below the national average. Captive monkeys, in fact captive mammals of virtually all species including mice, are typically obese compared to animals in the wild [57, 58]. So how to think about the results in terms of which animals should be considered obese, which normal, which restricted, is anything but obvious. One approach, implemented in fruit flies, could be to observe the reproductive effects of different dietary regimes and designate a regime as DR only if it is accompanied by a reduction in reproductive output due to nutrient limitation [59]. Again though, how to translate this to humans remains problematic.

Because calorie intake per se was assumed to be the key to the DR effect, considerable research effort went into exploring the energetic consequences of DR. Particularly after the damaging effects of reactive oxygen species (ROS) became known [60], a clean and neat mechanistic hypothesis to explain the DR effect was developed. This hypothesis proposed that restricting energy intake reduced daily metabolic rate, which in turn reduced ROS production and its consequent tissue damage. This led *ipso facto* to increased health and longevity. Like many a clean and neat hypothesis before it, the reduced metabolism hypothesis crashed on the rocks of experimentation and biological complexity. Both mice and rats experiencing DR have an initial transitory period where metabolic rate really did fall; however, after a few weeks energy consumption stabilizes at about the same rate in ad lib-fed and DR animals when measured as metabolic rate per gram of lean body mass [61, 62]. While these observations seemingly killed the reduced metabolism hypothesis, the reduced ROS hypothesis was left largely unscathed, as it became clear that there is not a simple one-to-one relationship between metabolic rate and ROS production. DR mice and rats in fact were consistently found to produce lower levels of ROS than controls [63]. The ROS hypothesis was seriously challenged, though, by a series of genetic manipulations of mouse cellular antioxidants, both under- and overexpressing them, which in turn reduced or increased ROS damage to tissues, yet which produced no consistent changes in lifespan [64].

A number of other simple physiological hypotheses were advanced by logic and slain by experimentation, including the glucocorticoid cascade hypothesis [65, 66] and the advanced glycation end-products hypothesis [67, 68]. This simple dietary manipulation, reducing food intake, alters so many cellular and physiological aspects – hormones, growth factors, inflammation, cell and protein turnover, body temperature, etc. – of rodent biology that no single factor could be determined to completely explain the DR effect. One notable discovery has been that chronic DR consistently enhances a range of protective responses from apoptosis of damaged cells to increased expression of DNA repair enzymes to elevated xenobiotic detoxification processes to the activation of the proteostasis network [69–74]. Indeed, DR was found to protect rats and mice from a broad range of carcinogens, cardiotoxins, hepatotoxins, and neurotoxins [75].

By the early 2000s the consensus on the primacy of reduced calories over specific dietary macronutrients and the universality of the DR effect began to come apart. That consensus, particularly as it related to reduced caloric intake per se, had been based on one study on one sex, male, of one inbred rat strain, F344, fed a diet that made that sex/strain particularly prone to age-related nephropathy [76, 77]. Several reports that restriction of sulfhydryl-containing amino acids in otherwise isocaloric diets extended mean and maximum longevity in the F344 rats were published in the 1990s, but in the early 2000s these studies were extended to several rat genotypes [78] and then to mice [79]. More recently, a thorough "nutritional geometry" approach to mouse diet composition and caloric content – this approach systematically varies macronutrient ratios and caloric density – has called into question several aspects of the standard rodent DR paradigm [80]. The results of this study are complex, but two basic themes emerge. First, reduced caloric intake does not necessarily increase longevity. Unlike previous studies, caloric intake was manipulated by diluting individual diets with indigestible cellulose, which is more like the typical DR paradigm used in invertebrate models. Second, median longevity, though unrelated to total caloric intake, was related to the balance of macronutrients. Specifically, mice lived longer when fed diets that were low in protein and high in carbohydrates. Whether the lengthened lifespan associated with reduced protein consumption was due to reduced consumption of specific amino acids as the earlier studies might suggest, was not investigated. It should be noted that this was a single study with a single mouse genotype with modest sample sizes for each of the 25 dietary/survival groups. Nevertheless when combined with the methionine restriction studies published earlier, and the invertebrate studies discussed below, it certainly calls into question the traditional wisdom that calories – not specific macronutrients – are what modulate lifespan.

Also called into question has been the conventional wisdom that the DR effect is close to universal. Despite the fact that laboratory mouse or rat strains, due to the random fixation of alleles during inbreeding, are expected to have their idiosyncrasies, DR had been reported to extend life in a wide variety of strains and outbred stocks of laboratory rodents [54]. Even exceptionally long-lived Ames dwarf mice were observed to live longer with DR [81]. However, recent work on a panel of 41 recombinant inbred (RI) mouse lines at two independent animal facilities found that slightly more of the strains had their lives shortened by 40 % DR – the standard mouse protocol – as had their lives lengthened [82, 83]. This was true for both sexes at one facility (Fig. 3). Only females were examined at the other. The results were strikingly similar to findings with respect to variation in replicative lifespan in 166 single gene deletion strains in yeast [84]. The mouse results should be considered preliminary, as the number of individual lifespans per treatment and strain was small (N = 5 in one study, N = 10–12 in the other) and the strain results varied somewhat between facilities. Yet the indication of substantial genetic variability in the response to DR offers a promising tool for the genetic dissection of the rodent DR effect and its physiological mechanisms the same way it has proven to be a useful tool in yeast.

4.1.2 Dietary Restriction in Flies

The lack of success in discovering mechanisms of the DR effect in rodents even after decades of intensive study led by the 2000s to exploration of its effects in worms and flies, the hope being that the genetic power available in these model invertebrates would provide insight into DR not easily available in mice. In addition, because fly and worm lifespans are short and large numbers of animals can be

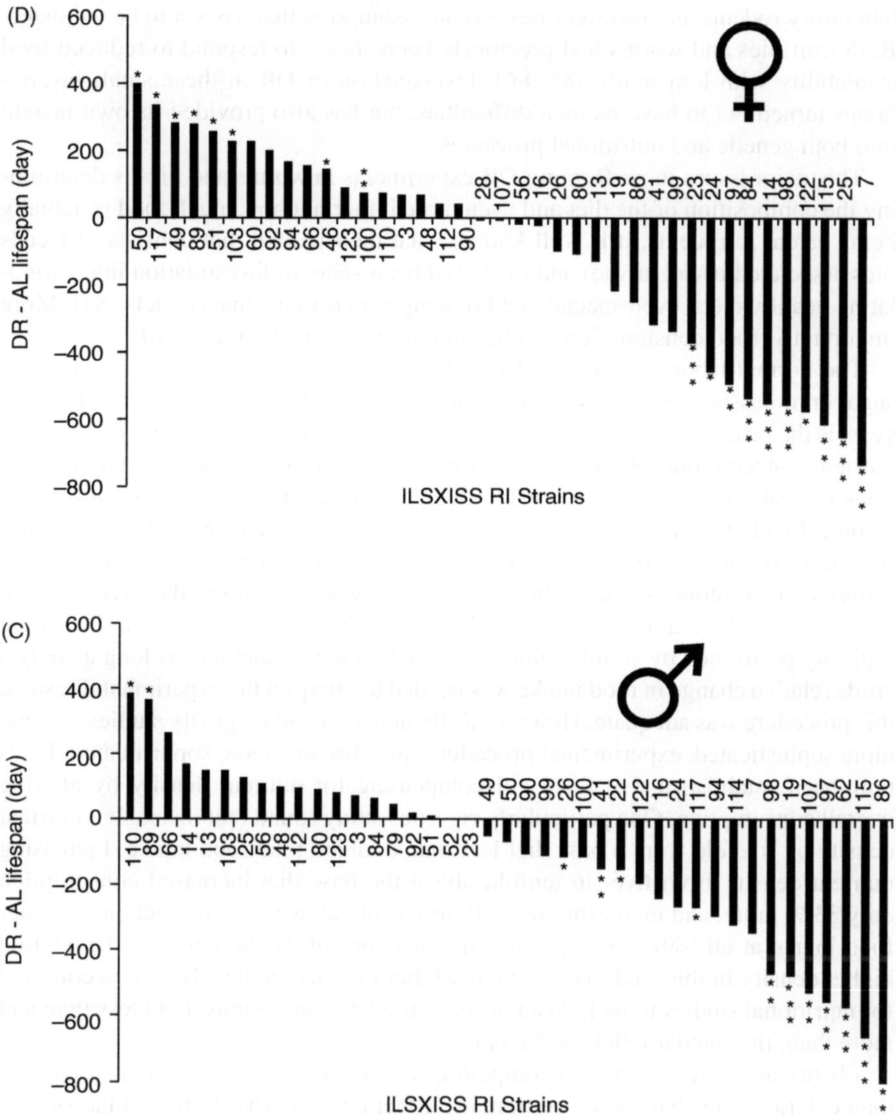

Fig. 3 Variation in the impact of 40 % dietary restriction depending upon mouse genotype. The data are from 41 recombinant inbred lines derived from an original cross of 8 laboratory strains followed by divergent selection for alcohol sensitivity (Reproduced with permission from Liao et al. [82]. *Aging Cell.* ©Blackwell Publishing Ltd/Anatomical Society of Great Britain and Ireland 2010)

easily maintained, experimental refinements such as including many dietary treatments within a single experiment can be routinely done. An implicit assumption of these studies – that DR affected longevity and aging via similar mechanisms in laboratory rodents and invertebrates – is an assumption that has yet to be validated. Both fruit flies and worms had previously been shown to respond to reduced food availability with longer life [85, 86]. Investigation of DR in these small invertebrates turned out to have its own difficulties, but has also provided its own insight into both genetic and nutritional processes.

The major issues in performing DR experiments in worms and flies is determining the composition of the diet and even more important how much food is actually being eaten. In rodents, it is well-known what their natural diet consists of (seeds plus associated insect larvae) and there had been years of investigation into formulating healthy diets, even specialized breeding versus maintenance diets [87]. More importantly, food consumption can be, and often is, directly measured.

The normal laboratory diet of fruit flies is an agar-base combination of yeast, sugar or molasses, cornmeal, and other carbohydrates [88]. In most laboratory diets, yeast is the primary protein source. There is no true standard laboratory diet and the nutrient concentration of the food can vary as much as tenfold among laboratories. Flies typically eat only 1–2 μg of food daily, so quantifying food consumption is technically challenging, although possible [89]. As a consequence, food consumption in fly studies is rarely assessed. Given the different nutrient concentrations of different laboratories' standard diets, nutrient consumption of standard-fed flies can vary dramatically among laboratories. Until recently, DR experiments in flies were typically performed by simply diluting the lab's standard diet and as long as only a crude relative change in food intake was needed to interpret the experimental results, this procedure was adequate. However as fly nutrition and longevity studies became more sophisticated, experimental procedures also became more sophisticated. It has now been established that flies will compensate for nutrient density by altering overall consumption. One particularly rigorous study found that increasing nutrient density by fivefold from a base diet less than doubled total food intake. Increasing nutrient density from five- to tenfold above the base diet increased consumption only 33 % more, and increasing from 10 to 15-fold above the base diet did not alter food intake at all [89]. To emphasize the diversity of fly diets in use, the 15-fold higher density in this study is the standard diet in other studies. Now it is common for nutritional studies to include an array of food concentrations, both less than and more than, the standard diet for the lab.

Fly research has shown most compellingly that macronutrient composition rather than calories alone has the most dramatic effect on longevity. In particular, several studies have concluded that reduced protein rather than calories is the key determinant of fly longevity [90, 91, 21, 92]. The role of specific amino acids has not yet been completely clarified, although as with laboratory rodents methionine appears to be a particularly important amino acid [91]. Recently Piper formulated a chemically defined diet for flies that should allow further refinement of the relationship between nutrition and aging in this species [93]. One particularly interesting discovery is that even the aroma of extra yeast is enough to shorten fly lifespan [94].

Nutritional research in flies has also illuminated a potentially serious confound in assessing genetic or even the pharmacological influence on aging and longevity in studies that rely on ad lib feeding as virtually all do. The impact of a gene or a drug may be sensitive to dietary factors or may affect the amount of food eaten. For

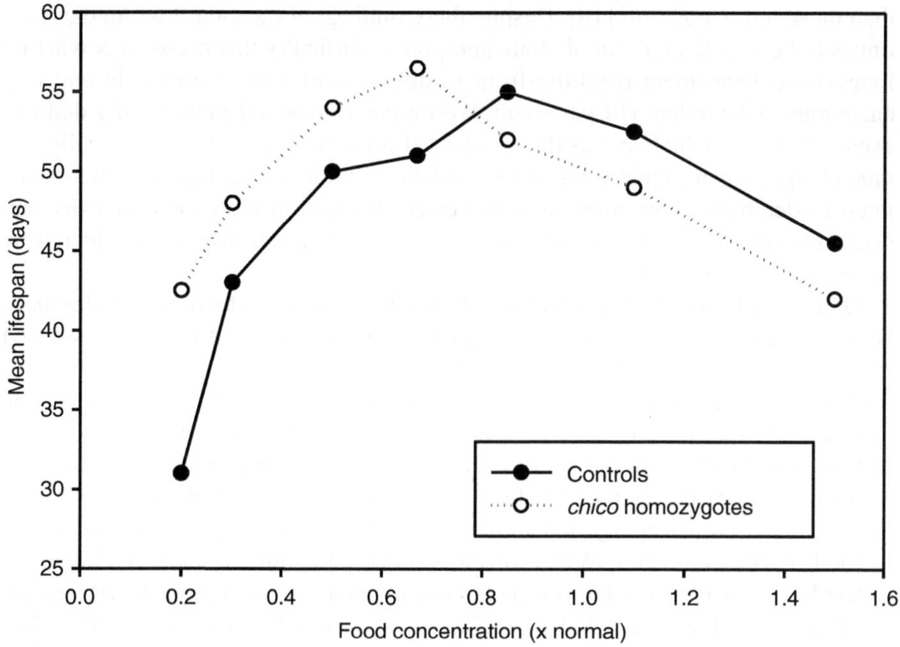

Fig. 4 Impact of nutritional density on lifespan in flies. The *chico* mutation is reported to make flies live longer (Clancy et al. 2001). Note, however, that the nutritional density of a "normal" diet varies dramatically among laboratories, such that what is considered a "normal" diet could affect whether the same mutation is a "short-gevity" or a longevity mutation and that at no food concentration is *chico* significantly longer-lived than control flies on their optimal longevity diet (Redrawn from Clancy et al. [96]. With permission from AAAS)

instance, certain mutations in the *Drosophila* gene *chico* have been reported to extend life [95]. However, this life extension was seen only at some food densities [96]. At other densities *chico* is shorter-lived than controls (Fig. 4) and at no density is *chico* longer-lived than the control flies on their own optimal-longevity diet. As the food density of a "normal" diet in flies is completely arbitrary, the finding that this *chico* mutation extends life is a happenstance of a particular standard diet. Other labs would have observed the same mutation to be life-shortening under their standard conditions. As long as studies are performed over a broad range of food densities, this should not be a problem. However, genetic studies of longevity rarely examine a range of nutrient conditions. In mouse studies, for instance, it is virtually never done.

4.1.3 Dietary Restriction in Worms

Turning to worms, their normal diet is unknown. Furthermore, the standard laboratory diet, live *E. coli* OP50, while having numerous advantages in terms of convenience and standardization, is clearly not a healthy worm diet. In fact, it is toxic. Worms fed killed *E. coli* live 16–40 % longer than worms fed live *E. coli* [97, 98] and worms fed a different bacterial species, *Bacillus subtilis*, live about 40 % longer than those fed live *E. coli* [18]. Despite these findings, the standard worm diet continues to be live *E. coli*. An obvious interpretive difficulty this presents is whether longevity enhancement resulting from feeding less of a toxic diet is in any way analogous to the rodent DR paradigm. Worms can also be fed a chemically-defined, axenic diet [99], which avoids the problem of frank toxicity – worms typically live much longer on an axenic versus an *E. coli* diet – but for either type of diet it is difficult to determine how much is being eaten. Because it increases longevity and decreases fertility relative to standard worm diet, axenic rearing is considered by some a form of DR [100].

Greer and Brunet (2011) identified 12 methods of life-extending DR that have been used in worms, including two synthetic liquid diets and one mutation that reduces pharyngeal pumping rate. Surprisingly few worm studies use more than two feeding levels (control vs restricted) even though considerably more information can emerge from multiple feeding level studies [101]. Any hope that the powerful genetic tools available in the worm would quickly clarify the molecular pathway or pathways responsible for the DR effect in other species has been dashed however as the genes that are necessary and sufficient for life extension by DR depend on the method of DR employed. Which of the various paradigms is most analogous to rodent DR is not obvious. One of the favored hypotheses among rodent DR specialists, that reduced insulin or IGF signaling is responsible for much of the DR effect, does not appear to be the case for worms.

So, as these examples illustrate, after 80 years of research the key mechanisms by which DR lengthens life remains an unsolved puzzle.

4.2 *Genetic Approaches to Retarding Aging*

If worm biology has not yet helped solve the mechanistic mystery of the DR effect, it has been instrumental in dissecting the genetics of longevity. Several hundred worm genes significantly extend life when wholly or partially inactivated. Given that the active forms of these genes were selected over millions of years of evolution, this large number is surprising to say the least. It will be interesting if anything like this turns out to be true of other model organisms or whether this is a quirk of worm biology, perhaps due to the centrality of the dauer larval stage in its life history. Still, some of the largest effects on worm longevity are still due to some of the earliest genes discovered to affect aging. Given that a complete review of the numerous genetic influences on aging and longevity is beyond the scope of this chapter, I will focus on just the two that seem at this juncture to be the most robust.

4.2.1 Insulin/IGF Signaling

The first two longevity mutants discovered in worms were both in the insulin/IGF signaling pathway, a phosphatidylinositol-3-kinase (*age-1*) [8] and the insulin/IGF receptor (*daf-2*) [9]. They lengthened worm life by as much as 75–100 %. Insulin and IGF (insulin-like growth factor) signaling are combined in this section because in worms and flies there is a single receptor that mediates the activity of multiple ligands. Subsequent mutations that reduce signaling through homologous pathways causing both dwarfing and lengthened life were discovered in flies [95, 102] and mice [103–105]. There is even evidence that reduced IGF-1 signaling may play a role in exceptional human longevity [106]. Thus signaling via insulin and/or IGF appears to be a highly conserved modulator of growth, metabolism, reproduction, and longevity – at least under laboratory conditions. Some evidence suggests that in more challenging circumstances the longevity effect may disappear or even be reversed [107, 108].

The evidence in mice deserves some special attention because in some ways it is weaker than in other model systems perhaps because the genetic tools are less robust but also possibly because the effect is less significant in mice or in mammals. It is important to note that too great a reduction in insulin or IGF signaling is life-shortening or lethal. It is also important to note that most of the mouse evidence for involvement of the insulin/IGF networks in longevity is indirect. For instance, possibly the most robust longevity-enhancing genetic treatment in mice is to reduce or eliminate signaling, not through insulin or IGF, but through the growth hormone receptor. The impact on longevity is substantial in both sexes (25–50 % increase) and unlike most mouse genetic longevity enhancements, it has been replicated in multiple genetic backgrounds, in multiple laboratories, and multiple times in the same laboratory [105, 109–111]. Although eliminating growth hormone signaling dramatically reduces circulating (but not locally produced) IGF-1, growth hormone also has effects independent of circulating IGF-1. Directly reducing IGF-1 activity by using mice haploinsufficient for the IGF-1 receptor has a substantially smaller longevity effect (5–11 %) in females and no effect in males [112, 113]. Haploinsufficiency of other members of the IGF signaling cascade such as insulin receptor substrate 1 or 2 also have small or sex-specific effects on longevity [114, 115]. Knocking out insulin – as opposed to IGF – signaling specifically in adipose tissue also modestly extends life (~18 %) in both sexes in mice [116]. Another effect of disruption of the growth hormone receptor is reduced plasma insulin. So potentially the large longevity increase with growth hormone receptor disruption is a combination of alterations to both insulin and IGF. On the other hand, researchers may be overlooking a key component of growth hormone signaling that is independent of insulin or IGF or it may be that haploinsufficiency or complete inactivation of activity in specific tissues are too crude manipulations to understand the context-specific roles of the signaling activity Both worm and fly mutations in insulin/IGF signaling that most effectively extend life are modulated reductions in signaling, rather than its complete ablation.

4.2.2 mTOR

The mechanistic Target Of Rapamycin (mTOR) is a highly conserved serine/threonine kinase that lies at the hub of a complex cellular signaling network that is centrally involved in metabolic processes, many diseases, and even aging itself.

Integrating environmental information on nutrient availability and a variety of stressors, mTOR activation promotes anabolism. Its suppression does the reverse – plus it modulates various stress responses [117]. First described in yeast, mTOR is a component of two distinct complexes, mTOR Complex 1 (mTORC1) which mediates most of the processes mentioned previously, and is directly inhibited by the drug rapamycin, and mTOR Complex 2 (mTORC2), which functions in the modulation of metabolism and various aspects of the cell's cytoskeleton and is not directly responsive to rapamycin. It attracted the attention of aging researchers due to its nutrient responsiveness and the known effects of DR on longevity [118]. Genetic inhibition of mTOR or its downstream mediators increases longevity in worms [119], flies [120], and mice [121]. Some types of DR in worms are ineffective when components of the mTOR network are suppressed. In flies, lifespan extension by reduced essential amino acid availability is also modulated by mTOR and not by insulin/IGF signaling [122]. DR in mice inhibits mTOR activity and pharmacological inhibition of mTOR extends life and health (see below). Therefore, it seems reasonable to speculate that inhibition of mTOR may play a role in DR's effects on health and longevity. Now that mouse genotypes that fail to respond to DR with increased life and health have been identified [82, 83], the role of mTOR and insulin/IGF signaling in the mouse DR response should be accessible.

4.2.3 Healthspan Versus Lifespan

What if an apparently frail and feeble mouse genotype seems to live on and on compared to more robust and vigorous control mice? To a number of researchers, the Ames or Snell dwarf mice appeared to be just such mice. Although they live dramatically longer than littermate controls (24–60 % longer), they are tiny, and because of their small size, they are particularly sensitive to cold. Females are sterile, males sub-fertile, and, when young, animals of neither sex moved around in their cages as much as controls. However, some aspects of their aging process appear to correlate with better health, rather than simply increased longevity. For example, some cognitive abilities appear better preserved with age [123], and neurogenesis continues later in life. Specifically, while early in life their spontaneous neuronal activity rate is lower than controls, Ames dwarf mice maintain this activity rate with age, such that later in life, it surpasses that of old control mice. Nevertheless, their small size and seeming frailty, I believe, led many researchers to raise questions about the quality of life associated with the longer lives Ames or Snell dwarf mice lived – particularly as people begin to consider the possibility of translating these successes from laboratory species to humans. These are legitimate questions with enormous societal implications. The simple assumptions that extended life equals extended health or that extended life will reduce the period of debility near life's end need to be critically evaluated and more difficult questions may follow. For instance, if we medically retard aging, giving most of us an extra 10 years of healthy life, but an additional 10 years of unhealthy life as well, is this worthwhile? Philosophers and economists might well differ in their opinions, but without evidence, these will be sterile debates. We need information. The Geroscience Hypothesis dictates that basic aging researchers need to define and evaluate healthspan as well as lifespan in their experiments.

Assessing healthspan in laboratory animals turns out to be considerably more difficult – and prone to interpretation – than assessing lifespan. The easiest case to evaluate would seem to be rodent DR, because researchers have been investigating its impact on a spectrum of physiological variables throughout life for decades [54]. There is consensus that many rodent genotypes on DR are still physically fit at ages when their ad lib-fed brethren are all dead and that DR animals are more likely to die without detectable pathological lesions than fully-fed animals. Whether the known downsides of DR – reduced libido, muscle mass, and bone density, slower wound healing and some immunosuppression – would be worth the extended lifespan, assuming that the health benefits in humans were the same, would be a matter of personal psychology. However, as obesity rates are rising world-wide, this is one treatment that even if proven to effectively extend health and reduce morbidity, would not likely be adopted *en masse*.

However, rodent DR is an exception to the rule. Generally, we know relatively little about the health consequences of our various life-extending treatments, particularly with invertebrate species – and what we do know is not necessarily reassuring. For instance, one reasonable metric of health might be resilience in the face of physiological stress, an increase in which often accompanies increased longevity [124, 125]. Yet, genetic variation for reduced mortality in ten inbred lines of *Drosophila* failed to exhibit any correlation with genetic variation for resilience to cold-stress, even though both traits varied [126]. Only recently has substantial effort gone into assessing the health consequences of some of the many longevity-enhancing mutations found in *C. elegans* [127]. A recent examination of four different worm longevity mutations, including the most robust of the known mutations, *daf-2(e1370)*, employing four carefully thought out metrics of worm health (heat and oxidative stress resistance plus activity in liquid or solid media) found that none of the mutations compressed morbidity (defined as a loss of ≥50 % the capacity of a young adult) relative to wild-type by any metric. Moreover, in only 5 of 16 possible cases (4 longevity mutations × 4 health measures) was healthy life extended, and in all of these the unhealthy period of life was also extended. In 7 of 16 possible cases, the period of healthy life was actually shortened compared to wild-type and the unhealthy period extended. While these results should be considered provisional as the investigation of *C. elegans* health is in its infancy, they are clearly provocative.

Research into laboratory rodent health has a much longer history, is considerably easier to assess, and is undoubtedly more relevant to what we might expect in humans. Moreover, what is known about the health consequences of life-extending interventions in mice is considerably more promising than evidence to date from the invertebrates (see below). However, functional metrics are performed (or at least reported) haphazardly and it is never clear whether all investigated metrics have been reported or there was a selection for reporting those that improved. What is needed in rodent aging research is a widely-agreed upon panel of functional indicators that is implemented for any new life-extending interventions. Such a panel is currently under development (Richardson, et al., in press).

4.3 Pharmacological Approaches to Retarding Aging

Arguably the biggest and most exciting advance in basic aging research recently is success in extending mouse life pharmacologically. Although various drugs and supplements have been reported to extend the lives of laboratory rodents at least since the early 1960s [128–132], until recently none had withstood the test of independent replication. A difficult problem with rodent longevity studies is that the cost and time involved in doing them makes replication rare. However, this is essential, so that scientists do not spend years pursuing dead ends arising from anomalous experimental results. A recent example is the 1999 report that a targeted mutation in the mouse $p66^{shc}$ gene increased lifespan by 30 % [133], a finding that was never independently validated until 15 years later when it could not be replicated [134].

For this reason and others, the National Institute of Aging's (NIA) Interventions Testing Program (ITP) represents a major advance in the implementation of drug testing protocols for mouse life extension. A key feature of the ITP is that compounds are tested at three independent research sites (University of Michigan, Jackson Laboratories, University of Texas Health Science Center San Antonio) using identical experimental protocols at the three sites [34]. As a consequence, a positive hit for longevity is immediately confirmed. A second key feature is the use of both sexes of genetically heterogeneous mice – the previously mentioned UM HET3 mouse stock – which prevents a positive hit because of a quirk of an individual inbred strain like C57BL/6 or because of a difference between the sexes.

To date, results from 16 different compounds have been published (Table 3). Some of the compounds have been tested at several doses and/or initiated at several different ages. One major result is that none of the compounds tested to date has significantly shortened mouse lifespan. Another rather astonishing result is that so far 5 of the 16 compounds have significantly extended life in either males alone or in both sexes.

Each of these successful interventions deserves a few words. Aspirin is a familiar nonsteroidal anti-inflammatory drug with anti-thrombotic and antioxidant properties. As chronic, low level inflammation is a hallmark of aging itself and damage from ROS have been implicated in numerous disease processes, aspirin had multiple rationales for being tested. At the single dose tested, there was a small (8 %) but statistically significant lengthening of median lifespan in males but no significant effect in females and no effect on maximum longevity (last surviving 10 %) in either sex [135]. Further investigation indicated that females had less bioactive plasma levels of aspirin and its metabolites. As a large (19,000 participants), multinational randomized clinical trial of low dose aspirin, looking at its impact on variety later life maladies is now underway [136], there is no follow-up to the aspirin study planned for the NIA ITP at this point.

Table 3 Compounds tested and results from the NIA Interventions Testing Program

Compound	Concentration in food	Age at initiation (months)	Effect in males	Effect in females
Aspirin	20 ppm	4	8 %	0
NFP	200 ppm	4	0	0
NDGA	2500 ppm	9	12 %	0
NDGA	800 ppm	6	8 %[a]	–
NDGA	2500 ppm	6	10 %[a]	–
NDGA	5000 ppm	6	9 %[a]	0
4-OH-PBN	315 ppm	4	0	0
CAPE	30 ppm	4	0	0
CAPE	300 ppm	4	0	0
Enalapril maleate	120 ppm	4	0	0
Rapamycin	14 ppm	20	9(28)%[b]	13(38)%[b]
Rapamycin	14 ppm	9	10 %	18 %
Rapamycin	4.9 ppm	9	3 %	16 %
Rapamycin	14 ppm	9	13 %	21 %
Rapamycin	42 ppm	9	23 %	26 %
Simvastatin	12 ppm	10	0	0
Simvastatin	120 ppm	10	0	0
Resveratrol	300 ppm	12	0	0
Resveratrol	1200 ppm	12	0	0
Resveratrol	300 ppm	4	0	0
Oxaloacetic acid	2200 ppm	4	0	0
Green tea extract	2000 ppm	4	0	0
Curcumin	2000 ppm	4	0	0
Medium chain triglyceride oil	60,000 ppm	4	0	0
17α-estradiol	4.8 ppm	10	12 %	0
Acarbose	1000 ppm	4	22 %	5 %

Effect = % median lifespan increase (all statistically significant)
[a]Survival curves not yet complete, results provisional
[b]Numbers are percent life extension from the time (20 months) the mice began receiving rapamycin

NDGA (nordihydroguaiaretic acid) is a natural product produced by the creosote bush with both anti-inflammatory and antioxidant properties, which is used in some traditional medicine pharmacopeias. NDGA had a slightly larger (12 %) effect on median lifespan than aspirin also in males only [137] with no statistical effect on females and no effect on maximum longevity in either sex (Table 3). Males appear to metabolize NDGA considerably slower than females and so maintain its levels in their blood considerably longer [137]. Although complete survival data are not yet available, NDGA follow-up studies begun a little earlier in life (6 months versus 9 months for the original study) using a lower dose, the same dose, and twice as high a dose as the original study appear to have very similar results. At all doses median male survival of those getting the drug is 8–10 % greater than controls and at no dose do females appear to be living longer [138].

A nonfeminizing estrogen with little or no affinity for the classical estrogen receptors, 17α-estradiol has been repeatedly reported to have neuroprotective and antioxidant properties [139]. It was hypothesized that males in particular might benefit from 17α-estradiol. Males did. Median longevity of males increased by a statistically significant 12 %, but as with aspirin and NDGA there was no significant effect on median longevity of females or on maximum longevity of either sex. An interesting aspect of the 17α-estradiol studies is that there was a dramatically larger effect (28 % increase in median longevity) at one facility than either of the others (a nonsignificant 3 % at each). The larger effect was not due to longer-lived treated mice at that site but to shorter-lived controls [138]. So this general result should be treated somewhat cautiously until more research is done.

The fourth compound found to extend mouse longevity is acarbose, commonly used clinically particularly in Europe in the treatment of type II diabetes. It ameliorates the postprandial spike in plasma glucose. Acarbose is not metabolized, it inhibits α-glucosidases in the intestines and therefore slows the breakdown of dietary carbohydrates into glucose. For that reason, it was hypothesized that it might act as a DR mimetic. Median male longevity was increased at all sites by 22 % for the pooled data. Maximum male longevity was also increased significantly (by 11 %). For females, the data are more complex. The pooled data showed a 5 % increase in median female longevity, which reached statistical significance ($p=0.01$). However, there was essentially no absolute median difference at one site, a nonsignificant 7 % increase at another site, and a marginally significant ($p=0.04$) 7 % difference at the third. As no statistical corrections were done for multiple comparisons, this result, like the female 17α-estradiol results, should be interpreted with caution. Surprisingly, maximum longevity of females was increased at all sites in absolute terms, by 9 % overall, a highly significant ($p=0.001$) result. Recall that these are genetically heterogeneous mice, so the puzzling results where there are large inter-site differences or a marginal result in median longevity but a more significant result in maximum longevity could be a consequence of that genetic heterogeneity. The drugs could be beneficial for some genotypes and not for others.

To summarize the ITP results to this point. Four of 16 drugs tested have shown highly statistically significant effects on median male longevity, only one of those four (acarbose) showed a statistically significant increase in median female longevity and there is reason to interpret that one result cautiously. Only one of the four drugs – again, acarbose – increased maximum longevity significantly, and it did so in both sexes. Follow-up studies on acarbose are currently underway.

By far the biggest success of the NIA ITP has been the discovery that rapamycin treatment increases longevity robustly in both mouse sexes even when initiated relatively late in life [140]. Used clinically as part of an antirejection cocktail primarily for people receiving kidney transplant [141, 142], for some cancer chemotherapeutic regimes [143], and to inhibit overgrowth of coronary artery stents [144, 145], rapamycin and its close chemical relatives (“rapalogs”) are well-known to directly inhibit mTOR kinase [146]. The biology of mTOR and the effects of its inhibition are too complex to cover here except cursorily as above, but many fine reviews are available [118, 147, 117]. In 2009 the NIA ITP reported that rapamycin when initiated at 20 months of age (the demographic equivalent of about 60 years old in human terms) lengthened life 9 % in male mice and 14 % in female mice as well as significantly extending maximum lifespan in both sexes [140]. Measured from the time at which rapamycin feeding had begun though, males lived 28 % longer than controls and females lived 38 % longer than controls. Follow-up studies with mice begun on the same dose of rapamycin at 9 months of age found a highly significant 10 % increase in median lifespan in males and an 18 % increase in females as well as 16 and 13 % increases in maximum longevity, respectively.

The longevity enhancing impact of rapamycin has now been replicated in both sexes, in both genetically heterogeneous and inbred mice, in a variety of labs, when administered in food or injected, at a variety of doses, and when started in adolescence, adulthood, or later in life [148–152]. In addition, it has now been shown to delay or even reverse cognitive deficits in mouse models of Alzheimer disease [153–155] and Huntington's disease [156], attenuate the progression of atherosclerotic plaques in atherosclerosis models [157, 158], prevent or delay both spontaneous and genetically-induced mouse cancers [149, 159, 160], and delays the onset of symptoms in a progeria model [161].

On top of these improvements in mouse models of various human diseases, rapamycin also improves a number of mouse health indicators. For instance, it enhances and broadens vaccine response [148, 162], delays normal cognitive decline during aging, reduces anxiety and depression [155, 151], increases spontaneous physical activity, improves cardiac function, and enhances sleep consolidation in older mice [163, 164, 151]. In sum, rapamycin administered to mice increases longevity, prevents or delays many diseases, and preserves many aspects of health. It might almost be said to retard aging itself.

All drugs have side-effects, rapamycin is no exception. Are any of these side-effects severe enough to eliminate it from consideration as a potential senescence-retarding intervention in humans? Because it has been in clinical use for years already, we know quite a bit about rapamycin's side-effects in people with various serious diseases. However because it is typically used in combination with other drugs and never given to completely healthy people, we know little about its side-effects in healthy people. From what is known, however, the side effects of most concern are immunosuppression, hyperglycemia, glucose intolerance, and insulin resistance [165–167]. However, in a genetically heterogeneous mouse stock, these effects were seen in young male mice during the first 6 weeks of rapamycin treatment but were substantially diminished and even reversed in some cases by 5 months of treatment [168]. Similar metabolic dysregulation was seen in both genetically heterogeneous and C57BL/6 J male mice for the first 4 months of treatment whether the mice were fed a normal or high-fat diet, but these effects disappeared within 2 weeks after cessation of treatment [167]. So at least in male mice, metabolic changes produced by chronic rapamycin treatment disappear quickly when treatment is halted and may be transient even with continued treatment. It will be enlightening to see whether these effects also occur in female mice and in both sexes of other species.

The use of rapamycin as a component of anti-rejection therapy following organ transplant suggests that if used chronically it may enhance susceptibility of infectious diseases. Without question rapamycin suppresses aspects of immune function [169–171]. However, it enhances other aspects, and consequently has been termed an immunomodulator rather an immunosuppressant [148, 172]. Chronic enteric rapamycin administration has been found to enhance resistance to pneumococcal pneumonia in elderly mice [173], although no such protection – and possibly reduced protection – was found against West Nile virus [174]. Moreover, a 6 week course of injected rapamycin prior to influenza vaccination has been found to enhance protection again influenza in both mice and humans [148, 172]. Therefore, the impact of chronic rapamycin on disease susceptibility in healthy humans is far from clear and should not by itself discourage trials in species other than mice.

5 Future Directions

As previously shown, dietary, genetic, and pharmacological interventions have been found that robustly extend life in model organisms and some have also been shown to extend multiple features of health. Where do we go from here if we are serious about ultimately discovering new ways to prolong human health? One obvious way forward is to solidify our knowledge base. That means replicating and optimizing successful interventions for both health and longevity in both sexes in other genotypes and other species. A surprising number of successful longevity interventions, both genetic and pharmacological, have had sex-specific effects (Austad & Bartke, *Gerontology*, in press). That also means evaluating interventions that have not already been approved for human use in other mammal species. Mice, particularly laboratory mice, are not an acceptable stand-in for all mammals. They have displayed a notable lack of success in predicting therapeutic efficacy in human diseases such as Alzheimer's disease, stroke, or even cancer. Mice have their obvious quirks such as their extreme susceptibility to cancer and limited cognitive sophistication. Their robust longevity response to constitutively-reduced growth hormone signaling has never been seen in another species and has failed to be observed even in their close relative, the laboratory rat [175]. Fortunately, additional mammal species such as the domestic dog and the small, short-lived primate, the common marmoset, appear to be excellent candidates for such studies.

Geroscience, as I hope this chapter has shown, is advancing more rapidly than almost anyone supposed. Its promise to enhance and extend human health could transform not only human health in the twenty-first century but also all the social institutions that depend on human health. In the year 2100, we may look back at the year 2000 and consider it as medically unsophisticated as we now think of the year 1900.

Acknowledgements The author is grateful for the advice and ideas shared by members of the Geroscience Network supported by NIH grant R24 AG044396.

Editor: Leslie Frieden, National Institute of Dental and Craniofacial Research (NIDCR), NIH.

References

1. Oeppen J, Vaupel JW (2002) Demography. Broken limits to life expectancy. Science 296(5570):1029–1031
2. Goldman DP, Cutler D, Rowe JW, Michaud PC, Sullivan J, Peneva D, Olshansky SJ (2013) Substantial health and economic returns from delayed aging may warrant a new focus for medical research. Health Aff (Millwood) 32(10):1698–1705. doi:10.1377/hlthaff.2013.0052
3. Murphy SL, Xu J, Kochanek KD (2013) Deaths: final data for 2010. Natl Vital Stat Rep 61(4):1–117
4. Raber J, Huang Y, Ashford JW (2004) ApoE genotype accounts for the vast majority of AD risk and AD pathology. Neurobiol Aging 25(5):641–650. doi:10.1016/j.neurobiolaging.2003.12.023
5. Miniño AM, Arias E, Kochanek KD, Murphy SL, Smith BL (2002) Deaths: final data for 2000. Natl Vital Stat Rep 50(15):1–120
6. Holloszy JO (1997) Mortality rate and longevity of food-restricted exercising male rats: a reevaluation. J Appl Physiol 82(2):399–403
7. Rajpathak SN, Liu Y, Ben-David O, Reddy S, Atzmon G, Crandall J, Barzilai N (2011) Lifestyle factors of people with exceptional longevity. J Am Geriatr Soc 59(8):1509–1512. doi:10.1111/j.1532-5415.2011.03498.x
8. Friedman DB, Johnson TE (1988) A mutation in the age-1 gene in Caenorhabditis elegans lengthens life and reduces hermaphrodite fertility. Genetics 118(1):75–86
9. Kenyon C, Chang J, Gensch E, Rudner A, Tabtiang R (1993) A C. elegans mutant that lives twice as long as wild type. Nature 366(6454):461–464. doi:10.1038/366461a0
10. Kimura KD, Tissenbaum HA, Liu Y, Ruvkun G (1997) daf-2, an insulin receptor-like gene that regulates longevity and diapause in Caenorhabditis elegans. Science 277(5328):942–946
11. Henderson ST, Rea SL, Johnson TE (2006) Dissecting the processes of aging using the nematode *Caenorhabditis elegans*. In: Masoro EJ, Austad SN (eds) Handbook of the biology of aging, 6th edn. Academic, San Diego, pp 360–399
12. Chalfie M, Tu Y, Euskirchen G, Ward WW, Prasher DC (1994) Green fluorescent protein as a marker for gene expression. Science 263(5148):802–805
13. Klass M, Hirsh D (1976) Non-ageing developmental variant of Caenorhabditis elegans. Nature 260(5551):523–525
14. Felix MA, Braendle C (2010) The natural history of Caenorhabditis elegans. Curr Biol 20(22):R965–R969. doi:10.1016/j.cub.2010.09.050
15. Riddle DL, Albert PS (1997) Genetic and environmental regulation of dauer larva development. In: Riddle DL, Blumenthal T, Meyer BJ, Priess JR (eds) C. elegans II, 2nd edn. Cold Spring Harbor Laboratory Press, Cold Spring Harbor, New York
16. Ayyadevara S, Alla R, Thaden JJ, Shmookler Reis RJ (2008) Remarkable longevity and stress resistance of nematode PI3K-null mutants. Aging Cell 7(1):13–22
17. Chen J, Senturk D, Wang JL, Muller HG, Carey JR, Caswell H, Caswell-Chen EP (2007) A demographic analysis of the fitness cost of extended longevity in Caenorhabditis elegans. J Gerontol A Biol Sci Med Sci 62(2):126–135
18. Garsin DA, Villanueva JM, Begun J, Kim DH, Sifri CD, Calderwood SB, Ruvkun G, Ausubel FM (2003) Long-lived C. elegans daf-2 mutants are resistant to bacterial pathogens. Science 300(5627):1921

19. Burger JM, Buechel SD, Kawecki TJ (2010) Dietary restriction affects lifespan but not cognitive aging in Drosophila melanogaster. Aging Cell 9(3):327–335. doi:10.1111/j.1474-9726.2010.00560.x
20. Bass TM, Grandison RC, Wong R, Martinez P, Partridge L, Piper MD (2007) Optimization of dietary restriction protocols in Drosophila. J Gerontol A Biol Sci Med Sci 62(10):1071–1081
21. Lee KP, Simpson SJ, Clissold FJ, Brooks R, Ballard JW, Taylor PW, Soran N, Raubenheimer D (2008) Lifespan and reproduction in Drosophila: new insights from nutritional geometry. Proc Natl Acad Sci U S A 105(7):2498–2503. doi:10.1073/pnas.0710787105
22. Deshpande SA, Carvalho GB, Amador A, Phillips AM, Hoxha S, Lizotte KJ, Ja WW (2014) Quantifying Drosophila food intake: comparative analysis of current methodology. Nat Methods 11(5):535–540. doi:10.1038/nmeth.2899
23. Wang L, Karpac J, Jasper H (2014) Promoting longevity by maintaining metabolic and proliferative homeostasis. J Exp Biol 217(Pt 1):109–118. doi:10.1242/jeb.089920
24. Resende LP, Jones DL (2012) Local signaling within stem cell niches: insights from Drosophila. Curr Opin Cell Biol 24(2):225–231. doi:10.1016/j.ceb.2012.01.004
25. Losick VP, Morris LX, Fox DT, Spradling A (2011) Drosophila stem cell niches: a decade of discovery suggests a unified view of stem cell regulation. Dev Cell 21(1):159–171. doi:10.1016/j.devcel.2011.06.018
26. Hedrich HJ (ed) (2012) The laboratory mouse. Handbook of experimental animals, 2nd edn. Academic, New York
27. Yuan R, Tsaih S-W, Petkova SB, de Evsikova CM, Xing S, Marion MA, Bogue MA, Mills KD, Peters LL, Bult CJ, Rosen CJ, Sundberg JP, Harrison DE, Churchill GA, Paigen B (2009) Aging in inbred strains of mice: study design and interim report on median lifespans and circulating IGF1 levels. Aging Cell 8:277–287
28. Miller RA, Nadon NL (2000) Principles of animal use for gerontological research. J Gerontol A Biol Sci Med Sci 55(3):B117–B123
29. Liang H, Masoro EJ, Nelson JF, Strong R, McMahan CA, Richardson A (2003) Genetic mouse models of extended lifespan. Exp Gerontol 38(11–12):1353–1364
30. Kastenmayer RJ, Fain MA, Perdue KA (2006) A retrospective study of idiopathic ulcerative dermatitis in mice with a C57BL/6 background. J Am Assoc Lab Anim Sci 45(6):8–12
31. Turturro A, Duffy P, Hass B, Kodell R, Hart R (2002) Survival characteristics and age-adjusted disease incidences in C57BL/6 mice fed a commonly used cereal-based diet modulated by dietary restriction. J Gerontol A Biol Sci Med Sci 57(11):B379–B389
32. Kasahara T, Abe K, Mekada K, Yoshiki A, Kato T (2010) Genetic variation of melatonin productivity in laboratory mice under domestication. Proc Natl Acad Sci U S A 107(14):6412–6417. doi:10.1073/pnas.0914399107
33. Ishimura R, Nagy G, Dotu I, Zhou H, Yang XL, Schimmel P, Senju S, Nishimura Y, Chuang JH, Ackerman SL (2014) RNA function. Ribosome stalling induced by mutation of a CNS-specific tRNA causes neurodegeneration. Science 345(6195):455–459. doi:10.1126/science.1249749
34. Miller RA, Harrison DE, Astle CM, Floyd RA, Flurkey K, Hensley KL, Javors MA, Leeuwenburgh C, Nelson JF, Ongini E, Nadon NL, Warner HR, Strong R (2007) An aging interventions testing program: study design and interim report. Aging Cell 6(4):565–575
35. Nadon NL, Strong R, Miller RA, Nelson J, Javors M, Sharp ZD, Peralba JM, Harrison DE (2008) Design of aging intervention studies: the NIA interventions testing program. Age 30(4):187–199
36. Spencer CC, Howell CE, Wright AR, Promislow DE (2003) Testing an 'aging gene' in long-lived drosophila strains: increased longevity depends on sex and genetic background. Aging Cell 2(2):123–130
37. Kortschak RD, Samuel G, Saint R, Miller DJ (2003) EST analysis of the cnidarian Acropora millepora reveals extensive gene loss and rapid sequence divergence in the model invertebrates. Curr Biol 13(24):2190–2195
38. Austad SN (2009) Is there a role for new invertebrate models for aging research? J Gerontol A Biol Sci Med Sci 64(2):192–194

39. Harel I, Benayoun BA, Machado B, Singh PP, Hu CK, Pech MF, Valenzano DR, Zhang E, Sharp SC, Artandi SE, Brunet A (2015) A platform for rapid exploration of aging and diseases in a naturally short-lived vertebrate. Cell 160(5):1013–1026. doi:10.1016/j.cell.2015.01.038
40. Shineman DW, Basi GS, Bizon JL, Colton CA, Greenberg BD, Hollister BA, Lincecum J, Leblanc GG, Lee LB, Luo F, Morgan D, Morse I, Refolo LM, Riddell DR, Scearce-Levie K, Sweeney P, Yrjanheikki J, Fillit HM (2011) Accelerating drug discovery for Alzheimer's disease: best practices for preclinical animal studies. Alzheimers Res Ther 3(5):28. doi:10.1186/alzrt90
41. Hall AM, Roberson ED (2012) Mouse models of Alzheimer's disease. Brain Res Bull 88(1):3–12. doi:10.1016/j.brainresbull.2011.11.017
42. Janssen JC, Beck JA, Campbell TA, Dickinson A, Fox NC, Harvey RJ, Houlden H, Rossor MN, Collinge J (2003) Early onset familial Alzheimer's disease: Mutation frequency in 31 families. Neurology 60(2):235–239
43. Costa DA, Cracchiolo JR, Bachstetter AD, Hughes TF, Bales KR, Paul SM, Mervis RF, Arendash GW, Potter H (2007) Enrichment improves cognition in AD mice by amyloid-related and unrelated mechanisms. Neurobiol Aging 28(6):831–844. doi:10.1016/j.neurobiolaging.2006.04.009
44. Berardi N, Braschi C, Capsoni S, Cattaneo A, Maffei L (2007) Environmental enrichment delays the onset of memory deficits and reduces neuropathological hallmarks in a mouse model of Alzheimer-like neurodegeneration. J Alzheimers Dis 11(3):359–370
45. Garcia-Mesa Y, Lopez-Ramos JC, Gimenez-Llort L, Revilla S, Guerra R, Gruart A, Laferla FM, Cristofol R, Delgado-Garcia JM, Sanfeliu C (2011) Physical exercise protects against Alzheimer's disease in 3xTg-AD mice. J Alzheimers Dis 24(3):421–454. doi:10.3233/JAD-2011-101635
46. Leon WC, Canneva F, Partridge V, Allard S, Ferretti MT, DeWilde A, Vercauteren F, Atifeh R, Ducatenzeiler A, Klein W, Szyf M, Alhonen L, Cuello AC (2010) A novel transgenic rat model with a full Alzheimer's-like amyloid pathology displays pre-plaque intracellular amyloid-beta-associated cognitive impairment. J Alzheimers Dis 20(1):113–126. doi:10.3233/JAD-2010-1349
47. Hsu PD, Lander ES, Zhang F (2014) Development and applications of CRISPR-Cas9 for genome engineering. Cell 157(6):1262–1278. doi:10.1016/j.cell.2014.05.010
48. Finch CE, Austad SN (2015) Commentary: is Alzheimer's disease uniquely human? Neurobiol Aging 36(2):553–555. doi:10.1016/j.neurobiolaging.2014.10.025
49. Austad SN (2009) Comparative biology of aging. J Gerontol A Biol Sci Med Sci 64(2):199–201
50. Tian X, Azpurua J, Hine C, Vaidya A, Myakishev-Rempel M, Ablaeva J, Mao Z, Nevo E, Gorbunova V, Seluanov A (2013) High-molecular-mass hyaluronan mediates the cancer resistance of the naked mole rat. Nature. nature12234 [pii]. doi:10.1038/nature12234
51. Ungvari Z, Buffenstein R, Austad SN, Podlutsky A, Kaley G, Csiszar A (2008) Oxidative stress in vascular senescence: lessons from successfully aging species. Front Biosci 13:5056–5070
52. Austad SN (2010) Methusaleh's Zoo: how nature provides us with clues for extending human health span. J Comp Pathol 142(Suppl 1):S10–S21
53. McCay CM, Crowell MF, Maynard LA (1935) The effect of retarded growth upon the length of the life span and upon ultimate body size. J Nutr 13:669–679
54. Weindruch R, Walford RL (1988) The retardation of aging and disease by dietary restriction. Charles C. Thomas, Springfield
55. Mattison JA, Roth GS, Beasley TM, Tilmont EM, Handy AM, Herbert RL, Longo DL, Allison DB, Young JE, Bryant M, Barnard D, Ward WF, Qi W, Ingram DK, de Cabo R (2012) Impact of caloric restriction on health and survival in rhesus monkeys from the NIA study. Nature 489(7415):318–321. nature11432 [pii]. doi:10.1038/nature11432
56. Colman RJ, Beasley TM, Kemnitz JW, Johnson SC, Weindruch R, Anderson RM (2014) Caloric restriction reduces age-related and all-cause mortality in rhesus monkeys. Nat Commun 5:3557. doi:10.1038/ncomms4557

57. Austad SN (2012) Ageing: mixed results for dieting monkeys. Nature 489(7415):210–211. nature11484 [pii]. doi:10.1038/nature11484
58. Austad SN, Kristan DM (2003) Are mice calorically restricted in nature? Aging Cell 2(4):201–207
59. Piper MD, Partridge L (2007) Dietary restriction in Drosophila: delayed aging or experimental artefact? PLoS Genet 3(4):e57. doi:10.1371/journal.pgen.0030057
60. Harman D (1956) Aging: a theory based on free radical and radiation chemistry. J Gerontol 11(3):298–300
61. Duffy PH, Feuers RJ, Hart RW (1990) Effect of chronic caloric restriction on the circadian regulation of physiological and behavioral variables in old male B6C3F1 mice. Chronobiol Int 7(4):291–303
62. McCarter RJ, Palmer J (1992) Energy metabolism and aging: a lifelong study of Fischer 344 rats. Am J Physiol 263(3 Pt 1):E448–E452
63. Sohal RS, Weindruch R (1996) Oxidative stress, caloric restriction, and aging. Science 273(5271):59–63
64. Perez VI, Bokov A, Remmen HV, Mele J, Ran Q, Ikeno Y, Richardson A (2009) Is the oxidative stress theory of aging dead? Biochim Biophys Acta 1790(10):1005–1014
65. Sapolsky RM, Krey LC, McEwen BS (1986) The neuroendocrinology of stress and aging: the glucocorticoid cascade hypothesis. Endocr Rev 7(3):284–301. doi:10.1210/edrv-7-3-284
66. Sabatino F, Masoro EJ, McMahan CA, Kuhn RW (1991) Assessment of the role of the glucocorticoid system in aging processes and in the action of food restriction. J Gerontol 46(B171):B179
67. Cerami A (1985) Hypothesis. Glucose as a mediator of aging. J Am Geriatr Soc 33(9):626–634
68. Masoro EJ (1996) Possible mechanisms underlying the antiaging actions of caloric restriction. Toxicol Pathol 24(6):738–741
69. Cabelof DC, Yanamadala S, Raffoul JJ, Guo Z, Soofi A, Heydari AR (2003) Caloric restriction promotes genomic stability by induction of base excision repair and reversal of its age-related decline. DNA Repair (Amst) 2(3):295–307
70. Heydari AR, You S, Takahashi R, Gutsmann A, Sarge KD, Richardson A (1996) Effect of caloric restriction on the expression of heat shock protein 70 and the activation of heat shock transcription factor 1. Dev Genet 18(2):114–124
71. Rao KS (2003) Dietary calorie restriction, DNA-repair and brain aging. Mol Cell Biochem 253(1–2):313–318
72. Frier B, Locke M (2005) Preservation of heat stress induced myocardial hsp 72 in aged animals following caloric restriction. Exp Gerontol 40(7):615–617
73. Hart RW, Leakey JE, Chou M, Duffy PH, Allaben WT, Feuers RJ (1992) Modulation of chemical toxicity by modification of caloric intake. Adv Exp Med Biol 322:73–81
74. Maglich JM, Watson J, McMillen PJ, Goodwin B, Willson TM, Moore JT (2004) The nuclear receptor CAR is a regulator of thyroid hormone metabolism during caloric restriction. J Biol Chem 279(19):19832–19838
75. Hart RW, Keenan K, Turturro A, Abdo KM, Leakey J, Lyn-Cook B (1995) Caloric restriction and toxicity. Fundam Appl Toxicol 25(2):184–195
76. Masoro EJ, Iwasaki K, Gleiser CA, McMahan CA, Seo EJ, Yu BP (1989) Dietary modulation of the progression of nephropathy in aging rats: an evaluation of the importance of protein. Am J Clin Nutr 49(6):1217–1227
77. Shimokawa I, Higami Y, Hubbard GB, McMahan CA, Masoro EJ, Yu BP (1993) Diet and the suitability of the male Fischer 344 rat as a model for aging research. J Gerontol 48(1):B27–B32
78. Zimmerman JA, Malloy V, Krajcik R, Orentreich N (2003) Nutritional control of aging. Exp Gerontol 38(1–2):47–52
79. Miller RA, Buehner G, Chang Y, Harper JM, Sigler R, Smith-Wheelock M (2005) Methionine-deficient diet extends mouse lifespan, slows immune and lens aging, alters glucose, T4, IGF-I and insulin levels, and increases hepatocyte MIF levels and stress resistance. Aging Cell 4(3):119–125

80. Solon-Biet SM, McMahon AC, Ballard JW, Ruohonen K, Wu LE, Cogger VC, Warren A, Huang X, Pichaud N, Melvin RG, Gokarn R, Khalil M, Turner N, Cooney GJ, Sinclair DA, Raubenheimer D, Le Couteur DG, Simpson SJ (2014) The ratio of macronutrients, not caloric intake, dictates cardiometabolic health, aging, and longevity in ad libitum-fed mice. Cell Metab 19(3):418–430. doi:10.1016/j.cmet.2014.02.009
81. Bartke A, Wright JC, Mattison JA, Ingram DK, Miller RA, Roth GS (2001) Extending the lifespan of long-lived mice. Nature 414(6862):412
82. Liao CY, Rikke BA, Johnson TE, Diaz V, Nelson JF (2010) Genetic variation in the murine lifespan response to dietary restriction: from life extension to life shortening. Aging Cell 9(1):92–95. ACE533 [pii]. doi:10.1111/j.1474-9726.2009.00533.x
83. Rikke BA, Liao CY, McQueen MB, Nelson JF, Johnson TE (2010) Genetic dissection of dietary restriction in mice supports the metabolic efficiency model of life extension. Exp Gerontol 45(9):691–701. doi:10.1016/j.exger.2010.04.008
84. Schleit J, Johnson SC, Bennett CF, Simko M, Trongtham N, Castanza A, Hsieh EJ, Moller RM, Wasko BM, Delaney JR, Sutphin GL, Carr D, Murakami CJ, Tocchi A, Xian B, Chen W, Yu T, Goswami S, Higgins S, Holmberg M, Jeong KS, Kim JR, Klum S, Liao E, Lin MS, Lo W, Miller H, Olsen B, Peng ZJ, Pollard T, Pradeep P, Pruett D, Rai D, Ros V, Singh M, Spector BL, Vander Wende H, An EH, Fletcher M, Jelic M, Rabinovitch PS, MacCoss MJ, Han JD, Kennedy BK, Kaeberlein M (2013) Molecular mechanisms underlying genotype-dependent responses to dietary restriction. Aging Cell 12(6):1050–1061. doi:10.1111/acel.12130
85. Klass MR (1977) Aging in the nematode Caenorhabditis elegans: major biological and environmental factors influencing life span. Mech Ageing Dev 6(6):413–429
86. Kopec S (1928) On the influence of intermittent starvation on the longevity of the imaginal stage of *Drosophila melanogaster*. Brit J Exp Biol 5:8
87. Turturro A, Witt WW, Lewis S, Hass BS, Lipman RD, Hart RW (1999) Growth curves and survival characteristics of the animals used in the Biomarkers of Aging Program. J Gerontol A Biol Sci Med Sci 54(11):B492–B501
88. Tatar M, Post S, Yu K (2014) Nutrient control of Drosophila longevity. Trends Endocrinol Metab 25(10):509–517. doi:10.1016/j.tem.2014.02.006
89. Carvalho GB, Kapahi P, Benzer S (2005) Compensatory ingestion upon dietary restriction in Drosophila melanogaster. Nat Methods 2(11):813–815
90. Min KJ, Tatar M (2006) Restriction of amino acids extends lifespan in Drosophila melanogaster. Mech Ageing Dev 127(7):643–646
91. Grandison RC, Piper MD, Partridge L (2009) Amino-acid imbalance explains extension of lifespan by dietary restriction in Drosophila. Nature 462(7276):1061–1064. doi:10.1038/nature08619
92. Piper MD, Mair W, Partridge L (2005) Counting the calories: the role of specific nutrients in extension of life span by food restriction. J Gerontol A Biol Sci Med Sci 60(5):549–555
93. Piper MD, Blanc E, Leitao-Goncalves R, Yang M, He X, Linford NJ, Hoddinott MP, Hopfen C, Soultoukis GA, Niemeyer C, Kerr F, Pletcher SD, Ribeiro C, Partridge L (2014) A holidic medium for Drosophila melanogaster. Nat Methods 11(1):100–105. doi:10.1038/nmeth.2731
94. Libert S, Zwiener J, Chu X, Vanvoorhies W, Roman G, Pletcher SD (2007) Regulation of Drosophila life span by olfaction and food-derived odors. Science 315(5815):1133–1137. doi:10.1126/science.1136610
95. Clancy DJ, Gems D, Harshman LG, Oldham S, Stocker H, Hafen E, Leevers SJ, Partridge L (2001) Extension of life-span by loss of CHICO, a Drosophila insulin receptor substrate protein. Science 292(5514):104–106
96. Clancy DJ, Gems D, Hafen E, Leevers SJ, Partridge L (2002) Dietary restriction in long-lived dwarf flies. Science 296(5566):319. doi:10.1126/science.1069366
97. Gems D, Riddle DL (2000) Defining wild-type life span in Caenorhabditis elegans. J Gerontol A Biol Sci Med Sci 55(5):B215–B219
98. Garigan D, Hsu AL, Fraser AG, Kamath RS, Ahringer J, Kenyon C (2002) Genetic analysis of tissue aging in Caenorhabditis elegans: a role for heat-shock factor and bacterial proliferation. Genetics 161(3):1101–1112

99. Houthoofd K, Gems D, Johnson TE, Vanfleteren JR (2007) Dietary restriction in the nematode Caenorhabditis elegans. Interdiscip Top Gerontol 35:98–114
100. Greer E, Brunet A (2011) The genetic network of life-span extension by dietary restriction. In: Masoro EJ, Austad SN (eds) Handbook of the biology of aging, 7th edn. Academic, San Diego, pp 3–24
101. Mair W, Panowski SH, Shaw RJ, Dillin A (2009) Optimizing dietary restriction for genetic epistasis analysis and gene discovery in C. elegans. PLoS One 4(2):e4535. doi:10.1371/journal.pone.0004535
102. Tatar M, Kopelman A, Epstein D, Tu MP, Yin CM, Garofalo RS (2001) A mutant Drosophila insulin receptor homolog that extends life-span and impairs neuroendocrine function. Science 292(5514):107–110
103. Brown-Borg HM, Borg KE, Meliska CJ, Bartke A (1996) Dwarf mice and the ageing process. Nature 384(6604):33
104. Flurkey K, Papaconstantinou J, Miller RA, Harrison DE (2001) Lifespan extension and delayed immune and collagen aging in mutant mice with defects in growth hormone production. Proc Natl Acad Sci USA 98(12):6736–6741
105. Coschigano KT, Clemmons D, Bellush LL, Kopchick JJ (2000) Assessment of growth parameters and life span of GHR/BP gene-disrupted mice. Endocrinology 141(7):2608–2613
106. Suh Y, Atzmon G, Cho MO, Hwang D, Liu B, Leahy DJ, Barzilai N, Cohen P (2008) Functionally significant insulin-like growth factor I receptor mutations in centenarians. Proc Natl Acad Sci USA 105(9):3438–3442
107. Van Voorhies WA, Fuchs J, Thomas S (2005) The longevity of Caenorhabditis elegans in soil. Biol Lett 1(2):247–249
108. Fabris N, Pierpaoli W, Sorkin E (1972) Lymphocytes, hormones and ageing. Nature 240(5383):557–559
109. Coschigano KT, Holland AN, Riders ME, List EO, Flyvbjerg A, Kopchick JJ (2003) Deletion, but not antagonism, of the mouse growth hormone receptor results in severely decreased body weights, insulin, and insulin-like growth factor I levels and increased life span. Endocrinology 144(9):3799–3810
110. Bonkowski MS, Rocha JS, Masternak MM, Al Regaiey KA, Bartke A (2006) Targeted disruption of growth hormone receptor interferes with the beneficial actions of calorie restriction. Proc Natl Acad Sci USA 103(20):7901–7905
111. Sun LY, Spong A, Swindell WR, Fang Y, Hill C, Huber JA, Boehm JD, Westbrook R, Salvatori R, Bartke A (2013) Growth hormone-releasing hormone disruption extends lifespan and regulates response to caloric restriction in mice. Elife 2:e01098. doi:10.7554/eLife.01098
112. Bokov AF, Garg N, Ikeno Y, Thakur S, Musi N, DeFronzo RA, Zhang N, Erickson RC, Gelfond J, Hubbard GB, Adamo ML, Richardson A (2011) Does reduced IGF-1R signaling in Igf1r+/– mice alter aging? PLoS One 6(11):e26891. PONE-D-11-07865 [pii]. doi:10.1371/journal.pone.0026891
113. Xu J, Gontier G, Chaker Z, Lacube P, Dupont J, Holzenberger M (2014) Longevity effect of IGF-1R(+/–) mutation depends on genetic background-specific receptor activation. Aging Cell 13(1):19–28. doi:10.1111/acel.12145
114. Selman C, Lingard S, Choudhury AI, Batterham RL, Claret M, Clements M, Ramadani F, Okkenhaug K, Schuster E, Blanc E, Piper MD, Al-Qassab H, Speakman JR, Carmignac D, Robinson IC, Thornton JM, Gems D, Partridge L, Withers DJ (2008) Evidence for lifespan extension and delayed age-related biomarkers in insulin receptor substrate 1 null mice. FASEB J 22(3):807–818
115. Taguchi A, Wartschow LM, White MF (2007) Brain IRS2 signaling coordinates life span and nutrient homeostasis. Science 317(5836):369–372
116. Bluher M, Kahn BB, Kahn CR (2003) Extended longevity in mice lacking the insulin receptor in adipose tissue. Science 299(5606):572–574
117. Dazert E, Hall MN (2011) mTOR signaling in disease. Curr Opin Cell Biol 23(6):744–755. doi:10.1016/j.ceb.2011.09.003
118. Kapahi P, Chen D, Rogers AN, Katewa SD, Li PW, Thomas EL, Kockel L (2010) With TOR, less is more: a key role for the conserved nutrient-sensing TOR pathway in aging. Cell Metab 11(6):453–465. S1550-4131(10)00153-1 [pii]. doi:10.1016/j.cmet.2010.05.001

119. Vellai T, Takacs-Vellai K, Zhang Y, Kovacs AL, Orosz L, Muller F (2003) Genetics: influence of TOR kinase on lifespan in C. elegans. Nature 426(6967):620
120. Kapahi P, Zid BM, Harper T, Koslover D, Sapin V, Benzer S (2004) Regulation of lifespan in Drosophila by modulation of genes in the TOR signaling pathway. Curr Biol 14(10):885–890
121. Selman C, Tullet JM, Wieser D, Irvine E, Lingard SJ, Choudhury AI, Claret M, Al-Qassab H, Carmignac D, Ramadani F, Woods A, Robinson IC, Schuster E, Batterham RL, Kozma SC, Thomas G, Carling D, Okkenhaug K, Thornton JM, Partridge L, Gems D, Withers DJ (2009) Ribosomal protein S6 kinase 1 signaling regulates mammalian life span. Science 326(5949):140–144
122. Emran S, Yang M, He X, Zandveld J, Piper MD (2014) Target of rapamycin signalling mediates the lifespan-extending effects of dietary restriction by essential amino acid alteration. Aging 6(5):390–398
123. Sun LY, Bartke A (2007) Adult neurogenesis in the hippocampus of long-lived mice during aging. J Gerontol A Biol Sci Med Sci 62(2):117–125
124. Martin GM, Austad SN, Johnson TE (1996) Genetic analysis of ageing: role of oxidative damage and environmental stresses. Nat Genet 13(1):25–34
125. Benedetti MG, Foster AL, Vantipalli MC, White MP, Sampayo JN, Gill MS, Olsen A, Lithgow GJ (2008) Compounds that confer thermal stress resistance and extended lifespan. Exp Gerontol 43(10):882–891. S0531-5565(08)00289-1 [pii]. doi:10.1016/j.exger.2008.08.049
126. Burger JM, Promislow DE (2006) Are functional and demographic senescence genetically independent? Exp Gerontol 41(11):1108–1116
127. Bansal A, Zhu LJ, Yen K, Tissenbaum HA (2015) Uncoupling lifespan and healthspan in Caenorhabditis elegans longevity mutants. Proc Natl Acad Sci U S A 112(3):E277–E286. doi:10.1073/pnas.1412192112
128. Harman D (1961) Prolongation of the normal lifespan and inhibition of spontaneous cancer by antioxidants. J Gerontol 16:247–254
129. Milgram NW, Racine RJ, Nellis P, Mendonca A, Ivy GO (1990) Maintenance on L-deprenyl prolongs life in aged male rats. Life Sci 47(5):415–420
130. Carney JM, Starke-Reed PE, Oliver CN, Landum RW, Cheng MS, Wu JF, Floyd RA (1991) Reversal of age-related increase in brain protein oxidation, decrease in enzyme activity, and loss in temporal and spatial memory by chronic administration of the spin-trapping compound N-tert-butyl-alpha-phenylnitrone. Proc Natl Acad Sci U S A 88(9):3633–3636
131. Edamatsu R, Mori A, Packer L (1995) The spin-trap N-tert-alpha-phenyl-butylnitrone prolongs the life span of the senescence accelerated mouse. Biochem Biophys Res Commun 211(3):847–849. doi:10.1006/bbrc.1995.1889
132. Baur JA, Pearson KJ, Price NL, Jamieson HA, Lerin C, Kalra A, Prabhu VV, Allard JS, Lopez-Lluch G, Lewis K, Pistell PJ, Poosala S, Becker KG, Boss O, Gwinn D, Wang M, Ramaswamy S, Fishbein KW, Spencer RG, Lakatta EG, Le CD, Shaw RJ, Navas P, Puigserver P, Ingram DK, de Cabo R, Sinclair DA (2006) Resveratrol improves health and survival of mice on a high-calorie diet. Nature 444(7117):337–342
133. Migliaccio E, Giorgio M, Mele S, Pelicci G, Reboldi P, Pandolfi PP, Lanfrancone L, Pelicci PG (1999) The p66shc adaptor protein controls oxidative stress response and life span in mammals. Nature 402(6759):309–313. doi:10.1038/46311
134. Ramsey JJ, Tran D, Giorgio M, Griffey SM, Koehne A, Laing ST, Taylor SL, Kim K, Cortopassi GA, Lloyd KC, Hagopian K, Tomilov AA, Migliaccio E, Pelicci PG, McDonald RB (2014) The influence of Shc proteins on life span in mice. J Gerontol A Biol Sci Med Sci 69(10):1177–1185. doi:10.1093/gerona/glt198
135. Strong R, Miller RA, Astle CM, Floyd RA, Flurkey K, Hensley KL, Javors MA, Leeuwenburgh C, Nelson JF, Ongini E (2008) Nordihydroguaiaretic acid and aspirin increase lifespan of genetically heterogeneous male mice. Aging Cell 7(5):641–650
136. Group AI (2013) Study design of ASPirin in Reducing Events in the Elderly (ASPREE): a randomized, controlled trial. Contemp Clin Trials 36(2):555–564. doi:10.1016/j.cct.2013.09.014

137. Strong R, Miller RA, Astle CM, Floyd RA, Flurkey K, Hensley KL, Javors MA, Leeuwenburgh C, Nelson JF, Ongini E, Nadon NL, Warner HR, Harrison DE (2008) Nordihydroguaiaretic acid and aspirin increase lifespan of genetically heterogeneous male mice. Aging Cell 7(5):641–650. doi:10.1111/j.1474-9726.2008.00414.x
138. Harrison DE, Strong R, Allison DB, Ames BN, Astle CM, Atamna H, Fernandez E, Flurkey K, Javors MA, Nadon NL, Nelson JF, Pletcher S, Simpkins JW, Smith D, Wilkinson JE, Miller RA (2014) Acarbose, 17-alpha-estradiol, and nordihydroguaiaretic acid extend mouse lifespan preferentially in males. Aging Cell 13(2):273–282. doi:10.1111/acel.12170
139. Moos W, Dykens JA, Nohynek D, Rubinchik E, Howell N (2009) Review of the effects of 17α−estrakiol in humans: a less feminizing estrogen with neuroprotective potential. Drug Dev Res 70:20
140. Harrison DE, Strong R, Sharp ZD, Nelson JF, Astle CM, Flurkey K, Nadon NL, Wilkinson JE, Frenkel K, Carter CS, Pahor M, Javors MA, Fernandez E, Miller RA (2009) Rapamycin fed late in life extends lifespan in genetically heterogeneous mice. Nature 460(7253):392–395
141. Charbonnier LM, Le MA (2012) Rapamycin as immunosuppressant in murine transplantation model. Methods Mol Biol 821:435–445. doi:10.1007/978-1-61779-430-8_28
142. Liu JY, Song M, Guo M, Huang F, Ma BJ, Zhu L, Xu G, Li J, You RX (2015) Sirolimus versus tacrolimus as primary immunosuppressant after renal transplantation: a meta-analysis and economics evaluation. Am J Ther. doi:10.1097/MJT.0000000000000186
143. Hall MN (2008) mTOR-what does it do? Transplant Proc 40(10 Suppl):S5–S8. S0041-1345(08)01361-4 [pii]. doi:10.1016/j.transproceed.2008.10.009
144. Adelman SJ (2010) Sirolimus and its analogs and its effects on vascular diseases. Curr Pharm Des 16(36):4002–4011, BSP/CPD/E-Pub/000283 [pii]
145. SoRelle R (2004) Everolimus stent proves effective. Circulation 109(18):e9041–e9042. doi:10.1161/01.CIR.0000132284.04047.55
146. Wullschleger S, Loewith R, Hall MN (2006) TOR signaling in growth and metabolism. Cell 124(3):471–484. S0092-8674(06)00108-5 [pii]. doi:10.1016/j.cell.2006.01.016
147. Thoreen CC, Sabatini DM (2009) Rapamycin inhibits mTORC1, but not completely. Autophagy 5(5):725–726
148. Chen C, Liu Y, Liu Y, Zheng P (2009) mTOR regulation and therapeutic rejuvenation of aging hematopoietic stem cells. Sci Signal 2(98):ra75
149. Anisimov VN, Zabezhinski MA, Popovich IG, Piskunova TS, Semenchenko AV, Tyndyk ML, Yurova MN, Rosenfeld SV, Blagosklonny MV (2011) Rapamycin increases lifespan and inhibits spontaneous tumorigenesis in inbred female mice. Cell Cycle 10(24):4230–4236. 18486 [pii]. doi:10.4161/cc.10.24.18486
150. Fok WC, Chen Y, Bokov A, Zhang Y, Salmon AB, Diaz V, Javors M, Wood WH 3rd, Zhang Y, Becker KG, Perez VI, Richardson A (2014) Mice fed rapamycin have an increase in lifespan associated with major changes in the liver transcriptome. PLoS One 9(1):e83988. doi:10.1371/journal.pone.0083988
151. Neff F, Flores-Dominguez D, Ryan DP, Horsch M, Schroder S, Adler T, Afonso LC, Aguilar-Pimentel JA, Becker L, Garrett L, Hans W, Hettich MM, Holtmeier R, Holter SM, Moreth K, Prehn C, Puk O, Racz I, Rathkolb B, Rozman J, Naton B, Ordemann R, Adamski J, Beckers J, Bekeredjian R, Busch DH, Ehninger G, Graw J, Hofler H, Klingenspor M, Klopstock T, Ollert M, Stypmann J, Wolf E, Wurst W, Zimmer A, Fuchs H, Gailus-Durner V, Hrabe de Angelis M, Ehninger D (2013) Rapamycin extends murine lifespan but has limited effects on aging. J Clin Invest 123(8):3272–3291. doi:10.1172/JCI67674
152. Miller RA, Harrison DE, Astle CM, Fernandez E, Flurkey K, Han M, Javors MA, Li X, Nadon NL, Nelson JF, Pletcher S, Salmon AB, Sharp ZD, Van Roekel S, Winkleman L, Strong R (2014) Rapamycin-mediated lifespan increase in mice is dose and sex dependent and metabolically distinct from dietary restriction. Aging Cell 13(3):468–477. doi:10.1111/acel.12194
153. Spilman P, Podlutskaya N, Hart MJ, Debnath J, Gorostiza O, Bredesen D, Richardson A, Strong R, Galvan V (2010) Inhibition of mTOR by rapamycin abolishes cognitive deficits and reduces amyloid-beta levels in a mouse model of Alzheimer's disease. PLoS One 5(4):e9979

154. Caccamo A, Majumder S, Richardson A, Strong R, Oddo S (2010) Molecular interplay between mammalian target of rapamycin (mTOR), amyloid-beta, and Tau: effects on cognitive impairments. J Biol Chem 285(17):13107–13120. M110.100420 [pii]. doi:10.1074/jbc.M110.100420
155. Halloran J, Hussong SA, Burbank R, Podlutskaya N, Fischer KE, Sloane LB, Austad SN, Strong R, Richardson A, Hart MJ, Galvan V (2012) Chronic inhibition of mammalian target of rapamycin by rapamycin modulates cognitive and non-cognitive components of behavior throughout lifespan in mice. Neuroscience 223C:102–113. S0306-4522(12)00672-0 [pii]. doi:10.1016/j.neuroscience.2012.06.054
156. Ravikumar B, Vacher C, Berger Z, Davies JE, Luo S, Oroz LG, Scaravilli F, Easton DF, Duden R, O'Kane CJ, Rubinsztein DC (2004) Inhibition of mTOR induces autophagy and reduces toxicity of polyglutamine expansions in fly and mouse models of Huntington disease. Nat Genet 36(6):585–595. doi:10.1038/ng1362
157. Pakala R, Stabile E, Jang GJ, Clavijo L, Waksman R (2005) Rapamycin attenuates atherosclerotic plaque progression in apolipoprotein E knockout mice: inhibitory effect on monocyte chemotaxis. J Cardiovasc Pharmacol 46(4):481–486
158. Beutner F, Brendel D, Teupser D, Sass K, Baber R, Mueller M, Ceglarek U, Thiery J (2012) Effect of everolimus on pre-existing atherosclerosis in LDL-receptor deficient mice. Atherosclerosis 222(2):337–343. doi:10.1016/j.atherosclerosis.2012.03.003
159. Livi CB, Hardman RL, Christy BA, Dodds SG, Jones D, Williams C, Strong R, Bokov A, Javors MA, Ikeno Y, Hubbard G, Hasty P, Sharp ZD (2013) Rapamycin extends life span of Rb1+/− mice by inhibiting neuroendocrine tumors. Aging 5(2):100–110
160. Hasty P, Livi CB, Dodds SG, Jones D, Strong R, Javors M, Fischer KE, Sloane L, Murthy K, Hubbard G, Sun L, Hurez V, Curiel TJ, Sharp ZD (2014) eRapa restores a normal life span in a FAP mouse model. Cancer Prev Res (Phila) 7(1):169–178. doi:10.1158/1940-6207.CAPR-13-0299
161. Ramos FJ, Chen SC, Garelick MG, Dai DF, Liao CY, Schreiber KH, MacKay VL, An EH, Strong R, Ladiges WC, Rabinovitch PS, Kaeberlein M, Kennedy BK (2012) Rapamycin reverses elevated mTORC1 signaling in lamin A/C-deficient mice, rescues cardiac and skeletal muscle function, and extends survival. Sci Transl Med 4(144):144ra103. 4/144/144ra103 [pii]. doi:10.1126/scitranslmed.3003802
162. Keating R, Hertz T, Wehenkel M, Harris TL, Edwards BA, McClaren JL, Brown SA, Surman S, Wilson ZS, Bradley P, Hurwitz J, Chi H, Doherty PC, Thomas PG, McGargill MA (2013) The kinase mTOR modulates the antibody response to provide cross-protective immunity to lethal infection with influenza virus. Nat Immunol 14(12):1266–1276. doi:10.1038/ni.2741
163. Zhang Y, Bokov A, Gelfond J, Soto V, Ikeno Y, Hubbard G, Diaz V, Sloane L, Maslin K, Treaster S, Rendon S, van Remmen H, Ward W, Javors M, Richardson A, Austad SN, Fischer K (2014) Rapamycin extends life and health in C57BL/6 mice. J Gerontol A Biol Sci Med Sci 69(2):119–130. doi:10.1093/gerona/glt056
164. Flynn JM, O'Leary MN, Zambataro CA, Academia EC, Presley MP, Garrett BJ, Zykovich A, Mooney SD, Strong R, Rosen CJ, Kapahi P, Nelson MD, Kennedy BK, Melov S (2013) Late-life rapamycin treatment reverses age-related heart dysfunction. Aging Cell 12(5):851–862. doi:10.1111/acel.12109
165. Fraenkel M, Ketzinel-Gilad M, Ariav Y, Pappo O, Karaca M, Castel J, Berthault MF, Magnan C, Cerasi E, Kaiser N, Leibowitz G (2008) mTOR inhibition by rapamycin prevents beta-cell adaptation to hyperglycemia and exacerbates the metabolic state in type 2 diabetes. Diabetes 57(4):945–957. doi:10.2337/db07-0922
166. Houde VP, Brule S, Festuccia WT, Blanchard PG, Bellmann K, Deshaies Y, Marette A (2010) Chronic rapamycin treatment causes glucose intolerance and hyperlipidemia by upregulating hepatic gluconeogenesis and impairing lipid deposition in adipose tissue. Diabetes 59(6):1338–1348. doi:10.2337/db09-1324
167. Liu Y, Diaz V, Fernandez E, Strong R, Ye L, Baur JA, Lamming DW, Richardson A, Salmon AB (2014) Rapamycin-induced metabolic defects are reversible in both lean and obese mice. Aging 6(9):742–754

168. Fang Y, Westbrook R, Hill C, Boparai RK, Arum O, Spong A, Wang F, Javors MA, Chen J, Sun LY, Bartke A (2013) Duration of rapamycin treatment has differential effects on metabolism in mice. Cell Metab 17(3):456–462. doi:10.1016/j.cmet.2013.02.008
169. Bruno L, Merkenschlager M (2008) Directing T cell differentiation and function with small molecule inhibitors. Cell Cycle 7(15):2296–2298
170. Kim KW, Chung BH, Kim BM, Cho ML, Yang CW (2015) The effect of mammalian target of rapamycin inhibition on T helper type 17 and regulatory T cell differentiation in vitro and in vivo in kidney transplant recipients. Immunology 144(1):68–78. doi:10.1111/imm.12351
171. Goldberg EL, Smithey MJ, Lutes LK, Uhrlaub JL, Nikolich-Zugich J (2014) Immune memory-boosting dose of rapamycin impairs macrophage vesicle acidification and curtails glycolysis in effector CD8 cells, impairing defense against acute infections. J Immunol 193(2):757–763. doi:10.4049/jimmunol.1400188
172. Mannick JB, Del Giudice G, Lattanzi M, Valiante NM, Praestgaard J, Huang B, Lonetto MA, Maecker HT, Kovarik J, Carson S, Glass DJ, Klickstein LB (2014) mTOR inhibition improves immune function in the elderly. Sci Transl Med 6(268):268ra179. doi:10.1126/scitranslmed.3009892
173. Hinojosa CA, Mgbemena V, Van RS, Austad SN, Miller RA, Bose S, Orihuela CJ (2012) Enteric-delivered rapamycin enhances resistance of aged mice to pneumococcal pneumonia through reduced cellular senescence. Exp Gerontol 47(12):958–965. S0531-5565(12)00250-1 [pii]. doi:10.1016/j.exger.2012.08.013
174. Goldberg EL, Romero-Aleshire MJ, Renkema KR, Ventevogel MS, Chew WM, Uhrlaub JL, Smithey MJ, Limesand KH, Sempowski GD, Brooks HL, Nikolich-Zugich J (2015) Lifespan-extending caloric restriction or mTOR inhibition impair adaptive immunity of old mice by distinct mechanisms. Aging Cell 14(1):130–138. doi:10.1111/acel.12280
175. Sonntag WE, Carter CS, Ikeno Y, Ekenstedt K, Carlson CS, Loeser RF, Chakrabarty S, Lee S, Bennett C, Ingram R, Moore T, Ramsey M (2005) Adult-onset growth hormone and insulin-like growth factor I deficiency reduces neoplastic disease, modifies age-related pathology, and increases life span. Endocrinology 146(7):2920–2932. doi:10.1210/en.2005-0058

Etiological Role of Aging in Chronic Diseases: From Epidemiological Evidence to the New Geroscience

Linda P. Fried and Luigi Ferrucci

Contents

1 Premise – Evolution of the Science of Chronic Diseases and Current State of the Field

1.1 Introduction

According to Wikipedia, "*Epidemiology is the science that studies the patterns, causes, and effects of health and disease conditions in defined populations. It is the cornerstone of public health, and informs policy decisions and*

L.P. Fried
Mailman School of Public Health, Columbia University Medical Center, New York, NY, USA
e-mail: lpfried@columbia.edu

L. Ferrucci (✉)
Intramural Research Program, National Institute on Aging, Baltimore, MD, USA
e-mail: Ferruccilu@mail.nih.gov

F. Sierra, R. Kohanski (eds.), *Advances in Geroscience*,
DOI 10.1007/978-3-319-23246-1_2

evidence-based practice by identifying risk factors for disease and targets for preventive healthcare". Consistent with this definition, over the last few decades, epidemiological studies identified a number of genetic and environmental risk factors for the majority of chronic diseases. There is no doubt that epidemiology has contributed tremendously to both the science of understanding of disease and to the science of prevention, both of which are necessary to achieve population health. It is currently believed that the increased longevity in the population and the decline in cardiovascular morbidity and mortality resulted from interventions on risk targets that were first identified in epidemiological studies.

Notably, however, the research on etiology of chronic diseases, which initially was mostly on cardiovascular disease and cancer, has been carried out in ways that, inadvertently, precluded understanding their relationship with aging. Since age and sex were considered unchangeable risk factors, they were generally factored out from all analyses as "potential confounders". "Adjusting for age and sex" was almost a *mantra* for seasoned epidemiologists; this, unfortunately, translated into a lost opportunity. Indeed, age is by far the strongest and most pervasive risk factor for almost all chronic diseases and medical conditions. The effect of aging on the risk of developing cardiovascular disease, cancer, diabetes, osteoarthritis and dementia, just to cite a few diseases, is several magnitudes higher than the risk attributed to all other known risk factors. The idea of "adjusting for age" obscures consideration of the effect of age, and also overlooks the critical nuance that chronological age is a poor approximation of biological aging. There is increasing heterogeneity with age between individuals in the physical and functional consequences of the aging process, which probably results from differentials in exposures across the life course and the intrinsic rate of biological aging. Understanding how the intrinsic biological mechanisms of aging affect most aspects of health in humans is a fascinating scientific challenge that has captured the attention of the greatest scientific minds over the centuries. However, with the current aging of the population, estimating biological aging is now also recognized as important for practical clinical purposes. To some extent, geriatricians and gerontologists have approached this problem through the conceptualization and operational definition of frailty as a diagnosable clinical syndrome that is a hallmark of the aging process and is marked by susceptibility to stress, definable biology, underlying loss of resiliency and diminished functional reserve. However, as research on the biology of aging in animal models progresses, it complements the work on mechanisms of aging-related dysregulation in humans; the two lines of investigation together suggest that a core set of mechanisms may reside at the basis of aging and resulting frailty. These same mechanisms may also contribute to disease and may be modifiable with appropriate interventions. In this chapter, we propose the idea that many chronic diseases in older age and frailty both originate, at least in part, from accelerated aging, and may mutually precipitate or exacerbate one another. We propose that this concept has enormous translational potential and is consistent with the new evidence emerging from the fields of Geroscience and Precision Medicine.

1.2 *Traditional Approach to Chronic Diseases*

The study of chronic diseases, etiological mechanisms, signs, symptoms and consequences, as well as potential treatments, has been focused primarily on analysis of organ-specific diseases and their clinical manifestations. Over the last 60 years, science has gone through a number of stages of such analysis and evidence. This progression began with population-based epidemiological studies that described the prevalence and incidence of chronic diseases, identified their etiologic risk factors and mechanisms, and led to the development – and evaluation – of clinical and population-based interventions, from Coronary Care Units to behavioral and pharmacologic therapies and primary prevention initiatives. Further, epidemiologic investigation led to evidence that there were independent predictors, namely environmental and behavioral risk factors, for specific chronic diseases that were potentially modifiable. Randomized controlled trials have shown that modification of such risk factors resulted in substantial primary prevention of morbidity and mortality. At the same time, intrinsic risk factors, such as the presence of hypertension, diabetes or elevated LDL cholesterol, have been shown to be predictive of subsequent cardiovascular disease and that screening for early identification of these conditions can lead to behavioral change or treatments that are effective in primary or secondary prevention of subsequent disease. Clinical and community-based guidelines, as well as health policies, have gone on to implement these recommendations on a population scale. These advances in knowledge and delivery of public health and medical science have been followed by a dramatic decline in cardiovascular morbidity and mortality. Age-adjusted death rates per 100,000 persons in the U.S. decreased from a peak of 307.4 in 1950 to 134.6 in 1996, an overall decline of 56 %; and they are continuing to decline today [1].

Overall, much population-based and clinical research has demonstrated that significant portions of chronic disease mortality and even the incidence of morbidity and resulting disability are either preventable, or can be delayed in onset. Following this line of research, geriatricians and gerontologists hypothesized that interventions could be developed to promote healthy and active aging, and that those interventions would include – but not be limited to – the primary and secondary prevention of chronic diseases [2]. The ultimate aim of those interventions is the "compression of morbidity" to the latest years in the human life span, including the delay of chronic disease morbidity and the onset of physical and cognitive disability. There is now a substantial literature to support the effectiveness of prevention of chronic diseases into the oldest ages [3], while the possibility to ultimately prevent physical and cognitive disability is still unanswered.

The many decades of science briefly summarized above have followed two pathways of reasoning. The traditional medical approach to chronic disease is to accomplish a diagnostic classification that is as precise as possible, based on symptoms, signs, clinical tests and other clinical elements. A correct and precise diagnosis allows access to the wealth of experience acquired in clinical medicine, including prognosis, and effective disease-modifying and symptomatic treatment of a specific disease. Physicians tend to work backwards in the etiologic pathway from making a disease diagnosis based on external clinical elements to generating hypotheses about pathophysiological mechanisms and the risk factors. Treatments aimed at prevention and cure are then administered that work 'forward' in the etiologic pathway, thereby correcting the clinical manifestation of diseases. A corollary of this method is the assumption that each syndromic manifestation has an underlying specific pathophysiology.

The traditional medical approach to human diseases has been quite successful in the care of young and middle-aged patients and prevention in these age groups. However, it has substantial limitations in the care of older patients for several reasons. First, the signs, symptoms, clinical manifestations, prognosis and response to treatment for many chronic diseases vary with age, and the specificity of symptoms decreases with increasing age. For example, in older persons episodes of hypoglycemia are often asymptomatic and signs of a previous acute myocardial infarction are often found in people with no history of symptoms. Second, the high likelihood of geriatric conditions and multimorbidity in older ages blurs diagnostic boundaries between diseases and complicates treatment choices. Third, in many older adults, the manifestations and clinical course of diseases are strongly affected by the underlying status of the "host" as well as by other coexisting diseases. Because of these reasons, considering aging as a 'confounder' in the study of chronic diseases ignores the complexity of the interactions between aging, disease and frailty. We now know that aging plays a central role in the pathogenesis, clinical presentation and response to treatment of many chronic diseases. Therefore, the patient's age (both biological and chronological) should be a primary clinical element that should affect choices of diagnostic, preventing and therapeutic interventions.

Emerging evidence on multi-morbidity and the frailty syndrome lays out the basis for making substantial progress in translating these concepts into improved care of older patients. Promising developments are coming, as well, from the rising interest in Geroscience and Precision Medicine [4]. The convergence of these scientific disciplines can be transformative in our understanding of the interplay between aging, frailty and disease, with the potential of producing dramatic improvements in public health.

In this chapter, we explore the evolution and current state of the science pertaining to possible links between aging and chronic disease(s), with a specific focus on the epidemiological evidence that such association is robust and not exclusively explained or sustained by a stochastic process. We seek to link together the mounting evidence that biological mechanisms that underlie aging lead to dysregulation of multiple physiological systems, loss of homeostatic capabilities and increased susceptibility to stress, and that these changes facilitate the emergence of both multimorbidity and clinically apparent frailty. Then, we consider whether the epidemiological literature is consistent with the stated hypothesis. Finally, we examine our current understanding of the biology of frailty as a basis for generating hypotheses about the biological mechanisms that link aging and chronic diseases.

1.3 From Clinical Presentation to Etiology and Biology

For over a thousand years, chronic diseases have been studied from the starting point of their externally apparent clinical manifestations. This approach led to the development of a number of classification systems, some relatively simple (such as the now disproven distinction between inflammatory and degenerative diseases) and some extremely precise and sophisticated (such as classification of lymphomas based on histological characteristics). While the ability to recognize specific diseases and to treat them successfully has increased tremendously, the limitations of these approaches have also become apparent. These are evident, in part, from the advent of applications such as high throughput genetic, genomic, proteomic and imaging biomarkers to the study of chronic diseases, in combination with the evidence from prospective/longitudinal epidemiological, population-based studies. Biomarker studies – perhaps better than any other scientific approaches – have offered evidence that, in many cases, diseases that are driven by different mechanisms converge into the same pathological and clinical manifestations. For example, it is now widely accepted that under the label of "Alzheimer's disease" exist a number of conditions with different underlying mechanisms [5]. Conversely, diseases that appear quite different from the perspective of phenotypic and end-organ manifestations are now known to have shared etiologic biomarkers (e.g., inflammation); this suggests that they are driven by the same pathophysiologic mechanisms, and that multiple, seemingly unrelated, chronic diseases share biomarker signatures. Interestingly, such biomarkers are often also related to aging itself and predict the development of frailty, a major adverse health outcome associated with aging. This is consistent with evidence that the biology of aging is associated with chronic disease development through mechanisms beyond the length of time for exposure and cumulative risk from external risk factors; rather, the evidence actually points to aging as playing a powerful causal role in development of chronic diseases.

2 Epidemiologic Interrogation of Chronic Disease

Over the last 60 years, we have learned – in large part through epidemiologic investigations - how the causes of morbidity have dramatically changed in our population. With the increase in longevity due to the demographic transition, chronic diseases have become the dominant causes of morbidity and mortality in the developed world, and are rapidly reaching that dominance globally. According to a recent report, over 85 % of US Medicare beneficiaries have at least one of nine chronic diseases [6].

The frequency of most of the major chronic diseases rises with age. However, we should not assume that the relationship between aging and disease is monotonic; in fact, it is quite complex. Some chronic diseases, such as those due to genetic defects, exposures during gestation, or environmental perturbation may become clinically evident early in life and are unlikely to emerge after a certain age. Some other diseases have a typical age of emergence and only rarely occur outside a certain time window (e.g., rosacea or lupus). However, the majority of chronic diseases such as dementia, cardiovascular diseases, cancer or osteoarthritis, show increased incidence and prevalence with aging, although many individuals never develop those diseases during their lifespan. Finally, some pathological conditions occur in everyone with aging (e.g., decline in renal function, decline in lean body mass) and, because they occur in everyone, we assume they are not a disease but rather part of "normal aging".

Given the extreme variability of human pathology, making a generalization is difficult. However, it is plausible that biological aging plays an important pathophysiologic role in diseases whose incidence and prevalence increase with aging. To offer the counterarguments first, there are basically two objections to this theory. The first is that not everyone develops those diseases as they age, although for some diseases, such as Alzheimer's disease, it has been argued that if they lived long enough, everyone would eventually develop the disease. The second argument is that not everyone develops diseases in the same sequence. In both cases, objections can be easily overcome by hypothesizing that the clinical emergence of disease results from a tradeoff between organ or tissue-specific susceptibility, the rate of progression by which subclinical processes become clinical, and the overall dysregulation induced by the aging process. As an example of the latter, aging can facilitate an imbalance in cholesterol metabolism, but such imbalance may never emerge clinically in individuals who do not have a certain genetic susceptibility and maintain a healthy diet and weight. Further, the clinical presentation of disease may be delayed by behavioral compensations, such as walking more slowly in patients with pulmonary diseases – so as not to experience the symptoms – or increased walking to improve muscle efficiency in utilizing oxygen in patients with peripheral artery disease – and thus decreasing the symptoms.

Unfortunately, modern medicine, public health and much of science has focused almost exclusively on mechanisms that create susceptibility to a single disease, and have substantially ignored the clues as to the potential direct contribution of aging and related biology to health as well as to chronic diseases. In this context, it is understandable why age has been considered merely a 'confounder'. In surveying the literature on chronic disease, the attributable fraction of the burden of chronic diseases to the health burden of aging appears quite high across diseases, even when the effect of powerful risk factors such as hypertension or smoking are factored out. However, such analyses have rarely been conducted comprehensively, because most studies have focused on one disease outcome and ignored the effect of competing risk or aggregate impact. Future studies are needed to estimate in large, representative cohorts the population-attributable fraction to aging and health of multiple chronic diseases after adjusting for known risk factors and using a multivariate approach that addresses competing risk and selective mortality. Such studies would help to estimate the extent to which the burden of morbidity in older persons is attributable to aging per se.

3 Where Does Aging Fit in the Study of the Etiology of Chronic Disease?

An interpretation of the epidemiologic literature strongly suggests that aging contributes independently to the pathogenesis of many chronic diseases, and there are truly very few exceptions, largely in the form of rare diseases. Almost counterintuitively, the only diseases that have been interrogated to reveal aging effects are the differential mechanisms for congestive heart failure by age [7], and the progeroid syndromes. Whether progeroid syndromes truly recapitulate the effect of aging in their early emergence of the aging phenotypes is unclear [8]. In general, very few of the findings of these studies have been translated into better understanding of the interface between aging and disease in people who have an average lifespan.

To go beyond "age as a confounder" or "aging as a process" independent of – and unrelated to – chronic diseases in the study of human pathology requires a shift in perspective. It is essential to go from the study of the clinical manifestations of a given disease and then its proximal drivers, to an agnostic consideration of the pathophysiologic events that precede the development of single and multiple disease(s) over the aging process, and to relate these outcomes to the processes and outcomes of aging including, but well beyond, disease(s). Theoretically and ideally, such a shift would require:

1. A basic understanding of the biological mechanisms of aging in humans. While we are far from having this knowledge in its entirety, some of the basic theories of aging have been developed and, to some extent, supported experimentally in model organisms (mostly yeast and worms, and in some cases, mice). Technology is currently available or in an advanced stage of development that should allow the testing of some of these theories in humans. Fully developing and testing such technology – and then the theories themselves – is clearly an important priority in aging research.
2. Criteria for disease diagnosis and multimorbidity classification should be developed that are not exclusively based on clinical manifestations and that do not ignore the role of the aging process or novel insights about causal pathways [9].
3. A list of potential biomarkers that change with aging and may be modified by chronic diseases – or that themselves modify disease – should be developed, and their ability to predict the decline in physical and cognitive function that occurs with aging should be evaluated. Ideally, these biomarkers should belong to pathways that have been demonstrated to be altered in animal models of aging and frailty.

3.1 *Multi-morbidity and Aging*

As outlined above, one of the landmarks of the interface between aging and diseases is multi-morbidity. In the U.S., nearly 80 % of Medicare beneficiaries 65 and older have at least 2 chronic conditions, and more than 60 % have at least 3 chronic conditions [10]. In fact, the condition of multi-morbidity is the most frequent medical condition that affects individuals 65 years and older. In a cross-sectional study that included 1.7 million patients in Scotland, UK, Barnett and colleagues found that 30.4 % of the population aged 45–64 years, 64.9 % of people ages 65–84 years, and 81.5 % of people aged 85 years and over reported at least two chronic conditions. Data from the InCHIANTI study, a longitudinal study of aging performed in Italy in a population-based sample, suggest that the average number of chronic diseases present in a person increases exponentially with aging, and the longitudinal rise in multi-morbidity is even higher than what can be estimated from a cross-sectional survey because of selective mortality [10]. Of course, the scenario of morbidity is more complex than just the counts of specific chronic diseases or considerations about their individual etiology. For example, in most studies on multimorbidity only a limited number of diseases are considered, typically 10–20, which is only a fraction of the diseases that affect aging humans. Prior reviews have discussed the significance of multimorbidity in the context of aging [11–17].

The development of multiple diseases in the same person occurs, in part, by chance alone, as a result of the length of exposure and cumulative development of the resulting pathophysiologic alterations. However, recent data suggest that the development of multi-morbidity is not merely the result of multiple independent, disease-specific pathophysiological processes. Their development can derive from shared etiologic factors, which predict many different chronic diseases that are age-related. That is, there are many environmental and behavioral risk factors (e.g., smoking) and shared intrinsic mechanisms, such as chronic inflammation, which drive development of many seemingly unrelated end-organ manifestations and diseases, from cardiovascular disease to diabetes to cancers and COPD [11]. Further, a number of studies have recently suggested that this is only part of the story. A landmark paper that tested the hypothesis that certain diseases tend to appear in clusters in some specific individuals was published by van den Akker and colleagues in 1998 [15]. Working in a large general practice, they used data on disease prevalence and co-prevalence to calculate the expected prevalence of multi-morbidity under the assumption of association by pure chance. The authors found that more people than expected did not have any disease or had four or more diseases, while less than expected had one or two diseases. This suggests that some individuals develop a global susceptibility to multiple diseases, while others appear to be unusually resistant, and that the co-occurrence of some specific diseases is not due to chance alone. These authors and others have suggested that "the force of aging" is different in different individuals and makes them more or less susceptible to multi-morbidity. This hypothesis has never been empirically evaluated, due to lack of data. However, since 1998, numerous studies, largely heterogeneous for sample size, age, settings, indices of multi-morbidity and statistical methods (observed-to-expected ratio, cluster analyses, factor analyses and multiple correspondence analyses), have been trying to describe possible patterns of associative multi-morbidity in order to generate helpful evidence for actual clinical practice and management [10, 17]. Compromising this research is the absence of standardization of methods, which has resulted in high variability among studies and inconsistency in results, making it difficult to compare them. Interestingly, chronological aging appears to be a strong predictor of mortality even in analyses that take into account the contributions of multiple diseases and risk factors [18, 19]. While this is not a surprising finding to clinicians, it clearly underlines the fact that that the parameters that are usually observed and considered in medical care do not exhaust all of the changes in health status that occur with aging. In other words, there are biological processes that occur with aging that are not revealed by clinically evident disease. It is quite possible that the excess mortality associated "purely" with old age may be due to subclinical disease, unobserved confounding and/or measurement error in assessing the full effect of diseases on prognosis. However, an alternative hypothesis is that some fundamental biological processes essential for life across cells, tissue and the entire organism deteriorate with the aging process and facilitate the emergence of both diseases and a general decline in health status that is not classified as "disease" but increase the risk of mortality.

An example based on the mitochondrial theory of aging may clarify this idea. A number of studies have found that aging is associated with loss of mitochondrial function; the mitochondrion being the key organelle responsible for cellular energy production in tissues and organs. Although mitochondrial volume appears to increase with aging, the electrical and chemical trans-membrane potential of the inner mitochondrial membrane is generally reduced and there are also functional defects in the electron transport chain that impair the efficiency of oxidative phosphorylation and limit the production of adenosine-5′-triphosphate (ATP) [20]. In addition, impairment in mitochondria may affect apoptosis as well as autophagy mechanisms.

When mitochondria are not functioning properly, tissues that have high energy demands suffer first. This is manifest in genetic mitochondrial disorders, which mostly affect the nervous system, heart and skeletal muscle [21]. A decline in mitochondrial function, regardless of its origin, has also been associated with central obesity, sarcopenia, fatigue, loss of physical fitness and diabetes [22]. Studies in animal models and in humans have demonstrated that the macroautophagy mechanisms aimed at removing and recycling defective mitochondria (mitophagy) become defective with aging, leading to the persistence of damaged mitochondria that generate excessive Radical Oxygen Species (ROS) and trigger a chronic inflammatory response. It is notable that inadequate energy support, excessive and unopposed oxidative stress as well as chronic inflammation have all been associated with the syndrome of frailty and with almost all chronic diseases whose incidence and prevalence increase with aging as well as with sarcopenia, gait disorders, and brain dysfunction. The translation of these biological events into both multi-morbidity and frailty is direct and intuitive. Further, it is easy to see how age-associated mitochondrial dysfunction can produce a number of vicious cycles that further impair mitochondrial physiology, for example, by causing ischemia or metabolic derangement. There are now many other examples implicating a single biological mechanism that, while driving the development of aging as a phenotype and aging-related frailty, also causes multi-morbidity. What is important here is that the triangle of aging-multimorbidity-and frailty may derive from the same root cause and, plausibly, they may all respond to the same interventions.

3.2 *Frailty*

We suggest above that the process of aging, and in particular "accelerated aging", may be best approximated by the concept of frailty. We develop further here the concept and the evidence that underlying mechanism(s) of aging – as exemplified above by the example of mitochondrial changes - may lead to this syndrome of frailty. Frailty is associated with a dysregulation of the complex adaptive mechanisms that support organismal resilience; the dysregulation leading to loss of homeostatic capabilities and increased susceptibility to stress. The net result is the emergence of a distinct clinical syndrome with a characteristic phenotype that is predictive of a range of adverse clinical outcomes.

We will first describe frailty as a prototypical constellation of signs and symptoms that allows a clinical diagnosis and then, working backwards in the causal pathway to etiology, we will consider how what we currently know about frailty informs understanding of aging and "accelerated aging".

There is now strong evidence to support frailty as being a clinical syndrome characterized by a constellation of symptoms and signs: muscle weakness, slowed gait, low physical activity, perceived low energy or "exhaustion", and weight loss that is not explained by a distinct disease process [23]. When a critical number of such signs and symptoms occur in the same individual they identify the frailty syndrome [23]. It is noteworthy that studies have demonstrated that the phenotypic criteria of frailty co-occur in ways that are consistent with the definition of a medical syndrome [24]. That they predict the clinical outcomes associated with being frail provides criterion validity.

The clinical presentation of frailty is consistent with dysregulation in multiple systems and organs [23], although it is, as yet, unknown whether such dysregulation is the primary causal event, a collateral, or a consequence. Operationally, when multiple of these criteria are present (three or more) the individual is considered "frail" and consistently, "frail" individuals have high risks – over 3 years – of mortality, disability, falls, hospitalization, loss of independence, and slower recovery from illness or surgery and greater postoperative adverse outcomes [23, 25]. Further, those who meet these criteria for frailty show decreased responsiveness to treatments, for example for HIV or renal failure [26]. This susceptibility to adverse outcomes occurs frequently in the context of a stressor, such as illness, hospitalization or surgery.

Clinical frailty develops progressively, so that testing positive for one or two criteria predicts the development of the full syndrome, with weakness and slowed gait being the most common earliest predictors [27]. Studies have found that the greater frailty is associated with greater risk for disability and loss of independence, for example, in the absence of an acute precipitant [28].

Clinical frailty is also associated with the presence of specific chronic diseases, particularly those with an inflammatory etiology, and the risk of frailty rises with the number of such diseases present [29]. While frailty incidence rises with increasing age, independently of ongoing chronic diseases, the association with the *subsequent* appearance of chronic diseases, including cardiovascular, kidney and rheumatologic diseases, suggests that there may be both a primary, aging-related phenotype of frailty, and a phenotype of frailty that is secondary to chronic disease. Further, new evidence indicates that obesity and aggregate risk for coronary artery disease in midlife predicts the development of pre-frailty and frailty 26 years later. Together these observations may implicate a shared etiology of aging (the process) and frailty (the clinical syndrome) [29].

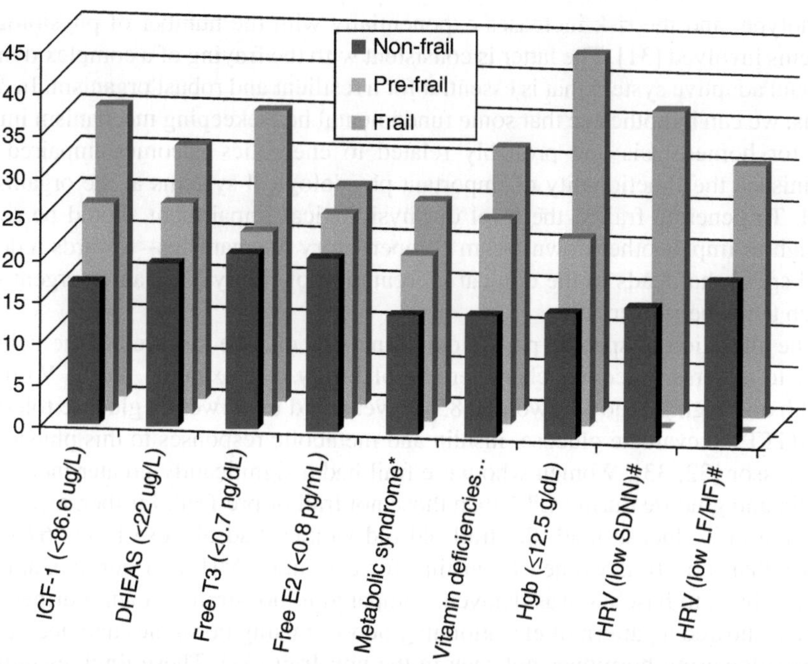

Fig. 1 Abnormal levels of biomarkers of multiple physiologic systems with frailty status (frail v. pre-frail vs. non-frail) in older adults. * Based on a 9-year incidence in the Cardiovascular Health Study. # Data only available for Frail versus non-frail. All other data from the Women's Health and Aging Study I and II, baseline (women 70–79 years old)

Although epidemiological research suggests that frailty is likely to cause an irreversible deterioration of health, there is some initial evidence that frailty can be prevented. For example, as shown by the InCHIANTI study, following a Mediterranean diet was protective against development of frailty as well as the development of cardiovascular diseases over a 6-year follow-up [30].

A large body of evidence indicates that frailty in all its clinical manifestations could be driven by a specific, although complex, pathophysiology that leads to dysregulation of multiple physiologic systems. Since aging is pervasive across the entire body, the more systems that are dysregulated, the greater the likelihood that the clinical manifestation is the result of accelerated aging or frailty rather than to specific disease. Longitudinal studies have found that dysregulation and loss of function tends to occur harmonically across multiple systems (Fig. 1), thus suggesting the existence of an underlying biological alteration, such as mitochondrial dysfunction, that may be altering energy availability and function for many systems. However, many of the systems affected (hormonal, inflammatory, muscle, anemia, autonomic) also mutually regulate each other, suggesting interrelatedness driving mutual dysregulation and consistent with one of the principles of a complex adaptive system. Further, the relationship goes beyond any particular dysregulated system: simply counting the number of systems dysregulated predicts the frailty phenotype, and the risk increases exponentially with the number of physiological systems involved [31]. The latter is consistent with the fraying of a complex dynamical and adaptive system that is essential for a resilient and robust organism. In other terms, we can hypothesize that some fundamental housekeeping mechanism important for homeostasis and probably related to energetics becomes impaired and diminishes the functionality of important physiological systems at the organismal level. To generate frailty, the level of physiological impairment should be severe enough to impair other downstream compensatory mechanisms– towards a downward spiral that leads to the clinical presentation of frailty [23], an emergent state which tends be irreversible.

The alteration of specific physiologic functions may be involved in the vulnerability to adverse outcomes characteristic of frailty. For example, in the Women's Health and Aging Study II, women 85–95 years old underwent a glucose tolerance test (GTT) to evaluate glucose, insulin and metabolic responses to this physiological stressor [32, 33]. Women who were frail had a significantly greater increase in insulin and glucose during GTT than those not frail or pre-frail. Further, at 120 min after the oral glucose load, the frail old-old women had glucose levels 67 mg/dl higher than non-frail women (adjusting for age and BMI), and normalization of glucose back to baseline was delayed, compared to non-frail women. Further, frail women showed a pattern of elevation in glucose- raising hormones and decrease in glucose-lowering hormones not seen in the non-frail [33]. These findings indicate that the entire physiological network of signals that regulate glucose homeostasis tends to be altered in frailty. Interestingly, most of the genetic mutations that have been associated with longevity in animal models are related to the insulin/IGF-1 pathway, which is involved in carbohydrate metabolism. Importantly, the diminished regulation of physiological responses to a stressor identifies the frail. However, while the frail were highly dysregulated compared to pre-frail and non-frail individuals, dysregulation was a generalized state for these women 85 and older, with the difference being only one of degree between the non-frail, pre-frail and frail. Specifically, only 27 % of the entire group of old-old women evaluated in this study had normal fasting glucose and normal OGTT, while 71 and 78 % of the full group met criteria for prediabetes and diabetes, respectively [34]. This suggests that findings of frailty are at the more severe end of dysregulation associated broadly with aging, and that the dysregulation of aging is interpretable, in this case, as disease [4].

The findings summarized above suggest that some specific cellular alterations might be key to maintaining the robust complex dynamic system of the human organism which is essential for health and resilience. The case of mitochondrial dysfunction described above is just an example of the many potential mechanisms that could be involved. The true underlying mechanisms of biological alteration that lead to frailty, and aging itself, remain unknown. The existence of a common causal pathway between aging and frailty could explain why the prevalence of frailty increases geometrically with aging and why the criteria used to define the frailty syndrome clinically include dimensions, such as sarcopenia and mobility, that are strongly modified by aging in all individuals and across species.

4 Conclusion

The analogies between aging and frailty may also explain why the clinical manifestations, evolution, prognostic implications and response to treatment of many chronic diseases are substantially different according to the age of the affected individuals. The multisystem nature of frailty, and of aging itself, may further offer insights into why many trials of single-agent replacement therapies in older adults have failed to improve targeted health outcomes. Thus, an important area of future research is to determine the physiological links that explain such age-associated differences in the manifestations of chronic morbidity – incorporating appreciation of multisystem phenomena and underlying drivers - so that this knowledge can be inserted into new clinical guidelines for the diagnosis and treatment of a series of pathologies in older, complex patients. Knowledge of the change in disease mechanisms, as well as the development of multimorbidity and frailty in association with age, offer the framing for breakthroughs in these research areas.

In summary, a focus on disease that avoids considering the biological processes of aging can lead to misunderstanding of the full scope of disease etiology in older patients and miss new therapeutic opportunities. Considering aging illuminates the question of how it is that biologic drivers of disease differ in the old compared to the young. Ultimately, it is possible that, as our understanding of the biology of aging grows, a new chapter of geriatric medicine will open. Perhaps the next generation of precision medicine is a disruptive vessel through which to accomplish some of this goal. Precision medicine is an emerging approach for disease treatment and prevention that takes into account individual variability in genes, environment, and lifestyle for each person. The extension of this method to aging and frailty appears to be the natural evolution of this idea and one that synergizes well with the new impetus of the Geroscience initiative. Placing this information in the context of the full human population experience can lead to understanding of the place of aging itself and the generalizable import of processes observed.

Acknowledgments Supported in part by the Intramural Research Program of the National Institute on Aging, National Institute on Health. The authors would like to thank Elisa Fabbri, Nida Raja and Maria O'Brien for help with finalizing the text.

Editor: Johanna Dwyer, Office of Dietary Supplements, Office of the Director, NIH.

References

1. National Heart, Lung and Blood Institute (1998) Morbidity & mortality: 1998 chartbook on cardiovascular, lung, and blood diseases. US Department of Health and Human Services, National Institutes of Health, Rockville
2. Hazzard WR (2000) Preventive gerontology: optimizing health and longevity for men and women across the lifespan. J Gend Specif Med 3(3):28–34
3. Fries JF, Bruce B, Chakravarty E (2011) Compression of morbidity 1980-2011: a focused review of paradigms and progress. J Aging Res 2011:261702. doi:10.4061/2011/261702, Epub 2011 Aug 23

4. Kennedy BK, Berger SL, Brunet A, Campisi J, Cuervo AM, Epel ES, Franceschi C, Lithgow GJ, Morimoto RI, Pessin JE, Rando TA, Richardson A, Schadt EE, Wyss-Coray T, Sierra F (2014) Geroscience: linking aging to chronic disease. Cell 159(4):709–713. doi:10.1016/j.cell.2014.10.039
5. Herrup K (2010) Reimagining Alzheimer's disease--an age-based hypothesis. J Neurosci 30(50):16755–16762. doi:10.1523/JNEUROSCI.4521-10.2010, Review
6. DuGoff EH, Canudas-Romo V, Buttorff C, Leff B, Anderson GF (2014) Multiple chronic conditions and life expectancy: a life table analysis. Med Care 52(8):688–694. doi:10.1097/MLR.0000000000000166
7. Rich MW (1997) Epidemiology, pathophysiology, and etiology of congestive heart failure in older adults. J Am Geriatr Soc 45(8):968–974
8. Arboleda G, Ramírez N, Arboleda H (2007) The neonatal progeroid syndrome (Wiedemann-Rautenstrauch): a model for the study of human aging? Exp Gerontol 42(10):939–943, Epub 2007 Jul 19. Review
9. Tinetti ME, Fried T (2004) The end of the disease era. Am J Med 116(3):179–185
10. Rocca WA, Boyd CM, Grossardt BR, Bobo WV, Finney Rutten LJ, Roger VL, Ebbert JO, Therneau TM, Yawn BP, St Sauver JL (2014) Prevalence of multimorbidity in a geographically defined American population: patterns by age, sex, and race/ethnicity. Mayo Clin Proc 89(10):1336–1349. doi:10.1016/j.mayocp.2014.07.010, Epub 2014 Sep 11
11. Fabbri E, An Y, Zoli M, Simonsick EM, Guralnik JM, Bandinelli S, Boyd CM, Ferrucci L (2015) Aging and the burden of multimorbidity: associations with inflammatory and anabolic hormonal biomarkers. J Gerontol A Biol Sci Med Sci 70(1):63–70
12. Fried LP, Ferrucci L, Darer J, Williamson JD, Anderson G (2004) Untangling the concepts of disability, frailty, and comorbidity: implications for improved targeting and care. J Gerontol A Biol Sci Med Sci 59(3):255–263
13. Boyd CM, Darer J, Boult C, Fried LP, Boult L, Wu AW (2005) Clinical practice guidelines and quality of care for older patients with multiple comorbid diseases: implications for pay for performance. JAMA 294(6):716–724
14. Tinetti ME, Fried TR, Boyd CM (2012) Designing health care for the most common chronic condition--multimorbidity. JAMA 307(23):2493–2494
15. van den Akker M, Buntinx F, Metsemakers JF, Roos S, Knottnerus JA (1998) Multimorbidity in general practice: prevalence, incidence, and determinants of co-occurring chronic and recurrent diseases. J Clin Epidemiol 51(5):367–375
16. St Sauver JL, Boyd CM, Grossardt BR, Bobo WV, Finney Rutten LJ, Roger VL, Ebbert JO, Therneau TM, Yawn BP, Rocca WA (2015) Risk of developing multimorbidity across all ages in an historical cohort study: differences by sex and ethnicity. BMJ Open 5(2), e006413. doi:10.1136/bmjopen-2014-006413
17. Barnett K, Mercer SW, Norbury M, Watt G, Wyke S, Guthrie B (2012) Epidemiology of multimorbidity and implications for health care, research, and medical education: a cross-sectional study. Lancet 380(9836):37–43
18. Fried LP, Kronmal RA, Newman AB, Bild DE, Mittelmark MB, Polak JF, Robbins JA, Gardin JM (1998) Risk factors for 5-year mortality in older adults: the Cardiovascular Health Study. JAMA 279(8):585–592
19. Newman AB, Boudreau RM, Naydeck BL, Fried LF, Harris TB (2008) A physiologic index of comorbidity: relationship to mortality and disability. J Gerontol A Biol Sci Med Sci 63(6):603–609
20. Chistiakov DA, Sobenin IA, Revin VV, Orekhov AN, Bobryshev YV (2014) Mitochondrial aging and age-related dysfunction of mitochondria. Biomed Res Int 2014:238463. doi:10.1155/2014/238463, Epub 2014 Apr 10. Review
21. Taylor RW, Turnbull DM (2005) Mitochondrial DNA mutations in human disease. Nat Rev Genet 6(5):389–402, Review
22. Coen PM, Jubrias SA, Distefano G, Amati F, Mackey DC, Glynn NW, Manini TM, Wohlgemuth SE, Leeuwenburgh C, Cummings SR, Newman AB, Ferrucci L, Toledo FG, Shankland E, Conley KE, Goodpaster BH (2013) Skeletal muscle mitochondrial energetics are associated

with maximal aerobic capacity and walking speed in older adults. J Gerontol A Biol Sci Med Sci 68(4):447–455. doi:10.1093/gerona/gls196, Epub 2012 Oct 9
23. Fried LP, Tangen CM, Walston J, Newman AB, Hirsch C, Gottdiener J, Seeman T, Tracy R, Kop WJ, Burke G, McBurnie MA, Cardiovascular Health Study Collaborative Research Group (2001) Frailty in older adults: evidence for a phenotype. J Gerontol A Biol Sci Med Sci 56(3):M146–M156
24. Bandeen-Roche K, Xue QL, Ferrucci L, Walston J, Guralnik JM, Chaves P, Zeger SL, Fried LP (2006) Phenotype of frailty: characterization in the women's health and aging studies. J Gerontol A Biol Sci Med Sci 61(3):262–266
25. Makary MA, Segev DL, Pronovost PJ, Syin D, Bandeen-Roche K, Patel P, Takenaga R, Devgan L, Holzmueller CG, Tian J, Fried LP (2010) Frailty as a predictor of surgical outcomes in older patients. J Am Coll Surg 210(6):901–908. doi:10.1016/j.jamcollsurg.2010.01.028, Epub 2010 Apr 28
26. Desquilbet L, Jacobson LP, Fried LP, Phair JP, Jamieson BD, Holloway M, Margolick JB (2011) A frailty-related phenotype before HAART initiation as an independent risk factor for AIDS or death after HAART among HIV-infected men. J Gerontol A Biol Sci Med Sci 66(9):1030–1038. doi:10.1093/gerona/glr097, Epub 2011 Jun 30
27. Xue QL, Bandeen-Roche K, Varadhan R, Zhou J, Fried LP (2008) Initial manifestations of frailty criteria and the development of frailty phenotype in the Women's Health and Aging Study II. J Gerontol A Biol Sci Med Sci 63(9):984–990
28. Boyd CM, Xue QL, Simpson CF, Guralnik JM, Fried LP (2005) Frailty, hospitalization, and progression of disability in a cohort of disabled older women. Am J Med 118(11):1225–1231
29. Strandberg TE, Sirola J, Pitkälä KH, Tilvis RS, Strandberg AY, Stenholm S (2012) Association of midlife obesity and cardiovascular risk with old age frailty: a 26-year follow-up of initially healthy men. Int J Obes (Lond) 36(9):1153–1157. doi:10.1038/ijo.2012.83
30. Talegawkar SA, Bandinelli S, Bandeen-Roche K, Chen P, Milaneschi Y, Tanaka T, Semba RD, Guralnik JM, Ferrucci L (2012) A higher adherence to a Mediterranean-style diet is inversely associated with the development of frailty in community-dwelling elderly men and women. J Nutr 142(12):2161–2166. doi:10.3945/jn.112.165498, Epub 2012 Oct 24
31. Fried LP, Xue Q-L, Cappola AR, Ferrucci L, Chaves P, Varadhan R, Guralnik JM, Leng SX, Semba RD, Walston JD (2009) Nonlinear multisystem physiological dysregulation associated with frailty in older women: implications for etiology and treatment. J Gerontol A Biol Sci Med Sci. 64(10):1049–1057. doi:10.1093/gerona/glp076
32. Kalyani RR, Varadhan R, Weiss CO, Fried LP, Cappola AR (2012) Frailty status and altered dynamics of circulating energy metabolism hormones after oral glucose in older women. J Nutr Health Aging 16(8):679–686. doi:10.1007/s12603-012-0369-5
33. Kalyani RR, Tian J, Xue QL, Walston J, Cappola AR, Fried LP, Brancati FL, Blaum CS (2012) Hyperglycemia and incidence of frailty and lower extremity mobility limitations in older women. J Am Geriatr Soc 60(9):1701–1707. doi:10.1111/j.1532-5415.2012.04099.x, Epub 2012 Aug 10
34. Barzilay JI, Blaum C, Moore T, Xue QL, Hirsch CH, Walston JD, Fried LP (2007) Insulin resistance and inflammation as precursors of frailty: the Cardiovascular Health Study. Arch Intern Med 167(7):635–641 [PMID: 17420420]

[illegible]

The Impact of Aging on Cancer Progression and Treatment

Shenghui He and Norman E. Sharpless

Contents

S. He, Ph.D. • N.E. Sharpless, M.D. (✉)
Department of Genetics, The Lineberger Comprehensive Cancer Center, University of North Carolina School Medicine CB #7295, Chapel Hill, NC, 27599, USA
e-mail: shenghui@email.unc.edu; Norman_Sharpless@med.unc.edu

F. Sierra, R. Kohanski (eds.), *Advances in Geroscience*,
DOI 10.1007/978-3-319-23246-1_3

1 Introduction

In higher organisms, cancer reflects the cost of the need for long-lived self-renewing somatic stem cells in proliferative tissues functioning throughout the lifespan. Such cells are necessary for the constant production of new cells to replace damaged or shed effector cells, thereby maintaining tissue and organ homeostasis. Somatic stem cells and their replicating progeny exhibit a staggering capacity for proliferation, but they also can undergo malignant transformation. DNA mutations in these cells along with changes of their epigenetic state almost inevitably allow for the emergence of aberrant self-renewing clones on their way to full neoplastic conversion. Mutations occur in these self-renewing compartments as a result of exposure to external carcinogens and genotoxic agents, but also occur solely through bad luck; for example, as an error of DNA replication [1]. These mutations provide the finite number of genetic alterations required for malignant transformation. Given the daily production of immense numbers of new cells, it is actually remarkable that highly replicating tissues only very rarely undergo neoplastic conversion. The finding that oncogenic events that characterize malignancy are very common, even present at birth [2], whereas cancer is an unusual disease mainly affecting the elderly, demonstrates the existence of very effective tumor suppression mechanisms.

In this Chapter, we will discuss the molecular basis of cancer and aging, with particular emphasis on the mechanisms that associate increased incidence of cancer with aging. Specifically, we will discuss how time-dependent accumulation of genetic and epigenetic alterations in self-renewing cells as a result of imperfect homeostatic mechanisms can act as a common molecular basis for aging and cancer; how tumor suppression mechanisms that have evolved to prevent cancer can cause age-related functional attrition of self-renewing cells, which in turn contributes to certain aging-related pathologies; and how aging-associated physiological changes can contribute to cancer initiation and progression. From these discussions, we hope to identify preventive measures that can minimize the risk of cancer while slowing the rate of aging.

2 Molecular Basis of Cancer

Cancer can largely be considered a genetic disease. That is, aberrant expression of proteins that normally regulate cell growth and proliferation cause cancer, either by over- or under-expression of normal versions of cellular proteins, or expression of mutant proteins that acquire de novo oncogenic functions. These changes in protein expression largely arise from fixed DNA mutations that occur as a result of replication error, carcinogen exposure and other genotoxic stresses. Cancer-causing mutations are grouped into two classes: activating events on *oncogenes* and inactivating events on *tumor suppressor genes* (Table 1). Ample evidence suggests that both

Table 1 The role of oncogenes and tumor suppressors in cancer and aging

	Cellular function	Role in cancer	Role in aging
Oncogenes	↑Cell cycle ↑Cell growth ↑Survival ↓Differentiation ↓Apoptosis	Gain-of-function mutation "One-hit" model Commonly initiating event "Oncogene addiction"	Loss-of-function impairs stem cell maintenance (pro-aging) Gain-of-function enhances stem cell self-renewal potential (anti-aging), but can activate tumor-suppressor response to induce senescence or apoptosis, or lead to clonal dominance of stem cells with defective differentiation potential (pro-aging)
Gatekeeper Tumor Suppressor	↓Cell cycle ↑Differentiation ↑Apoptosis	Loss-of-function mutation "Two-hits" Model Inactivation is required for tumor maintenance	Loss-of-function enhances stem cell function (anti-aging) but may induce stem cell exhaustion (pro-aging) Aberrant activation impairs stem cell function (pro-aging), but physiologically regulated increase in gene dose can in some cases extend lifespan by preventing cancer.
Caretaker Tumor Suppressor	DNA repair Telomere maintenance ROS detoxification Epigenetic maintenance	Loss-of-function mutation "Two-hits" Model Inactivation promotes (epi-) genomic instability Inactivation is not required for tumor maintenance	Loss-of-function impairs stem cell maintenance (pro-aging) Gain-of-function effects on aging are currently unclear

classes of mutations are needed for full malignant transformation, and generally multiple (>5) such events are required in adult cancers.

2.1 *Oncogenes*

Oncogenes are genes whose protein products promote neoplastic transformation of normal cells. They are often mutated or over-expressed forms of normal cellular genes (sometimes termed "proto-oncogenes"), but can also be encoded by certain strains of oncogenic viruses and acquired by normal cells following viral infection (e.g. SV40 large T antigen). Cellular proto-oncogenes encode proteins that play essential roles in regulating cell growth, survival, proliferation and differentiation. As a result, many proto-oncogenes are important regulators of embryonic development [3, 4], while some are specifically required for somatic stem cell maintenance and tissue homeostasis in adult mammals [5]. Given their unique ability to regulate cell growth and survival, proto-oncogenes are the targets of oncogenic mutations.

Oncogenic mutations either increase the gene's normal activity, or confer de novo oncogenic function to the mutated genes. Upon activation, a proto-oncogene becomes an oncogene, gaining the ability to confer growth and survival advantage to normal cells and promote cancer development. Since only one copy of the proto-oncogene needs to be mutated to exert its oncogenic function, activating mutations of proto-oncogenes can follow the "one-hit" model and often occur early during cancer development.

Oncogenes can be activated in a number of ways. For example, the mutant protein product possessing enhanced enzymatic activity or resistance to degradation/inactivation can be generated as a result of DNA mutations in the coding region of the oncogene (e.g. BRAF-V600E). Alternatively, increased expression of the oncogene protein product can occur as a result of gene amplification or promoter mutation (e.g. TERT), or DNA mutations affecting transcription efficiency or transcript stability. Lastly, chromosome rearrangement events involving one or more oncogenes can generate fusion proteins that acquire increased transcript stability or de novo oncogenic function (e.g. BCR-ABL, MLL-AF9).

An important concern is whether a given oncogene contributes only to cancer initiation or is it also required for the continued survival and expansion of cancer cells (termed 'tumor maintenance' or 'oncogene addiction'). Inhibiting oncogenic pathways involved in tumor maintenance from cancer cells causes tumor regression through increased cell death and/or cell cycle arrest [6]. Clearly, oncogenes to which a cancer is 'addicted' make better targets for cancer therapy. For example, the development of inhibitors against mutant BRAF (Vemurafenib) or the BCR-ABL fusion protein kinase (Imatinib) have substantially improved the treatment of BRAF-mutant melanoma and Philadelphia chromosome-positive chronic myelogenous leukemia (CML) patients, respectively. However, as with any disease based on clonal evolution, drug resistance frequently emerges in cancer cells. Thus a better understanding of the molecular function of oncogenes and their normal cellular counterparts may help to identify cooperating pathways which, when inhibited, can cause synthetic lethality of the drug resistant cancer cells.

While oncogenes are best known for their roles in cancer, dysregulation in their activity may also contribute to aging under physiological or pathological conditions. Because many proto-oncogenes are critical regulators of somatic stem cell function and maintenance in adult tissues, insufficient proto-oncogene activity may contribute to age-related functional attrition of somatic stem cells and aging of self-renewing tissues. This is best illustrated by studies of telomerase reverse transcriptase (*TERT*). Telomerase is activated in human cancers through several genetic mechanisms and is critical to transformation in some tissues (e.g. melanocytes). Loss of telomerase activity with attendant telomere shortening and dysfunction causes attrition of certain self-renewing cells and the manifestation of certain aspects of aging (e.g. pulmonary fibrosis, hair greying, cirrhosis, and bone marrow failure) in humans [7]. Another potential example is EZH2, a histone methyltransferase that is activated in a variety of cancers including lymphoma, melanoma and certain forms of lung cancer. EZH2 is a critical mediator of stem cell epigenetic state and self-renewal capacity [8], and EZH2 inhibitors are a promising new class of anti-cancer agent undergoing active clinical testing. However, a concern is that long-term use of EZH2 inhibitors may promote aging by augmenting somatic stem cell attrition.

A second way proto-oncogenes can contribute to aging is through the activation of tumor suppressor genes. Oncogene activation can trigger tumor suppressor responses in host cells, resulting in oncogene-induced apoptosis, senescence or differentiation. As activating mutations of oncogenes can accumulate in the stem cell pool over time, this can cause functional attrition of stem cells with age, resulting in reduced regenerative potential of aging tissues (discussed in detail in the next section).

A third and related way whereby oncogenic events can contribute to tissue aging is through the induction of advantaged clones that are defective for normal stem cell function. For example, work in human hematopoietic stem cells (HSCs) has shown that clones harboring a small number of oncogenic events may exhibit a modest increase in replicative capacity, in the absence of frank transformation [9–11]. In this case, these pre-neoplastic HSCs can out-compete normal HSCs, thereby "taking over" the bone marrow compartment. Although these pre-neoplastic HSCs may have an advantage in terms of proliferation and self-renewal compared to normal HSCs, they may also be defective in the production of differentiated blood elements, and thereby contribute to anemia and leukopenia, which are not uncommon features of hematopoietic aging in humans. It is likely that this type of stem cell competition resulting from oncogenic events that cause tissue dysplasia contributes to aging phenotypes in tissues beyond the bone marrow.

2.2 *Tumor Suppressor Genes*

Given the danger posed by stochastic oncogene activation or acquisition of viral oncogenes, mammalian cells are equipped with a variety of tumor suppression mechanisms to prevent and counter oncogene activation. These tasks are performed by so-called tumor suppressor genes. Tumor suppressors function to either prevent the emergence of neoplastic cells by maintaining genomic stability or to restrict the growth and proliferation of already damaged cells. For malignant transformation to succeed, tumor suppressor genes must be inactivated either by DNA mutations or epigenetic silencing. As both copies of tumor suppressor genes usually need to be inactivated to abolish their function, mutations of tumor suppressor genes in cancer cells often follow the "two-hit" model. There are some exceptions, as tumor suppressor genes can be 'haploinsufficient', meaning the loss of a single copy of the gene is sufficient to confer growth advantage to mutant cells; in other cases, mutation of the tumor suppressor proteins can create dominant negative or a gain-of-function form of the protein that is sufficient to drive neoplastic transformation. In these cases, mutations of tumor suppressor genes can act as an initiating event in malignant transformation.

Tumor suppressor genes have been functionally divided into at least two major categories (Table 1). One group of tumor suppressor genes are constitutive housekeeping genes that act as "caretakers" in the cells to maintain genomic and epigenetic stability thereby minimizing the risk of oncogene activation. Examples of this are genes such as *BRCA1* or *ATM* whose protein products prevent cancer by promoting genomic stability. In contrast, a second group of tumor suppressor genes are thought to be activated by characteristic features of early neoplasms, such as DNA damage or uncontrolled mitogenic signaling. Such tumor suppressor genes have been termed "gatekeepers" whose function is to restrict aberrant cell growth and proliferation [12]. While this classic distinction has intellectual appeal, limitations of this conceptual framework have emerged with a modern understanding of the cellular mechanisms of tumor suppression. For example, certain tumor suppressor genes such as *TP53* perform both caretaker and gatekeeper types of functions. Moreover, several recently appreciated tumor suppressor genes (e.g. *LKB1* or *PTEN*) are potent regulators of intracellular signaling, and their loss augments metabolism and growth in a way that does not obviously fit into either category. These limitations notwithstanding, we will discuss these tumor suppressor mechanisms and their relation to aging as grouped through this framework.

2.2.1 Caretaker Tumor Suppressor Genes

Caretaker tumor suppressors act to maintain genomic integrity or epigenetic stability, and do not directly regulate cell growth or proliferation [13]. The best characterized examples of caretaker tumor suppressors are genes involved in DNA damage response and repair.

DNA lesions occur constantly in all cells as a result of the intrinsic chemical instability of DNA, errors in DNA replication, or exposure to cell intrinsic (such as reactive oxygen species (ROS) generated from cellular metabolism) or extrinsic genotoxic agents (such as UV, radiation or chemical carcinogens). To deal with the many types of DNA lesions, mammalian cells have evolved diverse DNA repair mechanisms. When the lesion only occurs on one strand of the DNA double helix, it can be faithfully repaired using information from the complementary strand through mechanisms such as base excision repair (BER) or nucleotide excision repair (NER). In the case of DNA double strand breaks (DSBs), the lesions can either be repaired by the homologous recombination (HR) mechanism that uses the information from the homologous sister chromatid to faithfully repair the damage, or by the more efficient but error-prone non-homologous end joining (NHEJ) mechanism. In each case, the cell detects DNA lesions through sensor proteins, which recruit the appropriate DNA repair machineries; if the lesions occur during the cell cycle, additional checkpoint proteins are activated to arrest the cell cycle, allowing enough time to repair the lesion before resuming division. If the repair is not successful, more familiar gatekeeper tumor suppressors such as $p16^{INK4a}$ and p53 are activated to induce programmed cell death (apoptosis) or permanent cell cycle arrest (e.g. cellular senescence or terminal differentiation) to prevent the damaged cells from further propagation.

Given their critical role in maintaining genomic integrity, mutations in DNA repair genes are frequently associated with an increased incidence of cancer [14].

For example, women with a single copy of a germline mutation in *BRCA1* or *BRCA2*, which contribute to HR-mediated DSB repair, are at significantly increased risk of developing cancers of the breast, ovary and other tissues; while patients with *ATM* inactivation are most predisposed to leukemia and lymphoma. Interestingly, many congenital DNA repair deficiency diseases show a varying degree of accelerated aging in addition to their cancer-prone phenotype [13, 14]. Likewise, with the prominent exception of Hutchinson-Gilford progeria, nearly all the genetic segmental progerias result from mutations in genes associated with DNA metabolism (e.g. WRN), and generally lead to a cancer-prone phenotype. This relationship between DNA metabolism and aging is not entirely due to the secondary effects of cancer as patients of DNA repair deficiency diseases such as Cockayne Syndrome and Trichothiodystrophy do not have increased risk of cancer while still exhibiting some features of premature aging [13]. Likewise, external exposures that cause DNA damage also induce aspects of premature aging, providing an additional link between DNA metabolism and life-long somatic stem cell function. These observations, together with the findings that DNA repair genes are essential for maintaining long-term self-renewal potential of somatic stem cells throughout life at least in mouse [15], suggest DNA repair genes are not only essential for preventing neoplastic transformation, but may also play a role in regulating the rate of physiological aging. It is however unclear whether enhancing the activities of caretaker genes could delay the onset of stem cell aging.

While mutations of DNA repair genes can promote cancer initiation and clonal evolution, their continued inactivation is generally not required for cancer cell growth and proliferation. In fact, due to frequent defects in various DNA repair mechanisms, many types of cancer cells are critically dependent on specific DNA repair pathways for survival. This cancer cell specific dependency can be exploited therapeutically to target cancer cells without harming normal cells with a wholly intact repertoire of DNA repair activities [16, 17]. For example, cancer cells with *BRCA1/2* gene mutations are defective in the repair of double strand breaks via homologous recombination, and treating these cells with PARP1 inhibitors that cause accumulation of single strand breaks, which are later converted into DSBs, results in synthetic lethality in the mutant cells but not wild type cells [18, 19]. These findings of in vitro synthetic lethality in turn translate into significant clinical antitumor activity of PARP inhibitors in *BRCA1/2*-deficient malignancies. Several other defects in DNA repair and/or DNA damage-induced checkpoints are exploited therapeutically to allow for enhanced killing of cancer cells versus non-malignant tissues.

DNA repair genes are not the only caretaker tumor suppressors; genes involved in telomere maintenance, reactive oxygen species (ROS) detoxification and epigenetic modification are often also characterized as caretaker genes due to their ability to maintain genomic and epigenetic stability of the cell. However, their roles in cancer are more context-dependent. For example, telomerase is essential for maintaining genomic integrity by protecting chromosomes from fusing to each other, therefore acting as a caretaker against neoplastic transformation. However, several lines of data now show that telomerase activity is also oncogenic, presumably by enabling the immortalization of neoplastic cells. Similarly, while DNA and histone modification enzymes are important for maintaining the integrity of epigenetic information in the cell, aberrant activation of these genes is an emerging theme now noted in many types of cancer. A final example is the surprising role of NRF2 in squamous cancers of the aerodigestive tract. NRF2 regulates many genes involved in the detoxification of ROS. Excess ROS has traditionally been considered oncogenic, through effects on intra-cellular signaling as well as by inducing macromolecular damage. Large tumor sequencing studies such as the Cancer Genome Atlas, however, have firmly established that excess NRF2 activity is potently oncogenic in squamous cancers of the lung, oropharynx and esophagus, challenging the long held view of ROS as being oncogenic [20]. It is perhaps not a coincidence that the notion that ROS promotes physiological aging, as in the "free radical theory of aging", is also being seriously questioned in modern gerontology [21, 22]. Therefore, unlike gatekeeper tumor suppressors (discussed next), the activity of caretaker tumor suppressors does not directly inhibit cancer cell growth, and depending on the cellular context, can actually promote cancer progression.

2.2.2 Gatekeeper Tumor Suppressor Genes

Gatekeeper tumor suppressor genes are negative regulators of cell proliferation. Unlike the caretaker genes, gatekeeper genes are generally not essential for routine cellular function; instead they are only activated upon sensing persistent cellular stress such as chronic DNA damage or oncogene activation.

Perhaps the best known gatekeeper tumor suppressor gene is *TP53*, the so-called "guardian of the genome". *TP53* is the most highly mutated gene across all human cancers, and mice with heterozygous or homozygous loss of *TP53* are highly prone to sporadic cancers. *TP53* encodes a multifunctional transcription factor (p53) that, in response to a variety of stresses, controls the expression of genes involved in a wide range of cellular activities including DNA damage repair, cell cycle control, apoptosis and autophagy. Critically shortened telomeres and double strand DNA breaks can activate p53 through the checkpoint kinases ATM and CHK2, which then upregulate the expression of the cyclin-dependent kinase (CDK) inhibitor $p21^{CIP1}$ to induce cell cycle arrest to facilitate DNA repair. Moreover, p53 can also respond, in a poorly understood manner, to excess mitogenic signaling (e.g. "oncogenic stress") mainly through activation of the ARF (a.k.a. $p19^{ARF}$ in mouse or $p14^{ARF}$ in human) tumor suppressor protein (Fig. 2). ARF is encoded by the *CDKN2A* locus, which also encodes another important gatekeeper tumor suppressor, $p16^{INK4a}$, and *CDKN2A* expression is induced by a wide-variety of oncogenic signals [23, 24]. ARF binds to and inhibits the activity of MDM2, a potent negative regulator of p53, thereby leading to p53 protein stabilization and p53-mediated cell cycle arrest or apoptosis. Similar to p53, mice lacking one or both alleles of *Arf* are cancer-prone, and deletion of *ARF* is common in human malignancies. The cellular response to p53 activation depends on the nature and duration of the cellular stress, as well as the cellular context [25]. While strong, persistent signals are likely to induce apoptosis or senescence; low intensity, transient signals tend to induce reversible cell cycle arrest [26].

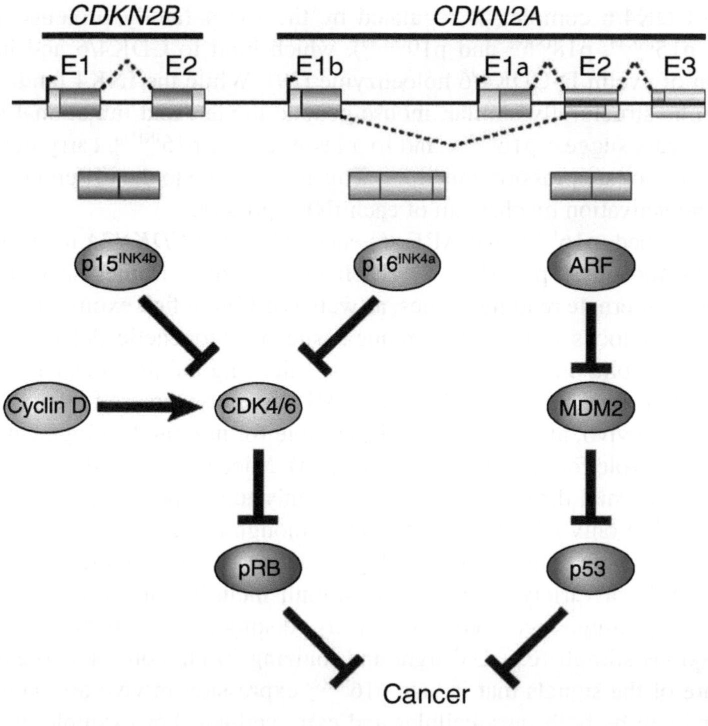

Fig. 1 The p16[INK4a]-RB and ARF-p53 tumor suppressor pathways. The *CDKN2A/B* locus on human chromosome 9p21 encodes three archetypical tumor suppressor proteins: p16[INK4a], ARF and p15[INK4b]. p16[INK4a] and ARF share common second and third exons, but are transcribed from separate promoters and first exons in alternative reading frames, thus yielding different proteins with distinct cellular functions. p16[INK4a] and p15[INK4b] belong to the INK4 family of cyclin dependent kinase inhibitors, which bind to CDK4/6 and inhibit the formation of cyclin D-CDK4/6 holoenzyme, leading to the stabilization of the RB family transcription factors and blocking cell cycle entry. ARF binds to and inhibits the activity of MDM2, a potent negative regulator of p53, leading to p53 protein stabilization. Activation of the p16[INK4a]-RB and ARF-p53 pathways upon oncogenic stress or persistent DNA damage can lead to cellular senescence or apoptosis, which serve as important tumor suppression mechanisms to prevent malignant transformation

A second group of important gatekeeper tumor suppressors involves the control of mammalian cell cycle progression, including the Retinoblastoma (RB) family proteins and several types of cyclin-dependent kinase inhibitors (Fig. 1). RB family proteins (RB, p107, and p130) are important regulators of the G1-S cell cycle transition, and they play a redundant role in binding and inhibiting the activity of E2F family transcription factors, which control the expression of genes important for DNA synthesis and cell division. Among the RB-family proteins, RB in particular is important for the induction of cellular senescence and for tumor suppression [27, 28]. RB protein activity is regulated by various cyclin/CDK protein complexes. Upon sensing mitogenic signals, activated cyclin-CDK complexes promote cell cycle entry by phosphorylating and inactivating RB proteins. The activity of the cyclin D-CDK4/6 complex is regulated by the INK4 family proteins (including $p16^{INK4a}$, $p15^{INK4b}$, $p18^{INK4c}$ and $p19^{INK4d}$), which bind to CDK4/6 and inhibit the formation of cyclin D-CDK4/6 holoenzyme [29]. While the INK4 family proteins appear to be structurally similar, mouse genetic models and mutational studies in human cancers suggest $p16^{INK4a}$, and to a lesser extent $p15^{INK4b}$, carry out the most important tumor suppression functions. This may be due to the different expression pattern and activation mechanism of each INK4 protein.

As mentioned, $p16^{INK4a}$ and ARF are encoded by the *CDKN2A* locus located on human chromosome 9p21 (Fig. 1). Their open reading frames share two exons translated in alternate reading frames, as well as different first exons and promoters. The *CDKN2A* locus is the most frequent site of cytogenetic deletion in human cancers, and point mutation or epigenetic silencing of the locus is also well-described. For the most part, $p16^{INK4a}$ and ARF are not expressed by normal proliferating cells in vivo, and are largely dispensable for normal development (with the exception of a role for Arf in eye development). Mice lacking both $p16^{INK4a}$ and *Arf* demonstrate normal development, but are highly tumor prone, even more so than animals lacking only $p16^{INK4a}$ or only *Arf*. Although these genes do not participate in normal cell cycle control, they are highly activated, albeit with delayed kinetics (1+ weeks), by a variety of oncogenic stimuli including in vitro culture, excess mitogenic signaling, increased cell density, disordered chromatin structure and many noxious stimuli (e.g. UV light and ionizing irradiation). The precise origin and nature of the signals that induce $p16^{INK4a}$ expression in vivo are poorly understood, and can be both intra-cellular and extra-cellular. For example, we recently showed that the presence of a small number of neoplastic cells within a given tissue will rapidly induce $p16^{INK4a}$ expression in nearby non-malignant cells [30]. Unlike $p21^{CIP1}$, $p16^{INK4a}$ is not acutely induced by DNA damage and does not play a role in the initial DNA damage-induced cell cycle arrest. Instead, the main function of $p16^{INK4a}$ appears to be to initiate and/or maintain a highly durable if not permanent cell cycle arrest (called 'cellular senescence') to prevent malignant transformation. In addition to the INK4 family proteins, another group of CDK inhibitors, the CIP/KIP family proteins (including $p21^{CIP1}$, $p27^{KIP1}$ and $p57^{KIP2}$), also represses the cell cycle by inhibiting the activity of the cyclin E-CDK2 complex. Their role in tumor suppression appears to be more limited. With the exception of bladder cancer where $p21^{CIP1}$ mutation is seen in 15 % of tumors, mutations of these proteins are unusual in most human malignancies, and mice deficient for $p21^{CIP1}$ or $p27^{KIP1}$ only exhibit a mildly increased risk of cancer.

Some evidence suggests the ARF-p53 and $p16^{INK4a}$-Rb tumor suppressor pathways are inactivated, through a wide variety of mechanisms, in almost all human cancers. Unlike caretaker tumor suppressors, the continued inactivation of gatekeeper tumor suppressors is required for cancer cell survival and proliferation, as re-expression of key gatekeeper genes such as p53 is sufficient to cause cancer regression [31, 32]. Activation of these tumor suppressor pathways may contribute to the therapeutic response to certain classes of anti-cancer agents; e.g. p53-mediated apoptosis may be important in the response to certain DNA damaging chemotherapy drugs. To date, however, re-activation or re-introduction of tumor suppressor genes in established cancers has proven technically difficult, and is currently not achievable in most cancers.

In addition to its critical role in cancer prevention, there is a large literature linking p53 with physiological aging (reviewed in [33]), but this relationship is complex. For example, as a critical regulator of genomic stability, p53 has been suggested to play a critical anti-aging role by minimizing the accumulation of macromolecular damage. Alternatively, excess p53 activation has been suggested to be age-promoting, perhaps through the induction of decreased function in somatic stem cell compartments. In support of this notion, mice expressing constitutively activated p53 exhibit a premature aging phenotype and have reduced life spans [34, 35], and hematopoietic stem cells isolated from these mice are functionally deficient in competitive transplantation settings [36]. However, mice engineered to carry an extra copy of the entire *Cdkn2a* locus and/or *p53*, whose activities are under physiological regulation, exhibit enhanced cancer resistance without detectable negative effects on aging and lifespan [37–39]. Therefore, excess, unbridled p53 may promote aging, whereas physiological p53 activation is likely anti-aging.

There are even stronger data linking $p16^{INK4a}$, and to a lesser degree ARF, to mammalian aging. The expression of both products of the *CDKN2a* locus increases exponentially with aging in all mammalian species studied [40–42], and this is thought to represent the accumulation of senescent cells with aging (covered below). The deletion of $p16^{INK4a}$ in mice appears to partially rescue the functional defects of old somatic stem cells in the nervous and hematopoietic system [43, 44], as well as improve the regenerative potential of aged lymphocytes and pancreatic beta cells [45–48]. Therefore, activation of $p16^{INK4a}$ appears to lead to a reduced replicative capacity in several somatic, self-renewing compartments, with an attendant decrease in physiological reserve. Alternatively, $p16^{INK4a}$ expression also appears to restrain forms of excess proliferation, malignant and non-malignant, that are pathogenic. For example, aberrant macrophage proliferation in atherogenic plaques appears to be restrained by $p16^{INK4a}$ and Arf expression, with reduced atherogenesis [49]. Beyond a wide range of murine studies supporting these relationships, there are also compelling genome-wide unbiased evidence for this model in humans. Genome wide association studies (GWAS) have linked regulation of the *CDKN2A* locus to many age-associated conditions such as cancer susceptibility, stroke, myocardial infarction, aortic aneurysm, glaucoma, and type 2 diabetes [50]. For cancer and atherosclerotic disease, it appears increased expression of *CDKN2A/B* transcripts is protective, but it is believed that excess expression of these transcripts may alternatively underlie the susceptibility to some of the other linked phenotypes.

2.2.3 Other Important Tumor Suppressors

As mentioned, there are additional genes that perform important tumor suppressor functions, but which do not fit neatly into either the caretaker or gatekeeper category. The Phosphatase and Tensin homolog (PTEN) protein converts phosphatidylinositol (3,4,5)-trisphosphate to phosphatidylinositol (4,5)-bisphosphate to inhibit downstream AKT signaling, thereby negatively regulating cell growth and proliferation [51]. PTEN mutations are frequently found in human brain and epithelial cancers, and mice engineered to delete *Pten* in somatic cells are cancer-prone. The human *STK11/LKB1* gene encodes a serine/threonine kinase that regulates energy metabolism and cell polarity through activation of a wide range of downstream kinases including the adenine monophosphate-activated protein kinase (AMPK) family of kinases. Somatic mutations of *LKB1* are found in a large fraction of human lung adenocarcinomas and cutaneous melanomas, and murine studies suggest loss of *LKB1* promotes cancer development, with a particularly strong association with distant metastases [52, 53]. The *NF1* gene encodes neurofibromin, a RAS GTPase-activating protein that negatively regulates RAS signaling. Loss of *NF1* causes hyperactive RAS signaling, leading to enhanced growth and decreased apoptosis of the mutant cell. Germ line mutation of *NF1* causes the tumor predisposition syndrome neurofibromatosis type I in humans, and somatic mutations of *NF1* are found in a wide range of human cancers. In all cases, these genes function as negative regulators of growth factor signaling and clearly act as tumor suppressors. Certainly the pathways regulated by these proteins, e.g. cellular growth and metabolism, have been linked to physiological aging through a variety of approaches, but the role of these specific proteins in human aging is unclear.

3 Cellular Mechanisms for Tumor Suppression

Upon detecting oncogenic stress, a proliferating cell can activate three major cellular tumor suppression mechanisms to prevent malignant transformation: permanent cell cycle arrest (cellular senescence), programmed cell death (e.g. through apoptosis, necroptosis or autophagy), or terminal differentiation (in the case of somatic stem and progenitor cells) (Fig. 2). Although much is known about the respective molecular biology of these processes, under which circumstances a damaged cell chooses each of these potential fates is not well-understood. Failure to activate the tumor suppression mechanisms, however, is a cardinal feature of all cancers.

3.1 Cellular Senescence

Among cellular tumor suppression mechanisms, the most important appears to be "cellular senescence", an irreversible form of cell cycle arrest, which is also intimately associated with aging. Senescence occurs in response to cellular stresses and displays features that distinguish it from other forms of growth arrest such as "quiescence" or "terminal differentiation" (reviewed in [54]). Several lines of evidence suggest that the senescence mechanism has beneficial features such as preventing cancer as well as non-neoplastic diseases characterized by pathogenic proliferation, and may also play an important role in wound healing. Importantly, however,

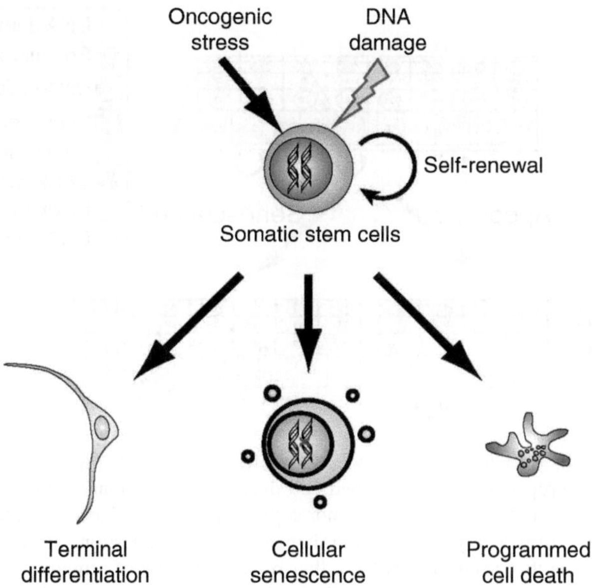

Fig. 2 Cellular tumor suppression mechanisms. Upon sensing oncogenic stress or persistent DNA damage, somatic stem cells and proliferating progenitors can activate three major cellular tumor suppression mechanisms to prevent malignant transformation: terminal differentiation, permanent cell cycle arrest (cellular senescence), or programmed cell death (e.g. through apoptosis, necroptosis, or autophagy). The choice to activate which specific mechanism is cell type and developmental stage dependent, and can lead to distinct physiological consequences

significant human and murine data suggest that senescence plays a causal role in some aspects of aging.

With regard to tumor suppression, senescence appears to be more critical than apoptosis or other forms of regulated cell death such as Autophagy. Mice with even mild defects in senescence activation are highly tumor-prone. In contrast, animals with near-complete deficiency in apoptosis (e.g. lacking Bax and Bak) display several developmental phenotypes and excess numbers of certain classes of hematopoietic cells, but are not generally cancer-prone [55]. In humans, unbiased genome-wide screens have demonstrated that inactivation of the senescence-promoting *CDKN2A* and *TP53* loci are common in familial cancer syndromes, and the Cancer Genome Atlas has established that loss of p16^{INK4a} and p53 functions are the most common somatic genetic events in human cancers, with possibly all tumors harboring one or multiple events targeting senescence (reviewed in [56, 57]). Even for a gene like *TP53*, which modulates both apoptosis and senescence, the available data suggest that its ability to promote cell cycle arrest is considerably more important to its ability to repress tumors than its effects on cell death [58, 59]. Therefore, not only is p53- and p16^{INK4a}–mediated cell cycle arrest (e.g. senescence) important for preventing tumorigenesis, it appears to have evolved in mammals to be the *most*

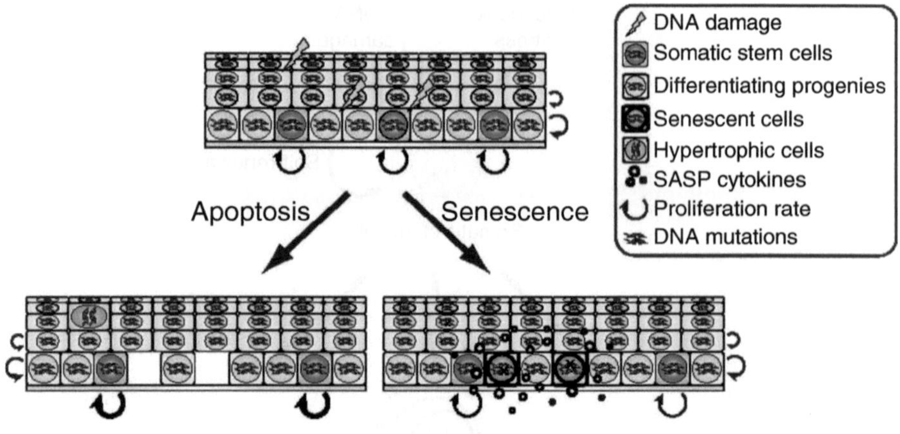

Fig. 3 Problems of Apoptosis vs. Senescence. While both are effective at preventing damaged cells from further propagation, programmed cell death (apoptosis) and cellular senescence have distinct advantages and disadvantages as tumor suppression mechanisms. The major advantage of apoptosis is that it irrevocably removes the damaged cells in question, however, this can leave "holes" in the tissue that need to be filled either by hypertrophy of remaining cells or increased proliferation of somatic stem and/or progenitor cells, which can compromise stem cell or tissue function. In contrast, while cellular senescence may have the advantage of preserving semi-functional cells after damage thus requiring less compensatory proliferation of the stem and progenitor cells, senescent cells can secrete harmful cytokines that interfere with normal tissue function, and senescent stem cells may occupy limited "niche" space leading to decreased overall function of the stem cell pool

critical anti-cancer mechanism. This fact suggests that senescence possesses some evolutionary advantage over apoptosis as a tumor suppressor mechanism.

One potential explanation for this observation is teleological: senescence did not evolve to suppress tumors, but rather was perfected through evolution to enhance species fitness by playing a role in physiologic processes such as the stress response or wound healing. By this argument, tumor suppression is merely a collateral benefit of the senescence mechanism, in the same way that decreased cancer is a benefit of more faithful DNA repair. This argument may be correct, but it has been hard to demonstrate what other function certain components of the senescence machinery (e.g. $p16^{INK4a}$) perform beyond limiting aberrant proliferation. For example, mice (or even in some rare cases, humans [60]) lacking $p16^{INK4a}$ appear phenotypically normal until the development of a wide variety of malignancies at an early age. So, even if the senescence response developed for additional reasons, the main phenotypic readout in mammals appears to be tumor predisposition.

Another possible reason for an evolutionary preference of senescence over apoptosis to censor tumors might be that apoptosis can produce new problems (Fig. 3). Since organ size and function of many tissues is regulated throughout the mammalian lifespan, a damaged cell censored by apoptosis would leave a niche "hole" that would then be filled either by hypertrophy of the remaining cells, or through increased proliferation of somatic stem cells. Hypertrophy is associated with tissue dysfunction in many organs (e.g. the heart), and would not solve the problem where diversity through cell number is important (e.g. the cellular immune system). Additional replication, however, may be undesirable as S-phase and mitosis provide opportunities for further errors of replication. Furthermore, the loss of certain cells to apoptosis in response to cellular stresses might, in particular, be functionally non-advantageous. For example, senescence, not apoptosis, is clearly the principal barrier to melanocyte transformation. As the primary function of melanocytes is to provide protection from damaging UV light through the synthesis of melanin, it could be evolutionarily undesirable for melanocytes, damaged as a result of UV exposure, to undergo apoptosis. Melanocyte senescence leaves a semi-functional, possibly protective melanocyte while diminishing the risk of malignant transformation. Therefore, senescence may be evolutionarily preferable to apoptosis as it may lead to a reduced lifetime proliferative burden on stem cells by leaving resident cells able to perform some desirable functions of related non-senescent cells (e.g. melanin synthesis). If correct, this model predicts that the clearance of senescence cells, for example through the use of 'senolytic' small molecules selectively toxic to senescent cells, might have unwanted toxicities. In contrast to this line of reasoning, Van Deursen and colleagues have reported that a genetic depletion of senescent cells is beneficial with regard to some aspects of aging in a strain of progeroid mice [61], although these experiments did not extend longevity or address the role of senescent cells in physiological aging. Ongoing experiments will address the consequences of senescence ablation in an aged individual.

3.2 *Programmed Cell Death*

As a tumor suppression mechanism, the major advantage of programmed cell death (such as apoptosis) over cellular senescence, is that it irrevocably removes damaged cells, thereby eliminating the potential for detrimental mutations to further propagate. This may be desirable under certain cellular contexts; for example, during early embryonic development it may be advantageous to eliminate cells with DNA damage instead of attempting repair, to prevent mutations from being transmitted into a large population of the cells. A similar situation may also apply to germ cells where the maintenance of genomic integrity is of utmost importance. Consistent with this notion, embryonic stem cells, early developing embryos and germ cells are all exquisitely sensitive to ionizing radiation-induced DNA damage, and they rapidly induce apoptosis without undergoing cell cycle arrest and DNA repair [62, 63]. These observations support the hypothesis that apoptosis may have evolved primarily as a mechanism to safeguard the genomic integrity against genotoxic stress, and the modest tumor suppression afforded by apoptosis in some tissues is only a collateral benefit. Nonetheless, apoptosis does play important roles in tumor suppression. Somatic mutations that affect genes regulating cell survival and apoptosis have been found in human cancers, particularly in tissues that have high intrinsic sensitivity to apoptosis such as lymphocytes; and pharmacological or biological means to induce cancer cell apoptosis are important therapeutic strategies to treat cancer.

3.3 Terminal Differentiation

Terminal differentiation is another form of permanent cell cycle arrest distinct from cellular senescence in that it is a developmentally programmed process for stem cells to generate normal effector cells, whereas senescence is stochastically induced by cellular stresses that distort homeostasis. The notion that terminal differentiation prevents oncogenic transformation is not new, as it has been demonstrated that the tumor suppressor protein APC prevents colon cancer by balancing the self-renewal and differentiation of intestinal stem cells through the regulation of cellular beta-catenin levels. However, the activity of APC is constitutive and not influenced by oncogenic stimuli. The first evidence suggesting terminal differentiation may act as a checkpoint against oncogenic stress comes from the finding that oncogene activation can induce differentiation. Over-expression of the *MYC* oncogene in the epidermis results in a long-term depletion of stem cells while accumulating differentiated keratinocytes [64]. Later, it was shown that DNA damage by ionizing radiation can directly trigger the differentiation of melanocyte stem cells and hematopoietic stem cells [65, 66]. These observations support the hypothesis that induced differentiation may be a conserved checkpoint mechanism to limit the self-renewal potential of somatic stem cells upon DNA damage or oncogene activation to prevent neoplastic transformation. However, it remains to be determined whether this mechanism applies to additional stem cell populations, and whether any cancer develops as a result of escaping the differentiation checkpoint mechanism.

4 Oncogenic Mutations Require a Proper Cellular Context for Malignant Transformation

While activating mutations of oncogenes and inactivating mutations of tumor suppressors are required by nearly every cancer, the particular genes that are preferentially mutated in each type of cancer are highly cancer-type specific. This suggests the non-genetic cellular context, that is the cellular epigenetic state, also plays a critical role in the impact of oncogenic mutations on cell biology. For example, activating mutations of the *H-RAS* oncogene cannot promote cancer in cells that do not express the *H-RAS* mRNA. Likewise, mutations that occur in replication incompetent cells (e.g. that are terminally differentiated) also do not drive oncogenesis. More subtly however, there are many examples in which the oncogenic potential of a mutant protein is restricted to a specific cell type or a particular developmental stage. For example, the loss of the VHL tumor suppressor is largely restricted to one type of advanced human cancer: renal carcinoma. VHL plays an important role in oxygen sensing, and VHL-inactivated cells act as if they are hypoxic. This observation can be interpreted as a demonstration that hypoxia does not promote cellular growth in most contexts, but is a stimulus to grow and divide in specific renal cells. Therefore, while cancer is a genetic disease, the specific cellular context determines the ability of oncogenic events to drive particular types of cancers.

5 How Does Cancer Relate to Aging?

Broadly defined, aging is the process of age-dependent decline in body function and increase in mortality that affect most living organisms. Aging is not a disease, rather, it is a collection of symptoms that reflect progressively decreased cellular function and disrupted tissue homeostasis over time. On the cellular level, aging is characterized by the time-dependent functional attrition of non-dividing (terminally differentiated) cells and impairment in somatic stem cell self-renewal and differentiation as a result of cell intrinsic and extrinsic stresses. These lead to impaired tissue homeostasis, disrupted tissue architecture, reduced organ function and altered energy metabolism, all of which contribute to the reduced fitness of aged organisms. Since aging is not caused by a single factor, no single intervention can be expected to treat all aging symptoms. The process of aging is not restricted to the aged. Mammalian cells experience DNA damage and other forms of age-promoting stresses on a daily basis. We are equipped with sophisticated repair mechanisms to deal with these challenges. Only when the extent of cellular and tissue damage exceeds the ability of our body to cope with these forms of damage do the symptoms of aging start to appear. Therefore, while completely preventing or reversing aging may be unlikely, minimizing these forms of daily damage may slow the rate of aging, while also reducing the incidence of cancer.

5.1 Aging as a Result of Time-Dependent Accumulation of Cellular Damage

One prominent feature of aging is the age-dependent accumulation of damage in various cellular components. Terminally differentiated cells such as neurons are sensitive to the accumulation of damaged lipids and proteins over time, largely due to their inability to dilute such molecules through successive cell divisions. In contrast, accumulation of genetic damage, including unrepaired DNA damage, is particularly detrimental to the function of long-term self-renewing cells including somatic stem cells. It is unclear if increases in macromolecular damage that are characteristic of aging solely reflect a lifetime's accumulation of toxic events, or also result from a reduced capacity to clear or repair damaged proteins and DNA with aging.

DNA damage impairs somatic stem cell function in several ways: it may lead to the generation of defective proteins that directly interfere with cellular function, or it may induce tumor suppression mechanisms as a result of persistent DNA damage or oncogene activation, which then limit the self-renewal potential of somatic stem cells through mechanisms such as senescence, cell death or differentiation. Therefore, the efficiency of somatic stem cells to successfully repair various types of DNA damage is a critical determinant of the rate of aging. As mentioned, several human progeroid syndromes (e.g. Bloom, Werner, Cockayne, Fanconi Anemia, etc.) are associated with mutations in DNA repair genes [67]. Telomere dysfunction is a specialized form of DNA damage, and mice deficient in telomerase have reduced somatic stem cells function and exhibit features of premature aging [15].

Therefore, despite their striking differences in presentation, aging and cancer seem to share a common origin: that is, the time-dependent accumulation of cellular macromolecular damage. When oncogenic mutations occur in a dividing cell as the result of DNA damage, the outcomes between cancer and aging depend on whether tumor suppression mechanisms can be successfully activated to prevent the damaged cell from further propagation: cancer arises when tumor suppression mechanisms fail, while aging results, at least partially, from the progressive functional attrition of self-renewing cells due to tumor suppression.

5.2 *How Is Cellular Senescence Related to Aging?*

Cellular senescence is thought to promote aging by reducing the regenerative potential of self-renewing cells and/or by leading to the production of detrimental cytokines and other biomolecules; however, some components of the senescence machinery (e.g. $p16^{INK4a}$) also appears to play a beneficial anti-aging role in some tissues by preventing the development of certain aging-associated pathologies (such as cancer and atherosclerosis). Therefore, while it may seem attractive to attempt to delay aging by attenuating tumor suppressor function or clearing senescent cells, more research is required on this topic to prevent unintended consequences.

5.2.1 Adverse Effects of Senescence with Regards to Aging: Loss of Self-Renewal

Although the expression of senescence markers is associated with aging, this observation does not establish a causal relationship between senescence and the loss of tissue replicative capacity associated with aging. To address this issue, a number of studies have been performed examining the effect of genetic ablation or overexpression of $p16^{INK4a}$ in self-renewing tissues and cell types (e.g. beta-cells, neural stem cells, HSCs, lymphocytes) on aging and age-related stresses. In mice, these studies have consistently shown that loss of $p16^{INK4a}$ can partially ameliorate the loss of tissue-specific replicative capacity with aging [43–48, 68], indicating that increasing levels of $p16^{INK4a}$ are not only associated with aging, but in part play a causal role in these tissues. In each of these compartments, $p16^{INK4a}$ deficiency attenuated the decline in proliferation and function as a function of advancing age, and the effects of $p16^{INK4a}$ loss are consistent across disparate self-renewing tissues. In no organ or tissue studied, however, does $p16^{INK4a}$ loss completely abrogate the effects of aging, indicating that $p16^{INK4a}$-independent aging also occurs in these compartments. Moreover, tissue specific ablation of $p16^{INK4a}$ in B-lymphocytes, as opposed to T-lymphocytes, was strongly associated with the development of cancer [46], making explicit the organismal costs of diminished $p16^{INK4a}$ function in some tissues.

Related observations have likewise suggested that activation of p53 or its effectors may compromise self-renewal to promote aging in mice [34, 35, 69] and humans [70]. The case for p53, however, is more complicated because it and its downstream effectors such as $p21^{CIP1}$ play important roles in regulating the DNA damage response, explaining why moderately increased gene dosage of *Cdkn2a* and *p53* in mice is associated with reduced aging [35, 37]. Additionally, HSCs from $p21^{CIP1}$–deficient mice demonstrate premature exhaustion [71], consistent with the notion that a p53- and $p21^{CIP1}$-dependent cell cycle pause in response to DNA damage or other stressors may be important for enhanced self-renewal and stem cell longevity in vivo. These results suggest that p53 activation can be both pro-aging and anti-aging depending on the nature and duration of the stress behind its activation.

Several lines of evidence have suggested that telomere dysfunction may contribute to mammalian aging by attenuating self-renewal and replicative capacity. Telomerase-deficient mice that have been serially backcrossed to harbor human length telomeres demonstrate regenerative failure in multiple organs due to a decline in stem cell proliferative capacity and tissue repair ability [72, 73]. In addition, human CD28 negative T cells, which have lost optimal telomerase activity, accumulate in vivo with age and display impaired function [74], and patients with congenital defects in telomerase activity display lymphopenia and lymphocyte hypo-proliferation [75, 76]. Together, these results support the notion that age-dependent activation of the cellular senescence program in proliferating cells impairs their self-renewal potential and promote aging.

5.2.2 Adverse Effects of Senescence with Regards to Aging: Gain of the Senescence Associated Secretory Phenotype (SASP)

In addition to a loss of replicative capacity, the accumulation of senescent cells with aging appears detrimental because of newly acquired functions of senescent cells. In particular, senescent cells display certain cell surface molecules and produce a raft of secreted SASP molecules associated with the 'sterile inflammation' characteristic of aging. While the production of SASP factors may be beneficial during transient episodes such as wound healing [77–79], the exponential accumulation of senescent cells with aging produces a monotonic increase in levels of SASP cytokines. Although numerous inflammatory markers increase with aging, the best studied with aging is IL-6, which is a reasonable aging biomarker whose expression is associated with worsening of age-associated phenotypes (reviewed in [80]).

Although perhaps beneficial in the short-term (e.g. in wound repair), elaboration of SASP cytokines appears detrimental in the long-term for several reasons. SASP factors may impair cellular differentiation and disrupt tissue homeostasis in otherwise undamaged tissues. Additionally, SASP factors appear to promote tumor growth in a paracrine manner [81–83]. Several cytokines associated with increased tumor progression are produced by senescent cells including IL-6, IL-8, GRO-alpha and VEGF; each of which could potentially promote tumor progression in nearby cells harboring oncogenic events. Finally, the chronic activation of immunity and associated immune dysfunction resulting from the potent pro-inflammatory cytokines associated with the SASP is associated with many aspects of aging (reviewed in [84]).

5.2.3 Benefits of Senescence with Regard to Aging

Although the field has perhaps focused more recently on the age-promoting roles of senescence, it is now clear that the senescence machinery also contributes important anti-aging roles. As mentioned, the clearest benefit of *p53* and *CDKN2A* activation is tumor suppression, and since neoplastic disease is strongly associated with aging, the anti-cancer functions of senescence attenuate this important age-associated phenotype. Animals lacking these pathways rapidly succumb to cancer, whereas mice engineered to exhibit a physiologically regulated increase in *Cdkn2a* and *p53* function exhibit a reduction in tumorigenesis with attendant longevity extension [37]. More recently, it has also become clear that the senescence machinery may limit non-malignant, but pathogenic proliferation in other age-associated settings. For example, work on humans and mice has suggested that *CDKN2A* expression plays an important role in preventing atherosclerosis by limiting disease-causing proliferation of monocyte/macrophage or smooth muscle progenitors [49, 85, 86]. It is also likely that activation of proteins associated with senescence may play beneficial roles in the prevention of other non-malignant diseases associated with aging (e.g. autoimmune conditions). While the notion that cancer and some aspects of aging are opposite outcomes based on the failure or success of senescence may be generally correct, these observations suggest that one should be careful as to how a given age-related phenotype is classified (i.e. as excess, aberrant proliferation, or replicative failure). Moreover, while activation of senescence may be "antagonistically pleiotropic" (that is, beneficial in young age but adverse in old age), the data seem clear with regard to senescence that some of the beneficial functions in youth actually provide a lifelong benefit, manifesting as a decreased incidence of certain classic age-associated conditions such as cancer and atherosclerosis.

6 Why Does the Incidence of Cancer Increase with Age?

The incidence of cancer increases sharply with aging (Fig. 4). For example, common epithelial malignancies of the breast and colon are highly unusual in individuals below the age of 40, with an exponential increase in incidence with aging such that such tumors are highly common in adults over the age of 65. Several reasons have been suggested for this intimate relationship between aging and cancer (Fig. 5):

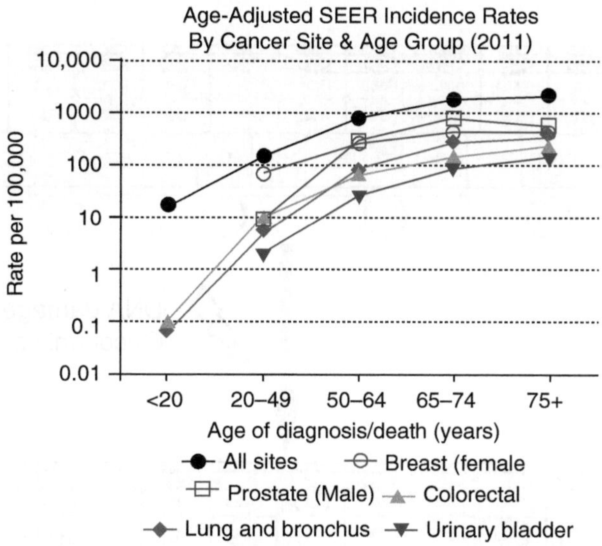

Fig. 4 Cancer incidence increases with aging. The incidence of most common types of human cancers, including lung, colorectal, urinary bladder, breast (female) and prostate (male), increases exponentially with age. While most types of cancers are extremely rare in young people, they are highly common in adults over the age of 65. Cancer incidence data are from the 2011 SEER (Surveillance, Epidemiology, and End Results Program) 18 areas. Rates are per 100,000 and are age-adjusted to the 2000 US Std Population (19 age groups – Census P25-1130). Cancer incidence data include all races and both sexes unless otherwise noted

6.1 Accumulation of DNA Mutations

Cancer requires the serial accumulation of a requisite number of DNA mutations in a long-lived self-renewing cell. One major source of time-dependent accumulation of DNA mutations likely comes from the intrinsic chemical instability of the DNA molecules. This is supported by the recent findings that most types of human cancer exhibit age-dependent enrichment of C>T substitutions at NpCpG trinucleotides, which is likely the result of the relatively elevated rate of spontaneous deamination of 5-methyl-cytocine at these locations [87]. The fact that this particular mutational signature is the only type that is strongly correlated with age across many cancer types suggests its accumulation is predominantly time dependent.

Another major source of DNA mutations likely occurs as accidents during DNA replication as cells divide throughout life. Since many of the self-renewing cells divide infrequently, the requisite number of mutations needed for malignant conversion may necessarily accumulate over years or decades. For example, HSCs represent the best understood system of human stem cell transformation. An adult human has fewer than 10^5 such cells that divide very rarely, approximately once every 40 weeks [88]. It is thought that these cells make roughly one mistake per replicative event, with the vast majority of errors in DNA replication being of no significance. As this small number of cells divides over decades, rare DNA mutations

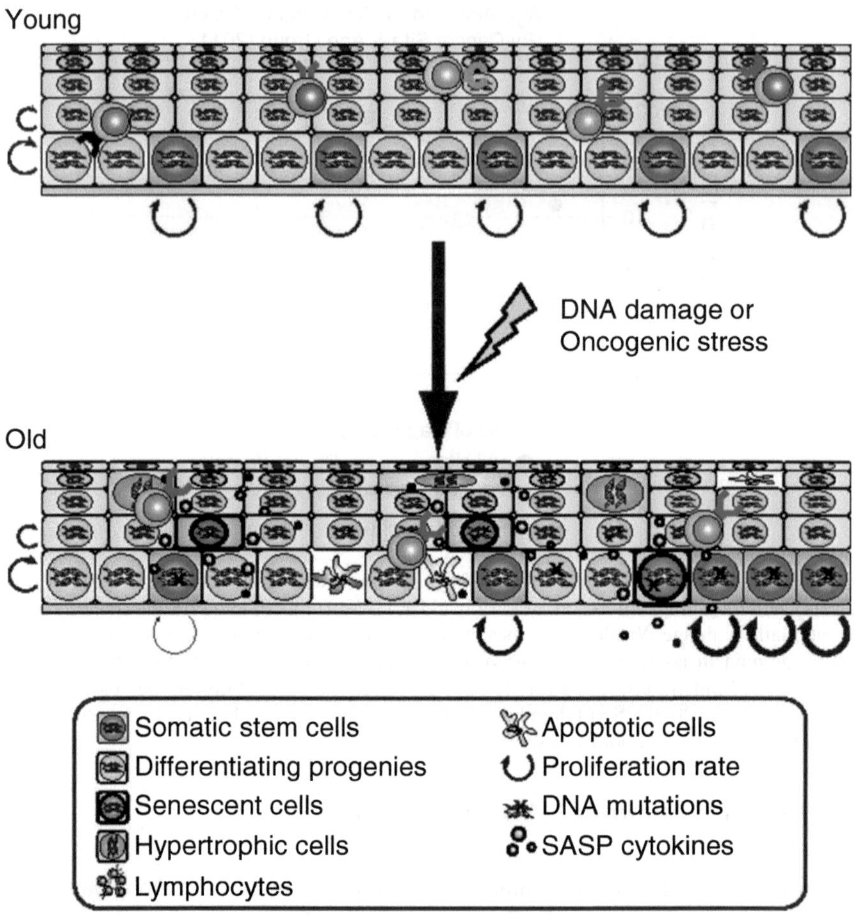

Fig. 5 Aging promotes cancer initiation and progression. Aging promotes neoplastic transformation and cancer development in several ways: (1) The time-dependent accumulation of DNA mutations in somatic stem cells can lead to the emergence of hyper-proliferative clones that may progress to full-blown cancer. (2) The functional attrition of somatic stem cells during aging as a result of genetic mutations and/or activation of cellular tumor suppression mechanisms can induce compensatory proliferation of remaining stem cells, which may lead to accumulation of new oncogenic mutations. (3) The age-dependent decline in adaptive immune system function as a result of reduced number, diversity, and functional competence of lymphocytes can lead to impaired immune surveillance and clearance of damaged or senescent cells from the body, which may contribute to increased cancer incidence in the elderly. (4) Accumulation of damaged and senescent cells in old tissue can disrupt tissue architecture and secrete pleotropic pro-inflammatory cytokines that contribute to cancer initiation and/or progression

occur stochastically which provide a selective advantage to a specific HSC, which then begins to overgrow the bone marrow. The increased replicative rate of this clone allows for further DNA mutations, and the serial accumulation of oncogenic driver events with inactivation of tumor suppressor mechanisms eventually leading to HSC cancer; that is, leukemia [89]. Presumably, cancer arises through similar mechanisms in other tissues, and in all cases, significant evidence suggests the period from the earliest stem oncogenic events to full malignant conversion can in some tissues last several decades. The more rapid induction of the same cancer type (leukemia) in mice over a much compressed period (~2–3 years) likely reflects several species differences including murine HSC dynamics (e.g. more rapid HSC proliferation) and reduced tumor suppressor barriers (e.g. differences in DNA repair, telomere biology and p53/p16^{INK4a} function).

Besides stochastic DNA mutations, genomic instability can also occur as a result of telomere shortening. In cells with intact DNA damage response machineries, critically shortened telomeres can trigger p53–mediated cellular senescence or apoptosis. In the absence of p53 however, telomere shortening instead causes chromosome end-to-end fusions, leading to chromosome breakage during the subsequent round of cell division, which can result in chromosomal abnormality and aneuploidy [90]. Human somatic cells start with approximately 15 kb of telomere sequence during early embryonic development [91]. In the absence of telomere maintenance mechanisms, each cell division is accompanied by a net loss of 100–200 bp of telomeric sequence. Even in stem cells such as HSCs that possess telomerase activity, telomere length decreases with age in vivo. Therefore, age-associated telomere shortening can contribute to increased cancer incidence in the elderly by inducing genomic instability and mutagenesis. Indeed, short telomere length has been associated with significantly increased risk of several types of human cancers [92, 93], and people with genetic deficiencies in telomere maintenance are predisposed to the development of certain cancers such as leukemia and keratinocytic cancers in addition to exhibiting features of premature aging [88]. Moreover, the human *TERT* locus on chromosome 5p has been associated through genome-wide association studies with a phenomenal range of human cancers, but in a complex manner. For example, common alleles with reduced telomerase activity (and presumably shorter telomeres) appear to increase human risk for the same cancers noted in the syndromes of congenital telomerase deficiency (e.g. leukemia), whereas the same alleles appear to decrease the risk for other cancer types such as melanoma, where telomerase clearly appears to be oncogenic [94].

6.2 Loss of Stem Cell Function with Compensatory Hyper-Proliferation

Given this model of how serial mutations occur in self renewing cells to yield malignant clones, it becomes obvious that anything which increases the chances of DNA mutation or which increases the rate of proliferation of tissue-specific stem cells will also increase the risk of malignant conversion. Carcinogens such as UV light and tobacco smoke are well known mutagens, which promote cancer by damaging DNA. Likewise, however, it is becoming clear that somatic stem cell function is compromised with aging in many tissues. This may occur as a result of tumor suppressor mechanisms such as senescence, which in turn occurs as a result of telomere shortening or other intrinsic and extrinsic cellular stresses. Importantly, however, the loss of even a fraction of a tissue's somatic stem cell compartment imposes a significant replicative burden on the remaining stem cells of that tissue. This compensatory increase in replication in turn translates into a greater risk of cancer, as well as additional possible stem cell attrition. Familiar examples of this process in humans are bone marrow failure syndromes such as aplastic anemia and myelodysplasia, where HSC numbers are sharply compromised, and malignant transformation to acute leukemia is highly enhanced.

6.3 Waning Immunity

Indubitably, the cellular immune system plays an important role in the surveillance and prevention of neoplastic clones. In fact, it appears that a substantial fraction of human cancers develop specific mechanisms to evade the immune system, thereby allowing for progression into later stages [95]. The ability of the immune system to repress malignancy may be particularly important in certain kinds of cancers that are very immunogenic; for example, those resulting from viral infections or which have tumor neo-antigens. Since the immune system plays an active role in cancer repression throughout life, the waning of immunity with aging, and in particular cellular immunity as mediated by T cells, likely plays an important role in tumorigenesis. Loss of immune function has been suggested to explain an observation in human melanoma: tumors driven by mutations of the *NRAS* oncogene (~25 % of melanoma) occur in patients significantly older than in those with tumors driven by mutations of the *BRAF* oncogene (~50 % of human melanoma). NRAS tumors are more strongly associated with chronic sun damage, and exhibit a larger number of antigenic mutations than *BRAF* tumors. Therefore, one hypothesis is that *NRAS* tumors are more efficiently cleared by younger individuals, but become more common with aging as cellular immunity decreases.

6.4 Changes in the Cellular Milieu

Beyond cell intrinsic events, it is also likely that extracellular changes that occur with aging also promote malignancy. For example, an increase in the number of senescent cells in aging tissues is thought to cause a significant increase in the local concentrations of pro-inflammatory cytokines secreted by senescent cells. The molecules elaborated by senescent cells are highly pleiotropic, causing many effects such as disruption of cellular differentiation, increasing proliferation, degradation of tissue matrix and increased cellular survival, which may in turn promote cancer initiation and/or progression. It should be noted that this model is somewhat controversial, with many authors suggesting senescence-related cytokines can also deter tumorigenesis, and therefore the effect of senescent cells with regard to cancer are likely tissue- and context-specific.

7 Preventing Cancer and Aging

Clearly, cancer and aging are intimately linked, such that cancer to some extent may be an almost inevitable consequence of advanced age in the setting of the accumulation of macromolecular damage, loss of somatic stem cell function, waning immunity, etc. While our understanding of the molecular basis of cancer and aging continues to grow, it is possible to design strategies to both decrease the incidence of cancer as well as slow the rate of aging.

7.1 Minimizing Cellular Damage

As it becomes increasingly clear that a time-dependent accumulation of cellular damage (particularly DNA damage and mutations) serves as a common molecular basis for both cancer and aging, reducing the total amount of damage our cells experience throughout life can be expected to protect against both cancer and aging. One way to achieve this is to minimize the exposure to environmental mutagens such as smoking, UV radiation and chemical carcinogens that can cause direct DNA damage and mutations. Indeed, it has been shown that quitting smoking has a myriad of health benefits in both target tissues such as the lung (e.g. reduced risk of lung cancer, emphysema and pulmonary fibrosis), as well as tissues less directly affected by tobacco inhalation (e.g. a reduction of cancer in many tissues, enhanced immunity, less vascular disease). Evidence suggests that tobacco exposure widely promotes aging-like phenotypes in many tissues, and therefore has been argued to represent the prototypical human 'gerontogen' [96, 97]; that is, an environmental exposure that promotes physiological aging. Similarly, avoiding UV exposure can dramatically decrease the incidence or cutaneous melanoma and slow skin aging. In addition to naturally occurring mutagens, cytotoxic agents that directly damage DNA are commonly used for treating a wide range of human malignancies. As a result, patients, particularly young children, who received cytotoxic cancer therapy, are at increased risk of developing secondary malignancies [35], and often display features of accelerated aging affecting a wide range of tissues [98]. The latter is believed to be partly due to reduced somatic stem cell function, either as a direct result of therapy-induced stem cell attrition (through cell death, senescence, or differentiation), or as the result of compensatory proliferation-induced exhaustion. Therefore, finding ways to protect stem cell function in cancer patients may help to reduce the aging-promoting effects of cytotoxic therapy in this special scenario. This issue is addressed in more detail in Chap. 4. The Impact of Cancer Treatments on Aging. Another potential way to reduce cellular damage is through modulating energy metabolism. It has long been observed that interventions such as dietary restrict (DR) and pharmacological inhibition of the mTOR pathway, which slow cell growth and metabolism, reduce the incidence of cancer and extend life span. One possible explanation for this is that decreased cellular energy metabolism may reduce the generation of harmful byproducts that cause cellular damage including DNA mutations. Alternatively, lowered cellular metabolism may slow cell proliferation, resulting in reduced replication-associated DNA damage and mutations.

7.2 Enhancing Immune System Function

The immune system plays an important role during both aging and tumorigenesis. As humans age, the function of the immune system declines sharply, resulting in increased susceptibility of the elderly to external infection. This aging-associated decline in adaptive immune response may also contribute to impaired immune surveillance and clearance of damaged or senescent cells from the body, which can contribute to aging and cancer development. On the other hand, an aberrant immune response such as autoimmune diseases also increases with age. Therefore, finding ways to augment the normal immune system function without over-activation in old individuals may ameliorate certain aspects of aging while providing an important anti-cancer defense. One suggested approach has been to augment thymic function, therefore delaying T cell aging, through supplementation with cytokines such as IL-7 [99].

7.3 Maintaining a Functional Stem Cell Pool

A key feature of aging is the loss of regenerative potential in self-renewing tissues as a result of functional attrition of somatic stem cells. Therefore, an important goal of aging research is to find ways to maintain functional stem cells throughout life. Functional attrition of stem cells with age occurs in part as the result of DNA damage accumulation that subsequently activates tumor suppression mechanisms. As DNA damage primarily occurs during DNA replication, the number of cell divisions a given stem cell goes through is expected to be negatively associated with their function. In support of this notion, many types of somatic stem cells including HSCs are found to be highly quiescent, and loss of quiescence is associated with premature stem cell exhaustion. Therefore, reducing the rate of stem cell proliferation may provide a way to slow stem cell aging. Alternatively, providing an exogenous source of healthy somatic stem cells, for example through a regenerative medicine approach, could provide a means to reduce both aging and cancer.

8 Summary

Serial accumulation of mutations in long-lived somatic stem cells as a result of imperfect DNA repair mechanism contributes to both cancer and aging. The choice between these two seemingly distinct outcomes depends on the effectiveness of cellular tumor suppression mechanisms against oncogenic mutation – failed tumor suppression results in cancer, while successful tumor suppression causes functional attrition of self-renewing cells, contributing to physiological aging. The increased incidence of cancer with aging reflects the time-dependent accumulation of mutations in somatic self-renewing cells, as well as waning immune surveillance and possibly pro-oncogenic changes to the tissue milieu with aging. While preventing aging or curing cancer is unlikely, minimizing cellular damage, enhancing immune system function, and maintaining a functional stem cell pool may slow the rate of aging while simultaneously reducing the risk of cancer.

Acknowledgments NES and SH are supported by grants from the NIA and NCI.
Editor: Kevin Howcroft (National Cancer Institute, NCI), NIH.

References

1. Tomasetti C, Vogelstein B (2015) Cancer etiology. Variation in cancer risk among tissues can be explained by the number of stem cell divisions. Science 347(6217):78–81. doi:10.1126/science.1260825
2. Mori H, Colman SM, Xiao Z, Ford AM, Healy LE, Donaldson C et al (2002) Chromosome translocations and covert leukemic clones are generated during normal fetal development. Proc Natl Acad Sci U S A 4:4
3. Davis AC, Wims M, Spotts GD, Hann SR, Bradley A (1993) A null c-myc mutation causes lethality before 10.5 days of gestation in homozygotes and reduced fertility in heterozygous female mice. Genes Dev 7(4):671–682
4. Johnson L, Greenbaum D, Cichowski K, Mercer K, Murphy E, Schmitt E et al (1997) K-ras is an essential gene in the mouse with partial functional overlap with N-ras. Genes Dev 11(19):2468–2481
5. Park IK, Qian D, Kiel M, Becker MW, Pihalja M, Weissman IL et al (2003) Bmi-1 is required for maintenance of adult self-renewing haematopoietic stem cells. Nature 423(6937):302–305. doi:10.1038/nature01587
6. Pagliarini R, Shao W, Sellers WR. Oncogene addiction: pathways of therapeutic response, resistance, and road maps toward a cure. EMBO Rep. 2015;16(3):280–296. doi:10.15252/embr.201439949.
7. Armanios M, Blackburn EH (2012) The telomere syndromes. Nat Rev Genet 13(10):693–704. doi:10.1038/nrg3246
8. Lund K, Adams PD, Copland M (2014) EZH2 in normal and malignant hematopoiesis. Leukemia 28(1):44–49. doi:10.1038/leu.2013.288
9. Genovese G, Kahler AK, Handsaker RE, Lindberg J, Rose SA, Bakhoum SF et al (2014) Clonal hematopoiesis and blood-cancer risk inferred from blood DNA sequence. N Engl J Med 371(26):2477–2487. doi:10.1056/NEJMoa1409405
10. Jaiswal S, Fontanillas P, Flannick J, Manning A, Grauman PV, Mar BG et al (2014) Age-related clonal hematopoiesis associated with adverse outcomes. N Engl J Med 371(26):2488–2498. doi:10.1056/NEJMoa1408617
11. Xie M, Lu C, Wang J, McLellan MD, Johnson KJ, Wendl MC et al (2014) Age-related mutations associated with clonal hematopoietic expansion and malignancies. Nat Med 20(12):1472–1478. doi:10.1038/nm.3733
12. Kinzler KW, Vogelstein B (1997) Cancer-susceptibility genes. Gatekeepers and caretakers. Nature 386(6627):761–763. doi:10.1038/386761a0
13. van Heemst D, den Reijer PM, Westendorp RG (2007) Ageing or cancer: a review on the role of caretakers and gatekeepers. Eur J Cancer 43(15):2144–2152. doi:10.1016/j.ejca.2007.07.011

14. Lombard DB, Chua KF, Mostoslavsky R, Franco S, Gostissa M, Alt FW (2005) DNA repair, genome stability, and aging. Cell 120(4):497–512. doi:10.1016/j.cell.2005.01.028
15. He S, Nakada D, Morrison SJ (2009) Mechanisms of stem cell self-renewal. Annu Rev Cell Dev Biol 25:377–406. doi:10.1146/annurev.cellbio.042308.113248
16. Hosoya N, Miyagawa K (2014) Targeting DNA damage response in cancer therapy. Cancer Sci 105(4):370–388. doi:10.1111/cas.12366
17. Dietlein F, Thelen L, Reinhardt HC (2014) Cancer-specific defects in DNA repair pathways as targets for personalized therapeutic approaches. Trends Genet 30(8):326–339. doi:10.1016/j.tig.2014.06.003
18. Bryant HE, Schultz N, Thomas HD, Parker KM, Flower D, Lopez E et al (2005) Specific killing of BRCA2-deficient tumours with inhibitors of poly(ADP-ribose) polymerase. Nature 434(7035):913–917. doi:10.1038/nature03443
19. Farmer H, McCabe N, Lord CJ, Tutt AN, Johnson DA, Richardson TB et al (2005) Targeting the DNA repair defect in BRCA mutant cells as a therapeutic strategy. Nature 434(7035):917–921. doi:10.1038/nature03445
20. Jaramillo MC, Zhang DD (2013) The emerging role of the Nrf2-Keap1 signaling pathway in cancer. Genes Dev 27(20):2179–2191. doi:10.1101/gad.225680.113
21. Salmon AB, Richardson A, Perez VI (2010) Update on the oxidative stress theory of aging: does oxidative stress play a role in aging or healthy aging? Free Radic Biol Med 48(5):642–655. doi:10.1016/j.freeradbiomed.2009.12.015
22. Perez VI, Bokov A, Van Remmen H, Mele J, Ran Q, Ikeno Y et al (2009) Is the oxidative stress theory of aging dead? Biochim Biophys Acta 1790(10):1005–1014. doi:10.1016/j.bbagen.2009.06.003
23. Sherr CJ (2001) The INK4a/ARF network in tumour suppression. Nat Rev Mol Cell Biol 2(10):731–737. doi:10.1038/35096061
24. Kim WY, Sharpless NE (2006) The regulation of INK4/ARF in cancer and aging. Cell 127(2):265–275. doi:10.1016/j.cell.2006.10.003
25. Fridman JS, Lowe SW (2003) Control of apoptosis by p53. Oncogene 22(56):9030–9040. doi:10.1038/sj.onc.1207116
26. Purvis JE, Karhohs KW, Mock C, Batchelor E, Loewer A, Lahav G (2012) p53 dynamics control cell fate. Science 336(6087):1440–1444. doi:10.1126/science.1218351
27. Burd CE, Sharpless NE (2010) What's so special about RB? Cancer Cell 17(4):313–314. doi:10.1016/j.ccr.2010.03.010
28. Chicas A, Wang X, Zhang C, McCurrach M, Zhao Z, Mert O et al (2010) Dissecting the unique role of the retinoblastoma tumor suppressor during cellular senescence. Cancer Cell 17(4):376–387. doi:10.1016/j.ccr.2010.01.023
29. Roussel MF (1999) The INK4 family of cell cycle inhibitors in cancer. Oncogene 18(38):5311–5317. doi:10.1038/sj.onc.1202998
30. Burd CE, Sorrentino JA, Clark KS, Darr DB, Krishnamurthy J, Deal AM et al (2013) Monitoring tumorigenesis and senescence in vivo with a p16(INK4a)-luciferase model. Cell 152(1–2):340–351. doi:10.1016/j.cell.2012.12.010
31. Ventura A, Kirsch DG, McLaughlin ME, Tuveson DA, Grimm J, Lintault L et al (2007) Restoration of p53 function leads to tumour regression in vivo. Nature 445(7128):661–665. doi:10.1038/nature05541
32. Martins CP, Brown-Swigart L, Evan GI (2006) Modeling the therapeutic efficacy of p53 restoration in tumors. Cell 127(7):1323–1334. doi:10.1016/j.cell.2006.12.007
33. Rufini A, Tucci P, Celardo I, Melino G (2013) Senescence and aging: the critical roles of p53. Oncogene 32(43):5129–5143. doi:10.1038/onc.2012.640
34. Maier B, Gluba W, Bernier B, Turner T, Mohammad K, Guise T et al (2004) Modulation of mammalian life span by the short isoform of p53. Genes Dev 18(3):306–319
35. Tyner SD, Venkatachalam S, Choi J, Jones S, Ghebranious N, Igelmann H et al (2002) p53 mutant mice that display early ageing-associated phenotypes. Nature 415(6867):45–53
36. Dumble M, Moore L, Chambers SM, Geiger H, Van Zant G, Goodell MA et al (2007) The impact of altered p53 dosage on hematopoietic stem cell dynamics during aging. Blood 109(4):1736–1742. doi:10.1182/blood-2006-03-010413

37. Matheu A, Maraver A, Klatt P, Flores I, Garcia-Cao I, Borras C et al (2007) Delayed ageing through damage protection by the Arf/p53 pathway. Nature 448(7151):375–379
38. Garcia-Cao I, Garcia-Cao M, Martin-Caballero J, Criado LM, Klatt P, Flores JM et al (2002) "Super p53" mice exhibit enhanced DNA damage response, are tumor resistant and age normally. EMBO J 21(22):6225–6235
39. Matheu A, Pantoja C, Efeyan A, Criado LM, Martin-Caballero J, Flores JM et al (2004) Increased gene dosage of Ink4a/Arf results in cancer resistance and normal aging. Genes Dev 18(22):2736–2746. doi:10.1101/gad.310304
40. Krishnamurthy J, Torrice C, Ramsey MR, Kovalev GI, Al-Regaiey K, Su L et al (2004) Ink4a/Arf expression is a biomarker of aging. J Clin Invest 114(9):1299–1307. doi:10.1172/JCI22475
41. Zindy F, Quelle DE, Roussel MF, Sherr CJ (1997) Expression of the p16INK4a tumor suppressor versus other INK4 family members during mouse development and aging. Oncogene 15(2):203–211. doi:10.1038/sj.onc.1201178
42. Liu Y, Sanoff HK, Cho H, Burd CE, Torrice C, Ibrahim JG et al (2009) Expression of p16(INK4a) in peripheral blood T-cells is a biomarker of human aging. Aging Cell 8(4):439–448. doi:10.1111/j.1474-9726.2009.00489.x
43. Molofsky AV, Slutsky SG, Joseph NM, He S, Pardal R, Krishnamurthy J et al (2006) Increasing p16INK4a expression decreases forebrain progenitors and neurogenesis during ageing. Nature 443(7110):448–452. doi:10.1038/nature05091
44. Janzen V, Forkert R, Fleming HE, Saito Y, Waring MT, Dombkowski DM et al (2006) Stem-cell ageing modified by the cyclin-dependent kinase inhibitor p16INK4a. Nature 443(7110):421–426
45. Krishnamurthy J, Ramsey MR, Ligon KL, Torrice C, Koh A, Bonner-Weir S et al (2006) p16INK4a induces an age-dependent decline in islet regenerative potential. Nature 443(7110):453–457. doi:10.1038/nature05092
46. Liu Y, Johnson SM, Fedoriw Y, Rogers AB, Yuan H, Krishnamurthy J et al (2011) Expression of p16(INK4a) prevents cancer and promotes aging in lymphocytes. Blood 117(12):3257–3267. doi:10.1182/blood-2010-09-304402
47. Signer RA, Montecino-Rodriguez E, Witte ON, Dorshkind K (2008) Aging and cancer resistance in lymphoid progenitors are linked processes conferred by p16Ink4a and Arf. Genes Dev 22(22):3115–3120. doi:10.1101/gad.1715808
48. Chen H, Gu X, Su IH, Bottino R, Contreras JL, Tarakhovsky A et al (2009) Polycomb protein Ezh2 regulates pancreatic beta-cell Ink4a/Arf expression and regeneration in diabetes mellitus. Genes Dev 23(8):975–985. doi:10.1101/gad.1742509
49. Kuo CL, Murphy AJ, Sayers S, Li R, Yvan-Charvet L, Davis JZ et al (2011) Cdkn2a is an atherosclerosis modifier locus that regulates monocyte/macrophage proliferation. Arterioscler Thromb Vasc Biol 31(11):2483–2492. doi:10.1161/ATVBAHA.111.234492
50. Jeck WR, Siebold AP, Sharpless NE (2012) Review: a meta-analysis of GWAS and age-associated diseases. Aging Cell 11(5):727–731. doi:10.1111/j.1474-9726.2012.00871.x
51. Song MS, Salmena L, Pandolfi PP (2012) The functions and regulation of the PTEN tumour suppressor. Nat Rev Mol Cell Biol 13(5):283–296. doi:10.1038/nrm3330
52. Liu W, Monahan KB, Pfefferle AD, Shimamura T, Sorrentino J, Chan KT et al (2012) LKB1/STK11 inactivation leads to expansion of a prometastatic tumor subpopulation in melanoma. Cancer Cell 21(6):751–764. doi:10.1016/j.ccr.2012.03.048
53. Ji H, Ramsey MR, Hayes DN, Fan C, McNamara K, Kozlowski P et al (2007) LKB1 modulates lung cancer differentiation and metastasis. Nature 448(7155):807–810. doi:10.1038/nature06030
54. Sharpless NE, Sherr CJ (2015) Forging a signature of in vivo senescence. [Review]. Nat Rev Cancer 15(7):397–408. doi:10.1038/nrc3960
55. Lindsten T, Ross AJ, King A, Zong WX, Rathmell JC, Shiels HA et al (2000) The combined functions of proapoptotic Bcl-2 family members bak and bax are essential for normal development of multiple tissues. Mol Cell 6(6):1389–1399
56. Sharpless NE, DePinho RA (2004) Telomeres, stem cells, senescence, and cancer. J Clin Invest 113(2):160–168
57. Campisi J, d'Adda di Fagagna F (2007) Cellular senescence: when bad things happen to good cells. Nat Rev Mol Cell Biol 8(9):729–740

58. Liu G, Terzian T, Xiong S, Van Pelt CS, Audiffred A, Box NF et al (2007) The p53-Mdm2 network in progenitor cell expansion during mouse postnatal development. J Pathol 213(4):360–368. doi:10.1002/path.2238
59. Barboza JA, Liu G, Ju Z, El-Naggar AK, Lozano G (2006) p21 delays tumor onset by preservation of chromosomal stability. Proc Natl Acad Sci U S A 103(52):19842–19847. doi:10.1073/pnas.0606343104
60. Gruis NA, van der Velden PA, Sandkuijl LA, Prins DE, Weaver-Feldhaus J, Kamb A et al (1995) Homozygotes for CDKN2 (p16) germline mutation in Dutch familial melanoma kindreds. Nat Genet 10(3):351–353. doi:10.1038/ng0795-351
61. Baker DJ, Wijshake T, Tchkonia T, LeBrasseur NK, Childs BG, van de Sluis B et al (2011) Clearance of p16Ink4a-positive senescent cells delays ageing-associated disorders. Nature 479(7372):232–236. doi:10.1038/nature10600
62. Liu JC, Lerou PH, Lahav G (2014) Stem cells: balancing resistance and sensitivity to DNA damage. Trends Cell Biol 24(5):268–274. doi:10.1016/j.tcb.2014.03.002
63. Heyer BS, MacAuley A, Behrendtsen O, Werb Z (2000) Hypersensitivity to DNA damage leads to increased apoptosis during early mouse development. Genes Dev 14(16):2072–2084
64. Eilers M, Eisenman RN (2008) Myc's broad reach. Genes Dev 22(20):2755–2766. doi:10.1101/gad.1712408
65. Wang J, Sun Q, Morita Y, Jiang H, Gross A, Lechel A et al (2012) A differentiation checkpoint limits hematopoietic stem cell self-renewal in response to DNA damage. Cell 148(5):1001–1014. doi:10.1016/j.cell.2012.01.040
66. Inomata K, Aoto T, Binh NT, Okamoto N, Tanimura S, Wakayama T et al (2009) Genotoxic stress abrogates renewal of melanocyte stem cells by triggering their differentiation. Cell 137(6):1088–1099. doi:10.1016/j.cell.2009.03.037
67. Lopez-Otin C, Blasco MA, Partridge L, Serrano M, Kroemer G (2013) The hallmarks of aging. Cell 153(6):1194–1217. doi:10.1016/j.cell.2013.05.039
68. Chen H, Gu X, Liu Y, Wang J, Wirt SE, Bottino R et al (2011) PDGF signalling controls age-dependent proliferation in pancreatic beta-cells. Nature 478(7369):349–355. doi:10.1038/nature10502
69. Choudhury AR, Ju Z, Djojosubroto MW, Schienke A, Lechel A, Schaetzlein S et al (2007) Cdkn1a deletion improves stem cell function and lifespan of mice with dysfunctional telomeres without accelerating cancer formation. Nat Genet 39(1):99–105. doi:10.1038/ng1937
70. Orsted DD, Bojesen SE, Tybjaerg-Hansen A, Nordestgaard BG (2007) Tumor suppressor p53 Arg72Pro polymorphism and longevity, cancer survival, and risk of cancer in the general population. J Exp Med 204(6):1295–1301
71. Cheng T, Rodrigues N, Shen H, Yang Y, Dombkowski D, Sykes M et al (2000) Hematopoietic stem cell quiescence maintained by p21cip1/waf1. Science 287(5459):1804–1808
72. Rudolph KL, Chang S, Lee HW, Blasco M, Gottlieb GJ, Greider C et al (1999) Longevity, stress response, and cancer in aging telomerase-deficient mice. Cell 96(5):701–712
73. Ju Z, Jiang H, Jaworski M, Rathinam C, Gompf A, Klein C, Ju Z, Jiang H, Jaworski M, Rathinam C, Gompf A, Klein C et al (2007) Telomere dysfunction induces environmental alterations limiting hematopoietic stem cell function and engraftment. Nat Med 13(6):742–747
74. Effros RB, Dagarag M, Spaulding C, Man J (2005) The role of CD8+ T-cell replicative senescence in human aging. Immunol Rev 205:147–157
75. Knudson M, Kulkarni S, Ballas ZK, Bessler M, Goldman F (2005) Association of immune abnormalities with telomere shortening in autosomal-dominant dyskeratosis congenita. Blood 105(2):682–688. doi:10.1182/blood-2004-04-1673
76. Jyonouchi S, Forbes L, Ruchelli E, Sullivan KE (2011) Dyskeratosis congenita: a combined immunodeficiency with broad clinical spectrum--a single-center pediatric experience. [Review]. Pediatr Allergy Immunol 22(3):313–319. doi:10.1111/j.1399-3038.2010.01136.x
77. Rodier F, Campisi J (2011) Four faces of cellular senescence. [Review]. J Cell Biol 192(4):547–556. doi:10.1083/jcb.201009094
78. Demaria M, Ohtani N, Youssef SA, Rodier F, Toussaint W, Mitchell JR et al (2014) An essential role for senescent cells in optimal wound healing through secretion of PDGF-AA. Dev Cell 31(6):722–733. doi:10.1016/j.devcel.2014.11.012

79. Neves J, Demaria M, Campisi J, Jasper H (2015) Of flies, mice, and men: evolutionarily conserved tissue damage responses and aging. Dev Cell 32(1):9–18. doi:10.1016/j.devcel.2014.11.028
80. Ershler WB, Keller ET (2000) Age-associated increased interleukin-6 gene expression, late-life diseases, and frailty. Annu Rev Med 51:245–270
81. Krtolica A, Parrinello S, Lockett S, Desprez PY, Campisi J (2001) Senescent fibroblasts promote epithelial cell growth and tumorigenesis: a link between cancer and aging. Proc Natl Acad Sci U S A 98(21):12072–12077
82. Parrinello S, Coppe JP, Krtolica A, Campisi J (2005) Stromal-epithelial interactions in aging and cancer: senescent fibroblasts alter epithelial cell differentiation. J Cell Sci 118(Pt 3):485–496
83. Liu D, Hornsby PJ (2007) Senescent human fibroblasts increase the early growth of xenograft tumors via matrix metalloproteinase secretion. Cancer Res 67(7):3117–3126. doi:10.1158/0008-5472.CAN-06-3452
84. Franceschi C, Capri M, Monti D, Giunta S, Olivieri F, Sevini F et al (2007) Inflammaging and anti-inflammaging: a systemic perspective on aging and longevity emerged from studies in humans. Mech Ageing Dev 128(1):92–105. doi:10.1016/j.mad.2006.11.016
85. Liu Y, Sanoff HK, Cho H, Burd CE, Torrice C, Mohlke KL et al (2009) INK4/ARF transcript expression is associated with chromosome 9p21 variants linked to atherosclerosis. PLoS One 4(4):e5027. doi:10.1371/journal.pone.0005027
86. Gonzalez-Navarro H, Abu Nabah YN, Vinue A, Andres-Manzano MJ, Collado M, Serrano M et al (2010) p19(ARF) deficiency reduces macrophage and vascular smooth muscle cell apoptosis and aggravates atherosclerosis. J Am Coll Cardiol 55(20):2258–2268
87. Alexandrov LB, Nik-Zainal S, Wedge DC, Aparicio SA, Behjati S, Biankin AV et al (2013) Signatures of mutational processes in human cancer. Nature 500(7463):415–421. doi:10.1038/nature12477
88. Catlin SN, Busque L, Gale RE, Guttorp P, Abkowitz JL (2011) The replication rate of human hematopoietic stem cells in vivo. Blood 117(17):4460–4466. doi:10.1182/blood-2010-08-303537
89. Jan M, Snyder TM, Corces-Zimmerman MR, Vyas P, Weissman IL, Quake SR et al (2012) Clonal evolution of preleukemic hematopoietic stem cells precedes human acute myeloid leukemia. Sci Transl Med 4(149):149ra118. doi:10.1126/scitranslmed.3004315
90. O'Sullivan RJ, Karlseder J (2010) Telomeres: protecting chromosomes against genome instability. Nat Rev Mol Cell Biol 11(3):171–181. doi:10.1038/nrm2848
91. Samassekou O, Gadji M, Drouin R, Yan J (2010) Sizing the ends: normal length of human telomeres. Ann Anat 192(5):284–291. doi:10.1016/j.aanat.2010.07.005
92. Prescott J, Wentzensen IM, Savage SA, De Vivo I (2012) Epidemiologic evidence for a role of telomere dysfunction in cancer etiology. Mutat Res 730(1–2):75–84. doi:10.1016/j.mrfmmm.2011.06.009
93. Hou L, Zhang X, Gawron AJ, Liu J (2012) Surrogate tissue telomere length and cancer risk: shorter or longer? Cancer Lett 319(2):130–135. doi:10.1016/j.canlet.2012.01.028
94. Huang FW, Hodis E, Xu MJ, Kryukov GV, Chin L, Garraway LA (2013) Highly recurrent TERT promoter mutations in human melanoma. Science 339(6122):957–959. doi:10.1126/science.1229259
95. Vesely MD, Kershaw MH, Schreiber RD, Smyth MJ (2011) Natural innate and adaptive immunity to cancer. Annu Rev Immunol 29:235–271. doi:10.1146/annurev-immunol-031210-101324
96. Martin GM (1987) Interactions of aging and environmental agents: the gerontological perspective. Prog Clin Biol Res 228:25–80
97. Sorrentino JA, Krishnamurthy J, Tilley S, Alb JG Jr, Burd CE, Sharpless NE (2014) p16INK4a reporter mice reveal age-promoting effects of environmental toxicants. J Clin Invest 124(1):169–173. doi:10.1172/JCI70960
98. Hudson MM, Ness KK, Gurney JG, Mulrooney DA, Chemaitilly W, Krull KR et al (2013) Clinical ascertainment of health outcomes among adults treated for childhood cancer. JAMA 309(22):2371–2381. doi:10.1001/jama.2013.6296
99. Holland AM, van den Brink MR (2009) Rejuvenation of the aging T cell compartment. Curr Opin Immunol 21(4):454–459. doi:10.1016/j.coi.2009.06.002

The Impact of Cancer Treatments on Aging

Changhan Lee and Valter Longo

Contents

C. Lee, Ph.D.
Davis School of Gerontology, University of Southern California, 3715 McClintock Avenue, Los Angeles, CA, 90089, USA
e-mail: changhan.lee@usc.edu

V. Longo, Ph.D. (✉)
Davis School of Gerontology, University of Southern California, 3715 McClintock Avenue, Los Angeles, CA, 90089, USA

IFOM, FIRC Institute of Molecular Oncology, Via Adamello 16, Milan 20139, Italy
e-mail: vlongo@usc.edu

F. Sierra, R. Kohanski (eds.), *Advances in Geroscience*,
DOI 10.1007/978-3-319-23246-1_4

1 Introduction

Owing to early detection and improved treatment, cancer survival rates in the US have recently been increasing [59]. This opens a new era in cancer care that is challenged with not only enhancing treatment efficacy but also improving the quality of the clinical outcome. There is an urgent need to address the highly complex and multifaceted problems that arise as a result of the therapy itself.

Modern cancer therapy evolved from a cytotoxic approach, using chemotherapeutics and ionizing radiation that are limited in their ability to distinguish healthy and cancerous cells, to drugs with specific molecular targets, such as tyrosine kinase inhibitors or antibodies recognizing specific receptors. Despite the leap in sophistication achieved through research and development, toxicity still remains the major limiting factor for many types of therapy, including the more targeted ones. This is especially relevant considering that many of the patients receiving these toxic therapies are older and are likely to have co-morbidities. More importantly, undesired side effects pose serious clinical challenges, including a negative impact on the quality of life and healthspan of cancer survivors.

We are now faced with an unprecedented range of dysfunctions and diseases resulting from the very interventions that have significantly improved the chance and length of survival after cancer diagnosis. This is an issue with multifaceted roots which requires a multilevel approach. There is much to be considered when addressing adverse health issues secondary to cancer treatment: (1) age-dependent co-morbidities affecting the great majority of patients who are older adults, (2) the potential compromise to health- span and quality of life, especially among childhood cancer survivors, (3) the incomplete understanding of the complex molecular pathways that can sensitize cancer cells while protecting healthy cells, (4) the multimodality nature of cancer treatments that involve a mix of molecular targets, and (5) a relatively short time frame to eliminate side effects compared to other chronic diseases.

The emerging interdisciplinary field of geroscience may hold the key to understanding the mechanisms involved in preventing and treating many of these side effects. This is especially important as the side effects involve damage to a wide-range of tissues, cells, organelles, and molecules that manifest over time and which can resemble phenotypes observed during aging. For example, cancer treatments such as chemotherapy or radiation can cause DNA alterations and mutations overlapping with those caused by oxidative damage and/or replication errors observed during aging.

2 Cancer Therapies

Traditionally, surgery has been a key treatment for solid tumors (e.g. breast, prostate, colorectal, gynecologic, and lung). The modern cancer therapy arsenal has greatly benefitted by the addition of therapies including chemotherapy and radiotherapy which have provided a broader control of the disease locally and systemically. Unfortunately, some of these therapies themselves can be toxic to the patient and also promote secondary tumors. Here we discuss the adverse effects that are known to affect patients during and long-after cancer therapy, ranging from mild side-effects, to long-term adverse secondary health problems and secondary cancers.

2.1 Chemotherapy

The beginning of modern chemotherapy began with the 1942 toxic mustard nitrogen gas studies by Goodman and Gillman, commissioned by the US Department of Defense in search for therapeutic properties of chemicals developed for chemical warfare during World War II [259]. After that, it took over 30 years for legislation by the US Congress to create the first federal program for anticancer drug discovery, leading to major improvements in the efficacy and reduction of the toxicity of chemo drugs. However, chemotherapy remains largely based on semi-targeted cytotoxity to rapidly proliferating cells [50, 143, 200]. Although chemotherapy was first thought to be quite selective, we now know normal cells including non-dividing cells also experience severe toxicity, leading to dose-limiting side-effects such as cardiotoxicity, myelosuppression, gastrointestinal damage, and fatigue. Although all chemotherapy agents produce reactive oxygen species (ROS) during apoptosis, some generate particularly high levels of ROS, such as the anthracyclines (doxorubicin, daunorubicin, and epirubicin), alkylating agents (cyclophosphamide and busulfan), platinum-based drugs (cisplatin, carboplatin, and oxaliplatin), podophyllotoxins (etoposide and teniposide), and camptothecins (topotecan and irinotecan) [50]. The US Food and Drug Administration (FDA) has approved 132 cancer chemotherapy drugs, 56 of which have been reported to cause oxidative stress, which leads to collateral damage to normal cells which is likely to overlap with that caused by ROS endogenously generated during mitochondrial respiration, and believed to contribute to aging phenotypes [41].

Anthracyclines produce the highest level of ROS by redox cycling and iron chelation, leading to mtDNA lesions and respiratory chain dysfunction, as well as disruption of antioxidant defense and repair systems [114, 115, 130, 219, 244]. Doxorubicin (a.k.a. Adriamycin) is an anthracycline widely used for the treatment of many cancers which exerts its antineoplastic effects by DNA intercalation, topoisomerase II inhibition, and oxidative stress. Its clinical use is limited due to common side-effects including myelosuppression, nausea, fatigue, but above all, cardiotoxicity [13, 137]. Cardiotoxicity is exponentially dose-dependent, and secondary to doxorubicin-induced oxidative stress with an estimated 7 % of patients developing congestive heart failure (CHF) after a cumulative dose of 550 mg/m^2 [218]. Although rare and generally transient, acute toxicity can be evident within a few minutes following administration [78]. Chronic cardiomyopathy usually occurs during the first year following treatment, but can manifest up to 20 years later and is irreversible. Doxorubicin-related cardiotoxicity is probably multifactorial, but most of the damage can be attributed to oxidative stress and lipid peroxidation that leads to mitochondrial DNA damage and dysfunction, mirroring what is observed during cardiac aging [62, 89, 165, 168, 214].

Cisplatin (cis-diamminedichloroplatinum(II)) is another chemotherapy drug widely used in many cancers, which suffers from oxidative stress-induced side effects, prominently in the kidneys and ear [53, 58, 194, 195, 202, 220].

2.2 *Radiotherapy*

Radiation-based treatment is a major arm of cancer therapy. Irradiating biological material leads to a rapid burst of ROS generated primarily because of the ionization of water molecules and direct ionization of target molecules [72, 189]. Ionizing radiation (IR) such as X-rays and γ-rays can cause direct macromolecular damage by energy deposition, but 60 % of damage is caused by ROS which is produced through the interaction of energy with water [11]. Although radiation-dependent ROS are effective in killing cancer cells, they also damage normal cells leading to dysfunction [18, 173, 256, 270]. It has been reported that even localized small field radiotherapy of the head and neck in patients can cause oxidative damage at the organismal level, and may lead to secondary mutations [193]. The use of the free-radical scavenger amifostine significantly reduced the side effects caused by radiotherapy [21, 28, 30], but additional studies are needed to understand how or if it affected the toxicity to the cancer cells.

Other studies also indicate that administering antioxidant enzymes with radiotherapy reduces the toxic side effects. For example, SOD was found to protect against radiation induced cystitis in bladder carcinoma without reduction in antitumor efficacy [92]. However, the use of antioxidants such as vitamin C is not expected to provide differential protection effects to normal and cancer cells, although it may do so with specific types of tumors.

2.3 *Target-Based Therapeutics*

Beyond the more traditional cytotoxic therapies, a new generation of target-based drugs has greatly enhanced the anti-cancer arsenal. There are various approaches to interfere with specific molecular targets that have been largely executed by antibodies and small molecules [121]. Although these inflict much less damage compared to the more traditional interventions, they can also promote damage in multiple systems, which could affect either the patient's aging process or cause significant damage resulting in loss of redundancy and frailty. For example, a kinase inhibitor such as rapamycin, well established to extend longevity in model organisms [19, 97, 190], is also effective in treating kidney cancer, but can contribute to either immunosuppression or immunostimulation, depending on the dose and mode of administration, which can have profound implications for the functionality of the immune system during aging [144, 145].

The concept of using immunity in cancer therapy was proposed in the late nineteenth century [47, 65] but it only progressed thanks to the introduction of hybridoma technology that enabled targeted antibody design and engineering [121, 204]. Targeted monoclonal antibodies can attach to cancer cells, based on the presence of an antigen on certain tumor cells, and can act as immunomodulatory agents [152, 175, 251], inhibit specific receptors or ligands [33, 135, 204, 261], or deliver cytotoxic agents such as radioisotopes, genes, and toxic drugs [25, 259]. Small molecules designed to target specific molecular pathways have also been actively developed recently, including various kinase inhibitors [102, 138, 271], cyclin-dependent kinase (CDK) inhibitors [163], drugs acting on hypoxia-related pathways [151, 257], and those targeting the PI3K/AKT/mTOR pathway [60, 79, 107, 198, 241, 242].

With the increasing appreciation of the role of metabolism in cancer, the concept of metabolic targets for cancer therapy has gained traction [81]. Metabolic dysfunction and mitochondrial alterations are common underlying characteristics of various age-dependent diseases, and several of the drugs have been proposed/tested off-label as cancer treatments or have multiple indications including those currently approved/developed for diabetes [66, 182], autoimmunity [29, 234], mitochondrial function [69, 80, 227], and other scopes [225, 238]. Because aging is a leading risk factor for most major diseases, older individuals often suffer from multiple conditions. Each disease can affect the other and accelerate the process. For instance, obesity and diabetes often develop with age, and a recent study with 5.24 million participants showed a strong association between being overweight/obese (BMI $\geq$25 kg/m^2) and 17 of 22 frequent cancers. Each 5 kg/m^2 increase in BMI was associated with higher risks of cancers of the uterus, gallbladder, kidney, liver, etc. [16]. However, obesity is not significantly linked to all types of cancer and the correlation shows clear sex-specific differences [229]. Also, diabetes is associated with an increased risk of some cancers, including liver, pancreas, endometrium, colorectal, breast, and bladder. This could be directly affected by hyperinsulinemia, hyperglycemia, and inflammation [84].

Although targeted therapeutics cause significantly lower toxicity than the traditional chemotherapeutics or ionizing radiation, they are not free of side effects [9, 85, 147, 187, 205, 224, 238]. Careful long-term follow-up studies on these newer therapeutics should be encouraged to better understand their immediate and long-term impact on cancer survivors.

3 The Clinical Impact of Cancer Therapy on Aging

Despite the fact that cancer therapeutics enter the clinic after tremendous scrutiny on safety, the side effects are still the major limiting factor. It is estimated that at least 50 % of cancer survivors will suffer from treatment-related side effects at the physical and psychosocial level [231]. Furthermore, subtler and less visible alterations secondary to cancer therapy may have profound effects later in life and affect the way we age. Importantly, these late-life effects are not scrutinized within the FDA process for approval of the drugs. This section discusses the potential impact these therapies have on lifespan/healthspan (including secondary tumors arising from therapy) and quality of life. The following section will discuss the molecular aspects of these events in detail.

Clinical and epidemiological studies have shown that long-term cancer survivors are at risk for late effects of their cancer treatment including secondary malignancies, cardiac and vascular abnormalities, pulmonary complications, infertility, endocrinopathies and other chronic conditions. Understanding the mechanistic details of damage, protection and death in both normal and malignant cells will be critical to develop adjuvant therapeutics to selectively protect the patient, but not the tumor. Because of its interdisciplinary and comprehensive nature, Geroscience is well-positioned to address these issues.

The treatment of pediatric cancers has witnessed remarkable improvements leading to a 5-year survival rate nearing 80 % [191]. Because most pediatric cancers are now curable, the adverse effects of treatment as patients age, including growth/development, fertility, secondary cancer, organ failure, and psychosocial challenges, are emerging issues for long-term childhood cancer survivors [191]. However, the more subtle but wider effects of a variety of treatments on cellular and organismal aging are likely to be underestimated. Based on a recent study of 10,397 adult survivors of childhood cancer, it is estimated that by 30 years from their cancer diagnosis, 73 % of them develop at least one chronic health condition, and 42 % exhibit a severe, life-threatening, or disabling condition or die from a chronic condition [170]. This is considerably higher than chronologically comparable controls [170]. Also, treatment-related secondary malignancies are a serious concern [17, 169, 255]. Ionizing radiation has been favored for the treatment of childhood cancers, leaving the survivors at increased risk of secondary malignancies, including those arising from the skin, breast, thyroid, and brain [8, 24, 110, 166, 176, 213]. Selected chemotherapeutics have also been successful for childhood cancers, but they can also cause adverse consequences. These include alkylating agents [117, 154], anthracyclines [125, 232], antimetabolites [7, 35, 185], corticosteroids [51, 70], platinum-based [253], and vinca alkaloids [82, 112]. Notably, long-term childhood cancer survivors also experience increased risk for metabolic dysfunctions, including obesity, diabetes, and hypertension, that are known to bring about additional major age-related diseases.

4 Geroscience and Cancer Treatment: Current Status

Current interventions to treat cancer affect several pathways that regulate the aging process. The undesired off-target events and their consequences may cause collateral damage to normal cells, alter the course of aging and also cause secondary damage to the patient as mentioned in the previous section. Here, the latest research on aging will be reconciled with some of the major molecular pathways affected by cancer therapies. Aging involves complex and interdependent biological processes many of which are affected by chemotherapy and radiotherapy and could contribute to accelerated aging.

4.1 Genomic Instability

Both chemo- and radiation therapies are aimed at destroying rapidly dividing cells, primarily by inducing damage to their replicating DNA. The lack of specificity in their actions, as discussed above, means that adjacent, non-cancerous cells, are also damaged. Thus, therapy-induced DNA damage may impede our ability to maintain genomic integrity and hasten the aging process. Chemotherapeutics and therapeutic irradiation are well known to cause DNA damage to the somatic nuclear and mitochondrial genomes, leading to functional decline of multiple organs and various diseases [26, 239]. Accumulation and incorporation of these DNA lesions can cause genomic instability, a hallmark of aging. Therefore, these therapies can be considered unintentional accelerators of aging, resulting in several phenotypes observed in older individuals; e.g. cardiomyopathies, and secondary malignancies.

4.1.1 Nuclear DNA

DNA is subject to frequent chemical alterations and breaks on a daily basis that requires sophisticated surveillance and repair to prevent genomic instability. The first line of defense is the chromatin structure where DNA is wrapped around a protein core (histones) in units known as nucleosomes. Then, there are highly specialized repair mechanisms that mend damaged DNA lesions ranging from error-free base excision repair, to error-prone translesion repair and non-homologous end-joining. Nevertheless, none of these systems are able to prevent the accumulation of age-dependent modifications and errors, and in fact, the error-prone systems can even promote point mutations and small DNA insertions and deletions, in an attempt to avoid the more problematic chromosome breaks and rearrangements. Genomic instability can arise from direct damage to the DNA, including single- and double-strand breaks, inter- and intra-strand base cross-linking, and depurination or depyrimidination, leading to transition or transversion mutations, or by progressive mutations, such as additions, deletions, frame-shifts, or substitutions. Chemotherapeutics and ionizing radiation can promote virtually all of the lesions listed above, leading to an accelerated genomic aging [5]. Therefore, further studies are necessary to: (1) assess the role of DNA damaging cancer therapy on the level and type of DNA damage and mutations, (2) determine how/if these change/increase during aging, (3) determine whether in addition to promoting secondary tumors they may play a role in accelerating multi-system aging and/or specific lesions/diseases. In addition to providing important clues that can help us redesign cancer treatment and improve the information given to patients in the informed consent, these studies can help test the genomic instability hypothesis of aging. Because each treatment causes a different profile of DNA mutations/lesions, these analyses could link a type of DNA damage (for example, point mutations) to specific age-dependent phenotypes (ex. secondary tumors and their type). The collaborative involvement of basic scientists, computational biologists, and clinicians will be necessary to achieve these goals.

4.1.2 Mitochondrial DNA

Chemotherapeutics and radiation can also damage mitochondrial DNA and cause dysfunction. The less sophisticated DNA repair mechanisms present in mitochondria may also contribute to enhancing this toxicity. There have been many reports of increased mtDNA mutations during aging, but the exact degree of causal contribution of mtDNA damage to human aging is still not entirely clear. Oxidative damage to the mtDNA has long been proposed to be involved in the aging process, but recent studies show evidence that replicative infidelity and spontaneous mutations may be more pertinent [210]. Comparison of mitochondria from the pre-frontal cortex of young (<1 year) or old (>75 years) brains showed a non-significant increase of oxidative DNA damage (8-oxodG), but a significant increase in DNA scars resulting from replication infidelity and spontaneous base hydrolysis, indicating that age-related mtDNA damage by ROS may be less than previously thought [116]. In fact, the mtDNA mutator mouse, which expresses a proofreading-deficient DNA polymerase γ ($PolgA^{mut}$), has increased levels of mtDNA mutations including deletions, and shows a premature aging phenotype, although the contribution of small mutations to aging phenotypes in these mice is an area of debate [3, 64, 226, 236, 237, 254]. Aging mitochondria have also been reported to carry large deletions in their DNA with a concurrent decline of mitochondrial energy production [15, 106, 124, 139]. One of the biggest hurdles of understanding mtDNA mutations is the lack of technology that is sensitive and accurate enough to detect single mutations with low background error frequencies.

Although currently less understood than nuclear DNA, mtDNA is also subject to damage by chemotherapy and radiation. Doxorubicin-induced mtDNA damage is proposed to occur via oxidative stress, as 8-oxoG adducts were increasingly detected preferentially in mtDNA over nuclear DNA following treatment in rodents and humans [129, 208]. Also, large mtDNA deletions in cardiomyocytes of mice chronically treated with doxorubicin have been reported [1]. Further, doxorubicin-induced cardiotoxicity is thought to result from the combined oxidative damage to mitochondrial DNA and lipids [129]. Additionally, platinum-based drugs [161, 180] and other chemotherapeutics [266] can damage mtDNA. Furthermore, ionizing radiation causes significant mtDNA damage, including a common deletion (~5 kb) that is more prominent in normal cells compared to cancer cells, and also found in aging mitochondria [181, 274]. Also, γ-radiation causes point mutations in the D-loop region of the mtDNA in an inconsistent manner [162]. As for nuclear DNA damage, the role of various chemotherapy drugs in causing different types of mtDNA damage can serve as an invaluable tool to determine the role of different categories of mtDNA damage on aging and age-related diseases. Again, it will be first essential to determine in mice and possibly human samples, what are the short-term and long-term effects of each chemotherapy drug on the mtDNA of different cell types.

4.1.3 Telomere Attrition

Telomeres are protective sequences that define and protect the ends of linear chromosomes, consisting of long double-stranded TTAGGG repeats that can reach 9–15 kb in humans and 100 kb in rodents [168]. Telomere length shortens in primary fibroblasts from older humans and also with serial passaging in vitro [89, 95]. Critical telomere shortening can signal cells to enter an irreversible proliferative arrest [20, 235], a condition known as replicative senescence [100]. Notably, ectopic expression of telomerase led to the immortalization of human fibroblasts [20]. Thus, it is not too surprising that telomeres have a critical role in cancer biology, Telomere-shortening and cellular senescence are thought to be important to protect against cancer [233]. However, telomere attrition can also promote oncogenesis by causing unregulated chromosomal rearrangements and genomic instability, as discussed above [260, 263].

There are numerous reports of telomere shortening after chemotherapy and ionizing radiation treatment, with older patients being at a greater risk. An accelerated rate of telomere attrition was found after various cancer chemotherapy treatments, including cisplatin and irradiation (>100-fold) [230], two doses of high-dose cyclophosphamide and Ara-C [186], six to eight cycles of CHOP (cyclophosphamide, Adriamycin, vincristine, and prednisone) [136], in a dose-dependent manner after CHOP, 5-FU, and fludarabine treatments [61]. Furthermore, in long-term survivors of childhood cancers, lower telomerase content was significantly related to treatment-related secondary thyroid cancer, resulting from intensive chemotherapy and ionizing radiation [87]. Also, therapy-related damage to hematopoietic stem cells (HSC) can cause myelodysplasia or acute myelogenous leukemia in recipients of autologous bone marrow and HSC transplant, posing a lethal complication [39]. Further studies should investigate both the potentially negative and positive effects of telomere shortening on aging and diseases, and particularly cancer. In fact, telomere shortening may both play an anti-cancer role by preventing the division of old and damaged cells but could also contribute to cell senescence and the pro-inflammatory effects of senescent cells [32, 223].

4.2 Epigenetic Alterations

In addition to mutations, cancer therapy can also cause epigenetic changes in normal and malignant cells. Epigenetics refers to genetic regulation without alterations of the DNA sequence. The term was first coined to connect genotype to phenotype and describe the deeply complex processes between them [243]. The currently understood molecular mediators of epigenetics are DNA methylation on CpG sites, histone modification (methylation and acetylation), chromatin remodeling, and non-coding RNAs (such as microRNAs and long non-coding RNAs), all processes involved in regulating gene expression. Unlike genetic information, epigenetic changes are plastic, dynamic, and diverse, making these changes a moving target during aging [73, 104]. There have been recent reports suggesting an age-specific CpG methylation pattern. Horvath reported 353 CpG sites pertinent to cell death/survival, cellular growth/proliferation, organismal/tissue development, and cancer, that can accurately predict chronological age across tissue and cell types to within a few years, starting from newborn and including induced pluripotent stem cells [103]. Hannum et al. built a quantitative model of aging based on >450,000 CpG markers from the blood of 656 individuals (19–101 years old) and showed that one's aging rate and chronological age can be predicted from this methylome. Also, the methylome may help explain the age-dependent epigenetic drift that may impact transcriptional patterns over time [94]. Also, a novel integrative epigenome-interactome approach identified tissue independent age-associated methylation interactome hotspots targeting stem-cell differentiation pathways with validation with independent DNA methylation data sets, encompassing over 1000 samples from different tissue types [252]. Another recent report examined more than 480,000 CpG sites from 965 individual samples and found 162 CpGs significantly associated with age, of which 65 were novel sites [39, 73].

Cytotoxic drugs and radiation can affect the DNA methylation profile and lead to epigenetic side effects that can persist much after treatment, including secondary cancer [52, 240]. Chemotherapy can affect DNA methylation and ionizing radiation can cause stable DNA hypomethylation in both target and bystander tissues. Breast cancer patients who received chemotherapy showed hypomethylation in 8 CpG sites in their blood cells that persisted 6-months after treatment [215]. DNA hypermethylation has also been shown in at least 1 gene in more than half of a set of leukemia patients that was associated with shortened average time to develop secondary therapy-related leukemia (49.3 vs. 133.2 months) [228]. Unlike chemotherapy that is usually delivered systemically, radiotherapy is generally focused on the tumor with precision. However, the bystander effect, which refers to indirect exposure and damage to neighboring tissues, in part by causing persistent epigenetic changes, still poses a serious clinical problem for cancer survivors [109, 123, 247]. In summary, epigenetic changes could contribute to the long-term effects of cancer therapy although they are not expected to promote the severe aging and disease phenotypes that mutations and gross chromosomal rearrangements are likely to promote.

4.3 Mitochondrial Function

Mitochondria are a major target for many cancer therapeutics, and are also strongly implicated in aging and age-related diseases. As mentioned above, mtDNA damage has been long thought to cause mitochondrial dysfunction.

Although the exact mechanisms by which mitochondria contribute to aging are still largely unclear, there are several promising theories. Because many oxygen species are so reactive, the mitochondrial free radical theory of aging has been one of the most cited theories for how mitochondrial and extra-mitochondrial damage occurs during aging. Briefly, the theory states that the superoxide and other free radicals that arise as by-products of mitochondrial respiration can cause oxidative damage to macromolecules including nucleic acids, proteins, and lipids. Indeed, ROS production increases with age and the cellular defense mechanisms against it declines. Recently, additional mechanisms of mitochondrial contribution to aging have been proposed including mitochondria as a signaling organelle [26, 40]. Notably, in worms, inducing a mitochondrial unfolded protein response (mtUPR) in neurons has been shown to transmit a mitochondrial signal to the gut that in turn leads to extended lifespan [63, 258]. The exact identity of the signal is currently under investigation. The emerging biology of mitochondrial-derived peptides (MDP) has also been shown to be involved in the regulation of aging and stress responses [134]. Humanin is a MDP encoded within the mitochondrial DNA that was first discovered through a search for neuroprotective factors from a cDNA library constructed from an unaffected brain fraction of an Alzheimer's patient [98]. Humanin levels decline with age and its expression is regulated by the growth hormone and insulin-like growth factor 1 axis (GH/IGF-I) [61]. Long-lived GH-deficient mice and the GH-deficient Ecuadorian cohort who are immune to cancer and diabetes have higher levels of humanin compared to their normal counterparts [61]. Notably, as discussed in the next section, humanin has recently been shown to protect against cancer treatment-related toxicity. Because cancer survivors experience delayed treatment-induced mitochondrial dysfunction that manifests as skeletal muscle dysfunction [203] and cardiac failure [129, 141], it would be of great benefit to be able to identify early markers of mitochondrial damage at the genetic, functional and signaling levels, and also interventions that can selectively protect mitochondria in normal cells during cancer treatment, such as humanin.

4.4 Cellular Senescence

Cellular senescence can result from different types of cancer treatments. As discussed above, senescence is defined by as an irreversible growth arrest that can be induced by telomere shortening. However, stress-induced premature senescence in the absence of reduced telomere length can occur in response to DNA damage (especially double strand breaks) [126], strong mitogenic stimuli (e.g. oncogenes), and mitochondrial dysfunction, including reactive oxygen species [233, 276]. The molecular mechanisms behind cellular senescence involve a network of cell cycle regulators, including cyclin-dependent kinase inhibitors (CDKis), notably p21 that is activated by the ATM/p53 pathway, and $p16^{INK4a}$, both of which converge on the tumor suppressor RB to arrest cell cycle [233]. Currently, the "senescent" state of a cell is assessed based on several characteristics including (i) enlarged morphology (e.g. twice the size of non-senescent cells) [99]; (ii) senescence associated β-galactosidase activity that reflects increased lysosomal biogenesis [128, 131]; (iii) senescence associated secretory phenotype (SASP), (iv) DNA-SCARS (DNA segments with chromatin alterations reinforcing senescence) [192], and (v) chromatin remodeling, including senescence-associated heterochromatin foci (SAHF) [4, 272]. More recently, the concept of 'deep' or 'late' senescence is emerging where senescent cells continue to evolve with time [233], including the senescence-associated opening of gene-poor heterochromatic regions where major retrotransposon families, L1, ALU and SVA reside, leading to increased transposition in senescent cells [54]. There are several ways senescence is thought to contribute to organismal aging and age-related diseases including cancer, inflammation, development and tissue/organ deterioration [32, 223, 233].

Traditional genotoxic cancer treatments (i.e. chemotherapy and ionizing radiation) can cause off-target cellular senescence of normal cells. Breast cancer patients who have received anthracycline-based chemotherapy showed elevated levels of $p16^{INK4a}$ in their blood cells immediately and 12 months after treatment [201]. Premature senescence by ionizing radiation and/or chemotherapy in hematopoietic stem cells has been reported in mice [153, 245]. Notably, the number of studies focusing on the role of cancer therapies on cell senescence and its consequent role on organismal aging and diseases is very limited, pointing to the need of additional studies which will also be useful to understand the role of senescence in aging and diseases. Studies could be designed to lead to demonstrations of the link between therapy, the presence of senescent cells and diseases. For example, fibroblasts or lymphocytes could be treated *in vitro* with chemotherapy drugs, taken through a sufficient number of population doublings to induce senescence and then injected into middle aged mice to determine whether the chemotherapy-treated, but also the untreated, senescent cells have effects on aging and age-related diseases.

4.5 Stem Cell Aging

The negative side effects of cancer therapy are often associated with toxicity to stem cells and, in some cases, their exhaustion. Aging is partially attributed to the loss of regenerative capacity with time, resulting in suboptimal tissue maintenance and repair. There are several proposed mechanisms that lead to stem cell aging and deficiency or dysfunction. Age-dependent accumulation of DNA damage and telomere shortening are thought to play a major role in stem cell damage, manifesting in replicative senescence [212]. Epigenetic changes representing proliferation history, independent of telomere length, have also been shown to cause hematopoietic stem cell aging [14]. Stem cells favor a specific microenvironment supported by certain types of cells, called niches, that provide factors such as TGF-β [246], bone morphogenetic protein (BMP) [55], Wnt [44], Notch [2], and cyclic ADP ribose [267]. The age-dependent decline of these cells is also thought to contribute to stem cell exhaustion with time and loss of activity in those remaining [171]. More recently, metabolism has emerged as an important factor in maintaining stem cell pluripotency. Proliferating cells, e.g. active stem cells and cancer cells but particularly cancer stem cells, require biosynthetic building blocks to synthesize DNA, protein, and lipid membranes to replicate. Rather than relying on increased consumption of extracellular nutrients, highly proliferating cells rely partially on de novo synthesis from glucose and glutamine. This high anabolic demand makes these dividing cells rely on aerobic glycolysis [74, 111]. Mitochondria, being the single most important metabolic organelle, have also risen as a key regulator of cell stemness [74, 111].

Traditional cytotoxic chemotherapy and ionizing radiation target rapidly proliferating cells with little distinction between normal and malignant cells. Therefore, quickly dividing normal stem cells are also targeted by these interventions, leading to undesired side effects that limit the efficacy of chemotherapy. For example, both cyclophosphamide and 5-FU, but also many other chemotherapy drugs targeting DNA, can cause severe damage to bone marrow stem cells leading to immunosuppression which is often severe enough to require bone marrow transplants. Furthermore, stem cells may be subject to therapy-induced DNA damage leading to senescence, and alterations in mitochondrial health that can also reduce their stemness. While the damage to the rapidly dividing hematopoietic stem cells is well documented, even non-dividing cells including muscle cells and neurons can also be severely damaged by several chemotherapy drugs. Therefore it will be important to understand how these drugs affect quiescent stem cells or those that are dividing very slowly such as the resident satellite cells in muscles.

5 Geroscience and Cancer Treatment: Filling in the Gaps

Geroscience is an emerging field that embraces and encompasses a wide range of disciplines, reflecting the complexity of aging. This also provides an entirely new opportunity to approach diseases with a wider net. For example, the toxic side-effects of cancer treatment arise because of the inability of treatments to fully distinguish normal and malignant cells. Therefore, the ability to separate the two cell types, to a degree similar to that attained by antibiotics, which can readily distinguish bacterial cells from ours, would undoubtedly improve our aim at cancer cells and dramatically increase the therapeutic window. Because geroscience is focused on the protection of normal cells from any type of damage, the field is well positioned to identify ways to protect normal and not cancer cells from toxins, not by simply screening for drugs with those properties but by understanding the fundamental mechanisms of protection in healthy and malignant cells. Ideally, normal cells (and the patient) could be protected while malignant cells could be sensitized to treatment. Chemoprotectants such as amifostine, glutathione, mesna, and dexrazoxane have been investigated and shown to provide drug-dependent protection to specific tissues, but the use of these compounds has not been shown to increase disease-free or overall survival [140]. One of the hallmarks of longevity interventions is increased stress resistance of the whole organism (and normal cells).

However, because of acquired mutations that render cells self-sufficient in growth and insensitive to anti-growth signals, cancer cells act autonomously and independently, providing an opportunity for geroscience to exploit the difference by selectively protecting the normal tissue without interfering, or even enhancing, treatment efficacy [93] (Fig. 1). Another opportunity that geroscience may provide is the identification of novel drug targets for cancer and/or the scientific basis for an oncocentric use of already FDA-approved pharmaceuticals.

The following are some geroscience-based interventions that deserve further investigation with the potential to provide a different approach to cancer treatment, whereby the focus not only continues to be on improving the killing of cancer cells, but also on the preservation of the patient's healthspan post-treatment.

5.1 Metabolic Interventions

5.1.1 Dietary Interventions

Dietary interventions have contributed greatly to our understanding of lifespan and stress resistance regulation. Dietary restriction (DR) is the most effective and reproducible intervention to decelerate the rate of aging and increase healthspan in various model organisms ranging from the simple yeast to worms, flies, rodents, and possibly non-human primates [48, 90, 118, 142]. In 1934, Crowell and McCay

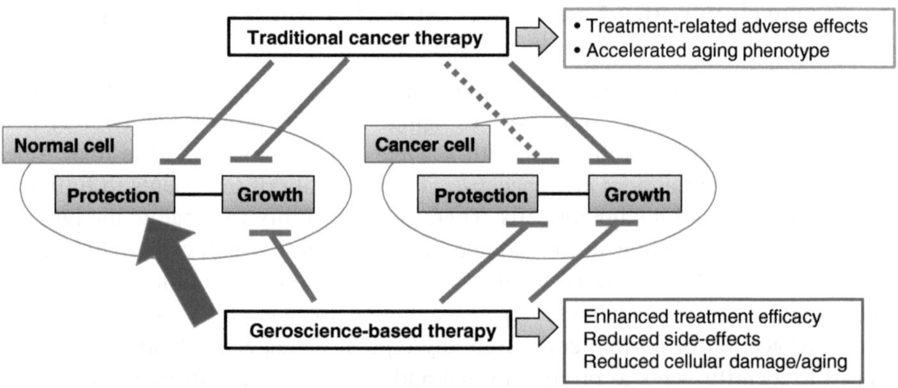

Fig. 1 Geroscience-based approaches have resulted in interventions aimed at protecting normal cells while instead sensitizing cancer cells to therapy. The Differential Stress Resistance and Sensitization strategies to treat cancer have resulted from basic and translation geroscience research that first described how starvation can protect normal cells from a wide variety of toxins and then identified proto-oncogenes as the negative regulators of the resistance. Basic geroscience research also described how cells expressing oncogene analogs have an adaptive disadvantage when expose to extreme and complex environments such as starvation in addition to toxins. These findings were exploited to protect mice and possibly humans from chemotherapy while rendering cancer cells more sensitive to the treatment

reported that rats fed a calorie restricted diet with sufficient nutrients starting after weaning extended lifespan nearly twice [150]. Following this seminal discovery, Walford and Weindruch reported that in mice, adult-initiated DR (undernutrition without malnutrition) which began at 12-months of age [248, 249], and DR initiated at weaning both increased lifespan and reduced tumor incidence [250], confirming the DR studies in rats. A >30-year longitudinal adult-onset DR study in rhesus monkeys performed at the Wisconsin National Primate Research Center (WNPRC) shows that DR (30 %) delayed disease onset and mortality, with a 50 % decrease in cancer incidence [48, 49]. However, a comparable >20-year study performed at the National Institute on Aging (NIA) shows a slightly different result where DR did not extend lifespan, but successfully improved health span, including reduced incidents of cancer and diabetes [149]. The disparities between the WNPRC and NIA studies were largely attributed to differences in diet composition and the genetic origin of the monkeys, suggesting that DR may need to be carefully considered if it is to be practiced in humans.

The mechanisms underlying DR are still unclear, but energy allocation appears to be fundamental. Because a cell can only retain a limited amount of energy at any given time, the cellular energetic network must economically balance the finite energy reserve between reproduction/growth and repair/maintenance [119]. However, under starvation or DR, the favored survival strategy is to discourage reproduction/growth and invest the remaining energy in repair/maintenance [119]. In a way, this switch of energy encourages the entrance into a maintenance mode, which could explain why DR or fasting reduces size and fertility, and increases lifespan and stress resistance [211].

Notably, short-term fasting selectively protects normal cells and mice while sensitizing malignant cells to chemotherapeutics and ionizing radiation, providing a means to induce differential stress resistance (DSR) [133, 183, 196]. Furthermore, short-term fasting may reduce chemotherapy-induced side effects such as nausea and vomiting in cancer patients [197]. Multiple fasting cycles protect hematopoietic cells from chemotoxicity and also promote their self-renewal in mice [42], and has the potential to prevent long-term treatment-related stem cell exhaustion. Short-term DR also enhances skeletal muscle stem cell function [38].

In addition to DR, simply restricting a single essential amino acid increases lifespan and stress resistance [57, 174, 188, 206, 277]. In flies, restoring essential amino acids to DR was sufficient to reverse lifespan extension [88]. Rodents fed a methionine or tryptophan restricted diet (MR and TR, respectively) lived longer and showed significantly less age-dependent diseases and enhanced resistance to oxidative stress [188], including tumor growth retardation in animals fed a MR diet [27, 222], and shows much potential in humans [67]. MR has been shown to have marked negative effects on cancer cells [36], whereas normal cells are relatively resistant to methionine restriction [37]. However, the use of specific amino acid restrictions for cancer treatment in vivo has limited applications since cancer cells can obtain methionine from other cells and tissues, therefore causing potentially more harm to normal cells than cancer cells. As has been shown for fasting, MR can also sensitize cancer cells to chemotherapeutics [86, 105, 221, 122, 268], suggesting its potential to induce differential stress sensitization (DSS). On the other hand, TR also provides longevity and reduced age-dependent deterioration [57, 174, 207], but has mainly been explored for neurological benefits due to its role in serotonin synthesis. Interventions that can induce DSR may increase the therapeutic window of available cancer treatments, while at the same time protecting normal cells from treatment and preventing long-term secondary adverse health issues in cancer survivors. Fasting and DR are candidate interventions with similar and different characteristics, but short-term fasting may be more clinically feasible because of its brevity, wide range, and efficacy [132]. Nonetheless, fasting is not a trivial intervention, especially for cancer cachectic patients, and fasting-mimicking diets providing high nourishment and a relatively high calorie content are likely to prove more beneficial.

5.1.2 Mitochondria Interventions

The recent identification of short open reading frames (sORFs) in the mtDNA that yields bioactive peptides such as humanin and MOTS-c, represent an entirely novel layer of signals that are inherently mitochondrial [87, 134]. The levels of these mitochondrial hormones decline with age both in the circulation and in relevant tissues in laboratory rodents [87, 136]. Humanin expression is regulated by the GH/IGF-I axis and its levels are positively correlated with longevity in long-lived mouse models [61]. Humanin is a protective factor against various types of stress, chiefly those that are related to oxidative stress. Many labs have reported on the protective effects of humanin in various disease models including Alzheimer's disease [98, 156, 167, 273], atherosclerosis [172, 269], and ischemic injury [164, 264, 265]. A recent report on bortezomib, a drug in clinical trial for childhood cancers, shows that humanin treatment successfully prevented bortezomib-induced toxicity to growth plate chondrocytes that lead to growth arrest without interfering with its anti-cancer effects [46, 68].

5.1.3 Glycolysis Blockade

Glucose is a major source of energy and carbon for mammalian cells. Inhibitors of glucose catabolism or generation, also considered calorie restriction mimetics, have been shown to increase lifespan in mice. Acarbose reduces the breakdown of starches and disaccharides to glucose by inhibiting α-glucosidases in the intestine, and thus limits glucose supply to cells. Acarbose treatment increased the median lifespan of male and female mice by 22 % and 5 %, respectively [96]. Acarbose is currently used to treat type 2 diabetes, but in addition it has been shown to also exhibit cardio-protective benefits [43]. In flies, feeding acarbose reduced tumor growth and improved survival [101].

2-deoxyglucose (2-DG) is a glucose analog that is phosphorylated by hexokinase to 2-DG-phosphate, which cannot be further metabolized and therefore blocks glycolysis. Rats fed 2-DG showed similarity to those under CR with improved glucose and insulin regulation [263], and increased recovery from stress [170]. 2-DG is actively being investigated for cancer treatment and although its use as a single modality drug in still debated [230], combination treatments with chemotherapy and radiotherapy are very promising [148, 186].

5.1.4 Intervening in the GH/IGF-I Axis

Dietary interventions alter a wide range of processes. However, there are certain key pathways that have been elucidated to mediate and/or mimic their effects. DR and fasting both reduce circulating levels of growth hormone (GH) and its downstream effector insulin-like growth factor 1 (IGF-1) [132]. In fact, it is suggested that the GH/IGF-I axis is a major mediator of the beneficial effects of DR [22], and a major regulator of lifespan and stress resistance [90, 118, 142]. GH deficient mice are resistant to stress, are smaller in size (dwarf), have reduced fertility and reduced levels of circulating GH/IGF-I, insulin, and glucose [113, 142, 160]. Conversely, mice overexpressing GH have a shortened lifespan [12]. Ecuadorian individuals with GH receptor (GHR) mutations coupled with severe GHR and IGF-1 deficiencies (Laron syndrome) are immune to diabetes and dramatically reduced cancer incidence [91]. Serum from Laron subjects protected human mammary epithelial cells from oxidative stress, concurrently reducing the expression of downstream signaling elements including RAS, protein kinase A (PKA), and mTOR. In addition, cells of long-lived mice with GH/IGF-I axis deficiencies are more resistant to oxidative stress (H_2O_2, paraquat), UV, genotoxins (methylmethanesulfonate, MMS), heat, and cadmium [160, 199], suggesting that enhanced stress resistance is at least partially responsible for longevity, and the possibility to enhance protection by interventions such as DR or down-regulation of the GH/IGF-I axis.

All things considered, the GH, IGF-I and insulin pathways appear to be major mediators/regulators of aging and stress resistance [22]. Therefore, interventions targeting the GH/IGF-1 axis may provide normal cells with increased stress resistance against cancer treatment and thus prevent secondary adverse health problems in cancer survivors. For instance, octreotide, a somatostatin analog that antagonizes GH production, has been shown to ameliorate chemoradiotherapy-induced diarrhea [216, 275]. Pegvisomant is a dominant negative GH mimetic that is FDA-approved for the treatment of acromegaly. It would be of interest to investigate how octreotide and Pegvisomant affect both the short-term and long-term effects of chemotherapy and other cancer treatments.

5.1.5 Intervening in the Nutrient Sensing Pathways

Nutrient sensing pathways are involved in longevity regulation and thought to partially mediate DR [75]. Many of these pathways are downstream of the GH/IGF-1 axis, including mTOR/S6K, PI3K/AKT, RAS, and AC/PKA and are highly conserved from single-celled yeast to humans [75]. In yeast, deleting human homologs of RAS (*RAS2*) and/or AKT (*SCH9/S6K*) increased lifespan to more than twofold while providing increased stress resistance against oxidants, genotoxins, and heat-shock [142]. Similarly, in *C. elegans*, mutations in the human homologs of insulin/IGF-1 receptor (*daf-2*) and PI3K (*age-1*) extended lifespan by twofold and increased resistance to thermal and oxidative stress [120]. In *D. melanogaster*, mutations in the insulin receptor substrate (*chico*) led to a 50 % lifespan extension [83].

Pharmacological manipulations can also increase lifespan, as shown in mice fed rapamycin [23], metformin [146], or SRT1720 [157, 158]. Rapamycin inhibits the mammalian target of rapamycin (mTOR), which can function downstream of IGF-I but can also be activated independently of IGF-1, and acts as a major regulator of cellular proliferation, metabolism, and stress [184, 262]. Rapamycin, the most experimentally successful longevity agent tested in model organisms, has recently been shown to extend both mean and maximum life span of both male and female mice, and in several genetic backgrounds [23]. Although rapamycin was initially used in the clinic as an immunosuppressant for organ transplants, its potential for cancer treatment was recognized more recently [34]. Analogs of rapamycin (rapalogs) with improved pharmacokinetics and solubility, including temsirolimus and everolimus, are being developed. Temsirolimus has been approved by the FDA for treating renal cell carcinoma, and clinical trials to test its efficacy in other cancers are underway [45, 209]. Everolimus is an oral rapalog that is also FDA-approved for various cancer treatments, including advanced renal carcinoma [155]. In addition to its anticancer effects, rapamycin has been shown to prevent stem cell senescence, protect mice from ionizing radiation-induced loss of proliferative basal epithelial stem cells [108], and enhance stem cell niche support [267].

Metformin is a front-line drug of choice for the treatment of type 2 diabetes, with several proposed mechanisms of action [76], that has recently gained much attention in cancer therapy [177]. An early report suggested that diabetes patients that received metformin as part of their treatment had a 23 % reduction in the risk for cancer [71]. A meta-analysis on 25 studies recruiting 579,621 patients reported that metformin use was associated with an overall 27 % reduction in the risk of developing any malignancy [77]. In particular, breast cancer has received much attention with promising results supporting the efficacy of metformin use in cancer [179]. Much of these studies involve diabetes patients who are at a higher risk for cancer [84], thus further randomized controlled clinical trials are needed to evaluate the efficacy of metformin in non-diabetic cancer patients. Furthermore, the use of metformin as a preventive measure of cancer should be considered [159]. In worms, metformin has been shown to extend lifespan through mitohormesis via the peroxiredoxin PRDX-2 [56], and by targeting the folate cycle in their bacterial feed, causing a methionine-restricted diet [31]. In mice, metformin (0.1 % w/w in diet) starting at middle age extended healthspan and lifespan in male mice, while a higher dose (1 % w/w) was toxic [127]. Notably, the combination or rapamycin and metformin may successfully antagonize cancer cells while protecting normal fibroblasts or epithelial cells, and thus prevent secondary health problems in cancer survivors [6].

Resveratrol activates Sirt1, an NAD^+-dependent deacetylase that has been shown to increase lifespan in lower organisms [251, 259]. In mammals, resveratrol treatment improved lifespan and healthspan in mice on a high-fat diet [65], and transgenic mice with moderate over-expression of Sirt1 showed an improved metabolic profile in multiple models of insulin resistance and diabetes [10, 178]. In addition to Sirt1, resveratrol is thought to act through multiple additional targets. In contrast, SRT1720 is a specific activator of SIRT1 that improves lifespan and healthspan in mice on both standard and high-fat diets [121, 157]. The effect of SRT1720 on cancer is mixed, with one study showing promotion of breast cancer cell migration and metastasis [217], and another showing increased apoptosis of breast cancer [135] and multiple myeloma cells [47].

6 Future Prospects for Geroscience and Cancer Treatments

Interventions aimed at affecting the vastly complex nature of the aging process require the understanding of its effect on a wide range of biological processes. However, the rate of aging is also affected by environmental factors such as chemotherapy which can profoundly alter its course. The topics discussed above have been categorized to help us understand this vast biology, but in fact they cover overlapping components of a biological network, each one influencing the other. For instance, telomere attrition can cause genomic instability that can lead to senescence, which in turn can affect inflammation (SASP) and stem cell exhaustion. Similarly, mitochondrial damage and increased ROS generation can damage DNA and cause genomic instability and senescence. The cytotoxic ripples inflicted by cancer chemotherapy and radiotherapy are wide and are among the most impactful interventions affecting aging and age-related diseases. Because many types of cellular damage caused by cancer therapy seem to accelerate those that occur naturally with age, a gero-centric approach may provide a more comprehensive solution both at the level of prevention and treatment. Geroscience can largely contribute to cancer therapy in at least three ways: (1) provide novel interventions and/or targets; (2) provide a method or intervention to selectively protect the patient, based on the stress-resistant phenotype of many long-lived model organisms, and (3) both enhance the killing of cancer cells while protecting the patient. On this line, geroscience-based intervention(s) should be promoted and urgently investigated to grasp the broader landscape of healthspan and quality of life of cancer-survivors. Lastly, there is a need to educate the patients of long-term consequences of cancer therapy and how geroscience can contribute to their decision-making and their post-treatment choices aimed at optimizing healthspan and quality of life.

Acknowledgments This work was funded in part by the National Institutes of Health (NIH); National Institutes of Aging (NIA) grant AG034906.

Editor: Julia Rowland, National Cancer Institute (NCI), NIH.

Author information VDL has equity interest in L-Nutra, a company that develops medical food.

References

1. Adachi K, Fujiura Y, Mayumi F, Nozuhara A, Sugiu Y, Sakanashi T et al (1993) A deletion of mitochondrial DNA in murine doxorubicin-induced cardiotoxicity. Biochem Biophys Res Commun 195(2):945–951. doi:10.1006/bbrc.1993.2135
2. Aguirre A, Rubio ME, Gallo V (2010) Notch and EGFR pathway interaction regulates neural stem cell number and self-renewal. Nature 467(7313):323–327. doi:10.1038/nature09347. [Research Support, N.I.H., Extramural Research Support, U.S. Gov't, Non-P.H.S.]
3. Ahlqvist KJ, Hamalainen RH, Yatsuga S, Uutela M, Terzioglu M, Gotz A et al (2012) Somatic progenitor cell vulnerability to mitochondrial DNA mutagenesis underlies progeroid phenotypes in Polg mutator mice. Cell Metab 15(1):100–109. doi:10.1016/j.cmet.2011.11.012. [Research Support, Non-U.S. Gov't]
4. Aird KM, Zhang R (2013) Detection of senescence-associated heterochromatin foci (SAHF). Methods Mol Biol 965:185–196. doi:10.1007/978-1-62703-239-1_12. [Research Support, N.I.H., Extramural. Research Support, Non-U.S. Gov't. Research Support, U.S. Gov't, Non-P.H.S.]
5. Alexeyev M, Shokolenko I, Wilson G, LeDoux S (2013) The maintenance of mitochondrial DNA integrity – critical analysis and update. Cold Spring Harb Perspect Biol 5(5):a012641. doi:10.1101/cshperspect.a012641. [Research Support, N.I.H., Extramural. Review]
6. Apontes P, Leontieva OV, Demidenko ZN, Li F, Blagosklonny MV (2011) Exploring long-term protection of normal human fibroblasts and epithelial cells from chemotherapy in cell culture. Oncotarget 2(3):222–233
7. Arikoski P, Voutilainen R, Kroger H (2003) Bone mineral density in long-term survivors of childhood cancer. J Pediatr Endocrinol Metab 16 Suppl 2:343–353. [Research Support, Non--U.S. Gov't Review]
8. Armstrong GT, Liu W, Leisenring W, Yasui Y, Hammond S, Bhatia S et al (2011) Occurrence of multiple subsequent neoplasms in long-term survivors of childhood cancer: a report from the childhood cancer survivor study. J Clin Oncol 29(22):3056–3064. doi:10.1200/JCO.2011.34.6585. [Research Support, N.I.H., Extramural Research Support, Non-U.S. Gov't]
9. Azad NS, Posadas EM, Kwitkowski VE, Steinberg SM, Jain L, Annunziata CM et al (2008) Combination targeted therapy with sorafenib and bevacizumab results in enhanced toxicity and antitumor activity. J Clin Oncol 26(22):3709–3714. doi:10.1200/JCO.2007.10.8332. [Clinical Trial, Phase I Research Support, N.I.H., Intramural]
10. Banks AS, Kon N, Knight C, Matsumoto M, Gutierrez-Juarez R, Rossetti L et al (2008) SirT1 gain of function increases energy efficiency and prevents diabetes in mice. Cell Metab 8(4):333–341. doi:10.1016/j.cmet.2008.08.014
11. Barcellos-Hoff MH, Park C, Wright EG (2005) Radiation and the microenvironment – tumorigenesis and therapy. Nat Rev Cancer 5(11):867–875. doi:10.1038/nrc1735. [Review]
12. Bartke A, Chandrashekar V, Bailey B, Zaczek D, Turyn D (2002) Consequences of growth hormone (GH) overexpression and GH resistance. Neuropeptides 36(2–3):201–208, S0143417902908899 [pii]
13. Bast A, Kaiserova H, den Hartog GJ, Haenen GR, van der Vijgh WJ (2007) Protectors against doxorubicin-induced cardiotoxicity: flavonoids. Cell Biol Toxicol 23(1):39–47. doi:10.1007/s10565-006-0139-4
14. Beerman I, Bock C, Garrison BS, Smith ZD, Gu H, Meissner A et al (2013) Proliferation-dependent alterations of the DNA methylation landscape underlie hematopoietic stem cell aging. Cell Stem Cell 12(4):413–425. doi:10.1016/j.stem.2013.01.017. [Research Support, Non-U.S. Gov't]
15. Bender A, Krishnan KJ, Morris CM, Taylor GA, Reeve AK, Perry RH et al (2006) High levels of mitochondrial DNA deletions in substantia nigra neurons in aging and Parkinson disease. Nat Genet 38(5):515–517. doi:10.1038/ng1769. [Research Support, Non-U.S. Gov't]

16. Bhaskaran K, Douglas I, Forbes H, dos-Santos-Silva I, Leon DA, Smeeth L (2014) Body-mass index and risk of 22 specific cancers: a population-based cohort study of 5.24 million UK adults. Lancet 384(9945):755–765. doi:10.1016/S0140-6736(14)60892-8
17. Bhatia S, Sklar C (2002) Second cancers in survivors of childhood cancer. Nat Rev Cancer 2(2):124–132. doi:10.1038/nrc722. [Review]
18. Bialkowski K, Kowara R, Windorbska W, Olinski R (1996) 8-Oxo-2′-deoxyguanosine level in lymphocytes DNA of cancer patients undergoing radiotherapy. Cancer Lett 99(1):93–97. [Research Support, Non-U.S. Gov't]
19. Bjedov I, Toivonen JM, Kerr F, Slack C, Jacobson J, Foley A et al (2010) Mechanisms of life span extension by rapamycin in the fruit fly Drosophila melanogaster. Cell Metab 11(1):35–46. doi:10.1016/j.cmet.2009.11.010. [Research Support, Non-U.S. Gov't]
20. Bodnar AG, Ouellette M, Frolkis M, Holt SE, Chiu CP, Morin GB et al (1998) Extension of life-span by introduction of telomerase into normal human cells. Science 279(5349):349–352. [Research Support, U.S. Gov't, P.H.S.]
21. Bohuslavizki KH, Klutmann S, Jenicke L, Kroger S, Buchert R, Mester J et al (1999) Salivary gland protection by S-2-(3-aminopropylamino)-ethylphosphorothioic acid (amifostine) in high-dose radioiodine treatment: results obtained in a rabbit animal model and in a double-blind multi-arm trial. Cancer Biother Radiopharm 14(5):337–347. [Clinical Trial Controlled Clinical Trial Research Support, Non-U.S. Gov't]
22. Bonkowski MS, Dominici FP, Arum O, Rocha JS, Al Regaiey KA, Westbrook R et al (2009) Disruption of growth hormone receptor prevents calorie restriction from improving insulin action and longevity. PLoS One 4(2), e4567. doi:10.1371/journal.pone.0004567
23. Bower JE, Bak K, Berger A, Breitbart W, Escalante CP, Ganz PA et al (2014) Screening, assessment, and management of fatigue in adult survivors of cancer: an American Society of Clinical oncology clinical practice guideline adaptation. J Clin Oncol 32(17):1840–1850. doi:10.1200/JCO.2013.53.4495. [Practice Guideline]
24. Bowers DC, Nathan PC, Constine L, Woodman C, Bhatia S, Keller K et al (2013) Subsequent neoplasms of the CNS among survivors of childhood cancer: a systematic review. Lancet Oncol 14(8):e321–e328. doi:10.1016/S1470-2045(13)70107-4. [Review]
25. Brannon-Peppas L, Blanchette JO (2004) Nanoparticle and targeted systems for cancer therapy. Adv Drug Deliv Rev 56(11):1649–1659. doi:10.1016/j.addr.2004.02.014. [Review]
26. Bratic A, Larsson NG (2013) The role of mitochondria in aging. J Clin Invest 123(3):951–957. doi:10.1172/JCI64125
27. Breillout F, Hadida F, Echinard-Garin P, Lascaux V, Poupon MF (1987) Decreased rat rhabdomyosarcoma pulmonary metastases in response to a low methionine diet. Anticancer Res 7(4B):861–867
28. Brenner W, Kampen WU, von Forstner C, Brummer C, Zuhayra M, Muhle C et al (2001) High-dose treatment with (186)Re-HEDP or (153)Sm-EDTMP combined with amifostine in a rabbit model. J Nucl Med 42(10):1545–1550. [Comparative Study Research Support, Non-U.S. Gov't]
29. Bryan J (2014) From cancer to rheumatoid arthritis treatment: the story of methotrexate. Diabetes 10:59
30. Buntzel J, Kuttner K, Frohlich D, Glatzel M (1998) Selective cytoprotection with amifostine in concurrent radiochemotherapy for head and neck cancer. Ann Oncol 9(5):505–509. [Clinical Trial Clinical Trial, Phase II Randomized Controlled Trial Research Support, Non-U.S. Gov't]
31. Cabreiro F, Au C, Leung KY, Vergara-Irigaray N, Cocheme HM, Noori T et al (2013) Metformin retards aging in C. elegans by altering microbial folate and methionine metabolism. Cell 153(1):228–239. doi:10.1016/j.cell.2013.02.035
32. Campisi J (2013) Aging, cellular senescence, and cancer. Annu Rev Physiol 75:685–705. doi:10.1146/annurev-physiol-030212-183653. [Review]
33. Carmeliet P, Jain RK (2011) Molecular mechanisms and clinical applications of angiogenesis. Nature 473(7347):298–307. doi:10.1038/nature10144. [Research Support, N.I.H.,

Extramural Research Support, Non-U.S. Gov't Research Support, U.S. Gov't, Non-P.H.S. Review]
34. Carver JR, Shapiro CL, Ng A, Jacobs L, Schwartz C, Virgo KS et al (2007) American Society of Clinical Oncology clinical evidence review on the ongoing care of adult cancer survivors: cardiac and pulmonary late effects. J Clin Oncol 25(25):3991–4008. doi:10.1200/JCO.2007.10.9777. [Meta-Analysis Review]
35. Castellino S, Lensing S, Riely C, Rai SN, Davila R, Hayden RT et al (2004) The epidemiology of chronic hepatitis C infection in survivors of childhood cancer: an update of the St Jude Children's Research Hospital hepatitis C seropositive cohort. Blood 103(7):2460–2466. doi:10.1182/blood-2003-07-2565. [Comparative Study Research Support, Non-U.S. Gov't Research Support, U.S. Gov't, P.H.S.]
36. Cavuoto P, Fenech MF (2012) A review of methionine dependency and the role of methionine restriction in cancer growth control and life-span extension. Cancer Treat Rev 38(6):726–736. doi:10.1016/j.ctrv.2012.01.004. [Review]
37. Cellarier E, Durando X, Vasson MP, Farges MC, Demiden A, Maurizis JC et al (2003) Methionine dependency and cancer treatment. Cancer Treat Rev 29(6):489–499, S030573720300118X [pii]
38. Cerletti M, Jang YC, Finley LW, Haigis MC, Wagers AJ (2012) Short-term calorie restriction enhances skeletal muscle stem cell function. Cell Stem Cell 10(5):515–519. doi:10.1016/j.stem.2012.04.002. [Research Support, N.I.H., Extramural Research Support, Non-U.S. Gov't Research Support, U.S. Gov't, Non-P.H.S.]
39. Chakraborty S, Sun CL, Francisco L, Sabado M, Li L, Chang KL et al (2009) Accelerated telomere shortening precedes development of therapy-related myelodysplasia or acute myelogenous leukemia after autologous transplantation for lymphoma. J Clin Oncol 27(5):791–798. doi:10.1200/JCO.2008.17.1033. [Research Support, N.I.H., Extramural]
40. Chandel NS (2014) Mitochondria as signaling organelles. BMC Biol 12:34. doi:10.1186/1741-7007-12-34. [Research Support, N.I.H., Extramural Review]
41. Chen Y, Jungsuwadee P, Vore M, Butterfield DA, St Clair DK (2007) Collateral damage in cancer chemotherapy: oxidative stress in nontargeted tissues. Mol Interv 7(3):147–156. doi:10.1124/mi.7.3.6. [Research Support, N.I.H., Extramural Review]
42. Cheng CW, Adams GB, Perin L, Wei M, Zhou X, Lam BS et al (2014) Prolonged fasting reduces IGF-1/PKA to promote hematopoietic-stem-cell-based regeneration and reverse immunosuppression. Cell Stem Cell 14(6):810–823. doi:10.1016/j.stem.2014.04.014. [Research Support, N.I.H., Extramural]
43. Chiasson JL, Josse RG, Gomis R, Hanefeld M, Karasik A, Laakso M et al (2003) Acarbose treatment and the risk of cardiovascular disease and hypertension in patients with impaired glucose tolerance: the STOP-NIDDM trial. JAMA 290(4):486–494. doi:10.1001/jama.290.4.486
44. Clevers H, Loh KM, Nusse R (2014) Stem cell signaling. An integral program for tissue renewal and regeneration: Wnt signaling and stem cell control. Science 346(6205):1248012. doi:10.1126/science.1248012. [Research Support, Non-U.S. Gov't Research Support, U.S. Gov't, Non-P.H.S. Review]
45. Cloughesy TF, Yoshimoto K, Nghiemphu P, Brown K, Dang J, Zhu S et al (2008) Antitumor activity of rapamycin in a Phase I trial for patients with recurrent PTEN-deficient glioblastoma. PLoS Med 5(1):e8. doi:10.1371/journal.pmed.0050008. [Clinical Trial, Phase I Research Support, N.I.H., Extramural Research Support, Non-U.S. Gov't]
46. Cohen P (2014) New role for the mitochondrial peptide humanin: protective agent against chemotherapy-induced side effects. J Natl Cancer Inst 106(3):dju006. doi:10.1093/jnci/dju006. [Comment Editorial Research Support, N.I.H., Extramural]
47. Coley WB (1898) The treatment of inoperable sarcoma with the mixed toxins of erysipelas and bacillus prodigiosus. JAMA 31(8):389–395
48. Colman RJ, Anderson RM, Johnson SC, Kastman EK, Kosmatka KJ, Beasley TM et al (2009) Caloric restriction delays disease onset and mortality in rhesus monkeys. Science 325(5937):201–204. doi:10.1126/science.1173635, 325/5937/201 [pii]

49. Colman RJ, Beasley TM, Kemnitz JW, Johnson SC, Weindruch R, Anderson RM (2014) Caloric restriction reduces age-related and all-cause mortality in rhesus monkeys. Nat Commun 5:3557. doi:10.1038/ncomms4557. [Research Support, N.I.H., Extramural Research Support, Non-U.S. Gov't]
50. Conklin KA (2004) Chemotherapy-associated oxidative stress: impact on chemotherapeutic effectiveness. Integr Cancer Ther 3(4):294–300. doi:10.1177/1534735404270335. [Comparative Study Review]
51. Cox CL, Zhu L, Hudson MM, Steen BD, Robison LL, Oeffinger KC (2013) Survivor typologies predict medical surveillance participation: the childhood cancer survivor study. Psychooncology 22(7):1534–1542. doi:10.1002/pon.3167. [Research Support, N.I.H., Extramural Research Support, Non-U.S. Gov't]
52. Csoka AB, Szyf M (2009) Epigenetic side-effects of common pharmaceuticals: a potential new field in medicine and pharmacology. Med Hypotheses 73(5):770–780. doi:10.1016/j.mehy.2008.10.039
53. Davis CA, Nick HS, Agarwal A (2001) Manganese superoxide dismutase attenuates Cisplatin-induced renal injury: importance of superoxide. J Am Soc Nephrol 12(12):2683–2690. [Research Support, U.S. Gov't, P.H.S.]
54. De Cecco M, Criscione SW, Peckham EJ, Hillenmeyer S, Hamm EA, Manivannan J et al (2013) Genomes of replicatively senescent cells undergo global epigenetic changes leading to gene silencing and activation of transposable elements. Aging Cell 12(2):247–256. doi:10.1111/acel.12047. [Research Support, N.I.H., Extramural Research Support, Non-U.S. Gov't]
55. de Cuevas M, Matunis EL (2011) The stem cell niche: lessons from the Drosophila testis. Development 138(14):2861–2869. doi:10.1242/dev.056242. [Research Support, N.I.H., Extramural Review]
56. De Haes W, Frooninckx L, Van Assche R, Smolders A, Depuydt G, Billen J et al (2014) Metformin promotes lifespan through mitohormesis via the peroxiredoxin PRDX-2. Proc Natl Acad Sci U S A 111(24):E2501–E2509. doi:10.1073/pnas.1321776111
57. De Marte ML, Enesco HE (1986) Influence of low tryptophan diet on survival and organ growth in mice. Mech Ageing Dev 36(2):161–171, 0047-6374(86)90017-5 [pii]
58. Dehne N, Lautermann J, Petrat F, Rauen U, de Groot H (2001) Cisplatin ototoxicity: involvement of iron and enhanced formation of superoxide anion radicals. Toxicol Appl Pharmacol 174(1):27–34. doi:10.1006/taap.2001.9171. [Research Support, Non-U.S. Gov't]
59. DeSantis CE, Lin CC, Mariotto AB, Siegel RL, Stein KD, Kramer JL et al (2014) Cancer treatment and survivorship statistics, 2014. CA Cancer J Clin 64(4):252–271. doi:10.3322/caac.21235
60. Dienstmann R, Rodon J, Serra V, Tabernero J (2014) Picking the point of inhibition: a comparative review of PI3K/AKT/mTOR pathway inhibitors. Mol Cancer Ther 13(5):1021–1031. doi:10.1158/1535-7163.MCT-13-0639. [Research Support, Non-U.S. Gov't]
61. Diker-Cohen T, Uziel O, Szyper-Kravitz M, Shapira H, Natur A, Lahav M (2013) The effect of chemotherapy on telomere dynamics: clinical results and possible mechanisms. Leuk Lymphoma 54(9):2023–2029. doi:10.3109/10428194.2012.757765
62. Doroshow JH (1983) Effect of anthracycline antibiotics on oxygen radical formation in rat heart. Cancer Res 43(2):460–472. [Research Support, U.S. Gov't, P.H.S.]
63. Durieux J, Wolff S, Dillin A (2011) The cell-non-autonomous nature of electron transport chain-mediated longevity. Cell 144(1):79–91. doi:10.1016/j.cell.2010.12.016. [Research Support, N.I.H., Extramural Research Support, Non-U.S. Gov't]
64. Edgar D, Shabalina I, Camara Y, Wredenberg A, Calvaruso MA, Nijtmans L et al (2009) Random point mutations with major effects on protein-coding genes are the driving force behind premature aging in mtDNA mutator mice. Cell Metab 10(2):131–138. doi:10.1016/j.cmet.2009.06.010. [Research Support, Non-U.S. Gov't]
65. Ehrlich P, Bolduan C (1906) Collected studies on immunity. Wiley, New York
66. Emami Riedmaier A, Fisel P, Nies AT, Schaeffeler E, Schwab M (2013) Metformin and cancer: from the old medicine cabinet to pharmacological pitfalls and prospects. Trends

Pharmacol Sci 34(2):126–135. doi:10.1016/j.tips.2012.11.005. [Research Support, Non-- U.S. Gov't Review]
67. Epner DE (2001) Can dietary methionine restriction increase the effectiveness of chemotherapy in treatment of advanced cancer? J Am Coll Nutr 20(5 Suppl):443S–449S; discussion 473S–475S
68. Eriksson E, Wickstrom M, Perup LS, Johnsen JI, Eksborg S, Kogner P et al (2014) Protective role of humanin on bortezomib-induced bone growth impairment in anticancer treatment. J Natl Cancer Inst 106(3):djt459. doi:10.1093/jnci/djt459. [Research Support, Non-U.S. Gov't]
69. Errico A (2014) Targeted therapy: Targeting mitochondria in pancreatic cancer. Nat Rev Clin Oncol 11(10):562. doi:10.1038/nrclinonc.2014.143. [Comment]
70. Essig S, Li Q, Chen Y, Hitzler J, Leisenring W, Greenberg M et al (2014) Risk of late effects of treatment in children newly diagnosed with standard-risk acute lymphoblastic leukaemia: a report from the Childhood Cancer Survivor Study cohort. Lancet Oncol 15(8):841–851. doi:10.1016/S1470-2045(14)70265-7. [Multicenter Study Research Support, N.I.H., Extramural Research Support, Non-U.S. Gov't]
71. Evans JM, Donnelly LA, Emslie-Smith AM, Alessi DR, Morris AD (2005) Metformin and reduced risk of cancer in diabetic patients. BMJ 330(7503):1304–1305. doi:10.1136/bmj.38415.708634.F7. [Research Support, Non-U.S. Gov't]
72. Fischer-Nielsen A, Jeding IB, Loft S (1994) Radiation-induced formation of 8-hydroxy-2′-deoxyguanosine and its prevention by scavengers. Carcinogenesis 15(8):1609–1612
73. Florath I, Butterbach K, Muller H, Bewerunge-Hudler M, Brenner H (2014) Cross-sectional and longitudinal changes in DNA methylation with age: an epigenome-wide analysis revealing over 60 novel age-associated CpG sites. Hum Mol Genet 23(5):1186–1201. doi:10.1093/hmg/ddt531. [Research Support, Non-U.S. Gov't]
74. Folmes CD, Dzeja PP, Nelson TJ, Terzic A (2012) Metabolic plasticity in stem cell homeostasis and differentiation. Cell Stem Cell 11(5):596–606. doi:10.1016/j.stem.2012.10.002. [Research Support, N.I.H., Extramural Research Support, Non-U.S. Gov't]
75. Fontana L, Partridge L, Longo VD (2010) Extending healthy life span – from yeast to humans. Science 328(5976):321–326. doi:10.1126/science.1172539
76. Foretz M, Guigas B, Bertrand L, Pollak M, Viollet B (2014) Metformin: from mechanisms of action to therapies. Cell Metab 20(6):953–966. doi:10.1016/j.cmet.2014.09.018. [Review]
77. Franciosi M, Lucisano G, Lapice E, Strippoli GF, Pellegrini F, Nicolucci A (2013) Metformin therapy and risk of cancer in patients with type 2 diabetes: systematic review. PLoS One 8(8):e71583. doi:10.1371/journal.pone.0071583. [Review]
78. Frishman WH, Sung HM, Yee HC, Liu LL, Keefe D, Einzig AI et al (1997) Cardiovascular toxicity with cancer chemotherapy. Curr Probl Cancer 21(6):301–360. [Review]
79. Fruman DA, Rommel C (2014) PI3K and cancer: lessons, challenges and opportunities. Nat Rev Drug Discov 13(2):140–156. doi:10.1038/nrd4204. [Research Support, N.I.H., Extramural Review]
80. Fulda S, Galluzzi L, Kroemer G (2010) Targeting mitochondria for cancer therapy. Nat Rev Drug Discov 9(6):447–464. doi:10.1038/nrd3137. [Research Support, Non-U.S. Gov't Review]
81. Galluzzi L, Kepp O, Kroemer G (2012) Mitochondria: master regulators of danger signalling. Nat Rev Mol Cell Biol 13(12):780–788. doi:10.1038/nrm3479
82. Gawade PL, Hudson MM, Kaste SC, Neglia JP, Wasilewski-Masker K, Constine LS et al (2014) A systematic review of selected musculoskeletal late effects in survivors of childhood cancer. Curr Pediatr Rev 10:249–262
83. Giannakou ME, Partridge L (2007) Role of insulin-like signalling in Drosophila lifespan. Trends Biochem Sci 32(4):180–188. doi:10.1016/j.tibs.2007.02.007, S0968-0004(07)00064-3 [pii]
84. Giovannucci E, Harlan DM, Archer MC, Bergenstal RM, Gapstur SM, Habel LA et al (2010) Diabetes and cancer: a consensus report. CA Cancer J Clin 60(4):207–221. doi:10.3322/caac.20078
85. Goldenberg DM (2002) Targeted therapy of cancer with radiolabeled antibodies. J Nucl Med 43(5):693–713. [Research Support, U.S. Gov't, P.H.S. Review]

86. Goseki N, Yamazaki S, Endo M, Onodera T, Kosaki G, Hibino Y et al (1992) Antitumor effect of methionine-depleting total parenteral nutrition with doxorubicin administration on Yoshida sarcoma-bearing rats. Cancer 69(7):1865–1872
87. Gramatges MM, Liu Q, Yasui Y, Okcu MF, Neglia JP, Strong LC et al (2014) Telomere content and risk of second malignant neoplasm in survivors of childhood cancer: a report from the Childhood Cancer Survivor Study. Clin Cancer Res 20(4):904–911. doi:10.1158/1078-0432.CCR-13-2076. [Research Support, N.I.H., Extramural Research Support, Non-U.S. Gov't]
88. Grandison RC, Piper MD, Partridge L (2009) Amino-acid imbalance explains extension of lifespan by dietary restriction in Drosophila. Nature 462(7276):1061–1064. doi:10.1038/nature08619, nature08619 [pii]
89. Greider CW (1998) Telomeres and senescence: the history, the experiment, the future. Curr Biol 8(5):R178–R181. [Research Support, U.S. Gov't, P.H.S. Review]
90. Guarente L, Kenyon C (2000) Genetic pathways that regulate ageing in model organisms. Nature 408(6809):255–262. doi:10.1038/35041700
91. Guevara-Aguirre J, Balasubramanian P, Guevara-Aguirre M, Wei M, Madia F, Cheng CW et al (2011) Growth hormone receptor deficiency is associated with a major reduction in pro-aging signaling, cancer, and diabetes in humans. Sci Transl Med 3(70):70ra13. doi:10.1126/scitranslmed.3001845
92. Gupta S, Singh KK, Vyas VJ, Chaturvedi VN, Reddy MV, Harinath BC (2000) Assessment of oxidative stress and effect of antioxidant supplementation during radiotherapy in carcinoma of upper digestive tract. Indian J Clin Biochem 15(1):52–55. doi:10.1007/BF02873549
93. Hanahan D, Weinberg RA (2011) Hallmarks of cancer: the next generation. Cell 144(5):646–674. doi:10.1016/j.cell.2011.02.013. [Research Support, N.I.H., Extramural Review]
94. Hannum G, Guinney J, Zhao L, Zhang L, Hughes G, Sadda S et al (2013) Genome-wide methylation profiles reveal quantitative views of human aging rates. Mol Cell 49(2):359–367. doi:10.1016/j.molcel.2012.10.016. [Research Support, N.I.H., Extramural Research Support, Non-U.S. Gov't]
95. Harley CB, Futcher AB, Greider CW (1990) Telomeres shorten during ageing of human fibroblasts. Nature 345(6274):458–460. doi:10.1038/345458a0. [Research Support, Non-U.S. Gov't]
96. Harrison DE, Strong R, Allison DB, Ames BN, Astle CM, Atamna H et al (2014) Acarbose, 17-alpha-estradiol, and nordihydroguaiaretic acid extend mouse lifespan preferentially in males. Aging Cell 13(2):273–282. doi:10.1111/acel.12170
97. Harrison DE, Strong R, Sharp ZD, Nelson JF, Astle CM, Flurkey K et al (2009) Rapamycin fed late in life extends lifespan in genetically heterogeneous mice. Nature 460(7253):392–395. doi:10.1038/nature08221. [Research Support, N.I.H., Extramural Research Support, U.S. Gov't, Non-P.H.S.]
98. Hashimoto Y, Niikura T, Tajima H, Yasukawa T, Sudo H, Ito Y et al (2001) A rescue factor abolishing neuronal cell death by a wide spectrum of familial Alzheimer's disease genes and Abeta. Proc Natl Acad Sci U S A 98(11):6336–6341. doi:10.1073/pnas.101133498. [Research Support, Non-U.S. Gov't]
99. Hayflick L (1965) The limited in vitro lifetime of human diploid cell strains. Exp Cell Res 37:614–636
100. Hayflick L, Moorhead PS (1961) The serial cultivation of human diploid cell strains. Exp Cell Res 25:585–621
101. Hirabayashi S, Baranski TJ, Cagan RL (2013) Transformed Drosophila cells evade diet-mediated insulin resistance through wingless signaling. Cell 154(3):664–675. doi:10.1016/j.cell.2013.06.030
102. Holderfield M, Deuker MM, McCormick F, McMahon M (2014) Targeting RAF kinases for cancer therapy: BRAF-mutated melanoma and beyond. Nat Rev Cancer 14(7):455–467. doi:10.1038/nrc3760. [Review]
103. Horvath S (2013) DNA methylation age of human tissues and cell types. Genome Biol 14(10):R115. doi:10.1186/gb-2013-14-10-r115

104. Horvath S, Zhang Y, Langfelder P, Kahn RS, Boks MP, van Eijk K et al (2012) Aging effects on DNA methylation modules in human brain and blood tissue. Genome Biol 13(10):R97. doi:10.1186/gb-2012-13-10-r97. [Research Support, N.I.H., Extramural]
105. Hoshiya Y, Kubota T, Matsuzaki SW, Kitajima M, Hoffman RM (1996) Methionine starvation modulates the efficacy of cisplatin on human breast cancer in nude mice. Anticancer Res 16(6B):3515–3517
106. Hsieh RH, Hou JH, Hsu HS, Wei YH (1994) Age-dependent respiratory function decline and DNA deletions in human muscle mitochondria. Biochem Mol Biol Int 32(6):1009–1022. [Research Support, Non-U.S. Gov't]
107. Husseinzadeh N, Husseinzadeh HD (2014) mTOR inhibitors and their clinical application in cervical, endometrial and ovarian cancers: a critical review. Gynecol Oncol 133(2):375–381. doi:10.1016/j.ygyno.2014.02.017. [Review]
108. Iglesias-Bartolome R, Patel V, Cotrim A, Leelahavanichkul K, Molinolo AA, Mitchell JB et al (2012) mTOR inhibition prevents epithelial stem cell senescence and protects from radiation-induced mucositis. Cell Stem Cell 11(3):401–414. doi:10.1016/j.stem.2012.06.007. [Research Support, N.I.H., Intramural Research Support, Non-U.S. Gov't]
109. Ilnytskyy Y, Koturbash I, Kovalchuk O (2009) Radiation-induced bystander effects in vivo are epigenetically regulated in a tissue-specific manner. Environ Mol Mutagen 50(2):105–113. doi:10.1002/em.20440. [Research Support, Non-U.S. Gov't]
110. Inskip PD, Robison LL, Stovall M, Smith SA, Hammond S, Mertens AC et al (2009) Radiation dose and breast cancer risk in the childhood cancer survivor study. J Clin Oncol 27(24):3901–3907. doi:10.1200/JCO.2008.20.7738
111. Ito K, Suda T (2014) Metabolic requirements for the maintenance of self-renewing stem cells. Nat Rev Mol Cell Biol 15(4):243–256. doi:10.1038/nrm3772. [Research Support, N.I.H., Extramural Research Support, Non-U.S. Gov't Review]
112. Jain P, Gulati S, Seth R, Bakhshi S, Toteja GS, Pandey RM (2013) Vincristine-induced neuropathy in childhood ALL (acute lymphoblastic leukemia) survivors: prevalence and electrophysiological characteristics. J Child Neurol 29(7):932–937. doi:10.1177/0883073813491829
113. Junnila RK, List EO, Berryman DE, Murrey JW, Kopchick JJ (2013) The GH/IGF-1 axis in ageing and longevity. Nat Rev Endocrinol 9(6):366–376. doi:10.1038/nrendo.2013.67
114. Kalyanaraman B (2007) Iron signaling and oxidant damage. Cardiovasc Toxicol 7(2):92–94. doi:10.1007/s12012-007-0025-1. [Review]
115. Kang YJ (2007) Antioxidant defense against anthracycline cardiotoxicity by metallothionein. Cardiovasc Toxicol 7(2):95–100. doi:10.1007/s12012-007-0007-3. [Research Support, N.I.H., Extramural Review]
116. Kennedy SR, Salk JJ, Schmitt MW, Loeb LA (2013) Ultra-sensitive sequencing reveals an age-related increase in somatic mitochondrial mutations that are inconsistent with oxidative damage. PLoS Genet 9(9), e1003794. doi:10.1371/journal.pgen.1003794
117. Kenney LB, Cohen LE, Shnorhavorian M, Metzger ML, Lockart B, Hijiya N et al (2012) Male reproductive health after childhood, adolescent, and young adult cancers: a report from the Children's Oncology Group. J Clin Oncol 30(27):3408–3416. doi:10.1200/JCO.2011.38.6938
118. Kenyon C (2001) A conserved regulatory system for aging. Cell 105(2):165–168, S0092-8674(01)00306-3 [pii]
119. Kirkwood TL, Kapahi P, Shanley DP (2000) Evolution, stress, and longevity. J Anat 197(Pt 4):587–590
120. Kleemann GA, Murphy CT (2009) The endocrine regulation of aging in Caenorhabditis elegans. Mol Cell Endocrinol 299(1):51–57. doi:10.1016/j.mce.2008.10.048, S0303-7207(08)00519-4 [pii]
121. Kohler G, Milstein C (1975) Continuous cultures of fused cells secreting antibody of predefined specificity. Nature 256(5517):495–497
122. Kokkinakis DM, Hoffman RM, Frenkel EP, Wick JB, Han Q, Xu M et al (2001) Synergy between methionine stress and chemotherapy in the treatment of brain tumor xenografts in athymic mice. Cancer Res 61(10):4017–4023

123. Koturbash I, Rugo RE, Hendricks CA, Loree J, Thibault B, Kutanzi K et al (2006) Irradiation induces DNA damage and modulates epigenetic effectors in distant bystander tissue in vivo. Oncogene 25(31):4267–4275. doi:10.1038/sj.onc.1209467. [Research Support, N.I.H., Extramural Research Support, Non-U.S. Gov't Research Support, U.S. Gov't, Non-P.H.S.]
124. Kraytsberg Y, Kudryavtseva E, McKee AC, Geula C, Kowall NW, Khrapko K (2006) Mitochondrial DNA deletions are abundant and cause functional impairment in aged human substantia nigra neurons. Nat Genet 38(5):518–520. doi:10.1038/ng1778. [Research Support, N.I.H., Extramural Research Support, U.S. Gov't, Non-P.H.S.]
125. Kremer LC, van Dalen EC, Offringa M, Ottenkamp J, Voute PA (2001) Anthracycline-induced clinical heart failure in a cohort of 607 children: long-term follow-up study. J Clin Oncol 19(1):191–196
126. Kuilman T, Michaloglou C, Mooi WJ, Peeper DS (2010) The essence of senescence. Genes Dev 24(22):2463–2479. doi:10.1101/gad.1971610. [Research Support, Non-U.S. Gov't Review]
127. Kuliawat R, Klein L, Gong Z, Nicoletta-Gentile M, Nemkal A, Cui L et al (2013) Potent humanin analog increases glucose-stimulated insulin secretion through enhanced metabolism in the beta cell. FASEB J 27(12):4890–4898. doi:10.1096/fj.13-231092. [Research Support, N.I.H., Extramural]
128. Kurz DJ, Decary S, Hong Y, Erusalimsky JD (2000) Senescence-associated (beta)-galactosidase reflects an increase in lysosomal mass during replicative ageing of human endothelial cells. J Cell Sci 113(Pt 20):3613–3622. [Comparative Study Research Support, Non-U.S. Gov't]
129. Lebrecht D, Kokkori A, Ketelsen UP, Setzer B, Walker UA (2005) Tissue-specific mtDNA lesions and radical-associated mitochondrial dysfunction in human hearts exposed to doxorubicin. J Pathol 207(4):436–444. doi:10.1002/path.1863. [Research Support, Non-U.S. Gov't]
130. Lebrecht D, Walker UA (2007) Role of mtDNA lesions in anthracycline cardiotoxicity. Cardiovasc Toxicol 7(2):108–113. doi:10.1007/s12012-007-0009-1. [Review]
131. Lee BY, Han JA, Im JS, Morrone A, Johung K, Goodwin EC et al (2006) Senescence-associated beta-galactosidase is lysosomal beta-galactosidase. Aging Cell 5(2):187–195. doi:10.1111/j.1474-9726.2006.00199.x. [Research Support, N.I.H., Extramural Research Support, Non-U.S. Gov't]
132. Lee C, Longo VD (2011) Fasting vs dietary restriction in cellular protection and cancer treatment: from model organisms to patients. Oncogene 30(30):3305–3316. doi:10.1038/onc.2011.91. [Comparative Study Review]
133. Lee C, Raffaghello L, Brandhorst S, Safdie FM, Bianchi G, Martin-Montalvo A et al (2012) Fasting cycles retard growth of tumors and sensitize a range of cancer cell types to chemotherapy. Sci Transl Med 4(124):124ra127, 10.1126/scitranslmed.3003293
134. Lee C, Yen K, Cohen P (2013) Humanin: a harbinger of mitochondrial-derived peptides? Trends Endocrinol Metab. doi:10.1016/j.tem.2013.01.005
135. Lee CS, Cragg M, Glennie M, Johnson P (2013) Novel antibodies targeting immune regulatory checkpoints for cancer therapy. Br J Clin Pharmacol 76(2):233–247. doi:10.1111/bcp.12164. [Research Support, Non-U.S. Gov't Review]
136. Lee JJ, Nam CE, Cho SH, Park KS, Chung IJ, Kim HJ (2003) Telomere length shortening in non-Hodgkin's lymphoma patients undergoing chemotherapy. Ann Hematol 82(8):492–495. doi:10.1007/s00277-003-0691-4. [Research Support, Non-U.S. Gov't]
137. Lefrak EA, Pitha J, Rosenheim S, Gottlieb JA (1973) A clinicopathologic analysis of adriamycin cardiotoxicity. Cancer 32(2):302–314
138. Levitzki A (2013) Tyrosine kinase inhibitors: views of selectivity, sensitivity, and clinical performance. Annu Rev Pharmacol Toxicol 53:161–185. doi:10.1146/annurev-pharmtox-011112-140341. [Review]
139. Lezza AM, Boffoli D, Scacco S, Cantatore P, Gadaleta MN (1994) Correlation between mitochondrial DNA 4977-bp deletion and respiratory chain enzyme activities in aging human skeletal muscles. Biochem Biophys Res Commun 205(1):772–779. doi:10.1006/bbrc.1994.2732. [Research Support, Non-U.S. Gov't]

140. Links M, Lewis C (1999) Chemoprotectants: a review of their clinical pharmacology and therapeutic efficacy. Drugs 57(3):293–308
141. Lipshultz SE, Adams MJ, Colan SD, Constine LS, Herman EH, Hsu DT et al (2013) Long-term cardiovascular toxicity in children, adolescents, and young adults who receive cancer therapy: pathophysiology, course, monitoring, management, prevention, and research directions: a scientific statement from the American Heart Association. Circulation 128(17):1927–1995. doi:10.1161/CIR.0b013e3182a88099. [Practice Guideline]
142. Longo VD, Finch CE (2003) Evolutionary medicine: from dwarf model systems to healthy centenarians? Science 299(5611):1342–1346. doi:10.1126/science.1077991. [Research Support, Non-U.S. Gov't Research Support, U.S. Gov't, P.H.S. Review]
143. Look MP, Musch E (1994) Lipid peroxides in the polychemotherapy of cancer patients. Chemotherapy 40(1):8–15. [Comparative Study]
144. Luan FL, Ding R, Sharma VK, Chon WJ, Lagman M, Suthanthiran M (2003) Rapamycin is an effective inhibitor of human renal cancer metastasis. Kidney Int 63(3):917–926
145. Mannick JB, Del Giudice G, Lattanzi M, Valiante NM, Praestgaard J, Huang B et al (2014) mTOR inhibition improves immune function in the elderly. Sci Transl Med 6(268):268ra179. doi:10.1126/scitranslmed.3009892. [Research Support, Non-U.S. Gov't]
146. Martin-Montalvo A, Mercken EM, Mitchell SJ, Palacios HH, Mote PL, Scheibye-Knudsen M et al (2013) Metformin improves healthspan and lifespan in mice. Nat Commun 4:2192. doi:10.1038/ncomms3192
147. Martins F, de Oliveira MA, Wang Q, Sonis S, Gallottini M, George S et al (2013) A review of oral toxicity associated with mTOR inhibitor therapy in cancer patients. Oral Oncol 49(4):293–298. doi:10.1016/j.oraloncology.2012.11.008. [Review]
148. Maschek G, Savaraj N, Priebe W, Braunschweiger P, Hamilton K, Tidmarsh GF et al (2004) 2-deoxy-D-glucose increases the efficacy of adriamycin and paclitaxel in human osteosarcoma and non-small cell lung cancers in vivo. Cancer Res 64(1):31–34
149. Mattison JA, Roth GS, Beasley TM, Tilmont EM, Handy AM, Herbert RL et al (2012) Impact of caloric restriction on health and survival in rhesus monkeys from the NIA study. Nature 489(7415):318–321. doi:10.1038/nature11432. [Research Support, N.I.H., Intramural]
150. McCay CM, Crowell MF, Maynard LA (1989) The effect of retarded growth upon the length of life span and upon the ultimate body size. 1935. Nutrition 5(3):155–171; discussion 172
151. Meadows KL, Hurwitz HI (2012) Anti-VEGF therapies in the clinic. Cold Spring Harb Perspect Med 2(10). doi:10.1101/cshperspect.a006577. [Review].
152. Mellman I, Coukos G, Dranoff G (2011) Cancer immunotherapy comes of age. Nature 480(7378):480–489. doi:10.1038/nature10673. [Review]
153. Meng A, Wang Y, Van Zant G, Zhou D (2003) Ionizing radiation and busulfan induce premature senescence in murine bone marrow hematopoietic cells. Cancer Res 63(17):5414–5419. [Research Support, U.S. Gov't, P.H.S.]
154. Metzger ML, Meacham LR, Patterson B, Casillas JS, Constine LS, Hijiya N et al (2013) Female reproductive health after childhood, adolescent, and young adult cancers: guidelines for the assessment and management of female reproductive complications. J Clin Oncol 31(9):1239–1247. doi:10.1200/JCO.2012.43.5511. [Review]
155. Meyerhardt JA, Mangu PB, Flynn PJ, Korde L, Loprinzi CL, Minsky BD et al (2013) Follow-up care, surveillance protocol, and secondary prevention measures for survivors of colorectal cancer: American Society of Clinical Oncology clinical practice guideline endorsement. J Clin Oncol 31(35):4465–4470. doi:10.1200/JCO.2013.50.7442. [Practice Guideline]
156. Miao J, Zhang W, Yin R, Liu R, Su C, Lei G et al (2008) S14G-Humanin ameliorates Abeta25-35-induced behavioral deficits by reducing neuroinflammatory responses and apoptosis in mice. Neuropeptides 42(5–6):557–567. doi:10.1016/j.npep.2008.08.004. [Research Support, Non-U.S. Gov't]
157. Minor RK, Baur JA, Gomes AP, Ward TM, Csiszar A, Mercken EM et al (2011) SRT1720 improves survival and healthspan of obese mice. Sci Rep 1:70. doi:10.1038/srep00070. [Research Support, N.I.H., Extramural Research Support, Non-U.S. Gov't]

158. Mitchell SJ, Martin-Montalvo A, Mercken EM, Palacios HH, Ward TM, Abulwerdi G et al (2014) The SIRT1 activator SRT1720 extends lifespan and improves health of mice fed a standard diet. Cell Rep 6(5):836–843. doi:10.1016/j.celrep.2014.01.031. [Research Support, N.I.H., Extramural Research Support, Non-U.S. Gov't]
159. Morales DR, Morris AD (2015) Metformin in cancer treatment and prevention. Annu Rev Med 66:17–29. doi:10.1146/annurev-med-062613-093128. [Research Support, Non-U.S. Gov't]
160. Murakami S (2006) Stress resistance in long-lived mouse models. Exp Gerontol 41(10):1014–1019. doi:10.1016/j.exger.2006.06.061, S0531-5565(06)00236-1 [pii]
161. Murata T, Hibasami H, Maekawa S, Tagawa T, Nakashima K (1990) Preferential binding of cisplatin to mitochondrial DNA and suppression of ATP generation in human malignant melanoma cells. Biochem Int 20(5):949–955. [Comparative Study]
162. Murphy JE, Nugent S, Seymour C, Mothersill C (2005) Mitochondrial DNA point mutations and a novel deletion induced by direct low-LET radiation and by medium from irradiated cells. Mutat Res 585(1–2):127–136. doi:10.1016/j.mrgentox.2005.04.011
163. Musgrove EA, Caldon CE, Barraclough J, Stone A, Sutherland RL (2011) Cyclin D as a therapeutic target in cancer. Nat Rev Cancer 11(8):558–572. doi:10.1038/nrc3090. [Research Support, Non-U.S. Gov't Review]
164. Muzumdar RH, Huffman DM, Calvert JW, Jha S, Weinberg Y, Cui L et al (2010) Acute humanin therapy attenuates myocardial ischemia and reperfusion injury in mice. Arterioscler Thromb Vasc Biol 30(10):1940–1948. doi:10.1161/ATVBAHA.110.205997. [Research Support, N.I.H., Extramural Research Support, Non-U.S. Gov't]
165. Myers CE, McGuire WP, Liss RH, Ifrim I, Grotzinger K, Young RC (1977) Adriamycin: the role of lipid peroxidation in cardiac toxicity and tumor response. Science 197(4299):165–167
166. Neglia JP, Robison LL, Stovall M, Liu Y, Packer RJ, Hammond S et al (2006) New primary neoplasms of the central nervous system in survivors of childhood cancer: a report from the Childhood Cancer Survivor Study. J Natl Cancer Inst 98(21):1528–1537. doi:10.1093/jnci/djj411. [Multicenter Study Research Support, N.I.H., Extramural Research Support, N.I.H., Intramural Research Support, Non-U.S. Gov't]
167. Niikura T, Sidahmed E, Hirata-Fukae C, Aisen PS, Matsuoka Y (2011) A humanin derivative reduces amyloid beta accumulation and ameliorates memory deficit in triple transgenic mice. PLoS One 6(1):e16259. doi:10.1371/journal.pone.0016259. [Research Support, N.I.H., Extramural Research Support, Non-U.S. Gov't]
168. O'Sullivan RJ, Karlseder J (2010) Telomeres: protecting chromosomes against genome instability. Nat Rev Mol Cell Biol 11(3):171–181. doi:10.1038/nrm2848. [Research Support, N.I.H., Extramural Research Support, Non-U.S. Gov't Review]
169. Oeffinger KC, Bhatia S (2009) Second primary cancers in survivors of childhood cancer. Lancet 374(9700):1484–1485. doi:10.1016/S0140-6736(09)61885-7. [Review]
170. Oeffinger KC, Mertens AC, Sklar CA, Kawashima T, Hudson MM, Meadows AT et al (2006) Chronic health conditions in adult survivors of childhood cancer. N Engl J Med 355(15):1572–1582. doi:10.1056/NEJMsa060185. [Comparative Study Research Support, N.I.H., Extramural Research Support, Non-U.S. Gov't]
171. Oh J, Lee YD, Wagers AJ (2014) Stem cell aging: mechanisms, regulators and therapeutic opportunities. Nat Med 20(8):870–880. doi:10.1038/nm.3651. [Research Support, N.I.H., Extramural Research Support, Non-U.S. Gov't Review]
172. Oh YK, Bachar AR, Zacharias DG, Kim SG, Wan J, Cobb LJ et al (2011) Humanin preserves endothelial function and prevents atherosclerotic plaque progression in hypercholesterolemic ApoE deficient mice. Atherosclerosis 219(1):65–73. doi:10.1016/j.atherosclerosis.2011.06.038
173. Olinski R, Zastawny TH, Foksinski M, Windorbska W, Jaruga P, Dizdaroglu M (1996) DNA base damage in lymphocytes of cancer patients undergoing radiation therapy. Cancer Lett 106(2):207–215. [Research Support, Non-U.S. Gov't]

174. Ooka H, Segall PE, Timiras PS (1988) Histology and survival in age-delayed low-tryptophan-fed rats. Mech Ageing Dev 43(1):79–98, 0047-6374(88)90099-1 [pii]
175. Page DB, Postow MA, Callahan MK, Allison JP, Wolchok JD (2014) Immune modulation in cancer with antibodies. Annu Rev Med 65:185–202. doi:10.1146/annurev-med-092012-112807. [Review]
176. Pappo AS, Armstrong GT, Liu W, Srivastava DK, McDonald A, Leisenring WM et al (2013) Melanoma as a subsequent neoplasm in adult survivors of childhood cancer: a report from the childhood cancer survivor study. Pediatr Blood Cancer 60(3):461–466. doi:10.1002/pbc.24266. [Multicenter Study Research Support, N.I.H., Extramural Research Support, Non-U.S. Gov't]
177. Pernicova I, Korbonits M (2014) Metformin – mode of action and clinical implications for diabetes and cancer. Nat Rev Endocrinol 10(3):143–156. doi:10.1038/nrendo.2013.256. [Research Support, Non-U.S. Gov't Review]
178. Pfluger PT, Herranz D, Velasco-Miguel S, Serrano M, Tschop MH (2008) Sirt1 protects against high-fat diet-induced metabolic damage. Proc Natl Acad Sci U S A 105(28):9793–9798. doi:10.1073/pnas.0802917105
179. Pizzuti L, Vici P, Di Lauro L, Sergi D, Della Giulia M, Marchetti P et al (2015) Metformin and breast cancer: basic knowledge in clinical context. Cancer Treat Rev 41(5):441–447. doi:10.1016/j.ctrv.2015.03.002. [Review]
180. Podratz JL, Knight AM, Ta LE, Staff NP, Gass JM, Genelin K et al (2011) Cisplatin induced mitochondrial DNA damage in dorsal root ganglion neurons. Neurobiol Dis 41(3):661–668. doi:10.1016/j.nbd.2010.11.017. [Comparative Study Research Support, N.I.H., Extramural]
181. Prithivirajsingh S, Story MD, Bergh SA, Geara FB, Ang KK, Ismail SM et al (2004) Accumulation of the common mitochondrial DNA deletion induced by ionizing radiation. FEBS Lett 571(1–3):227–232. doi:10.1016/j.febslet.2004.06.078. [Research Support, Non--U.S. Gov't Research Support, U.S. Gov't, P.H.S.]
182. Quinn BJ, Kitagawa H, Memmott RM, Gills JJ, Dennis PA (2013) Repositioning metformin for cancer prevention and treatment. Trends Endocrinol Metab 24(9):469–480. doi:10.1016/j.tem.2013.05.004
183. Raffaghello L, Lee C, Safdie FM, Wei M, Madia F, Bianchi G et al (2008) Starvation-dependent differential stress resistance protects normal but not cancer cells against high-dose chemotherapy. Proc Natl Acad Sci U S A 105(24):8215–8220. doi:10.1073/pnas.0708100105. [Research Support, N.I.H., Extramural Research Support, Non-U.S. Gov't]
184. Reiling JH, Sabatini DM (2006) Stress and mTORture signaling. Oncogene 25(48):6373–6383. doi:10.1038/sj.onc.1209889, 1209889 [pii]
185. Relling MV, Rubnitz JE, Rivera GK, Boyett JM, Hancock ML, Felix CA et al (1999) High incidence of secondary brain tumours after radiotherapy and antimetabolites. Lancet 354(9172):34–39. doi:10.1016/S0140-6736(98)11079-6. [Clinical Trial Research Support, Non-U.S. Gov't Research Support, U.S. Gov't, P.H.S.]
186. Ricca I, Compagno M, Ladetto M, Rocci A, Dell'Aquila M, Omede P et al (2005) Marked telomere shortening in mobilized peripheral blood progenitor cells (PBPC) following two tightly spaced high-dose chemotherapy courses with G-CSF. Leukemia 19(4):644–651. doi:10.1038/sj.leu.2403652. [Research Support, Non-U.S. Gov't]
187. Ricciardi S, Tomao S, de Marinis F (2009) Toxicity of targeted therapy in non-small-cell lung cancer management. Clin Lung Cancer 10(1):28–35. doi:10.3816/CLC.2009.n.004. [Review]
188. Richie JP Jr, Leutzinger Y, Parthasarathy S, Malloy V, Orentreich N, Zimmerman JA (1994) Methionine restriction increases blood glutathione and longevity in F344 rats. FASEB J 8(15):1302–1307
189. Riley PA (1994) Free radicals in biology: oxidative stress and the effects of ionizing radiation. Int J Radiat Biol 65(1):27–33. [Research Support, Non-U.S. Gov't Review]
190. Robida-Stubbs S, Glover-Cutter K, Lamming DW, Mizunuma M, Narasimhan SD, Neumann-Haefelin E et al (2012) TOR signaling and rapamycin influence longevity by regulating SKN-1/Nrf and DAF-16/FoxO. Cell Metab 15(5):713–724. doi:10.1016/j.cmet.2012.04.007. [Research Support, N.I.H., Extramural Research Support, Non-U.S. Gov't]

191. Robison LL, Hudson MM (2014) Survivors of childhood and adolescent cancer: life-long risks and responsibilities. Nat Rev Cancer 14(1):61–70. doi:10.1038/nrc3634. [Research Support, N.I.H., Extramural Research Support, Non-U.S. Gov't]
192. Rodier F, Munoz DP, Teachenor R, Chu V, Le O, Bhaumik D et al (2011) DNA-SCARS: distinct nuclear structures that sustain damage-induced senescence growth arrest and inflammatory cytokine secretion. J Cell Sci 124(Pt 1):68–81. doi:10.1242/jcs.071340. [Research Support, N.I.H., Extramural Research Support, Non-U.S. Gov't Research Support, U.S. Gov't, Non-P.H.S.]
193. Roszkowski K, Gackowski D, Rozalski R, Dziaman T, Siomek A, Guz J et al (2008) Small field radiotherapy of head and neck cancer patients is responsible for oxidatively damaged DNA/oxidative stress on the level of a whole organism. Int J Cancer 123(8):1964–1967. doi:10.1002/ijc.23700. [Research Support, Non-U.S. Gov't]
194. Rybak LP, Ravi R, Somani SM (1995) Mechanism of protection by diethyldithiocarbamate against cisplatin ototoxicity: antioxidant system. Fundam Appl Toxicol 26(2):293–300. [Research Support, U.S. Gov't, P.H.S.]
195. Rybak LP, Whitworth CA, Mukherjea D, Ramkumar V (2007) Mechanisms of cisplatin-induced ototoxicity and prevention. Hear Res 226(1–2):157–167. doi:10.1016/j.heares.2006.09.015. [Research Support, N.I.H., Extramural Review]
196. Safdie F, Brandhorst S, Wei M, Wang W, Lee C, Hwang S et al (2012) Fasting enhances the response of glioma to chemo- and radiotherapy. PLoS One 7(9), e44603. doi:10.1371/journal.pone.0044603
197. Safdie FM, Dorff T, Quinn D, Fontana L, Wei M, Lee C et al (2009) Fasting and cancer treatment in humans: a case series report. Aging (Albany NY) 1(12):988–1007. [Case Reports Research Support, Non-U.S. Gov't]
198. Saini KS, Loi S, de Azambuja E, Metzger-Filho O, Saini ML, Ignatiadis M et al (2013) Targeting the PI3K/AKT/mTOR and Raf/MEK/ERK pathways in the treatment of breast cancer. Cancer Treat Rev 39(8):935–946. doi:10.1016/j.ctrv.2013.03.009
199. Salmon AB, Murakami S, Bartke A, Kopchick J, Yasumura K, Miller RA (2005) Fibroblast cell lines from young adult mice of long-lived mutant strains are resistant to multiple forms of stress. Am J Physiol Endocrinol Metab 289(1):E23–E29. doi:10.1152/ajpendo.00575.2004, 00575.2004 [pii]
200. Sangeetha P, Das UN, Koratkar R (1990) Free radical generation in human leukocytes by CIS-unsaturated fatty acids is a calmodulin dependent process. Prostaglandins Leukot Essent Fatty Acids 39(1):27–30. [Comparative Study Research Support, Non-U.S. Gov't]
201. Sanoff HK, Deal AM, Krishnamurthy J, Torrice C, Dillon P, Sorrentino J et al (2014) Effect of cytotoxic chemotherapy on markers of molecular age in patients with breast cancer. J Natl Cancer Inst 106(4):dju057. doi:10.1093/jnci/dju057. [Research Support, N.I.H., Extramural Research Support, Non-U.S. Gov't]
202. Santos NA, Catao CS, Martins NM, Curti C, Bianchi ML, Santos AC (2007) Cisplatin-induced nephrotoxicity is associated with oxidative stress, redox state unbalance, impairment of energetic metabolism and apoptosis in rat kidney mitochondria. Arch Toxicol 81(7):495–504. doi:10.1007/s00204-006-0173-2
203. Scheede-Bergdahl C, Jagoe RT (2013) After the chemotherapy: potential mechanisms for chemotherapy-induced delayed skeletal muscle dysfunction in survivors of acute lymphoblastic leukaemia in childhood. Front Pharmacol 4:49. doi:10.3389/fphar.2013.00049
204. Scott AM, Wolchok JD, Old LJ (2012) Antibody therapy of cancer. Nat Rev Cancer 12(4):278–287. doi:10.1038/nrc3236. [Research Support, Non-U.S. Gov't Review]
205. Sebolt-Leopold JS, Herrera R (2004) Targeting the mitogen-activated protein kinase cascade to treat cancer. Nat Rev Cancer 4(12):937–947. doi:10.1038/nrc1503. [Review]
206. Segall P (1977) Long-term tryptophan restriction and aging in the rat. Aktuelle Gerontol 7(10):535–538
207. Segall PE, Timiras PS (1976) Patho-physiologic findings after chronic tryptophan deficiency in rats: a model for delayed growth and aging. Mech Ageing Dev 5(2):109–124

208. Serrano J, Palmeira CM, Kuehl DW, Wallace KB (1999) Cardioselective and cumulative oxidation of mitochondrial DNA following subchronic doxorubicin administration. Biochim Biophys Acta 1411(1):201–205. [Comparative Study Research Support, U.S. Gov't, P.H.S.]
209. Seto B (2012) Rapamycin and mTOR: a serendipitous discovery and implications for breast cancer. Clin Transl Med 1(1):29. doi:10.1186/2001-1326-1-29
210. Sevini F, Giuliani C, Vianello D, Giampieri E, Santoro A, Biondi F et al (2014) mtDNA mutations in human aging and longevity: controversies and new perspectives opened by high-throughput technologies. Exp Gerontol 56:234–244. doi:10.1016/j.exger.2014.03.022. [Research Support, Non-U.S. Gov't]
211. Shanley DP, Kirkwood TB (2000) Calorie restriction and aging: a life-history analysis. Evol Int J Org Evol 54(3):740–750
212. Sharpless NE, DePinho RA (2007) How stem cells age and why this makes us grow old. Nat Rev Mol Cell Biol 8(9):703–713
213. Sigurdson AJ, Ronckers CM, Mertens AC, Stovall M, Smith SA, Liu Y et al (2005) Primary thyroid cancer after a first tumour in childhood (the Childhood Cancer Survivor Study): a nested case–control study. Lancet 365(9476):2014–2023. doi:10.1016/S0140-6736(05)66695-0. [Research Support, N.I.H., Extramural Research Support, Non-U.S. Gov't Research Support, U.S. Gov't, P.H.S.]
214. Singal PK, Li T, Kumar D, Danelisen I, Iliskovic N (2000) Adriamycin-induced heart failure: mechanism and modulation. Mol Cell Biochem 207(1–2):77–86. [Research Support, Non--U.S. Gov't Review]
215. Smith AK, Conneely KN, Pace TW, Mister D, Felger JC, Kilaru V et al (2014) Epigenetic changes associated with inflammation in breast cancer patients treated with chemotherapy. Brain Behav Immun 38:227–236. doi:10.1016/j.bbi.2014.02.010. [Research Support, N.I.H., Extramural]
216. Sun JX, Yang N (2014) Role of octreotide in post chemotherapy and/or radiotherapy diarrhea: prophylaxis or therapy? Asia Pac J Clin Oncol 10(2):e108–e113. doi:10.1111/ajco.12055
217. Suzuki K, Hayashi R, Ichikawa T, Imanishi S, Yamada T, Inomata M et al (2012) SRT1720, a SIRT1 activator, promotes tumor cell migration, and lung metastasis of breast cancer in mice. Oncol Rep 27(6):1726–1732. doi:10.3892/or.2012.1750
218. Swain SM, Whaley FS, Ewer MS (2003) Congestive heart failure in patients treated with doxorubicin: a retrospective analysis of three trials. Cancer 97(11):2869–2879. doi:10.1002/cncr.11407. [Clinical Trial Clinical Trial, Phase III Randomized Controlled Trial Research Support, Non-U.S. Gov't]
219. Swift L, McHowat J, Sarvazyan N (2007) Anthracycline-induced phospholipase A2 inhibition. Cardiovasc Toxicol 7(2):86–91. doi:10.1007/s12012-007-0012-6. [Research Support, N.I.H., Extramural Review]
220. Taguchi T, Nazneen A, Abid MR, Razzaque MS (2005) Cisplatin-associated nephrotoxicity and pathological events. Contrib Nephrol 148:107–121. doi:10.1159/000086055. [Review]
221. Tan Y, Sun X, Xu M, Tan X, Sasson A, Rashidi B et al (1999) Efficacy of recombinant methioninase in combination with cisplatin on human colon tumors in nude mice. Clin Cancer Res 5(8):2157–2163
222. Tan Y, Xu M, Guo H, Sun X, Kubota T, Hoffman RM (1996) Anticancer efficacy of methioninase in vivo. Anticancer Res 16(6C):3931–3936
223. Tchkonia T, Zhu Y, van Deursen J, Campisi J, Kirkland JL (2013) Cellular senescence and the senescent secretory phenotype: therapeutic opportunities. J Clin Invest 123(3):966–972. doi:10.1172/JCI64098. [Research Support, N.I.H., Extramural Research Support, Non--U.S. Gov't Review]
224. Tebbutt N, Pedersen MW, Johns TG (2013) Targeting the ERBB family in cancer: couples therapy. Nat Rev Cancer 13(9):663–673. doi:10.1038/nrc3559. [Research Support, Non--U.S. Gov't Review]
225. Tennant DA, Duran RV, Gottlieb E (2010) Targeting metabolic transformation for cancer therapy. Nat Rev Cancer 10(4):267–277. doi:10.1038/nrc2817. [Review]

226. Trifunovic A, Wredenberg A, Falkenberg M, Spelbrink JN, Rovio AT, Bruder CE et al (2004) Premature ageing in mice expressing defective mitochondrial DNA polymerase. Nature 429(6990):417–423. doi:10.1038/nature02517. [Research Support, Non-U.S. Gov't]
227. Ubah OC, Wallace HM (2014) Cancer therapy: Targeting mitochondria and other sub-cellular organelles. Curr Pharm Des 20(2):201–222. [Review]
228. Uehara E, Takeuchi S, Tasaka T, Matsuhashi Y, Yang Y, Fujita M et al (2003) Aberrant methylation in promoter-associated CpG islands of multiple genes in therapy-related leukemia. Int J Oncol 23(3):693–696
229. Ungefroren H, Gieseler F, Fliedner S, Lehnert H (2015) Obesity and cancer. Horm Mol Biol Clin Investig 21:5–15. doi:10.1515/hmbci-2014-0046
230. Unryn BM, Hao D, Gluck S, Riabowol KT (2006) Acceleration of telomere loss by chemotherapy is greater in older patients with locally advanced head and neck cancer. Clin Cancer Res 12(21):6345–6350. doi:10.1158/1078-0432.CCR-06-0486. [Research Support, Non-U.S. Gov't]
231. Valdivieso M, Kujawa AM, Jones T, Baker LH (2012) Cancer survivors in the United States: a review of the literature and a call to action. Int J Med Sci 9(2):163–173. doi:10.7150/ijms.3827. [Research Support, N.I.H., Extramural Review]
232. van der Pal HJ, van Dalen EC, van Delden E, van Dijk IW, Kok WE, Geskus RB et al (2012) High risk of symptomatic cardiac events in childhood cancer survivors. J Clin Oncol 30(13):1429–1437. doi:10.1200/JCO.2010.33.4730. [Research Support, Non-U.S. Gov't Webcasts]
233. van Deursen JM (2014) The role of senescent cells in ageing. Nature 509(7501):439–446. doi:10.1038/nature13193. [Research Support, N.I.H., Extramural Research Support, Non--U.S. Gov't Review]
234. Vander Heiden MG (2011) Targeting cancer metabolism: a therapeutic window opens. Nat Rev Drug Discov 10(9):671–684. doi:10.1038/nrd3504. [Research Support, N.I.H., Extramural Research Support, Non-U.S. Gov't Review]
235. Vaziri H, Benchimol S (1998) Reconstitution of telomerase activity in normal human cells leads to elongation of telomeres and extended replicative life span. Curr Biol 8(5):279–282. [Research Support, Non-U.S. Gov't]
236. Vermulst M, Bielas JH, Kujoth GC, Ladiges WC, Rabinovitch PS, Prolla TA et al (2007) Mitochondrial point mutations do not limit the natural lifespan of mice. Nat Genet 39(4):540–543. doi:10.1038/ng1988. [Research Support, N.I.H., Extramural Research Support, Non-U.S. Gov't]
237. Vermulst M, Wanagat J, Kujoth GC, Bielas JH, Rabinovitch PS, Prolla TA et al (2008) DNA deletions and clonal mutations drive premature aging in mitochondrial mutator mice. Nat Genet 40(4):392–394. doi:10.1038/ng.95. [Research Support, N.I.H., Extramural Research Support, Non-U.S. Gov't]
238. Vignot S, Faivre S, Aguirre D, Raymond E (2005) mTOR-targeted therapy of cancer with rapamycin derivatives. Ann Oncol 16(4):525–537. doi:10.1093/annonc/mdi113. [Review]
239. Vijg J, Suh Y (2013) Genome instability and aging. Annu Rev Physiol 75:645–668. doi:10.1146/annurev-physiol-030212-183715. [Research Support, N.I.H., Extramural Research Support, Non-U.S. Gov't Review]
240. Voso MT, D'Alo F, Greco M, Fabiani E, Criscuolo M, Migliara G et al (2010) Epigenetic changes in therapy-related MDS/AML. Chem Biol Interact 184(1–2):46–49. doi:10.1016/j.cbi.2009.10.013. [Research Support, Non-U.S. Gov't]
241. Voss MH, Bastos DA, Karlo CA, Ajeti A, Hakimi AA, Feldman DR et al (2014) Treatment outcome with mTOR inhibitors for metastatic renal cell carcinoma with nonclear and sarcomatoid histologies. Ann Oncol 25(3):663–668. doi:10.1093/annonc/mdt578. [Research Support, Non-U.S. Gov't]
242. Voss MH, Hakimi AA, Pham CG, Brannon AR, Chen YB, Cunha LF et al (2014) Tumor genetic analyses of patients with metastatic renal cell carcinoma and extended benefit from mTOR inhibitor therapy. Clin Cancer Res 20(7):1955–1964. doi:10.1158/1078-0432.CCR-13-2345. [Research Support, N.I.H., Extramural Research Support, Non-U.S. Gov't]

243. Waddington CH (2012) The epigenotype. 1942. Int J Epidemiol 41(1):10–13. doi:10.1093/ije/dyr184. [Biography Classical Article Historical Article]
244. Wallace KB (2007) Adriamycin-induced interference with cardiac mitochondrial calcium homeostasis. Cardiovasc Toxicol 7(2):101–107. doi:10.1007/s12012-007-0008-2. [Research Support, N.I.H., Extramural Review]
245. Wang Y, Schulte BA, LaRue AC, Ogawa M, Zhou D (2006) Total body irradiation selectively induces murine hematopoietic stem cell senescence. Blood 107(1):358–366. doi:10.1182/blood-2005-04-1418. [Research Support, N.I.H., Extramural Research Support, U.S. Gov't, Non-P.H.S.]
246. Watabe T, Miyazono K (2009) Roles of TGF-beta family signaling in stem cell renewal and differentiation. Cell Res 19(1):103–115. doi:10.1038/cr.2008.323. [Review]
247. Weigel C, Schmezer P, Plass C, Popanda O (2014) Epigenetics in radiation-induced fibrosis. Oncogene. doi:10.1038/onc.2014.145
248. Weindruch R, Gottesman SR, Walford RL (1982) Modification of age-related immune decline in mice dietarily restricted from or after midadulthood. Proc Natl Acad Sci U S A 79(3):898–902
249. Weindruch R, Walford RL (1982) Dietary restriction in mice beginning at 1 year of age: effect on life-span and spontaneous cancer incidence. Science 215(4538):1415–1418
250. Weindruch R, Walford RL, Fligiel S, Guthrie D (1986) The retardation of aging in mice by dietary restriction: longevity, cancer, immunity and lifetime energy intake. J Nutr 116(4):641–654
251. Weiner LM, Surana R, Wang S (2010) Monoclonal antibodies: versatile platforms for cancer immunotherapy. Nat Rev Immunol 10(5):317–327. doi:10.1038/nri2744. [Research Support, N.I.H., Extramural Review]
252. West J, Beck S, Wang X, Teschendorff AE (2013) An integrative network algorithm identifies age-associated differential methylation interactome hotspots targeting stem-cell differentiation pathways. Sci Rep 3:1630. doi:10.1038/srep01630. [Research Support, Non-U.S. Gov't]
253. Whelan K, Stratton K, Kawashima T, Leisenring W, Hayashi S, Waterbor J et al (2011) Auditory complications in childhood cancer survivors: a report from the childhood cancer survivor study. Pediatr Blood Cancer 57(1):126–134. doi:10.1002/pbc.23025. [Comparative Study Multicenter Study Research Support, N.I.H., Extramural Research Support, Non-U.S. Gov't]
254. Williams SL, Huang J, Edwards YJ, Ulloa RH, Dillon LM, Prolla TA et al (2010) The mtDNA mutation spectrum of the progeroid Polg mutator mouse includes abundant control region multimers. Cell Metab 12(6):675–682. doi:10.1016/j.cmet.2010.11.012. [Research Support, Non-U.S. Gov't Research Support, U.S. Gov't, P.H.S.]
255. Wilson CL, Cohn RJ, Johnston KA, Ashton LJ (2010) Late mortality and second cancers in an Australian cohort of childhood cancer survivors. Med J Aust 193(5):258–261. [Research Support, Non-U.S. Gov't]
256. Wilson VL, Taffe BG, Shields PG, Povey AC, Harris CC (1993) Detection and quantification of 8-hydroxydeoxyguanosine adducts in peripheral blood of people exposed to ionizing radiation. Environ Health Perspect 99:261–263
257. Wilson WR, Hay MP (2011) Targeting hypoxia in cancer therapy. Nat Rev Cancer 11(6):393–410. doi:10.1038/nrc3064. [Research Support, Non-U.S. Gov't Review]
258. Woo DK, Shadel GS (2011) Mitochondrial stress signals revise an old aging theory. Cell 144(1):11–12. doi:10.1016/j.cell.2010.12.023. [Comment Research Support, N.I.H., Extramural]
259. Wu AM, Senter PD (2005) Arming antibodies: prospects and challenges for immunoconjugates. Nat Biotechnol 23(9):1137–1146. doi:10.1038/nbt1141. [Research Support, N.I.H., Extramural Research Support, U.S. Gov't, Non-P.H.S. Research Support, U.S. Gov't, P.H.S. Review]
260. Wu X, Amos CI, Zhu Y, Zhao H, Grossman BH, Shay JW et al (2003) Telomere dysfunction: a potential cancer predisposition factor. J Natl Cancer Inst 95(16):1211–1218. [Research Support, U.S. Gov't, P.H.S.]

261. Wu Y, Cain-Hom C, Choy L, Hagenbeek TJ, de Leon GP, Chen Y et al (2010) Therapeutic antibody targeting of individual Notch receptors. Nature 464(7291):1052–1057. doi:10.1038/nature08878. [Research Support, N.I.H., Extramural Research Support, U.S. Gov't, Non-P.H.S.]
262. Wullschleger S, Loewith R, Hall MN (2006) TOR signaling in growth and metabolism. Cell 124(3):471–484. doi:10.1016/j.cell.2006.01.016, S0092-8674(06)00108-5 [pii]
263. Xu L, Li S, Stohr BA (2013) The role of telomere biology in cancer. Annu Rev Pathol 8:49–78. doi:10.1146/annurev-pathol-020712-164030. [Research Support, N.I.H., Extramural Research Support, Non-U.S. Gov't Review]
264. Xu X, Chua CC, Gao J, Chua KW, Wang H, Hamdy RC et al (2008) Neuroprotective effect of humanin on cerebral ischemia/reperfusion injury is mediated by a PI3K/Akt pathway. Brain Res 1227:12–18. doi:10.1016/j.brainres.2008.06.018. [Research Support, N.I.H., Extramural Research Support, Non-U.S. Gov't Research Support, U.S. Gov't, Non-P.H.S.]
265. Xu X, Chua CC, Gao J, Hamdy RC, Chua BH (2006) Humanin is a novel neuroprotective agent against stroke. Stroke 37(10):2613–2619. doi:10.1161/01.STR.0000242772.94277.1f. [Comparative Study Research Support, Non-U.S. Gov't Research Support, U.S. Gov't, Non-P.H.S.]
266. Yanez JA, Teng XW, Roupe KA, Fariss MW, Davies NM (2003) Chemotherapy induced gastrointestinal toxicity in rats: involvement of mitochondrial DNA, gastrointestinal permeability and cyclooxygenase -2. J Pharm Pharm Sci 6(3):308–314. [Research Support, Non-U.S. Gov't]
267. Yilmaz OH, Katajisto P, Lamming DW, Gultekin Y, Bauer-Rowe KE, Sengupta S et al (2012) mTORC1 in the Paneth cell niche couples intestinal stem-cell function to calorie intake. Nature 486(7404):490–495. doi:10.1038/nature11163. [Research Support, N.I.H., Extramural Research Support, Non-U.S. Gov't]
268. Yoshioka T, Wada T, Uchida N, Maki H, Yoshida H, Ide N et al (1998) Anticancer efficacy in vivo and in vitro, synergy with 5-fluorouracil, and safety of recombinant methioninase. Cancer Res 58(12):2583–2587
269. Zacharias DG, Kim SG, Massat AE, Bachar AR, Oh YK, Herrmann J et al (2012) Humanin, a cytoprotective peptide, is expressed in carotid artherosclerotic plaques in humans. PLoS One 7(2):e31065. doi:10.1371/journal.pone.0031065. [Research Support, N.I.H., Extramural]
270. Zastawny TH, Czerwinska B, Drzewiecka B, Olinski R (1996) Radiation induced oxidative DNA base damage and its repair in liver chromatin DNA of rats upon whole body gamma-irradiation. Acta Biochim Pol 43(3):579–582. [Research Support, Non-U.S. Gov't]
271. Zhang J, Yang PL, Gray NS (2009) Targeting cancer with small molecule kinase inhibitors. Nat Rev Cancer 9(1):28–39. doi:10.1038/nrc2559. [Review]
272. Zhang R, Poustovoitov MV, Ye X, Santos HA, Chen W, Daganzo SM et al (2005) Formation of MacroH2A-containing senescence-associated heterochromatin foci and senescence driven by ASF1a and HIRA. Dev Cell 8(1):19–30. doi:10.1016/j.devcel.2004.10.019. [Comparative Study Research Support, Non-U.S. Gov't Research Support, U.S. Gov't, Non-P.H.S. Research Support, U.S. Gov't, P.H.S.]
273. Zhang W, Li Z, Hao J, Zhang Z, Liu L, Mao N et al (2012) S14G-humanin improves cognitive deficits and reduces amyloid pathology in the middle-aged APPswe/PS1dE9 mice. Pharmacol Biochem Behav 100(3):361–369. doi:10.1016/j.pbb.2011.09.012. [Research Support, Non-U.S. Gov't]
274. Zhao XT, Feng JB, Li YW, Luo Q, Yang XC, Lu X et al (2012) Identification of two novel mitochondrial DNA deletions induced by ionizing radiation. Biomed Environ Sci 25(5):533–541. doi:10.3967/0895-3988.2012.05.006. [Research Support, Non-U.S. Gov't]
275. Zidan J, Haim N, Beny A, Stein M, Gez E, Kuten A (2001) Octreotide in the treatment of severe chemotherapy-induced diarrhea. Ann Oncol 12(2):227–229. [Clinical Trial]
276. Ziegler DV, Wiley CD, Velarde MC (2014) Mitochondrial effectors of cellular senescence: beyond the free radical theory of aging. Aging Cell. doi:10.1111/acel.12287
277. Zimmerman JA, Malloy V, Krajcik R, Orentreich N (2003) Nutritional control of aging. Exp Gerontol 38(1–2):47–52, S0531556502001493 [pii]

Cardiovascular Disease and Aging

Ying Ann Chiao, Edward Lakatta, Zoltan Ungvari, Dao-Fu Dai, and Peter Rabinovitch

Contents

Y.A. Chiao, Ph.D. • D.-F. Dai, M.D., Ph.D. • P. Rabinovitch, M.D., Ph.D. (✉)
Department of Pathology, University of Washington, Seattle, WA, 98195, USA
e-mail: annchiao@uw.edu; ddaofu@uw.edu; petersr@uw.edu

E. Lakatta, M.D.
Laboratory of Cardiovascular Science, Biomedical Research Center, National Institute of Aging, Baltimore, MD, USA
e-mail: lakattae@mail.nih.gov

Z. Ungvari, M.D., Ph.D.
Department of Geriatric Medicine, Reynolds Oklahoma Center on Aging, University of Oklahoma Health Science Center, Oklahoma City, OK, USA
e-mail: Zoltan-Ungvari@ouhsc.edu

F. Sierra, R. Kohanski (eds.), *Advances in Geroscience*,
DOI 10.1007/978-3-319-23246-1_5

1 Introduction

Aging impairs the cardiovascular system and is the dominant risk factor for cardiovascular disease (CVD). The incidence and prevalence of most CVD increase with advancing age, and CVD are the leading cause of death for populations over 65 years of age. According to the heart disease and stroke statistics 2014 update published by the American Heart Association, Americans 60–79 years of age have over 70 % prevalence rate and Americans over 80 years of age have over 80 % prevalence rate of cardiovascular diseases [1]. However, most of the research efforts on prevention of these diseases have ignored age and have focused instead on development of interventions that target "traditional" CV risk factors such as hypertension, high blood cholesterol and triglycerides. In addition to the increased prevalence of cardiovascular diseases, aging is also associated with impaired responses to cardiovascular diseases. Aging also leads to deterioration of the structure and function of the heart and vasculature in individuals without overt cardiovascular disease, as reflected in cardiovascular measurements that are made in healthy individuals at rest and during exercise, both of which reveal aging-related changes. Therefore, it is important to understand the molecular mechanism of cardiovascular aging and how the age-related changes in the cardiovascular system interact with the pathophysiological mechanisms that lead to cardiovascular disease. In this chapter, we will describe the characteristics of cardiovascular aging in humans and in mammalian models, and review the roles of different hallmarks of aging in cardiovascular aging. Valvular degeneration in the aging heart is not addressed in this review, and the reader is referred to other sources [2, 3].

2 Cardiovascular Aging in Humans

2.1 Cardiac Aging

2.1.1 Cardiac Structure/Function at Rest

A continuum of expression of cardiac structural and functional alterations occurs with age in healthy humans (Fig. 1). On the one hand, a unified interpretation of identified cardiac aging changes at rest in otherwise healthy persons (Fig. 1) suggests that these changes are at least in part adaptive, occurring to some extent in response to changes that occur within the arterial tree that result in increased load on left ventricular ejection [4]. On the other hand, these age-associated cardiac changes also have direct relevance to the steep age-associated increases in left ventricular hypertrophy (LVH), chronic heart failure, and atrial fibrillation (AF).

2.1.2 Cardiac Structure

Cross-sectional studies of subjects without hypertension or clinically apparent CVD indicate that with advancing age the walls of the left ventricle increase in thickness, largely because of an increase in ventricular myocyte size and an increase in vascular

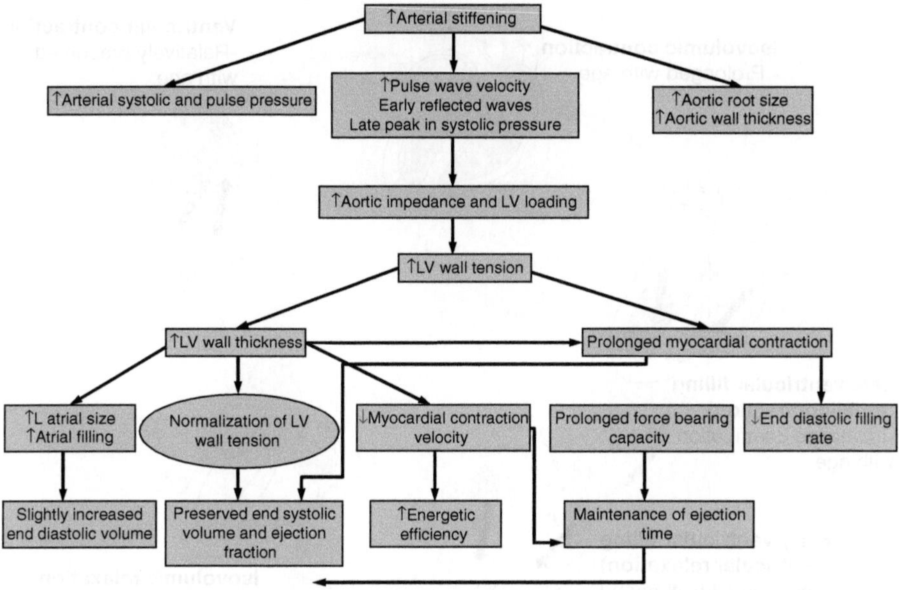

Fig. 1 Arterial and cardiac changes that occur with aging in healthy humans (Adapted from Lakatta [252])

impedance, and this helps moderate the increase in LV wall tension (Fig. 1). Modest increases in collagen levels and non-enzymatic cross linking, rendering collagen stiffer, also occur with aging. In older hospitalized patients without apparent cardiovascular disease, however, the cardiac myocyte-to-collagen mass ratio in the older heart either remains constant or increases because of an increase in myocyte size.

2.1.3 Left Ventricular Function at Rest

Increased central arterial stiffness is a factor that leads to an increase in the afterload on the left ventricle, leading to increased left ventricular wall thickness (Fig. 1). The left ventricle and the central arteries have bidirectional constant interactions. One useful index of this interaction is termed arterial-ventricular coupling. This tight heart-arterial coupling is thought to allow the cardiovascular system to optimize energetic efficiency. The LV ejection fraction (EF) at rest, the most commonly used clinical measure of heart-arterial crosstalk, is preserved during aging (Fig. 2). The average value of resting EF is approximately 65 %, and very few healthy, sedentary, community-dwelling older individuals have EF <50 % (which would indicate impaired LV-arterial crosstalk [5].)

Prolonged contraction of the thickened LV wall maintains a normal ejection time in older persons in the presence of the increased resistance to pulsatile blood flow from the heart. This preserves the systolic cardiac pumping function at rest. One disadvantage of prolonged contraction is that at the time of the mitral valve opening, myocardial relaxation is relatively more incomplete in older than in younger individuals. The LV early diastolic filling rate progressively slows after the age of 20

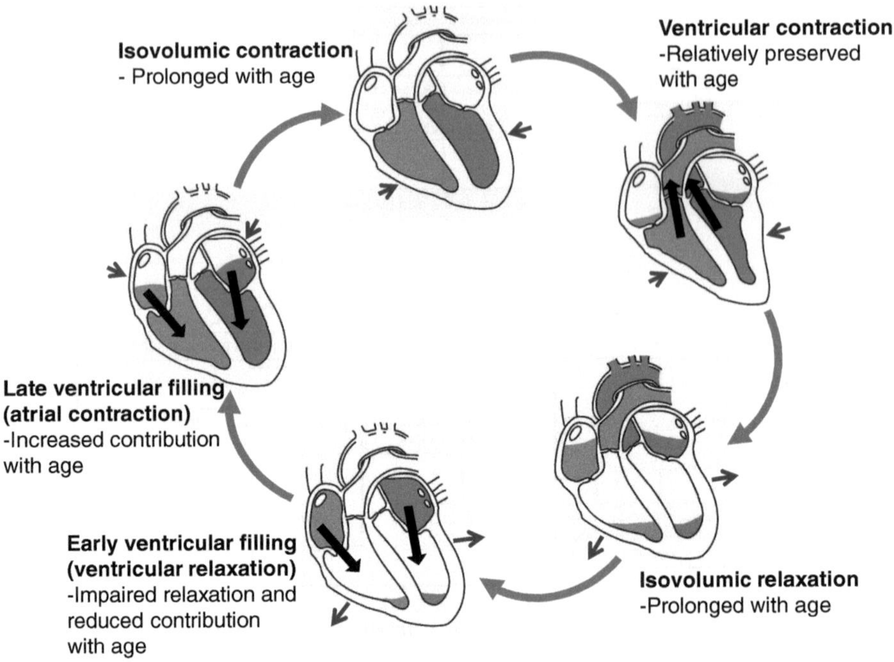

Fig. 2 The age-related changes in different phases of the cardiac cycle. With age, ventricular systole remains relative preserved while isovolumic relaxation and contraction are prolonged. Ventricular relaxation is impaired and the relative contribution of early ventricular filling is reduced, with an increased contribution of atrial contraction (late filling)

years [6–8], so that by 80 years the rate is reduced, on average, up to 50 %. Structural (fibrous) changes within the LV myocardium or residual myofilament Ca^{2+} activation from the preceding systole are putative mechanisms for a reduced early diastolic LV filling rate. Structural changes and functional heterogeneity occurring within the left ventricle with aging may also contribute to this reduction in peak LV filling rate. However, concomitant adaptations – left atrial enlargement and an enhanced atrial contribution to ventricular filling (Fig. 1) – compensate for the reduced early filling and prevent a reduction of the end diastolic volume. Thus, despite the age-associated changes in the diastolic filling pattern in older, healthy persons, their left ventricular end-diastolic volume in the supine position is not compromised and does not substantially differ from that of their younger counterparts [9, 10]. Thus, neither EF nor cardiac output at rest declines with advancing age.

2.1.4 Cardiovascular Reserve

During exercise, impaired heart rate acceleration and impaired augmentation of blood ejection from the left ventricle, accompanied by a modest increase in LV end-diastolic volume are the most dramatic changes in cardiac reserve capacity that occur with aging in healthy, community-dwelling persons [11]. Mechanisms that underlie the age-associated reduction in maximum ejection fraction are multifactorial and include a reduction in intrinsic myocardial contractility, an increase in vascular afterload, and arterial-ventricular load mismatching. Although these age-associated changes in cardiovascular reserve, per se, are usually insufficient to cause clinical heart failure, they do affect its clinical presentation, the threshold for symptoms and signs, and the severity and prognosis of heart failure secondary to any level of disease burden. Notably, this loss of reserve limits exercise capacity and contributes to frailty in the elderly.

Sympathetic neural impulses to the heart via beta-adrenergic receptor stimulation of heart and vascular cells elicit the recruitment of cardiovascular reserve capacity during stress. The essence of sympathetic modulation of the cardiovascular system is to ensure that the heart beats faster, to ensure that it retains a small size by reducing the diastolic filling period, reducing LV afterload, to augment myocardial contractility and relaxation, and to redistribute blood to working muscles and to skin to dissipate heat. A sizeable component of the age-associated deficit in cardiovascular reserve is composed of diminished effectiveness of the autonomic modulation of heart rate, LV contractility, and arterial afterload.

Multiple lines of evidence support the idea that the efficiency of postsynaptic beta-adrenergic signaling declines with aging in numerous species [4]. One line of evidence stems from the observation that cardiovascular responses to beta-adrenergic agonist infusions at rest decrease with age in humans and mammalian models. A second type of evidence is that acute beta-adrenergic receptor blockade changes the exercise hemodynamic profile of younger persons to make it resemble that of older individuals. Significant beta blockade–induced LV dilation occurs only in younger subjects. The heart rate reduction during exercise in the presence of acute beta-adrenergic blockade is greater in younger versus older human subjects, as are the age-associated deficits in LV early diastolic filling rate, both at rest and during exercise. In older dogs it has also been observed that an age-associated increase in aortic impedance during exercise is abolished by acute beta-adrenergic blockade [12].

Apparent deficits in sympathetic modulation of cardiac and arterial functions with aging occur in the presence of exaggerated neurotransmitter levels. Plasma levels of norepinephrine and epinephrine, during any perturbation from the supine basal state, increase to a greater extent in older compared with younger healthy humans. The age-associated increase in plasma levels of norepinephrine results from an increased spillover into the circulation and, to a lesser extent, reduced plasma clearance. Deficient norepinephrine reuptake at nerve endings is a primary mechanism for increased spillover into the circulation of older persons during acute, graded exercise (exercise intensity is progressively increased until the subject reaches a self-imposed fatigue level). During prolonged exercise, however, diminished neurotransmitter reuptake might also be associated with depletion and reduced release and spillover in older persons.

2.1.5 Heart Rhythm

Beat-to-beat fluctuation of heart rate (HR) at rest is known as HR variability. It is a physiological phenomenon that reflects different inputs to the cardiac conduction system and variation results from normal fluctuation in these inputs (not to be confused with arrhythmia, described below). However, HR variability declines steadily with age and this reduction is an indicator of cardiac autonomic dysregulation commonly found in older people; it has been linked to increased risk for morbid and fatal outcomes [13].

An increase in the prevalence and complexity of both supraventricular and ventricular premature beats and arrhythmias, whether detected by resting ECG, ambulatory monitoring, or exercise testing, occurs in otherwise healthy older patients and are generally not associated with heart disease, even in older persons. Over a 10-year mean follow-up period, isolated atrial premature beats, even if frequent, were not predictive of increased cardiac risk in these individuals [14].

Short bursts of paroxysmal supraventricular tachycardia (PSVT) are observed in 1–2 % of apparently healthy individuals older than 65 years who were rigorously screened to exclude disease. PSVT at rest or induced by exercise is an early clue that some healthy individuals are at an increased risk for future atrial fibrillation (AF). Another risk factor for AF may be the increase in left atrial size (Fig. 1) that accompanies advancing age in otherwise healthy persons [4].

Limited data available in older persons without apparent heart disease support a marked age-associated increase in the prevalence and complexity of ventricular ectopy (VE); i.e., premature heart beats originating from the ventricular myocardium, both at rest and during exercise, at least in men. A steep increase in the prevalence of VE with advancing age occurs in both clinically free of heart disease individuals and in unselected populations [15]. Neither the prevalence nor the complexity of VE at rest was a determinant of future coronary events over a 10-year mean follow-up period [14].

2.2 *Arterial Aging*

Intima-media thickening increases with advancing age in animal models that are devoid of atherosclerosis [16], thus indicating that these age-associated alterations are due to the aging process and not to superimposed atherosclerosis. Aging is also associated with thickening of the walls of the central arteries. Intima-media thickness (IMT) rises nearly threefold between the third and ninth decades of life, which is mainly attributable to an increase in intimal thickness in the context of inflammation [17]. A growing body of literature has shown that traditional cardiovascular risk factors and prevalent CVD are associated with increased IMT, and that IMT is itself a potent and independent predictor of adverse CV events. Diffuse intima-media thickening, however, should not be construed as synonymous with "subclinical atherosclerosis," particularly in the absence of plaques. This thickening is due to myriad age-associated biochemical, cellular, and morphologic changes in the arterial wall, which are modulated by the same factors that have been implicated in the genesis of various CVD [18]. Thus, IMT remains a useful marker of an arterial risk factor associated with aging [19]. In contrast to *central* arteries, IMT is only modestly correlated with *coronary* artery disease [20].

One of the hallmarks of central arterial aging is an age-associated increase in arterial wall stiffness. The age-associated increase in stiffness has frequently been attributed to the fraying and breakdown of elastin due to the lifelong repeated cycles of distention and recoil of the central aorta as well as the increased deposition and covalent cross-linking of collagen molecules. It is now recognized that arterial stiffening can be modulated by several factors besides aging, including lifestyle (e.g., salt intake, exercise, or weight loss) [21], signaling molecules (e.g., nitric oxide) [22], inflammation, and genetics [23]. Manifestations of arterial aging vary among the different vascular beds, reflecting differences in the structural compositions of the arteries and perhaps differences in the age-associated signaling cascades that modulate the arterial properties, or differences in the response to these signals across the arterial tree. For example, in contrast to the central elastic arteries the stiffness of the muscular arteries does not increase with age (e.g., brachial and femoral arteries).

Aortic pulse wave velocity has been anointed the "gold standard" for the noninvasive assessment of central arterial stiffness [24]. Pulse wave velocity has been shown to be an independent predictor of morbidity and mortality in healthy subjects and in individuals with various levels of cardiovascular risk. It is likely that arterial stiffness is not only a risk marker but also a risk factor for cardiovascular diseases.

Increased central arterial stiffening is a likely explanation of the age-associated changes in blood pressures, whereby systolic blood pressure continues to increase with advancing age, and diastolic blood pressure increases until the fifth decade, then levels off and starts to decrease after the age of 60 years [25]. The decrease in diastolic blood pressure may compromise coronary blood flow, which occurs predominantly in diastole, and a further increase in pulse pressure, which can be twice as high in older vs. younger persons. Numerous clinical and epidemiologic studies in several different populations with varying prevalence of cardiovascular diseases have demonstrated that central pulse pressure is an important predictor of adverse outcomes, often more potent than systolic or diastolic blood pressures. Increased central arterial pulse pressure is transmitted to small arteries of the kidney and heart, damaging these vessels and organs, often resulting in stroke, myocardial infarction and chronic renal disease which increase exponentially with advancing age. Both animal and clinical studies have recently demonstrated that arterial stiffness precedes the development of hypertension [26–28]. Interventions to prevent or to delay arterial stiffening have predominantly focused on pharmacologic antihypertensive therapies. However, these strategies are aimed at lowering blood pressure, whereby the reduction in stiffness is a secondary effect due to reverse remodeling of the arterial wall in response to the lower pressures. Because central arterial stiffness is a potent predictor of mortality and morbidity independent of blood pressure, a more direct approach that would target the stiffening process is desirable.

2.3 Clinical Implications of Cardiovascular Aging

In summary, the most dramatic changes in cardiac function that occur with aging in healthy, adult, community-dwelling subjects, ranging in age from 20 to 85 years, are increased LV wall thickness, alterations in the diastolic filling pattern, impaired LV ejection and HR reserve capacity, and altered heart rhythm. Although these age-associated changes do not usually result in clinical heart disease per se, they do compromise the cardiac reserve capacity and affect the threshold for symptoms and signs, as well as the severity and prognosis of heart failure secondary to any given disease-related challenge. This is true for both systolic and diastolic heart failure. Thus, age-associated changes in the heart structure and function that occur in the absence of a clinical diagnosis of heart disease explain the increased risk for left ventricular hypertrophy, atrial fibrillation, congestive heart failure and diastolic heart failure, all of which occur at markedly higher rates in older persons than in younger persons. These three cardiac diagnoses become interrelated in older persons, in part because of this link with age-associated cardiac changes. An age-dependent increase in left ventricular mass increases the stiffness of the left ventricle and promotes an increase in end diastolic filling pressure, which is an important contributor to diastolic heart failure in older persons. In addition, increased diastolic filling pressure results in left atrial dilation, predisposes the heart to AF. When associated with tachycardia and loss of atrioventricular coupling, AF reduces diastolic filling time and eliminates atrial systolic contribution to left ventricular filling, thereby compounding the predisposition to diastolic heart failure.

Age-associated changes are increasingly recognized as risk factors for CVD. Because many of the age-associated alterations in CV structure and function, at both the cellular and molecular levels, are specific risk factors for cardiovascular diseases, there is an urgency to incorporate cardiovascular aging into clinical medicine to preserve the healthspan of older persons. In spite of the fact that CV aging is a major risk factor for CV disease, it has remained for the most part outside of mainstream clinical medicine because the pathophysiologic implications of these changes are largely underappreciated and are not well disseminated in the medical community. However, an understanding of the age-associated alterations in cardiac and arterial structure and function at both the cellular and molecular levels [29, 30] provides valuable clues that may assist in the development of effective therapies to prevent, to delay, or to attenuate the CV changes that accompany aging.

3 Cardiovascular Aging in Mammalian Models

3.1 The Cardiac Aging Phenotype in Mammalian Models

Most of what we know about the molecular basis of cardiovascular aging comes from animal models of human aging and heart function. Cardiac aging responses have been characterized in multiple animal models, including nonhuman primates, dogs, rats and mice. However, due to the different species, strains and definitions of age groups used by different studies, the results should be interpreted cautiously.

Rodents, particularly mice, are widely used in cardiac aging studies. While the rodent heart is different from those of primates and other larger mammals (particularly the electrical conduction system), in general, cardiac aging in rodents closely recapitulates the cardiac aging phenotypes seen in humans without overt cardiovascular diseases [31]. Dai and colleagues showed in a mouse longevity cohort that there were significant age-dependent linear trends for several cardiac parameters [32]. They showed by echocardiography that left ventricular mass index and left atrial dimension significantly increased with age. Diastolic function measured by tissue Doppler echocardiography revealed an age-dependent decline in the ratio of early to late diastolic mitral annular velocity (Ea/Aa) and the frequency of diastolic dysfunction [33], defined as Ea/Aa <1, was increased in C57B6 mice over 24 months of age. The proportion of mice with atrial dilation also significantly increased with age [32]. Systolic function, measured by ejection fraction (EF) and fractional shortening (FS) remain relatively unchanged in middle-age and old mice but decreased in 32 month-old senescent mice [34, 35]. The myocardial performance index was significantly worsened with age [36], indicating that a greater fraction of systole is spent to overcome the pressure changes during isovolumic phases; this has been shown to reflect both LV systolic and/or diastolic dysfunction [37]. The aging-associated changes in different phases of the cardiac cycle are summarized in Fig. 2.

At the cellular level, reduced myocyte number and hypertrophy of remaining myocytes have been demonstrated in aging rats and mice [34, 38]. Structurally, LV wall thickness increases in middle-age and old mice, but decreases in senescent mice [34, 35, 39]. This suggests that the hypertrophic growth of the myocytes in middle-age and old mice becomes decompensated in senescent mice, consistent with myocyte loss [34]. At a histopathological level, the aged hearts display interstitial fibrosis, cytoplasmic vacuolization and hyalinization, increased variation in myocyte fiber size, collapse of sarcomeres, mineralization, arteriosclerosis and arterio*lo*sclerosis [40].

The relatively short lifespan and the availability of genetically modified mice also make mouse models useful tools for study of the molecular mechanisms of cardiac aging. In addition, the general absence of common cardiovascular risk factors such as diabetes and hypertension [32, 41], allows intrinsic cardiac aging changes to be distinguished from disease-induced changes.

3.2 The Vascular Aging Phenotype in Mammalian Models

Aged laboratory rodents exhibit a range of vascular changes also observed in humans. These include endothelial dysfunction [42], structural remodeling, arterial stiffening, vascular oxidative stress and inflammation [43], vascular calcification, microvascular rarefaction [44], autoregulatory dysfunction and impaired functional adaptation to hypertension [45, 46], blood brain barrier disruption [45], neurovascular uncoupling [47], impaired cellular stress resistance [48], increased susceptibility for vascular injury, and mitochondrial dysregulation [49, 50]. Thus, laboratory rodents are well-suited models for studying these aspects of aging-induced vascular pathologies. In contrast, laboratory rodents are not ideal models for age-related increases in blood pressure [51]. Aged wild type mice and rats also do not develop atherosclerotic plaques spontaneously (in the absence of genetic depletion of *Ldlr* or *Apoe*) similar to those observed in aged primates.

The age-related damage of the arterial tree in non-human primates is of particular interest, as these species are physiologically and phylogenetically closer to humans than the more commonly studied rodent models. With age, non-human primates exhibit increased arterial stiffness [52], increased central arterial pulse pressure, increased carotid intima-media thickness, hypertension [53], activation of processes involved in atherogenesis [54], development of aneurysms [55], vascular oxidative stress and inflammation [56, 57], endothelial dysfunction and apoptosis [16], alterations of the blood brain barrier [58] and cellular mitochondrial content [59] and impaired cellular stress resistance [57], mimicking the human vascular aging phenotype.

4 Molecular Mechanisms of Cardiovascular Aging

4.1 Mechanisms of Cardiac Aging

While the phenotypes of cardiac aging have been well-characterized for decades, the molecular mechanisms of cardiac aging are just beginning to be revealed. Using genetic models and interventions targeting different pillars of geroscience, researchers have revealed several critical molecular mechanisms of cardiac aging in the past decade that have the potential to translate into cardiovascular healthspan interventions (summarized in Fig. 3).

4.1.1 Macromolecular Damage and Mitochondrial ROS

There are multiple sources of reactive oxygen species (ROS) in the cells but mitochondria, which generate ROS during oxidative phosphorylation, are the origin of most ROS. The heart has a high metabolic demand and is rich in mitochondria and

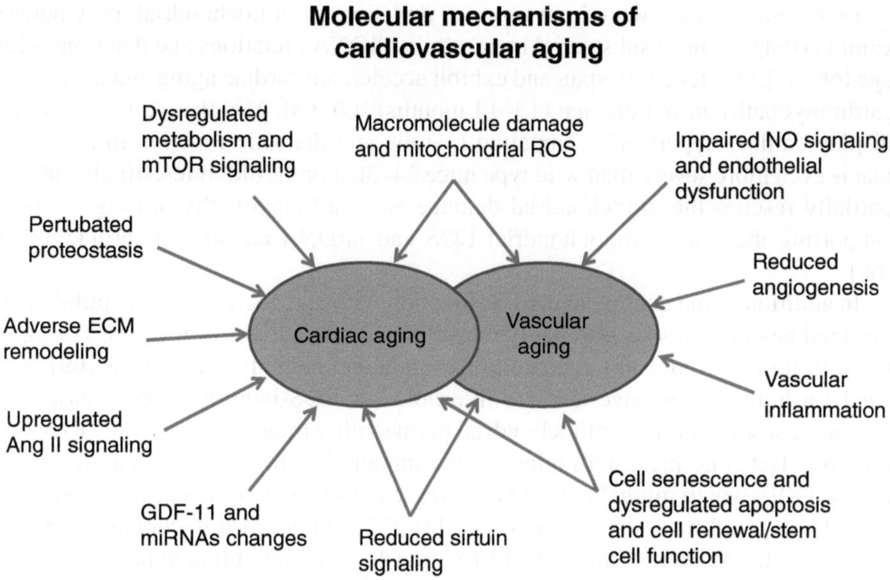

Fig. 3 A diagram summarizing the molecular mechanisms that mediate cardiovascular aging. Multiple pathways and molecules have been demonstrated to play critical roles in cardiac and vascular aging. The details are discussed in Sect. 4

therefore may be especially prone to oxidative damage. Increasing evidence suggests the role of abnormal mitochondrial ROS (mtROS) production and detoxification in mitochondrial dysfunction and cardiomyopathy in old age (reviewed in [60–62]). It has been shown that mitochondrial production of ROS significantly increases in the aging heart [63]. With advanced age, oxidative phosphorylation and mitochondrial state 3 respiration are significantly reduced due to diminished complex I and complex IV activities (see review [64]). The impairment in electron transport chain function may lead to elevated electron leakage and increased mitochondrial ROS production.

Direct evidence of the role of mitochondrial ROS in aging was provided by mice overexpressing catalase targeted to the mitochondria (mCAT). The mCAT mice, but not mice overexpressing wild-type human catalase (naturally delivered to peroxisomes, pCAT), showed an 18 % extension of mean and maximal lifespan [65]. Although cardiac failure was not the major cause of death in these mice, they exhibited great attenuation of many cardiac aging phenotypes, including left ventricular hypertrophy, diastolic dysfunction and impaired myocardial performance, as well as significant amelioration of age-dependent cardiomyocyte hypertrophy, interstitial fibrosis and mitochondrial ultrastructural changes [32, 65]. These cardiac aging benefits in mCAT mice were accompanied by significantly reduced mitochondrial protein oxidative damage and mitochondrial DNA mutation and deletion frequencies [32, 65].

In contrast, mice with homozygous mutation of mitochondrial polymerase gamma (Polg$^{m/m}$) have substantial increases in mtDNA mutations and deletions with age [66, 67], shortened lifespan and exhibit accelerated cardiac aging and developed cardiomyopathy in middle age (13–14 months) [66, 68]. Middle age Polg$^{m/m}$ mice display cardiac hypertrophy, impaired systolic and diastolic function to an extent that is even more severe than wild type mice 24–30 months old. Interestingly, mCAT partially rescues the mitochondrial damage and cardiomyopathy in Polg$^{m/m}$ mice, supporting the role of mitochondrial ROS and mtDNA damage in cardiac aging [68].

In addition to the decline in cardiac function with age, increased susceptibility of the aged heart to stress is also likely related to mitochondrial dysfunction. The aged myocardium has reduced tolerance to ischemia and hemodynamic stress [69], and aged cardiomyocytes have a lower threshold for ROS-induced ROS release and increased susceptibility to mitochondrial permeability transition pore (mPTP) opening [70]. Ischemic preconditioning is also impaired in the aged myocardium (see review [70]), which might be due to a decrease in mitochondrial heat shock protein-70 [71], reduced nitric oxide bioavailability [72], damaged mitochondria that are vulnerable to stress, and diminished PKC translocation into mitochondria [73, 74].

4.1.2 Metabolism and mTOR Signaling

Mechanistic target of rapamycin (mTOR) integrates nutrient and hormonal cues to regulate growth and longevity and is an important modulator of aging and age-related disease [75]. The nutrient sensitive TORC1 branch of the mTOR pathway is a complex that includes mTOR and Raptor (regulatory associated protein of mTOR) and activity of this complex is inhibited by rapamycin. Active TORC1 phosphorylates p70S6K, which accelerates ribosome biogenesis. TORC1 also phosphorylates 4E binding protein 1 (4EBP1) which results in its release from the inactive 4EBP1/eukaryotic initiation factor 4E (eIF4E) complex to allow cap-dependent translation initiation (see review [76]). Studies have shown that increased mTOR signaling impairs – and reduced mTOR signaling improves – resistance to cardiac aging.

In a *Drosophila* model of aging related attenuation of cardiac function, Bodmer's laboratory initially showed that inhibition of the mTOR pathway could attenuate the aging-related decline [77]. Subsequently, they determined that overexpression of 4EBP prevented the aging cardiac decline to the same extent as overexpression of the TOR antagonist TSC, and that, conversely, overexpression of eIF4E leads to a more rapid decline of myocardial function with age [78]. These findings suggest that mTOR signaling through eIF4E plays a primary role in cardiac aging in Drosophila.

In the mouse model, cardiac-specific gene ablation of TSC1, which is an engineered model for increased mTOR signaling, develop cardiomyopathy with the occurrence of scattered foci of enlarged ventricular myocytes and have a median survival of only 6 months [79]. Direct genetic evidence is not yet available in mice to document that decreased mTOR activity is associated with improved cardiac aging; however data from caloric restriction (CR) and rapamycin support this conclusion.

CR has been shown to extend longevity in numerous animal models and a large body of evidence supports a major role of inhibition of mTOR signaling in mediating the CR benefit [80]. Literature from both rhesus monkey studies and short-duration CR in humans provides strong evidence that CR reduces aging- and/or diet-dependent risk factors associated with heart disease: resting heart rate and blood pressure are decreased, insulin sensitivity is enhanced, lipid profiles are improved, and inflammatory processes that likely contribute to atherosclerosis are reduced [81]. Taffett et al. found that CR of mice had a large positive effect on age-related impaired diastolic function [82]. A later study suggested that in humans that had undertaken CR for a mean of 6.5 years, there was lower blood pressure, lower systemic oxidative stress, and improved diastolic function [83]. In the Dahl salt-sensitive rat (with high-salt induced hypertension), moderate calorie restriction markedly attenuated changes in heart weight, left ventricular mass; wall thickness and echocardiography demonstrated that CR reduced cardiac diastolic dysfunction in this model [84].

The role of mTOR is further supported by the beneficial effects of rapamycin on cardiac aging. Rapamycin has been demonstrated to extend lifespan in model organisms including yeast, fly, and mouse [85–87]. Long-term rapamycin treatment for 1 year initiated at mid-life reduced hypertrophy but failed to restore systolic function in aged male mice [88]. Flynn et al. showed that short-term treatment for 12 weeks initiated at late life can attenuate age-related cardiac hypertrophy and marginally improve systolic function in female mice, and was accompanied by a reduction in age-related inflammation [89]. Our recent study demonstrated that short-term rapamycin (10 weeks) recapitulated the effect of CR to improve diastolic function and LV hypertrophy in old female mice [90]. This was accompanied by restoration of proteomic and metabolic profiles to more youthful phenotypes.

4.1.3 Proteostasis

Protein homeostasis (proteostasis) is a steady state between protein synthesis and degradation. It functions to maintain protein abundance and quality and supports normal physiological function. Failure to maintain protein homeostasis, leading to accumulation of defective proteins, has been observed in several age-related diseases [91], including neurodegenerative diseases [92], cardiac dysfunction [93, 94], cataracts [95], and sarcopenia [96, 97]. Several protein degradation pathways have been implicated in these failures, including autophagy and the ubiquitin-proteosome system.

Perturbation of proteostasis has also been observed in normal aging [98]. One of the hallmarks of aging hearts is the accumulation of myocardial lipofuscin. This "wear and tear" pigment is membrane-bound cellular waste that can neither be degraded nor ejected from the cell and is composed of incomplete lysosomal degradation products, predominantly from damaged mitochondria [99]. This accumulation is attributable to increased protein oxidation and damage and age-dependent declines in autophagy and ubiquitin-mediated degradation [76, 100]. In normal physiology, removal of damaged mitochondria occurs primarily through fusion and fission, autophagy and lysosomal degradation. When mitochondrial turnover is perturbed by changes in the rates of mitochondrial fission or fusion or alterations in autophagy, the result is an accumulation of damaged and dysfunctional mitochondria. Dysfunctional mitochondria produce high levels of ROS, have impaired ATP production capacity, and likely participate in aberrant signaling [101, 102]. Tissues with accumulated damaged mitochondria may become senescent and cells may undergo apoptosis when a critical threshold of dysfunctional mitochondria have accumulated or when the affected tissue is challenged with an external stress exceeding the cellular functional reserve. Accumulation of damaged, high ROS-producing mitochondria have been demonstrated in aged hearts [101, 102] and are shown to play a role in cardiac hypertrophy [103].

Age dependent cardiac hypertrophy and diastolic dysfunction are also accompanied by cardiac proteomic changes [90]. Briefly, the levels of several mitochondrial proteins, including those involved in tricyclic acid cycle and electron transport chains as well as major metabolic pathways, such as fatty acid beta oxidation, amino acid metabolism, and ketogenesis are significantly reduced in the aged heart, with concurrent increase of proteins involved in glycolysis and gluconeogenesis [90]. Normal adult hearts preferentially utilize fatty acids as the main energy source, while diseased and failing heart use glucose as the main energy source. In addition, consistent with age-dependent cardiac hypertrophy, there are increased levels of extracellular structural proteins (and their associated signaling pathways) with age [90]. These changes may be the result of an underlying decline in protein quality control systems, which in turn lead to the accumulation of damaged proteins that are unable to efficiently perform their biological roles.

As the efficiency of protein degradation decreases in old age, it is generally expected that the overall protein turnover rate should be slower in tissues from older individuals. This was supported by early studies performed using the classical method of measuring the bulk rate of incorporation or wash-out of radioactive label in total protein [104, 105]. However, using a novel non-radioisotope deuterated leucine labeling method followed by proteomics analysis, recent studies demonstrate that the average proteome turnover rate is not significantly different in the aged mouse heart [90]. Similar observations have been made in other tissues, including liver [106] and aged skeletal muscle (Kruse and Marcinek, ms. submitted). These studies examine the individual protein turnover rates simultaneously for hundreds of leucine containing proteins and are independent of differences in amino acid precursor pool sizes [107]. The discrepancy between these recent data and earlier findings might be explained by the greater influence of abundant proteins in the old "bulk" protein measurements, or may be due to precursor pool differences. Other recent studies utilizing a similar metabolic labeling-based mass spectrometry approach to assess in vivo protein turnover have observed turnover rates consistent with our observations in aging mice [108, 109]. Thus, while the overall turnover rates of proteins are not significantly different or only slightly slower in the aged heart, the increased prevalence of damaged proteins and decreased efficiency of proteostatic maintenance in old age may have a balanced effect on turnover that mediates cardiac aging.

4.1.4 Stem Cells

Despite extensive investigation since the initial reports in 2003, the existence, identities and possible roles of adult cardiac stem cells in cardiac physiology and disease remains under debate. Earlier studies suggest the existence of multipotent populations of cells in the heart, such as c-kit+and sca1+/c-kit- cells, that are capable of differentiating into cardiomyocytes following isolation and culture [110, 111]. In contrast, a recent study demonstrates that c-kit+cells only minimally contribute to cardiomyocytes regeneration during development, aging or in response to injury [112]. Regardless of their role during normal physiology, these cells are apparently insufficient to prevent the progression of cardiovascular aging or to spontaneously regenerate the heart following acute ischemic events. Possible explanations include limited capacity of these cells to regenerate myocardium in the presence of continuous stress (such as pressure overload and ischemia), and the intrinsic aging of cardiac stem cells. Evidence of the latter was provided by experiments in rodents, which revealed that cardiac c-kit+stem cells in older animals had a higher rate of apoptosis and shorter telomeres [113]. In rodent model of diabetic cardiomyopathy, c-kit+cells demonstrate telomere shortening, increased expression of senescence markers p53 and p16INK4a, and an increase in apoptosis. Interestingly, in diabetic cardiomyopathy all of the above changes were attenuated by the ablation of the p66Shc gene [114]. p66Shc is activated by oxidative stress and translocated to the mitochondrial intermembrane space where it binds and oxidizes cytochrome c, producing H_2O_2 [115, 116], suggesting a central role of mitochondrial ROS in aging of cardiac stem (c-kit+) cells. Studies using cardiosphere-derived cells, another type of cardiac stem cell, also demonstrate a significant age-dependent decline in the number and function of stem cells derived from mouse atrial explant [117]. Bergmann et al. used data from environmental ^{14}C in human tissues, and mathematical modeling, to estimate that cardiomyocyte turnover decreased from 1 % per year at the age of 25 to 0.45 % per year by the age of 75 in adult human hearts [118]. Using multi-isotope imaging mass spectrometry to detect ^{15}N thymidine, Senyo et al. reported that the estimated annual rate of cardiomyocytes DNA synthesis is 5.5 % in the young adult and 2.6 % in the old mice [119]. Taking into account the multinucleation and polyploidization of adult cardiomyocytes, the estimated turnover rate of cardiomyocytes is 0.76 % per year in the young adult mice, and the rate further declines with age [119]. At this time, the prevailing view is that cardiac stem cells are present and function during heart development, that few new cardiomyocytes are formed in the adult heart, but these arise from existing cardiomyocytes, whether during normal aging or in response to injury [120].

4.1.5 Extracellular Matrix Remodeling

The extracellular matrix (ECM) provides structural support to the heart, and its composition is a major determinant of the stiffness of the myocardium, which is a major regulator of diastolic function [121]. ECM remodeling is a dynamic process and ECM composition is tightly regulated by the balance of the synthesis and degradation of ECM by matrix metalloproteinases (MMPs) and other proteases. Myocardial fibrosis is a hallmark of cardiac aging and deregulations of ECM synthesis and degradation have both been implicated in cardiac aging and pathology. Cardiac fibroblasts are the primary sources of cardiac ECM proteins, and it has been shown that properties of cardiac fibroblasts and their progenitors are altered with aging [122]. Studies have shown that fibroblasts isolated from old hearts have a lower proliferative capacity and have impaired differentiation into myofibroblasts in response to injury [35, 123]. Cardiac fibroblasts can be derived from several lineages including mesenchymal and myeloid origins, and the Entman group has demonstrated increased differentiation of these progenitors into mesenchymal fibroblasts and myeloid fibroblasts which contributes in increased cardiac fibrosis in aged hearts [122, 124, 125].

Transforming growth factor-β (TGF-β), a family of profibrotic cytokines, induce the expression of ECM proteins and suppress matrix degradation by MMPs [126]. TGF-β1 heterozygous mice showed reduced myocardial fibrosis and stiffness and increased LV compliance at 24 months of age [127]. Connective tissue growth factor (CTGF) is a downstream mediator of TGF-β and its expression increases with aging [128]. Cardiomyocyte-specific CTGF overexpressing mice display accelerated cardiac aging and develop age-related cardiac dysfunction as early as 7 months of age [129]. Consistent with this, Reed and colleagues demonstrated diastolic dysfunction in senescence-accelerated mice at 6 months of age, accompanied by increased LV fibrosis and increased TGF-β and CTGF expression [130]. In another study, Bradshaw and colleagues showed that deletion of secreted protein acidic and rich in cysteine (SPARC) resulted in reduced fibrillar collagen content in the LV and decreased LV diastolic stiffness [131]. This evidence suggests that increased ECM synthesis is an important mediator in diastolic dysfunction with age and reduced ECM synthesis can improve cardiac aging.

MMPs are a family of 25 zinc-dependent enzymes that regulate the degradation of ECM proteins. MMP proteolytic activity is inhibited specifically in the tissue by the tissue inhibitors of matrix metalloproteinase (TIMPs), a family composed TIMP-1, -2, -3, and -4 [132]. The expression levels of MMPs and TIMPs change differentially with age but their roles in cardiac aging have not been well established. In human plasma, the levels of MMP-2, MMP-7, TIMP-1, TIMP-2 and TIMP-4 increase but levels of MMP-9 decrease with age [133]. In CB6F1 mice, the levels of MMPs -3, -9, -14 and TIMP-4 increase from middle-age to old-age [35]. Spinale et al. showed that cardiac-specific MT1-MMP overexpressing mice display increased myocardial collagen deposition and LV dysfunction in middle-age, suggesting accelerated cardiac aging [134]. In a recent study, Chiao et al. showed that MMP-9 levels increase in the LV and plasma of aged C57Bl6 mice [135]. They later demonstrated that aged MMP-9 null mice exhibit reduced collagen deposition and preserved diastolic function and these attenuated cardiac aging phenotypes are accompanied by reduced expression of pro-fibrotic proteins, periostin and CTGF, and a compensatory increase in MMP-8 levels in the left ventricle [136]. These findings suggest a role of ECM remodeling, under complex regulation by MMPs, in cardiac aging.

4.1.6 Angiotensin Signaling

The renin-angiotensin aldosterone system (RAAS) is a major regulatory system of cardiovascular function. RAAS blockade demonstrates tremendous positive effects on the cardiovascular system, including antihypertensive, anti-inflammatory, reduction of oxidative stress and antiproliferative effects on vascular smooth muscle cells. These beneficial effects clinically translate into protection from hypertensive target organ damage, improvement of chronic heart failure, reduction of atherosclerosis as well as decreased frequency of atrial fibrillation and stroke. The reduction of these aging-related cardiovascular diseases is greater than expected by the effect of blood pressure lowering alone, and hence suggests that RAAS blockade may have a direct role in cardiovascular aging. In the heart, this is supported by the findings that intracardiac Angiotensin II concentrations are significantly increased with age, and many structural, functional and molecular changes found in aged hearts are consistent with the effects of Angiotensin II [32, 137].

Direct evidence for the role of angiotensin II in cardiac aging was provided by the studies showing that inhibition of RAAS by either angiotensin converting enzyme inhibitor enalapril, angiotensin receptor blocker losartan or genetic disruption of Angiotensin receptor type I extended the lifespan of normal rodents and slowed the onset of age related cardiovascular pathologies [138, 139], including reduction of age-related myocardial fibrosis and fibrosis-related arrhythmias [140]. Protection by RAAS blockade is not specific to cardiovascular system, but is also evident in other angiotensin responsive tissues, such as the aging kidney [141].

The mechanism of RAAS protection involves a reduction of ROS. Angiotensin II binds to ATR1, a Gαq protein coupled-receptor (GPCR), which activates NADPH oxidase (NOX2) on the cell membrane and/or NOX4 on the mitochondrial membrane [142]. ROS from NOX2 and/or NOX4 lead to increase mitochondrial ROS production [143, 144]. The fact that mCAT (mitochondrial) but not pCAT (peroxisomal) catalase attenuates Angiotensin II- and Gαq-induced cardiac hypertrophy and failure emphasize the central role of ROS amplification within mitochondria [145]. Mechanisms of ROS amplification may include ROS-induced ROS release: ROS within mitochondria could lead to electron leakage from the electron transport chain, which might further stimulate ROS production. In addition, a ROS-mtDNA damage "vicious cycle" has been postulated in which ROS cause mtDNA mutation, which produce defective respiratory proteins resulting in more ROS production. The involvement of mitochondrial DNA mutations/deletions in this vicious cycle is supported by the observation that primary damage to mtDNA in $Polg^{m/m}$ mice is sufficient to increase mitochondrial ROS, induce cardiac hypertrophy and systolic dysfunction [68, 145]. Breaking the ROS vicious cycle within mitochondria by transgenic expression of mCAT or mitochondrial targeted antioxidants effectively attenuates both cardiac hypertrophy and failure [146, 147].

4.1.7 Other Factors: Sirtuins, GDF-11 and miRNA

Sirtuin 1 (SIRT1) is a NAD+–dependent protein deacetylase that affects aging and lifespan in model organisms. The Sadoshima lab showed that expression of SIRT1 increases in aged heart, and that low to moderate overexpression (2.5–7.5 fold) of SIRT1 can attenuate cardiac aging responses, while high level overexpression (12.5 fold) resulted in cardiac hypertrophy and myocardial fibrosis at young age and death within a year [148]. In a recent study, mice deficient in the mitochondrial deacetylase SIRT3 displayed accelerated cardiac aging changes, including early age onset of hypertrophy associated with fibrosis, age-dependent increased mitochondrial swelling due to increased mPTP opening, and increased mortality after transverse aortic constriction [149].

Using heterochronic parabiosis, Loffredo et al. recently demonstrated that factors present in the circulation of young mice can reverse cardiac hypertrophy in aged mouse hearts [150]. Using an aptomer-based proteomic approach they identified growth differentiation factor 11(GDF-11) as a circulating factor that declines with age and can reverse age-related cardiac hypertrophy. Recapitulating the effect of parabiosis, treatment with GDF-11 also reduced expression of hypertrophic markers (ANP and BNP) and increased SERCA-2 expression in old hearts [150]. However, a later study has shown that the reagents for GDF-11 detection used in Loffredo's study cross-react with myostatin, and suggested that GDF-11 levels actually tend to increase with age in rat and human sera [151]. Further study will be required to determine the precise role of GDF-11 in cardiac aging.

Increasing evidence suggest that microRNAs (miRNAs) are important regulators of cardiovascular aging and diseases [152, 153]. In heart failure-prone C57Bl6×129Sv mice, van Almen et al. demonstrated that the expression of the miR-17-92 cluster (consisting of miR-18a, miR19a and miR-19b) decreases while the expression of their targets, CTGF and ECM protein thrombospondin-1 (TSP-1) increases. They showed that miR-18a and miR-19b regulated expression of CTGF, TSP-1 and collagen in *in vitro* aged cardiomyocyte cultures, and suggested that these miRNAs mediate age-related ECM remodeling in the heart [154]. In C57Bl6 mice, Jazbutyte et al. demonstrated an age-related increase in miR-22 in hearts and that miR-22 regulates cardiac fibroblast senescence [155]. In a recent study, Boon et al. demonstrated that expression of miR-34a increased in aged mouse hearts and *in vivo* silencing of miR-34a for 1 week can rescue the increase in cardiomyocyte cell death in aged mice [156]. They also showed that aged miR-34a knockout mice have improved contractile function and reduced cardiac hypertrophy compared to wild-type littermates. In the same study, they demonstrated that inhibition of miR-34a can also improve contractile function in Ku80 knockout mice (a mouse model of accelerated aging). These observations suggest increased miR-34a expression in the aged heart contributes to cardiac aging. However, whether the impact of miR-34a on cardiac aging is mediated by the newly identified target phosphatase 1 nuclear targeting subunit (PNUTS) and/ or other miR-34a targets (e.g. SIRT1) remains to be elucidated.

4.2 Mechanisms of Vascular Aging

The vasculature is a pervasive system whose age-related alterations fundamentally impact the function of every organ. As atherosclerotic diseases are a leading cause for mortality and morbidity, the mechanisms of vascular aging that have direct relevance for atherogenesis are considered, focusing on the role of oxidative stress and chronic low-grade inflammation. Importantly, the microcirculation, with a total length of ~100,000 km, interacts with virtually every cell in the human body. In the past decade a growing number of publications have revised our understanding of the important role of age-related functional and phenotypic alterations of microvascular endothelial cells, both in the aging process and the development of multiple diseases of aging. Thus, we also review recent insights into the mechanisms of microvascular dysfunction in aging and how these might contribute to age-related functional decline of multiple organ systems.

4.2.1 Oxidative-Nitrative Stress and Endothelial Dysfunction in Vascular Aging

It is well-established that nitric oxide (NO) is a crucial factor for the health and function of endothelial cells, regulating vascular tone, structural remodeling, cell proliferation, angiogenic processes, inflammation, hemostasis and barrier function [43]. Considerable evidence has been published that endothelial function is impaired in aging due to an increased production of ROS [42, 50, 157–163], which contributes to the development of a wide range of age-related pathologies, including coronary artery disease and stroke (recently reviewed [43]). Impaired bioavailability of NO due to the functional inactivation of NO by increased O_2^- was shown to result in a severe impairment of flow/shear stress-induced vasodilation in the coronary circulation [42] and other vascular beds [157], compromising adjustment of blood flow to tissue oxygen demand. Impaired flow-induced vasodilation likely contributes to decreased exercise capacity and myocardial ischemia in the elderly. Impaired endothelial release of NO is also responsible for erectile dysfunction in older men. In addition, endothelial cell-derived NO confers significant vasoprotective and cardioprotective effects, including inhibition of inflammatory cell adhesion to endothelial cells and thrombocyte aggregation, disruption of inflammatory processes, inhibition of apoptosis and preservation of endothelial progenitor cell function [43]. Endothelium-derived NO was also shown to regulate mitochondrial biogenesis and tissue energy metabolism [164, 165]. Thus, the significant age-related impairment of NO bioavailability [166], exacerbated by an age-related decline in eNOS expression [42, 167–170] and availability of tetrahydrobiopterin [171], is likely to promote both vascular inflammation and atherogenesis and lead to cellular energetic imbalance. The key role of endothelium-derived NO in protecting the cardiovascular system during aging is underscored by the finding that eNOS knockout mice exhibit a premature cardiac aging phenotype associated with early mortality [172]. Decreased endothelial NO production has been linked to increased apoptosis of endothelial cells in aging [170, 173], which likely represents an important mechanism involved in age-related microvascular rarefaction (see below). In addition to inactivating NO and eliciting oxidative macromolecular damage, increased production of ROS in the aged vasculature has important signaling roles both in vascular endothelial and smooth muscle cells. Importantly, increased levels of ROS were shown to activate redox-sensitive cellular signaling pathways implicated in inflammatory processes in the aged vasculature [174]. Increased oxidative stress also activates MMPs and promotes pathological vascular remodeling and vascular injury (reviewed in [31, 174, 175]). Caloric restriction was shown to up-regulate eNOS and increase NO bioavailability [165, 176], which contribute both to its vasoprotective and metabolic effects.

One of the major molecular mechanisms underlying vascular oxidative stress in aging is an increased expression and activity of NAD(P)H oxidases (NOX) [42, 159, 163, 177–179]. Previous studies show that inhibition of NOX exerts vasoprotective effects in aging, improving microvascular function [179], restoring penile erection [180] and inhibiting the progression of atherosclerosis [181]. There are multiple pathways which are likely involved in age-related activation and up-regulation of NOX, including hypertension [178] and up-regulation of the local renin-angiotensin II system and TNFα in the vascular wall [182, 183]. Hypertension also elicits increased mitochondrial production of ROS in the vasculature [184] and this effect is exacerbated in aging [185]. Mitochondrial-located Nox4 is a major source of pressure overload-induced oxidative stress in the heart [186] and its expression is up-regulated in the vasculature of hypertensive aged mice [46]. Moreover, angiotensin II induces expression of Nox4 both in smooth muscle cells [187] and in cardiac myocytes [146]. Thus, the effects of oxidative and nitrative stresses in aging are observed primarily in the vascular endothelium, but also have effects in the vascular smooth muscle cells.

4.2.2 Sirtuins in Vascular Aging

There is evidence that SIRT1 is also dysregulated in the aged vasculature [188] and previous studies demonstrated that SIRT1 is a key mediator of the anti-aging effects of caloric restriction [189], including its anti-oxidant and anti-inflammatory vascular effects [176]. The first pharmacological activator of SIRT1 to be widely studied was resveratrol [189]. Resveratrol was shown to confer multifaceted vasoprotective effects in aging, including attenuation of oxidative stress, increases in NO bioavailability, anti-apoptotic and anti-inflammatory effects [47, 56, 190–192], mimicking many of the effects of caloric restriction. Resveratrol was shown to improve endothelial function in hypertensive patients [193] and to prevent arterial wall inflammation and stiffening in nonhuman primates [190]. Recent data demonstrate that SIRT1 overexpression in vascular smooth muscle cells decreases blood pressure and inhibits vascular remodeling in angiotensin II-treated mice [194] thereby supporting antihypertensive effects. However, resveratrol is not specific to SIRT1 and it was shown to have many other targets in the vasculature, including Nrf2 [195]. More recently, treatment with SRT1720, a synthetic sirtuin activating compound with improved bioavailability and specificity for SIRT1 activation, was also reported to normalize aortic superoxide production, decrease NF-κB activation and down-regulate TNFα in old mice [196]. Several lines of evidence support the concept that activation of SIRT1 confers anti-atherogenic effects. First, there are studies demonstrating an association of genetic variations at the SIRT1 locus with carotid atherosclerosis [197]. Second, $Apoe^{-/-}$ mice expressing inactive truncated SIRT1 (Δex4) in smooth muscle cells demonstrate increased DNA damage-response and apoptosis, increased atherosclerosis and medial degeneration [198]. Third, treatment with the pharmacological SIRT1 activator SRT3025 inhibits atherogenesis in $Apoe^{-/-}$ mice fed a high-cholesterol diet (although it seems to be ineffective in $Ldlr^{-/-}$ mice) [199]. Recent studies suggest that the mitochondrial sirtuins (including SIRT3), which are in part responsible for the regulation of ATP production, metabolism, apoptosis and cell signaling, are also dysregulated in the aged vasculature [188]. However, the role of SIRT3 in atheroprotection is controversial [200].

4.2.3 Vascular Inflammation in Aging

There are substantial experimental and clinical data demonstrating that aging is associated with chronic low-grade inflammation [201], which promotes the development of age-related vascular diseases, including atherosclerosis [174]. Even in normal, healthy aging there is a pro-inflammatory shift in the gene expression profile of vascular endothelial and smooth muscle cells, including an up-regulation of inflammatory cytokines, chemokines, adhesion molecules and iNOS both in laboratory rodents and primates [42, 50, 173, 192, 202–204]. Vascular inflammation in aging contributes to the development of vascular dysfunction [182, 205] and promotes endothelial apoptosis in aging [173, 182]. Secretion of inflammatory mediators from microvascular endothelial cells is also likely to affect the function of cells in the parenchyma of the supplied organs. For example, neural stem cells were shown to lie close to blood vessels, and their function is likely directly affected by pro-inflammatory changes in the specialized microenvironment of this vascular niche [206]. In this regard it should be noted that age-related functional and phenotypic alterations of the microcirculation also promote chronic inflammation indirectly in the brain and other organs. Accordingly, aging is associated with significant blood brain barrier disruption in the hippocampus and other brain regions, which is exacerbated by hypertension [46]. Passing through the damaged blood brain barrier, plasma constituents, including IgG and fibrinogen, can enter the brain promoting neuroinflammation (e.g. activation of microglia by IgG via the IgG Fc receptors) [207, 208]. Thus, microvascular aging (via endothelial activation and extravasation of leukocytes, secretion of inflammatory mediators and disruption of barrier function) likely contributes to a wide range of age-related chronic diseases.

There is ample evidence that increased NF-κB activation in the aging vascular cells contributes to endothelial activation and pro-inflammatory gene expression [50, 163, 209, 210]. Age-related mechanisms that contribute to NF-κB activation include increased production of mitochondria-derived H_2O_2 [50], Nrf2 dysfunction [56], up-regulation of tissue renin-angiotensin system (RAS) [204, 211–213] and paracrine TNFα signaling [182]. There is strong evidence that age-related vascular inflammation is reversible both by dietary interventions (caloric restriction [176, 214]) and pharmacological treatments that inhibit NF-κB activation [192].

4.2.4 Vascular Cell Senescence, Endothelial Cell Renewal and Apoptosis in Vascular Aging

Role of Cellular Senescence in Vascular Aging

Mitotically competent vascular cells, including endothelial cells, smooth muscle cells, adventitial fibroblasts and pericytes, can react to diverse endogenous and exogenous stressors (e.g. paracrine signals, DNA damage, dysfunctional telomeres) by permanently withdrawing from the cell cycle, a response termed "cellular senescence" [215]. Experiments on endothelial cells *in vitro* suggest that oxidative and nitrative stress are an important stimuli for the induction of senescence [215]. There is increasing evidence that senescent cells accumulate with age in the cardiovascular system. Yet, a controversy exists regarding the exact biological role of senescent cells and the relationship between cellular senescence and vascular aging. Apart from the alterations related to permanent cell-cycle exit, senescent cells acquire distinct phenotypic changes, including the "senescence-associated secretory phenotype" (SASP) [216], which likely contribute to the development of age-related vascular diseases by altering the tissue microenvironment, impairing the function of neighboring cells via the secretion of paracrine mediators and changing the composition of the extracellular matrix. Some of these phenotypic changes are potentially important in altering the regenerative and angiogenic capacity of the vascular endothelium and promoting inflammatory processes and atherogenesis during aging [215].

Increased Endothelial Apoptosis in Aging

Programed cell death might account for some aging phenotypes in various organs [217], as well as the genesis of age-related cardiovascular pathologies. While an attractive hypothesis, the relationship between vascular aging and apoptosis remains unclear. Aging is associated with increased apoptosis of endothelial cells in the vasculature of non-human primates [16]. The percentage of apoptotic endothelial cells also increases with age in the vasculature of laboratory rodents [173, 182, 192]. The available data suggest that multiple age-related mechanisms, including impaired bioavailability of NO, circulating IGF-1 deficiency, impaired cellular oxidative stress resistance, up-regulation of TNFα and/or mitochondrial oxidative stress are likely to contribute to this increase in apoptotic endothelial cells [173, 182, 218]. Factors present in the circulation of patients with peripheral artery disease were shown to confer significant pro-apoptotic effects in cultured endothelial cells derived from aged rodents [219]. Yet, in human patients no significant correlation was found between patient's age and the number of apoptotic cells in the coronary circulation [220]. In addition to large vessel pathologies, increased apoptotic cell death at the level of the capillaries is also likely to contribute to microvascular rarefaction (see below) and, concomitantly, to the declines in muscle mass [221] and organ function during aging.

Impaired Angiogenesis and Microvascular Rarefaction in Aging

The process of angiogenesis is critical for maintenance of the microvasculature and cardiovascular homeostasis. Previous studies demonstrate that aging is associated with a progressive deterioration of microvascular homeostasis due to age-related impairment of angiogenic processes [43, 222–226]. It is assumed that these changes have a key role in the age-related decline in microvascular density (microvascular rarefaction) [227] that has been observed in multiple organ systems with age, including the heart [228], kidney [229] and skin [230]. Microvascular rarefaction is thought to decrease tissue blood supply, contribute to the development of hypertension and impair adaptation to hypoxia [231–233]. There is also an age-related rarefaction of the cerebral microvasculature, which likely contributes to a decline in cerebral blood flow that reduces metabolic support for neural signaling, promoting cognitive dysfunction in the absence of or preceding neurodegeneration in the elderly [44, 234, 235]. However, the age-related loss of microvascular plasticity has significance beyond metabolic support for neuronal signaling, since neurogenesis in the adult brain is regulated coordinately with capillary growth [44].

The mechanisms by which aging impairs angiogenesis and promotes microvascular rarefaction includes an age-related impairment of the growth hormone/IGF-1 axis (reviewed recently [236]). Previous studies demonstrate that growth hormone supplementation substantially increases cortical vascular density in older rats [234], which was accompanied by a significant improvement of cognitive function. Other factors appear to be involved: pituitary adenylate cyclase-activating polypeptide (PACAP) is an evolutionarily conserved neuropeptide secreted by endothelial cells and neurons that confers important anti-aging effects. Recent studies show that secretion of PACAP by cerebral microvessels also significantly declines with age [237, 238]. In vitro evidence suggests that age-related decline in autocrine PACAP signaling contributes to impairment of endothelial angiogenic capacity with age

[238]. There is also evidence that Dicer1 (ribonuclease III, a key enzyme of the miRNA processing machinery) is essential for normal endothelial angiogenic processes, and that age-related dysregulation of Dicer1-dependent miRNA expression is a potential mechanism underlying impaired angiogenesis and cerebromicrovascular rarefaction in aging [226]. Interestingly, lifelong caloric restriction restores the angiogenic capacity of aged endothelial cells, promoting endothelial cell proliferation and capillary morphogenesis and increasing capillary density in the brain of aged laboratory rodents [214, 239, 240]. SIRT1 also seems to regulate angiogenesis, as suggested by studies on mice with endothelium-specific knockout of SIRT1 [241]. Importantly, Nrf2 has also been implicated in regulation of endothelial angiogenic capacity [242]. In that regard it is significant that cerebral capillary density can be increased in aged mice by resveratrol treatment [191].

Impaired Endothelial Progenitor Cell Function in Aging

The link between vascular aging and a decline in the replicative function of endothelial progenitor cells (EPCs) is controversial. It is still not well established whether aging affects total EPC number [243–245], however, previous studies provided evidence that the function of circulating EPCs is impaired with age [244, 246]. Aging was shown to impair neovascularization, which depends on the intact function of highly proliferative EPCs. A role for age-related alterations of circulating factors in EPC dysfunction is suggested by the findings that the presence of sera from young rats in the culture medium improves the function of EPCs isolated from aged rats [247]. Importantly, IGF-1 was shown to exert beneficial effects on the function of progenitor cells in the cardiovascular system, including antioxidant effects, upregulation of telomerase activity, delaying replicative senescence, and increasing the pool of functionally competent progenitor cells [248]. Both circulating IGF-1 levels and paracrine IGF-1 decrease in aging, which likely exert deleterious effects in EPCs [249]. Indeed, there are studies demonstrating that in human patients age-dependent impairment of EPCs is corrected by GH-mediated increase of IGF-1 [250]. Interestingly, regular aerobic exercise was also shown to increase both the number and migratory activity of EPCs in previously sedentary older men [251].

5 Future Prospects and Interventions

Age-associated alterations in arterial, microvascular and cardiac structure and function represent links that explain, at least in part, the reason why aging is by far the greatest risk factor for cardiovascular disease. It is critical to understand the molecular mechanisms of cardiovascular aging, their interactions with both cardiovascular disease pathogenesis and systemic aging processes, and identify novel pathways that could be targeted for interventions aiming at retardation or attenuation of these age-associated alterations. The recent studies on the roles of different hallmarks of aging have advanced our understanding of cardiovascular aging and shed light on potential therapeutic strategies. Several examples of such potential therapies are indicated in the sections above, including mitochondrial protective agents, rapamycin and GDF-11. Further understanding of the mechanisms of cardiovascular aging will guide the future translational studies on novel therapeutics to treat age-related cardiovascular disease and to improve healthy cardiovascular aging. Cardiovascular aging is a promising frontier that is ripe for, and in dire need of, attention to prevent age-associated deterioration of healthspan.

Acknowledgments Y.A.C is a Glenn/AFAR postdoctoral fellow for Translational Research on Aging. We acknowledge support from NIH intramural funding for E.G.L., and the Ellison Medical Foundation and the American Federation for Aging Research, as well as NIH grants AG001751, AG038550 and HL101186 for P.S.R.

Editor: Youngsuk Oh, National Heart, Lung and Blood Institute (NHLBI), NIH.

References

1. Go AS, Mozaffarian D, Roger VL, Benjamin EJ, Berry JD, Blaha MJ et al (2014) Heart disease and stroke statistics--2014 update: a report from the American Heart Association. Circulation 129(3):e28–e292. doi:10.1161/01.cir.0000441139.02102.80
2. Hakuno D, Kimura N, Yoshioka M, Fukuda K (2009) Molecular mechanisms underlying the onset of degenerative aortic valve disease. J Mol Med (Berl) 87(1):17–24. doi:10.1007/s00109-008-0400-9
3. Connell PS, Han RI, Grande-Allen KJ (2012) Differentiating the aging of the mitral valve from human and canine myxomatous degeneration. J Vet Cardiol 14(1):31–45. doi:10.1016/j.jvc.2011.11.003, S1760-2734(12)00004-5 [pii]
4. Gerstenblith G, Frederiksen J, Yin FC, Fortuin NJ, Lakatta EG, Weisfeldt ML (1977) Echocardiographic assessment of a normal adult aging population. Circulation 56(2):273–278
5. Vasan RS, Levy D (2000) Defining diastolic heart failure: a call for standardized diagnostic criteria. Circulation 101(17):2118–2121
6. Benjamin EJ, Levy D, Anderson KM, Wolf PA, Plehn JF, Evans JC et al (1992) Determinants of Doppler indexes of left ventricular diastolic function in normal subjects (the Framingham Heart Study). Am J Cardiol 70(4):508–515 [Research Support, Non-U.S. Gov't Research Support, U.S. Gov't, P.H.S.]
7. Schulman SP, Lakatta EG, Fleg JL, Lakatta L, Becker LC, Gerstenblith G (1992) Age-related decline in left ventricular filling at rest and exercise. Am J Physiol 263(6 Pt 2):H1932–H1938 [Research Support, U.S. Gov't, P.H.S.]
8. Swinne CJ, Shapiro EP, Lima SD, Fleg JL (1992) Age-associated changes in left ventricular diastolic performance during isometric exercise in normal subjects. Am J Cardiol 69(8):823–826
9. Fleg JL, O'Connor F, Gerstenblith G, Becker LC, Clulow J, Schulman SP et al (1995) Impact of age on the cardiovascular response to dynamic upright exercise in healthy men and women. J Appl Physiol 78(3):890–900 [Research Support, U.S. Gov't, P.H.S.]
10. Rodeheffer RJ, Gerstenblith G, Beard E, Fleg JL, Becker LC, Weisfeldt ML et al (1986) Postural changes in cardiac volumes in men in relation to adult age. Exp Gerontol 21(4–5):367–378 [Research Support, U.S. Gov't, P.H.S.]

11. Najjar SS, LE, Gerstenblith G (2011) Cardiovascular aging: the next frontier in cardiovascular prevention. In: Foody JM, Blumenthal R, Wong NA (eds) Preventive cardiology: a companion to Braunwald's heart disease. Elsevier/Saunders, Philadelphia, pp 415–432
12. Yin FC, Weisfeldt ML, Milnor WR (1981) Role of aortic input impedance in the decreased cardiovascular response to exercise with aging in dogs. J Clin Invest 68(1):28–38 [Research Support, Non-U.S. Gov't Research Support, U.S. Gov't, P.H.S.]
13. Tsuji H, Larson MG, Venditti FJ Jr, Manders ES, Evans JC, Feldman CL et al (1996) Impact of reduced heart rate variability on risk for cardiac events. The Framingham Heart Study. Circulation 94(11):2850–2855 [Research Support, Non-U.S. Gov't]
14. Fleg JL, Kennedy HL (1992) Long-term prognostic significance of ambulatory electrocardiographic findings in apparently healthy subjects greater than or equal to 60 years of age. Am J Cardiol 70(7):748–751
15. Fleg JL, Kennedy HL (1982) Cardiac arrhythmias in a healthy elderly population: detection by 24-hour ambulatory electrocardiography. Chest 81(3):302–307
16. Asai K, Kudej RK, Shen YT, Yang GP, Takagi G, Kudej AB et al (2000) Peripheral vascular endothelial dysfunction and apoptosis in old monkeys. Arterioscler Thromb Vasc Biol 20(6):1493–1499 [Research Support, U.S. Gov't, P.H.S.]
17. Virmani R, Avolio AP, Mergner WJ, Robinowitz M, Herderick EE, Cornhill JF et al (1991) Effect of aging on aortic morphology in populations with high and low prevalence of hypertension and atherosclerosis. Comparison between occidental and Chinese communities. Am J Pathol 139(5):1119–1129 [Comparative Study Research Support, U.S. Gov't, P.H.S.]
18. Lakatta EG, Wang M, Najjar SS (2009) Arterial aging and subclinical arterial disease are fundamentally intertwined at macroscopic and molecular levels. Med Clin North Am 93(3):583–604. doi:10.1016/j.mcna.2009.02.008, [Research Support, N.I.H., Intramural Review] Table of Contents
19. Najjar SS (2009) IMT as a risk marker: the plot thickens. J Am Soc Echocardiogr 22(5):505–507. doi:10.1016/j.echo.2009.03.019 [Comment Editorial Research Support, N.I.H., Intramural]
20. Adams MR, Nakagomi A, Keech A, Robinson J, McCredie R, Bailey BP et al (1995) Carotid intima-media thickness is only weakly correlated with the extent and severity of coronary artery disease. Circulation 92(8):2127–2134 [Research Support, Non-U.S. Gov't]
21. Wildman RP, Farhat GN, Patel AS, Mackey RH, Brockwell S, Thompson T et al (2005) Weight change is associated with change in arterial stiffness among healthy young adults. Hypertension 45(2):187–192. doi:10.1161/01.HYP.0000152200.10578.5d [Research Support, N.I.H., Extramural Research Support, U.S. Gov't, P.H.S.]
22. Wilkinson IB, Franklin SS, Cockcroft JR (2004) Nitric oxide and the regulation of large artery stiffness: from physiology to pharmacology. Hypertension 44(2):112–116. doi:10.1161/01.HYP.0000138068.03893.40 [Review]
23. Tarasov KV, Sanna S, Scuteri A, Strait JB, Orru M, Parsa A et al (2009) COL4A1 is associated with arterial stiffness by genome-wide association scan. Circ Cardiovasc Genet 2(2):151–158. doi:10.1161/CIRCGENETICS.108.823245 [Research Support, N.I.H., Extramural Research Support, N.I.H., Intramural Research Support, Non-U.S. Gov't]
24. Laurent S, Cockcroft J, Van Bortel L, Boutouyrie P, Giannattasio C, Hayoz D et al (2006) Expert consensus document on arterial stiffness: methodological issues and clinical applications. Eur Heart J 27(21):2588–2605. doi:10.1093/eurheartj/ehl254
25. Franklin SS, Gustin WT, Wong ND, Larson MG, Weber MA, Kannel WB et al (1997) Hemodynamic patterns of age-related changes in blood pressure. The Framingham Heart Study. Circulation 96(1):308–315 [Research Support, Non-U.S. Gov't Research Support, U.S. Gov't, P.H.S.]
26. AlGhatrif M, Strait JB, Morrell CH, Canepa M, Wright J, Elango P et al (2013) Longitudinal trajectories of arterial stiffness and the role of blood pressure: the Baltimore Longitudinal Study of Aging. Hypertension 62(5):934–941. doi:10.1161/HYPERTENSIONAHA.113.01445
27. Kaess BM, Rong J, Larson MG, Hamburg NM, Vita JA, Levy D et al (2012) Aortic stiffness, blood pressure progression, and incident hypertension. JAMA 308(9):875–881. doi:10.1001/2012.jama.10503

28. Weisbrod RM, Shiang T, Al Sayah L, Fry JL, Bajpai S, Reinhart-King CA et al (2013) Arterial stiffening precedes systolic hypertension in diet-induced obesity. Hypertension 62(6):1105–1110. doi:10.1161/HYPERTENSIONAHA.113.01744
29. Feridooni HA, Dibb KM, Howlett SE (2014) How cardiomyocyte excitation, calcium release and contraction become altered with age. J Mol Cell Cardiol. doi:10.1016/j.yjmcc.2014.12.004 [Review]
30. Wang M, Jiang L, Monticone RE, Lakatta EG (2014) Proinflammation: the key to arterial aging. Trends Endocrinol Metab 25(2):72–79. doi:10.1016/j.tem.2013.10.002 [Research Support, N.I.H., Intramural Review]
31. Lakatta EG, Levy D (2003) Arterial and cardiac aging: major shareholders in cardiovascular disease enterprises: Part I: aging arteries: a "set up" for vascular disease. Circulation 107(1):139–146
32. Dai DF, Santana LF, Vermulst M, Tomazela DM, Emonds MJ, Maccoss MJ, Gollahon K, Martin GM, Loeb LA, Ladiges WC, Rabinovitch PS (2009) Overexpression of catalase targeted to mitochondria attenuates murine cardiac aging. Circulation 119(21):2789–2797. doi:10.1161/CIRCULATIONAHA.108.822403 [Research Support, N.I.H., Extramural]
33. Khouri SJ, Maly GT, Suh DD, Walsh TE (2004) A practical approach to the echocardiographic evaluation of diastolic function. J Am Soc Echocardiogr 17(3):290–297
34. Lin J, Lopez EF, Jin Y, Van Remmen H, Bauch T, Han HC et al (2008) Age-related cardiac muscle sarcopenia: combining experimental and mathematical modeling to identify mechanisms. Exp Gerontol 43(4):296–306
35. Lindsey ML, Goshorn DK, Squires CE, Escobar GP, Hendrick JW, Mingoia JT et al (2005) Age-dependent changes in myocardial matrix metalloproteinase/tissue inhibitor of metalloproteinase profiles and fibroblast function. Cardiovasc Res 66(2):410–419. doi:10.1016/j.cardiores.2004.11.029 [Research Support, N.I.H., Extramural Research Support, U.S. Gov't, Non-P.H.S. Research Support, U.S. Gov't, P.H.S.]
36. Barger JL, Kayo T, Vann JM, Arias EB, Wang J, Hacker TA et al (2008) A low dose of dietary resveratrol partially mimics caloric restriction and retards aging parameters in mice. PLoS One 3(6), e2264
37. Tei C, Nishimura RA, Seward JB, Tajik AJ (1997) Noninvasive Doppler-derived myocardial performance index: correlation with simultaneous measurements of cardiac catheterization measurements. J Am Soc Echocardiogr 10(2):169–178
38. Anversa P, Palackal T, Sonnenblick EH, Olivetti G, Meggs LG, Capasso JM (1990) Myocyte cell loss and myocyte cellular hyperplasia in the hypertrophied aging rat heart. Circ Res 67(4):871–885 [Research Support, U.S. Gov't, P.H.S.]
39. Burgess ML, McCrea JC, Hedrick HL (2001) Age-associated changes in cardiac matrix and integrins. Mech Ageing Dev 122(15):1739–1756
40. Treuting PM, Linford NJ, Knoblaugh SE, Emond MJ, Morton JF, Martin GM et al (2008) Reduction of age-associated pathology in old mice by overexpression of catalase in mitochondria. J Gerontol A Biol Sci Med Sci 63(8):813–822
41. Zheng F, Plati AR, Potier M, Schulman Y, Berho M, Banerjee A et al (2003) Resistance to glomerulosclerosis in B6 mice disappears after menopause. Am J Pathol 162(4):1339–1348
42. Csiszar A, Ungvari Z, Edwards JG, Kaminski PM, Wolin MS, Koller A et al (2002) Aging-induced phenotypic changes and oxidative stress impair coronary arteriolar function. Circ Res 90(11):1159–1166
43. Ungvari Z, Kaley G, de Cabo R, Sonntag WE, Csiszar A (2010) Mechanisms of vascular aging: new perspectives. J Gerontol A Biol Sci Med Sci 65(10):1028–1041. doi:10.1093/gerona/glq113, glq113 [pii]
44. Riddle DR, Sonntag WE, Lichtenwalner RJ (2003) Microvascular plasticity in aging. Ageing Res Rev 2(2):149–168, S1568163702000648 [pii]
45. Toth P, Csiszar A, Tucsek Z, Sosnowska D, Gautam T, Koller A et al (2013) Role of 20-HETE, TRPC channels, and BKCa in dysregulation of pressure-induced Ca2+ signaling and myogenic constriction of cerebral arteries in aged hypertensive mice. Am J Physiol Heart Circ Physiol 305(12):H1698–H1708. doi:10.1152/ajpheart.00377.2013, ajpheart.00377.2013 [pii]

46. Toth P, Tucsek Z, Sosnowska D, Gautam T, Mitschelen M, Tarantini S et al (2013) Age-related autoregulatory dysfunction and cerebromicrovascular injury in mice with angiotensin II-induced hypertension. J Cereb Blood Flow Metab 33(11):1732–1742. doi:10.1038/jcbfm.2013.143, jcbfm2013143 [pii]
47. Toth P, Tarantini S, Tucsek Z, Ashpole NM, Sosnowska D, Gautam T et al (2014) Resveratrol treatment rescues neurovascular coupling in aged mice: role of improved cerebromicrovascular endothelial function and down-regulation of NADPH oxidas. Am J Physiol Heart Circ Physiol 306(3):H299–H308. doi:10.1152/ajpheart.00744.2013, ajpheart.00744.2013 [pii]
48. Ungvari Z, Bailey-Downs L, Sosnowska D, Gautam T, Koncz P, Losonczy G et al (2011) Vascular oxidative stress in aging: a homeostatic failure due to dysregulation of Nrf2-mediated antioxidant response. Am J Physiol Heart Circ Physiol 301(2):H363–H372
49. Ungvari ZI, Labinskyy N, Gupte SA, Chander PN, Edwards JG, Csiszar A (2008) Dysregulation of mitochondrial biogenesis in vascular endothelial and smooth muscle cells of aged rats. Am J Physiol Heart Circ Physiol 294(5):H2121–H2128
50. Ungvari ZI, Orosz Z, Labinskyy N, Rivera A, Xiangmin Z, Smith KE et al (2007) Increased mitochondrial H2O2 production promotes endothelial NF-kB activation in aged rat arteries. Am J Physiol Heart Circ Physiol 293(1):H37–H47
51. Han J, Hosokawa M, Umezawa M, Yagi H, Matsushita T, Higuchi K et al (1998) Age-related changes in blood pressure in the senescence-accelerated mouse (SAM): aged SAMP1 mice manifest hypertensive vascular disease. Lab Anim Sci 48(3):256–263
52. Qiu H, Zhu Y, Sun Z, Trzeciakowski JP, Gansner M, Depre C et al (2010) Short communication: vascular smooth muscle cell stiffness as a mechanism for increased aortic stiffness with aging. Circ Res 107(5):615–619. doi:10.1161/CIRCRESAHA.110.221846, CIRCRESAHA.110.221846 [pii]
53. Eichberg JW, Shade RE (1987) "Normal" blood pressure in chimpanzees. J Med Primatol 16(5):317–321
54. Strong JP (1976) Atherosclerosis in primates. Introduction and overview. Primates Med 9:1–15
55. Zarins CK, Glagov S, Vesselinovitch D, Wissler RW (1990) Aneurysm formation in experimental atherosclerosis: relationship to plaque evolution. J Vasc Surg 12(3):246–256, 0741-5214(90)90144-Y [pii]
56. Csiszar A, Sosnowska D, Wang M, Lakatta EG, Sonntag WE, Ungvari Z (2012) Age-associated proinflammatory secretory phenotype in vascular smooth muscle cells from the non-human primate Macaca mulatta: reversal by resveratrol treatment. J Gerontol A Biol Sci Med Sci 67(8):811–820. doi:10.1093/gerona/glr228, glr228 [pii]
57. Ungvari Z, Bailey-Downs L, Gautam T, Sosnowska D, Wang M, Monticone RE et al (2011) Age-associated vascular oxidative stress, Nrf2 dysfunction and NF-kB activation in the non-human primate Macaca mulatta. J Gerontol A Biol Sci Med Sci 66(8):866–875
58. Mackic JB, Bading J, Ghiso J, Walker L, Wisniewski T, Frangione B et al (2002) Circulating amyloid-beta peptide crosses the blood-brain barrier in aged monkeys and contributes to Alzheimer's disease lesions. Vascul Pharmacol 38(6):303–313, S1537189102001982 [pii]
59. Burns EM, Kruckeberg TW, Comerford LE, Buschmann MT (1979) Thinning of capillary walls and declining numbers of endothelial mitochondria in the cerebral cortex of the aging primate, Macaca nemestrina. J Gerontol 34(5):642–650
60. Mammucari C, Rizzuto R (2010) Signaling pathways in mitochondrial dysfunction and aging. Mech Ageing Dev 131(7–8):536–543. doi:10.1016/j.mad.2010.07.003, S0047-6374(10)00135-1 [pii]
61. Trifunovic A, Larsson NG (2008) Mitochondrial dysfunction as a cause of ageing. J Intern Med 263(2):167–178. doi:10.1111/j.1365-2796.2007.01905.x, JIM1905 [pii]
62. Terzioglu M, Larsson NG (2007) Mitochondrial dysfunction in mammalian ageing. Novartis Found Symp 287:197–208; discussion 208–113
63. Judge S, Jang YM, Smith A, Hagen T, Leeuwenburgh C (2005) Age-associated increases in oxidative stress and antioxidant enzyme activities in cardiac interfibrillar mitochondria: implications for the mitochondrial theory of aging. FASEB J 19(3):419–421

64. Navarro A, Boveris A (2007) The mitochondrial energy transduction system and the aging process. Am J Physiol Cell Physiol 292(2):C670–C686
65. Schriner SE, Linford NJ, Martin GM, Treuting P, Ogburn CE, Emond M et al (2005) Extension of murine life span by overexpression of catalase targeted to mitochondria. Science 308(5730):1909–1911
66. Trifunovic A, Wredenberg A, Falkenberg M, Spelbrink JN, Rovio AT, Bruder CE et al (2004) Premature ageing in mice expressing defective mitochondrial DNA polymerase. Nature 429(6990):417–423
67. Kujoth GC, Hiona A, Pugh TD, Someya S, Panzer K, Wohlgemuth SE et al (2005) Mitochondrial DNA mutations, oxidative stress, and apoptosis in mammalian aging. Science 309(5733):481–484
68. Dai DF, Chen T, Wanagat J, Laflamme M, Marcinek DJ, Emond MJ et al (2010) Age-dependent cardiomyopathy in mitochondrial mutator mice is attenuated by overexpression of catalase targeted to mitochondria. Aging Cell 9(4):536–544. doi:10.1111/j.1474-9726.2010.00581.x, ACE581 [pii]
69. Isoyama S, Nitta-Komatsubara Y (2002) Acute and chronic adaptation to hemodynamic overload and ischemia in the aged heart. Heart Fail Rev 7(1):63–69
70. Juhaszova M, Rabuel C, Zorov DB, Lakatta EG, Sollott SJ (2005) Protection in the aged heart: preventing the heart-break of old age? Cardiovasc Res 66(2):233–244
71. Nitta Y, Abe K, Aoki M, Ohno I, Isoyama S (1994) Diminished heat shock protein 70 mRNA induction in aged rat hearts after ischemia. Am J Physiol 267(5 Pt 2):H1795–H1803
72. Chou TC, Yen MH, Li CY, Ding YA (1998) Alterations of nitric oxide synthase expression with aging and hypertension in rats. Hypertension 31(2):643–648
73. Korzick DH, Holiman DA, Boluyt MO, Laughlin MH, Lakatta EG (2001) Diminished alpha1-adrenergic-mediated contraction and translocation of PKC in senescent rat heart. Am J Physiol Heart Circ Physiol 281(2):H581–H589
74. Tani M, Honma Y, Hasegawa H, Tamaki K (2001) Direct activation of mitochondrial K(ATP) channels mimics preconditioning but protein kinase C activation is less effective in middle-aged rat hearts. Cardiovasc Res 49(1):56–68
75. Kennedy BK, Steffen KK, Kaeberlein M (2007) Ruminations on dietary restriction and aging. Cell Mol Life Sci 64(11):1323–1328. doi:10.1007/s00018-007-6470-y
76. Johnson SC, Rabinovitch PS, Kaeberlein M (2013) mTOR is a key modulator of ageing and age-related disease. Nature 493(7432):338–345. doi:10.1038/nature11861, nature11861 [pii]
77. Luong N, Davies CR, Wessells RJ, Graham SM, King MT, Veech R et al (2006) Activated FOXO-mediated insulin resistance is blocked by reduction of TOR activity. Cell Metab 4(2):133–142. doi:10.1016/j.cmet.2006.05.013
78. Wessells R, Fitzgerald E, Piazza N, Ocorr K, Morley S, Davies C et al (2009) d4eBP acts downstream of both dTOR and dFoxo to modulate cardiac functional aging in Drosophila. Aging Cell 8(5):542–552. doi:10.1111/j.1474-9726.2009.00504.x
79. Meikle L, McMullen JR, Sherwood MC, Lader AS, Walker V, Chan JA et al (2005) A mouse model of cardiac rhabdomyoma generated by loss of Tsc1 in ventricular myocytes. Hum Mol Genet 14(3):429–435. doi:10.1093/hmg/ddi039 [Research Support, N.I.H., Extramural Research Support, U.S. Gov't, P.H.S.]
80. Blagosklonny MV (2010) Calorie restriction: decelerating mTOR-driven aging from cells to organisms (including humans). Cell Cycle 9(4):683–688
81. Mattson MP, Wan R (2005) Beneficial effects of intermittent fasting and caloric restriction on the cardiovascular and cerebrovascular systems. J Nutr Biochem 16(3):129–137. doi:10.1016/j.jnutbio.2004.12.007 [Review]
82. Taffet GE, Pham TT, Hartley CJ (1997) The age-associated alterations in late diastolic function in mice are improved by caloric restriction. J Gerontol A Biol Sci Med Sci 52(6):B285–B290 [Research Support, Non-U.S. Gov't Research Support, U.S. Gov't, P.H.S.]
83. Meyer TE, Kovacs SJ, Ehsani AA, Klein S, Holloszy JO, Fontana L (2006) Long-term caloric restriction ameliorates the decline in diastolic function in humans. J Am Coll Cardiol 47(2):398–402. doi:10.1016/j.jacc.2005.08.069 [Research Support, N.I.H., Extramural Research Support, Non-U.S. Gov't]

84. Seymour EM, Parikh RV, Singer AA, Bolling SF (2006) Moderate calorie restriction improves cardiac remodeling and diastolic dysfunction in the Dahl-SS rat. J Mol Cell Cardiol 41(4):661–668. doi:10.1016/j.yjmcc.2006.07.012 [Research Support, N.I.H., Extramural Research Support, Non-U.S. Gov't]
85. Bjedov I, Toivonen JM, Kerr F, Slack C, Jacobson J, Foley A et al (2010) Mechanisms of life span extension by rapamycin in the fruit fly Drosophila melanogaster. Cell Metab 11(1):35–46. doi:10.1016/j.cmet.2009.11.010
86. Harrison DE, Strong R, Sharp ZD, Nelson JF, Astle CM, Flurkey K et al (2009) Rapamycin fed late in life extends lifespan in genetically heterogeneous mice. Nature 460(7253):392–395. doi:10.1038/nature08221
87. Powers RW 3rd, Kaeberlein M, Caldwell SD, Kennedy BK, Fields S (2006) Extension of chronological life span in yeast by decreased TOR pathway signaling. Genes Dev 20(2):174–184. doi:10.1101/gad.1381406
88. Neff F, Flores-Dominguez D, Ryan DP, Horsch M, Schroder S, Adler T et al (2013) Rapamycin extends murine lifespan but has limited effects on aging. J Clin Invest 123(8):3272–3291. doi:10.1172/JCI67674, 67674 [pii]
89. Flynn JM, O'Leary MN, Zambataro CA, Academia EC, Presley MP, Garrett BJ et al (2013) Late-life rapamycin treatment reverses age-related heart dysfunction. Aging Cell. doi:10.1111/acel.12109
90. Dai DF, Karunadharma PP, Chiao YA, Basisty N, Crispin D, Hsieh EJ et al (2014) Altered proteome turnover and remodeling by short-term caloric restriction or rapamycin rejuvenate the aging heart. Aging Cell 13(3):529–539. doi:10.1111/acel.12203
91. Bedford L, Hay D, Devoy A, Paine S, Powe DG, Seth R et al (2008) Depletion of 26S proteasomes in mouse brain neurons causes neurodegeneration and Lewy-like inclusions resembling human pale bodies. J Neurosci 28(33):8189–8198. doi:10.1523/JNEUROSCI.2218-08.2008
92. Douglas PM, Dillin A (2010) Protein homeostasis and aging in neurodegeneration. J Cell Biol 190(5):719–729. doi:10.1083/jcb.201005144
93. Christians ES, Benjamin IJ (2012) Proteostasis and REDOX state in the heart. Am J Physiol Heart Circ Physiol 302(1):H24–H37. doi:10.1152/ajpheart.00903.2011
94. Hedhli N, Pelat M, Depre C (2005) Protein turnover in cardiac cell growth and survival. Cardiovasc Res 68(2):186–196. doi:10.1016/j.cardiores.2005.06.025
95. Surguchev A, Surguchov A (2010) Conformational diseases: looking into the eyes. Brain Res Bull 81(1):12–24. doi:10.1016/j.brainresbull.2009.09.015
96. Vinciguerra M, Musaro A, Rosenthal N (2010) Regulation of muscle atrophy in aging and disease. Adv Exp Med Biol 694:211–233
97. Marzetti E, Calvani R, Bernabei R, Leeuwenburgh C (2012) Apoptosis in skeletal myocytes: a potential target for interventions against sarcopenia and physical frailty – a mini-review. Gerontology 58(2):99–106. doi:10.1159/000330064
98. de Magalhães JP (2004) From cells to ageing: a review of models and mechanisms of cellular senescence and their impact on human ageing. Exp Cell Res 300(1):1–10. doi:10.1016/j.yexcr.2004.07.006
99. Gray DA, Woulfe J (2005) Lipofuscin and aging: a matter of toxic waste. Sci Aging Knowledge Environ 2005(5):re1. doi:10.1126/sageke.2005.5.re1, 2005/5/re1 [pii]
100. Morimoto RI, Cuervo AM (2009) Protein homeostasis and aging: taking care of proteins from the cradle to the grave. J Gerontol A Biol Sci Med Sci 64(2):167–170. doi:10.1093/gerona/gln071
101. Muscari C, Caldarera CM, Guarnieri C (1990) Age-dependent production of mitochondrial hydrogen peroxide, lipid peroxides and fluorescent pigments in the rat heart. Basic Res Cardiol 85(2):172–178
102. Sohal RS, Sohal BH (1991) Hydrogen peroxide release by mitochondria increases during aging. Mech Ageing Dev 57(2):187–202, 0047-6374(91)90034-W [pii]

103. Brunk UT, Terman A (2002) The mitochondrial-lysosomal axis theory of aging: accumulation of damaged mitochondria as a result of imperfect autophagocytosis. Eur J Biochem 269(8):1996–2002, 2869 [pii]
104. Lewis SE, Goldspink DF, Phillips JG, Merry BJ, Holehan AM (1985) The effects of aging and chronic dietary restriction on whole body growth and protein turnover in the rat. Exp Gerontol 20(5):253–263 [Research Support, Non-U.S. Gov't]
105. Merry BJ, Holehan AM, Lewis SE, Goldspink DF (1987) The effects of ageing and chronic dietary restriction on in vivo hepatic protein synthesis in the rat. Mech Ageing Dev 39(2):189–199 [Research Support, Non-U.S. Gov't]
106. Karunadharma PP, Basisty N, Dai D-F, Chiao YA, Quarles EK, Hsieh EJ et al (2015) Subacute calorie restriction and rapamycin discordantly alter mouse liver proteome homeostasis and reverse aging effects. Aging Cell. doi:10.1111/acel.12317
107. Hsieh EJ, Shulman NJ, Dai DF, Vincow ES, Karunadharma PP, Pallanck L et al (2012) Topograph, a software platform for precursor enrichment corrected global protein turnover measurements. Mol Cell Proteomics 11(11):1468–1474. doi:10.1074/mcp.O112.017699
108. Miller BF, Robinson MM, Bruss MD, Hellerstein M, Hamilton KL (2012) A comprehensive assessment of mitochondrial protein synthesis and cellular proliferation with age and caloric restriction. Aging Cell 11(1):150–161. doi:10.1111/j.1474-9726.2011.00769.x
109. Price JC, Guan S, Burlingame A, Prusiner SB, Ghaemmaghami S (2010) Analysis of proteome dynamics in the mouse brain. Proc Natl Acad Sci U S A 107(32):14508–14513. doi:10.1073/pnas.1006551107
110. Oh H, Bradfute SB, Gallardo TD, Nakamura T, Gaussin V, Mishina Y et al (2003) Cardiac progenitor cells from adult myocardium: homing, differentiation, and fusion after infarction. Proc Natl Acad Sci U S A 100(21):12313–12318
111. Beltrami AP, Barlucchi L, Torella D, Baker M, Limana F, Chimenti S et al (2003) Adult cardiac stem cells are multipotent and support myocardial regeneration. Cell 114(6):763–776, S0092867403006871 [pii]
112. van Berlo JH, Kanisicak O, Maillet M, Vagnozzi RJ, Karch J, Lin SC et al (2014) c-kit+cells minimally contribute cardiomyocytes to the heart. Nature 509(7500):337–341. doi:10.1038/nature13309, nature13309 [pii]
113. Anversa P, Rota M, Urbanek K, Hosoda T, Sonnenblick EH, Leri A et al (2005) Myocardial aging--a stem cell problem. Basic Res Cardiol 100(6):482–493
114. Rota M, LeCapitaine N, Hosoda T, Boni A, De Angelis A, Padin-Iruegas ME et al (2006) Diabetes promotes cardiac stem cell aging and heart failure, which are prevented by deletion of the p66shc gene. Circ Res 99(1):42–52
115. Orsini F, Migliaccio E, Moroni M, Contursi C, Raker VA, Piccini D et al (2004) The life span determinant p66Shc localizes to mitochondria where it associates with mitochondrial heat shock protein 70 and regulates trans-membrane potential. J Biol Chem 279(24):25689–25695
116. Giorgio M, Migliaccio E, Orsini F, Paolucci D, Moroni M, Contursi C et al (2005) Electron transfer between cytochrome c and p66Shc generates reactive oxygen species that trigger mitochondrial apoptosis. Cell 122(2):221–233. doi:10.1016/j.cell.2005.05.011 [Research Support, Non-U.S. Gov't]
117. Hsiao LC, Perbellini F, Gomes RS, Tan JJ, Vieira S, Faggian G et al (2014) Murine cardiosphere-derived cells are impaired by age but not by cardiac dystrophic dysfunction. Stem Cells Dev 23(9):1027–1036. doi:10.1089/scd.2013.0388
118. Bergmann O, Bhardwaj RD, Bernard S, Zdunek S, Barnabe-Heider F, Walsh S et al (2009) Evidence for cardiomyocyte renewal in humans. Science 324(5923):98–102. doi:10.1126/science.1164680, 324/5923/98 [pii]
119. Senyo SE, Steinhauser ML, Pizzimenti CL, Yang VK, Cai L, Wang M et al (2013) Mammalian heart renewal by pre-existing cardiomyocytes. Nature 493(7432):433–436. doi:10.1038/nature11682, nature11682 [pii]

120. Senyo SE, Lee RT, Kuhn B (2014) Cardiac regeneration based on mechanisms of cardiomyocyte proliferation and differentiation. Stem Cell Res 13(3 Pt B):532–541. doi:10.1016/j.scr.2014.09.003
121. Ouzounian M, Lee DS, Liu PP (2008) Diastolic heart failure: mechanisms and controversies. Nat Clin Pract Cardiovasc Med 5(7):375–386. doi:10.1038/ncpcardio1245 [Research Support, Non-U.S. Gov't Review]
122. Cieslik KA, Trial J, Crawford JR, Taffet GE, Entman ML (2014) Adverse fibrosis in the aging heart depends on signaling between myeloid and mesenchymal cells; role of inflammatory fibroblasts. J Mol Cell Cardiol 70:56–63. doi:10.1016/j.yjmcc.2013.10.017
123. Cieslik KA, Trial J, Entman ML (2011) Defective myofibroblast formation from mesenchymal stem cells in the aging murine heart rescue by activation of the AMPK pathway. Am J Pathol 179(4):1792–1806. doi:10.1016/j.ajpath.2011.06.022
124. Cieslik KA, Taffet GE, Carlson S, Hermosillo J, Trial J, Entman ML (2011) Immune-inflammatory dysregulation modulates the incidence of progressive fibrosis and diastolic stiffness in the aging heart. J Mol Cell Cardiol 50(1):248–256. doi:10.1016/j.yjmcc.2010.10.019
125. Cieslik KA, Trial J, Carlson S, Taffet GE, Entman ML (2013) Aberrant differentiation of fibroblast progenitors contributes to fibrosis in the aged murine heart: role of elevated circulating insulin levels. FASEB J 27(4):1761–1771. doi:10.1096/fj.12-220145
126. Bujak M, Frangogiannis NG (2007) The role of TGF-beta signaling in myocardial infarction and cardiac remodeling. Cardiovasc Res 74(2):184–195. doi:10.1016/j.cardiores.2006.10.002 [Research Support, N.I.H., Extramural Research Support, Non-U.S. Gov't Review]
127. Brooks WW, Conrad CH (2000) Myocardial fibrosis in transforming growth factor beta(1) heterozygous mice. J Mol Cell Cardiol 32(2):187–195. doi:10.1006/jmcc.1999.1065 [Research Support, U.S. Gov't, Non-P.H.S.]
128. Wang M, Zhang J, Walker SJ, Dworakowski R, Lakatta EG, Shah AM (2010) Involvement of NADPH oxidase in age-associated cardiac remodeling. J Mol Cell Cardiol 48(4):765–772. doi:10.1016/j.yjmcc.2010.01.006 [Research Support, N.I.H., Intramural Research Support, Non-U.S. Gov't]
129. Panek AN, Posch MG, Alenina N, Ghadge SK, Erdmann B, Popova E et al (2009) Connective tissue growth factor overexpression in cardiomyocytes promotes cardiac hypertrophy and protection against pressure overload. PLoS One 4(8):e6743. doi:10.1371/journal.pone.0006743 [Research Support, Non-U.S. Gov't]
130. Reed AL, Tanaka A, Sorescu D, Liu H, Jeong EM, Sturdy M et al (2011) Diastolic dysfunction is associated with cardiac fibrosis in the senescence-accelerated mouse. Am J Physiol Heart Circ Physiol 301(3):H824–H831. doi:10.1152/ajpheart.00407.2010
131. Bradshaw AD, Baicu CF, Rentz TJ, Van Laer AO, Bonnema DD, Zile MR (2010) Age-dependent alterations in fibrillar collagen content and myocardial diastolic function: role of SPARC in post-synthetic procollagen processing. Am J Physiol Heart Circ Physiol 298(2):H614–H622. doi:10.1152/ajpheart.00474.2009 [Research Support, N.I.H., Extramural Research Support, U.S. Gov't, Non-P.H.S.]
132. Tayebjee MH, Lip GYH, Blann AD, MacFadyen RJ (2005) Effects of age, gender, ethnicity, diurnal variation and exercise on circulating levels of matrix metalloproteinases (MMP)-2 and –9, and their inhibitors, tissue inhibitors of matrix metalloproteinases (TIMP)-1 and –2. Thromb Res 115(3):205–210
133. Bonnema DD, Webb CS, Pennington WR, Stroud RE, Leonardi AE, Clark LL et al (2007) Effects of age on plasma matrix metalloproteinases (MMPs) and tissue inhibitor of metalloproteinases (TIMPs). J Card Fail 13(7):530–540. doi:10.1016/j.cardfail.2007.04.010 [Research Support, N.I.H., Extramural]
134. Spinale FG, Escobar GP, Mukherjee R, Zavadzkas JA, Saunders SM, Jeffords LB et al (2009) Cardiac-restricted overexpression of membrane type-1 matrix metalloproteinase in mice: effects on myocardial remodeling with aging. Circ Heart Fail 2(4):351–360. doi:10.1161/

CIRCHEARTFAILURE.108.844845 [Research Support, N.I.H., Extramural Research Support, U.S. Gov't, Non-P.H.S.]
135. Chiao YA, Dai Q, Zhang J, Lin J, Lopez EF, Ahuja SS et al (2011) Multi-analyte profiling reveals matrix metalloproteinase-9 and monocyte chemotactic protein-1 as plasma biomarkers of cardiac aging. Circ Cardiovasc Genet 4(4):455–462. doi:10.1161/CIRCGENETICS.111.959981 [Research Support, N.I.H., Extramural Research Support, Non-U.S. Gov't Research Support, U.S. Gov't, Non-P.H.S.]
136. Chiao YA, Ramirez TA, Zamilpa R, Okoronkwo SM, Dai Q, Zhang J et al (2012) Matrix metalloproteinase-9 deletion attenuates myocardial fibrosis and diastolic dysfunction in ageing mice. Cardiovasc Res 96(3):444–455. doi:10.1093/cvr/cvs275 [Research Support, N.I.H., Extramural Research Support, Non-U.S. Gov't Research Support, U.S. Gov't, Non-P.H.S.]
137. Groban L, Pailes NA, Bennett CD, Carter CS, Chappell MC, Kitzman DW et al (2006) Growth hormone replacement attenuates diastolic dysfunction and cardiac angiotensin II expression in senescent rats. J Gerontol A Biol Sci Med Sci 61(1):28–35
138. Basso N, Cini R, Pietrelli A, Ferder L, Terragno NA, Inserra F (2007) Protective effect of long-term angiotensin II inhibition. Am J Physiol Heart Circ Physiol 293(3):H1351–H1358. doi:10.1152/ajpheart.00393.2007, 00393.2007 [pii]
139. Benigni A, Corna D, Zoja C, Sonzogni A, Latini R, Salio M et al (2009) Disruption of the Ang II type 1 receptor promotes longevity in mice. J Clin Invest 119(3):524–530
140. Stein M, Boulaksil M, Jansen JA, Herold E, Noorman M, Joles JA et al (2010) Reduction of fibrosis-related arrhythmias by chronic renin-angiotensin-aldosterone system inhibitors in an aged mouse model. Am J Physiol Heart Circ Physiol 299(2):H310–H321. doi:10.1152/ajpheart.01137.2009, ajpheart.01137.2009 [pii]
141. Inserra F, Basso N, Ferder M, Userpater M, Stella I, Paglia N et al (2009) Changes seen in the aging kidney and the effect of blocking the renin-angiotensin system. Ther Adv Cardiovasc Dis 3(5):341–346. doi:10.1177/1753944709339195, 1753944709339195 [pii]
142. Mollnau H, Wendt M, Szocs K, Lassegue B, Schulz E, Oelze M et al (2002) Effects of angiotensin II infusion on the expression and function of NAD(P)H oxidase and components of nitric oxide/cGMP signaling. Circ Res 90(4):E58–E65
143. Doughan AK, Harrison DG, Dikalov SI (2008) Molecular mechanisms of angiotensin II-mediated mitochondrial dysfunction: linking mitochondrial oxidative damage and vascular endothelial dysfunction. Circ Res 102(4):488–496. doi:10.1161/CIRCRESAHA.107.162800, CIRCRESAHA.107.162800 [pii]
144. Kimura S, Zhang GX, Nishiyama A, Shokoji T, Yao L, Fan YY et al (2005) Mitochondria-derived reactive oxygen species and vascular MAP kinases: comparison of angiotensin II and diazoxide. Hypertension 45(3):438–444
145. Dai DF, Johnson SC, Villarin JJ, Chin MT, Nieves-Cintron M, Chen T et al (2011) Mitochondrial oxidative stress mediates angiotensin II-induced cardiac hypertrophy and G{alpha}q overexpression-induced heart failure. Circ Res 108(7):837–846. doi:10.1161/CIRCRESAHA.110.232306, CIRCRESAHA.110.232306 [pii]
146. Dai DF, Chen T, Szeto H, Nieves-Cintron M, Kutyavin V, Santana LF et al (2011) Mitochondrial targeted antioxidant peptide ameliorates hypertensive cardiomyopathy. J Am Coll Cardiol 58(1):73–82. doi:10.1016/j.jacc.2010.12.044, S0735-1097(11)01177-6 [pii]
147. Dai DF, Hsieh EJ, Liu Y, Chen T, Beyer RP, Chin MT et al (2012) Mitochondrial proteome remodelling in pressure overload-induced heart failure: the role of mitochondrial oxidative stress. Cardiovasc Res 93(1):79–88. doi:10.1093/cvr/cvr274 [Research Support, N.I.H., Extramural]
148. Alcendor RR, Gao S, Zhai P, Zablocki D, Holle E, Yu X et al (2007) Sirt1 regulates aging and resistance to oxidative stress in the heart. Circ Res 100(10):1512–1521. doi:10.1161/01.RES.0000267723.65696.4a [Research Support, N.I.H., Extramural Research Support, Non-U.S. Gov't]

149. Hafner AV, Dai J, Gomes AP, Xiao CY, Palmeira CM, Rosenzweig A et al (2010) Regulation of the mPTP by SIRT3-mediated deacetylation of CypD at lysine 166 suppresses age-related cardiac hypertrophy. Aging (Albany NY) 2(12):914–923, 100252 [pii]
150. Loffredo FS, Steinhauser ML, Jay SM, Gannon J, Pancoast JR, Yalamanchi P et al (2013) Growth differentiation factor 11 is a circulating factor that reverses age-related cardiac hypertrophy. Cell 153(4):828–839. doi:10.1016/j.cell.2013.04.015 [Research Support, N.I.H., Extramural Research Support, Non-U.S. Gov't]
151. Egerman MA, Cadena SM, Gilbert JA, Meyer A, Nelson HN, Swalley SE et al (2015) GDF11 increases with age and inhibits skeletal muscle regeneration. Cell Metab. doi:10.1016/j.cmet.2015.05.010
152. Quiat D, Olson EN (2013) MicroRNAs in cardiovascular disease: from pathogenesis to prevention and treatment. J Clin Invest 123(1):11–18. doi:10.1172/JCI62876 [Research Support, N.I.H., Extramural Research Support, Non-U.S. Gov't Review]
153. Smith-Vikos T, Slack FJ (2012) MicroRNAs and their roles in aging. J Cell Sci 125(Pt 1):7–17. doi:10.1242/jcs.099200 [Research Support, N.I.H., Extramural Review]
154. van Almen GC, Verhesen W, van Leeuwen RE, van de Vrie M, Eurlings C, Schellings MW et al (2011) MicroRNA-18 and microRNA-19 regulate CTGF and TSP-1 expression in age-related heart failure. Aging Cell 10(5):769–779. doi:10.1111/j.1474-9726.2011.00714.x [Research Support, Non-U.S. Gov't]
155. Jazbutyte V, Fiedler J, Kneitz S, Galuppo P, Just A, Holzmann A et al (2013) MicroRNA-22 increases senescence and activates cardiac fibroblasts in the aging heart. Age 35(3):747–762. doi:10.1007/s11357-012-9407-9 [Comparative Study Research Support, Non-U.S. Gov't]
156. Boon RA, Iekushi K, Lechner S, Seeger T, Fischer A, Heydt S et al (2013) MicroRNA-34a regulates cardiac ageing and function. Nature 495(7439):107–110. doi:10.1038/nature11919 [Research Support, Non-U.S. Gov't]
157. Sun D, Huang A, Yan EH, Wu Z, Yan C, Kaminski PM et al (2004) Reduced release of nitric oxide to shear stress in mesenteric arteries of aged rats. Am J Physiol Heart Circ Physiol 286(6):H2249–H2256
158. Hamilton CA, Brosnan MJ, McIntyre M, Graham D, Dominiczak AF (2001) Superoxide excess in hypertension and aging: a common cause of endothelial dysfunction. Hypertension 37(2):529–534
159. van der Loo B, Labugger R, Skepper JN, Bachschmid M, Kilo J, Powell JM et al (2000) Enhanced peroxynitrite formation is associated with vascular aging. J Exp Med 192(12):1731–1744
160. Francia P, Delli Gatti C, Bachschmid M, Martin-Padura I, Savoia C, Migliaccio E et al (2004) Deletion of p66shc gene protects against age-related endothelial dysfunction. Circulation 110(18):2889–2895
161. Csiszar A, Labinskyy N, Orosz Z, Xiangmin Z, Buffenstein R, Ungvari Z (2007) Vascular aging in the longest-living rodent, the naked mole-rat. Am J Physiol 293:H919–H927
162. Jablonski KL, Seals DR, Eskurza I, Monahan KD, Donato AJ (2007) High-dose ascorbic acid infusion abolishes chronic vasoconstriction and restores resting leg blood flow in healthy older men. J Appl Physiol 103:1715–1721
163. Donato AJ, Eskurza I, Silver AE, Levy AS, Pierce GL, Gates PE et al (2007) Direct evidence of endothelial oxidative stress with aging in humans: relation to impaired endothelium-dependent dilation and upregulation of nuclear factor-kappaB. Circ Res 100(11):1659–1666
164. Nisoli E, Clementi E, Paolucci C, Cozzi V, Tonello C, Sciorati C et al (2003) Mitochondrial biogenesis in mammals: the role of endogenous nitric oxide. Science 299(5608):896–899
165. Nisoli E, Tonello C, Cardile A, Cozzi V, Bracale R, Tedesco L et al (2005) Calorie restriction promotes mitochondrial biogenesis by inducing the expression of eNOS. Science 310(5746):314–317
166. Tschudi MR, Barton M, Bersinger NA, Moreau P, Cosentino F, Noll G et al (1996) Effect of age on kinetics of nitric oxide release in rat aorta and pulmonary artery. J Clin Invest 98(4):899–905

167. Tanabe T, Maeda S, Miyauchi T, Iemitsu M, Takanashi M, Irukayama-Tomobe Y et al (2003) Exercise training improves ageing-induced decrease in eNOS expression of the aorta. Acta Physiol Scand 178(1):3–10
168. Woodman CR, Price EM, Laughlin MH (2002) Aging induces muscle-specific impairment of endothelium-dependent dilation in skeletal muscle feed arteries. J Appl Physiol 93(5):1685–1690
169. Matsushita H, Chang E, Glassford AJ, Cooke JP, Chiu CP, Tsao PS (2001) eNOS activity is reduced in senescent human endothelial cells: preservation by hTERT immortalization. Circ Res 89(9):793–798
170. Hoffmann J, Haendeler J, Aicher A, Rossig L, Vasa M, Zeiher AM et al (2001) Aging enhances the sensitivity of endothelial cells toward apoptotic stimuli: important role of nitric oxide. Circ Res 89(8):709–715
171. Sindler AL, Delp MD, Reyes R, Wu G, Muller-Delp JM (2009) Effects of aging and exercise training on eNOS uncoupling in skeletal muscle resistance arterioles. J Physiol. doi:10.1113/jphysiol.2009.172221, jphysiol.2009.172221 [pii]
172. Li W, Mital S, Ojaimi C, Csiszar A, Kaley G, Hintze TH (2004) Premature death and age-related cardiac dysfunction in male eNOS-knockout mice. J Mol Cell Cardiol 37(3):671–680
173. Csiszar A, Ungvari Z, Koller A, Edwards JG, Kaley G (2004) Proinflammatory phenotype of coronary arteries promotes endothelial apoptosis in aging. Physiol Genomics 17(1):21–30
174. Csiszar A, Wang M, Lakatta EG, Ungvari ZI (2008) Inflammation and endothelial dysfunction during aging: role of NF-{kappa}B. J Appl Physiol 105(4):1333–1341. doi:10.1152/japplphysiol.90470.2008, 90470.2008 [pii]
175. Dai DF, Rabinovitch PS, Ungvari Z (2012) Mitochondria and cardiovascular aging. Circ Res 110(8):1109–1124. doi:10.1161/CIRCRESAHA.111.246140, 110/8/1109 [pii]
176. Csiszar A, Podlutsky A, Wolin MS, Losonczy G, Pacher P, Ungvari Z (2009) Oxidative stress and accelerated vascular aging: implications for cigarette smoking. Front Biosci 14:3128–3144, 3440 [pii]
177. Adler A, Messina E, Sherman B, Wang Z, Huang H, Linke A et al (2003) NAD(P)H oxidase-generated superoxide anion accounts for reduced control of myocardial O2 consumption by NO in old Fischer 344 rats. Am J Physiol Heart Circ Physiol 285(3):H1015–H1022
178. Jacobson A, Yan C, Gao Q, Rincon-Skinner T, Rivera A, Edwards J et al (2007) Aging enhances pressure-induced arterial superoxide formation. Am J Physiol Heart Circ Physiol 293(3):H1344–H1350. doi:10.1152/ajpheart.00413.2007, 00413.2007 [pii]
179. Park L, Anrather J, Girouard H, Zhou P, Iadecola C (2007) Nox2-derived reactive oxygen species mediate neurovascular dysregulation in the aging mouse brain. J Cereb Blood Flow Metab 27(12):1908–1918. doi:10.1038/sj.jcbfm.9600491, 9600491 [pii]
180. Silva FH, Monica FZ, Bau FR, Brugnerotto AF, Priviero FB, Toque HA et al (2013) Superoxide anion production by NADPH oxidase plays a major role in erectile dysfunction in middle-aged rats: prevention by antioxidant therapy. J Sex Med 10(4):960–971. doi:10.1111/jsm.12063
181. Kinkade K, Streeter J, Miller FJ (2013) Inhibition of NADPH oxidase by apocynin attenuates progression of atherosclerosis. Int J Mol Sci 14(8):17017–17028. doi:10.3390/ijms140817017, ijms140817017 [pii]
182. Csiszar A, Labinskyy N, Smith K, Rivera A, Orosz Z, Ungvari Z (2007) Vasculoprotective effects of anti-TNFalfa treatment in aging. Am J Pathol 170(1):388–698
183. Oudot A, Martin C, Busseuil D, Vergely C, Demaison L, Rochette L (2006) NADPH oxidases are in part responsible for increased cardiovascular superoxide production during aging. Free Radic Biol Med 40(12):2214–2222. doi:10.1016/j.freeradbiomed.2006.02.020, S0891-5849(06)00157-2 [pii]
184. Dikalov SI, Ungvari Z (2013) Role of mitochondrial oxidative stress in hypertension. Am J Physiol Heart Circ Physiol 305(10):H1417–H1427. doi:10.1152/ajpheart.00089.2013, ajpheart.00089.2013 [pii]

185. Springo Z, Tarantini S, Toth P, Tucsek Z, Tarantini S, Koller A, et al (2015) Aging exacerbates pressureinduced mitochondrial oxidative stress in mouse cerebral arteries. J Gerontol Biol Med Sci. doi:10.1093/gerona/glu244
186. Kuroda J, Ago T, Matsushima S, Zhai P, Schneider MD, Sadoshima J (2010) NADPH oxidase 4 (Nox4) is a major source of oxidative stress in the failing heart. Proc Natl Acad Sci U S A 107(35):15565–15570. doi:10.1073/pnas.1002178107, 1002178107 [pii]
187. Briones AM, Tabet F, Callera GE, Montezano AC, Yogi A, He Y et al (2011) Differential regulation of Nox1, Nox2 and Nox4 in vascular smooth muscle cells from WKY and SHR. J Am Soc Hypertens 5(3):137–153. doi:10.1016/j.jash.2011.02.001, S1933-1711(11)00041-6 [pii]
188. Roos CM, Hagler M, Zhang B, Oehler EA, Arghami A, Miller JD (2013) Transcriptional and phenotypic changes in aorta and aortic valve with aging and MnSOD deficiency in mice. Am J Physiol Heart Circ Physiol 305(10):H1428–H1439. doi:10.1152/ajpheart.00735.2012, ajpheart.00735.2012 [pii]
189. Baur JA, Ungvari Z, Minor RK, Le Couteur DG, de Cabo R (2012) Are sirtuins viable targets for improving healthspan and lifespan? Nat Rev Drug Discov 11(6):443–461. doi:10.1038/nrd3738, nrd3738 [pii]
190. Mattison JA, Wang M, Bernier M, Zhang J, Park SS, Maudsley S et al (2014) Resveratrol prevents high fat/sucrose diet-induced central arterial wall inflammation and stiffening in nonhuman primates. Cell Metab 20(1):183–190. doi:10.1016/j.cmet.2014.04.018, S1550-4131(14)00208-3 [pii]
191. Oomen CA, Farkas E, Roman V, van der Beek EM, Luiten PG, Meerlo P (2009) Resveratrol preserves cerebrovascular density and cognitive function in aging mice. Front Aging Neurosci 1:4. doi:10.3389/neuro.24.004.2009
192. Pearson KJ, Baur JA, Lewis KN, Peshkin L, Price NL, Labinskyy N et al (2008) Resveratrol delays age-related deterioration and mimics transcriptional aspects of dietary restriction without extending life span. Cell Metab 8(2):157–168
193. Carrizzo A, Puca A, Damato A, Marino M, Franco E, Pompeo F et al (2013) Resveratrol improves vascular function in patients with hypertension and dyslipidemia by modulating NO metabolism. Hypertension 62(2):359–366. doi:10.1161/HYPERTENSIONAHA.111.01009, HYPERTENSIONAHA.111.01009 [pii]
194. Gao P, Xu TT, Lu J, Li L, Xu J, Hao DL et al (2014) Overexpression of SIRT1 in vascular smooth muscle cells attenuates angiotensin II-induced vascular remodeling and hypertension in mice. J Mol Med (Berl) 92(4):347–357. doi:10.1007/s00109-013-1111-4
195. Ungvari Z, Bagi Z, Feher A, Recchia FA, Sonntag WE, Pearson K et al (2010) Resveratrol confers endothelial protection via activation of the antioxidant transcription factor Nrf2. Am J Physiol Heart Circ Physiol 299(1):H18–H24
196. Gano LB, Donato AJ, Pasha HM, Hearon CM Jr, Sindler AL, Seals DR (2014) The SIRT1 activator SRT1720 reverses vascular endothelial dysfunction, excessive superoxide production, and inflammation with aging in mice. Am J Physiol Heart Circ Physiol 307(12):H1754–H1763. doi:10.1152/ajpheart.00377.2014, ajpheart.00377.2014 [pii]
197. Kedenko L, Lamina C, Kedenko I, Kollerits B, Kiesslich T, Iglseder B et al (2014) Genetic polymorphisms at SIRT1 and FOXO1 are associated with carotid atherosclerosis in the SAPHIR cohort. BMC Med Genet 15(1):112. doi:10.1186/s12881-014-0112-7, s12881-014-0112-7 [pii]
198. Gorenne I, Kumar S, Gray K, Figg N, Yu H, Mercer J et al (2013) Vascular smooth muscle cell sirtuin 1 protects against DNA damage and inhibits atherosclerosis. Circulation 127(3):386–396. doi:10.1161/CIRCULATIONAHA.112.124404, CIRCULATIONAHA.112.124404 [pii]
199. Miranda MX, van Tits LJ, Lohmann C, Arsiwala T, Winnik S, Tailleux A et al (2014) The Sirtl activator SRT3025 provides atheroprotection in Apoe–/– mice by reducing hepatic Pcsk9 secretion and enhancing Ldlr expression. Eur Heart J. doi:10.1093/eurheartj/ehu095, ehu095 [pii]

200. Winnik S, Gaul DS, Preitner F, Lohmann C, Weber J, Miranda MX et al (2014) Deletion of Sirt3 does not affect atherosclerosis but accelerates weight gain and impairs rapid metabolic adaptation in LDL receptor knockout mice: implications for cardiovascular risk factor development. Basic Res Cardiol 109(1):399. doi:10.1007/s00395-013-0399-0
201. Franceschi C, Bonafe M, Valensin S, Olivieri F, De Luca M, Ottaviani E et al (2000) Inflamm-aging. An evolutionary perspective on immunosenescence. Ann N Y Acad Sci 908:244–254
202. Csiszar A, Ungvari Z, Koller A, Edwards JG, Kaley G (2003) Aging-induced proinflammatory shift in cytokine expression profile in rat coronary arteries. FASEB J 17(9):1183–1185
203. Cernadas MR, Sanchez de Miguel L, Garcia-Duran M, Gonzalez-Fernandez F, Millas I, Monton M et al (1998) Expression of constitutive and inducible nitric oxide synthases in the vascular wall of young and aging rats. Circ Res 83(3):279–286
204. Wang M, Zhang J, Jiang LQ, Spinetti G, Pintus G, Monticone R et al (2007) Proinflammatory profile within the grossly normal aged human aortic wall. Hypertension 50(1):219–227
205. Arenas IA, Xu Y, Davidge ST (2006) Age-associated impairment in vasorelaxation to fluid shear stress in the female vasculature is improved by TNF-{alpha} antagonism. Am J Physiol Heart Circ Physiol 290(3):H1259–H1263
206. Shen Q, Wang Y, Kokovay E, Lin G, Chuang SM, Goderie SK et al (2008) Adult SVZ stem cells lie in a vascular niche: a quantitative analysis of niche cell-cell interactions. Cell Stem Cell 3(3):289–300. doi:10.1016/j.stem.2008.07.026, S1934-5909(08)00397-4 [pii]
207. Zlokovic BV (2008) The blood-brain barrier in health and chronic neurodegenerative disorders. Neuron 57(2):178–201. doi:10.1016/j.neuron.2008.01.003, S0896-6273(08)00034-2 [pii]
208. Fernandez-Vizarra P, Lopez-Franco O, Mallavia B, Higuera-Matas A, Lopez-Parra V, Ortiz-Munoz G et al (2012) Immunoglobulin G Fc receptor deficiency prevents Alzheimer-like pathology and cognitive impairment in mice. Brain 135(Pt 9):2826–2837. doi:10.1093/brain/aws195, aws195 [pii]
209. Zou Y, Yoon S, Jung KJ, Kim CH, Son TG, Kim MS et al (2006) Upregulation of aortic adhesion molecules during aging. J Gerontol A Biol Sci Med Sci 61(3):232–244
210. Donato AJ, Black AD, Jablonski KL, Gano LB, Seals DR (2008) Aging is associated with greater nuclear NFkappaB, reduced IkappaBalpha, and increased expression of proinflammatory cytokines in vascular endothelial cells of healthy humans. Aging Cell 7(6):805–812. doi:10.1111/j.1474-9726.2008.00438.x, ACE438 [pii]
211. Spinetti G, Wang M, Monticone R, Zhang J, Zhao D, Lakatta EG (2004) Rat aortic MCP-1 and its receptor CCR2 increase with age and alter vascular smooth muscle cell function. Arterioscler Thromb Vasc Biol 24(8):1397–1402
212. Wang M, Takagi G, Asai K, Resuello RG, Natividad FF, Vatner DE et al (2003) Aging increases aortic MMP-2 activity and angiotensin II in nonhuman primates. Hypertension 41(6):1308–1316
213. Wang M, Zhang J, Spinetti G, Jiang LQ, Monticone R, Zhao D et al (2005) Angiotensin II activates matrix metalloproteinase type II and mimics age-associated carotid arterial remodeling in young rats. Am J Pathol 167(5):1429–1442
214. Csiszar A, Gautam T, Sosnowska D, Tarantini S, Banki E, Tucsek Z et al (2014) Caloric restriction confers persistent anti-oxidative, pro-angiogenic, and anti-inflammatory effects and promotes anti-aging miRNA expression profile in cerebromicrovascular endothelial cells of aged rats. Am J Physiol Heart Circ Physiol 307(3):H292–H306. doi:10.1152/ajpheart.00307.2014, ajpheart.00307.2014 [pii]
215. Erusalimsky JD (2009) Vascular endothelial senescence: from mechanisms to pathophysiology. J Appl Physiol 106(1):326–332. doi:10.1152/japplphysiol.91353.2008, 91353.2008 [pii]
216. Coppe JP, Patil CK, Rodier F, Sun Y, Munoz DP, Goldstein J et al (2008) Senescence-associated secretory phenotypes reveal cell-nonautonomous functions of oncogenic RAS and the p53 tumor suppressor. PLoS Biol 6(12):2853–2868. doi:10.1371/journal.pbio.0060301, 08-PLBI-RA-2566 [pii]

217. Kajstura J, Cheng W, Sarangarajan R, Li P, Li B, Nitahara JA et al (1996) Necrotic and apoptotic myocyte cell death in the aging heart of Fischer 344 rats. Am J Physiol 271(3 Pt 2):H1215–H1228
218. Bailey-Downs LC, Mitschelen M, Sosnowska D, Toth P, Pinto JT, Ballabh P et al (2012) Liver-specific knockdown of IGF-1 decreases vascular oxidative stress resistance by impairing the Nrf2-dependent antioxidant response: a novel model of vascular aging. J Gerontol Biol Med Sci 67(4):313–329
219. Gardner AW, Parker DE, Montgomery PS, Sosnowska D, Casanegra AI, Ungvari Z et al (2014) Greater endothelial apoptosis and oxidative stress in patients with peripheral artery disease. Int J Vasc Med 2014:160534. doi:10.1155/2014/160534
220. Boddaert J, Mallat Z, Fornes P, Esposito B, Lecomte D, Verny M et al (2005) Age and gender effects on apoptosis in the human coronary arterial wall. Mech Ageing Dev 126(6–7):678–684. doi:10.1016/j.mad.2005.01.001, S0047-6374(05)00017-5 [pii]
221. Wang H, Listrat A, Meunier B, Gueugneau M, Coudy-Gandilhon C, Combaret L et al (2014) Apoptosis in capillary endothelial cells in ageing skeletal muscle. Aging Cell 13(2):254–262. doi:10.1111/acel.12169
222. Tarnawski AS, Pai R, Tanigawa T, Matysiak-Budnik T, Ahluwalia A (2010) PTEN silencing reverses aging-related impairment of angiogenesis in microvascular endothelial cells. Biochem Biophys Res Commun 394(2):291–296. doi:10.1016/j.bbrc.2010.02.161, S0006-291X(10)00402-X [pii]
223. Bach MH, Sadoun E, Reed MJ (2005) Defects in activation of nitric oxide synthases occur during delayed angiogenesis in aging. Mech Ageing Dev 126(4):467–473. doi:10.1016/j.mad.2004.10.005, S0047-6374(04)00256-8 [pii]
224. Sadoun E, Reed MJ (2003) Impaired angiogenesis in aging is associated with alterations in vessel density, matrix composition, inflammatory response, and growth factor expression. J Histochem Cytochem 51(9):1119–1130
225. Ahluwalia A, Tarnawski AS (2011) Activation of the metabolic sensor – AMP activated protein kinase reverses impairment of angiogenesis in aging myocardial microvascular endothelial cells. Implications for the aging heart. J Physiol Pharmacol 62(5):583–587
226. Ungvari Z, Tucsek Z, Sosnowska D, Toth P, Gautam T, Podlutsky A et al (2013) Aging-induced dysregulation of dicer1-dependent MicroRNA expression impairs angiogenic capacity of rat cerebromicrovascular endothelial cells. J Gerontol A Biol Sci Med Sci 68(8):877–891. doi:10.1093/gerona/gls242, gls242 [pii]
227. Ungvari Z, Csiszar A (2012) The emerging role of IGF-1 deficiency in cardiovascular aging: recent advances. J Gerontol A Biol Sci Med Sci. doi:10.1093/gerona/gls072, gls072 [pii]
228. Anversa P, Li P, Sonnenblick EH, Olivetti G (1994) Effects of aging on quantitative structural properties of coronary vasculature and microvasculature in rats. Am J Physiol 267(3 Pt 2):H1062–H1073
229. Kang DH, Anderson S, Kim YG, Mazzalli M, Suga S, Jefferson JA et al (2001) Impaired angiogenesis in the aging kidney: vascular endothelial growth factor and thrombospondin-1 in renal disease. Am J Kidney Dis 37(3):601–611, S0272638601863678 [pii]
230. Montagna W, Carlisle K (1979) Structural changes in aging human skin. J Invest Dermatol 73(1):47–53
231. Murugesan N, Demarest TG, Madri JA, Pachter JS (2011) Brain regional angiogenic potential at the neurovascular unit during normal aging. Neurobiol Aging. doi:10.1016/j.neurobiolaging.2011.09.022, S0197-4580(11)00374-5 [pii]
232. Benderro GF, Lamanna JC (2011) Hypoxia-induced angiogenesis is delayed in aging mouse brain. Brain Res 1389:50–60. doi:10.1016/j.brainres.2011.03.016, S0006-8993(11)00513-0 [pii]
233. Ingraham JP, Forbes ME, Riddle DR, Sonntag WE (2008) Aging reduces hypoxia-induced microvascular growth in the rodent hippocampus. J Gerontol A Biol Sci Med Sci 63(1):12–20, 63/1/12 [pii]

234. Sonntag WE, Lynch CD, Cooney PT, Hutchins PM (1997) Decreases in cerebral microvasculature with age are associated with the decline in growth hormone and insulin-like growth factor 1. Endocrinology 138(8):3515–3520
235. Khan AS, Sane DC, Wannenburg T, Sonntag WE (2002) Growth hormone, insulin-like growth factor-1 and the aging cardiovascular system. Cardiovasc Res 54(1):25–35, S0008636301005338 [pii]
236. Sonntag WE, Deak F, Ashpole N, Toth P, Csiszar A, Freeman W et al (2013) Insulin-like growth factor-1 in CNS and cerebrovascular aging. Front Aging Neurosci 5:27. doi:10.3389/fnagi.2013.00027
237. Tripathy D, Sanchez A, Yin X, Martinez J, Grammas P (2012) Age-related decrease in cerebrovascular-derived neuroprotective proteins: effect of acetaminophen. Microvasc Res 84(3):278–285. doi:10.1016/j.mvr.2012.08.004, S0026-2862(12)00153-7 [pii]
238. Banki E, Sosnowska D, Tucsek Z, Gautam T, Toth P, Tarantini S et al (2014) Age-related decline of autocrine PACAP impairs angiogenic capacity of rat cerebromicrovascular endothelial cells. J Gerontol Biol Med Sci. doi:10.1093/gerona/glu116
239. Csiszar A, Sosnowska D, Tucsek Z, Gautam T, Toth P, Losonczy G et al (2013) Circulating factors induced by caloric restriction in the nonhuman primate Macaca mulatta activate angiogenic processes in endothelial cells. J Gerontol A Biol Sci Med Sci 68(3):235–249. doi:10.1093/gerona/gls158, gls158 [pii]
240. Lynch CD, Cooney PT, Bennett SA, Thornton PL, Khan AS, Ingram RL et al (1999) Effects of moderate caloric restriction on cortical microvascular density and local cerebral blood flow in aged rats. Neurobiol Aging 20(2):191–200, S0197458099000329 [pii]
241. Maizel J, Xavier S, Chen J, Lin CH, Vasko R, Goligorsky MS (2014) Sirtuin 1 ablation in endothelial cells is associated with impaired angiogenesis and diastolic dysfunction. Am J Physiol Heart Circ Physiol 307(12):H1691–H1704. doi:10.1152/ajpheart.00281.2014, ajpheart.00281.2014 [pii]
242. Valcarcel-Ares MN, Gautam T, Warrington JP, Bailey-Downs L, Sosnowska D, de Cabo R et al (2012) Disruption of Nrf2 signaling impairs angiogenic capacity of endothelial cells: implications for microvascular aging. J Gerontol A Biol Sci Med Sci 67(8):821–829
243. Chang EI, Loh SA, Ceradini DJ, Chang EI, Lin SE, Bastidas N et al (2007) Age decreases endothelial progenitor cell recruitment through decreases in hypoxia-inducible factor 1alpha stabilization during ischemia. Circulation 116(24):2818–2829. doi:10.1161/CIRCULATIONAHA.107.715847, CIRCULATIONAHA.107.715847 [pii]
244. Heiss C, Keymel S, Niesler U, Ziemann J, Kelm M, Kalka C (2005) Impaired progenitor cell activity in age-related endothelial dysfunction. J Am Coll Cardiol 45(9):1441–1448. doi:10.1016/j.jacc.2004.12.074, S0735-1097(05)00357-8 [pii]
245. Thijssen DH, Vos JB, Verseyden C, van Zonneveld AJ, Smits P, Sweep FC et al (2006) Haematopoietic stem cells and endothelial progenitor cells in healthy men: effect of aging and training. Aging Cell 5(6):495–503. doi:10.1111/j.1474-9726.2006.00242.x, ACE242 [pii]
246. Keymel S, Kalka C, Rassaf T, Yeghiazarians Y, Kelm M, Heiss C (2008) Impaired endothelial progenitor cell function predicts age-dependent carotid intimal thickening. Basic Res Cardiol 103(6):582–586. doi:10.1007/s00395-008-0742-z
247. Zhu G, Song M, Wang H, Zhao G, Yu Z, Yin Y et al (2009) Young environment reverses the declined activity of aged rat-derived endothelial progenitor cells: involvement of the phosphatidylinositol 3-kinase/Akt signaling pathway. Ann Vasc Surg 23(4):519–534. doi:10.1016/j.avsg.2008.11.013, S0890-5096(09)00039-9 [pii]
248. Torella D, Rota M, Nurzynska D, Musso E, Monsen A, Shiraishi I et al (2004) Cardiac stem cell and myocyte aging, heart failure, and insulin-like growth factor-1 overexpression. Circ Res 94(4):514–524
249. Humpert PM, Djuric Z, Zeuge U, Oikonomou D, Seregin Y, Laine K et al (2008) Insulin stimulates the clonogenic potential of angiogenic endothelial progenitor cells by IGF-1 receptor-dependent signaling. Mol Med 14(5–6):301–308. doi:10.2119/2007-00052.Humpert

250. Thum T, Hoeber S, Froese S, Klink I, Stichtenoth DO, Galuppo P et al (2007) Age-dependent impairment of endothelial progenitor cells is corrected by growth-hormone-mediated increase of insulin-like growth-factor-1. Circ Res 100(3):434–443
251. Hoetzer GL, Van Guilder GP, Irmiger HM, Keith RS, Stauffer BL, DeSouza CA (2007) Aging, exercise, and endothelial progenitor cell clonogenic and migratory capacity in men. J Appl Physiol 102(3):847–852. doi:10.1152/japplphysiol.01183.2006, 01183.2006 [pii]
252. Lakatta EG (1993) Cardiovascular regulatory mechanisms in advanced age. Physiol Rev 73(2):413–467

The Impact of Aging on Ischemic Stroke

Farida Sohrabji

Contents

1 Introduction

The high incidence of stroke worldwide is considered a global epidemic. There are two broad types of stroke: hemorrhagic stroke and ischemic stroke. Hemorrhagic stroke is due to weakening of the vessel wall and eventual rupture and spillage of blood in the brain parenchyma. Hemorrhagic strokes are not as common (about 80,000 cases per year in the US) but are more likely to be fatal as compared to ischemic strokes. Ischemic strokes result from blockage or constriction of a cerebral vessel

F. Sohrabji, Ph.D.
Women's Health in Neuroscience Program, Department of Neuroscience and Experimental Therapeutics, Texas A&M Health Science Center, 8447 State Hwy 47, MREB 4108, Bryan, TX, 77807, USA
e-mail: Sohrabji@medicine.tamhsc.edu

F. Sierra, R. Kohanski (eds.), *Advances in Geroscience*,
DOI 10.1007/978-3-319-23246-1_6

resulting in rapid neuronal death due to loss of oxygen and glucose. Ischemic stroke is more common and can run the gamut from mild symptoms to chronic disability and death. Stroke symptoms can vary dramatically from person to person, based in part on the specific blood vessel, the brain region, the duration of ischemia and, crucially, on the general health of the individual. Not surprisingly, stroke is the leading non-martial cause of serious long-term disability and a stroke-related death occurs every 47 s in the US, underscoring the enormous medical, financial and societal burden of this disease. Few therapies are available for stroke patients outside of rehabilitative therapy.

2 Stroke and Aging

In the US, ischemic stroke is considered a disease of the elderly. Thus, while stroke incidence is low among younger demographics, the prevalence of stroke in the sixth–seventh decade of life (60–79) is 6.2 % for men and 6.9 % for women, which then doubles to 13.9 % for males and 13.8 % for females in the 80+ years age group [102]. The increased risk for stroke with age coupled with a growing aging population will lead to an additional 3.4 million people affected by stroke in the next 15 years [209].

Besides elevating the risk for stroke, age also adversely affects stroke outcomes [128]. Stroke outcomes can be assessed by several measures including survival, functional recovery, and length of hospitalization. An early study of 415 patients with transient ischemic attack (TIAs) reported that 1 and 5 years survival was negatively correlated with advanced age [121]. A prospective analysis of 515 stroke patients (Copenhagen Stroke Study) showed that outcome measures such as activities of daily living (measured by the Barthel Index [BI]) were significantly worse in older patients, although neurological outcomes as measured by the Scandinavian Stroke Score were not affected by age [192]. Since ADL status reflects not only recovery from the stroke but also compensation from the non-affected side, it suggests a poorer ability to compensate among this older population. Furthermore, hospitalization length was significantly increased in older patients (>65 years) with stroke [225]. Observational studies in university hospital settings reported that age was a highly significant predictor of poor functional outcome [1, 66, 139]. Moreover, a small study of centenarians also confirmed that strokes were much more severe in this population than in other age groups [207]. Morphologically, aging is associated with decreased salvage of penumbral tissue, and leptomeningeal collateral circulation is reduced in aging stroke patients, which associated with a poorer outcome [15].

2.1 Is Age a Non-modifiable Risk Factor?

Based on the evidence above, age is often referred to as a 'non-modifiable' risk factor for stroke. In evaluating this statement, it should be recognized that the aging demographic is a highly variable one. Sex/gender differences, life style factors, and ethnicity all impact stroke incidence. Additionally, the incidence of hypertension, metabolic disease, hyper-cholesteronemia, that are co-morbid for various neurologic diseases including stroke, all contribute to the variability in the elderly population. Accordingly, while the prevalence of stroke is one-sixth of the 85+ age group, Alzheimer's disease, another neurologic illness, occurs in one-third of the 80+ age group. Thus while age is a contributing factor for many illness, the relatively lower incidence of stroke in the aging population suggest that other factors may also modulate age to elevate the risk for this disease. Some of these factors are considered below.

2.1.1 Sex Differences in Stroke

A principal variable affecting stroke incidence in aging is the sex of the patient. Most ischemic strokes occur in the elderly and among this elderly demographic, women are more likely to get a stroke [217]. In addition to a higher incidence of ischemic stroke at older ages, females display more non-classical stroke symptoms and tend to have worse outcomes from stroke. Thus while stroke is the 4th leading cause of death overall, it is the 3rd leading cause of death in women, and the 5th leading cause of death in men [193]. In fact, the rates of stroke-related death have declined over the last 25 years for men but not women [232]. Furthermore, since women live longer than men, it is projected that stroke-related disability and institutionalization is likely to affect women more than men [145]. Women account for 60 % of stroke-related deaths [169], even after normalization for age. The 5 y stroke recurrence is also disproportionately higher in females (20 %) as compared to males (10 %) in the 45–64 years age range [232]. A Canadian stroke registry study reported that 10 % of women stroke patients were discharged to long term care as compared to 5 % of men [133], despite the observation that stroke size tends not to be different in males and females [248]. In the Danish National Registry analysis, women were reported to have more severe strokes than men although they exhibited a survival advantage compared to men, especially at advanced ages [206].

The increased incidence of stroke among older women, especially when compared to the relative low risk among younger demographics has led to the hypothesis that the loss of ovarian hormones, principally estrogen, at menopause may be a contributory factor. However, analysis of hormone use and stroke incidence in postmenopausal women does not support this conclusion. An early case–control study reported no increased risk for stroke in postmenopausal women who took hormone therapy, relative to those not taking hormones [216]. In a different multicenter case-controlled study, increased lifetime exposure to estrogen was associated with a lower risk of stroke, but interestingly, a lower age at menarche increased the odds of stroke [63]. The influential Women's Health Initiative (WHI) study indicated that hormone use actually increased stroke risk. The WHI study was a randomized, double blind, placebo-controlled multicenter trial, which compared the risk of myocardial infarction, stroke and dementia in women who consumed daily conjugated equine estrogens (CEE) [115]. CEE + progestins [279] or placebo. Hormone therapy groups showed an increased risk for stroke; however, subgroup analyses indicated that most of this risk was seen in the older age groups. In the CEE trial, increased risk for stroke was statistically significant for the 60–69 years old group but not the 50–59 years old group [115]. In an observational analysis of postmenopausal women in the Nurse's Health Study, estrogen and estrogen + progestin use increased the risk of stroke irrespective of the age of the user or time since menopause [107]. However, the observational arm of the WHI study showed no increased risk for stroke in the CEE or CEE + progestin arm [222, 223]. A possible factor in the discrepancy between the WHI trial and the WHI observational study was that the initiation of hormones was much earlier in the latter study. However other health characteristics among this group can also impact stroke risk in conjunction with hormone therapy (HT). In the observational trial (SHOW study) there were no differences in stroke risk due to HT use; however, HT users were more likely to be normotensive and lean as compared to non-users in this study [42] which was not the case in the WHI study, where hypertension incidence was similar in CEE users and non-users [115]. A similar interaction between HT and hypertension was seen in the Danish Nurses study, where normotensive women who used hormone therapy were not different from controls, while the risk for stroke was elevated among hypertensive women who used hormone therapy [170].

Another modifier of the effectiveness of HT for stroke is the concept of timing. The timing hypothesis postulates that hormone treatment is likely to be cardio- and stroke-protective only if taken during the perimenopause or early after menopause. Subgroup analyses of the WHI study (described above) support this idea, as well as data from a prospective study of Swedish women, where stroke risk was significantly decreased in women who initiated hormone treatment prior to menopause [157]. In a population-based nested case–control study of 50–69 year old women, HT did not significantly elevate ischemic stroke risk [8], further supporting the idea that HT at ages closer to the menopause may be harmless for stroke. Coronary artery calcification, a surrogate marker of cardiac disease, was reduced by estrogen in the youngest cohort of WHI (50–59 years) [177], also signifying that estrogen's effects can be modulated by the age of the user. Finally, a study of non-users found that stroke-related mortality in women 65 and older was higher in women with higher levels of endogenous estrogen [176], implying that elevated levels of hormones in late life, whether exogenous or endogenous, may exert a deleterious effect on stroke.

The accumulated evidence of sex differences in the incidence, mortality and outcome for stroke prompted the recent American Heart Association/American Stroke Association guidelines for sex-specific recommendations for the female stroke patient [41]. This recommendation underscored stroke risk factors that were stronger or more prevalent in females than males, such as migraine, atrial fibrillation, Type 2 diabetes, and hypertension, as well as risk factors specific to women such as pregnancy, gestational diabetes, preeclampsia, oral contraceptives and estrogen therapy. Thus the aging woman may represent a specifically vulnerable population for stroke.

2.1.2 Hypertension and Aging in the Context of Stroke

Hypertension is a significant risk factor for global diseases such as cardiovascular disease, congestive heart failure, stroke and end-stage renal disease. It is a leading risk factor of stroke among the elderly. Hypertension is highly prevalent in the aging population, estimated at 69.6 % of all women aged 65–74, and 64.7 % of all men aged 65–74 [102]. One report estimates that non-hypertensive individuals aged 55–65 are virtually all (90 %) likely to develop stage 1 hypertension and are at a 40 % risk of developing stage 2 hypertension [271], underscoring the close link between aging and hypertension. Systolic blood pressure rises gradually between 30 and 80 years of age, and systolic hypertension is very common after age 50 [219]. Hypertension is seen in 77 % of all stroke patients [169], and the Framingham study reported that hypertension was the factor most strongly associated with stroke in elderly males and females, increasing the odds ratio of stroke 1.9 in aging men and 2.3 in aging women [132]. A similar finding has been reported for both men and women aged 80+ years [13].

While hypertension increases with age, an increase in systolic blood pressure (SBP) was frequently thought to be a normal, and therefore inconsequential, part of the aging process, while increases in diastolic BP (DBP) were seen as the greater threat to cardiovascular and stroke incidence. However, in the US, more than 3/4 of all untreated hypertensive patients 50 years and older are of the "isolated systolic hypertension" (ISH) subtype [90], suggesting that hypertension among the elderly is of a different type [91]. Aging affects hemodynamics, with increases in systolic BP, diastolic BP and mean arterial pressure (MAP) starting at 50–55 years of age. Thereafter, with age there is a leveling off of DBP and MAP, with a steady increase in SBP. This pattern is thought to reflect a transition from age-related changes in peripheral vascular resistance to large artery stiffness [89]. Additionally, the difference between SBP and DBP, or pulse pressure, is currently considered to be a more predictive risk factor for cardiovascular disease. Both aging and hypertension can impair endothelial, and therefore, microvessel function. With age, vascular tone is affected as well as arterial remodeling, resulting in increased pulse wave velocity, a phenomenon referred to as vascular aging [196], a determining factor for cardiovascular disease.

A series of studies examining the effect of anti-hypertensive therapy in elderly patients showed that such interventions reduced all strokes 30–57 % compared to placebo, and reduced fatal strokes by 39–76 % (reviewed in [14]). Thus aggressive management of blood pressure in the elderly may dilute the impact of age on stroke. This may underlie the steady decline in stroke mortality and stroke-associated disability among the elderly that has been reported between 1967 and 1985, although the incidence of stroke remains high in this group [24]. While control of risk factors including hypertension, hyperlipidemia, diabetes and cigarette smoking are necessary at all ages, they are most likely to be beneficial in the elderly [25]. For every 5 mmHg reduction in BP there is a 14 % decline in stroke mortality [51] and a 35–40 % reduction of stroke occurrence [53]. Taken together, the evidence indicates that there appears to be no age threshold where treatment for hypertension is not likely to be beneficial for stroke [16].

2.1.3 Other Comorbidities and Stroke

Aging-related changes in blood pressure also intersect with dyslipidemia in this group, possibly due to lipid action/accumulation on the arterial wall. Elevated cholesterol and lower levels of high-density lipoproteins are associated with stroke in aging and lipid-lowering drugs reduce the incidence of myocardial infarction [6, 237]. While major trials have indicated that statins are well tolerated in the elderly, the association between hyperlipidemia and stroke is not strong in this group. In a study of high-risk elderly patients (< or = 82 years of age), HMG-CoA reductase inhibitors (statins) appear to lower stroke risk [6]. However, this may also result from other protective actions of statins on the endothelium, including anti-oxidant, anti-inflammatory effects and stabilization of plaques [179]. After age 65, blood lipid levels are less prominent risk factors for cardiovascular diseases and by age 75, blood lipids have little predictive value [29]. In fact, specific lipids may be associated with longevity in the elderly population, for example sphingomyelin in women [103, 238]. Thus among the elderly the risk imposed by hypertension is likely more severe than hyperlipidemia. By contrast, cigarette smoking, whether past or current, also impacts vascular disease [56], and is a risk factor for heart failure even in older adults. In current smokers this risk is elevated irrespective of the 'pack-years' of smoking exposure [104]. Remarkably, prior to 75 years of age, hypertension and diabetes are much less important risk factors as compared to heavy (>2 drinks/day) alcohol consumption at midlife [129].

2.1.4 Diabetes

Cardiovascular complications are the most common non-fatal complication of diabetes among older adults [122]. The prevalence of type 2 diabetes is 16.5 % in men and 12.7 % in women in the 75–84 age range where strokes are common, and glucose intolerance was found in more than 1/3 of all participants in the Framingham study that were 65 or older [281]. Altered glucose metabolism is not necessarily a component of aging, and may represent a sub population that is generally at higher risk for other adverse geriatric processes [131]. Diabetes is frequently associated with cognitive impairment, dementia, depression, and stroke [28]; however, in some cases, it is the co-occurrence of hypertension in diabetic populations that may increase the risk for stroke. Some support for this idea comes from the fact that vascular disease increases before the elevation of glucose levels and more than 25 % of newly diagnosed diabetic patients already have cardiovascular disease [281].

The convergence of comorbid disease and sociocultural stressors during aging as risk factors for stroke fits well with the concept of an 'allostatic load' [182]. Allostatic load refers to the cumulative lifespan exposure to adverse circumstances, and integrates with the 3-hit hypothesis where disease susceptibility is thought to result from genetic predisposition, early life events and later-life events [60]. Saban and colleagues propose that early life events may predispose an inflammatory epigenetic signature, which is made worse with chronic stressors such as social disadvantage [48] and psychological stress, culminating in increased risk for CVD and stroke [235].

3 Impact of Aging on Conventional Stroke Therapy

Tissue plasminogen activator (tPA; Alteplase) is the only FDA-approved therapy for stroke, and its mode of action consists of proteolytic degradation of the clot, with the goal of re-establishing circulation. tPA has also been shown to increase the risk for hemorrhagic transformation, which occurs subsequent to ischemic stroke and cerebral infarction. Although hemorrhagic transformation (intracerebral hemorrhage) may occur spontaneously after ischemic stroke, thrombolytic therapy occasionally leads to this complication, possibly due to the actions of tPA on matrix metalloproteinases [148, 268]. In animal models, tPA increases permeability of the blood brain barrier in aged (18–20 month old) male Wistar rats as compared to young (3–4 months old) males, and this is related to disassembly of tight junction proteins such as claudin and occludin [134].

The actions of tPA are poorly studied in the elderly, despite the fact that people aged 85 and over are the fastest growing stroke demographic in the US. In the landmark tPA study at NINDS, tPA was noted to cause intracerebral hemorrhage (ICH) in a small proportion of patients and this was embodied in the recommendation that tPA after 3 h was not advised. In this study, the conversion to ICH was 2.87 times greater in patients older than 80 as compared to patients younger than 80 [264], although age was not an independent predictor of ICH. In contrast, age was an independent predictor of hemorrhage in the European Acute Stroke Study [149]. In a small retrospective study of 22 stroke patients who were 90 years or older, most patients had poor outcomes at 30 days post stroke and many died [180]. Similarly, a report of patients from a German registry indicated that the rate of ICH was much higher in older patients (10.3 %; 75 or older) [116]. Although the explanation for this age response is not clear, the greater incidence of underlying vascular pathology including cerebral amyloid angiopathy [105] as well as deficient clearance of tPA in the elderly have been implicated as factors. In fact, in carefully selected older populations, with treatment by stroke specialists and careful adherence to the NINDS guidelines, there does not appear to be an increased risk of ICH in elderly patients treated with tPA [260].

Sex differences in treatment among patients that receive tPA may also factor into the sex differences in stroke outcomes. In a study spanning stroke patients over a decade (1997–2006), men were more likely than women to receive IV tPA, angioplasty/stents, carotid endarterectomy, or cardiac reperfusion. However towards the end of the study period, sex differences in the use of IV tPA were eliminated [267], which suggests that greater overall tPA use and emphasis on early time-to-treatment may decrease sex differences in acute stroke care. More recently, a comparison of white and black male and female stroke patients found no differences in the outcome of tPA administration in men, but reported that black women were less likely to get tPA than white women [36]. In a regional study, women were more likely to be excluded from tPA for hypertension as compared to men, suggesting that under-treatment of stroke risk factors in women may further impact stroke therapies as well [175]. Increased assignment of tPA therapy to women should be encouraged as treatment outcomes do not differ between tPA treated men and women, while in non-tPA administered groups, males were more likely to have a better neurologic score as compared to women [247].

3.1 Failed Stroke Trials

Although several drugs have been identified in preclinical studies, only a few of these have made it to clinical trials and none have succeeded [44]. These include the SAINT trials that tested the free radical scavenger NXY-059; the RANTTAS trials for Tirilazad mesylate, a lipid peroxidation inhibitor; the INWEST trials using nimodipine, a calcium channel blocker and the Selfotel investigations using an NMDA blocker. While several reasons may explain why the preclinical promise of these drugs was not borne out in clinical trials, an important consideration is the lack of aging animals used in the preclinical studies [166]. Most preclinical studies used healthy young animals as test subjects, which clearly does not approximate the human population [185]. In fact, irrespective of the treatment, stroke outcome was significantly affected by the age of the patient in the combined SAINT trial analyses [70]. Preclinical studies with these drugs routinely failed to use clinically relevant animal models, such as the aged and those with comorbid diseases. A comprehensive review of preclinical studies that lay the groundwork for these failed drugs found that virtually all studies (43/45) used only younger animals [270]. These and other studies provided the impetus for the STAIR recommendations, which specifically included recommendations for clinically relevant animal models [87].

3.2 Preclinical Therapies

Emerging therapies are focusing on interventions that lead to long-term brain repair and plasticity, using cell based therapies and pharmacological therapies.

The use of cell therapies and grafts in stroke has focused on adult stem cells or induced pluripotent stem cells [23, 110]. Both human and animal stroke brains show signs of proliferation, including the aging human [174]. Intra-parenchymal [158] and intra-arterial [159] delivery of bone marrow derived mesenchymal cells is reported to improve neurological outcomes and functional performance when delivered post stroke. Interestingly, grafts of human umbilical tissue-derived cells have also been shown effective for neural recovery in aged animals [294]. Environmental enrichment appears to improve neurologic function in both young and old animals [38], and further enhances functional recovery when combined with stem cell therapy [117]. Although the mechanism of action is not well understood, several end points are improved including release of trophic factors, anti-inflammatory effects, angiogenesis and cell survival (reviewed in [47]).

Angiogenesis is considered critical to long term stroke recovery [11], and the low rate of vessel formation in the elderly is thought to be associated with low rates of functional recovery. Formation of new blood vessels is a desired therapeutic outcome, and ischemic events provide important signals for new vessel formation, such as secretion of angiogenic and matrix remodeling factors. Angiogenesis is also critical for providing a niche for neurogenesis [210]. Studies have shown that VEGF is angiogenic in the post stroke brain but may also cause blood brain barrier 'leakiness' and hemorrhagic transformation in the early acute phase of stroke [296]. Similarly, stem cell transplant-derived VEGF is also neuroprotective [118]. However, rAAV-mediated VEGF therapy showed that the angiogenic response to this growth factor is attenuated in aged animals [94], while sildenafil treatment post stroke enhanced recovery and angiogenesis in both young and aged animals [293], and adenoviral transfer of adiponectin was more effective in old than young stroke animals [184]. These studies, while promising, underscore the need for preclinical studies to mimic clinically valid aspects of the patient population, including old age and comorbidities.

4 Animal Models to Gauge the Impact of Aging on Stroke

While preclinical models do not measure risk, they can be useful in assessing stroke severity (outcomes) in terms of infarct volume, neurological scores, the impact of comorbid disease and response to therapies. Age differences in stroke outcomes in preclinical models tend to mirror the findings of the clinical studies. With increasing age, there is a greater conversion of ischemic tissue to infarcted tissue in stroke patients as detected by MRI [17]. Similarly, aged animals show more infarcted damage than young animals, and young animals are likely to show spontaneous recovery soon after the ischemic event, while in aged animals, such recovery is usually longer and never reaches the same recovery levels as young animals [38].

Young and aged male and female mice subject to a middle cerebral artery occlusion (MCAo) show that young females sustain a smaller infarct as compared to young males or aged female mice [164] although by 2 weeks post stroke, both young and aged groups were comparable [178]. In a study of neonates (10 day), and adult animals at 2 and 6 months, functional recovery was best in the neonate and was impaired in the older age groups, suggesting that the plastic environment of the immature brain is better suited for stroke recovery [284]. Similarly, post stroke epilepsy, a common complication in this disease, was more common in older animals as compared to younger ones [135]. In studies comparing young and middle aged animals, the latter group showed worse infarct volume in the spontaneously hypertensive rat strain (SHR) as compared to the normotensive Kyoto Wistar strain [152]. In fact, middle-aged SHR rats show spontaneous white matter disease, cognitive decline and heightened inflammation as compared to normotensive controls [130], accentuating their value as preclinical models for stroke.

Sex differences also modulate stroke outcome in the context of aging in animal models. Adult females have a smaller infarct and better cerebral blood flow than age-matched males both in normoglycemic [4] and diabetic [266] animals. However, although female mice sustain a much smaller infarct [178], they showed significantly more mortality and poorer stroke outcomes as compared to older males. These sex differences prompted several studies addressing the contribution of hormones to stroke outcomes, specifically estrogen. Using natural variations in circulating estrogen levels, Liao and colleagues [160] showed that the extent of ischemic damage was inversely related to circulating levels of estrogen [160]. In fact, replacement with 17β estradiol [77, 234, 243] and its inactive stereoisomer 17α estradiol [249] as well as the conjugate equine estrogen preparation [181] all reduce infarct volume in female animals. Exogenous estrogen replacement is neuroprotective when given prior [77] or subsequent to the injury [167, 286]. However, it should be noted that all these studies were done in young female animals that were ovariectomized to mimic a surgical menopause. In contrast, as mentioned earlier, elevated levels of sex hormones may have a negative effect on stroke in the aged.

Hormone treatment in studies using older female animals does not reliably result in stroke neuroprotection and may in fact exacerbate stroke recovery. While some studies show that estrogen treatment to middle-aged or older female animals is neuroprotective [76], a growing number of studies show that estrogen treatment to ovariectomized middle-aged female animals either has no protective effect [62, 154] or paradoxically, increases infarct volume [243, 244]. Besides gonadal steroids, other endocrine systems are also affected by aging and disease, and it has been proposed that these changes may impact the overall effectiveness of estrogen in an aging model [254]. Support for this idea comes from a study where post stroke IGF-1 treatment to estrogen-exposed middle aged female rats reversed the detrimental effects of estrogen. Conversely, IGF-1 receptor antagonist treatment of young females abrogated the protective effects of estrogen in this group, suggesting that cross talk between hormone systems (in this case estrogen and IGF-1) may be critical for neuroprotection [244]. Thus the loss of both IGF-1 and estrogen in aging females may be responsible for the more severe stroke outcomes seen in this group. Interestingly, IGF-1 levels are higher in young males, while their stroke outcomes are worse than young females, suggesting that the correlation between IGF-1 and neuroprotection may be more complex.

Overall, when key variables that affect stroke outcomes in patients, such as advanced age, sex and hypertension are included in preclinical studies in animal models, they appropriately reflect the outcomes seen in human disease.

4.1 *Pathophysiology of Stroke*

4.1.1 Ischemia Induced Cascade

Loss of glucose and oxygen to brain cells causes a series of events resulting in neuronal death either through apoptosis or necrosis. Cell death occurs not only in those areas directly affected by the ischemia, but also in neighboring cells as a result of an ischemic cascade initiated in proximal cells. Many of these processes occur simultaneously, beginning with a failure of ATP-dependent systems that result in unregulated calcium entry into the cell. A feed forward process then ensues, whereby calcium-induced release of the excitatory amino acid glutamate, further increases Ca^{+2} accumulation. Consequently, stimulation of calcium dependent enzymes initiate a wide variety of cellular reactions resulting in free radical formation and oxidative stress. Death of ischemic neurons causes toxicity in the local microenvironment, and activates local immune and inflammatory cells, thus amplifying the possibility of cell death (reviewed in [212]).

4.2 *Is the Cellular Response to Stroke Different in Aging Animals?*

At the cellular level, aged animals are able to mount a cytoprotective response to stroke but the timing of proliferation and activation of key support cells such as glia and endothelial cells is accelerated, resulting in rapid infarct development and poor prognosis in aged animals [221]. Endothelial cells, astrocytes and microglia are the major support cells of the brain and play a critical role in preserving neurons following ischemic injury. A critical way in which these cells interact is the neurovascular unit, where blood brain barrier components (endothelial cells, astrocytes and pericytes) form a functional unit with neighboring neurons. The blood brain barrier itself consists of endothelial cells and their intercellular tight junctions, supported by astrocytic end feet and pericytes [2, 113, 144, 230]. Paracellular transport between adjacent endothelial cells is restricted by the presence of tight junctions, composed of large transmembrane proteins such as claudins and occludins. Collectively, this structure maintains the homeostatic environment of the brain and excludes peripheral cells, proteins, and many molecules, including cytotoxic compounds. Functional changes in the blood brain barrier occur as a result of ischemia, including loss of endothelial tight junctions, the internalization of plasma proteins, and trafficking of peripheral immune cells into the brain parenchyma. Coupled with distress signals from local brain cells, this promotes the intercellular transfer of peripheral immune cells and transcytosis of plasma proteins, thus amplifying the inflammatory response in the ischemic brain. The aging blood brain barrier, and its cellular components, may well underlie the greater stroke severity seen in this group.

4.2.1 Aging and the Blood Brain Barrier

Among the many age-related changes in the brain, alterations in the blood brain barrier are most likely to elevate the risk and severity of stroke. Age-related changes in the microvasculature increase blood brain barrier permeability which is further increased in patients with vascular dementia or Alzheimer's disease [83]. There is growing evidence that hemorrhagic transformation after tPA treatment is due to its action on proteolysis in the neurovascular unit (reviewed in [278]). One possibility is that tPA may act via matrix metalloproteinases (MMP) which are known to promote barrier leakiness. Stroke patients who receive tPA have elevated MMP-9 [119] and MMP9 levels are highest in tPA patients with hemorrhagic conversion [187]. Furthermore, MMP inhibitors significantly reduce the severity of tPA induced cerebral hemorrhage [258]. Increased blood brain barrier permeability with age has been reported in both animals and humans (reviewed in [200]). Following stroke, blood brain permeability is enhanced in older animals and tPA action on the barrier is accompanied by activation of occludin and claudin-5 [134].

Sex differences and alterations in barrier function due to menopause or reproductive senescence are relatively understudied. Experimental studies evaluating the influence of estrogen on blood brain barrier permeability generally indicate a protective function [253]. 17β-estradiol also reduces edema in an experimental stroke model by reducing the activity [199] and abundance [46] of the Na-K-Cl cotransporter. However, the synthetic estrogen ethinyl estradiol has been shown to increase endothelial permeability to albumin [93].

In middle-aged female rats, there is increased permeability of the blood brain barrier in the hippocampus and olfactory bulb as compared to younger females [21]. At the molecular/cellular level, this is accompanied by increased perivascular IgG expression in the hippocampus, a marker commonly used to assess barrier integrity in aging and disease. Furthermore, constitutive expression of claudin-5 and occludin were not altered by age, however junctional localization of these proteins, which is critical for their barrier function was reduced in cerebral microvessels from middle aged reproductively senescent females [19]. Disrupted tight junctions are also seen in aging female hamsters [101]. In fact, cerebral microvessels from a small sample of pre and post-menopausal women also confirmed this reproductive age-related loss of junctional localization [19].

Consistent with the high demand for active transport in these cells, barrier-forming endothelial cells contain significantly greater mitochondrial content than non-barrier forming endothelial cells [205]. In both rats and monkeys, mitochondrial content of endothelial cells is reduced with age [40], while mitochondrial DNA mutations increase during aging and in age-related neurologic conditions such as Alzheimer's disease [190]. Mitochondrial oxidative stress is a leading cause of vascular/endothelial dysfunction in the aging population. With advancing age, the accumulation of reactive oxygen species increases in endothelial cells (reviewed in

[58, 269], resulting in inactivation of the vasodilator nitric oxide (NO), and consequently, reduced vasodilator capacity and perfusion of tissues [57]. Compensatory responses such as increased iNOS in microvessels are also seen in hypertensive subjects [250] and aging male rats [43]. Age-related elevation of oxidative stress is accompanied by a chronic low-grade inflammatory phenotype marked by NF-kB activation [269]. Furthermore, age-associated mitochondrial oxidative stress promotes mitochondrial protein oxidation and mitochondrial DNA mutations (reviewed in [59]), and it is speculated to cause endothelial apoptosis [171, 285]. Decreased density of cerebral arterioles in aging [256] is consistent with the idea of vascular deterioration.

4.2.2 Aging and Angiogenesis

In addition to their central role in the blood brain barrier, cerebrovascular endothelial cells play a critical role in stroke-associated angiogenesis and regulation of blood flow. Angiogenesis or formation of new vessels is an adaptive response to ischemic injury [64, 255]. Angiogenesis is stimulated by hypoxia, which upregulates angiogenic factors such as HIF-1 and VEGF [106, 224]. Post-stroke angiogenesis is closely associated with neurogenesis [7, 52] such that the angiogenic niche promotes neurogenesis [202]. Neurons and astrocytes within the neurovascular unit also secrete angiogenic factors, which in turn enhance proliferation and differentiation of neuronal precursor cells to promote neurogenesis [259]. Thus reduced functional capacity of endothelial cells with age will not only affect vascular repair but also neurogenesis.

The impact of aging on angiogenesis in the stroke brain is poorly studied and the results equivocal. Aged male F344 rats show a significant decrease in capillary angiogenesis compared to young animals following hypoxia [125], while in mice, both young and aged animals display similar microvessel densities 2–3 weeks after hypoxia, although hypoxia-induced upregulation of HIF-1α and Ang-2 was significantly reduced in the aged mice [30]. Aging also impairs the angiogenic potential of senescent human umbilical vein endothelial cells [97], which is associated with reduced VEGF levels [231]. Exogenous VEGF treatment of aging mice, however, did not improve the angiogenic response, suggesting that VEGF-activated downstream signaling pathways may be permanently changed with age [94]. Although HIF-1 activation is a primary step in angiogenesis, PPAR-gamma coactivator 1-alpha (PGC 1α) is also capable of inducing angiogenesis in aged F344 rats [194]. Using gene expression at the neurovascular unit as a marker for angiogenic capacity, resting gene expression of PGC 1α was lower in aging animals, and the angiogenic response to hypoxia was also weaker [191]. Estrogen promotes angiogenesis, decreases free radical production, increases cell survival, and stimulates angiogenesis in cerebral endothelial cells [141]. It also increased microvessel density prior to [10] and 10 days post stroke [9], indicating that the loss of estrogen in aging females may impair repair processes.

4.2.3 Aging and Cerebral Blood Flow

Reduced cerebral blood flow during aging appears to be universal, and has been documented in rats [33, 203], monkeys [197] and humans [240]. Interestingly, cerebral blood flow reduction may occur as early as midlife in humans (50 years) [240]), consistent with elevated stroke risk during this time frame. Impaired endothelial vasodilation, which is an early marker for arterial aging, has been attributed to the twin perils of oxidative stress and inflammation. Accordingly, the vasodilatory compound nitric oxide (NO) is reduced by an imbalance in pro and anti-oxidant systems in the aging vasculature, resulting in high levels of reactive oxygen species and low levels of antioxidant response due to decreases in critical proteins including manganese superoxide dismutase (mnSOD2) and nuclear factor-erythroid 2 p45-related factor 2factor-2 (Nrf2) (reviewed in [80]). This is believed to act as a stimulus for a low -grade inflammation [282] typically seen in aging vessels [290], resulting in a deleterious feed-forward amplification of oxidative stress.

One of the most prominent reasons for NO insufficiency is age, due in part to the age-related reduction of nitric oxide synthase [218] and an age-related increase in arginase, that degrades the natural substrate for NOS [32]. In fact age is the most significant predictor of endothelial-dependent vasodilation [100]. In coronary circulation, endothelial-derived NO decreases dramatically in 70–80 year old patients as compared to 20 year old patients [79], consistent with the observation that cerebral blood flow is negatively correlated with age in a study of 20–100 year old individuals [261]. In contrast, the renin angiotensin system is permissive for hypertension [183] and age-related imbalances in the renin-angiotensin system (RAS) is also a risk factor for cardiovascular disease [12]. In animal models, inhibition of RAS increases lifespan and reduces age-related hypertension (reviewed in [71]). Through production of vasodilatory and vasoconstrictive molecules, the cerebrovascular endothelium plays an important role in regulating blood flow, which is also critical during reperfusion following stroke [109]. Some evidence links the age response to an imbalance between vasoconstrictive and vasodilatory factors, with an elevation in the former [72]. Vascular reactivity is also altered with age, such that adenosine administration induced greater vasodilation in young animals as compared to older animals [127], while intravascular serotonin exacerbates vasoconstriction in older animals [111]. Thus, the aging brain is more likely to be subject to hypoperfusion and potentially, greater neuronal damage in response to ischemic stroke conditions.

4.2.4 The Aging Astrocyte

Astrocytes play an important role in the normal and pathologic brain. Specifically, astrocytes regulate synaptic activity, extracellular matrix secretion, blood–brain barrier integrity and the inflammatory response (reviewed in [252]). Following brain injury, astrocytes become reactive with increased expression of glial fibrillary acidic protein (GFAP) [215, 262]. Astrocytes offer trophic support to neurons through secretion and/or expression of several soluble factors including neurotrophins [172, 227] growth factors [226, 297] and by regulating local glutamate concentrations [239]. Astrocytes are also a source of inflammatory cytokines/chemokines [137, 150, 226] that can potentially ameliorate or exacerbate the injury response. However, astrocytic response to injury can be modulated by several factors including age and hormonal status.

The aging brain shows distinct changes in astrocyte morphology [272], GFAP expression [195, 233] and astrocyte numbers [112, 189]. Aging accelerates injury-induced astrocyte reactivity [18, 221], and after ischemic injury, this enhanced glial response accelerates glial scar formation [18]. Furthermore, reduced expression of astrocyte-derived SC1, an extracellular matrix-associated glycoprotein, indicates that matrix remodeling may also be impaired in aged rats following focal ischemia [168].

Reduced astrocyte function with age may also impact neurogenesis, a potential endogenous repair mechanism following brain injury. Decreased Wnt3 secretion (a neural stem cell differentiation factor) in astrocytes from middle aged (9 month old) mice resulted in impaired hippocampal neurogenesis in this age group as compared to younger (3 months.) mice [186, 204]. Reductions in astrocyte-derived growth factors and their receptors such as insulin-like growth factor 1 (IGF-1) [78, 155], fibroblast growth factor receptor 2 (FGFR-2) [45] and vascular endothelial growth factor (VEGF) [34] may also contribute to reduced adult neurogenesis.

Both constitutive and injury-induced functions of astrocytes are affected by age. Ex vivo cultures of astrocytes from the olfactory bulb of middle-aged female rats show several phenotypic changes such as increased stress fiber formation, reduced laminin deposition, increased BDNF expression and reduced TrkB expression compared to young astrocytes [156]. Furthermore, aging astrocytes show an impairment in their ability to promote neuronal differentiation of neural progenitor cells [156]. Moreover, ex vivo cultures of ischemia-activated astrocytes from aging females show reduced glutamate uptake as compared to astrocytes harvested from young adult females [155]. The impaired glutamate clearance and metabolic dysregulation observed in the aging astrocyte promotes a more toxic microenvironment in the older brain, thus probably contributing to the increased infarct size observed in old rats [221, 243].

4.2.5 Effects of Sex Hormones on Astrocyte Function

The sex (or gonadal) hormones androgen, estrogen and progesterone and their metabolites play a significant role in regulating astrocyte activity (reviewed in [251]), specifically by increasing expression of astrocyte-derived growth factors [92, 220] and regulating the glial glutamate transporters [213]. In the context of injury, reactive astrogliosis is attenuated following replacement with estrogen, testosterone [26, 61], dihyrotestosterone [61] progesterone as well as the neurosteroids dehydroepiandrosterone, pregnenolone and pregnenolone sulfate [96]. In addition, progesterone, estradiol [95] and the selective estrogen receptor modulators (SERMS) raloxifene and tamoxifen [27] reduced reactive gliosis in females. Ischemia-activated cortical astrocytes from acyclic middle aged females showed impairments in glutamate clearance and lower growth factor synthesis as compared to astrocyte cultures from young females [155], indicating that a constitutive loss of ovarian estrogen may affect astrocyte function.

Interestingly, neuroprotective interventions by sex hormones may originate from the astrocyte itself or may result from paracrine signaling. Aromatase expression, a key enzyme in estrogen biosynthesis, is enhanced following a sub-lethal pressure increase in astrocyte cultures [98] while estrogen treatment of astrocytes inhibits the steroid hydroxylase CYP7B1 which is responsible for metabolizing dehydroepiandrosterone (DHEA), an important precursor for both estrogen and testosterone [86]. Taken together these data suggest that sex hormones play a supportive role in astrocyte function, and further support the hypothesis that the loss of ovarian hormones at menopause may accelerate stroke severity in females.

4.2.6 Aging Microglia

Microglia are the main effectors of the innate response in the ischemic brain. Microglia are thought to contain multimolecular complexes called inflammasomes, which act as intracellular sensors for host-derived signals in cases of brain injury and stroke [276]. Activated microglia are responsible for phagocytosis of non-functioning cells and synthesis and release of cytokines that can result in cell death. Accordingly, minocycline, an anti-inflammatory compound that targets microglia, reduces microglial activation and improves function and survival rate [114, 165]. In a small clinical trial, minocycline use was reported to improve neurologic function in stroke patients [147]. On the other hand, microglia also promote a neurogenic environment after stroke, since they also produce growth factors that are neuroprotective, and can elevate growth factor synthesis in neighboring cells [287]. Mesenchymal stem cells cultured in microglial media exhibit elevated level of several growth factors including VEGF and IGF-1 [298]. Furthermore, selective ablation of proliferating microglia exacerbates ischemic injury [146] while exogenous application of microglia reduces ischemic injury [138].

Similar to macrophages, at least two activation states have been proposed for microglia. The classical M1 response is induced by microbial agents or T helper cell type 1 secretions and results in the production of pro-inflammatory cytokines (IL-1β, TNF-α, IL-6, and IL-12), which are cytotoxic. The M2 activation is induced by Th2 cytokines (IL-4, IL-13), and is characterized by anti-inflammatory responses and tissue repair. Both age [151] and mitochondrial dysfunction [85] impair IL-4 mediated alternate activation and produce a failure to elevate growth factor levels in microglia. With aging, the brain environment develops a more pro-inflammatory profile and aging microglia may be a crucial component of this process. Microglial priming, which occurs with age, is associated with increased expression of inflammatory cytokines and an enhanced activated morphological profile (reviewed in [198]). Oxidative stress and free radical accumulation with age is thought to increase inflammasome activity, and deletion of NLRP3 inflammasome in cells of myeloid origin reduces the inflammatory profile of microglia in a region specific manner [290].

4.2.7 Peripheral Immune Cells

One of the consequences of blood brain barrier dysregulation after stroke, coupled with elevated expression of specific receptors and adhesion factors by endothelial cells and astrocytes, is the influx of peripheral immune cells into the brain. Consequently, leukocytes including T cells, B cells, neutrophils and monocytes (macrophages) are mobilized to the ischemic region [81, 99], contributing to secondary inflammation and accelerating cell death [173]. The significant loss of splenic weight after transient [201] and permanent [274] ischemia suggests that immune cells stored in the spleen cells are mobilized into circulation after injury and recruited to the brain [241]. Accordingly, splenectomy [3] or treatments that maintain spleen mass such as infusion with human umbilical cord blood cells [274] reduced infarct volume. The latter treatment also reduced ischemia induced elevation of inflammatory cytokines in the spleen [273], suggesting that the spleen may be a good target for stroke therapy.

More recent work has shown that the type of immune cells recruited to the brain may also affect stroke outcomes. Macrophages are immunologically difficult to distinguish from microglia and the reduction of this cell type in the brain after splenectomy indicates that macrophages are mobilized to the ischemic brain [3]. Growing evidence indicates that T cell recruitment contributes to increased stroke pathology [108] and consistent with this evidence, T cell knock out mice have lower infarct volumes [288]. Additionally, specific cohorts of T cells have been shown to have cell protective or cell toxic effects. Thus, the gammadelta IL-17 producing cells (Th17) have been implicated in increased cell death [246], while regulatory T cells (Treg) and IL-10-producing Breg cells are thought to provide cell protection [35, 161]. Irrespective of age, young, middle-aged and aging mice subject to ischemia improved after treatment with recombinant T cell receptor ligand (RTL), which causes T cells to become non-pathogenic [75, 299]. This study supports the idea that different neuroprotective mechanisms may be activated with age. In this context, it is worth noting that while splenectomy is neuroprotective in males, it does not improve infarct volume or neuroinflammation in females, so it remains to be determined whether this surgical process would be equally effective in aging females [74].

The use of older animals in stroke research is particularly critical in light of the evidence that immune and other somatic responses will shape the ischemic response. Franceschi and colleagues have proposed the term inflamm-aging in recognition of the impact of aging on the immune system. Specifically, this denotes the upregulation of the innate immune response in the elderly, as well as the persistent low grade chronic inflammatory state seen in this group [88]. Not only does this condition underlie many neurologic diseases including Alzheimer's and stroke, but also contributes to a paradoxical condition of immunodeficiency, whereby hyperexcitability of the immune system contributes to immune-suppression and lymphopenia that is often seen in the elderly.

4.3 Global Gene Regulatory Mechanisms in Aging

The large number of gene families and disparate cell types that are affected during aging and ischemia suggest the involvement of global regulatory mechanisms. Aged animals not only exhibit altered levels of specific genes such as inflammatory genes following ischemia as compared to young animals [221], but also display unique patterns of gene expression, with adult (3–4 months old) male rats upregulating genes involved in oxidative stress and aged (19–20 months old) rats displaying increased expression of pro-apoptotic and phagocytosis-promoting genes [39]. During the last decade, significant research has focused on molecular processes capable of exerting widespread effects on the genome. These include both small (microRNA) non-coding RNA species that act as translational repressors, long non-coding RNA (lnc) that can act as transcriptional enhancers or repressors, and chemical modification of the genome.

4.3.1 Non-coding RNA

MicroRNAs (miRNAs) are 18–25 nucleotide-long, non-coding RNA molecules that are important regulators of mRNA transcript stability [65] and mRNA translation [5]. MicroRNAs can occur within exons, introns and polycistronic clusters in the genome [153].

Though relatively few in number, miRNAs are predicted to control a large proportion of the tissue- and cell-specific transcriptome [142, 162] regulating important biological processes including mitosis, tissue-specific cellular differentiation, and cell death [55], maintaining the pluripotent state of embryonic stem cells [120], delaying neuronal maturation [143], or promoting neuronal differentiation [54]. Montano and Long (2010) [188] have proposed that RNA surveillance by regulatory molecules such as miRNA influence life span and longevity. Clusters of miRNA increase or decrease with aging, although these alterations are not always predictive of either benign or maladaptive aging. For example, miR-1 declines with age (day 15) in *C. elegans* and its disappearance has been correlated with muscle aging [124]. However, only 7 % of *C. elegans* survive to day 15, which suggests alternatively that declining miR-1 may be a compensatory process that promotes survival.

Specific pathogenic processes that contribute to stroke are associated with miRNA species, such as miR33 with hyperlipidemia, miR155 with hypertension, miR21 and −126 with atherosclerosis and miR222 with plaque rupture (reviewed in [228]). MiRNA profiles are altered with stroke both in humans [163] and in experimental [67, 126] stroke models. MiR17 is reported to be significantly elevated in acute stroke patients as compared to controls (patients with vascular risk but no stroke) [136], while plasma levels of miR-210 are lowered in ischemic stroke patients as compared to controls, although stroke outcomes are better in patients where miR-210 levels are higher [292].

A recent study evaluated the association of microRNA polymorphisms with the risk of ischemic stroke in a Chinese population [123]. miR146a/rs2910164 C/G genotypes were significantly associated with increased risk of ischemic stroke, and the association was more pronounced in subjects over 60 years old, females, non-drinkers, and subjects without hypertension or diabetes mellitus. In animal models, miR15a has been specifically associated with endothelial cell loss and blood brain barrier breakdown in stroke [289]. MicroRNA manipulation has also been shown to mediate neuroprotection in stroke models. Specifically, antagomirs to Let7f [242] miR200c [257] mir29c [211] and mir181 [283] have been shown to reduce ischemic injury and improve behavioral function. High levels of miR1 are associated with oxidative stress [277] and anti-miR1 antagomirs reduce cortical infarction in experimental stroke [242]. IGF-1, which is neuroprotective for stroke in middle age females, regulates several microRNA that are implicated in barrier function [20].

Long Non Coding (lnc) RNA on the other hand are >200 nucleotides and can extend to several kilobases in length. LncRNA can enhance or repress transcription by epigenetic silencing or by serving as scaffolds for large protein complexes [229]. At the present time, very little is known about the involvement of lncRNA in stroke as compared to microRNA. In two recent studies, an experimental stroke model identified a large cohort of stroke-responsive lncRNA [68] and a subset of these stroke-responsive lncRNA were found to associate with chromatin modifying proteins including Sin3A and coREST, suggesting a mechanism by which non coding RNA can regulate the genome post-ischemia [69].

Little is known about non-coding RNA regulation following ischemia in aging populations, and this represents an important future direction for understanding stroke pathophysiology. A recent study comparing adult and middle-aged male and female rats showed that a small cohort of circulating miRNAs discriminate groups with small infarct volumes (adult females) versus groups with large infarcts (middle-aged females, young males) [245]. Such analyses may be useful in uncovering new mechanistic targets for understanding the impact of age in stroke outcomes.

4.3.2 Epigenenomic Modifications

Heritable, but potentially reversible, modifications to the genome represent the most fundamental aspect of gene regulation aside from the DNA sequence [140]. Epigenetic modifications can be classified into two categories: those that create an active chromatin state and positively regulate gene expression, or those that create a repressive chromatin state and negatively regulate gene expression [84]. Active/inactive states of the genome are regulated by chemical modifications of nucleotides and/or post-translational modification of histones in conjunction with chromatin compaction (euchromatin or heterochromatin) [49, 214]. Although as many as 60 covalent modifications of DNA and histones have been identified to date [140] most studies have focused on the methylation of cytosine located 5′ to guanine (CpG methylation), and the methylation of lysine residues within the N-terminal tails of histone H3. DNA methylation is typically associated with repressed chromatin at promoters and regulatory elements, while histone H3 methylation of lysines 4, 36 and 79 (H3K4, H3K36, and H3K79) are implicated in activation of transcription, while methylation of lysines 9 and 27 of histone H3 (H3K9 and H3K27) are implicated in transcriptional repression [140]. Similarly, histone acetylation is also associated with unwinding of chromatin.

4.3.3 Histone Acetylation and Aging

Histone acetylation typically results in the unwinding of chromatin, which facilitates gene expression, while histone deacetylation causes chromatin compaction. With age, changes in the activity and cellular localization of histone deacetylases (HDACs) can alter the careful balance of histone acetyltransferases (HATs) and HDACs required for maintaining histone acetylation [22, 73]. Decreased acetylation levels have been reported for animal models of neurodegenerative diseases [236]. Age-dependent changes in synaptic plasticity [291] and memory impairment [214] were also associated with lowered histone acetylation.

4.3.4 DNA Methylation in Aging

Both hypo- and hypermethylation of DNA have been reported in aging. One of the earliest studies linking epigenomic changes and aging reported that spawning fish have global decreases in DNA methylation with age [31]. Global hypomethylation was reported in human embryonic lung fibroblasts resulting from peroxide-induced senescence [295], and conversely, experimentally-induced hypomethylation reduced the lifespan of MRC-5 cells by 25 % [82]. A pervasive loss of methylation at Alu repetitive elements was also seen in a large scale study with humans ranging in age from 55 to 92 years [37]. In age-related dementias such as AD, overexpression of the amyloid precursor protein is associated with hypomethylation of the gene as well as APP promoter demethylation in the cortex [265, 280]. The mechanism underlying DNA hypomethylation is not well understood. One obvious mechanism is via altered expression of demethylases, but methylation can also be reduced due to insufficient methyl donors, or essential nutrients involved in metabolism of methyl groups such as folic acid, Vitamin B12 and choline, suggesting that nutritional deficiencies in aging may impact the epigenome. In the brain, increased DNA methylation but decreased histone methylation has been reported with aging [263].

In view of the crucial role that astrocytes play in the ischemic brain, age-related epigenomic modification in this cell type may critically affect stroke outcomes. Astrocytes isolated from adult mice showed higher expression of genes involved in hemoglobin synthesis and neuronal differentiation than aging astrocytes, while aged astrocytes showed higher expression of genes implicated in zinc ion binding and an increased inflammatory phenotype indicating that normal aging alters gene expression profiles in astrocytes [208]. Astrocytes harvested from the ischemic cortex of young adult female rats had greater H3K4 specific methyltransferase activity as compared to middle-aged females and consistent with this finding, astrocytes from young adult females displayed more H3K4me3 enriched peaks than middle-aged females [50]. H3K4me3 enriched peaks at the mir17-20 cluster and the VEGF gene were further confirmed by measuring elevated mir20 RNA expression and VEGF protein in astrocytes from young adult females relative to middle aged females [50]. The use of these new tools to address cell specific changes in aging and ischemia will be critical for the development of next generation drug therapies, where the focus will be on global regulatory mechanisms.

5 Conclusions

Stroke occurs more often in the elderly and the outcomes are more severe in this group, but it would be incorrect to call age a 'non-modifiable' risk factor. With greater recognition of stroke risk factors during aging, such as hypertension and diabetes, and aggressive management of these diseases and better access to health care among the elderly, the impact of age as a risk factor should be mitigated. Healthier life style choices have extended life span and the concern is that longevity may bring about greater disability. However, a study that measured cumulative risk (exercise, body mass and smoking) found that cumulative disability was reduced in the elderly with healthy habits and disability was postponed and compressed to the last few years of life [275].

Acknowledgments Supported by NS074895 and AG042189. The author thanks Tiffany Heard for her assistance in the preparation of this chapter.

Editor: Roderick Corriveau, National Institute of Neurological Disorders and Stroke (NINDS), NIH.

References

1. Abanto C, Ton TG, Tirschwell DL et al (2013) Predictors of functional outcome among stroke patients in Lima, Peru. J Stroke Cerebrovasc Dis 22:1156–1162
2. Abbott NJ, Rönnbäck L, Hansson E (2006) Astrocyte-endothelial interactions at the blood–brain barrier. Nat Rev Neurosci 7:41–53
3. Ajmo CT Jr, Vernon DO, Collier L, Hall AA, Garbuzova-Davis S, Willing A, Pennypacker KR (2008) The spleen contributes to stroke-induced neurodegeneration. J Neurosci Res 86:2227–2234
4. Alkayed NJ, Harukuni I, Kimes AS, London ED, Traystman RJ, Hurn PD (1998) Gender-linked brain injury in experimental stroke. Stroke 29:159–165; discussion 166
5. Ambros V (2001) MicroRNAs: tiny regulators with great potential. Cell 107(7):823–826

6. Andrawes WF, Bussy C, Belmin J (2005) Prevention of cardiovascular events in elderly people. Drugs Aging 22:859–876
7. Arai K, Lok J, Guo S, Hayakawa K, Xing C, Lo EH (2011) Cellular mechanisms of neurovascular damage and repair after stroke. J Child Neurol 26:1193–1198
8. Arana A, Varas C, Gonzalez-Perez A, Gutierrez L, Bjerrum L, Garcia Rodriguez LA (2006) Hormone therapy and cerebrovascular events: a population-based nested case–control study. Menopause 13:730–736
9. Ardelt AA, Anjum N, Rajneesh KF, Kulesza P, Koehler RC (2007) Estradiol augments peri-infarct cerebral vascular density in experimental stroke. Exp Neurol 206:95–100
10. Ardelt AA, McCullough LD, Korach KS, Wang MM, Munzenmaier DH, Hurn PD (2005) Estradiol regulates angiopoietin-1 mRNA expression through estrogen receptor-alpha in a rodent experimental stroke model. Stroke 36:337–341
11. Arenillas JF, Sobrino T, Castillo J, Dávalos A (2007) The role of angiogenesis in damage and recovery from ischemic stroke. Curr Treat Options Cardiovasc Med 9:205–212
12. Arnold AC, Gallagher PE, Diz DI (2013) Brain renin-angiotensin system in the nexus of hypertension and aging. Hypertens Res 36:5–13
13. Aronow WS, Ahn C, Gutstein H (1996) Risk factors for new atherothrombotic brain infarction in 664 older men and 1,488 older women. Am J Cardiol 77:1381–1383
14. Aronow WS (2013) Hypertension-related stroke prevention in the elderly. Curr Hypertens Rep 15:582–589
15. Arsava EM, Vural A, Akpinar E, Gocmen R, Akcalar S, Oguz KK, Topcuoglu MA (2014) The detrimental effect of aging on leptomeningeal collaterals in ischemic stroke. J Stroke Cerebrovasc Dis 23:421–426
16. Asmar R (2003) Benefits of blood pressure reduction in elderly patients. J Hypertens Suppl 21:25–30
17. Ay H, Koroshetz WJ, Vangel M, Benner T, Melinosky C, Zhu M et al (2005) Conversion of ischemic brain tissue into infarction increases with age. Stroke 36:2632–2636
18. Badan I, Buchhold B, Hamm A, Gratz M, Walker LC, Platt D, Kessler C, Popa-Wagner A (2003) Accelerated glial reactivity to stroke in aged rats correlates with reduced functional recovery. J Cereb Blood Flow Metab 23:845–854
19. Bake S, Friedman JA, Sohrabji F (2009) Reproductive age-related changes in the blood brain barrier: expression of IgG and tight junction proteins. Microvasc Res 78:413–424
20. Bake S, Selvamani A, Cherry J, Sohrabji F (2014) Blood brain barrier and neuroinflammation are critical targets of IGF-1-mediated neuroprotection in stroke for middle-aged female rats. PLoS One 9:e91427
21. Bake S, Sohrabji F (2004) 17beta-estradiol differentially regulates blood–brain barrier permeability in young and aging female rats. Endocrinology 145:5471–5475
22. Baltan S (2012) Histone deacetylase inhibitors preserve function in aging axons. J Neurochem 123(Suppl 2):108–115
23. Bang OY, Lee JS, Lee PH, Lee G (2005) Autologous mesenchymal stem cell transplantation in stroke patients. Ann Neurol 57:874–882
24. Barker WH, Mullooly JP (1997) Stroke in a defined elderly population. Stroke 28:284–290
25. Barnett HJ (2002) Stroke prevention in the elderly. Clin Exp Hypertens 24:563–571
26. Barreto G, Veiga S, Azcoitia I, Garcia-Segura LM, Garcia-Ovejero D (2007) Testosterone decreases reactive astroglia and reactive microglia after brain injury in male rats: role of its metabolites, oestradiol and dihydrotestosterone. Eur J Neurosci 25:3039–3046
27. Barreto G, Santos-Galindo M, Diz-Chaves Y, Pernía O, Carrero P, Azcoitia I, Garcia-Segura LM (2009) Selective estrogen receptor modulators decrease astrogliosis in the injured brain: effects of aging and prolonged depletion of ovarian hormones. Endocrinology 150:5010–5015
28. Bauduceau B, Bourdel-Marchasson I, Brocker P, Taillia H (2005) The brain of the elderly diabetic patient. Diabetes Metab 31:5S92–5S97

29. Beckett N, Nunes M, Bulpitt C (2000) Is it advantageous to lower cholesterol in the elderly hypertensive? Cardiovasc Drugs Ther 14:397–405
30. Benderro GF, Lamanna JC (2011) Hypoxia-induced angiogenesis is delayed in aging mouse brain. Brain Res 1389:50–60
31. Berdyshev GD, Korotaev GK, Boiarskikh GV, Vaniushin BF (1967) Nucleotide composition of DNA and RNA from somatic tissues of humpback and its changes during spawning. Biokhimiia 32:988–993
32. Berkowitz DE, White R, Li D et al (2003) Arginase reciprocally regulates nitric oxide synthase activity and contributes to endothelial dysfunction in aging blood vessels. Circulation 108:2000–2006
33. Berman RF, Goldman H, Altman HJ (1988) Age-related changes in regional cerebral blood flow and behavior in Sprague–Dawley rats. Neurobiol Aging 9:691–696
34. Bernal GM, Peterson DA (2011) Phenotypic and gene expression modification with normal brain aging in GFAP-positive astrocytes and neural stem cells. Aging Cell 10:466–482
35. Bodhankar S, Chen Y, Vandenbark AA, Murphy SJ, Offner H (2014) Treatment of experimental stroke with IL-10-producing B-cells reduces infarct size and peripheral and CNS inflammation in wild-type B-cell-sufficient mice. Metab Brain Dis 29:59–73
36. Boehme AK, Siegler JE, Mullen MT, Albright KC, Lyerly MJ, Monlezun DJ, Jones EM, Tanner R, Gonzales NR, Beasley TM, Grotta JC, Savitz SI, Martin-Schild S (2014) Racial and gender differences in stroke severity, outcomes, and treatment in patients with acute ischemic stroke. J Stroke Cerebrovasc Dis 23:e255–e261
37. Bollati V, Schwartz J, Wright R, Litonjua A, Tarantini L, Suh H, Sparrow D, Vokonas P, Baccarelli A (2009) Decline in genomic DNA methylation through aging in a cohort of elderly subjects. Mech Ageing Dev 130:234–239
38. Buchhold B, Mogoanta L, Suofu Y, Hamm A, Walker L, Kessler C et al (2007) Environmental enrichment improves functional and neuropathological indices following stroke in young and aged rats. Restor Neurol Neurosci 25:467–484
39. Buga AM, Scholz CJ, Kumar S, Herndon JG, Alexandru D, Cojocaru GR, Dandekar T, Popa-Wagner A (2012) Identification of new therapeutic targets by genome-wide analysis of gene expression in the ipsilateral cortex of aged rats after stroke. PLoS One 7, e50985
40. Burns EM, Kruckeberg TW, Gaetano PK (1981) Changes with age in cerebral capillary morphology. Neurobiol Aging 2:283–291
41. Bushnell C, McCullough LD, Awad IA, Chireau MV, Fedder WN, Furie KL, Howard VJ, Lichtman JH, Lisabeth LD, Pina IL, Reeves MJ, Rexrode KM, Saposnik G, Singh V, Towfighi A, Vaccarino V, Walters MR, American Heart Association Stroke Council; Council on Cardiovascular and Stroke Nursing; Council on Clinical Cardiology; Council on Epidemiology and Prevention; Council for High Blood Pressure Research, Guidelines for the prevention of stroke in women: a statement for healthcare professionals from the American Heart Association/American Stroke Association (2014) Guidelines for the prevention of stroke in women: a statement for healthcare professionals from the American Heart Association/ American Stroke Association. Stroke 45:1545–1588
42. Bushnell C (2009) Stroke Hormones and Outcomes in Women (SHOW) study: is the 'healthy-user effect' valid for women after stroke? Womens Health (Lond Engl) 5:485–496
43. Cernadas MR, Sánchez de Miguel L, García-Durán M, González-Fernández F, Millás I, Montón M, Rodrigo J, Rico L, Fernández P, de Frutos T, Rodríguez-Feo JA, Guerra J, Caramelo C, Casado S, López-Farré AJ (1998) Expression of constitutive and inducible nitric oxide synthases in the vascular wall of young and aging rats. Circ Res 83:279–286
44. Chacon MR, Jensen MB, Sattin JA, Zivin JA (2008) Neuroprotection in cerebral ischemia: emphasis on the SAINT trial. Curr Cardiol Rep 10:37–42
45. Chadashvili T, Peterson DA (2006) Cytoarchitecture of fibroblast growth factor receptor 2 (FGFR-2) immunoreactivity in astrocytes of neurogenic and non-neurogenic regions of the young adult and aged rat brain. J Comp Neurol 498:1–15

46. Chang E, O'Donnell ME, Barakat AI (2008) Shear stress and 17beta-estradiol modulate cerebral microvascular endothelial Na-K-Cl cotransporter and Na/H exchanger protein levels. Am J Physiol Cell Physiol 294:C363–C371
47. Chen J, Venkat P, Zacharek A, Chopp M (2014) Neurorestorative therapy for stroke. Front Hum Neurosci 8:382
48. Chen R, Crichton S, McKevitt C, Rudd AG, Sheldenkar A, Wolfe CD (2015) Association between socioeconomic deprivation and functional impairment after stroke: the South London stroke register. Stroke 46:800–805
49. Cheung P, Allis CD, Sassone-Corsi P (2000) Signaling to chromatin through histone modifications. Cell 103:263–271
50. Chisholm N, Henderson M, Selvamani A, Park MJ, Dindot S, Miranda RC, Sohrabji F (2015) Histone methylation patterns in astrocytes are influenced by age following ischemia. Epigenetics 10:142–152
51. Chobanian AV, Bakris GL, Black HR et al (2003) Seventh report of the joint national committee on prevention, detection, evaluation, and treatment of high blood pressure. Hypertension 42:1206–1252
52. Chopp M, Zhang ZG, Jiang Q (2007) Neurogenesis, angiogenesis, and MRI indices of functional recovery from stroke. Stroke 38:827–831
53. Collins R, Peto R, MacMahon S et al (1990) Blood pressure, stroke, and coronary heart disease. Part 2, short-term reductions in blood pressure: overview of randomised drug trials in their epidemiological context. Lancet 335:827
54. Conaco C, Otto S, Han J, Mandel G (2006) Reciprocal actions of REST and a microRNA promote neuronal identity. Proc Natl Acad Sci U S A 103(7):2422–2427
55. Croce CM, Calin GA (2005) miRNAs, cancer, and stem cell division. Cell 122(1):6–7
56. Csiszar A, Podlutsky A, Wolin MS, Losonczy G, Pacher P, Ungvari Z (2009) Oxidative stress and accelerated vascular aging: implications for cigarette smoking. Front Biosci (Landmark Ed) 14:3128–3144
57. Csiszar A, Pacher P, Kaley G, Ungvari Z (2005) Role of oxidative and nitrosative stress, longevity genes and poly(ADP-ribose) polymerase in cardiovascular dysfunction associated with aging. Curr Vasc Pharmacol 3:285–291
58. Csiszar A, Wang M, Lakatta EG, Ungvari Z (2008) Inflammation and endothelial dysfunction during aging: role of NF-kappaB. J Appl Physiol 105:1333–1341
59. Dai DF, Rabinovitch PS (2009) Cardiac aging in mice and humans: the role of mitochondrial oxidative stress. Trends Cardiovasc Med 19:213–220
60. Daskalakis NP, Bagot RC, Parker KJ, Vinkers CH, de Kloet ER (2013) The three-hit concept of vulnerability and resilience: toward understanding adaptation to early-life adversity outcome. Psychoneuroendocrinology 38:1858–1873
61. Day JR, Laping NJ, Lampert-Etchells M, Brown SA, O'Callaghan JP, McNeill TH, Finch CE (1993) Gonadal steroids regulate the expression of glial fibrillary acidic protein in the adult male rat hippocampus. Neuroscience 55:435–443
62. De Butte-Smith M, Nguyen AP, Zukin RS, Etgen AM, Colbourne F (2007) Failure of estradiol to ameliorate global ischemia-induced CA1 sector injury in middle-aged female gerbils. Brain Res 1153:214–220
63. de Lecinana MA, Egido JA, Fernandez C, Martinez-Vila E, Santos S, Morales A, Martinez E, Pareja A, Alvarez-Sabin J, Casado I (2007) Risk of ischemic stroke and lifetime estrogen exposure. Neurology 68:33–38
64. del Zoppo GJ, Milner R (2006) Integrin-matrix interactions in the cerebral microvasculature. Arterioscler Thromb Vasc Biol 26:1966–1975
65. Denli AM, Hannon GJ (2003) RNAi: an ever-growing puzzle. Trends Biochem Sci 28(4):196–201
66. Denti L, Scoditti U, Tonelli C, Saccavini M, Caminiti C, Valcavi R, Benatti M, Ceda GP (2010) The poor outcome of ischemic stroke in very old people: a cohort study of its determinants. J Am Geriatr Soc 58:12–17

67. Dharap A, Bowen K, Place R, Li LC, Vemuganti R (2009) Transient focal ischemia induces extensive temporal changes in rat cerebral microRNAome. J Cereb Blood Flow Metab 29:675–687
68. Dharap A, Nakka VP, Vemuganti R (2012) Effect of focal ischemia on long noncoding RNAs. Stroke 43:2800–2802
69. Dharap A, Pokrzywa C, Vemuganti R (2013) Increased binding of stroke-induced long non-coding RNAs to the transcriptional corepressors Sin3A and coREST. ASN Neuro 5:283–289
70. Diener HC, Lees KR, Lyden P, Grotta J, Davalos A, Davis SM, Shuaib A, Ashwood T, Wasiewski W, Alderfer V, Hårdemark HG, Rodichok L, SAINT I and II Investigators (2008) NXY-059 for the treatment of acute stroke: pooled analysis of the SAINT I and II Trials. Stroke 39:1751–1758
71. Diz DI (2008) Lewis K. Dahl memorial lecture: the Renin-Angiotensin system and aging hypertension. Hypertension 52:37–43
72. Donato AJ, Gano LB, Eskurza I, Silver AE, Gates PE, Jablonski K, Seals DR (2009) Vascular endothelial dysfunction with aging: endothelin-1 and endothelial nitric oxide synthase. Am J Physiol Heart Circ Physiol 297:H425–H432
73. Dos Santos Sant'Anna G, Rostirola Elsner V, Moyses F, Reck Cechinel L, Agustini Lovatel G, Rodrigues Siqueira I (2013) Histone deacetylase activity is altered in brain areas from aged rats. Neurosci Lett 556:152–154
74. Dotson AL, Wang J, Saugstad J, Murphy SJ, Offner H (2015) Splenectomy reduces infarct volume and neuroinflammation in male but not female mice in experimental stroke. J Neuroimmunol 278:289–298
75. Dotson AL, Zhu W, Libal N, Alkayed NJ, Offner H (2014) Different immunological mechanisms govern protection from experimental stroke in young and older mice with recombinant TCR ligand therapy. Front Cell Neurosci 8:284
76. Dubal DB, Wise PM (2001) Neuroprotective effects of estradiol in middle-aged female rats. Endocrinology 142:43–48
77. Dubal D, Kashon M, Pettigrew L, Ren J, Finklestein S, Rau S, Wise P (1998) Estradiol protects against ischemic injury. J Cereb Blood Flow Metab 18:1253–1258
78. Duenas M, Luquin S, Chowen JA, Torres-Aleman I, Naftolin F, Garcia-Segura LM (1994) Gonadal hormone regulation of insulin-like growth factor-I-like immunoreactivity in hypothalamic astroglia of developing and adult rats. Neuroendocrinology 59: 528–538
79. Egashira K, Inou T, Hirooka Y, Kai H, Sugimachi M, Suzuki S, Kuga T, Urabe Y, Takeshita A (1993) Effects of age on endothelium-dependent vasodilation of resistance coronary artery by acetylcholine in humans. Circulation 88:77–81
80. El Assar M, Angulo J, Rodríguez-Mañas L (2013) Oxidative stress and vascular inflammation in aging. Free Radic Biol Med 65:380–401
81. Emsley HC, Smith CJ, Gavin CM, Georgiou RF, Vail A, Barberan EM, Hallenbeck JM, del Zoppo GJ, Rothwell NJ, Tyrrell PJ, Hopkins SJ (2003) An early and sustained peripheral inflammatory response in acute ischaemic stroke: relationships with infection and atherosclerosis. J Neuroimmunol 139:93–101
82. Fairweather DS, Fox M, Margison GP (1987) The in vitro lifespan of MRC-5 cells is shortened by 5-azacytidine-induced demethylation. Exp Cell Res 168(1):153–159
83. Farrall AJ, Wardlaw JM (2009) Blood–brain barrier: ageing and microvascular disease – systematic review and meta-analysis. Neurobiol Aging 30:337–352
84. Feinberg AP (2007) Phenotypic plasticity and the epigenetics of human disease. Nature 447(7143):433–440
85. Ferger AI, Campanelli L, Reimer V, Muth KN, Merdian I, Ludolph AC, Witting A (2010) Effects of mitochondrial dysfunction on the immunological properties of microglia. J Neuroinflammation 7:45

86. Fex Svenningsen A, Wicher G, Lundqvist J, Pettersson H, Corell M, Norlin M (2011) Effects on DHEA levels by estrogen in rat astrocytes and CNS co-cultures via the regulation of CYP7B1-mediated metabolism. Neurochem Int 58:620–624
87. Fisher M, Feuerstein G, Howells DW, Hurn PD, Kent TA, Savitz SI, Lo EH, STAIR Group (2009) Update of the stroke therapy academic industry roundtable preclinical recommendations. Stroke 40:2244–2250
88. Franceschi C, Bonafè M, Valensin S, Olivieri F, De Luca M, Ottavini E, De Benedictis G (2000) Inflamm-aging. An evolutionary perspective on immunosenescence. Ann N Y Acad Sci 908:244–254
89. Franklin SS, Gustin W, Wong ND et al (1997) Hemodynamic patterns of age-related changes in blood pressure. The Framingham Heart Study. Circulation 96:308–315
90. Franklin SS, Jacobs MJ, Wong ND et al (2001) Predominance of systolic hypertension among middle-aged and elderly US hypertensives. Hypertension 37:869–874
91. Franklin SS (2012) Elderly hypertensives: how are they different? J Clin Hypertens (Greenwich) 14:779–786
92. Galbiati M, Saredi S, Melcangi RC (2003) Steroid hormones and growth factors act in an integrated manner at the levels of hypothalamic astrocytes: a role in the neuroendocrine control of reproduction. Ann N Y Acad Sci 1007:162–168
93. Gammal E, Zuk A (1980) Effect of ethinyl estradiol on endothelial permeability to 125I-labeled albumin in female rats. Exp Mol Pathol 32:91–101
94. Gao P, Shen F, Gabriel RA, Law D, Yang E, Yang GY, Young WL, Su H (2009) Attenuation of brain response to vascular endothelial growth factor-mediated angiogenesis and neurogenesis in aged mice. Stroke 40:3596–3600
95. Garcia-Estrada J, Del Rio JA, Luquin S, Soriano E, Garcia-Segura LM (1993) Gonadal hormones down-regulate reactive gliosis and astrocyte proliferation after a penetrating brain injury. Brain Res 628:271–278
96. Garcia-Estrada J, Luquin S, Fernandez AM, Garcia-Segura LM (1999) Dehydroepiandrosterone, pregnenolone and sex steroids down-regulate reactive astroglia in the male rat brain after a penetrating brain injury. Int J Dev Neurosci 17:145–151
97. Garfinkel S, Hu X, Prudovsky IA, McMahon GA, Kapnik EM, McDowell SD, Maciag T (1996) FGF-1-dependent proliferative and migratory responses are impaired in senescent human umbilical vein endothelial cells and correlate with the inability to signal tyrosine phosphorylation of fibroblast growth factor receptor-1 substrates. J Cell Biol 134:783–791
98. Gatson JW, Simpkins JW, Yi KD, Idris AH, Minei JP, Wigginton JG (2011) Aromatase is increased in astrocytes in the presence of elevated pressure. Endocrinology 152:207–213
99. Gendron A, Teitelbaum J, Cossette C, Nuara S, Dumont M, Geadah D, du Souich P, Kouassi E (2002) Temporal effects of left versus right middle cerebral artery occlusion on spleen lymphocyte subsets and mitogenic response in Wistar rats. Brain Res 955:85–97
100. Gerhard M, Roddy MA, Creager SJ, Creager MA (1996) Aging progressively impairs endothelium-dependent vasodilation in forearm resistance vessels of humans. Hypertension 27:849–853
101. Gerrits PO, de Weerd H, van der Want JJ, Kortekaas R, Luiten PG, Veening JG (2010) Microvascular changes in estrogen-alpha sensitive brainstem structures of aging female hamsters. Neurosci Res 67:267–274
102. Go AS, Mozaffarian D, Roger VL et al (2013) Executive summary: heart disease and stroke statistics–2013 update: a report from the American Heart Association. Circulation 127:143–152
103. Gonzalez-Covarrubias V (2013) Lipidomics in longevity and healthy aging. Biogerontology 14:663–672
104. Gopal DM, Kalogeropoulos AP, Georgiopoulou VV, Smith AL, Bauer DC, Newman AB, Kim L, Bibbins-Domingo K, Tindle H, Harris TB, Tang WW, Kritchevsky SB, Butler J (2012) Cigarette smoking exposure and heart failure risk in older adults: the Health, Aging, and Body Composition Study. Am Heart J 164:236–242

105. Greenberg SM, Vonsattel JP (1997) Diagnosis of cerebral amyloid angiopathy. Sensitivity and specificity of cortical biopsy. Stroke 28:1418–1422
106. Greenberg DA (1998) Angiogenesis and stroke. Drug News Perspect 11:265–270
107. Grodstein F, Manson JE, Stampfer MJ, Rexrode K (2008) Postmenopausal hormone therapy and stroke: role of time since menopause and age at initiation of hormone therapy. Arch Intern Med 168:861–866
108. Gronberg NV, Johansen FF, Kristiansen U, Hasseldam H (2013) Leukocyte infiltration in experimental stroke. J Neuroinflammation 10:115
109. Gursoy-Ozdemir Y, Yemisci M, Dalkara T (2012) Microvascular protection is essential for successful neuroprotection in stroke. J Neurochem 123(Suppl 2):2–11
110. Haas S, Weidner N, Winkler J (2005) Adult stem cell therapy in stroke. Curr Opin Neurol 18:59–64
111. Hajdu MA, McElmurry RT, Heistad DD, Baumbach GL (1993) Effects of aging on cerebral vascular responses to serotonin in rats. Am J Physiol 264:H2136–H2140
112. Hansen LA, Armstrong DM, Terry RD (1987) An immunohistochemical quantification of fibrous astrocytes in the aging human cerebral cortex. Neurobiol Aging 8:1–6
113. Haseloff RF, Blasig IE, Bauer HC, Bauer H (2005) In search of the astrocytic factor(s) modulating blood–brain barrier functions in brain capillary endothelial cells in vitro. Cell Mol Neurobiol 25:25–39
114. Hayakawa K, Mishima K, Nozako M, Hazekawa M, Mishima S, Fujioka M, Orito K, Egashira N, Iwasaki K, Fujiwara M (2008) Delayed treatment with minocycline ameliorates neurologic impairment through activated microglia expressing a high-mobility group box1-inhibiting mechanism. Stroke 39:951–958
115. Hendrix SL, Wassertheil-Smoller S, Johnson KC, Howard BV, Kooperberg C, Rossouw JE, Trevisan M, Aragaki A, Baird AE, Bray PF, Buring JE, Criqui MH, Herrington D, Lynch JK, Rapp SR, Torner J, WHI Investigators (2006) Effects of conjugated equine estrogen on stroke in the Women's Health Initiative. Circulation 113:2425–2434
116. Heuschmann PU, Kolominsky-Rabas PL, Roether J, Misselwitz B, Lowitzsch K, Heidrich J, Hermanek P, Leffmann C, Sitzer M, Biegler M, Buecker-Nott HJ, Berger K, German Stroke Registers Study Group (2004) Predictors of in-hospital mortality in patients with acute ischemic stroke treated with thrombolytic therapy. JAMA 292:1831–1838
117. Hicks AU, Hewlett K, Windle V, Chernenko G, Ploughman M, Jolkkonen J et al (2007) Enriched environment enhances transplanted subventricular zone stem cell migration and functional recovery after stroke. Neuroscience 146:31–40
118. Horie N, Pereira MP, Niizuma K et al (2011) Transplanted stem cell-secreted vascular endothelial growth factor effects poststroke recovery, inflammation, and vascular repair. Stem Cells 29:274–285
119. Horstmann S, Kalb P, Koziol J, Gardner H, Wagner S (2003) Profiles of matrix metalloproteinases, their inhibitors, and laminin in stroke patients: influence of different therapies. Stroke 34:2165–2170
120. Houbaviy HB, Murray MF, Sharp PA (2003) Embryonic stem cell-specific MicroRNAs. Dev Cell 5:351–358
121. Howard G, Toole JF, Frye-Pierson J, Hinshelwood LC (1987) Factors influencing the survival of 451 transient ischemic attack patients. Stroke 18:552–557
122. Huang ES, Laiteerapong N, Liu JY, John PM, Moffet HH, Karter AJ (2014) Rates of complications and mortality in older patients with diabetes mellitus: the diabetes and aging study. JAMA Intern Med 174:251–258
123. Huang S, Zhou S, Zhang Y, Lv Z, Li S, Xie C, Ke Y, Deng P, Geng Y, Zhang Q, Chu X, Yi Z, Zhang Y, Wu T, Cheng J (2015) Association of the genetic polymorphisms in pre-MicroRNAs with risk of ischemic stroke in a Chinese population. PLoS One 10, e0117007
124. Ibáñez-Ventoso C, Yang M, Guo S, Robins H, Padgett RW, Driscoll M (2006) Modulated microRNA expression during adult lifespan in Caenorhabditis elegans. Aging Cell 5:235–246

125. Ingraham JP, Forbes ME, Riddle DR, Sonntag WE (2008) Aging reduces hypoxia-induced microvascular growth in the rodent hippocampus. J Gerontol A Biol Sci Med Sci 63:12–20
126. Jeyaseelan K, Lim KY, Armugam A (2008) MicroRNA expression in the blood and brain of rats subjected to transient focal ischemia by middle cerebral artery occlusion. Stroke 39:959–966
127. Jiang HX, Chen PC, Sobin SS, Giannotta SL (1992) Age related alterations in the response of the pial arterioles to adenosine in the rat. Mech Ageing Dev 65:257–276
128. Jongbloed L (2006) Prediction of function after stroke: a critical review. Stroke 17:765–776
129. Kadlecová P, Andel R, Mikulík R, Handing EP, Pedersen NL (2015) Alcohol consumption at midlife and risk of stroke during 43 years of follow-up: cohort and twin analyses. Stroke 46:627–633
130. Kaiser D, Weise G, Möller K, Scheibe J, Pösel C, Baasch S, Gawlitza M, Lobsien D, Diederich K, Minnerup J, Kranz A, Boltze J, Wagner DC (2014) Spontaneous white matter damage, cognitive decline and neuroinflammation in middle-aged hypertensive rats: an animal model of early-stage cerebral small vessel disease. Acta Neuropathol Commun 2:169
131. Kalyani RR, Egan JM (2013) Diabetes and altered glucose metabolism with aging. Endocrinol Metab Clin North Am 42:333–347
132. Kannel WB (1993) Hypertension in the elderly: epidemiologic appraisal from the Framingham study. Cardiol Elderly 1:359–363
133. Kapral MK, Fang J, Hill MD, Silver F, Richards J, Jaigobin C, Cheung AM (2005) Sex differences in stroke care and outcomes: results from the Registry of the Canadian Stroke Network. Stroke 36:809–814
134. Kaur J, Tuor UI, Zhao Z, Barber PA (2011) Quantitative MRI reveals the elderly ischemic brain is susceptible to increased early blood–brain barrier permeability following tissue plasminogen activator related to claudin 5 and occludin disassembly. J Cereb Blood Flow Metab 31:1874–1885
135. Kelly KM, Kharlamov A, Hentosz TM, Kharlamova EA, Williamson JM, Bertram EH 3rd, Kapur J, Armstrong DM (2001) Photothrombotic brain infarction results in seizure activity in aging Fischer 344 and Sprague Dawley rats. Epilepsy Res 47:189–203
136. Kim JM, Jung KH, Chu K, Lee ST, Ban J, Moon J, Kim M, Lee SK, Roh JK (2015) Atherosclerosis-related circulating MicroRNAs as a predictor of stroke recurrence. Transl Stroke Res 6:191–197
137. Kim JS, Gautam SC, Chopp M, Zaloga C, Jones ML, Ward PA, Welch KM (1995) Expression of monocyte chemoattractant protein-1 and macrophage inflammatory protein-1 after focal cerebral ischemia in the rat. J Neuroimmunol 56:127–134
138. Kitamura Y, Takata K, Inden M, Tsuchiya D, Yanagisawa D, Nakata J, Taniguchi T (2004) Intracerebroventricular injection of microglia protects against focal brain ischemia. J Pharmacol Sci 94:203–206
139. Knoflach M, Matosevic B, Rucker M et al (2012) Functional recovery after ischemic stroke – a matter of age: data from the Austrian Stroke Unit Registry. Neurology 78:279–285
140. Kouzarides T (2007) Chromatin modifications and their function. Cell 128(4):693–705
141. Krause DN, Duckles SP, Pelligrino DA (2006) Influence of sex steroid hormones on cerebrovascular function. J Appl Physiol 101:1252–1261
142. Krek A, Grun D, Poy MN, Wolf R, Rosenberg L, Epstein EJ, MacMenamin P, da Piedade I, Gunsalus KC, Stoffel M et al (2005) Combinatorial microRNA target predictions. Nat Genet 37:495–500
143. Krichevsky AM, King KS, Donahue CP, Khrapko K, Kosik KS (2003) A microRNA array reveals extensive regulation of microRNAs during brain development. RNA 9:1274–1281
144. Kulik T, Kusano Y, Aronhime S, Sandler AL, Winn HR (2008) Regulation of cerebral vasculature in normal and ischemic brain. Neuropharmacology 55:281–288
145. Lai SM, Duncan PW, Dew P, Keighley J (2005) Sex differences in stroke recovery. Prev Chronic Dis 2:A13

146. Lalancette-Hébert M, Gowing G, Simard A, Weng YC, Kriz J (2007) Selective ablation of proliferating microglial cells exacerbates ischemic injury in the brain. J Neurosci 27:2596–2605
147. Lampl Y, Boaz M, Gilad R, Lorberboym M, Dabby R, Rapoport A, Anca-Hershkowitz M, Sadeh M (2007) Minocycline treatment in acute stroke: an open-label, evaluator-blinded study. Neurology 69:1404–1410
148. Lapchak PA, Chapman DF, Zivin JA (2000) Metalloproteinase inhibition reduces thrombolytic (tissue plasminogen activator)-induced hemorrhage after thromboembolic stroke. Stroke 31:3034–3040
149. Larrue V, von Kummer R, del Zoppo G, Bluhmki E (1997) Hemorrhagic transformation in acute ischemic stroke. Potential contributing factors in the European Cooperative Acute Stroke Study. Stroke 28:957–960
150. Lau LT, Yu AC (2001) Astrocytes produce and release interleukin-1, interleukin-6, tumor necrosis factor alpha and interferon-gamma following traumatic and metabolic injury. J Neurotrauma 18:351–359
151. Lee DC, Ruiz CR, Lebson L, Selenica ML, Rizer J, Hunt JB Jr, Rojiani R, Reid P, Kammath S, Nash K, Dickey CA, Gordon M, Morgan D (2013) Aging enhances classical activation but mitigates alternative activation in the central nervous system. Neurobiol Aging 34:1610–1620
152. Lee JT, Liu HL, Yang JT, Yang ST, Lin JR, Lee TH (2014) Longitudinal MR imaging study in the prediction of ischemic susceptibility after cerebral hypoperfusion in rats: influence of aging and hypertension. Neuroscience 257:31–40
153. Lee Y, Kim M, Han J, Yeom KH, Lee S, Baek SH, Kim VN (2004) MicroRNA genes are transcribed by RNA polymerase II. Embo J 23(20):4051–4060
154. Leon RL, Li X, Huber JD, Rosen CL (2012) Worsened outcome from middle cerebral artery occlusion in aged rats receiving 17beta-estradiol. Endocrinology 153:3386–3393
155. Lewis DK, Thomas KT, Selvamani A, Sohrabji F (2012) Age-related severity of focal ischemia in female rats is associated with impaired astrocyte function. Neurobiol Aging 33:1123.e1–1123.e16
156. Lewis DK, Johnson AB, Stoehlgren S, Harms A, Sohrabji F (2008) Effects of estrogen receptor agonists on regulation of the inflammatory response in astrocytes from young adult and middle-aged female rats. J Neuroimmunol 195:47–59
157. Li C, Engstrom G, Hedblad B, Berglund G, Janzon L (2006) Risk of stroke and hormone replacement therapy. A prospective cohort study. Maturitas 54:11–18
158. Li Y, Chen J, Chopp M (2001) Adult bone marrow transplantation after stroke in adult rats. Cell Transplant 10:31–40
159. Li Y, Chen J, Wang L, Lu M, Chopp M (2001) Treatment of stroke in rat with intracarotid administration of marrow stromal cells. Neurology 56:1666–1672
160. Liao S, Chen W, Kuo J, Chen C (2001) Association of serum estrogen level and ischemic neuroprotection in female rats. Neurosci Lett 297:159–162
161. Liesz A, Suri-Payer E, Veltkamp C, Doerr H, Sommer C, Rivest S, Giese T, Veltkamp R (2009) Regulatory T cells are key cerebroprotective immunomodulators in acute experimental stroke. Nat Med 15:192–199
162. Lim LP, Lau NC, Garrett-Engele P, Grimson A, Schelter JM, Castle J, Bartel DP, Linsley PS, Johnson JM (2005) Microarray analysis shows that some microRNAs downregulate large numbers of target mRNAs. Nature 433(7027):769–773
163. Liu DZ, Tian Y, Ander BP et al (2010) Brain and blood microRNA expression profiling of ischemic stroke, intracerebral hemorrhage, and kainate seizures. J Cereb Blood Flow Metab 30:92–101
164. Liu F, Yuan R, Benashski SE, McCullough LD (2009) Changes in experimental stroke outcome across the life span. J Cereb Blood Flow Metab 29:792–802

165. Liu Z, Fan Y, Won SJ, Neumann M, Hu D, Zhou L, Weinstein PR, Liu J (2007) Chronic treatment with minocycline preserves adult new neurons and reduces functional impairment after focal cerebral ischemia. Stroke 38:146–152
166. Liu F, McCullough LD (2011) Middle cerebral artery occlusion model in rodents: methods and potential pitfalls. J Biomed Biotechnol 2011:464701
167. Liu R, Wang X, Liu Q, Yang SH, Simpkins JW (2007) Dose dependence and therapeutic window for the neuroprotective effects of 17beta-estradiol when administered after cerebral ischemia. Neurosci Lett 415:237–241
168. Lively S, Moxon-Emre I, Schlichter LC (2011) SC1/hevin and reactive gliosis after transient ischemic stroke in young and aged rats. J Neuropathol Exp Neurol 70:913–929
169. Lloyd-Jones D, Adams RJ, Brown TM et al (2010) Heart disease and stroke statistics–2010 update: a report from the American Heart Association. Circulation 121:e46–e215
170. Lokkegaard E, Jovanovic Z, Heitmann BL, Keiding N, Ottesen B, Hundrup YA, Obel EB, Pedersen AT (2003) Increased risk of stroke in hypertensive women using hormone therapy: analyses based on the Danish Nurse Study. Arch Neurol 60:1379–1384
171. Lopez-Lopez C, Dietrich MO, Metzger F, Loetscher H, Torres-Aleman I (2007) Disturbed cross talk between insulin-like growth factor I and AMP-activated protein kinase as a possible cause of vascular dysfunction in the amyloid precursor protein/presenilin 2 mouse model of Alzheimer's disease. J Neurosci 27:824–831
172. Lu B, Yokoyama M, Dreyfus CF, Black IB (1991) NGF gene expression in actively growing brain glia. J Neurosci 11:318–326
173. Lucas SM, Rothwell NJ, Gibson RM (2006) The role of inflammation in CNS injury and disease. Br J Pharmacol 147(Suppl 1):S232–S240
174. Macas J, Nern C, Plate KH, Momma S (2006) Increased generation of neuronal progenitors after ischemic injury in the aged adult human forebrain. J Neurosci 26:13114–13119
175. Madsen TE, Khoury JC, Alwell KA, Moomaw CJ, Kissela BM, De Los Rios La Rosa F, Woo D, Adeoye O, Flaherty ML, Khatri P, Ferioli S, Kleindorfer D (2015) Analysis of tissue plasminogen activator eligibility by sex in the greater cincinnati/northern kentucky stroke study. Stroke 46:717–721
176. Maggio M, Ceda GP, Lauretani F, Bandinelli S, Ruggiero C, Guralnik JM, Metter EJ, Ling SM, Paolisso G, Valenti G, Cappola AR, Ferrucci L (2009) Relationship between higher estradiol levels and 9-year mortality in older women: the Invecchiare in Chianti study. J Am Geriatr Soc 57:1810–1815
177. Manson JE, Allison MA, Rossouw JE, Carr JJ, Langer RD, Hsia J, Kuller LH, Cochrane BB, Hunt JR, Ludlam SE, Pettinger MB, Gass M, Margolis KL, Nathan L, Ockene JK, Prentice RL, Robbins J, Stefanick ML (2007) Estrogen therapy and coronary-artery calcification. N Engl J Med 356:2591–2602
178. Manwani B, Liu F, Xu Y, Persky R, Li J, McCullough LD (2011) Functional recovery in aging mice after experimental stroke. Brain Behav Immun 25:1689–1700
179. Marzilli M (2010) Pleiotropic effects of statins: evidence for benefits beyond LDL-cholesterol lowering. Am J Cardiovasc Drugs 10(Suppl 1):3–9
180. Mateen FJ, Nasser M, Spencer BR, Freeman WD, Shuaib A, Demaerschalk BM, Wijdicks EF (2009) Outcomes of intravenous tissue plasminogen activator for acute ischemic stroke in patients aged 90 years or older. Mayo Clin Proc 84:334–338
181. McCullough LD, Alkayed NJ, Traystman RJ, Williams MJ, Hurn PD (2001) Postischemic estrogen reduces hypoperfusion and secondary ischemia after experimental stroke. Stroke 32:796–802
182. McEwen BS (1998) Stress, adaptation, and disease. Allostasis and allostatic load. Ann N Y Acad Sci 840:33–44
183. McKinley MJ, Albiston AL, Allen AM, Mathai ML, May CN, McAllen RM, Oldfield BJ, Mendelsohn FA, Chai SY (2003) The brain renin-angiotensin system: location and physiological roles. Int J Biochem Cell Biol 35:901–918

184. Miao J, Shen LH, Tang YH, Wang YT, Tao MX, Jin KL, Zhao YJ, Yang GY (2013) Overexpression of adiponectin improves neurobehavioral outcomes after focal cerebral ischemia in aged mice. CNS Neurosci Ther 19:969–977
185. Minnerup J, Sutherland BA, Buchan AM, Kleinschnitz C (2012) Neuroprotection for stroke: current status and future perspectives. Int J Mol Sci 13:11753–11772
186. Miranda CJ, Braun L, Jiang Y et al (2012) Aging brain microenvironment decreases hippocampal neurogenesis through Wnt-mediated survivin signaling. Aging Cell 11:542–552
187. Montaner J, Molina CA, Monasterio J, Abilleira S, Arenillas JF, Ribo M, Quintana M, Alvarez-Sabin J (2003) Matrix metalloproteinase-9 pretreatment level predicts intracranial hemorrhagic complications after thrombolysis in human stroke. Circulation 107:598–603
188. Montano M, Long K (2010) RNA surveillance-An emerging role for RNA regulatory networks in aging. Ageing Res Rev 10:216–224
189. Mouton PR, Long JM, Lei DL, Howard V, Jucker M, Calhoun ME, Ingram DK (2002) Age and gender effects on microglia and astrocyte numbers in brains of mice. Brain Res 956:30–35
190. Muller WE, Eckert A, Kurz C, Eckert GP, Leuner K (2010) Mitochondrial dysfunction: common final pathway in brain aging and Alzheimer's disease–therapeutic aspects. Mol Neurobiol 41:159–171
191. Murugesan N, Demarest TG, Madri JA, Pachter JS (2012) Brain regional angiogenic potential at the neurovascular unit during normal aging. Neurobiol Aging 33:1004.e1001–1004.e1016
192. Nakayama H, Jorgensen HS, Raaschou HO, Olsen TS (1994) The influence of age on stroke outcome. The Copenhagen Stroke Study. Stroke 25:808–813
193. National Center for Health Statistics. Health, United States (2011) US Dept of Health and Human Services, Centers for Disease Control and Prevention, National Center for Health Statistics, Hyattsville; 2012. http://www.cdc.gov/nchs/hus/contents2011.htm#031
194. Ndubuizu OI, Tsipis CP, Li A, LaManna JC (2010) Hypoxia-inducible factor-1 (HIF-1)-independent microvascular angiogenesis in the aged rat brain. Brain Res 1366:101–109
195. Nichols NR, Day JR, Laping NJ, Johnson SA, Finch CE (1993) GFAP mRNA increases with age in rat and human brain. Neurobiol Aging 14:421–429
196. Nilsson PM (2014) Hemodynamic aging as the consequence of structural changes associated with early vascular aging (EVA). Aging Dis 5:109–113
197. Noda A, Ohba H, Kakiuchi T, Futatsubashi M, Tsukada H, Nishimura S (2002) Age-related changes in cerebral blood flow and glucose metabolism in conscious rhesus monkeys. Brain Res 936:76–81
198. Norden DM, Muccigrosso MM, and Godbout JP (2015) Microglial priming and enhanced reactivity to secondary insult in aging, and traumatic CNS injury, and neurodegenerative disease. Neuropharmacology 29–41
199. O'Donnell ME, Lam TI, Tran LQ, Foroutan S, Anderson SE (2006) Estradiol reduces activity of the blood–brain barrier Na-K-Cl cotransporter and decreases edema formation in permanent middle cerebral artery occlusion. J Cereb Blood Flow Metab 26:1234–1249
200. Oakley R, Tharakan B (2014) Vascular hyperpermeability and aging. Aging Dis 5:114–125
201. Offner H, Subramanian S, Parker SM, Wang C, Afentoulis ME, Lewis A, Vandenbark AA, Hurn PD (2006) Splenic atrophy in experimental stroke is accompanied by increased regulatory T cells and circulating macrophages. J Immunol 176:6523–6531
202. Ohab JJ, Fleming S, Blesch A, Carmichael ST (2006) A neurovascular niche for neurogenesis after stroke. J Neurosci 26:13007–130016
203. Ohata M, Sundaram U, Fredericks WR, London ED, Rapoport SI (1981) Regional cerebral blood flow during development and ageing of the rat brain. Brain 104:319–332
204. Okamoto M, Inoue K, Iwamura H, Terashima K, Soya H, Asashima M, Kuwabara T (2011) Reduction in paracrine Wnt3 factors during aging causes impaired adult neurogenesis. FASEB J 25:3570–3582

205. Oldendorf WH, Cornford ME, Brown WJ (1977) The large apparent work capability of the blood–brain barrier: a study of the mitochondrial content of capillary endothelial cells in brain and other tissues of the rat. Ann Neurol 1:409–417
206. Olsen TS, Andersen KK (2010) Female survival advantage relates to male inferiority rather than female superiority: a hypothesis based on the impact of age and stroke severity on 1-week to 1-year case fatality in 40,155 men and women. Gend Med 7:284–295
207. Olsen TS, Andersen KK (2013) Stroke in centenarians. Geriatr Gerontol Int 14:84–88
208. Orre M, Kamphuis W, Osborn LM, Melief J, Kooijman L, Huitinga I, Klooster J, Bossers K, Hol EM (2014) Acute isolation and transcriptome characterization of cortical astrocytes and microglia from young and aged mice. Neurobiol Aging 35:1–14
209. Ovbiagele B, Goldstein LB, Higashida RT, Howard VJ, Johnston SC, Khavjou OA et al (2013) Forecasting the future of stroke in the United States: a policy statement from the American Heart Association and American Stroke Association. Stroke 44:2361–2375
210. Palmer TD, Willhoite AR, Gage FH (2000) Vascular niche for adult hippocampal neurogenesis. J Comp Neurol 425:479–494
211. Pandi G, Nakka VP, Dharap A, Roopra A, Vemuganti R (2013) MicroRNA miR-29c down-regulation leading to de-repression of its target DNA methyltransferase 3a promotes ischemic brain damage. PLoS One 8, e58039
212. Pandya RS, Mao L, Zhou H, Zhou S, Zeng J, Popp AJ, Wang X (2011) Central nervous system agents for ischemic stroke: neuroprotection mechanisms. Cent Nerv Syst Agents Med Chem 11:81–97
213. Pawlak J, Brito V, Kuppers E, Beyer C (2005) Regulation of glutamate transporter GLAST and GLT-1 expression in astrocytes by estrogen. Mol. Brain Res 138:1–7
214. Peleg S, Sananbenesi F, Zovoilis A, Burkhardt S, Bahari-Javan S, Agis-Balboa RC, Cota P, Wittnam JL, Gogol-Doering A, Opitz L, Salinas-Riester G, Dettenhofer M, Kang H, Farinelli L, Chen W, Fischer A (2010) Altered histone acetylation is associated with age-dependent memory impairment in mice. Science 328:753–756
215. Petito CK, Morgello S, Felix JC, Lesser ML (1990) The two patterns of reactive astrocytosis in postischemic rat brain. J Cereb Blood Flow Metab 10:850–859
216. Petitti DB, Sidney S, Quesenberry CP Jr, Bernstein A (1998) Ischemic stroke and use of estrogen and estrogen/progestogen as hormone replacement therapy. Stroke 29:23–28
217. Petrea RE, Beiser AS, Seshadri S, Kelly-Hayes M, Kase CS, Wolf PA (2009) Gender differences in stroke incidence and poststroke disability in the Framingham heart study. Stroke 40:1032–1037
218. Pie JE, Baek SY, Kim HP, Ryu SD, Chung WG, Cha YN, Park CS (2002) Age-related decline of inducible nitric oxide synthase gene expression in primary cultured rat hepatocytes. Mol Cells 13:399–406
219. Pinto E (2007) Blood pressure and ageing. Postgrad Med J 83:109–114
220. Platania P, Seminara G, Aronica E, Troost D, Vincenza C, Angela S (2005) 17beta-estradiol rescues spinal motoneurons from AMPA-induced toxicity: a role for glial cells. Neurobiol Dis 20:461–470
221. Popa-Wagner A, Badan I, Walker L, Groppa S, Patrana N, Kessler C (2007) Accelerated infarct development, cytogenesis and apoptosis following transient cerebral ischemia in aged rats. Acta Neuropathol 113:277–293
222. Prentice RL, Langer R, Stefanick ML, Howard BV, Pettinger M, Anderson G, Barad D, Curb JD, Kotchen J, Kuller L, Limacher M, Wactawski-Wende J, Women's Health Initiative Investigators (2005) Combined postmenopausal hormone therapy and cardiovascular disease: toward resolving the discrepancy between observational studies and the Women's Health Initiative clinical trial. Am J Epidemiol 162:404–414
223. Prentice RL, Langer RD, Stefanick ML, Howard BV, Pettinger M, Anderson GL, Barad D, Curb JD, Kotchen J, Kuller L, Limacher M, Wactawski-Wende J, Women's Health Initiative Investigators (2006) Combined analysis of Women's Health Initiative observational and clinical trial data on postmenopausal hormone treatment and cardiovascular disease. Am J Epidemiol 163:589–599

224. Pugh CW, Ratcliffe PJ (2003) Regulation of angiogenesis by hypoxia: role of the HIF system. Nat Med 9:677–684
225. Qureshi AI, Adil MM, Zacharatos H, Suri MF (2013) Factors associated with length of hospitalization in patients admitted with transient ischemic attack in United States. Stroke 44:1601–1605
226. Ridet J, Malhotra S, Privat A, Gage F (1997) Reactive astrocytes: cellular and molecular cues to biological function. Trends Neurosci 20:570–577
227. Riley C, Cope T, Buck C (2004) CNS neurotrophins are biologically active and expressed by multiple cell types. J Mol Histol 35:771–783
228. Rink C, Khanna S (2011) MicroRNA in ischemic stroke etiology and pathology. Physiol Genomics 43:521–528
229. Rinn JL, Chang HY (2012) Genome regulation by long noncoding RNAs. Annu Rev Biochem 81:145–166
230. Risau W, Wolburg H (1990) Development of the blood–brain barrier. Trends Neurosci 13:174–178
231. Rivard A, Fabre JE, Silver M, Chen D, Murohara T, Kearney M, Magner M, Asahara T, Isner JM (1999) Age-dependent impairment of angiogenesis. Circulation 99:111–120
232. Roger VL, Go AS, Lloyd-Jones DM et al (2011) Heart disease and stroke statistics–2011 update: a report from the American Heart Association. Circulation 123:e18–e209
233. Rozovsky I, Finch C, Morgan T (1998) Age-related activation of microglia and astrocytes: in vitro studies show persistent phenotypes of aging, increased proliferation, and resistance to down-regulation. Neurobiol Aging 19:97–103
234. Rusa R, Alkayed NJ, Crain BJ, Traystman RJ, Kimes AS, London ED, Klaus JA, Hurn PD (1999) 17beta-estradiol reduces stroke injury in estrogen-deficient female animals. Stroke 30:1665–1670
235. Saban KL, Mathews HL, DeVon HA, Janusek LW (2014) Epigenetics and social context: implications for disparity in cardiovascular disease. Aging Dis 5:346–355
236. Saha RN, Pahan K (2006) HATs and HDACs in neurodegeneration: a tale of disconcerted acetylation homeostasis. Cell Death Differ 13:539–550
237. Sarti C, Kaarisalo M, Tuomilehto J (2000) The relationship between cholesterol and stroke: implications for antihyperlipidaemic therapy in older patients. Drugs Aging 17:33–51
238. Schoenmaker M, de Craen AJM et al (2006) Evidence of genetic enrichment for exceptional survival using a family approach: the Leiden longevity study. Eur J Hum Genet 14:79–84
239. Schousboe A, Svenneby G, Hertz L (1977) Uptake and metabolism of glutamate in astrocytes cultured from dissociated mouse brain hemispheres. J Neurochem 29:999–1005
240. Schultz SK, O'Leary DS, Boles Ponto LL, Watkins GL, Hichwa RD, Andreasen NC (1999) Age-related changes in regional cerebral blood flow among young to mid-life adults. Neuroreport 10:2493–2496
241. Seifert HA, Hall AA, Chapman CB, Collier LA, Willing AE, Pennypacker KR (2012) A transient decrease in spleen size following stroke corresponds to splenocyte release into systemic circulation. J Neuroimmune Pharmacol 7:1017–1024
242. Selvamani A, Sathyan P, Miranda RC, Sohrabji F (2012) An antagomir to microRNA Let7f promotes neuroprotection in an ischemic stroke model. PLoS One 7(2), e32662
243. Selvamani A, Sohrabji F (2010) Reproductive age modulates the impact of focal ischemia on the forebrain as well as the effects of estrogen treatment in female rats. Neurobiol Aging 31:1618–1628
244. Selvamani A, Sohrabji F (2010) The neurotoxic effects of estrogen on ischemic stroke in older female rats is associated with age-dependent loss of insulin-like growth factor-1. J Neurosci 30:6852–6861
245. Selvamani A, Williams M, Miranda RC, Sohrabji F (2014) Circulating miRNA profiles provide a biomarker for severity of stroke outcomes associated with age and sex in a rat model. Clin Sci 127:77–89
246. Shichita T, Sugiyama Y, Ooboshi H, Sugimori H, Nakagawa R, Takada I, Iwaki T, Okada Y, Iida M, Cua DJ, Iwakura Y, Yoshimura A (2009) Pivotal role of cerebral interleukin-17-pro-

ducing gammadeltaT cells in the delayed phase of ischemic brain injury. Nat Med 15:946–950
247. Shobha N, Sylaja PN, Kapral MK, Fang J, Hill MD, Investigators of the Registry of the Canadian Stroke Network (2010) Differences in stroke outcome based on sex. Neurology 74:767–771
248. Silva GS, Lima FO, Camargo EC, Smith WS, Lev MH, Harris GJ, Halpern EF, Koroshetz W, Furie KL (2010) Gender differences in outcomes after ischemic stroke: role of ischemic lesion volume and intracranial large-artery occlusion. Cerebrovasc Dis 30:470–475
249. Simpkins JW, Rajakumar G, Zhang YQ, Simpkins CE, Greenwald D, Yu CJ, Bodor N, Day AL (1997) Estrogens may reduce mortality and ischemic damage caused by middle cerebral artery occlusion in the female rat. J Neurosurg 87:724–730
250. Smith CJ, Santhanam L, Bruning RS, Stanhewicz A, Berkowitz DE, Holowatz LA (2011) Upregulation of inducible nitric oxide synthase contributes to attenuated cutaneous vasodilation in essential hypertensive humans. Hypertension 58:935–942
251. Sohrabji F, Bake S, Lewis DK (2013) Age-related changes in brain support cells: implications for stroke severity. Neurochem Int 63:291–301
252. Sohrabji F, Selvamani A, Balden R (2013) Revisiting the timing hypothesis: biomarkers that define the therapeutic window of estrogen for stroke. Horm Behav 63:222–230
253. Sohrabji F (2007) Guarding the blood–brain barrier: a role for estrogen in the etiology of neurodegenerative disease. Gene Expr 13:311–319
254. Sohrabji F (2014) Estrogen-IGF-1 interactions in neuroprotection: ischemic stroke as a case study. Front Neuroendocrinol 36:1–14
255. Sonntag WE, Eckman DM, Ingraham J, Riddle DR (2007) Chapter 12. Regulation of cerebrovascular aging. In: Riddle DR (ed) Brain aging: models, methods, and mechanisms. CRC Press, Boca Raton
256. Sonntag WE, Lynch CD, Cooney PT, Hutchins PM (1997) Decreases in cerebral microvasculature with age are associated with the decline in growth hormone and insulin-like growth factor 1. Endocrinology 138:3515–3520
257. Stary CM, Xu L, Sun X, Ouyang YB, White RE, Leong J, Li J, Xiong X, Giffard RG (2015) MicroRNA-200c contributes to injury from transient focal cerebral ischemia by targeting reelin. Stroke 46:551–556
258. Sumii T, Lo EH (2002) Involvement of matrix metalloproteinase in thrombolysis-associated hemorrhagic transformation after embolic focal ischemia in rats. Stroke 33:831–836
259. Sun Y, Jin K, Xie L, Childs J, Mao XO, Logvinova A, Greenberg DA (2003) VEGF-induced neuroprotection, neurogenesis, and angiogenesis after focal cerebral ischemia. J Clin Invest 111:1843–1851
260. Sylaja PN, Cote R, Buchan AM, Hill MD, Canadian Alteplase for Stroke Effectiveness Study Inestigators (2006) Thrombolysis in patients older than 80 years with acute ischaemic stroke: Canadian Alteplase for Stroke Effectiveness Study. J Neurol Neurosurg Psychiatry 77:826–829
261. Tachibana H, Meyer JS, Okayasu H, Shaw TG, Kandula P, Rogers RL (1984) Xenon contrast CT-CBF scanning of the brain differentiates normal age-related changes from multi-infarct dementia and senile dementia of Alzheimer type. J Gerontol 39:415–423
262. Takamiya Y, Kohsaka S, Toya S, Otani M, Tsukada Y (1988) Immunohistochemical studies on the proliferation of reactive astrocytes and the expression of cytoskeletal proteins following brain injury in rats. Brain Res 466:201–210
263. Thakur MK, Kanungo MS (1981) Methylation of chromosomal proteins and DNA of rat brain and its modulation by estradiol and calcium during aging. Exp Gerontol 16:331–336
264. The National Institute of Neurological Disorders and Stroke rt-PA Stroke Study Group (1995) Tissue plasminogen activator for acute ischemic stroke. N Engl J Med 333: 1581–1587
265. Tohgi H, Utsugisawa K, Nagane Y, Yoshimura M, Genda Y, Ukitsu M (1999) Reduction with age in methylcytosine in the promoter region −224 approximately −101 of the amyloid precursor protein gene in autopsy human cortex. Brain Res Mol Brain Res 70:288–292

266. Toung TK, Hurn PD, Traystman RJ, Sieber FE (2000) Estrogen decreases infarct size after temporary focal ischemia in a genetic model of type 1 diabetes mellitus. Stroke 31:2701–2706
267. Towfighi A, Markovic D, Ovbiagele B (2013) Sex differences in revascularization interventions after acute ischemic stroke. J Stroke Cerebrovasc Dis 22:e347–e353
268. Tsuruoka A, Atsumi C, Mizukami H, Imai T, Hagiwara Y, Hasegawa Y (2014) Effects of edaravone, a free radical scavenger, on circulating levels of MMP-9 and hemorrhagic transformation in patients with intravenous thrombolysis using low-dose alteplase. J Stroke Cerebrovasc Dis 23:2894–2899
269. Ungvari Z, Orosz Z, Rivera A, Labinsky N, Xiangmin Z, Olson S, Podlutsky A, Csiszar A (2007) Resveratrol increases vascular oxidative stress resistance. Am J Physiol Heart Circ Physiol 292:H2417–H2424
270. van der Worp HB, de Haan P, Morrema E, Kalkman CJ (2005) Methodological quality of animal studies on neuroprotection in focal cerebral ischaemia. J Neurol 252:1108–1114
271. Vasan RS, Beiser A, Seshadri S et al (2002) Residual lifetime risk for developing hypertension in middle-aged women and men: the Framingham Heart Study. JAMA 287:1003
272. Vaughan DW, Peters A (1974) Neuroglial cells in the cerebral cortex of rats from young adulthood to old age: an electron microscope study. J Neurocytol 3:405–429
273. Vendrame M, Gemma C, de Mesquita D, Collier L, Bickford PC, Sanberg CD, Sanberg PR, Pennypacker KR, Willing AE (2005) Anti-inflammatory effects of human cord blood cells in a rat model of stroke. Stem Cells Dev 14:595–604
274. Vendrame M, Gemma C, Pennypacker KR, Bickford PC, Davis Sanberg C, Sanberg PR, Willing AE (2006) Cord blood rescues stroke-induced changes in splenocyte phenotype and function. Exp Neurol 199:191–200
275. Vita AJ, Terry RB, Hubert HB, Fries JF (1998) Aging, health risks, and cumulative disability. N Engl J Med 338:1035–1041
276. Walsh JG, Muruve DA, Power C (2014) Inflammasomes in the CNS. Nat Rev Neurosci 15:84–97
277. Wang L, Yuan Y, Li J, Ren H, Cai Q, Chen X, Liang H, Shan H, Fu ZD, Gao X, Lv Y, Yang B, Zhang Y (2015) MicroRNA-1 aggravates cardiac oxidative stress by post-transcriptional modification of the antioxidant network. Cell Stress Chaperones 20:411–420
278. Wang X, Tsuji K, Lee SR, Ning M, Furie KL, Buchan AM, Lo EH (2004) Mechanisms of hemorrhagic transformation after tissue plasminogen activator reperfusion therapy for ischemic stroke. Stroke 35:2726–2730
279. Wassertheil-Smoller S, Hendrix SL, Limacher M, Heiss G, Kooperberg C, Baird A, Kotchen T, Curb JD, Black H, Rossouw JE, Aragaki A, Safford M, Stein E, Laowattana S, Mysiw WJ, WHI Investigators (2003) Effect of estrogen plus progestin on stroke in postmenopausal women: the Women's Health Initiative: a randomized trial. JAMA 289:2673–2684
280. West RL, Lee JM, Maroun LE (1995) Hypomethylation of the amyloid precursor protein gene in the brain of an Alzheimer's disease patient. J Mol Neurosci 6:141–146
281. Wilson PW, Kannel WB (2002) Obesity, diabetes, and risk of cardiovascular disease in the elderly. Am J Geriatr Cardiol 11:119–123
282. Wu C, Hu Q, Chen J, Yan F, Li J, Wang L, Mo H, Gu C, Zhang P, Chen G (2013) Inhibiting HIF-1α by 2ME2 ameliorates early brain injury after experimental subarachnoid hemorrhage in rats. Biochem Biophys Res Commun 437:469–474
283. Xu LJ, Ouyang YB, Xiong X, Stary CM, Giffard RG (2015) Post-stroke treatment with miR-181 antagomir reduces injury and improves long-term behavioral recovery in mice after focal cerebral ischemia. Exp Neurol 264:1–7
284. Yager JY, Wright S, Armstrong EA, Jahraus CM, Saucier DM (2006) The influence of aging on recovery following ischemic brain damage. Behav Brain Res 173:171–180
285. Yamamori T, White AR, Mattagajasingh I et al (2005) P66shc regulates endothelial NO production and endothelium-dependent vasorelaxation: implications for age-associated vascular dysfunction. J Mol Cell Cardiol 39:992–995
286. Yang SH, He Z, Wu SS, He YJ, Cutright J, Millard WJ, Day AL, Simpkins JW (2001) 17-beta estradiol can reduce secondary ischemic damage and mortality of subarachnoid hemorrhage. J Cereb Blood Flow Metab 21:174–181

287. Yenari MA, Kauppinen TM, Swanson RA (2010) Microglial activation in stroke: therapeutic targets. Neurotherapeutics 7:378–391
288. Yilmaz G, Arumugam TV, Stokes KY, Granger DN (2006) Role of T lymphocytes and interferon-gamma in ischemic stroke. Circulation 113:2105–2112
289. Yin KJ, Deng Z, Hamblin M, Xiang Y, Huang H, Zhang J, Jiang X, Wang Y, Chen YE (2010) Peroxisome proliferator-activated receptor delta regulation of miR-15a in ischemia-induced cerebral vascular endothelial injury. J Neurosci 30(18):6398–6408
290. Youm YH, Grant RW, McCabe LR, Albarado DC, Nguyen KY, Ravussin A, Pistell P, Newman S, Carter R, Laque A, Münzberg H, Rosen CJ, Ingram DK, Salbaum JM, Dixit VD (2013) Canonical Nlrp3 inflammasome links systemic low-grade inflammation to functional decline in aging. Cell Metab 18:519–532
291. Zeng Y, Tan M, Kohyama J, Sneddon M, Watson JB, Sun YE, Xie CW (2011) Epigenetic enhancement of BDNF signaling rescues synaptic plasticity in aging. J Neurosci 31:17800–17810
292. Zeng L, Liu J, Wang Y, Wang L, Weng S, Tang Y, Zheng C, Cheng Q, Chen S, Yang GY (2011) MicroRNA-210 as a novel blood biomarker in acute cerebral ischemia. Front Biosci (Elite Ed) 3:1265–1272
293. Zhang L, Zhang RL, Wang Y, Zhang C, Zhang ZG, Meng H, Chopp M (2005) Functional recovery in aged and young rats after embolic stroke: treatment with a phosphodiesterase type 5 inhibitor. Stroke 36:847–852
294. Zhang L, Yi L, Chopp M, Kramer BC, Romanko M, Gosiewska A et al (2013) Intravenous administration of human umbilical tissue-derived cells improves neurological function in aged rats after embolic stroke. Cell Transplant 22:1569–1576
295. Zhang W, Ji W, Yang J, Yang L, Chen W, Zhuang Z (2008) Comparison of global DNA methylation profiles in replicative versus premature senescence. Life Sci 83:475–480
296. Zhang ZG, Zhang L, Jiang Q et al (2000) VEGF enhances angiogenesis and promotes blood–brain barrier leakage in the ischemic brain. J Clin Invest 106:829–838
297. Zhang Z, Trautmann K, Artelt M, Burnet M, Schluesener HJ (2006) Bone morphogenetic protein-6 is expressed early by activated astrocytes in lesions of rat traumatic brain injury. Neuroscience 138:47–53
298. Zhou C, Zhang C, Chi S, Xu Y, Teng J, Wang H, Song Y, Zhao R (2009) Effects of marrow stromal cells on activation of microglial cells and production of inflammatory factors induced by lipopolysaccharide. Brain Res 7(1269):23–30
299. Zhu W, Dotson AL, Libal NL, Lapato AS, Bodhankar S, Offner H, Alkayed NJ (2015) Recombinant T-cell receptor ligand RTL1000 limits inflammation and decreases infarct size after experimental ischemic stroke in middle-aged mice. Neuroscience 288:112–119

The Role of Aging in Alzheimer's Disease

Geoffrey A. Kerchner and Tony Wyss-Coray

Contents

G.A. Kerchner
Department of Neurology and Neurological Sciences, Stanford University School of Medicine, Stanford, CA, USA
e-mail: kerchner@stanford.edu

T. Wyss-Coray (✉)
Department of Neurology and Neurological Sciences, Stanford University School of Medicine, Stanford, CA, USA

Center for Tissue Regeneration, Repair and Restoration, Veterans Affairs Palo Alto Health Care System, Palo Alto, CA, USA
e-mail: twc@stanford.edu

F. Sierra, R. Kohanski (eds.), *Advances in Geroscience*,
DOI 10.1007/978-3-319-23246-1_7

1 Introduction

Aging changes the adult brain both structurally and functionally, facilitating and accelerating cognitive impairments and susceptibility to degenerative disorders even in healthy individuals [1, 2]. Indeed, aging remains the single most predominant risk factor for dementia and related neurodegenerative diseases, including Alzheimer's disease (AD). Considering the rate at which the human population is aging, it becomes imperative to identify means by which to maintain cognitive integrity by protecting against, or even counteracting, the effects of aging. Indeed, according to the Department of Health and Human Services, forty million Americans are older than 65 years of age and by 2030 this number will have doubled, accounting for one in five people in the United States. Hence, halting or reversing brain aging by only a small fraction could delay neurodegeneration and dementia, which would have a significant impact on the quality of life in old persons. Here, we will review the clinical distinctions between age-related cognitive decline and AD and explore the role of established hallmarks of aging in the aged brain and in AD.

2 Alzheimer's Disease as a Clinical Entity

AD may be suspected in a patient who exhibits gradual, progressive cognitive decline [3]. The most common symptom associated with AD is loss of recent memory—that is, difficulty forming new memories. People who know the patient well, like a spouse or other family member, may notice that the patient asks the same question more than once, repeats stories or conversations, or has trouble remembering recent events. By contrast, memory for events in the more distant past may be normal. At first, recent memory loss is typically mild, subtle, and in some cases evident only to the patient; as the amnesia gradually worsens over months to a year or so, it eventually catches the attention of family, friends, and clinicians. A clinician may test memory by providing a list of words or other new information, then distract the patient for a few minutes with other tasks or conversation, and finally ask the patient to recall.

Other cognitive symptoms may occur in addition to or instead of recent memory loss. Common symptoms that may emerge either early or late in the course of AD include word finding deficits, difficulty with navigation or decoding a complex visual scene, and trouble with executive function tasks that include planning, organizing, judgment, or problem solving. Motor deficits typically do not occur until late in the disease, except for apraxia, or loss of the ability to carry out complex motor tasks, which may occur earlier. In some cases of AD, language, visuospatial, or executive deficits may dominate in the relative absence of amnesia.

When the clinician discovers cognitive dysfunction that is objectively evident yet not severe enough to interfere with the patient's ability to perform typical, everyday activities, a diagnosis of mild cognitive impairment may be appropriate [4]. When cognitive dysfunction associates with difficulty performing everyday tasks (e.g., driving, balancing a checkbook, keeping track of appointments, etc.), then a diagnosis of dementia may be given. Dementia may be subcategorized as early- versus late-onset, based arbitrarily on whether symptoms emerge before or after age sixty-five. Importantly, mild cognitive impairment and dementia are umbrella terms that are used to describe how much a patient's cognitive dysfunction has impacted everyday life; however, because these terms do not convey the suspected cause of the cognitive dysfunction, they are not intended as stand-alone diagnoses. Patients with amnestic mild cognitive impairment (i.e., in whom episodic memory impairment is present) have a 10–15 % annual risk of progressing to dementia, often due to AD [5].

While there are many possible causes of mild cognitive impairment and dementia, AD accounts for approximately two thirds of cases among older adults. Other common causes, each with a slightly different but nevertheless overlapping spectrum of typical symptoms, include stroke (vascular dementia), Lewy body disease (closely associated with Parkinson's disease), and frontotemporal dementia. Many other possible causes, while less frequent, may be worthy of consideration because some are reversible or at least treatable; these include vitamin deficiencies (e.g., Vitamin B12), hormone imbalance (e.g., hypothyroidism), infections (e.g., HIV, syphilis, or Lyme disease), tumors or other cancer-related conditions, psychiatric disease (e.g., depression), medication side effect or toxic exposure, or inflammatory or autoimmune diseases (e.g., multiple sclerosis). A standard medical evaluation includes a detailed history, physical examination, neurological examination, objective cognitive testing (either a brief bedside test or a comprehensive neuropsychological evaluation that includes tests of memory, language, visuospatial function, and executive function), neuroimaging with either computed tomography (CT) or magnetic resonance imaging (MRI), and blood tests for thyroid hormone function and Vitamin B12 levels, among other tests as indicated by the patient's specific situation. While such tests cannot provide direct evidence for AD, they can help to raise or lower suspicion for other diagnoses.

AD is diagnosed when a patient exhibits typical symptoms of the illness, and no other cause is uncovered during the medical evaluation. Importantly, AD is defined by the presence of two distinct protein accumulations in the brain of a patient with progressive cognitive decline: beta-amyloid, which self-aggregates into extracellular plaques between cortical neurons, and the microtubule-associated protein tau, which self-aggregates into tangles within the neurites and cell bodies of cortical neurons [6]. Classically, because these protein aggregates may be observed directly only during microscopic evaluation of the post-mortem brain, a diagnosis of AD in life may only be suspected. However, newly emerging biomarker tests are changing this situation, by allowing clinicians to test for the presence of beta-amyloid or tau pathology in a living patient. These tests, which do not substitute for the certainty of an autopsy but nevertheless may increase or decrease suspicion for AD, include direct measurements of beta-amyloid and tau concentrations in the cerebrospinal fluid obtained by lumbar puncture [7], as well as positron emission tomography (PET) scan after intravenous administration of a radiolabeled tracer that selectively binds either to beta-amyloid [8] or tau deposits [9] in the brain.

Treatment options for AD are limited. An active lifestyle may reduce the rate of future cognitive decline and reduce dependence upon caregivers; regular aerobic exercise, social and cognitive stimulation, and a healthy diet are important lifestyle components. Four medications are currently approved for the treatment of AD: donepezil, rivastigmine, and galantamine are centrally-acting cholinesterase inhibitors that increase the half-life of acetylcholine at synapses in the brain, modestly boosting alertness and cognitive function for many patients with mild-to-moderate

AD dementia. Memantine is a *N*-methyl-D-aspartate-type (NMDA) glutamate receptor antagonist that exhibits similar modest benefits among patients with moderate-to-severe AD dementia. None of these medications exhibits any robust influence over the pace of disease progression, and none has any direct impact on the underlying molecular pathogenesis. Clinical trials of many other potential therapies are underway.

3 Spectrum of Normal and Abnormal Aging

Cognitive change occurs with aging in the absence of any disease [10]. Beginning gradually in middle age, cognitive processing speed slows, fluid intelligence declines, and episodic memory skills wane. Older adults exhibit changes in sustained attention, working memory, and distractibility compared to their younger counterparts [11]. Cognitive change attributed to age alone, while robust enough to register as slight changes over time on neuropsychological assessments, is never sufficiently severe to impact daily function or to merit a clinical designation of mild cognitive impairment or dementia. Moreover, isolated age-associated cognitive decline is very slow, with changes that are perceptible perhaps decade-to-decade, but not year-to-year [12].

It is not the case that cognitive aging is simply a slow-motion version of AD. Rather, there are distinct processes at play in the two conditions. Indeed, fundamental changes occur naturally throughout the brain over the human life span, from macroscopic loss of brain volume to microscopic reductions in neuron and synapse numbers. Altered gene expression affects neural signaling with age; for instance, NMDA receptors, the calcium-permeable ion channels that play a central role in the induction of synaptic plasticity, exhibit a robust age-dependent change in subunit composition that affects how new experiences sculpt synaptic strength [13, 14].

However, the distinction between pathological and non-pathological cognitive decline is generally evident only in hindsight, based on the retrospectively-observed pace and pattern of cognitive change, and the end result in terms of functional status. Prospectively, it is very difficult to know, among a group of similar individuals experiencing the sorts of typical, age-associated cognitive changes described here, who will go on to experience a benign course, versus those who are actually experiencing the first, subtle symptoms of a cognitive illness like AD. A major challenge for the field is the fact that the molecular changes associated with AD and other neurodegenerative illnesses begin years—likely decades—before symptoms become clinically apparent. When symptoms do appear, they are at first subtle and not easily differentiated from the typical effects of aging. Given the likelihood that newly-emerging drug therapies will be effective only at an early stage of the disease, this symptomatic overlap between typical aging and prodromal AD is problematic, as there is a need to identify patients before the damage is indelible.

Even at a molecular level, distinguishing typical aging from disease is not straightforward. Roughly a third of asymptomatic older adults exhibit signs of beta-amyloid plaque deposition on amyloid PET imaging [15]. Similarly, as many as 60 % of adults around age 60 years exhibit at least the first stages of tau protein deposits in the medial temporal lobe of the brain at autopsy [16], far exceeding the proportion who exhibit actual cognitive decline; newer data suggest that tau pathology may begin in the locus coeruleus, a brainstem nucleus containing noradrenergic neurons, as early as middle age [17]. The detection of molecular pathology in the brain, either at autopsy or with newly emerging molecular diagnostic tests, is not synonymous with the presence of a clinical disease. Certainly the presence of abnormal protein accumulations raises the risk of future cognitive decline, but the magnitude of that risk is not yet definitively established by available research. In other words, a cognitively healthy patient who tests positive on an amyloid PET scan has an unclear fate; she may be on the precipice of imminent cognitive decline or may survive many years with normal cognitive health. Even if we could peer into her brain with a high-powered microscope years before her death, we may still lack confidence about her cognitive fate.

It is worth mentioning that beta-amyloid and tau are not unique in their capacity to self-aggregate in the brains of older adults. Lewy bodies, the intracytoplasmic neuronal accumulations of alpha-synuclein protein associated with Parkinson disease and Lewy body dementia, frequently appear in the brains of cognitively healthy older adults. The same is true for TDP-43, a protein that self-aggregates in some forms of frontotemporal dementia. When are plaques, tangles, Lewy bodies, and TDP-43 deposits simply benign findings in an aging brain, and at what point do they become significant enough to be blamed for cognitive changes? To confuse the picture even more, it is rare to find, at autopsy, isolated evidence of a single neurodegenerative disease; rather, Lewy bodies and TDP-43 deposits frequently appear in the brains of patients with AD, and patients with Lewy body dementia often exhibit plaques and tangles. How do we know where one disease stops and the other starts?

The answers to these questions are not yet clear. Indeed, the very notion of what constitutes normality is open to debate. What is clear is that cognitive changes due to the inherent biology of aging versus superimposed neurodegenerative disease—while distinct processes—are not mutually exclusive, and probably exist on a spectrum in every aging adult.

4 Risk Factors for Alzheimer's Disease

Despite the uncertainties outlined above, there are some known risk factors for AD. These may be categorized by whether or not they are modifiable: Modifiable risk factors for AD, while less important than non-modifiable factors in the overall risk profile, deserve close attention because they may be mitigated through lifestyle choices. They include the presence of vascular risk factors and insulin-resistant diabetes mellitus [18]. Repeated concussions or other forms of head trauma associate with an increased chance of developing AD or other neurodegenerative illnesses later in life [19]. A sedentary lifestyle may increase one's chances of developing AD, whereas regular aerobic exercise is protective [20].

Less exposure to education is an additional modifiable risk factor for AD. Individuals with higher education and perhaps greater cognitive stimulation in midlife appear to be protected from developing symptoms of cognitive decline, even though, at autopsy, they may nevertheless exhibit the neuropathological signs of AD [21]. Such findings have given rise to the concept of cognitive reserve, or the buildup of resistance against the cognitively detrimental effects of damage to the brain. Although the mechanism of cognitive reserve is not known, it is hypothesized that education and ongoing learning in life lead to a greater number of synapses and perhaps even neurons in the brain that can act in a compensatory capacity in the face of a degenerative illness.

Non-modifiable risk factors dominate the overall risk for developing AD. Chronological age stands out as conferring by far the most risk. About a third of individuals aged 85 years and older suffer from AD dementia; many more have the pathological signs of the disease, in the absence of overt symptoms. The vast majority of patients are 75 years or older.

Genetics and family history are also non-modifiable risks. Of the many genes that influence overall risk for late-onset, sporadic AD, the *APOE* genotype has a greater effect than all of the others. The *APOE* gene encodes the apolipoprotein E protein, which plays an important role in cholesterol trafficking; it is expressed in astroglial cells and, to a small extent, in neurons in the brain. There are three *APOE* allelic isoforms: ε3, ε4, and ε2, in decreasing order of population prevalence. Carriers of the ε4 allele exhibit an increased risk of developing AD, such that heterozygotes and homozygotes have a 2–3-fold and 12-fold increased lifetime risk relative to noncarriers [22]. Carriers of the least common ε2 allele have a reduced risk relative to ε3 homozygotes, the most common genotype. Gender also plays a role; female *APOE* ε4 heterozygotes exhibit a greater chance of developing AD than their male counterparts [23]. A history of AD in multiple family members may increase suspicion for an *APOE* ε4 allele or other risk-conferring genetic polymorphisms.

It is important to point out that genetic *polymorphisms* that influence the risk of acquiring late-onset, sporadic AD are very different from genetic *mutations* associated with familial, early-onset AD. Mutations in the genes for amyloid precursor protein (APP), presenilin 1, and presenilin 2 lead to cognitive decline in the third or fourth decade of life, causing AD in an autosomal dominant, completely penetrant fashion. Down syndrome (trisomy 21) associates with early-onset AD, due to overexpression of the APP gene, which happens to localize to chromosome 21. Such genetic cases account for only a tiny fraction of overall AD burden worldwide and are not directly relevant to the vast majority of patients who experience sporadic disease.

While much is known about risk factors for AD, less is known about the predictors of better or worse phenotypes of normal aging in the absence of neurodegenerative disease. As described in the previous section, part of the difficulty with such research is a lack of clarity about the presence or absence of preclinical neurodegenerative disease in a living individual. For instance, whereas the *APOE* ε4 allele is not associated with any difference in baseline cognition among older adults, it does predict a faster rate of decline [24, 25]; however, given the tight association of *APOE* ε4 carrier status with beta-amyloid accumulation, this observation may be explained by the emergence of AD symptoms among carriers rather than an acceleration of AD-independent aspects of cognitive aging. Most likely, many lifestyle factors that associate with a reduced risk for AD, like exercise and education, also predict a more benign course of cognitive aging. Genetics influences the pace of cognitive aging, although the identity of specific genes is only beginning to emerge [25, 26]. As one example, a polymorphism in the gene encoding klotho, a transmembrane β-glucuronidase enzyme, promotes longevity and associates with improved cognition in older adults [27]. The gene for another protein, forkhead box O3, exhibits a polymorphism associated with longevity [28], and its possible association with cognitive aging should be examined.

5 Structural and Functional Changes in the Brain

A proliferation of advanced neuroimaging technologies and analytical techniques has broadened the perspective on how the brain changes in aging and AD. MRI reveals fine structural details of the brain and permits volumetric analysis and detection of atrophy in subjects who undergo repeated, longitudinal imaging. Indeed, perhaps not surprisingly, atrophy is a robust finding in aging adults. Not only do older adults exhibit smaller brains than younger adults, but also year-to-year volumetric decline occurs among cognitively healthy older adults who do not exhibit any concerning signs for preclinical AD [29–31]. Atrophy accelerates in AD, but this is observed only with serial imaging in the same patient and is not captured with a single scan at one point in time.

The hippocampus is a special area of focus for structural imaging, because postmortem pathology points to this structure as an early victim of neurodegeneration and atrophy in AD [32]. Indeed, hippocampal atrophy is a robust finding that differentiates patients with AD from age-matched, cognitively-normal controls [33]. However, the finding emerges only after averaging images from groups of patients and controls; there is so much variability in hippocampal size from individual to individual that hippocampal volume, whether in raw, unadjusted terms or normalized to overall brain volume, is not strongly predictive of diagnosis until rather late into the course of dementia. In other words, a single MRI in one individual is less sensitive to AD than traditional clinical data. The overlap between healthy aging and AD is evident even when using ultra-high field MRI to resolve a very tiny subregion of the hippocampus known to degenerate first in AD [34].

Functional imaging offers another set of perspectives on how the brain changes with age and with the onset of neurodegenerative disease. Whereas structural imaging yields a finding only after enough wholesale loss of neurons, neurites, glial cells, etc. occurs to result in macroscopic shrinkage of a brain region, functional imaging offers the ability to detect changes in patterns of neural activation that could predate involution. When subjects perform episodic memory tasks during functional MRI scanning, complex results emerge out of older adults who are cognitively normal or have mild cognitive impairment or AD dementia: compared to young adults, healthy older adults exhibit very similar patterns of hippocampal activation during memory tasks, although they show differential activation of frontal and parietal lobe areas [35]. In patients with mild cognitive impairment, hippocampal activity during episodic memory tasks actually increases relative to age-matched, cognitively healthy controls; as the clinical severity worsens towards dementia, hippocampal activation during memory tasks decreases below that seen in controls [36]. The mechanism and implications of this apparent hippocampal hyperactivity during prodromal AD are not known, including whether it represents a compensatory strategy or is instead an inherent part of the pathophysiology. Interestingly, network dysfunction, resulting in hippocampal overactivation and subclinical seizures, has been observed in mouse models of AD [37]. Antiepileptic therapy, specifically with levetiracetam, improves network dysfunction and cognitive deficits in the mouse models [38] and is now being explored in humans [39, 40].

High-resolution MRI, sufficient to resolve individual hippocampal subfields, yields additional insights. It is clear from post-mortem tissue examination that the hippocampal CA1 subfield is selectively vulnerable to degeneration in AD, and disproportionate CA1 atrophy is observed using a variety of advanced structural MRI techniques in vivo [34, 41, 42]. Functional imaging reveals a slightly different story: In healthy older adults and patients with mild cognitive impairment, it is the dentate gyrus and CA3 areas (which cannot be differentiated with in vivo imaging because of their close proximity) that appear abnormal, exhibiting hyperactivity in association with impairment of pattern separation, or the ability to discriminate novel stimuli from slightly different, previously-encountered stimuli [43, 44]. This finding of dentate gyrus/CA3 hyperactivity – which may possibly underlie the observation of global hippocampal hyperactivity observed with lower resolution functional MRI – differentiated patients with mild cognitive impairment from normal controls, and among healthy controls, the amount of activity in this region correlated negatively with cognitive performance. Together, these data suggest that dentate gyrus/CA3 hyperactivity, while more apparent in prodromal AD than in healthy aging, is nevertheless a feature in both. This contrasts with some other data, derived in part from animal models, suggesting that normal aging and AD are cleanly distinguished by dentate gyrus versus CA1 dysfunction [45]; however, these findings have not been readily reproduced in humans.

Functional MRI, in addition to showing patterns of regional brain activation while performing a task, also reveals areas of the brain that exhibit coordinated activity when the subject is at rest, in so-called resting-state networks. One resting-state network, the default mode network, includes the medial and dorsolateral parietal association cortex, medial prefrontal cortex, and the medial temporal lobe, and is more active than other networks at rest. The integrity of the default mode network deteriorates in AD [46], and it also deteriorates in healthy older adults at risk of AD because of *APOE* ε4 positivity or the presence of amyloid deposition [47].

In summary, and echoing the theme of earlier sections, brain structure and function clearly changes with aging and with AD, but there is a high degree of overlap between these two states. This overlap may reflect a limitation of available technology or may instead reflect a true overlap in underlying biology.

6 Biological Hallmarks in Brain Aging and Alzheimer's Disease

Given these astonishingly intricate connections between brain aging and AD, how can we possibly untangle and comprehend the extent to which aging is responsible for the development of the disease? In their landmark review, "the hallmarks of aging", López-Otín and colleagues describe nine cellular and molecular, age-dependent changes which have been observed across multiple species, tissues, and cells and can be considered common hallmarks of the aging process [48]. We will use these hallmarks as guides to compare normal brain aging with AD in humans and discuss animal and tissue model studies to support findings and arguments, if appropriate. A few hallmarks for which there is substantial evidence for involvement in AD are discussed in more detail: telomere attrition, epigenetic alterations, loss of proteostasis, stem cell dysfunction, altered intercellular communication and neuroinflammation. To what extent these processes underlie or promote AD is unclear but they provide possible therapeutic targets that might interfere with AD pathogenesis by impinging on the aging process directly. As well, we hope that future studies of these well-defined hallmarks in the aging and AD brain will provide additional insight into the mechanisms underlying AD.

6.1 Genomic Instability and Telomere Attrition

Aging results in the accumulation of damage in nuclear and mitochondrial DNA and interference with DNA repair processes can accelerate aging. Telomeres shorten with cell divisions and restoring telomere length can delay aging. While glial and endothelial cells in the brain have the capacity to divide the vast majority of neurons don't renew and some may live longer than a century. As such, they need to withstand stressors and employ repair mechanisms to maintain proper function. Of particular importance is the preservation of DNA, which can be damaged by reactive oxygen and nitrogen species, replication errors, and other mechanisms. Among other defects this leads to single base damage and single or double strand DNA breaks in neurons [49]. Pathways which increase damage or decrease repair could conceivably contribute to brain aging and AD and there is some evidence in support of this hypothesis [50, 51]. In a landmark study, Yankner and colleagues showed that with age DNA damage accumulates in the brain, specifically in promoter regions of genes involved in synaptic plasticity, vesicular transport, and mitochondrial function, and that levels of these genes were reduced with age [52]. Remarkably, neurons may regulate DNA errors to control plasticity and cellular diversity and to facilitate chromatin remodeling in these highly transcriptionally active cells. Thus, exploratory behavior and neuronal activity in mice led to the formation of transient double stranded DNA breaks. In APP transgenic mice more breaks were observed and these were maybe more permanent and damaging [53].

Telomere attrition contributes to the age-dependent loss of proliferative capacity in some cell types and provides a strong link to cellular senescence and organismal aging. The activity of telomerase, the enzyme that adds telomeric repeats to terminal DNA, decreases with age in mice [54, 55], and mice lacking telomerase show reduced neural stem cell activity in adults but not during development [54, 56]. This decrease in telomerase activity may in part be dependent on p53 as removal of p53 is sufficient to rescue defects in proliferation, self-renewal, and differentiation of neural stem cells in telomerase-deficient mice. Also, physical exercise, which is known to increase adult neurogenesis and cognitive function, increased telomerase activity in neural stem cells [55]. The role of telomeres in AD and AD mouse models is not well understood and controversial [57]; while there appears to be no significant telomere abnormality in AD brains, several studies reported accelerated telomere shortening in leukocytes from patients compared with age-matched controls [58]. However, among 10 studies investigating telomere length in leukocytes from AD patients, half observed a shortening of telomeres while the others found no change [57]. In a recent study of 274 individuals, leukocytes with either shorter or longer telomeres compared with those found in normal subjects were both associated with mild cognitive impairment, a major risk factor for development of AD [59]. Possibly explaining some of these discrepancies between studies or even within studies are observations that leukocyte telomere length, or better, the ratio between telomerase activity and telomere length, are inversely related to hippocampal volume in early aging [60]. In experimental models of AD, APP mice lacking telomerase (G3tert deficient) show reduced amyloidosis accompanied by microglial activation and impaired neurogenesis [61] suggesting independent effects of telomeres on different AD relevant pathways. Consistent with this interpretation, and this may relate to neuroinflammation and changes in intercellular communication, telomeres in microglia may be altered in the aging brain [62]. Microglia are phagocytic cells and the key representative of the immune system in the brain (see section 6.6). They maintain proliferative capacity in adulthood but seem to show signs of senescence in normal and AD brains [63]. Accordingly, mice lacking telomerase showed reduced dendrites and dendrite branching and increased expression of activation markers in microglia [61]. Adding complexity, telomerase (TERT) has recently been found to exert extranuclear functions beyond telomere maintenance, including protection of mitochondria from oxidative stress and prosurvival effects in mature neurons. Human TERT was detected in mature human neurons and while its total level was unchanged in AD brains, the enzyme colocalized with mitochondria in advanced stages of the disease [64]. Intriguingly, in these AD brains, TERT was not detected in neurons with tau pathology and in cultured neurons, TERT protected against the accumulation of tau lesions, supporting the idea that the protein confers a protective effect to neurons [64]. Together, these reports suggest that DNA damage and telomerase activity may have a role in both normal brain aging and AD but the exact mechanisms remain to be defined.

6.2 Epigenetic Alterations

Posttranslational modifications of histones and changes in DNA methylation patterns constitute important aspects of gene regulation and they have increasingly been implicated in aging. Epigenetic regulation of gene expression also has a critical role in memory and learning and may thus assume a particularly important role in the aging of this tissue [65]. Consequently, it is likely that age-related changes in epigenetic processes would contribute to cognitive dysfunction and AD. What's more, epigenetic mechanisms are key to neuronal fate determination and function with repressor element 1-silencing transcription factor (REST) playing a dominant role [66]. Together with HDAC, MECP2 and CoREST, this factor regulates developmental and adult neurogenesis as well as gene activity in mature neurons. Interestingly, as cellular stress increases with age, REST appears to be important in promoting expression of anti-apoptotic, anti-oxidant, and other protective genes including BCL2, SOD1, and FOXO [67]. In AD, REST is depleted from the nucleus and appears in autophagosomes instead; moreover, genetic ablation of REST in mice or the REST homologue SPR-4 in *C. elegans* increases neuronal susceptibility to oxidative stress and results in neurodegeneration and shortened lifespan, respectively. Epigenetic regulation of memory and hippocampal plasticity in particular is altered with age in mice undergoing specific learning tasks. Thus, while baseline hippocampal histone H3 and H4 acetylation as well as levels of HDACs and histone acetyltransferases (HATs) were unchanged with age, old mice failed to acetylate histone H4K12 in response to a learning task [68]. Consistent with this lack of H4K12 acetylation, only few genes were differentially expressed based on genome wide transcript analysis or ChIP sequencing in the aged hippocampus in response to the learning paradigm [68]. To identify genes involved in age-related memory loss, Pavlopoulos and colleagues analyzed gene expression in the dentate gyrus, a region of the hippocampus thought to show age- rather than AD-specific deterioration, in pathology-free subjects from 33 to 88 years of age. The histone binding protein RbAp48, which functions in histone acetylation and transcriptional regulation, showed the most prominent decline with age, and mice expressing an inhibitor of RbAp48 showed hippocampus dependent memory deficits associated with a regional decrease in histone acetylation. Upregulation of RbAp48 reversed age-related memory loss and normalized histone acetylation [69].

DNA methylation controls the expression of genes and, with age, these patterns change, although it has not been demonstrated that reversing age-related DNA methylation can extend lifespan. A study of close to 400 human brains aged 1–102 years showed a strong positive correlation between methylation and age across three brain regions [70]. Two independent, epigenome wide association studies (EWAS) of AD analyzed together multiple brain regions in close to 1,000 autopsied brains and correlated DNA methylation changes with pathology [71–73]. Remarkably, four methylated sites, close to *ANK1*, *RHBDF2*, *RPL13* and *CDH23*, were linked to AD pathology in both studies and all except CDH23 are biologically linked to *PTK2B*, a known AD-associated gene. In a more focused approach, analysis of 28 genetic loci associated with AD in 740 autopsied brains from AD cases and controls showed significant DNA methylation changes in SORL1, ABCA7, HLA-DRB5, SLC24A4 and BIN1, some of which correlated with brain pathology [74]. Using a genome wide approach, Gjoneska et al. recently compared the transcriptome and epigenome of a mouse model of AD-related neurodegeneration and brain tissue dissected from AD brains [75]. They observed a downregulation of synaptic plasticity genes and regulatory regions and a concomitant increase in immune response genes and regulatory regions such as the ETS transcription factor PU.1, a critical factor for the development of the myeloid lineage including microglia [76].

Remarkably, apart from DNA methylation, few studies assessed the role of epigenetic changes in the human brain in normal aging or AD thus far and there is little consensus at this point. Studies investigating miRNA changes in the disease focused frequently only on one or a few specific species and only one study carried out an unbiased screen [77]. Clearly, larger and broader screens will be necessary to gain an understanding of the role of these epigenetic regulators. Using selected reaction monitoring (SRM) proteomics and confirmation by immunoblot, Zhang et al. reported strongly reduced acetylation of histone H3 K18/23 in temporal lobes of AD brains compared with controls in a study of 15 brains in total [78]. Histone modifications have been implicated in AD pathogenesis in mouse models. For example, the deacetylase SIRT1 shows disease-dependent upregulation in neurons in a mouse model of AD and overexpression of SIRT1 reduced neurodegeneration dependent on its deacetylase activity [79]. Subsequent studies by the same group showed that DNA double strand breaks led to a recruitment of SIRT1 in injured neurons where it deaceylated HDAC1 and thus stimulated enzymatic activity, which was necessary for DNA repair [80]. While overall SIRT1 mRNA levels were not changed with age in the cortex of wild type mice [81], SIRT1 activity decreased with normal brain aging in microglia and resulted in increased production of IL1-beta and tau-dependent memory deficits [82]. It is thus possible that, in this fashion, loss of deacetylase activity promotes neuroinflammation and AD pathogenesis.

Overall, it is well known that epigenetic changes are key to learning and memory and some studies suggest that these pathways become dysfunctional during brain aging [65, 83, 84]. It seems clear from recent studies that the AD brain shows methylation changes which are linked to the genetic and biological understanding of the disease. As ChIP sequencing techniques become more reliable and amenable to high throughput studies, it is likely that additional epigenetic modification will be discovered in AD and aid in the understanding of the disease. In general, our knowledge of epigenetic changes with normal brain aging remains scant.

6.3 *Mitochondrial Dysfunction*

Mitochondria are the cell's purveyors of life and death by regulating energy metabolism and cell death pathways. They also regulate cellular redox potential, calcium levels, cell cycle and influence many other key pathways. It is thus not surprising that these organelles have a major role in aging in general, and in brain aging and AD in particular. For excellent general overviews about this large field of research we refer to [85–87]. Neurons have very high numbers of mitochondria and their dysfunction would be expected to have significant consequences but, at the same time, it could be expected that nature has built in extra protective mechanisms to avoid premature failure of the nervous system. One hypothesis proposes that mitochondrial DNA damage accumulates with age, leading to reduced mitochondrial enzymatic activity, a loss of mitochondrial bioenergetic function, and subsequent cellular degeneration [88]. There is indeed support for an accumulation of mitochondrial damage in the aging normal human brain and in AD [89–91]. Additionally, mitochondrial enzyme levels including the nuclear encoded F1 ATP synthase [52] or activity of complex I and complex IV [92] were reduced in the aging human or rodent brain, respectively. Likewise, it has been reported that pyruvate and ketoglutarate dehydrogenase as well as COX activities are reduced in mitochondria from AD brains compared with normal controls, and that the number of normal-appearing mitochondria is lower in AD [86]. Mitochondrial impairments have also been extensively documented in APP or tau transgenic mouse models for AD [85]. In summary, while it is clear that mitochondria are altered and fulfill a key role in both brain aging and AD, it is debated whether dysfunctional mitochondria are principal drivers of brain aging and AD or primarily a consequence of accumulating beta-amyloid and other brain lesions.

6.4 *Loss of Proteostasis*

The equilibrium between synthesis and degradation determines the levels of a given protein but many other regulatory pathways contribute to this process as well as the qualitative aspects of the product, including protein folding, polymerization/aggregation, tagging for degradation, etc. Changes in any of these have been implicated in aging and age-related loss of protein homeostasis, and would be expected to have particularly drastic consequences in a tissue with largely post-mitotic cells. Indeed, neurodegenerative diseases are largely characterized by accumulation of protein deposits and strong evidence points to a causal role for protein dyshomeostasis in many such diseases. AD is the archetype neurodegenerative disease where to this day, only the combined presence and abundance of amyloid plaques and neurofibrillary tangles allows for the unequivocal diagnosis of AD by the neuropathologist upon autopsy. We have learned that amyloid plaques are extracellular assemblies of highly ordered fibrils consisting predominantly of 40–42 amino acid fragments derived from APP, and that autosomal dominant mutations in APP result in amyloid accumulation and disease [93]. Likewise, we know that neurofibrillary tangles form inside neurons, are made of fibrils of the microtubule associated protein tau, and that autosomal dominant mutations in tau result in tangle formation and development of a related neurological disease called frontotemporal dementia [6]. Thus, AD is clearly a disease of proteostasis failure and much of the last two decades of research in the field has tried to understand the basis of this failure in both autosomal dominant and sporadic forms of the disease [93]. Ground-breaking measurements of beta-amyloid steady state levels in brains of living humans by Bateman and colleagues demonstrated that production of beta-amyloid is not increased in sporadic AD, but that instead, impaired clearance is chiefly responsible for its accumulation [94]. This clearance defect may include a combination of problems including impairments in phagocytic capacity of microglia [95] and astrocytes [96], transport of beta-amyloid across the blood–brain-barrier [97], or extracellular degradation of beta-amyloid [98]. Significant evidence exists that beta-amyloid may also accumulate inside neurons and contribute to their dysfunction [99].

Posttranslationally modified tau protein is a sensitive indicator of neuronal injury and forms increasingly more complex and insoluble aggregates in neurons in AD [6]. It has long been maintained that this pathology begins in the entorhinal cortex from which it spreads to hippocampal and cortical areas [100] but, in a herculean effort characterizing tau pathology in more than 2,300 postmortem brains aged 1–100 years old, the same authors reported that subtle tau abnormalities occur already in the youngest brains and that they are first observed in the locus caeruleus [17]. Braak concludes from these findings “that the pathologic process underlying AD is not age-dependent but an uncommonly slowly progressive one that frequently extends into old age”. While this interpretation is possible, the lack of fibrillar tau in younger brains could also indicate that age-dependent processes turn relatively harmless tau lesions into neurotoxic ones. Furthermore, the fact that tangle pathology can be replicated in mouse models in a matter of months – albeit with mutant forms of tau – argues for biological processes, rather than time alone, playing a key role in tauopathies.

Both beta-amyloid and tau accumulate in the brain decades before clinical symptoms manifest and changes in extracellular beta-amyloid and tau levels in the CSF appear to be reliable predictors of clinical disease onset [7]. As discussed above, new PET imaging probes, which allow for the detection of abnormal beta-amyloid [8] and tau [9] deposits in living people, seem to support these findings made in CSF. Because many brains from cognitively healthy people show abnormal beta-amyloid and tau deposits, the distinction between normal brain aging and slow disease progression becomes extremely difficult. Studies in mouse models expressing human APP or tau containing familial mutations form pathological protein deposits but, thus far, to our knowledge, no data exist whether such deposits develop with the same kinetics in a young or old mouse brain, within the same mouse model.

Interestingly, part of the machinery that controls protein folding and maintenance and elimination of abnormally folded proteins shows prominent changes in aging neurons and other brain cells. Proteomic studies of AD brains and mouse models of the disease showed significant oxidation and inactivation of various proteins and enzymes including UCHL1 [101], which is necessary for the degradation of misfolded or damaged proteins by the proteasome, the ubiquitin E3 ligase CHIP, hsp70, ubiquitin, etc. [102, 103]. In support for a central role of the ubiquitin-proteasome degradation system (UPS) in AD, unbiased iTRAQ mass spectrometry of tissue from AD and age-matched controls showed prominent changes of multiple proteins which interact with ubiquitin in the hippocampus and cortex [104]. Genetic approaches in mouse models of AD support the notion that beta-amyloid and phosphorylated tau impair the UPS and that changes in UPS function directly modulate the accumulation of these proteins in vivo [102, 105, 106].

The kinase mechanistic target of rapamycin (mTOR), which is a key regulator of protein homeostasis and inhibitor of autophagy (see below), is highly expressed together with PI3K and Akt in the developing brain, and tightly regulated in the adult brain during learning and memory processes. Sustained activation of this pathway together with an insulin and IGF-1 resistant state has been observed in AD and linked to abnormal phosphorylation of tau and, ultimately, desensitization to insulin receptor and IGF-1 receptor signaling [107]. Beta-amyloid is a strong candidate for activating this PI3K/Akt/mTOR pathway in a feed-forward loop [107]. Inhibition of mTOR by pharmacological or genetic means reduces neurodegeneration, tauopathy, and amyloidosis in mouse models of AD [108]. Inhibition of mTOR by rapamycin, which increased levels of autophagy, ameliorated cognitive deficits and lowered beta-amyloid brain concentrations in APP transgenic mice [109]; likewise, rapamycin mediated an increase in GSK-beta phosphorylation and autophagy induction was associated with fewer cognitive deficits and reduced tau pathology in tau transgenic mice [108]. While some of these effects of rapamycin target autophagy, other more general metabolic and anti-aging pathways will be activated as well.

Interestingly, human brains display a diversity of granular structures containing cross-linked proteins, lipids, and carbohydrates which accumulate prominently with age and occur alongside the classical beta-amyloid and tau lesions in AD. Some of these structures are extracellular, such as corporea amylacea. Others, including Hirano bodies, Marinesco bodies, granulovacuolar inclusions, stress granules, and lipofuscin are found inside neurons or glial cells; with the exception of lipofuscin granules, these structures remain poorly characterized and their origin and significance to neurodegeneration are still unclear [110]. Of particular interest to neurodegeneration are lipofuscin deposits which were first described in neurons by Hannover in 1842 and are now a well-established aging marker for post-mitotic cells [111]. Lipofuscin forms as undegradable material within lysosomes and can accumulate independent of age, e.g. in lysosomal storage diseases, as a result of abnormal autophagy or impaired lysosomal degradation. As proposed by Brunk and Terman [111], during normal aging, oxidative modifications to macromolecules make them undegradable in lysosomes, leading to the recruitment of large amounts of newly synthesized lysosomal enzymes which, however, still fail to degrade the material. Consequently, there will be an insufficient supply of lysosomal enzymes available for autophagy, leading to accumulation of aged mitochondria and other cellular organelles and material. The levels of many lysosomal proteins and enzymes are increased in the aging brain and more dramatically in AD and lysosomal dysfunction may have a major role in AD pathogenesis [112]. Along with lysosomal dysfunction, or possibly as a consequence, the AD brain is characterized by extensive endosomal abnormalities and the accumulation of autophagosomes [113]. Genetic manipulation of pathways that regulate autophagosomal pathways in APP or tau transgenic mice showed prominent effects on disease progression and accumulation of these proteins [108, 114, 115]. Stress granules, which contain proteins and RNAs that accumulate in response to stress, are gaining interest due to their possible role in neurodegenerative diseases [116]. Relevant to AD, granulovacuolar degeneration bodies were found to contain markers of stress granules, including pS6 and p54/Rck, and to be highly abundant in neurons from AD compared with control brains. Such neurons did not contain tau aggregates and displayed reduced oxidative damage, suggesting a potentially protective function of these granules [117]. In line with these findings, stress granules containing different types of proteins were associated with tau pathology in AD brains and tau transgenic mice and, based on experiments overexpressing the stress granule protein TIA-1, the authors proposed that some stress granules promote tauopathy [118].

In aggregate, these observations strongly support a major role for protein dyshomeostasis in both normal brain aging and AD. The relative importance of various aspects of protein maintenance and turnover in AD pathogenesis are unclear as they are unclear for aging in general. It is likely, however, that pathways that stabilize protein homeostasis will be beneficial for AD.

6.5 Stem Cell Exhaustion

Stem cells are critical to the regeneration of many tissues and with age, the capacity of these cells to self-renew and produce progeny declines. The adult brain was considered in the past to be a post-mitotic tissue without stem cell activity, and it took several decades from the first reports of adult neurogenesis in rats by Altman and colleagues [119] until it became accepted that several mammalian species, including primates, have the capability to generate new neurons in select brain regions [120]. Following a pioneering study by Gage and colleagues demonstrating the uptake of BrdU into dividing hippocampal neurons in cancer patients treated with this drug [121], Frisen and his team produced the most convincing evidence for neurogenesis in the human brain, thus far, by taking advantage of radioactive ^{14}C isotope released into the biosphere following atomic bomb tests to birth date neurons [122]. Their studies showed that a large portion of the human hippocampus is subject to neuronal turnover, that 700 new cells are added per day, on average, and that neurogenesis declines with age – intriguingly, that decline seems less pronounced than the one observed in rodents [122, 123]. Based on extensive studies of the functional relevance of adult neurogenesis in rodents (see below) it is likely that human neurogenesis contributes to cognitive function and, consequently, it is possible that the age-related decline in neurogenesis results in reduced function [120].

Adult neurogenesis in rodents occurs mainly in two distinct brain regions, the subventricular zone (SVZ) and the subgranular zone (SGZ) of the dentate gyrus, a subregion of the hippocampus. Adult neural stem cells are a relatively quiescent population that can both self-renew and give rise to more rapidly dividing progenitors which in turn produce neurons (neurogenesis), as well as astrocytes and oligodendrocytes (gliogenesis) [124]. Ultimately, newly born neurons in the SVZ migrate and incorporate into the olfactory bulb where they contribute to olfaction. In a similar fashion, neurons born in the SGZ become granule neurons that integrate into the existing circuitry of the hippocampus where they are involved in learning and memory. Consequently, the profound decline in neurogenesis with age is linked to functional declines in olfaction and spatial learning and memory [124, 125]. The cause for this decline and exhaustion of stem cell activity is unknown.

Theoretically, inhibition of adult neural stem cell activity can occur due to limits in the potential number of cellular divisions, through intra- or extracellular factors that induce cell cycle arrest to maintain a pool of viable quiescent stem cells, or the molecular composition of the neurogenic niche to either enhance or mitigate cellular proliferation [126, 127]. Indeed, recent studies in the aged brain have linked the decline in neural progenitor cell function and olfactory bulb neurogenesis to $p16^{INK4a}$ [128], the polycomb gene Bmi-1 signaling pathway [129], and FoxO3a [130], for example. In addition, telomere control seems to be important in neural stem cells as mice deficient in telomerase show exhaustion of SVZ stem cells [56] and telomerase expression decreases with age [54, 55] (see also Sect. 6.1). Furthermore, adult neurogenesis is controlled by a number of epigenetic mechanisms. For example, the immediate early gene Gadd45b, which is required for neuronal activity-induced DNA demethylation, is necessary for hippocampal neurogenesis and induction of BDNF and FGF [131]. In addition, valproic acid, which blocks histone deacetylases (HDAC), promotes neuronal differentiation [132] and REST controls neurogenesis by keeping stem cells in a quiescent state [133].

Besides these intracellular factors, a number of more global extracellular regulators have resulted in changes of neural progenitor cell proliferation and neurogenesis, including insulin-like growth factor 1 (IGF-1) and vascular endothelial growth factor (VEGF) [134, 135]. Importantly, the neurogenic niche is localized around blood vessels, which allows for the possible communication with the systemic environment [136–139]. Hence, perturbations to the systemic milieu of an organism, such as those induced by exercise or dietary restriction, can increase stem cell functionality and enhance learning and memory in aging mice [140, 141] and exercise in humans has similar beneficial effects [142–144]. Using heterochronic parabiosis, Villeda et al. showed that blood-borne factors present in the systemic milieu can inhibit or promote adult neurogenesis in an age-dependent fashion in mice [136]. Accordingly, exposing an old mouse to a young systemic environment or to plasma from young mice increased neurogenesis, synaptic plasticity, and improved contextual fear conditioning, spatial learning and memory [136, 145]. In support of the existence of pro-neurogenic factors in the young circulation, heterochronic parabiosis increased neurogenesis in the old SVZ, thus enhancing olfaction in mice, and growth and differentiation factor (GDF)-11 was sufficient to mimic at least some of these benefits [146]. Given the potent effects of growth factors on neurogenesis it is maybe not surprising that mTOR signalling holds a key in controlling proliferation of stem and progenitor cells in the brain [147]. Specifically, the mTOR complex-1 (mTORC1) is expressed in a unique pool of transiently amplifying progenitor cells and signalling via mTOR decreases with age. Surprisingly, inhibition of mTORC1 depletes this progenitor pool, inducing a quiescence-like phenotype which may be similar to the one observed with aging and which was reversible by EGF-induced upregulation of mTOR activity [147]. How the beneficial effects of rapamycin on brain function factor into this observation is currently unclear, but it is interesting to speculate whether factors associated with a young circulatory environment exert similar effects on quiescent cell populations.

Conversely, circulating factors associated with aging or inflammation are known to inhibit neurogenesis and cognitive function and non-steroidal anti-inflammatory drugs were able to prevent this [148]. For example, increased levels of TNFα [149] or IL1-β [150] resulted in reduced levels of adult neurogenesis and adult rats treated with IFN-α showed depression-like symptoms and reduced cell proliferation in the dentate gyrus [151]. Interestingly, patients treated with recombinant IFN-α as part of a therapy for hepatitis C virus infection or cancer frequently develop depression or cognitive deficits linked to the treatment [152, 153] which has been suggested to depend on binding of IFN-α to complement receptor 2 on neural stem cells [154]. In an attempt to identify age-related factors linked to reduced neurogenesis, Villeda and colleagues used a focused proteomic approach in the heterochronic parabiosis model [136]. They demonstrated that CCL11/eotaxin, a chemokine involved in allergic responses and not previously linked to aging, was sufficient to reduce neurogenesis and impair cognition when administered systemically to young mice. Interestingly, CCL11 expression is also increased in both the choroid plexus during aging [155] and in fat deposits with obesity, an important risk factor for cognitive impairment; conversely, CCL11 decreases following exercise in obese patients [156, 157].

The first study that explored neurogenesis in AD brains reported an increase in markers for immature neurons including doublecortin and NSA-NCAM, although no statistical analysis of these results was presented [158] and there is still no consensus as to the degree to which neurogenesis is altered in the disease. In mouse models overexpressing APP alone or in combination with presenilin containing familial AD mutations, neurogenesis was typically reduced when compared with wild type mice [159, 160] although some groups reported increased adult neurogenesis in certain strains [161]. Interestingly, while overexpression of wild type or mutant presenilin-1 reduced the number of neural progenitors in the mouse hippocampus, only mutant protein was sufficient to reduce the survival of newborn neurons [162]. Because APP and presenilin function in neurogenesis during development, discrepancies observed in adult neurogenesis between the various models could be a result of differences in temporal, spatial, or levels of expression of transgenes [124, 125].

In summary, there is overwhelming evidence that neural stem cells have key functions in mammalian learning and memory and that adult neurogenesis takes place in the human hippocampus. It is also clear that with age stem cell activity declines prominently. The mechanisms behind this exhaustion and the functional consequences to AD and human cognition in general need further investigation.

6.6 Altered Intercellular Communication and Neuroinflammation

Communication between cells of the same type, the same tissue, or across tissues is governed by thousands of secreted proteins, peptides, lipid mediators, and other molecules and must be under tight regulatory control. As the organism develops, matures, and ages, there are dramatic changes in the composition of this network, and with disease, additional adaptations will occur. Proteomic, lipidomic, and metabolomic technologies are trying to characterize these changes and a growing number of studies have applied them to brain aging and AD. Because brain tissue typically cannot be studied at the molecular level in living individuals, it has been particularly challenging to study physiological changes in the human brain or chronic neurodegenerative diseases such as AD, which develop over decades and have no strong genetic component or etiology. As an alternative, and taking advantage of the concept of intercellular communication between tissues, a growing number of studies are correlating changes in blood to normal brain aging or AD, with the hypothesis that such changes mirror, in part, changes in the brain.

Age-related changes in intercellular communication have been studied as a function of changes in secreted proteins including endocrine and neuroendocrine factors [48]. Quantifiable molecular markers of intercellular communication in the blood have indeed greatly advanced the understanding and diagnosis of human disease, and recent studies in blood suggest that aging is similarly associated with changes in intercellular communication factors. Targeted proteomic studies measuring concentrations of dozens to hundreds of intercellular communication factors in plasma from AD patients using antibody-based multiplex assays described protein signatures which may be specific to prodromal stages of the disease [163], or which characterize patients who progress from a prodromal stage to AD [164–166]. Other signatures appear to correlate with APOE genotype [167] or with pathological changes such as Aβ and tau protein levels in CSF or brains of AD patients [168, 169]. While there is some general overlap between these studies (e.g. apoE, clusterin, ICAM1, RANTES, complement) most signatures have not been independently validated and their biological significance or diagnostic value is unclear. More sophisticated and unbiased methods to study the plasma proteome use mass spectrometry, often in combination with initial fractionation or selection steps such as 2D gel electrophoresis, chromatography, or antibodies. One such study found complement factor H, alpha2-macroglobulin and other proteins to be associated with AD [170]. To our knowledge, unbiased plasma proteome approaches have not been used to study human brain aging but, as described above, Villeda and colleagues used multiplex assays to identify plasma communication factors that correlate with age-related changes in neurogenesis and which are altered in response to heterochronic parabiosis in mice [136]. They discovered proteins which increase with age in mice and humans, including in CSF, and showed that at least one of these factors, CCL11, is sufficient to induce changes in young mice that would normally occur at a much older age. Whether this represents "accelerated" aging of the brain remains to be investigated but the study demonstrated that intercellular communication factors in the circulation are not only correlated with, but also sufficient to modulate brain aging.

Interestingly, other proteins associated with brain aging and reduced neurogenesis in the above study, including the chemokines CCL2, CCL12, CCL19, as well as beta2-microglobulin and haptoglobin [136], are immune regulatory factors and are likely part of the concerted low-grade inflammatory response associated with aging, also known as "inflammaging" [171]. Searching for factors which decrease with aging, increase with heterochronic parabiosis, and benefit the old brain, Katsimpardi et al. reported that GDF-11 increases neurogenesis, improves olfaction, and exerts beneficial effects on the brain vasculature [146]. Other age-related circulatory factors with beneficial effects on the brain include Klotho, a pleiotropic protein which suppresses insulin and wnt signaling and has been shown to extend lifespan in mice. Recent studies show that a lifespan-extending version of the protein, which is expressed at increased levels in the circulation of carriers, is associated with enhanced cognition in humans, and that mice overexpressing klotho showed increased synaptic plasticity and memory function, possibly by increasing levels of the NMDA receptor subunit GluN2B [27]. Furthermore, overexpression of klotho reduced mortality and enhanced cognition in APP mice [172]. It is likely that other age-related circulatory factors with detrimental or beneficial effects on the aging brain will be discovered and it remains to be explored whether they have a role in AD.

Immune and inflammatory processes have long been suspected to have a role in AD [173, 174], and long-term use of NSAIDs, taken for several years before the onset of clinical symptoms, and ideally at younger age, is associated with reduced risk for AD [175–177]. In addition, studies in patients with mild to moderate AD have shown that systemic inflammation and the number of systemic inflammatory events (e.g. urinary tract infection) contribute to the progression and severity of AD [178, 179]. However, the link between inflammation and AD is still unclear and likely involves both systemic and central mechanisms. One idea about how brain inflammation ties in with AD is that astrocytes and microglia become senescent and assume a so-called senescence associated secretory phenotype (SASP; [180]). Indeed, some aging astrocytes appear to express the hallmarks of the SASP including increased expression of intermediate filament proteins, cytokines, and intracellular protein aggregates [181]. As discussed above, this aged phenotype could be the result of epigenetic changes or replicative senescence. In support of epigenetic factors, Gjoneska et al. find a coordinated increase in the expression of immune response genes targeted by the transcription factor PU.1, which is restricted to microglia in the brain [75]. Alternatively, these phagocytic cells may become activated and "inflamed" as a result of protein dyshomeostasis. For a detailed discussion of microglia in brain aging and AD see [63]. Additional support for a role of immune-based mechanisms in brain aging and AD comes from recent genome wide expression studies in multiple brain regions of >1,600 brains. One study showed that activation of inflammatory genes is a hallmark of normal brain aging that likely precedes the development of AD [182]. A gene network analysis has also suggested that TYROBP/DAP12, an adaptor protein expressed in microglia, is deregulated in AD patients compared to healthy controls [183]. DAP12 binds to various receptors including complement receptor 3, one of the major phagocytic receptors expressed on microglia, as well as TREM2 [184]. In fact, the most direct evidence that altered immune function and inflammation have a role in AD comes from genetic studies which show that rare polymorphisms in the microglial gene TREM2 increase AD risk by several fold [185, 186]. Genome wide association studies (GWAS) in AD also identified a number of single nucleotide polymorphisms in immune related genes, which modify the risk of developing the disease [187–189]. In the brain, most of these receptors are exclusively expressed on microglia.

In conclusion, there is very strong genetic, transcriptomic, and proteomic evidence that altered intercellular communication and inflammation are major components of both normal brain aging and AD. It is, however, not necessarily clear whether upregulation of immune and inflammatory pathways is a driver of aging and disease or, at least in part, a reparative or regenerative process [174]. Likewise, the interplay between central and systemic inflammation is complex as age-related changes in the hypothalamus, for example, may orchestrate organismal aging [190].

7 Conclusions and Future Prospects

In this brief and somewhat selective review of brain aging and AD we conclude that all aspects of organismal aging are observed in the brain and the boundaries between aging and disease are largely fluent. As observed in other tissues, defined hallmarks of aging are strongly inter-dependent and it remains elusive whether such hallmarks are a cause or consequence of the aging process. Clearly, they are not the result of AD unless the disease starts at birth, as has been postulated by some [17], in which case aging itself would need to be called a disease. Rather, we believe there is a continuum from brain aging to AD, which is susceptible to genetic, epigenetic, and environmental influences. An overwhelming number of individuals above age 80 show signs of AD pathology and cognitive decline; still, there are extremely rare, exceptional centenarians without pathology [191] indicating that AD is not entirely inevitable – but almost. Studying these rare subjects should be a high priority.

Genetic mouse models of AD recapitulate pathological abnormalities, synaptic loss, and cognitive deficits often during young adulthood while the same manifestations in humans are not observed until much later in life. If these models are to be relevant for the understanding of disease progression, particularly sporadic AD, the role of aging needs to be considered and aggressive models which develop disease before midlife are likely not very informative.

Given the profound role immune responses and inflammation seem to play in both brain aging and AD, it will be critical to determine which aspects of immunity are maladaptive and which ones are attempts to maintain or repair damage. Indeed, the oldest old may show gains in brain immunity [192] and rejuvenation of immunity has been proposed as an approach to treat brain aging [193]. In fact, a number of colony stimulating factors, which are mitogens for immune cells, show benefit in models of AD and some are being evaluated in patients [194].

Remarkable effects of heterochronic parabiosis in mice on multiple tissues including the brain [136, 145, 146] suggest that the brain is malleable and that a dysregulation of intercellular communication may underlie, in part, brain aging and cognitive decline. Understanding the molecular basis of these rejuvenation approaches will likely provide new insight into brain aging and plasticity.

In this vein, and maybe most importantly, neuroscientists have focused for too long on neurons alone and ignored glial cells and vascular cells, let alone systemic factors as important regulators of brain function and dysfunction. It is now evident that with age, every cell type shows signs of deterioration and dependent on a person's genetic and environmental exposure, the sequence and severity of cellular dysfunction may produce the particular manifestations of aging and neurodegeneration. By altering basic processes involved in aging, it might be possible to counteract the cellular dysfunction that leads to neurodegeneration, including AD.

Acknowledgments This work was supported by the Department of Veterans Affairs (T.W.-C.), and the NIA (AG045034 (T.W.-C.)).

Editor: Bradley Wise, National Institute on Aging (NIA), NIH.

References

1. Yankner BA, Lu T, Loerch P (2008) The aging brain. Ann Rev Pathol Mech Dis 3:41–66. doi:10.1146/annurev.pathmechdis.2.010506.092044
2. Bishop NA, Lu T, Yankner BA (2010) Neural mechanisms of ageing and cognitive decline. Nature 464:529–535. doi:10.1038/nature08983
3. McKhann GM, Knopman DS, Chertkow H et al (2011) The diagnosis of dementia due to Alzheimer's disease: recommendations from the National Institute on Aging-Alzheimer's Association workgroups on diagnostic guidelines for Alzheimer's disease. Alzheimers Dement 7:263–269. doi:10.1016/j.jalz.2011.03.005
4. Albert MS, DeKosky ST, Dickson D et al (2011) The diagnosis of mild cognitive impairment due to Alzheimer's disease: recommendations from the National Institute on Aging-Alzheimer's Association workgroups on diagnostic guidelines for Alzheimer's disease. Alzheimers Dement 7:270–279. doi:10.1016/j.jalz.2011.03.008
5. Vos SJ, Verhey F, Frolich L et al (2015) Prevalence and prognosis of Alzheimer's disease at the mild cognitive impairment stage. Brain. doi:10.1093/brain/awv029
6. Huang Y, Mucke L (2012) Alzheimer mechanisms and therapeutic strategies. Cell 148:1204–1222. doi:10.1016/j.cell.2012.02.040
7. Jack CR Jr, Holtzman DM (2013) Biomarker modeling of Alzheimer's disease. Neuron 80:1347–1358. doi:10.1016/j.neuron.2013.12.003

8. Cohen AD, Klunk WE (2014) Early detection of Alzheimer's disease using PiB and FDG PET. Neurobiol Dis 72:117–122. doi:10.1016/j.nbd.2014.05.001
9. Okamura N, Harada R, Furumoto S et al (2014) Tau PET imaging in Alzheimer's disease. Curr Neurol Neurosci Rep 14:500–507. doi:10.1007/s11910-014-0500-6
10. Deary IJ, Corley J, Gow AJ et al (2009) Age-associated cognitive decline. Br Med Bull 92:135–152. doi:10.1093/bmb/ldp033
11. Chadick JZ, Zanto TP, Gazzaley A (2014) Structural and functional differences in medial prefrontal cortex underlie distractibility and suppression deficits in ageing. Nat Commun 5:4223. doi:10.1038/ncomms5223
12. Hayden KM, Reed BR, Manly JJ et al (2011) Cognitive decline in the elderly: an analysis of population heterogeneity. Age Ageing 40:684–689. doi:10.1093/ageing/afr101
13. Tang YP, Shimizu E, Dube GR et al (1999) Genetic enhancement of learning and memory in mice. Nature 401:63–69. doi:10.1038/43432
14. Cui Z, Feng R, Jacobs S et al (2013) Increased NR2A: NR2B ratio compresses long-term depression range and constrains long-term memory. Sci Rep 3:1036. doi:10.1038/srep01036
15. Chetelat G, La Joie R, Villain N et al (2013) Amyloid imaging in cognitively normal individuals, at-risk populations and preclinical Alzheimer's disease. NeuroImage Clin 2:356–365. doi:10.1016/j.nicl.2013.02.006
16. Braak H, Braak E (1997) Frequency of stages of Alzheimer-related lesions in different age categories. NBA 18:351–357. doi:10.1016/S0197-4580(97)00056-0
17. Braak H, Thal DR, Ghebremedhin E, Del Tredici K (2011) Stages of the pathologic process in Alzheimer disease: age categories from 1 to 100 years. J Neuropathol Exp Neurol 70:960–969. doi:10.1097/NEN.0b013e318232a379
18. Barnes DE, Yaffe K (2011) The projected effect of risk factor reduction on Alzheimer's disease prevalence. Lancet Neurol 10:819–828. doi:10.1016/S1474-4422(11)70072-2
19. McKee AC, Daneshvar DH (2015) The neuropathology of traumatic brain injury. Handb Clin Neurol 127:45–66. doi:10.1016/B978-0-444-52892-6.00004-0
20. Erickson KI, Weinstein AM, Lopez OL (2012) Physical activity, brain plasticity, and Alzheimer's disease. Arch Med Res 43:615–621. doi:10.1016/j.arcmed.2012.09.008
21. Wilson RS, Boyle PA, Yu L et al (2013) Life-span cognitive activity, neuropathologic burden, and cognitive aging. Neurology 81:314–321. doi:10.1212/WNL.0b013e31829c5e8a
22. Michaelson DM (2014) APOE epsilon4: the most prevalent yet understudied risk factor for Alzheimer's disease. Alzheimers Dement 10:861–868. doi:10.1016/j.jalz.2014.06.015
23. Altmann A, Tian L, Henderson VW et al (2014) Sex modifies the APOE-related risk of developing Alzheimer disease. Ann Neurol 75:563–573. doi:10.1002/ana.24135
24. Yaffe K, Cauley J, Sands L, Browner W (1997) Apolipoprotein E phenotype and cognitive decline in a prospective study of elderly community women. Arch Neurol 54:1110–1114. doi:10.1001/archneur.1997.00550210044011
25. De Jager PL, Shulman JM, Chibnik LB et al (2012) A genome-wide scan for common variants affecting the rate of age-related cognitive decline. NBA 33:1017.e1–15. doi:10.1016/j.neurobiolaging.2011.09.033
26. Brooks-Wilson AR (2013) Genetics of healthy aging and longevity. Hum Genet 132:1323–1338. doi:10.1007/s00439-013-1342-z
27. Dubal DB, Yokoyama JS, Zhu L et al (2014) Life extension factor klotho enhances cognition. Cell Reports 7(4):1065–1076. doi:10.1016/j.celrep.2014.03.076
28. Willcox BJ, Donlon TA, He Q et al (2008) FOXO3A genotype is strongly associated with human longevity. Proc Natl Acad Sci 105:13987–13992. doi:10.1073/pnas.0801030105
29. Fjell AM, McEvoy L, Holland D et al (2014) What is normal in normal aging? Effects of aging, amyloid and Alzheimer's disease on the cerebral cortex and the hippocampus. Prog Neurobiol 117:20–40. doi:10.1016/j.pneurobio.2014.02.004
30. Fjell AM, Westlye LT, Grydeland H et al (2014) Accelerating cortical thinning: unique to dementia or universal in aging? Cereb Cortex 24:919–934. doi:10.1093/cercor/bhs379
31. Fraser MA, Shaw ME, Cherbuin N (2015) A systematic review and meta-analysis of longitudinal hippocampal atrophy in healthy human ageing. Neuroimage. doi:10.1016/j.neuroimage.2015.03.035

32. Braak H, Alafuzoff I, Arzberger T et al (2006) Staging of Alzheimer disease-associated neurofibrillary pathology using paraffin sections and immunocytochemistry. Acta Neuropathol 112:389–404. doi:10.1007/s00401-006-0127-z
33. Jack CRJ, Petersen RC, O'Brien PC, Tangalos EG (1992) MR-based hippocampal volumetry in the diagnosis of Alzheimer's disease. Neurology 42:183–188
34. Kerchner GA, Hess CP, Hammond-Rosenbluth KE et al (2010) Hippocampal CA1 apical neuropil atrophy in mild Alzheimer disease visualized with 7-T MRI. Neurology 75:1381–1387
35. Chhatwal JP, Sperling RA (2012) Functional MRI of mnemonic networks across the spectrum of normal aging, mild cognitive impairment, and Alzheimer's disease. J Alzheimers Dis 31:S155–S167. doi:10.3233/JAD-2012-120730
36. Dickerson BC, Sperling RA (2008) Functional abnormalities of the medial temporal lobe memory system in mild cognitive impairment and Alzheimer's disease: insights from functional MRI studies. Neuropsychologia 46:1624–1635. doi:10.1016/j.neuropsychologia.2007.11.030
37. Palop JJ, Mucke L (2010) Amyloid-β–induced neuronal dysfunction in Alzheimer's disease: from synapses toward neural networks. Nat Neurosci 13:812–818. doi:10.1038/nn.2583
38. Sanchez PE, Zhu L, Verret L et al (2012) Levetiracetam suppresses neuronal network dysfunction and reverses synaptic and cognitive deficits in an Alzheimer's disease model. Proc Natl Acad Sci 109:E2895–E2903. doi:10.1073/pnas.1121081109
39. Vossel KA, Beagle AJ, Rabinovici GD et al (2013) Seizures and epileptiform activity in the early stages of Alzheimer disease. JAMA Neurol 70:1158–1166. doi:10.1001/jamaneurol.2013.136
40. Bakker A, Krauss GL, Albert MS et al (2012) Reduction of hippocampal hyperactivity improves cognition in amnestic mild cognitive impairment. Neuron 74:467–474. doi:10.1016/j.neuron.2012.03.023
41. Apostolova LG, Mosconi L, Thompson PM et al (2010) Subregional hippocampal atrophy predicts Alzheimer's dementia in the cognitively normal. Neurobiol Aging 31:1077–1088. doi:10.1016/j.neurobiolaging.2008.08.008
42. La Joie R, Perrotin A, La Sayette de V et al (2013) Hippocampal subfield volumetry in mild cognitive impairment, Alzheimer's disease and semantic dementia. Neuroimage Clin 3:155–162. doi:10.1016/j.nicl.2013.08.007
43. Yassa MA, Stark SM, Bakker A et al (2010) High-resolution structural and functional MRI of hippocampal CA3 and dentate gyrus in patients with amnestic Mild Cognitive Impairment. Neuroimage 51:1242–1252. doi:10.1016/j.neuroimage.2010.03.040
44. Yassa MA, Lacy JW, Stark SM et al (2011) Pattern separation deficits associated with increased hippocampal CA3 and dentate gyrus activity in nondemented older adults. Hippocampus 21:968–979. doi:10.1002/hipo.20808
45. Small SA, Schobel SA, Buxton RB et al (2011) A pathophysiological framework of hippocampal dysfunction in ageing and disease. Nat Rev Neurosci. doi:10.1038/nrn3085
46. Greicius MD, Srivastava G, Reiss AL, Menon V (2004) Default-mode network activity distinguishes Alzheimer's disease from healthy aging: evidence from functional MRI. Proc Natl Acad Sci U S A 101:4637–4642. doi:10.1073/pnas.0308627101
47. Dennis EL, Thompson PM (2014) Functional brain connectivity using fMRI in aging and Alzheimer's disease. Neuropsychol Rev 24:49–62. doi:10.1007/s11065-014-9249-6
48. López-Otín C, Blasco MA, Partridge L et al (2013) The hallmarks of aging. Cell 153:1194–1217. doi:10.1016/j.cell.2013.05.039
49. Pan L, Penney J, Tsai L-H (2014) Chromatin regulation of DNA damage repair and genome integrity in the central nervous system. J Mol Biol 426:3376–3388. doi:10.1016/j.jmb.2014.08.001
50. Adamec E, Vonsattel JP, Nixon RA (1999) DNA strand breaks in Alzheimer's disease. Brain Res 849:67–77
51. Madabhushi R, Pan L, Tsai L-H (2014) DNA damage and its links to neurodegeneration. Neuron 83:266–282. doi:10.1016/j.neuron.2014.06.034

52. Lu T, Pan Y, Kao S-Y et al (2004) Gene regulation and DNA damage in the ageing human brain. Nature 429:883–891. doi:10.1038/nature02661
53. Suberbielle E, Sanchez PE, Kravitz AV et al (2013) Physiologic brain activity causes DNA double-strand breaks in neurons, with exacerbation by amyloid-β. Nat Neurosci 16:613–621. doi:10.1038/nn.3356
54. Ferron SR, Marques-Torrejon MA, Mira H et al (2009) Telomere shortening in neural stem cells disrupts neuronal differentiation and neuritogenesis. J Neurosci 29:14394–14407. doi:10.1523/JNEUROSCI.3836-09.2009
55. Wolf SA, Melnik A, Kempermann G (2011) Physical exercise increases adult neurogenesis and telomerase activity, and improves behavioral deficits in a mouse model of schizophrenia. Brain Behav Immun 25:971–980. doi:10.1016/j.bbi.2010.10.014
56. Ferron S (2004) Telomere shortening and chromosomal instability abrogates proliferation of adult but not embryonic neural stem cells. Development 131:4059–4070. doi:10.1242/dev.01215
57. Eitan E, Hutchison ER, Mattson MP (2014) Telomere shortening in neurological disorders: an abundance of unanswered questions. Trends Neurosci 37:256–263. doi:10.1016/j.tins.2014.02.010
58. Panossian LA, Porter VR, Valenzuela HF et al (2003) Telomere shortening in T cells correlates with Alzheimer's disease status. NBA 24:77–84
59. Roberts RO, Boardman LA, Cha RH et al (2014) Short and long telomeres increase risk of amnestic mild cognitive impairment. Mech Ageing Dev 141–142:64–69. doi:10.1016/j.mad.2014.10.002
60. Jacobs EG, Epel ES, Lin J et al (2014) Relationship between leukocyte telomere length, telomerase activity, and hippocampal volume in early aging. JAMA Neurol 71:921–923. doi:10.1001/jamaneurol.2014.870
61. Rolyan H, Scheffold A, Heinrich A et al (2011) Telomere shortening reduces Alzheimer's disease amyloid pathology in mice. Brain 134:2044–2056. doi:10.1093/brain/awr133
62. Flanary BE, Streit WJ (2004) Progressive teromere shortening occurs in cultured rat microglia, but not astrocytes. Glia 45(1):75–88. PMID 14648548
63. Mosher KI, Wyss-Coray T (2014) Microglial dysfunction in brain aging and Alzheimer's disease. Biochem Pharmacol 88:594–604. doi:10.1016/j.bcp.2014.01.008
64. Spilsbury A, Miwa S, Attems J, Saretzki G (2015) The role of telomerase protein TERT in Alzheimer's disease and in tau-related pathology in vitro. J Neurosci 35:1659–1674. doi:10.1523/JNEUROSCI.2925-14.2015
65. Kosik KS, Rapp PR, Raz N et al (2012) Mechanisms of age-related cognitive change and targets for intervention: epigenetics. J Gerontol A Biol Sci Med Sci 67:741–746. doi:10.1093/gerona/gls110
66. Ballas N, Mandel G (2005) The many faces of REST oversee epigenetic programming of neuronal genes. Curr Opin Neurobiol 15:500–506. doi:10.1016/j.conb.2005.08.015
67. Lu T, Aron L, Zullo J et al (2014) REST and stress resistance in ageing and Alzheimer's disease. Nature 507(7493):448–454. doi:10.1038/nature13163
68. Peleg S, Sananbenesi F, Zovoilis A et al (2010) Altered histone acetylation is associated with age-dependent memory impairment in mice. Science 328:753–756. doi:10.1126/science.1186088
69. Pavlopoulos E, Jones S, Kosmidis S et al (2013) Molecular mechanism for age-related memory loss: the histone-binding protein RbAp48. Sci Transl Med 5:200ra115. doi:10.1126/scitranslmed.3006373
70. Hernandez DG, Nalls MA, Gibbs JR et al (2011) Distinct DNA methylation changes highly correlated with chronological age in the human brain. Hum Mol Genet 20:1164–1172. doi:10.1093/hmg/ddq561
71. Lunnon K, Smith R, Hannon E et al (2014) Methylomic profiling implicates cortical deregulation of ANK1 in Alzheimer's disease. Nat Neurosci 17:1164–1170. doi:10.1038/nn.3782
72. De Jager PL, Srivastava G, Lunnon K et al (2014) Alzheimer's disease: early alterations in brain DNA methylation at ANK1, BIN1, RHBDF2 and other loci. Nat Neurosci 17:1156–1163. doi:10.1038/nn.3786

73. Lord J, Cruchaga C (2014) The epigenetic landscape of Alzheimer's disease. Nat Neurosci 17:1138–1140. doi:10.1038/nn.3792
74. Yu L, Chibnik LB, Srivastava GP et al (2015) Association of brain DNA methylation in SORL1, ABCA7, HLA-DRB5, SLC24A4, and BIN1With pathological diagnosis of Alzheimer disease. JAMA Neurol 72:15–24. doi:10.1001/jamaneurol.2014.3049
75. Gjoneska E, Pfenning AR, Mathys H et al (2015) Conserved epigenomic signals in mice and humans reveal immune basis of Alzheimer's disease. Nature 518:365–369. doi:10.1038/nature14252
76. Satoh J-I, Asahina N, Kitano S, Kino Y (2014) A comprehensive profile of ChIP-Seq-Based PU.1/Spi1 target genes in microglia. GRSB 8:127–13. doi:10.4137/GRSB.S19711
77. Bennett DA, Yu L, Yang J et al (2015) Epigenomics of Alzheimer's disease. Transl Res 165:200–220. doi:10.1016/j.trsl.2014.05.006
78. Zhang K, Schrag M, Crofton A et al (2012) Targeted proteomics for quantification of histone acetylation in Alzheimer's disease. Proteomics 12:1261–1268. doi:10.1002/pmic.201200010
79. Kim D, Nguyen MD, Dobbin MM et al (2007) SIRT1 deacetylase protects against neurodegeneration in models for Alzheimer's disease and amyotrophic lateral sclerosis. EMBO J 26:3169–3179. doi:10.1038/sj.emboj.7601758
80. Dobbin MM, Madabhushi R, Pan L et al (2013) SIRT1 collaborates with ATM and HDAC1 to maintain genomic stability in neurons. Nat Neurosci 16:1008–1015. doi:10.1038/nn.3460
81. Huang PS, Son JH, Abbott LC, Winzer-Serhan UH (2011) Regulated expression of neuronal SIRT1 and related genes by aging and neuronal β2-containing nicotinic cholinergic receptors. Neuroscience 196:189–202. doi:10.1016/j.neuroscience.2011.09.007
82. Cho SH, Chen JA, Sayed F et al (2015) SIRT1 deficiency in microglia contributes to cognitive decline in aging and neurodegeneration via epigenetic regulation of IL-1. J Neurosci 35:807–818. doi:10.1523/JNEUROSCI.2939-14.2015
83. Gapp K, Woldemichael BT, Bohacek J, Mansuy IM (2014) Reviewepigenetic regulation in neurodevelopment and neurodegenerative diseases. Neuroscience 264:99–111. doi:10.1016/j.neuroscience.2012.11.040
84. Penner MR, Roth TL, Barnes CA, Sweatt JD (2010) An epigenetic hypothesis of aging-related cognitive dysfunction. Front Aging Neurosci. doi:10.3389/fnagi.2010.00009
85. Eckert A, Schmitt K, Götz J (2011) Mitochondrial dysfunction – the beginning of the end in Alzheimer's disease? Separate and synergistic modes of tau and amyloid-β toxicity. Alzheimers Res Ther 3:15. doi:10.1186/alzrt74
86. Swerdlow RH (2011) Brain aging, Alzheimer's disease, and mitochondria. BBA Mol Basis Dis 1812:1630–1639. doi:10.1016/j.bbadis.2011.08.012
87. Santos RX, Correia SC, Zhu X et al (2013) Mitochondrial DNA oxidative damage and repair in aging and Alzheimer's disease. Antioxid Redox Signal 18:2444–2457. doi:10.1089/ars.2012.5039
88. Linnane AW, Marzuki S, Ozawa T, Tanaka M (1989) Mitochondrial DNA mutations as an important contributor to ageing and degenerative diseases. Lancet 1:642–645
89. Mecocci P, MacGarvey U, Kaufman AE et al (1993) Oxidative damage to mitochondrial DNA shows marked age-dependent increases in human brain. Ann Neurol 34:609–616. doi:10.1002/ana.410340416
90. Mecocci P, MacGarvey U, Beal MF (1994) Oxidative damage to mitochondrial DNA is increased in Alzheimer's disease. Ann Neurol 36:747–751. doi:10.1002/ana.410360510
91. Lin MT, Simon DK, Ahn CH et al (2002) High aggregate burden of somatic mtDNA point mutations in aging and Alzheimer's disease brain. Hum Mol Genet 11:133–145
92. Navarro A, Boveris A (2006) The mitochondrial energy transduction system and the aging process. AJP Cell Physiol 292:C670–C686. doi:10.1152/ajpcell.00213.2006
93. Selkoe DJ (2011) Alzheimer's disease. Cold Spring Harb Perspect Biol 3:a004457. doi:10.1101/cshperspect.a004457
94. Bateman RJ, Munsell LY, Morris JC et al (2006) Human amyloid-β synthesis and clearance rates as measured in cerebrospinal fluid in vivo. Nat Med 12:856–861. doi:10.1038/nm1438

95. Krabbe G, Halle A, Matyash V et al (2013) Functional impairment of microglia coincides with beta-amyloid deposition in mice with Alzheimer-like pathology. PLoS One 8:e60921. doi:10.1371/journal.pone.0060921.s005
96. Wyss-Coray T, Loike JD, Brionne TC et al (2003) Adult mouse astrocytes degrade amyloid-β in vitro and in situ. Nat Med 9:453–457. doi:10.1038/nm838
97. Sagare AP, Bell RD, Zlokovic BV (2012) Neurovascular dysfunction and faulty amyloid -peptide clearance in Alzheimer disease. Cold Spring Harb Perspect Med 2:a011452. doi:10.1101/cshperspect.a011452
98. Saido T, Leissring MA (2012) Proteolytic degradation of amyloid -protein. Cold Spring Harb Perspect Med 2:a006379. doi:10.1101/cshperspect.a006379
99. LaFerla FM (2002) Calcium dyshomeostasis and intracellular signalling in Alzheimer's disease. Nat Rev Neurosci 3:862–872. doi:10.1038/nrn960
100. Braak H, Braak E (1991) Neuropathological stageing of Alzheimer-related changes. Acta Neuropathol 82:239–259
101. Sultana R, Boyd-Kimball D, Poon HF et al (2006) Redox proteomics identification of oxidized proteins in Alzheimer's disease hippocampus and cerebellum: an approach to understand pathological and biochemical alterations in AD. Neurobiol Aging 27:1564–1576. doi:10.1016/j.neurobiolaging.2005.09.021
102. Petrucelli L, Dickson D, Kehoe K et al (2004) CHIP and Hsp70 regulate tau ubiquitination, degradation and aggregation. Hum Mol Genet 13:703–714. doi:10.1093/hmg/ddh083
103. Riederer BM, Leuba G, Vernay A, Riederer IM (2011) The role of the ubiquitin proteasome system in Alzheimer's disease. Exp Biol Med 236:268–276. doi:10.1258/ebm.2010.010327
104. Manavalan A, Mishra M, Feng L (2013) Brain site-specific proteome changes in aging-related dementia. Exp Mol Med 45:e39–17. doi:10.1038/emm.2013.76
105. Almeida CG (2006) Beta-amyloid accumulation impairs multivesicular body sorting by inhibiting the ubiquitin-proteasome system. J Neurosci 26:4277–4288. doi:10.1523/JNEUROSCI.5078-05.2006
106. Dickey CA, Kamal A, Lundgren K et al (2007) The high-affinity HSP90-CHIP complex recognizes and selectively degrades phosphorylated tau client proteins. J Clin Invest 117:648–658. doi:10.1172/JCI29715
107. O'Neill C (2013) PI3-kinase/Akt/mTOR signaling: impaired on/off switches in aging, cognitive decline and Alzheimer's disease. Exp Gerontol 48:647–653. doi:10.1016/j.exger.2013.02.025
108. Caccamo A, Magrì A, Medina DX et al (2013) mTOR regulates tau phosphorylation and degradation: implications for Alzheimer's disease and other tauopathies. Aging Cell 12:370–380. doi:10.1111/acel.12057
109. Spilman P, Podlutskaya N, Hart MJ et al (2010) Inhibition of mTOR by rapamycin abolishes cognitive deficits and reduces amyloid-β levels in a mouse model of Alzheimer's disease. PLoS One 5:e9979–8. doi:10.1371/journal.pone.0009979
110. Mrak R, Griffin WS, Graham DI (1997) Agin-associated changes in human brain. J Neuropathol Exp Neurol 56:1269–1275
111. Brunk UT, Terman A (2002) The mitochondrial-lysosomal axis theory of aging: accumulation of damaged mitochondria as a result of imperfect autophagocytosis. Eur J Biochem 269:1996–2002. doi:10.1046/j.1432-1033.2002.02869.x
112. Nixon RA, Cataldo AM, Mathews PM (2000) The endosomal-lysosomal system of neurons in Alzheimer's disease pathogenesis: a review. Neurochem Res 25:1161–1172
113. Nixon RA, Wegiel J, Kumar A et al (2005) Extensive involvement of autophagy in Alzheimer disease: an immuno-electron microscopy study. J Neuropathol Exp Neurol 64:113–122
114. Pickford F, Masliah E, Britschgi M et al (2008) The autophagy-related protein beclin 1 shows reduced expression in early Alzheimer disease and regulates amyloid beta accumulation in mice. J Clin Invest 118:2190–2199. doi:10.1172/JCI33585
115. Nilsson P, Loganathan K, Sekiguchi M et al (2013) Aβ secretion and plaque formation depend on autophagy. Cell Rep 5:61–69. doi:10.1016/j.celrep.2013.08.042

116. Ash PEA, Vanderweyde TE, Youmans KL et al (2014) Pathological stress granules in Alzheimer's disease. Brain Res 1584:52–58. doi:10.1016/j.brainres.2014.05.052
117. Castellani RJ, Gupta Y, Sheng B et al (2011) A novel origin for granulovacuolar degeneration in aging and Alzheimer's disease: parallels to stress granules. Lab Invest 91:1777–1786. doi:10.1038/labinvest.2011.149
118. Vanderweyde T, Yu H, Varnum M et al (2012) Contrasting pathology of the stress granule proteins TIA-1 and G3BP in tauopathies. J Neurosci 32:8270–8283. doi:10.1523/JNEUROSCI.1592-12.2012
119. Altman J, Das GD (1965) Autoradiographic and histological evidence of postnatal hippocampal neurogenesis in rats. J Comp Neurol 124:319–335
120. van Wijngaarden P, Franklin RJM (2013) Ageing stem and progenitor cells: implications for rejuvenation of the central nervous system. Development 140:2562–2575. doi:10.1242/dev.092262
121. Eriksson PS, Perfilieva E, Björk-Eriksson T et al (1998) Neurogenesis in the adult human hippocampus. Nat Med 4:1313–1317. doi:10.1038/3305
122. Spalding KL, Bergmann O, Alkass K et al (2013) Dynamics of hippocampal neurogenesis in adult humans. Cell 153:1219–1227. doi:10.1016/j.cell.2013.05.002
123. Göritz C, Frisén J (2012) Neural stem cells and neurogenesis in the adult. Cell Stem Cell 10:657–659. doi:10.1016/j.stem.2012.04.005
124. Lazarov O, Mattson MP, Peterson DA et al (2010) When neurogenesis encounters aging and disease. Trends Neurosci 33:569–579. doi:10.1016/j.tins.2010.09.003
125. Mu Y, Gage FH (2011) Adult hippocampal neurogenesis and its role in Alzheimer's disease. Mol Neurodeg 6:85. doi:10.1186/1750-1326-6-85
126. Ming G-L, Song H (2011) Adult neurogenesis in the mammalian brain: significant answers and significant questions. Neuron 70:687–702. doi:10.1016/j.neuron.2011.05.001
127. Fuentealba LC, Obernier K, Alvarez-Buylla A (2012) Adult neural stem cells bridge their niche. Stem Cell 10:698–708. doi:10.1016/j.stem.2012.05.012
128. Molofsky AV, Slutsky SG, Joseph NM et al (2006) Increasing p16INK4a expression decreases forebrain progenitors and neurogenesis during ageing. Nature 443:448–452. doi:10.1038/nature05091
129. Molofsky AV, Pardal R, Iwashita T et al (2003) Bmi-1 dependence distinguishes neural stem cell self-renewal from progenitor proliferation. Nature 425:962–967. doi:10.1038/nature02060
130. Renault VM, Rafalski VA, Morgan AA et al (2009) FoxO3 regulates neural stem cell homeostasis. Cell Stem Cell 5:527–539. doi:10.1016/j.stem.2009.09.014
131. Ma DK, Jang M-H, Guo JU et al (2009) Neuronal activity-induced Gadd45b promotes epigenetic DNA demethylation and adult neurogenesis. Science 323:1074–1077. doi:10.1126/science.1166859
132. Hsieh J, Nakashima K, Kuwabara T et al (2004) Histone deacetylase inhibition-mediated neuronal differentiation of multipotent adult neural progenitor cells. Proc Natl Acad Sci U S A 101:16659–16664. doi:10.1073/pnas.0407643101
133. Gao Z, Ure K, Ding P et al (2011) The master negative regulator REST/NRSF controls adult neurogenesis by restraining the neurogenic program in quiescent stem cells. J Neurosci 31:9772–9786. doi:10.1523/JNEUROSCI.1604-11.2011
134. Trejo JL, Carro E, Torres-Aleman I (2001) Circulating insulin-like growth factor I mediates exercise-induced increases in the number of new neurons in the adult hippocampus. J Neurosci 21:1628–1634
135. Fabel K, Fabel K, Tam B et al (2003) VEGF is necessary for exercise-induced adult hippocampal neurogenesis. Eur J Neurosci 18:2803–2812. doi:10.1046/j.1460-9568.2003.03041.x
136. Villeda SA, Luo J, Mosher KI et al (2011) The ageing systemic milieu negatively regulates neurogenesis and cognitive function. Nature 477:90–94. doi:10.1038/nature10357
137. Shen Q, Wang Y, Kokovay E et al (2008) Adult SVZ stem cells lie in a vascular niche: a quantitative analysis of niche cell-cell interactions. Cell Stem Cell 3:289–300. doi:10.1016/j.stem.2008.07.026

138. Tavazoie M, Van der Veken L, Silva-Vargas V et al (2008) A specialized vascular niche for adult neural stem cells. Cell Stem Cell 3:279–288. doi:10.1016/j.stem.2008.07.025
139. Palmer TD, Willhoite AR, Gage FH (2000) Vascular niche for adult hippocampal neurogenesis. J Comp Neurol 425:479–494
140. van Praag H, Kempermann G, Gage FH (1999) Running increases cell proliferation and neurogenesis in the adult mouse dentate gyrus. Nat Neurosci 2:266–270. doi:10.1038/6368
141. van Praag H, Kempermann G, Gage FH (2000) Neural consequences of environmental enrichment. Nat Rev Neurosci 1:191–198. doi:10.1038/35044558
142. Aberg MAI, Waern M, Nyberg J et al (2012) Cardiovascular fitness in males at age 18 and risk of serious depression in adulthood: Swedish prospective population-based study. Br J Psychiatry. doi:10.1192/bjp.bp.111.103416
143. Knöchel C, Oertel-Knöchel V, O'Dwyer L et al (2012) Cognitive and behavioural effects of physical exercise in psychiatric patients. Prog Neurobiol 96:46–68. doi:10.1016/j.pneurobio.2011.11.007
144. Larson EB, Wang L, Bowen JD et al (2006) Exercise is associated with reduced risk for incident dementia among persons 65 years of age and older. Ann Intern Med 144:73–81
145. Villeda SA, Plambeck KE, Middeldorp J et al (2014) Young blood reverses age-related impairments in cognitive function and synaptic plasticity in mice. Nat Med 20:659–663. doi:10.1038/nm.3569
146. Katsimpardi L, Litterman NK, Schein PA et al (2014) Vascular and neurogenic rejuvenation of the aging mouse brain by young systemic factors. Science 344:630–634. doi:10.1126/science.1251141
147. Paliouras GN, Hamilton LK, Aumont A et al (2012) Mammalian target of rapamycin signaling is a key regulator of the transit-amplifying progenitor pool in the adult and aging forebrain. J Neurosci 32:15012–15026. doi:10.1523/JNEUROSCI.2248-12.2012
148. Monje ML, Toda H, Palmer TD (2003) Inflammatory blockade restores adult hippocampal neurogenesis. Science 302:1760–1765. doi:10.1126/science.1088417
149. Iosif RE (2006) Tumor necrosis factor receptor 1 is a negative regulator of progenitor proliferation in adult hippocampal neurogenesis. J Neurosci 26:9703–9712. doi:10.1523/JNEUROSCI.2723-06.2006
150. Koo JW, Duman RS (2008) IL-1β is an essential mediator of the antineurogenic and anhedonic effects of stress. Proc Natl Acad Sci U S A 105:751–756
151. Kaneko N, Kudo K, Mabuchi T et al (2006) Suppression of cell proliferation by interferon-alpha through interleukin-1 production in adult rat dentate gyrus. Neuropsychopharmacology 31:2619–2626. doi:10.1038/sj.npp.1301137
152. Scheibel RS, Valentine AD, O'Brien S, Meyers CA (2004) Cognitive dysfunction and depression during treatment with interferon-alpha and chemotherapy. J Neuropsychiatry Clin Neurosci 16:185–191. doi:10.1176/appi.neuropsych.16.2.185
153. Hilsabeck RC, Hassanein TI, Ziegler EA et al (2005) Effect of interferon-alpha on cognitive functioning in patients with chronic hepatitis C. J Int Neuropsychol Soc 11:16–22. doi:10.1017/S1355617705050022
154. Moriyama M, Fukuhara T, Britschgi M et al (2011) Complement receptor 2 is expressed in neural progenitor cells and regulates adult hippocampal neurogenesis. J Neurosci 31:3981–3989. doi:10.1523/JNEUROSCI.3617-10.2011
155. Baruch K, Ron-Harel N, Gal H et al (2013) CNS-specific immunity at the choroid plexus shifts toward destructive Th2 inflammation in brain aging. Proc Natl Acad Sci. doi:10.1073/pnas.1211270110
156. Vasudevan AR (2006) Eotaxin and obesity. J Clin Endocrinol Metab 91:256–261. doi:10.1210/jc.2005-1280
157. Choi KM, Kim JH, Cho GJ et al (2007) Effect of exercise training on plasma visfatin and eotaxin levels. Eur J Endocrinol 157:437–442. doi:10.1530/EJE-07-0127
158. Jin K, Peel AL, Mao XO et al (2004) Increased hippocampal neurogenesis in Alzheimer's disease. Proc Natl Acad Sci U S A 101:343–347. doi:10.1073/pnas.2634794100

159. Haughey NJ, Nath A, Chan SL et al (2002) Disruption of neurogenesis by amyloid beta-peptide, and perturbed neural progenitor cell homeostasis, in models of Alzheimer's disease. J Neurochem 83:1509–1524
160. Verret L, Jankowsky JL, Xu GM et al (2007) Alzheimer's-type amyloidosis in transgenic mice impairs survival of newborn neurons derived from adult hippocampal neurogenesis. J Neurosci 27:6771–6780. doi:10.1523/JNEUROSCI.5564-06.2007
161. López-Toledano MA, Shelanski ML (2007) Increased neurogenesis in young transgenic mice overexpressing human APP(Sw, Ind). J Alzheimers Dis 12:229–240
162. Wen PH, Shao X, Shao Z et al (2002) Overexpression of wild type but not an FAD mutant presenilin-1 promotes neurogenesis in the hippocampus of adult mice. Neurobiol Dis 10: 8–19. doi:10.1006/nbdi.2002.0490
163. Hu WT, Holtzman DM, Fagan AM et al (2012) Plasma multianalyte profiling in mild cognitive impairment and Alzheimer disease. Neurology 79:897–905. doi:10.1212/WNL.0b013e318266fa70
164. Ray S, Britschgi M, Herbert C et al (2007) Classification and prediction of clinical Alzheimer's diagnosis based on plasma signaling proteins. Nat Med 13:1359–1362. doi:10.1038/nm1653
165. Johnstone D, Milward EA, Berretta R et al (2012) Multivariate protein signatures of pre-clinical Alzheimer's disease in the Alzheimer's disease neuroimaging initiative (ADNI) plasma proteome dataset. PLoS One 7:e34341. doi:10.1371/journal.pone.0034341.t009
166. Hye A, Riddoch-Contreras J, Baird AL et al (2014) Plasma proteins predict conversion to dementia from prodromal disease. Alzheimers Dement 10:799–807.e2. doi:10.1016/j.jalz.2014.05.1749
167. Soares HD, Potter WZ, Pickering E et al (2012) Plasma biomarkers associated with the apolipoprotein E genotype and Alzheimer disease. Arch Neurol 69(10):1310–1317. doi:10.1001/archneurol.2012.1070
168. Britschgi M, Rufibach K, Huang SL et al (2011) Modeling of pathological traits in Alzheimer's disease based on systemic extracellular signaling proteome. Mol Cell Proteomics 10(M111):008862
169. Kiddle SJ, Thambisetty M, Simmons A et al (2012) Plasma based markers of [11C] PiB-PET brain amyloid burden. PLoS One 7:e44260. doi:10.1371/journal.pone.0044260.t003
170. Hye A, Lynham S, Thambisetty M et al (2006) Proteome-based plasma biomarkers for Alzheimer's disease. Brain 129:3042–3050. doi:10.1093/brain/awl279
171. Franceschi C, Campisi J (2014) Chronic inflammation (inflammaging) and its potential contribution to age-associated diseases. J Gerontol A Biol Sci Med Sci 69:S4–S9. doi:10.1093/gerona/glu057
172. Dubal DB, Zhu L, Sanchez PE et al (2015) Life extension factor klotho prevents mortality and enhances cognition in hAPP transgenic mice. J Neurosci 35:2358–2371. doi:10.1523/JNEUROSCI.5791-12.2015
173. Wyss-Coray T (2006) Inflammation in Alzheimer disease: driving force, bystander or beneficial response? Nat Med 12:1005–1015. doi:10.1038/nm1484
174. Lucin KM, Wyss-Coray T (2009) Immune activation in brain aging and neurodegeneration: too much or too little? Neuron 64:110–122. doi:10.1016/j.neuron.2009.08.039
175. McGeer PL, McGeer E, Rogers J, Sibley J (1990) Anti-inflammatory drugs and Alzheimer disease. Lancet 335:1037
176. Côté S, Carmichael P-H, Verreault R et al (2012) Nonsteroidal anti-inflammatory drug use and the risk of cognitive impairment and Alzheimer's disease. Alzheimers Dement 8: 219–226. doi:10.1016/j.jalz.2011.03.012
177. Vlad SC, Miller DR, Kowall NW, Felson DT (2008) Protective effects of NSAIDs on the development of Alzheimer disease. Neurology 70:1672–1677. doi:10.1212/01.wnl.0000311269.57716.63
178. Holmes C, Cunningham C, Zotova E et al (2011) Proinflammatory cytokines, sickness behavior, and Alzheimer disease. Neurology 77:212–218. doi:10.1212/WNL.0b013e318225ae07

179. Holmes C, Cunningham C, Zotova E et al (2009) Systemic inflammation and disease progression in Alzheimer disease. Neurology 73:768–774. doi:10.1212/WNL.0b013e3181b6bb95
180. Coppé J-P, Patil CK, Rodier F et al (2008) Senescence-associated secretory phenotypes reveal cell-nonautonomous functions of oncogenic RAS and the p53 tumor suppressor. Plos Biol 6:e301. doi:10.1371/journal.pbio.0060301
181. Salminen A, Ojala J, Kaarniranta K et al (2011) Astrocytes in the aging brain express characteristics of senescence-associated secretory phenotype. Eur J Neurosci 34:3–11. doi:10.1111/j.1460-9568.2011.07738.x
182. Podtelezhnikov AA, Tanis KQ, Nebozhyn M et al (2011) Molecular insights into the pathogenesis of Alzheimer's disease and its relationship to normal aging. PLoS One 6:e29610. doi:10.1371/journal.pone.0029610
183. Zhang B, Gaiteri C, Bodea L-G et al (2013) Integrated systems approach identifies genetic nodes and networks in late-onset alzheimer's disease. Cell 153:707–720. doi:10.1016/j.cell.2013.03.030
184. Linnartz B, Neumann H (2012) Microglial activatory (immunoreceptor tyrosine-based activation motif)- and inhibitory (immunoreceptor tyrosine-based inhibition motif)-signaling receptors for recognition of the neuronal glycocalyx. Glia. doi:10.1002/glia.22359
185. Jonsson T, Stefansson H, Steinberg S et al (2013) Variant of TREM2 associated with the risk of Alzheimer's disease. N Engl J Med 368:107–116. doi:10.1056/NEJMoa1211103
186. Guerreiro R, Wojtas A, Bras J et al (2013) TREM2 variants in Alzheimer's disease. N Engl J Med 368:117–127. doi:10.1056/NEJMoa1211851
187. Naj AC, Jun G, Beecham GW et al (2011) Common variants at MS4A4/MS4A6E, CD2AP, CD33 and EPHA1 are associated with late-onset alzheimer's disease. Nat Genet 43:436–441. doi:10.1038/ng.801
188. Hollingworth P, Harold D, Sims R et al (2011) Common variants at ABCA7, MS4A6A/MS4A4E, EPHA1, CD33 and CD2AP are associated with Alzheimer's disease. Nat Genet 43:429–435. doi:10.1038/ng.803
189. Lambert J-C, Ibrahim-Verbaas CA, Harold D et al (2013) Meta-analysis of 74,046 individuals identifies 11 new susceptibility loci for Alzheimer's disease. Nat Genet 45(12):1452–1458. doi:10.1038/ng.2802
190. Zhang G, Li J, Purkayastha S et al (2013) Hypothalamic programming of systemic ageing involving IKK-β, NF-κB and GnRH. Nature 497:211–216. doi:10.1038/nature12143
191. den Dunnen WFA, Brouwer WH, Bijlard E et al (2008) No disease in the brain of a 115-year-old woman. Neurobiol Aging 29:1127–1132. doi:10.1016/j.neurobiolaging.2008.04.010
192. Katsel P, Tan W, Haroutunian V (2009) Gain in brain immunity in the oldest-old differentiates cognitively normal from demented individuals. PLoS One 4:e7642. doi:10.1371/journal.pone.0007642.t009
193. Ron-Harel N, Schwartz M (2009) Immune senescence and brain aging: can rejuvenation of immunity reverse memory loss? Trends Neurosci 32:367–375. doi:10.1016/j.tins.2009.03.003
194. Lampron A, Gosselin D, Rivest S (2011) Targeting the hematopoietic system for the treatment of Alzheimer's disease. Brain Behav Immun 25(Suppl 1):S71–S79. doi:10.1016/j.bbi.2010.12.018

Parkinson's Disease and Aging

Julie K. Andersen and Shankar Chinta

Contents

1 Clinical Symptoms

Parkinson's disease (PD) is the most common movement disease of older adults, with an average onset of around 60 years of age [1, 2]. Life expectancy is reduced amongst patients suffering from PD, with a mortality ratio twice that of unaffected

J.K. Andersen, Ph.D. (✉) • S. Chinta, Ph.D.
Buck Institute for Research on Aging, 8001 Redwood Blvd, Novato, CA 94945, USA
e-mail: jandersen@buckinstitute.org; schinta@buckinstitute.org

F. Sierra, R. Kohanski (eds.), *Advances in Geroscience*,
DOI 10.1007/978-3-319-23246-1_8

individuals [3]. Mortality risk factors associated with the disease include cognitive decline and age at time of onset [4]. The disease results in progressive deficits in motor movement due to the preferential loss of dopaminergic (DAergic) neurons in an area of the midbrain known as the substantia nigra pars compacta (SNpc) [5, 6]. Cardinal symptoms include resting tremor, delayed initiation and slowness of movement (bradykinesia), muscular rigidity, and impaired balance [7]. Onset of motor symptoms is often asymmetric, with the most common initial manifestation being a mild resting tremor in one arm [8]. As the disease progresses, the patient may note gait difficulties, micrographia (shrinkage in size of handwriting), and issues with axial posture [9]. Other changes related to motor function include decreased facial expression and laryngeal dysfunction resulting in difficulties with speech [10].

The involvement of several neurotransmitter systems beyond DAergic that may play a significant role in the disease has recently been recognized [11]. These include the serotonergic, cholinergic, and noradrenergic networks, effects on which may explain some of the non-motor symptoms associated with the disease [12–15]. These can occur prior to the onset of motor symptoms and are currently being tested clinically as early disease indicators, including hyposmia (reduction in olfactory function) and disturbances in REM sleep [16–18]. Other common non-motor symptoms accompanying the disease include constipation due to reduced intestinal motility, fatigue, pain, depression, anxiety, and cognitive issues [11, 19]. Patients with PD indeed have a two to six-fold increased risk for development of dementia than control populations [20]. Cause of death include pulmonary infections due to aspiration of saliva into the lungs, accidents associated with increased risks of falls due to loss in motor function, chocking due to difficulty swallowing, and blockage of blood vessels leading to pulmonary embolism and deep vein thrombosis. Individuals do not die of PD, but rather of the life-threatening complications associated with the condition [21, 22].

Progression of symptoms is also associated with the presence of abnormal intraneuronal proteinacious accumulations called Lewy bodies that occur in both neuronal cell bodies and in neurites [6, 23]. A major component of these inclusions is a protein known as α-synuclein. PD is considered part of a larger group of disorders known as synucleopathies that include other related diseases such as Dementia with Lewy Bodies (DLB) [24]. Extent and expression of the disease at defined sites within the nervous system has been proposed to track closely with the presence of Lewy bodies at these various locations and may in part explain patient-to-patient variability in symptoms [19]. According to the Braak staging scale, a means of classifying neuropathology associated with PD based on observations at autopsy, Lewy bodies first appear in enteric neurons of the gut then the olfactory bulb and lower brain stem, prior to individuals displaying any motor symptoms [25–27]. As the disease evolves, Lewy bodies spread to the SNpc and then to the neocortex. This has led to the prion hypothesis of PD that postulates that α-synuclein spreads transcellularly, temporally inducing progressive pathological aggregation within affected neurons in a pattern that tracks with resulting symptomology [28]. If correct, this could have major implications in terms of the success of proposed therapies for the disorder, including use of cellular transplantation (see below). However there are clearly caveats to the data supporting this theory that are under intense investigation [29].

2 Current Therapies

Currently, there is no cure for PD. Existing therapies are instead primarily directed towards alleviating the most disturbing symptoms in individuals with the disease, which can vary widely in terms of both presence of particular symptoms and rate of progression [7]. More than a third of individuals with PD, for example, do not display a tremor, therefore presence of tremor in an older individual alone is not necessarily a sign of PD per se. For this reason, there is no standard treatment for the disorder and treatments must be tailored to meet the needs of each individual patient.

Approaches to lessen PD symptoms include both medication and surgical therapies [10]. The most widely used therapy is Levo-Dopa (L-Dopa) [30]. L-dopa is a precursor of dopamine and acts to increase synthesis of this neurotransmitter in remaining DAergic neurons of the SNpc. It is particularly effective in reducing bradykinesia and rigidity in early stages of the disease. It is often given in conjunction with carbidopa, a peripheral dopa decarboxylase inhibitor that acts to lessen side effects due to conversion of L-dopa to dopamine outside of the brain, which can lead to nausea and vomiting [31]. Due to the rapid turnover of dopamine (DA), resulting in oscillations in its levels and difficulties in adjusting dosage to obtain ideal brain concentrations, patients often develop symptoms related to the drug itself, including dyskinesias (abnormal involuntary movements) and fluctuations in treatment efficacy causing transient freezing [32]. These side effects are controlled in many cases by addition of other drugs to the patient's medication regime [32]. While L-dopa can control PD symptoms, it does not however halt the progression of the underlying disease. As neurons continue to be lost, L-dopa efficacy becomes diminished due to loss of striatal DAergic terminals necessary for neurotransmitter storage and buffering [33]. An alternative or complement to treatment with L-dopa is the use of dopamine agonists, which act as a sort of pseudo-DA to activate DAergic receptors in the brain [34, 35]. These include Requip, Mirapex, and Neupro that are taken either alone or in combination with Sinemet (L-dopa/carbidopa) [36]. Due to better tolerance and reduced long-term risk of complications, these are now often the first choice of treatment for PD. They however have their own side effects, including increased risk for drug-associated psychosis [37]. Other alternatives include the use of anticholinergics (Artane, Cogentin) which balance levels of acetycholine in the brain and Eldepryl and Azilect (selegeline, rasagaline) which prevent breakdown of dopamine via inhibiting the enzyme monoamine oxidase B (MAO-B) [38–40]. A newly described MAO-B inhibitor, safinamide, is currently undergoing clinical trial [31, 41]. Another strategy to prolong DA response is extracellular inhibition of its degradation by the enzyme catechol-o-methyl-transferase (COMT) via administration of entacapone or related agents. Given that PD can occur in conjunction with several other age-related comorbidities, it is particularly important for patients to check with their physicians about potential negative interactions between drugs used to treat PD and other age-related conditions. Sometimes there are additionally dietary considerations due to interactions of medications with specific foods [38]. The American Academy of Neurology has recently released a series of recommendations for the treatment of non-motor symptoms including erectile dysfunction, constipation, night-time sleep disturbances due to uncontrolled limb movement, daytime somnolence, and fatigue, all of which may cause considerable pain and discomfort and affect quality of daily life (American Academy of Neurology, 2010).

At the point in time at which medicinal therapies lose their effectiveness, surgical interventions such as deep brain stimulation (DBS) may be tried [42, 43]. DBS involves electrical stimulation of the thalamus and subthalamic regions, aiding in the attenuation of tremors and freezing by blocking certain electrical signals within the brain contributing to these symptoms [9]. It is important to emphasize that although all of these therapies can act to reduce motor symptoms and therefore increase the longevity of PD patients, they do not result in a halting of the underlying neurodegeneration associated with the disease.

Other recommended evidence-based treatment approaches include general lifestyle modifications such as rest and exercise for improvement of motor function and reducing stress, and speech therapy to aid with losses in communication abilities [44–46]. Use of alternative medicinal therapies including herbs, vitamins, and supplements have in the majority of cases not been thoroughly studied scientifically in terms of either efficacy or safety nor vetted by the FDA for use in the disease. In the few cases in which such agents have undergone more rigorous study (vitamin E, coenzyme Q), their beneficial effects have not held out under clinical trial conditions, although they may be of benefit to certain patients [31, 47]. Creatine is currently under a phase III clinical trial for efficacy [48]. A meta-analysis of vitamin E intake in the food suggests a relationship between high intake and reduced risk for development of PD [49, 50]; however, vitamin E did not show efficacy in a previous clinical trial (DATATOP). Other over-the-counter nutriceuticals (low-dose lithium, vitamin D, nicotine, etc.) have shown efficacy either in pre-clinical or in epidemiological studies but have yet to be thoroughly assessed in human clinical trials, although some trials are being initiated or are ongoing [51]. Other clinical trials involve repositioning drugs already approved for treatment of other diseases for use in PD, including pioglitazone (BioPEP), normally prescribed for type II diabetes, and the calcium channel blocker isradipine (STEADY PD III), used to treat high blood pressure. Results from recent studies suggesting that statins may lower the risk for PD have also peaked interest in initiating longer-term studies to verify these effects [52].

Over the last decade, there has been a large amount of research put into novel restorative therapies including gene therapy and cellular replacement via transplantation. Based on pre-clinical data, for example, clinical trials were performed infusing the neurotrophic factor glial-derived neurotrophic factor (GDNF) directly into the brains of PD patients as a means of protecting against additional DAergic cell loss [53]. However, results from various clinical trials proved to be contradictory, perhaps due to differences in trial design. Other confounding factors include heterogeneity of the patient population and the difficulties inherent in teasing out symptomatic versus drug-mediated effects. Recruitment for phase I clinical trials is currently underway to test the safety and tolerability of adeno associated virus

(AAV)-mediated delivery of GDNF [54]. A previous AAV phase II trial delivering the GDNF analogue neurturin (NTN) showed some negative side effects and a lack of overall improvement in symptoms [55, 56]. A subsequent phase II trial utilizing higher viral dosages and longer follow-up times is currently underway [57]. Some success has been met in the use of gene therapy involving the AAV-mediated delivery of glutamic acid decarboxylase (GAD), an enzyme that acts to modulate GABA production in the subthalamic nucleus. This was reported to result in improved motor function in PD patients following a 6-month observational period, with minimal side effects (64, 65).

As with gene therapy, several clinical trials using cellular transplantation have been undertaken, as early as the 1980's. The goal has been to replace lost DAergic neurons, utilizing DAergic fetal cells [58, 59]. Study results have been variable, but more detailed placebo-controlled double-blind trials (versus open-labeled) generally report lack of significant overall improvement in motor function and the induction of drug-related side effects such as dyskinesia. Recent excitement has been engendered by the ability to produce inducible pluripotent stem cell (iPSC) populations from a patient's own tissues which can be differentiated towards a DAergic fate for transplant, particularly in combination with gene therapy to promote their survival in vivo [60–62]. However, excitement has been somewhat tempered based on evidence suggesting that transplanted cells have limited success in clinical trials and may actually take on the same fate as affected endogenous neurons [28]. More studies are undoubtedly required before moving such studies towards new clinical trials, including how to deal with reduced transplant efficiency in the older brain. In other words, not only do we need to consider the cells themselves, but also the environment into which they are placed.

3 Known Molecular or Cellular Underpinnings

Although familial forms of the disease account for roughly 5–10 % of PD, it is largely considered a sporadic condition arising from a combination of genetic susceptibility and environmental exposures across the lifespan [63]. These vary from individual to individual, which may to some degree help explain disparate disease presentation. This has led to the notion that PD may constitute not a single disease entity, but rather a syndrome.

Aging is the number one risk factor for the disorder and factors common to both normal brain aging and selective cell death in PD merit special consideration in terms of overlapping factors that may be causative in both conditions [64–66]. Is PD a case of accelerated aging in a specific population of cells particularly sensitive to certain genetically induced phenotypes and/or environmental exposures? Advanced age is certainly directly linked to a more rapid disease progression and older individuals are more refractory to medical treatments for the disorder, suggesting that there is an important interplay between the two [67].

Through studies of both the post-mortem PD brain and genetic pathways responsible for familial forms of the disease, much has been gleaned about the cause of cellular dysfunction in PD. Some of the pathways implicated so far include elements identified as also important for basic aging including mitochondrial dysfunction, oxidative stress, calcium mishandling, inflammation, and effects on proteostasis [68, 69]. Identification of these molecular targets has led to exploration of interventions designed to prevent or reverse their detrimental effects as a means of slowing or reversing the course of the disease.

3.1 Mitochondrial Dysfunction

Neurons are particularly dependent on mitochondrial oxidative phosphorylation as a source of ATP, with a very limited capacity for glycolysis [70, 71]. Since the discovery of defects in mitochondrial complex I activity in postmortem brain samples and increased levels of mtDNA mutations in patients with the disease, mitochondrial dysfunction has become a major focus of research [69, 72]. As a consequence, models based on both mitochondrial neurotoxins (MPTP, paraquat, rotenone) and gene defects that impact on mitochondrial function identified in rarer familial forms of the disease (α-synuclein, parkin, PINK, LRRK2) have been established [70]. These have been helpful in the segmental dissection of different aspects of disease pathology, including the role of mitochondrial defects in neuropathological features associated with the disease [73].

The master transcription factor peroxisome proliferator-activated receptor γ (PPARγ) coactivatory-1α (PGC-1α) is an important regulator of mitochondrial biogenesis and function. Its dysregulation has recently been identified as a key factor in PD (138). Activators of PGC-1α including resveratrol (acting via SIRT1) and diabetes drugs such as pioglitazone (currently in clinical trial for early PD, BioPEP) have been shown to be neuroprotective in animal models of the disease (139). However, use of agents that act to increase mitochondrial biogenesis will need to be balanced with those which increase lysosomal turnover of defective mitochondria so as to not increase the build-up of the latter. Interestingly, recent data in a Huntington mouse model has demonstrated that the neuroprotective effects of increased PGC-1α result in up-regulation of the master transcriptional regulator of lysosomal biogenesis, transcription factor EB (TFEB), suggesting the existence of a tight regulation between mitochondrial biogenesis and turnover [74]. Enhancement of fission-fusion events in early stages of the disease may also be effective in repairing damaged mitochondria. However, as levels of damaged mitochondria increase, these processes lose their effectiveness and are replaced by removal of dysfunctional mitochondrial via lysosomal degradation. Genes associated with familial forms of PD including parkin, PINK, and LRRK2 have suggested that defects in mitochondrial turnover or mitophagy contribute to neuropathology and interventions that up-regulate these processes are likely to improve neuronal function [69]. TFEB has itself been recently been identified as a potential target for intervention in PD, although primarily as a means of removing defective protein aggregates (see below) [75]. Ability to remove both defective proteins and organelles suggests that TFEB induction could have dual neuroprotective benefits [76].

3.2 Oxidative Stress

Mitochondrial dysfunction itself can result in production of ROS that in turn can cause further damage to the mitochondria, resulting in a destructive feed forward cycle, ultimately resulting in neuronal cell death. Oxidative stress is another feature of normal brain aging that has been widely implicated in select cell death associated with PD (86–87). Whether or not age-related oxidative stress actually contributes to brain aging is controversial; however, DAergic neurons are particularly sensitive to the effects of oxidative stress due to several factors including their high number of synapses, their undermyelinated axons requiring large numbers of mitochondria to support their activity and the ability for dopamine to be oxidized to toxic by-products that can impact on mitochondrial function [77, 78]. This can be exacerbated by the relatively low levels of antioxidants such as glutathione within these cells, high levels of the ROS-producing enzyme monoamine oxidase B (MAO-B) within neighboring astrocytes, and regionally high levels of reactive iron [79]. Due to its proximity to mitochondrial ROS-production, lack of protective histones, and its high replication rate, mitochondrial DNA (mtDNA) is particularly prone to the deleterious effects of oxidative modification [80]. Mitochondrial damage is another feature shared by the normal aging brain and DAergic cells within the PD SNpc [81, 82].

As mentioned previously, reductions in oxidative stress via vitamin E did not show efficacy in earlier clinical trials (DATATOP), although whether this was due to inappropriate dosing regimes or lack of targeting to the most vulnerable cellular locations is unclear. A recent clinical trial using mitochondrially targeted CoQ10 (MitoQ) also failed to demonstrate slowing of clinical progression of the disease [83].

3.3 Calcium Mishandling

A loss in cellular calcium homeostasis is another potential contributor to the pathogenesis of PD that is also altered in the normal aging brain [84]. Losses in mitochondrial function can affect the ability of the organelle to sequester calcium and this in turn can result in the generation of mitochondrial-mediated oxidative stress and subsequent damage to the organelle that can further affect its function [85].

A growing body of evidence suggests that with age, nigral DAergic neurons become more reliant on L-type calcium channels to maintain pace-making activity and that agents that block channel activity such as isradipine (currently under clinical trial, STEADY PD III) may slow or halt PD progression [79, 86]. A recent study reported that DAergic SNpc neurons in slices isolated from older wild type mice (25–30 months) show significantly lower spontaneous firing rates, shorter spike widths, and lower pacemaking ability, as compared to younger animals (2–7 months) [87]. This effect is due to the impact of L-type Cav1.2α and Cav1.3α channels on pacemaker firing. This was speculated to be due to changes in calcium stores resulting in effects on calcium-activated potassium channels, suggesting that age-related alterations in the function of these channels may be an important underlying cause of PD [86, 87].

3.4 *Inflammation*

Immune cell activation likely evolved as a means of removing foreign pathogens or damaged cells, the latter as a consequence of acute cellular stress or injury. However, in the face of chronic age-related neurodegeneration, inflammatory responses may actually contribute to neuronal loss via the ongoing release of damaging inflammatory agents, including reactive oxygen and nitrogen species (ROS/RNS), inflammatory cytokines and chemokines, and matrix metalloproteinases (MMPs) [88, 89]. Several lines of evidence suggest that inflammatory processes within the central nervous system (CNS) may contribute to neuronal cell loss associated with PD [89, 90]. The SNpc itself contains a particularly high density of immune cells such as microglia, even in the normal young brain. Post-mortem studies in the PD SNpc demonstrate the presence of markers of both innate and adaptive immunity. Activated microglia, astrocytes and $CD4^+$ and $CD8^+$ T lymphocytes have been detected in affected brain regions, along with increased expression of pro-inflammatory mediators [91–94]. In addition, preclinical studies in various animal models strongly suggest the involvement of inflammatory processes in associated neuronal cell death [95]. Neuroinflammatory processes may contribute to deleterious events leading to neuronal degeneration. Results from epidemiological studies in humans have suggested that use of anti-inflammatory drugs reduces the risk for development of PD, although reports have been somewhat mixed [96–99]. Possible factors involved in neuroprotection may include inhibition of cyclooxygenase 2 and reduced production of prostaglandins [100, 101]. Interestingly, ibuprofen, which shows the strongest effects in a prospective meta-analysis study, is a known inducer of PPARγ [102]. A genome-wide association study (GWAS) has confirmed the involvement of various immune-related factors [103]. Neuroinflammatory processes represent an attractive therapeutic target for slowing progression of the disorder.

During aging, astrocyte numbers increase and a greater proportion become activated (astrogliosis) [104, 105]. This is correlated with an increase in levels of glial fibrillary activated protein (GFAP). While microglial cell numbers do not appear to be significantly increased in normal aging as they are in PD, the cells that are present are more likely to be in an activated state, known as 'microglial priming', an enhancement of the innate immune response [106].

It is important to note that in addition to involvement in neuroinflammatory events, astrocytes also normally interact with endothelial cells to help maintain integrity of the blood–brain-barrier (BBB), remove ions and glutamate from the extracellular space following bursts of neuronal activity, and respond to the metabolic needs of neurons including via the release of supportive growth factors [107]. These physiological functions decrease with brain aging and may contribute to neuronal losses associated with PD.

3.5 *Defects in Proteostasis*

A common feature of neurodegenerative diseases including PD is accumulation of proteinaceous deposits in the CNS or peripheral nervous system [28]. The location of these deposits appears to be disease-specific. Unwanted proteins are cleared from the cell by the ubiquitin proteasome system (UPS) or lysosomal-mediated autophagy [108]. Defects in both of these systems have been associated with both aging and PD, and attempts to correct them by various means have been extensively studied [109, 110]. The PD-related protein α-synuclein, associated with both familial and sporadic PD, comprises roughly 1 % of proteins found within Lewy bodies and its toxicity is believed to be caused by accumulation of misfolded proteins and formation of toxic oligomers [111]. Inclusions containing α-synuclein aggregates have been found in transplanted fetal DAergic cells in PD patients. As cells were derived from unrelated donors, it is unlikely that they are the source of the pathology and this is cited as evidence, along with recent work in mouse models, backing the 'prion transmission theory' of PD progression [28, 112]. This is another potential target for therapy, but although prion-like disease spread is known to require endosomal release and uptake, much more about the molecular mechanisms involved in this process still needs to be elucidated. A recent publication suggested that one of the factors involved in removal of intracellular α-synuclein via secretion into the extracellular space is the protein ATP13A2 or PARK9 whose mutation is involved in a juvenile onset form of parkinsonism called Kufor-Rakeb syndrome (KRS) [113].

One interventional target that has been extensively explored experimentally in PD is the heat shock proteins (HSPs) [114]. This family of proteins is a group of molecular chaperones that promote either proper refolding of proteins or their transfer to the UPS and lysosome for degradation. Drugs such as the hsp70 inducer geldanamycin have shown efficacy in several animal models of PD, but unfortunately have side effects including liver damage [115]. Drugs that enhance clearance of defective proteins have also shown efficacy in animal PD models, including rapamycin that enhances lysosomal protein turnover [116]. Interventions that bind and prevent toxic oligomerization of α-synuclein including polyphenols like curcumin, immunotherapy or siRNA therapy have also been demonstrated to have some benefits in animal models and a clinical trial of the latter (AFFiRiS) has recently been initiated [117, 118]. There is however some concern that this may prevent potentially neuroprotective aggregation of the protein, and this may actually exacerbate the condition. It is also conceivable that post-translational modifications that drive aggregate formation may evade antibody binding and may increase neuroinflammation (see above).

4 What Have We Learned and Where Do We Go from Here?

Development of PD therapeutics that halt or slow disease progression and prevent disability is the highest priority for researchers in the field. Unfortunately, attempts to intervene have to date provided little to no benefit in human clinical trials [51]. There may be several contributing factors involved, including heterogeneity of the patient populations examined, interventions attempted too late in disease stage for effect, short durations of treatments, and small group sizes. It may also reflect lack of thoroughly vetted pre-clinical studies in order to better understand dose responses, treatment pharmacokinetics, and appropriate therapeutic windows. Identification of a disease-modifying neuroprotective intervention has remained elusive to date, likely due in part to these factors. It may however also imply that exploration of novel targets far afield from those conventionally studied is needed.

5 Possible Role of Major Hallmarks of Aging

A traditional approach in age-related disease research has been to investigate single disease conditions in isolation. While this has yielded important information and enormous efforts have been expended to develop therapies based on these findings, there have been few successful interventions as a consequence. This is likely due to an incomplete picture of the nature of complex chronic disease states. Preventive and therapeutic treatments for diseases like PD will likely require the discovery of unique targets far removed from those identified to date by conventional approaches.

Recently, it has begun to be appreciated that in order to fully understand these disorders, researchers need to better understand the underlying role of aging in conditions as disparate as atherosclerosis, diabetes, cancer, osteoporosis, and neurodegenerative diseases. Studying the role of aging mechanisms across a wide variety of disease states will allow scientists to broaden the scope of research beyond traditional disciplines, towards the central concept that these multiple human disease states likely arise from a common underlying cause: aging itself. The term "geroscience" was coined by scientists at the Buck Institute for Research on Aging in 2007 as an acknowledgement and organizing principle of this scientific concept. In 2011, the Geroscience Interest Group (GSIG) was formed, a trans-NIH team interested in this concept, and geroscience as a scientific discipline was formally recognized in the U.S. Senate Appropriations in 2013. The NIH, with support from the Alliance for Aging Research and the Gerontological Society of America, hosted a conference entitled "Advances in Geroscience: Impacts on Healthspan and Chronic Disease" in October of 2013 to examine the extent to which the physiological effects of aging represent a common major risk factor for chronic diseases. This led to a leading-edge commentary in the journal *Cell* entitled 'Geroscience: Linking Aging to Chronic Disease' [119].

PD itself falls under this rubric and potentially could be viewed as a case of accelerated aging in a particular population of cells in response to individual variation in genetic makeup and environmental exposures. Interestingly, the SNpc displays more age-related neuronal cell loss than any other region of the brain [65]. How does aging then contribute to the preferential death of these cells?

5.1 Age-Related Loss in Adaptation to Stress and Increased Macromolecular Damage

Aging leads to a decrease in the efficiency of many aspects of organismal stress-resistance [120]. This includes reductions in the ability to maintain proteostasis and function of organelles such as the mitochondria and endoplasmic reticulum (ER) due to losses in both the Unfolded Protein Stress (UPS) and autophagic functions. Cellular defenses against the effects of oxidative stress are also reduced. The decreased ability of aged organisms to deal with stressors can lead to damage to both proteins and organelles that, due to their unique cellular features, can have particularly devastating effects on DAergic SNpc neurons [121]. Many of the disease-related effects that occur within the PD SNpc also occur during normal aging, albeit to a lesser degree [87]. These include: (1) effects on mitochondrial function including reduced enzymatic activity of both mitochondrial complex I and α-ketoglutarate dehydrogenase (KDGH), reduced NAD^+ levels, and increased mtDNA deletions [122–125]; (2) reduced ability to deal with effects of increased cellular ROS including reductions in cellular glutathione content and reduced activity of the major anti-oxidant regulator nuclear factor erythroid-derived 2 (NRF2), coupled with increased levels of redox-available iron in this brain region [126–128]; and (3) increased levels of protein accumulation, including elevated levels of α-synuclein [129]. In the normal aging SNpc, α-synuclein appears to be primarily in a more soluble form rather than associated with Lewy bodies as in the PD SNpc, although in both aging and disease this appears to be associated with a decline in neuronal function [130]. This may be due to the presence of higher levels of ROS and RNS in the context of PD. Accumulation of α-synuclein has been linked to alterations in mitochondrial fission-fusion and function [131]. Taken together, these data suggests that PD may constitute acceleration of the aging process in the SNpc in susceptible individuals. Reductions in the deleterious effects of these age-related changes may therefore greatly reduce the risk of developing PD.

Conversion of α-ketoglutarate and NAD^+ to succinyl coA and NADH by KDGH is the rate-limiting step in the conversion of carbohydrates to energy as part of the mitochondrial TCA cycle. NADH and succinate feed into electron transport chain (ETC) complexes I and II, respectively, to support mitochondrial respiration. KDGH activity in the brain is reduced in both aging and in PD, likely as a consequence of increases in mitochondrial ROS and lipid peroxidation to which the enzyme is particularly susceptible [132, 133]. This is reversible by raising levels of thiols, particularly glutathione, which is normally found at millimolar levels in the cell and is reduced during normal aging and in PD [134]. Lipoic acid, via its ability to activate the NRF2 pathway, can induce phase II genes including those involved in glutathione synthesis and metabolism [135]. NRF2 is also involved in a pathway resulting in induction of the mitochondrial transcription factor A (TFAM) that controls mtDNA transcription, translation, and repair. Accumulation of mtDNA damage has been reported to be particularly high in the aged SNpc and conditionally knocking out TFAM selectively within DAergic neurons results in PD-like neuronal cell loss and formation of protein inclusions, suggesting age-related increases in mtDNA mutation load may be a contributing factor to the disease [136]. Lipoic acid has been reported to have restorative effects on mitochondrial complex I activity in both aging and PD models [66].

NRF2 is induced by PGC1α and its co-activators, coordinating in turn expression of several nuclear-encoded mitochondrial proteins including TFAM. Losses in PGC1α have been implicated in age-related reduction in muscular strength as well as cardiac and cognitive function, due to an increase in metabolic abnormalities and subsequent reduction in cellular, tissue, and organ function [137]. PGC1α has also recently been identified as a central therapeutic target for the treatment of PD [138].

In addition to PGC1α, mitochondrial biogenesis is also controlled by the NAD^+ dependent protein deacetylase sirtuin 1 (SIRT1). Sirtuins are a family of conserved enzymes whose modulation has been demonstrated to alter the course of aging in various model systems. Alterations in the activity of these enzymes have also recently been shown to have effects in neurodegenerative diseases, including PD [139]. Activation of SIRT1 by resveratrol, for example, has been shown to be neuroprotective in mouse PD models [140]. SIRT1 inactivation has been reported to result in increased activity of the heat shock factor I (HSF1) that drives transcription of a group of molecular chaperones that regulate protein homeostasis [141]. Recently, inactivation of SIRT1 via its nitrosylation, which prevents its ability to bind zinc, has been demonstrated to result in enhanced inflammation including in a mouse model of PD, contributing to neurodegenerative effects. This was discovered to be due to activation of p53 and subsequent induction of NFkappaB [142]. The authors point to inhibition of SIRT1 by nitrosylation as a possible major target for other major age-related diseases that involve inflammatory processes, including diabetes, atherosclerosis, and Alzheimer's disease (see below). In contrast to SIRT1, SIRT2 inhibition has been reported to be protective in the MPTP model of PD, by preventing deacetylation of Foxo3a and subsequent activation of apoptosis via the factor BIM [143]. Its inhibition may also prevent formation of α-synuclein aggregates.

Aging is correlated with loss in function of various molecular chaperones normally involved in repair of conformational alterations in cellular proteins in response to stress events. This is particularly relevant in post-mitotic cells such as neurons, and chaperone deficiency has been implicated in several neurodegenerative diseases, including PD [144–146]. Loss of the ability to respond to stress can result in accumulation of age- and disease-related misfolded proteins, protein aggregation, and disruption of cellular function. The capacity to induce heat shock proteins including HSP70 and HSP90 is compromised in both the aging brain and in PD and further exacerbated in the face of age-related increases in oxidative damage. Interventions towards replacing levels of chaperone function have been proposed as a potential therapeutic for brain aging.

5.2 *Systemic Inflammaging: 'Damage at a Distance'*

In addition to local inflammation within the brain itself, systemic inflammation has also been suggested to play a role in PD [147]. The brain is normally protected from the effects of systemic inflammation due to the presence of the blood–brain barrier (BBB). Age-related damage, however, contributes to BBB leakiness, allowing the influx of systemic pro-inflammatory factors into the brain and resulting in the activation of microglia and astrocytes that can exacerbate ongoing neurodegeneration [148]. Recent studies suggest that activation of the peripheral immune system can elicit a pro-inflammatory response in the brain of aged subjects that does not occur in younger cohorts. For example, peripheral lipopolysaccharide (LPS) challenge in older mice results in enhanced microglial secretion of the pro-inflammatory cytokines IL-1β and IL-1 [149]. It has been suggested that this may be due to age-related microglial priming resulting in enhanced activation following entrance of immune signals from the periphery, releasing elevated levels of pro-inflammatory cytokines. Infectious diseases are associated with increased risk for development of PD, particularly in the elderly [150].

In addition to systemic pro-inflammatory factors, damage to the BBB can also allow the entrance of peripheral immune cells into the brain, including neutrophils and macrophages [151–153]. Secretion of chemokines by activated microglia can attract neutrophils and monocytes from the bloodstream. In contrast, up-regulation of anti-inflammatory factors in the periphery may act to reduce glial cell activation in the brain and therefore neuropathology. In addition to entrance via a leaky BBB, inflammatory agents may also enter the CNS via the autonomic nervous system, particularly the vagal nerve afferents which lies close to the liver and lymphatic nodes [154, 155]. Conversely, damage within the brain may trigger inflammatory effects in the periphery. For example, brain injury has been reported to result in increases in pro-inflammatory cells in the liver, resulting in neutrophil translocation in the brain [156].

5.3 Cellular Senescence

Cellular senescence is a potent anti-cancer mechanism that occurs in most, if not all, dividing cell types [157]. The senescence response arrests cell proliferation, stably and essentially irreversibly, in response to stresses that puts cells at risk for malignant transformation. Senescent cells can secrete numerous pro-inflammatory cytokines, chemokines, growth factors and proteases, a feature termed the senescence-associated secretory phenotype (SASP) [158–162]. Many SASP factors have been shown, or are suspected, to cause or contribute to the loss of tissue structure and function that occurs with age, including those associated with neurodegenerative disease. A seminal publication showed that elimination of senescent cells that accumulate in a progeroid mouse model prevents the onset of three major aging phenotypes (cataracts, sarcopenia and loss of subcutaneous fat), providing the first evidence that senescent cells play a causal role in at least some age-related pathologies in vivo [163]. While cell senescence has been causally linked to age-related pathologies in peripheral tissues, its potential role in brain aging and neurodegenerative disease has just begun to be explored. Recent findings suggest that cultured glial cells are capable of undergoing senescence and developing a SASP. In response to exogenous H_2O_2, for example, cultured astrocytes displayed numerous senescent characteristics: arrested growth, an enlarged morphology, senescence-associated β-galactosidase (SA-βgal) activity, and increased expression of p21 and $p16^{INK4a}$ [164]. In vivo and in conjunction with age-related cognitive impairment, changes in the aging brain include increased expression of astrocytic GFAP in cells with a senescent-like morphology. Telomere shortening in rat microglia both in culture following repeated cell divisions and with advancing age in vivo has been reported to lead to cellular senescence that may impact cellular function [165, 166]. In response to repeated lipopolysaccharide administration, cultured microglial cells display growth arrest, enhanced SA-βgal activity, and senescence-associated heterochromatic foci [167]. This may be what primes microglia for enhanced activation in response to systemic inflammatory stimuli. Both normal brain aging and chronic age-related neurodegenerative diseases are associated with microglial-mediated increases in components that are associated with the SASP, including pro-inflammatory cytokines such as IL-1β and IL-6. Other non-neuronal cell types in the brain may also be capable of undergoing senescence including oligodendrocytes, which could affect neuronal myelination and subsequent signaling capabilities as well as endothelial cells, resulting in breakdown of the BBB and influx of peripheral inflammatory factors [168]. Cellular senescence has been reported to occur in the vascular endothelium in the periphery, suggesting that this same cell type may be vulnerable in the aging brain. Replication-competent cell types capable of undergoing senescence and expressing a SASP could potentially affect the function of neighboring neurons and contribute to their degeneration. They could also promote an amplification of glial cell activation. A recent publication showed that both $p16^{INK4a}$ and the SASP factor matrix metalloproteinase (MMP) 3 increase in brains from AD patients compared to age-matched controls; interestingly, this increase occurred primarily in cortical astrocytes [169]. This may be important not only in these disease states themselves, but in terms of the effective use of cellular transplantation as a therapy for these disorders (see below).

5.4 Reduced Adult Neurogenesis

Neurogenesis entails the production of new neurons from neural precursor cells (NPCs). In adult brains, NPCs proliferate in the subventricular zone (SVZ) of the lateral ventricle and the subgranular zone (SGZ) of the hippocampal dentate gyrus (DG) [170]. They then migrate to sites of injury (in the case of PD, the nigrostriata) and differentiate into functional neuronal and glial cell types. We and others showed that the ability of NPCs to proliferate, migrate, and differentiate is significantly blunted by advancing age and the degree of this decline is exacerbated by PD-inducing agents such as MPTP [171–175]. This may be due to increased oxidative stress or neuroinflammation, resulting in neural progenitor cell (NPC) senescence or loss. It is currently not clear that induction of endogenous NPCs alone is sufficient in the context of the aging brain to restore losses in DAergic SNpc neurons. In addition to neurons, it will also likely be of import to consider alterations in the regenerative capacity and function of non-neuronal brain cells during aging and in PD.

Cellular transplantation to replace lost or damaged neurons in patients with the disease is a therapeutic option that mimics what occurs to a lesser degree during endogenous adult neurogenesis. However as mentioned previously, there are potential roadblocks. Upon extensive culturing (2 months), DAergic neurons derived from patients with sporadic PD show characteristic morphological features found in the diseased brain, including reduced neurite number, accumulation of cellular α-synuclein, and increased numbers of autophagic vacuoles [176]. This suggests that long-term cell survival may be diminished, particularly in the environment of on aging brain. It would be of interest to know whether cell survival is increased in brains made more youthful, for example following removal of senescent cells.

5.5 Beyond Mendelian Genetics: Epigenetics and the Transcriptome

Increases in DNA methylation and alterations in post-translational histone modifications including changes in histone methylation and acetylation patterns have been reported in various models of PD and observed in patient samples [177]. For example, DNA methylation has been shown to regulate expression of the gene encoding α-synuclein, SNCA, as well as others. Masliah and colleagues have reported that α-synuclein can interact with the demethylase Dnmt1 in the cytoplasm, preventing it from translocating to the nucleus and resulting in elevations in DNA methylation [178, 179]. Alpha-synuclein has itself been reported to interact directly with histones and to inhibit histone H3 acetylation [180, 181]. Both of these can impact gene expression patterns. Our laboratory has shown that selective binding of α-synuclein to particular histone marks not only tracks with human disease and occurs in PD models, but also leads to functional effects on patterns of genome-wide transcription, including expression of PGC1alpha [181] (our unpublished data). HDAC inhibitors have been shown to be neuroprotective in various models of the disorder [182]. The histone deacetylase LSD-1 is inhibited by lithium, an agent shown by our laboratory to be neuroprotective in PD mouse models [183, 184]. LSD-1 is also inhibited by MAO-B inhibitors currently under use as treatment for the disorder [143]. A known target of transcriptional LSD-1 repression is *p57*^{kip2}, *a* gene that aids in the differentiation and maintenance of midbrain DAergic neurons [185]. Beyond its effects on histones, LSD-1 inhibition can also impact on p53's ability to interact with 53BP1 and to induce apoptosis [186]. Its other non-histone substrates include FOXO transcription factors and NFkappaB [187]. Lithium and LSD-1 inhibition both result in increased lifespan in *C. elegans*, suggesting that they may have impacts in terms of aging itself [188].

Aging is also associated with extensive remodeling of gene expression profiles in different tissues as a consequence of epigenetic alterations. The HSPs 22 and 70, implicated in PD as well as in lifespan determination, are strongly regulated by histone modification [189]. Changes in mitochondrial function that increase cellular ROS and affect NAD^+/NADH ratios in PD and aging may also impact nuclear DNA methylation and histone modifications [190, 191]. Alterations in epigenetics as a consequence of aging and PD are likely impacted by individual environmental factors over the lifespan, which may in part explain individual variability in the presentation of both conditions.

6 Needed Research in This Area

There are several areas of research needed in order for us to achieve a better understanding of the interplay between aging and PD. These include a better understanding of the dual protective roles of autophagy in turnover of damaged proteins and organelles like the mitochondria, the precise sources of inflammation (glial cell activation, cellular senescence), and causes of lost neurogenesis in adult neural stem cells (e.g. whether cellular senescence is a factor). We also need to better understand the difference between transient stressors that allow adaptation versus chronic stressors which result in damage—retrograde mitochondrial signaling including brief elevations in mitochondrial ROS is a good example of the former [192]. More work needs to be devoted to linking findings in cellular and animal models to humans. This include the development of working mouse models of PD that take aging into consideration, as well as the use of patient iPSC-derived DAergic neurons, human autopsy tissues, and epidemiological/clinical studies. Undoubtedly this would be aided by an increased use of non-human primate models in both PD and aging. Studies in rhesus monkeys show a strong correlation not only between the presence of α-synuclein in aged versus MPTP-treated monkeys, but also of other disease hallmarks including selective reductions in DAergic SNpc cell number, defects in UPS and lysosomal activities, markers of oxidative and nitrosative stress, and neuroinflammation [193]. Marmosets are shorter-lived and have also been used in PD research, so they represent an attractive alternative model [194].

An especially vital area of research is obtaining a better understanding of the interrelationship between genes and environment (the 'exposome') on PD and aging [195]. Dietary fat and exposure to heavy metals and pesticides are all examples of environmental factors linked to increased risk for PD [196–198]. The relationship between dietary restriction (DR), the gut microbiome, and aging may also have important implications for mechanisms underlying PD. Intermittent fasting (every other day fasting) has been proposed to have an effect on brain function [199]. DR has been linked to prevention of aging in several model systems as well as beneficial effects in PD mouse models [200]. PD patients show elevated heart rate and impaired cardiovascular stress response due to effects on the autonomic nervous system preceding DAergic SNpc loss correlating with α-synuclein accumulation [201]; DR resulted in prevention of these effects in a mouse model of the disease [202]. Production of ketone bodies during periods of fasting may also have beneficial effects in the brain due to inhibition of histone deacetylases, impacting on the transcription of genes like neurotrophic BDNF, reduction of pro-inflammatory cytokines, enhancement of neurogenesis, and perhaps even induction of mitochondrial biogenesis via SIRT1-related elevation in PGC1α levels [203].

The diversity and make-up of the gut microbiome has been shown to change with age, coinciding with inflammaging [204]. These alterations have been demonstrated to be involved in risk for chronic age-related diseases including cardiovascular disease, inflammatory bowel syndrome, metabolic disease, and cancer [205]. This is alterable for better or worse by lifestyle and diet, and as a consequence the gut microbiome has been identified as a target for improving overall health in the elderly population [206]. Scientific evidence for an involvement of the gut microbiome in brain function has recently begun to gain ground for disorders such as autism and depression [207]. Given its role in inflammaging, composition of the gut microbiome could potentially also play a role in neurodegenerative diseases like PD. The gut microbiome is responsible for the production and processing of micronutrients such as folate, thiamine, riboflavin, and biotin. Deficiencies in these have all been linked to PD as well as to aging in some cases [208]. Pyroxidine is also produced via activity of gut microbes and is known to accelerate the rate of conversion of L-Dopa in the periphery, which can be slowed by inclusion of carbidopa [209]. Disruptions in circadian rhythms have recently been linked to alterations in the gut microbiome [210]. Both are associated with similar chronic age-related diseases [205, 211]. Mice with genetically altered circadian rhythms were found to have significantly altered gut microbiota when fed a high-fat, high-sugar diet [212, 213]. These data suggest that age-related changes in the circadian clock can drive changes in the microbiome and impact chronic disease states, perhaps including PD. Indeed loss of midbrain DAergic neurons in the Mitopark mouse has recently been found to be associated with disruption of circadian rhythms [214].

Recent animal studies have also shown that gut microorganisms can activate the vagus nerve via immunomodulatory effects and that this plays a critical role in mediating brain function [215, 216]. The vagus nerve connects the enteric nervous system to the brain and is considered a possible pathway for transmission of α-synuclein [217]. Understanding the role of the vagus nerve may have important implications for the development of microbial- or nutrition-based therapeutic strategies for PD and related disorders.

7 Conclusions

The most recent scientific evidence in the field suggests that PD is one of several disorders for which aging is not merely a risk factor, but an underlying cause of disease. This suggests the exciting possibility that by seeking to prevent or slow basic processes that drive aging, we will uncover potent new therapies for PD and related neurodegenerative conditions as well as other age-related disorders. This enterprise will involve additional research in order to identify the most promising potential therapeutic directions.

Acknowledgments This work was funded by an Ellison Senior Research Fellowship and the Buck Impact Circle (JKA).

Editor: Jovier Evans, National Institute of Mental Health, NIMH (NIH)

References

1. de Lau LM, Breteler MM (2006) Epidemiology of Parkinson's disease. Lancet Neurol 5(6):525–535
2. Jankovic J (2008) Parkinson's disease: clinical features and diagnosis. J Neurol Neurosurg Psychiatry 79(4):368–376
3. Hoehn MM, Yahr MD (1967) Parkinsonism: onset, progression and mortality. Neurology 17(5):427–442
4. Macleod AD, Taylor KS, Counsell CE (2014) Mortality in Parkinson's disease: a systematic review and meta-analysis. Mov Disord 29(13):1615–1622
5. Davie CA (2008) A review of Parkinson's disease. Br Med Bull 86:109–127
6. Lees AJ, Hardy J, Revesz T (2009) Parkinson's disease. Lancet 373(9680):2055–2066
7. Jankovic J, Aguilar LG (2008) Current approaches to the treatment of Parkinson's disease. Neuropsychiatr Dis Treat 4(4):743–757
8. Cooper CA et al (2009) Does laterality of motor impairment tell us something about cognition in Parkinson disease? Parkinsonism Relat Disord 15(4):315–317
9. Olanow CW, Brin MF, Obeso JA (2000) The role of deep brain stimulation as a surgical treatment for Parkinson's disease. Neurology 55(12 Suppl 6):S60–S66
10. Weintraub D, Comella CL, Horn S (2008) Parkinson's disease – Part 2: treatment of motor symptoms. Am J Manag Care 14(2 Suppl):S49–S58

11. Rascol O et al (2003) Limitations of current Parkinson's disease therapy. Ann Neurol 53(Suppl 3):S3–S12; discussion S12–5
12. Pillon B et al (1989) Does cognitive impairment in Parkinson's disease result from non-dopaminergic lesions? J Neurol Neurosurg Psychiatry 52(2):201–206
13. Agid Y et al (1989) Biochemistry of Parkinson's disease 28 years later: a critical review. Mov Disord 4(Suppl 1):S126–S144
14. Wakabayashi K, Takahashi H (1997) Neuropathology of autonomic nervous system in Parkinson's disease. Eur Neurol 38(Suppl 2):2–7
15. Martignoni E et al (1995) Autonomic disorders in Parkinson's disease. J Neural Transm Suppl 45:11–19
16. Xiao Q, Chen S, Le W (2014) Hyposmia: a possible biomarker of Parkinson's disease. Neurosci Bull 30(1):134–140
17. Silveira-Moriyama L et al (2009) Olfaction in patients with suspected parkinsonism and scans without evidence of dopaminergic deficit (SWEDDs). J Neurol Neurosurg Psychiatry 80(7):744–748
18. Sixel-Doring F et al (2014) Rapid eye movement sleep behavioral events: a new marker for neurodegeneration in early Parkinson disease? Sleep 37(3):431–438
19. Truong DD, Bhidayasiri R, Wolters E (2008) Management of non-motor symptoms in advanced Parkinson disease. J Neurol Sci 266(1–2):216–228
20. Poewe W et al (2008) Diagnosis and management of Parkinson's disease dementia. Int J Clin Pract 62(10):1581–1587
21. Bloem BR, van Vugt JP, Beckley DJ (2001) Postural instability and falls in Parkinson's disease. Adv Neurol 87:209–223
22. Pinter B et al (2015) Mortality in Parkinson's disease: a 38-year follow-up study. Mov Disord 30(2):266–269
23. Recasens A, Dehay B (2014) Alpha-synuclein spreading in Parkinson's disease. Front Neuroanat 8:159
24. Eller M, Williams DR (2011) alpha-Synuclein in Parkinson disease and other neurodegenerative disorders. Clin Chem Lab Med 49(3):403–408
25. Goedert M et al (2013) 100 years of Lewy pathology. Nat Rev Neurol 9(1):13–24
26. Braak H et al (2013) Age-related appearance of dendritic inclusions in catecholaminergic brainstem neurons. Neurobiol Aging 34(1):286–297
27. Clairembault T et al (2015) Enteric glial cells: new players in Parkinson's disease? Mov Disord 30(4):494–498
28. Luk KC, Lee VM (2014) Modeling Lewy pathology propagation in Parkinson's disease. Parkinsonism Relat Disord 20(Suppl 1):S85–S87
29. Visanji NP et al (2014) Alimentary, my dear Watson? The challenges of enteric alpha-synuclein as a Parkinson's disease biomarker. Mov Disord 29(4):444–450
30. Jankovic J (2002) Levodopa strengths and weaknesses. Neurology 58(4 Suppl 1):S19–S32
31. Parkinson Study Group, Q.E.I et al (2014) A randomized clinical trial of high-dosage coenzyme Q10 in early Parkinson disease: no evidence of benefit. JAMA Neurol 71(5):543–552
32. Pilleri M, Antonini A (2015) Therapeutic strategies to prevent and manage dyskinesias in Parkinson's disease. Expert Opin Drug Saf 14(2):281–294
33. Wilson JM et al (1996) Differential changes in neurochemical markers of striatal dopamine nerve terminals in idiopathic Parkinson's disease. Neurology 47(3):718–726
34. Bodagh IY, Robertson DR (1994) A risk-benefit assessment of drugs used in the management of Parkinson's disease. Drug Saf 11(2):94–103
35. Stathis P, Konitsiotis S, Antonini A (2015) Dopamine agonists early monotherapy for the delay of development of levodopa-induced dyskinesias. Expert Rev Neurother 15(2):207–213
36. Tintner R, Jankovic J (2002) Treatment options for Parkinson's disease. Curr Opin Neurol 15(4):467–476

37. Zhang G et al (2014) Impulsive and compulsive behaviors in Parkinson's disease. Front Aging Neurosci 6:318
38. Rascol O et al (2011) A double-blind, delayed-start trial of rasagiline in Parkinson's disease (the ADAGIO study): prespecified and post-hoc analyses of the need for additional therapies, changes in UPDRS scores, and non-motor outcomes. Lancet Neurol 10(5):415–423
39. Pagano G et al (2015) Cholinesterase inhibitors for Parkinson's disease: a systematic review and meta-analysis. J Neurol Neurosurg Psychiatry 86(7):767–773
40. Follmer C (2014) Monoamine oxidase and alpha-synuclein as targets in Parkinson's disease therapy. Expert Rev Neurother 14(6):703–716
41. Dezsi L, Vecsei L (2014) Safinamide for the treatment of Parkinson's disease. Expert Opin Investig Drugs 23(5):729–742
42. Deuschl G et al (2006) A randomized trial of deep-brain stimulation for Parkinson's disease. N Engl J Med 355(9):896–908
43. Eller T (2011) Deep brain stimulation for Parkinson's disease, essential tremor, and dystonia. Dis Mon 57(10):638–646
44. Lamotte G et al (2015) Effects of endurance exercise training on the motor and Non-motor features of Parkinson's disease: a review. J Parkinsons Dis 5(1):21–41
45. Shanahan J et al (2015) Dance for people with Parkinson disease: what is the evidence telling Us? Arch Phys Med Rehabil 96(1):141–153
46. Suchowersky O et al (2006) Practice Parameter: neuroprotective strategies and alternative therapies for Parkinson disease (an evidence-based review): report of the Quality Standards Subcommittee of the American Academy of Neurology. Neurology 66(7):976–982
47. Storch A et al (2007) Randomized, double-blind, placebo-controlled trial on symptomatic effects of coenzyme Q(10) in Parkinson disease. Arch Neurol 64(7):938–944
48. Elm JJ, Investigators NN-P (2012) Design innovations and baseline findings in a long-term Parkinson's trial: the National Institute of Neurological Disorders and Stroke Exploratory Trials in Parkinson's Disease Long-Term Study-1. Mov Disord 27(12):1513–1521
49. Etminan M, Gill SS, Samii A (2005) Intake of vitamin E, vitamin C, and carotenoids and the risk of Parkinson's disease: a meta-analysis. Lancet Neurol 4(6):362–365
50. Zhang SM et al (2002) Intakes of vitamins E and C, carotenoids, vitamin supplements, and PD risk. Neurology 59(8):1161–1169
51. Athauda D, Foltynie T (2015) The ongoing pursuit of neuroprotective therapies in Parkinson disease. Nat Rev Neurol 11(1):25–40
52. Roy A, Pahan K (2011) Prospects of statins in Parkinson disease. Neuroscientist 17(3):244–255
53. Sherer TB et al (2006) Crossroads in GDNF therapy for Parkinson's disease. Mov Disord 21(2):136–141
54. Richardson RM et al (2011) Interventional MRI-guided putaminal delivery of AAV2-GDNF for a planned clinical trial in Parkinson's disease. Mol Ther 19(6):1048–1057
55. Bartus RT et al (2011) Bioactivity of AAV2-neurturin gene therapy (CERE-120): differences between Parkinson's disease and nonhuman primate brains. Mov Disord 26(1):27–36
56. Bartus RT et al (2011) Properly scaled and targeted AAV2-NRTN (neurturin) to the substantia nigra is safe, effective and causes no weight loss: support for nigral targeting in Parkinson's disease. Neurobiol Dis 44(1):38–52
57. Herzog CD et al (2013) Enhanced neurotrophic distribution, cell signaling and neuroprotection following substantia nigral versus striatal delivery of AAV2-NRTN (CERE-120). Neurobiol Dis 58:38–48
58. Freed CR et al (2003) Do patients with Parkinson's disease benefit from embryonic dopamine cell transplantation? J Neurol 250(Suppl 3):III44–III46
59. Bjorklund A et al (2003) Neural transplantation for the treatment of Parkinson's disease. Lancet Neurol 2(7):437–445

60. Ambasudhan R et al (2014) Potential for cell therapy in Parkinson's disease using genetically programmed human embryonic stem cell-derived neural progenitor cells. J Comp Neurol 522(12):2845–2856
61. Ray S et al (2014) The Parkinson disease-linked LRRK2 protein mutation I2020T stabilizes an active state conformation leading to increased kinase activity. J Biol Chem 289(19):13042–13053
62. Bjorklund A, Kordower JH (2013) Cell therapy for Parkinson's disease: what next? Mov Disord 28(1):110–115
63. Hardy J et al (2009) The genetics of Parkinson's syndromes: a critical review. Curr Opin Genet Dev 19(3):254–265
64. Collier TJ, Kanaan NM, Kordower JH (2011) Ageing as a primary risk factor for Parkinson's disease: evidence from studies of non-human primates. Nat Rev Neurosci 12(6):359–366
65. Reeve A, Simcox E, Turnbull D (2014) Ageing and Parkinson's disease: why is advancing age the biggest risk factor? Ageing Res Rev 14:19–30
66. Phillipson OT (2014) Management of the aging risk factor for Parkinson's disease. Neurobiol Aging 35(4):847–857
67. Hindle JV (2010) Ageing, neurodegeneration and Parkinson's disease. Age Ageing 39(2):156–161
68. Esposito E, Di Matteo V, Di Giovanni G (2007) Death in the substantia nigra: a motor tragedy. Expert Rev Neurother 7(6):677–697
69. Camilleri A, Vassallo N (2014) The centrality of mitochondria in the pathogenesis and treatment of Parkinson's disease. CNS Neurosci Ther 20(7):591–602
70. Nicholls DG, Budd SL (2000) Mitochondria and neuronal survival. Physiol Rev 80(1):315–360
71. Bolanos JP, Almeida A, Moncada S (2010) Glycolysis: a bioenergetic or a survival pathway? Trends Biochem Sci 35(3):145–149
72. Perier C, Vila M (2012) Mitochondrial biology and Parkinson's disease. Cold Spring Harb Perspect Med 2(2):a009332
73. Gu G et al (2002) Mitochondrial DNA deletions/rearrangements in parkinson disease and related neurodegenerative disorders. J Neuropathol Exp Neurol 61(7):634–639
74. Tsunemi T et al (2012) PGC-1alpha rescues Huntington's disease proteotoxicity by preventing oxidative stress and promoting TFEB function. Sci Transl Med 4(142):142ra97
75. Decressac M et al (2013) TFEB-mediated autophagy rescues midbrain dopamine neurons from alpha-synuclein toxicity. Proc Natl Acad Sci U S A 110(19):E1817–E1826
76. Decressac M, Bjorklund A (2013) TFEB: Pathogenic role and therapeutic target in Parkinson disease. Autophagy 9(8):1244–1246
77. Hirsch E, Graybiel AM, Agid YA (1988) Melanized dopaminergic neurons are differentially susceptible to degeneration in Parkinson's disease. Nature 334(6180):345–348
78. Segura-Aguilar J et al (2014) Protective and toxic roles of dopamine in Parkinson's disease. J Neurochem 129(6):898–915
79. Sulzer D (2007) Multiple hit hypotheses for dopamine neuron loss in Parkinson's disease. Trends Neurosci 30(5):244–250
80. Shokolenko IN, Wilson GL, Alexeyev MF (2014) Aging: A mitochondrial DNA perspective, critical analysis and an update. World J Exp Med 4(4):46–57
81. Santos RX et al (2013) Mitochondrial DNA oxidative damage and repair in aging and Alzheimer's disease. Antioxid Redox Signal 18(18):2444–2457
82. Sanders LH et al (2014) Mitochondrial DNA damage: molecular marker of vulnerable nigral neurons in Parkinson's disease. Neurobiol Dis 70:214–223
83. Snow BJ et al (2010) A double-blind, placebo-controlled study to assess the mitochondria-targeted antioxidant MitoQ as a disease-modifying therapy in Parkinson's disease. Mov Disord 25(11):1670–1674
84. Chan CS, Gertler TS, Surmeier DJ (2009) Calcium homeostasis, selective vulnerability and Parkinson's disease. Trends Neurosci 32(5):249–256

85. Zhu J, Chu CT (2010) Mitochondrial dysfunction in Parkinson's disease. J Alzheimers Dis 20(Suppl 2):S325–S334
86. Chan CS et al (2007) 'Rejuvenation' protects neurons in mouse models of Parkinson's disease. Nature 447(7148):1081–1086
87. Costa KM (2014) The effects of aging on substantia nigra dopamine neurons. J Neurosci 34(46):15133–15134
88. Su X, Federoff HJ (2014) Immune responses in Parkinson's disease: interplay between central and peripheral immune systems. Biomed Res Int 2014:275178
89. McGeer PL, McGeer EG (2004) Inflammation and the degenerative diseases of aging. Ann N Y Acad Sci 1035:104–116
90. Kannarkat GT, Boss JM, Tansey MG (2013) The role of innate and adaptive immunity in Parkinson's disease. J Parkinsons Dis 3(4):493–514
91. More SV et al (2013) Cellular and molecular mediators of neuroinflammation in the pathogenesis of Parkinson's disease. Mediators Inflamm 2013:952375
92. Mogi M et al (1994) Interleukin-1 beta, interleukin-6, epidermal growth factor and transforming growth factor-alpha are elevated in the brain from parkinsonian patients. Neurosci Lett 180(2):147–150
93. Mogi M et al (1994) Tumor necrosis factor-alpha (TNF-alpha) increases both in the brain and in the cerebrospinal fluid from parkinsonian patients. Neurosci Lett 165(1–2):208–210
94. Perry VH (2012) Innate inflammation in Parkinson's disease. Cold Spring Harb Perspect Med 2(9):a009373
95. Ton TG et al (2006) Nonsteroidal anti-inflammatory drugs and risk of Parkinson's disease. Mov Disord 21(7):964–969
96. Gao X et al (2011) Use of ibuprofen and risk of Parkinson disease. Neurology 76(10):863–869
97. Chen Y, Swanson RA (2003) Astrocytes and brain injury. J Cereb Blood Flow Metab 23(2):137–149
98. Hirsch EC, Hunot S (2009) Neuroinflammation in Parkinson's disease: a target for neuroprotection? Lancet Neurol 8(4):382–397
99. Esposito E et al (2007) Non-steroidal anti-inflammatory drugs in Parkinson's disease. Exp Neurol 205(2):295–312
100. Teismann P (2012) COX-2 in the neurodegenerative process of Parkinson's disease. Biofactors 38(6):395–397
101. Figueiredo-Pereira ME et al (2014) Neuroinflammation and J2 prostaglandins: linking impairment of the ubiquitin-proteasome pathway and mitochondria to neurodegeneration. Front Mol Neurosci 7:104
102. Gagne JJ, Power MC (2010) Anti-inflammatory drugs and risk of Parkinson disease: a meta-analysis. Neurology 74(12):995–1002
103. Lubbe S, Morris HR (2014) Recent advances in Parkinson's disease genetics. J Neurol 261(2):259–266
104. Cotrina ML et al (2001) Expression and function of astrocytic gap junctions in aging. Brain Res 901(1–2):55–61
105. Rodriguez-Arellano JJ et al (2015) Astrocytes in physiological aging and Alzheimer's disease. Neuroscience pii: S0306-4522(15)00031-7. doi:10.1016/j.neuroscience.2015.01.007. [Epub ahead of print]
106. Patterson SL (2015) Immune dysregulation and cognitive vulnerability in the aging brain: interactions of microglia, IL-1beta, BDNF and synaptic plasticity. Neuropharmacology 96(Pt A):11–18
107. Perea G, Sur M, Araque A (2014) Neuron-glia networks: integral gear of brain function. Front Cell Neurosci 8:378
108. Osellame LD, Duchen MR (2014) Quality control gone wrong: mitochondria, lysosomal storage disorders and neurodegeneration. Br J Pharmacol 171(8):1958–1972

109. Dehay B et al (2013) Lysosomal impairment in Parkinson's disease. Mov Disord 28(6):725–732
110. Atkin G, Paulson H (2014) Ubiquitin pathways in neurodegenerative disease. Front Mol Neurosci 7:63
111. Snead D, Eliezer D (2014) Alpha-synuclein function and dysfunction on cellular membranes. Exp Neurobiol 23(4):292–313
112. Herva ME, Spillantini MG (2015) Parkinson's disease as a member of Prion-like disorders. Virus Res 207:38–46.
113. Tsunemi T, Hamada K, Krainc D (2014) ATP13A2/PARK9 regulates secretion of exosomes and alpha-synuclein. J Neurosci 34(46):15281–15287
114. Cox D, Carver JA, Ecroyd H (2014) Preventing alpha-synuclein aggregation: the role of the small heat-shock molecular chaperone proteins. Biochim Biophys Acta 1842(9):1830–1843
115. Shen HY et al (2005) Geldanamycin induces heat shock protein 70 and protects against MPTP-induced dopaminergic neurotoxicity in mice. J Biol Chem 280(48):39962–39969
116. Procaccio V et al (2014) Perspectives of drug-based neuroprotection targeting mitochondria. Rev Neurol (Paris) 170(5):390–400
117. Schneeberger A et al (2012) Vaccination for Parkinson's disease. Parkinsonism Relat Disord 18(Suppl 1):S11–S13
118. Bourdenx M, Dehay B, Bezard E (2014) Down-regulating alpha-synuclein for treating synucleopathies. Mov Disord 29(12):1463–1465
119. Kennedy BK et al (2014) Geroscience: linking aging to chronic disease. Cell 159(4):709–713
120. Epel ES, Lithgow GJ (2014) Stress biology and aging mechanisms: toward understanding the deep connection between adaptation to stress and longevity. J Gerontol A Biol Sci Med Sci 69(Suppl 1):S10–S16
121. Nunomura A et al (2007) Neuronal death and survival under oxidative stress in Alzheimer and Parkinson diseases. CNS Neurol Disord Drug Targets 6(6):411–423
122. van Muiswinkel FL et al (2004) Expression of NAD(P)H: quinone oxidoreductase in the normal and Parkinsonian substantia nigra. Neurobiol Aging 25(9):1253–1262
123. Itoh K et al (1996) Cytochrome c oxidase defects of the human substantia nigra in normal aging. Neurobiol Aging 17(6):843–848
124. Bender A et al (2006) High levels of mitochondrial DNA deletions in substantia nigra neurons in aging and Parkinson disease. Nat Genet 38(5):515–517
125. Ali K, Morris HR (2015) Parkinson's disease: chameleons and mimics. Pract Neurol 15(1):14–25
126. Zhang J et al (2013) The small co-chaperone p23 overexpressing transgenic mouse. J Neurosci Methods 212(2):190–194
127. Sarbishegi M, Mehraein F, Soleimani M (2014) Antioxidant role of oleuropein on midbrain and dopaminergic neurons of substantia nigra in aged rats. Iran Biomed J 18(1):16–22
128. Connor JR et al (2009) Altered dopaminergic profile in the putamen and substantia nigra in restless leg syndrome. Brain 132(Pt 9):2403–2412
129. Double KL et al (2010) Selective cell death in neurodegeneration: why are some neurons spared in vulnerable regions? Prog Neurobiol 92(3):316–329
130. Yankner BA, Lu T, Loerch P (2008) The aging brain. Annu Rev Pathol 3:41–66
131. Xie W, Chung KK (2012) Alpha-synuclein impairs normal dynamics of mitochondria in cell and animal models of Parkinson's disease. J Neurochem 122(2):404–414
132. Kumar MJ, Nicholls DG, Andersen JK (2003) Oxidative alpha-ketoglutarate dehydrogenase inhibition via subtle elevations in monoamine oxidase B levels results in loss of spare respiratory capacity: implications for Parkinson's disease. J Biol Chem 278(47):46432–46439
133. Mallajosyula JK et al (2009) Metabolic control analysis in a cellular model of elevated MAO-B: relevance to Parkinson's disease. Neurotox Res 16(3):186–193
134. Smeyne M, Smeyne RJ (2013) Glutathione metabolism and Parkinson's disease. Free Radic Biol Med 62:13–25

135. Packer L, Cadenas E (2011) Lipoic acid: energy metabolism and redox regulation of transcription and cell signaling. J Clin Biochem Nutr 48(1):26–32
136. Belin AC et al (2011) Modeling Parkinson's disease genetics: altered function of the dopamine system in Adh4 knockout mice. Behav Brain Res 217(2):439–445
137. Austin S, St-Pierre J (2012) PGC1alpha and mitochondrial metabolism – emerging concepts and relevance in ageing and neurodegenerative disorders. J Cell Sci 125(Pt 21):4963–4971
138. Zheng B et al (2010) PGC-1alpha, a potential therapeutic target for early intervention in Parkinson's disease. Sci Transl Med 2(52):52ra73
139. Herskovits AZ, Guarente L (2013) Sirtuin deacetylases in neurodegenerative diseases of aging. Cell Res 23(6):746–758
140. Ferretta A et al (2014) Effect of resveratrol on mitochondrial function: implications in parkin-associated familiar Parkinson's disease. Biochim Biophys Acta 1842(7):902–915
141. Verma P et al (2014) HSF1 protects neurons through a novel trimerization- and HSP-independent mechanism. J Neurosci 34(5):1599–1612
142. Shinozaki S et al (2014) Inflammatory stimuli induce inhibitory S-nitrosylation of the deacetylase SIRT1 to increase acetylation and activation of p53 and p65. Sci Signal 7(351):ra106
143. Liu D, Zempleni J (2014) Transcriptional regulation of the albumin gene depends on the removal of histone methylation marks by the FAD-dependent monoamine oxidase lysine-specific demethylase 1 in HepG2 human hepatocarcinoma cells. J Nutr 144(7):997–1001
144. Dimant H, Ebrahimi-Fakhari D, McLean PJ (2012) Molecular chaperones and co-chaperones in Parkinson disease. Neuroscientist 18(6):589–601
145 Cortez L, Sim V (2014) The therapeutic potential of chemical chaperones in protein folding diseases. Prion 8(2). pii: 28938
146. Bolshette NB et al (2014) Protein folding and misfolding in the neurodegenerative disorders: a review. Rev Neurol (Paris) 170(3):151–161
147. Collins LM et al (2012) Contributions of central and systemic inflammation to the pathophysiology of Parkinson's disease. Neuropharmacology 62(7):2154–2168
148. Perry VH, Cunningham C, Holmes C (2007) Systemic infections and inflammation affect chronic neurodegeneration. Nat Rev Immunol 7(2):161–167
149. Cunningham C et al (2005) Central and systemic endotoxin challenges exacerbate the local inflammatory response and increase neuronal death during chronic neurodegeneration. J Neurosci 25(40):9275–9284
150. Zhang Y (2003) Transcriptional regulation by histone ubiquitination and deubiquitination. Genes Dev 17(22):2733–2740
151. Brochard V et al (2009) Infiltration of CD4+ lymphocytes into the brain contributes to neurodegeneration in a mouse model of Parkinson disease. J Clin Invest 119(1):182–192
152. Vitte J et al (2004) Oxidative stress level in circulating neutrophils is linked to neurodegenerative diseases. J Clin Immunol 24(6):683–692
153. Dantzer R et al (2008) From inflammation to sickness and depression: when the immune system subjugates the brain. Nat Rev Neurosci 9(1):46–56
154. Pavlov VA, Tracey KJ (2005) The cholinergic anti-inflammatory pathway. Brain Behav Immun 19(6):493–499
155. Maguire-Zeiss KA, Federoff HJ (2010) Future directions for immune modulation in neurodegenerative disorders: focus on Parkinson's disease. J Neural Transm 117(8):1019–1025
156. Morris AP et al (2012) Large-scale association analysis provides insights into the genetic architecture and pathophysiology of type 2 diabetes. Nat Genet 44(9):981–990
157. Campisi J (2001) Cellular senescence as a tumor-suppressor mechanism. Trends Cell Biol 11(11):S27–S31
158. Coppe JP et al (2008) Senescence-associated secretory phenotypes reveal cell-nonautonomous functions of oncogenic RAS and the p53 tumor suppressor. PLoS Biol 6(12):2853–2868
159. Coppe JP et al (2010) The senescence-associated secretory phenotype: the dark side of tumor suppression. Annu Rev Pathol 5:99–118

160. Campisi J et al (2011) Cellular senescence: a link between cancer and age-related degenerative disease? Semin Cancer Biol 21(6):354–359
161. Chinta SJ et al (2013) Environmental stress, ageing and glial cell senescence: a novel mechanistic link to Parkinson's disease? J Intern Med 273(5):429–436
162. Tchkonia T et al (2013) Cellular senescence and the senescent secretory phenotype: therapeutic opportunities. J Clin Invest 123(3):966–972
163. Baker DJ et al (2011) Clearance of p16Ink4a-positive senescent cells delays ageing-associated disorders. Nature 479(7372):232–236
164. Bitto A et al (2010) Stress-induced senescence in human and rodent astrocytes. Exp Cell Res 316(17):2961–2968
165. Flanary BE, Streit WJ (2003) Telomeres shorten with age in rat cerebellum and cortex in vivo. J Anti Aging Med 6(4):299–308
166. Flanary BE, Streit WJ (2004) Progressive telomere shortening occurs in cultured rat microglia, but not astrocytes. Glia 45(1):75–88
167. Yu HM et al (2012) Repeated lipopolysaccharide stimulation induces cellular senescence in BV2 cells. Neuroimmunomodulation 19(2):131–136
168. Chinta SJ et al (2015) Cellular senescence and the aging brain. Exp Gerontol 68:3–7
169. Bhat R et al (2012) Astrocyte senescence as a component of Alzheimer's disease. PLoS One 7(9), e45069
170. Fuchs E, Flugge G (2014) Adult neuroplasticity: more than 40 years of research. Neural Plast 2014:541870
171. Crews L et al (2008) Alpha-synuclein alters Notch-1 expression and neurogenesis in mouse embryonic stem cells and in the hippocampus of transgenic mice. J Neurosci 28(16):4250–4260
172. Winner B et al (2008) Mutant alpha-synuclein exacerbates age-related decrease of neurogenesis. Neurobiol Aging 29(6):913–925
173. Peng J et al (2008) Fibroblast growth factor 2 enhances striatal and nigral neurogenesis in the acute 1-methyl-4-phenyl-1,2,3,6-tetrahydropyridine model of Parkinson's disease. Neuroscience 153(3):664–670
174. Peng J, Andersen JK (2011) Mutant alpha-synuclein and aging reduce neurogenesis in the acute 1-methyl-4-phenyl-1,2,3,6-tetrahydropyridine model of Parkinson's disease. Aging Cell 10(2):255–262
175. Harvey K, Marchetti B (2014) Regulating Wnt signaling: a strategy to prevent neurodegeneration and induce regeneration. J Mol Cell Biol 6(1):1–2
176. Sanchez-Danes A et al (2012) Disease-specific phenotypes in dopamine neurons from human iPS-based models of genetic and sporadic Parkinson's disease. EMBO Mol Med 4(5):380–395
177. Feng Y, Jankovic J, Wu YC (2015) Epigenetic mechanisms in Parkinson's disease. J Neurol Sci 349(1–2):3–9
178. Desplats P et al (2011) Alpha-synuclein sequesters Dnmt1 from the nucleus: a novel mechanism for epigenetic alterations in Lewy body diseases. J Biol Chem 286(11):9031–9037
179. Masliah E et al (2013) Distinctive patterns of DNA methylation associated with Parkinson disease: identification of concordant epigenetic changes in brain and peripheral blood leukocytes. Epigenetics 8(10):1030–1038
180. Kontopoulos E, Parvin JD, Feany MB (2006) Alpha-synuclein acts in the nucleus to inhibit histone acetylation and promote neurotoxicity. Hum Mol Genet 15(20):3012–3023
181. Siddiqui A et al (2012) Selective binding of nuclear alpha-synuclein to the PGC1alpha promoter under conditions of oxidative stress may contribute to losses in mitochondrial function: implications for Parkinson's disease. Free Radic Biol Med 53(4):993–1003
182. Abel T, Zukin RS (2008) Epigenetic targets of HDAC inhibition in neurodegenerative and psychiatric disorders. Curr Opin Pharmacol 8(1):57–64

183. Kim YH et al (2011) Lithium protects against oxidative stress-mediated cell death in alpha-synuclein-overexpressing in vitro and in vivo models of Parkinson's disease. J Neurosci Res 89(10):1666–1675
184. Lieu CA et al (2014) Lithium prevents parkinsonian behavioral and striatal phenotypes in an aged parkin mutant transgenic mouse model. Brain Res 1591:111–117
185. Joseph B et al (2003) p57(Kip2) cooperates with Nurr1 in developing dopamine cells. Proc Natl Acad Sci U S A 100(26):15619–15624
186. Jin L et al (2013) Loss of LSD1 (lysine-specific demethylase 1) suppresses growth and alters gene expression of human colon cancer cells in a p53- and DNMT1(DNA methyltransferase 1)-independent manner. Biochem J 449(2):459–468
187. Lynch JT, Harris WJ, Somervaille TC (2012) LSD1 inhibition: a therapeutic strategy in cancer? Expert Opin Ther Targets 16(12):1239–1249
188. McColl G et al (2008) Pharmacogenetic analysis of lithium-induced delayed aging in Caenorhabditis elegans. J Biol Chem 283(1):350–357
189. Cloutier P et al (2013) A newly uncovered group of distantly related lysine methyltransferases preferentially interact with molecular chaperones to regulate their activity. PLoS Genet 9(1), e1003210
190. Ghosh S, Sengupta S, Scaria V (2014) Comparative analysis of human mitochondrial methylomes shows distinct patterns of epigenetic regulation in mitochondria. Mitochondrion 18:58–62
191. Philp A et al (2014) Understanding the acetylome: translating targeted proteomics into meaningful physiology. Am J Physiol Cell Physiol 307(9):C763–C773
192. Hill S, Van Remmen H (2014) Mitochondrial stress signaling in longevity: a new role for mitochondrial function in aging. Redox Biol 2:936–944
193. Potts LF et al (2014) Modeling Parkinson's disease in monkeys for translational studies, a critical analysis. Exp Neurol 256:133–143
194. Eslamboli A (2005) Marmoset monkey models of Parkinson's disease: which model, when and why? Brain Res Bull 68(3):140–149
195. Vrijheid M (2014) The exposome: a new paradigm to study the impact of environment on health. Thorax 69(9):876–878
196. Kamel F et al (2014) Dietary fat intake, pesticide use, and Parkinson's disease. Parkinsonism Relat Disord 20(1):82–87
197. Goldman SM (2014) Environmental toxins and Parkinson's disease. Annu Rev Pharmacol Toxicol 54:141–164
198. Kones R (2010) Low-fat versus low-carbohydrate diets, weight loss, vascular health, and prevention of coronary artery disease: the evidence, the reality, the challenge, and the hope. Nutr Clin Pract 25(5):528–541
199. Mattson MP (2014) Interventions that improve body and brain bioenergetics for Parkinson's disease risk reduction and therapy. J Parkinsons Dis 4(1):1–13
200. Mattson MP (2014) Challenging oneself intermittently to improve health. Dose Response 12(4):600–618
201. Goldstein DS et al (2007) Cardiac sympathetic denervation preceding motor signs in Parkinson disease. Clin Auton Res 17(2):118–121
202. Griffioen KJ et al (2013) Dietary energy intake modifies brainstem autonomic dysfunction caused by mutant alpha-synuclein. Neurobiol Aging 34(3):928–935
203. Mattson MP (2015) Lifelong brain health is a lifelong challenge: from evolutionary principles to empirical evidence. Ageing Res Rev 20:37–45
204. Santoro A et al (2014) Combating inflammaging through a Mediterranean whole diet approach: the NU-AGE project's conceptual framework and design. Mech Ageing Dev 136–137:3–13
205. Lakshminarayanan B et al (2014) Compositional dynamics of the human intestinal microbiota with aging: implications for health. J Nutr Health Aging 18(9):773–786

206. Kinross J, Nicholson JK (2012) Gut microbiota: dietary and social modulation of gut microbiota in the elderly. Nat Rev Gastroenterol Hepatol 9(10):563–564
207. Hsiao EY et al (2013) Microbiota modulate behavioral and physiological abnormalities associated with neurodevelopmental disorders. Cell 155(7):1451–1463
208. Murakami K et al (2010) Dietary intake of folate, vitamin B6, vitamin B12 and riboflavin and risk of Parkinson's disease: a case–control study in Japan. Br J Nutr 104(5):757–764
209. Mars H (1974) Levodopa, carbidopa, and pyridoxine in Parkinson disease. Metabolic interactions. Arch Neurol 30(6):444–447
210. Forsyth CB et al (2015) Circadian rhythms, alcohol and gut interactions. Alcohol 49(4):389–398
211. Arellanes-Licea E et al (2014) The circadian timing system: a recent addition in the physiological mechanisms underlying pathological and aging processes. Aging Dis 5(6):406–418
212. Laermans J et al (2014) Shifting the circadian rhythm of feeding in mice induces gastrointestinal, metabolic and immune alterations which are influenced by ghrelin and the core clock gene Bmal1. PLoS One 9(10), e110176
213. Coomans CP et al (2013) Detrimental effects of constant light exposure and high-fat diet on circadian energy metabolism and insulin sensitivity. FASEB J 27(4):1721–1732
214. Fifel K, Cooper HM (2014) Loss of dopamine disrupts circadian rhythms in a mouse model of Parkinson's disease. Neurobiol Dis 71:359–369
215. Matteoli G, Boeckxstaens GE (2013) The vagal innervation of the gut and immune homeostasis. Gut 62(8):1214–1222
216. Matteoli G et al (2014) A distinct vagal anti-inflammatory pathway modulates intestinal muscularis resident macrophages independent of the spleen. Gut 63(6):938–948
217. Holmqvist S et al (2014) Direct evidence of Parkinson pathology spread from the gastrointestinal tract to the brain in rats. Acta Neuropathol 128(6):805–820

[illegible]

Aging and the Bone-Muscle Interface

Simon Melov and Clifford J. Rosen

Contents

1 Skeletal and Muscle Compartments: Relationship to the Geroscience Pillars

The skeleton is one of the largest systems in the body consisting of a mineralized matrix and a highly active cellular remodeling unit, composed of osteoblasts, osteoclasts, osteocytes and bone lining cells [1]. The most obvious function of the skeleton is to provide structural integrity for the organism while maintaining a degree of

S. Melov, Ph.D.
Buck Institute for Research on Aging, Novato, CA, USA
e-mail: smelov@buckinstitute.org

C.J. Rosen, M.D. (✉)
Department of Medicine, Musculoskeletal Center,
Center for Clinical & Translational Research,
Maine Medical Center Research Institute, Scarborough, ME, USA
e-mail: cjrofen@gmail.com

F. Sierra, R. Kohanski (eds.), *Advances in Geroscience*,
DOI 10.1007/978-3-319-23246-1_9

elasticity that allows for a broad range of locomotor activities. In addition to its structural function, the skeleton also serves as a mineral depot insuring the maintenance of serum calcium and phosphate levels through normal remodeling processes and by secretion of peptide factors such as FGF-23. Every 10 years the human skeleton is completely remodeled. Just as importantly, the skeleton encases the bone marrow and maintains a niche along trabecular bone elements for active hematopoiesis. The marrow niche consists of osteoblasts, adipocytes, reticuloendothelial cells, sinusoids, and mesenchymal stromal and stem cells. These progenitors respond to injuries at various sites, not restricted to bone, providing new cells for critical repair processes [2]. Remarkably, the adult skeleton also harbors a huge adipose depot, comprising 10–15 % of all fatty tissues in the body [3]. With aging that percentage increases, particularly as peripheral adipose depots shrink. Whether this is a response to chronic inflammation, metabolic changes from insulin resistance, disuse because of muscle atrophy or molecular drivers related to senescence within the marrow is currently being debated. Regardless, alterations in either the structural or metabolic functions of the skeleton, a key component of the aging process, have tremendous implications for the overall health of the organism.

Muscle represents the single largest component of the body, representing ~40 % mass by weight in a typical individual. There are more than 600 muscles in the body, with distinct differences in metabolism, cellular makeup, the ability to generate force, as well as specificity for workload. Skeletal muscles are composed of individual fibers, anchored to the skeleton through tendons, and each fiber is essentially a single cell, containing multiple nuclei throughout its length. Muscle fibers can be generally broken down into two key types, fast twitch, and slow twitch, with a number of further subtypes dependent on species. Fast twitch fibers are primarily glycolytic, and generally are associated with explosive energy demands such as sprinting. Slow twitch fibers are primarily oxidative, and are typically associated with more sustained workloads such as long distance running, and certain muscles can be predominantly comprised of only one type, while other muscles have mixed fiber types. Muscle is one of the major sites of metabolism in the body, and is responsible for more energy consumption than fat and bone combined. Although we do not yet have a complete picture of age-related atrophy with regards to each distinct muscle in any species, it is generally agreed that there is a loss of muscle mass with age. The exact extent of loss is not yet clear, and may vary depending on environment, lifestyle, and genetics. Importantly, accompanying the loss of muscle, there is a corresponding decline in function. Initially, the loss of muscle mass with age was thought to be due to a loss of type II fibers, however recent studies have suggested that the loss in muscle mass can be almost entirely explained by atrophy, rather than loss, of type II fibers alone. Therefore, changes in metabolism, mitochondrial autophagy, enhanced reactive oxygen species and impaired stem cell regenerative capacity, all could lead to muscle atrophy, disuse and ultimately to a failure in loading [4]. Thus musculoskeletal aging must be considered within the context of extending healthspan since general mobility is critical for a good quality of life. Muscle and bone are intimately associated with each other, yet we typically study each in isolation, and there are few studies examining them together as a functional system.

There are three skeletal compartments with unique functional characteristics, trabecular bone, cortical bone and the periosteum. Each compartment is composed of nearly identical cells that arise from the same precursor but functionally may be very different. Not unexpectedly, these cells and their compartments differ in their response to the stressors of aging, including changes in metabolism, reactive oxygen species, and inflammation. Trabecular bone comprises about 20 % of the human skeleton in young adults and is present primarily in the axial bones. In rodents it comprises less than 5 % of bone mass and cross sectional studies to date have suggested that it is lost much earlier during the aging process relative to cortical bone. Its large surface area provides a framework for skeletal remodeling and calcium homeostasis, ensuring adequate calcium for the body while also providing the elasticity necessary for bi- and quadra-pedal locomotion.

It is estimated that every 10 years, the human skeleton is remodeled, with the greatest frequency of turnover in trabecular bone [1, 5]. There is no evidence that remodeling slows with age, and indeed, there may be increased bone resorption during the latter decades of life. This may be associated with enhanced reactive oxygen species generation, epigenetic changes, or enhanced expression of RANKL from osteocytes that leads to greater bone resorption.

The cortical compartment surrounds the trabecular elements and is composed of a solid layer of calcified matrix in lamellar bone. Cortical bone is a hard tissue with elastic properties that enhance its strength. It is also very dynamic. The endocortical surface of cortical bone is subject to remodeling due to the presence of osteoclasts and osteoblasts, as well as its proximity to the marrow space, where progenitor cells reside, although the rate of turnover is much lower than in the trabecular skeleton. The periosteum is the outer layer of the skeleton and serves several functions but does not remodel, even though it is a major source of progenitor cells for fracture healing and for endochondral bone formation. Importantly, the periosteum contributes to skeletal growth, the response to mechanical loading, the injury and fracture response, and the compensatory mechanisms associated with aging that attempt to preserve strength in the face of trabecular and cortical loss. Importantly, despite several determinants directly related to aging (i.e. geroscience pillars) including metabolic, inflammatory and epigenetic processes, the periosteum appears to respond appropriately to changes with the trabecular compartment even at advanced ages. However, few studies have addressed why a highly innervated and vascular tissue (i.e. the periosteum) can withstand the stressors of aging that have deleterious effects on other hard and soft tissues.

Like bone, there are substantial changes in the musculature with age. Between the forth and fifth decade of life, homeostatic control of muscle mass declines, resulting in an overall loss of muscle mass in later years of life – termed sarcopenia ("poverty of the flesh"). This loss of muscle mass due to intrinsic aging has been estimated to be of the order of 1 % a year from ~50 years of age, and can result in as much as a 30 % loss of muscle mass by the mid-80s. Although sarcopenia appears to be a universal attribute of aging, which has been documented in multiple mammalian species [6–9] as well as invertebrate models of aging [10, 11], comparatively little work has been done in determining how sarcopenia dynamically affects the more than 600 different muscle types in aging mammals [7, 9, 12–14]. Even less has been reported about the effect of sarcopenia on different compartments of the skeleton. Most studies to date have focused on large muscle groups where it is possible to obtain sufficient material to enumerate the number and type of fibers present at specific ages, and nearly all data collected to date is cross sectional in nature. Common methods to measure atrophy in the fibers and subsequent loss of muscle mass include non-invasive techniques such as MRI (for evaluating cross sectional area of the thigh for example), as well as muscle biopsies, used to enumerate the total number of fibers present in the biopsy as well as defining the level of atrophy and the corresponding fiber type (myosin heavy chain specific isoforms distinguish fiber types). The mechanisms through which muscle mass is lost with age are currently unknown, but are likely multifactorial. Factors that have been implicated in sarcopenia include lifestyle, hormonal milieu, mitochondrial dysfunction, nutritional status, protein synthesis, stem cell decline, loss of neuromuscular junctions, and fiber atrophy [4, 15–22]. Mitochondrial dysfunction could result from enhanced ROS production, epigenetic changes due to environmental determinants, metabolic load (particularly glucose) and inflammation. Since both bone and muscle share common regulatory features such as cytokines, neurotropic factors, and the sympathetic nervous system, it seems likely that with aging there are concomitant changes in both tissues that ultimately result in falls and fractures.

Skeletal aging begins after the rapid phases of bone modeling and growth during adolescence. The acquisition of peak bone mass is related to a complex series of hormonal changes associated with matrix biosynthesis and mineralization. This is followed in humans by a plateau phase of variable duration in early adulthood, followed by loss of bone mass from two of the three skeletal compartments. During growth, bone modeling occurs through endochondral bone formation (i.e. requiring cartilage) by longitudinal expansion of the growth plate from the secondary and primary spongiosa [23]. Membranous bone formation without the need for chondrocytes occurs primarily in the craniofacial bones. After peak bone acquisition (ages 12–16), skeletal remodeling balances resorption with formation by a general maintenance phase that may last from 5 to 25 years in humans. Ultimately bone loss occurs from the trabecular and cortical skeleton although there is debate about the timing and magnitude as well as the compartmental specific effects [24]. In general, trabecular bone is lost first and may begin as early as the third decade of life in humans [25]. By the nineth decade of life, there is a marked diminution in trabecular bone in the distal extremities with even less trabeculae in the vertebrae. In female C57BL/6J mice, trabecular bone from the distal femur and tibia begins an incessant decline by as early as 8 weeks of age, whereas spongy bone loss from the vertebrae doesn't start until nearly 16 weeks [26]. These changes are relatively similar with regards to the biological age of human bone loss. However, unlike humans, trabecular bone in the femur/tibia of mice is essentially absent after ~12 months of age, although further studies are needed to longitudinally follow bone with age. In addition, both in humans and mice there is a much slower rate of loss of trabecular bone in males than females. It is still uncertain how age-related trabecular bone changes impact, if at all the muscle bone interface.

Cortical bone remodels in humans and in rats but less so in mice. Conversely, bone loss from this compartment starts much later on the endocortical surface than on the endosteal perimeter, and is incessant [27]. Cortical thickness declines with aging at a relatively constant rate although it can be accelerated by hormonal imbalances (e.g. estrogen loss at menopause, hypogonadism in males, glucocorticoid excess in both sexes) [28]. The stressors that compose the pillars of Geroscience and impact soft tissue also play a major role in cortical bone changes. However, less work has been done in determining the relative magnitude or the sum of these effects. Traditionally, hormonal imbalances have been ascribed as the major mechanism for bone loss. But emerging studies suggest that accumulation of toxic substances, mitochondrial dysfunction, metabolic stress, and impaired stem cell responsiveness also play critical roles.

Cortical bone can become porous with age, although the mechanism for the development of these pores within the matrix remains uncertain. One possibility is that cortical bone becomes trabecularized by higher rates of resorption and this leads to areas of porosity that are imbedded within the cortex. Seeman has proposed the notion that most age-related osteoporosis represents disease of the cortical skeleton [29]. Hence around midlife, in women, remodeling balance becomes negative; less bone is deposited than it is resorbed by each bone's basic multicellular units (BMUs), and remodeling rate increases; there are more BMUs removing bone upon intracortical, endocortical, and trabecular surfaces. Canals enlarge and coalesce creating giant pores. Remodeling upon trabecular surfaces removes canals, whereas intracortical and endocortical remodeling fragments the cortex. Seeman proposes that bone loss becomes almost entirely cortical as trabeculae disappear [29]. Overall, remodeling removes more bone from a diminishing total mineralized bone matrix volume so that by old age, total mineralized bone matrix volume is halved; but 70 % of all bone loss is cortical because 80 % of the skeleton is cortical; a third of all the bone loss arises from the 20 % of the skeleton that is trabecular. Hence most of the fractures occurring with aging are non-vertebral (hip, humerus, tibia, fibula, radius) and predominantly cortical whereas 20 % are vertebral. If indeed, cortical bone changes are the major determinants of osteoporosis, then the impact on muscle, and vice versa must be significant.

As cortical bone thins, and porosity increases, structural fragility becomes more pronounced. Intriguingly, in mammals there is a compensatory mechanism in place during aging that is activated by the rapid loss of long bone. This is termed 'periosteal expansion' and it has a potential to increase bone area and partially buffer the higher rate of endosteal and endocortical resorption with age [30]. The unique capacity of the periosteum, which is the site of insertion of tendons from large muscles, to expand, improves skeletal properties such as the polar moment of inertia, and stiffness, thereby partially preserving bone strength and reducing the risk of fracture. Males tend to have a more vigorous periosteal response to aging and injury than females; this may be due to inherent cell autonomous differences by sex. The molecular drivers of this difference have not been elucidated nor is it clear that androgens direct this process. Moreover, the signals for periosteal compensation are also not known but it is this interface between bone and muscle that provides some fascinating insights into the physiology of aging and hence can shed light on the defects inherent in osteoporosis.

2 Unique Aspects of the Bone-Muscle Interface: Relationship to Geroscience

To understand the importance of the muscle-bone interface within the broader context of geroscience, it is critical to define the anatomical relationships. Muscles insert on bone via tendons that connect to a fibrous layer on the surface of bone. The periosteal layer or membrane is contiguous with this fibrous layer and covers all bones in the body except the joints. It is made up of sensory fibers, endothelial tissue, stromal elements including mesenchymal stromal cells (MSCs) that could become osteoblasts, vascular networks and some adipocytes. The periosteum is really a dense irregular connective tissue [31]. It can be divided into an outer "fibrous layer" and an inner "cambium layer" (or "osteogenic layer"). The fibrous layer contains fibroblasts, while the cambium layer contains progenitor cells that develop into osteoblasts. These osteoblasts are responsible for increasing the width of a long bone and the overall size of the other bone types. After a bone fracture the progenitor cells can develop into osteoblasts and chondroblasts, which are essential for fracture healing. Bone has very few long track sensory nerves beyond the innervations to osteoblasts whereas the periosteum has nociceptive nerve endings, making it very sensitive to manipulation. The nerve endings are accompanied by many blood vessels, branches of which penetrate the bone to supply the osteocytes, or older osteoblasts embedded within the cortex. These perpendicular branches pass into the bone along channels known as Volkmann canals to the vessels in the Haversian canals, which run the length of the bone.

Osteocytes are older osteoblasts that serve as mechano-sensors to modulate skeletal remodeling through the secretion of peptide factors such as sclerostin. This connection between cell surfaces (via the periosteum), which is activated by loading of the bone, can respond to fluid flux within the cortical lacunae and communicate with other cells via the canaliculi. Recent studies of the osteocyte further support its importance in regulating bone remodeling through several factors, including RANKL which can cause osteocytic osteolysis (i.e. dissolution of matrix around osteocytes) and sclerostin which by binding to Lrp5, a critical receptor for osteoblasts, can inhibit new bone formation [1, 32]. In addition, osteocytes also secrete endocrine factors such as FGF-23 which modulates phosphate homeostasis [1]. Aging bone is characterized by osteocytic drop out, or what is termed 'empty lacunae' [1, 25–27]. Apoptosis is the presumed mechanism, but the molecular drivers of that process are not known. In a recent study, Jilka et al. genetically deleted two osteocytic genes related to apoptosis and found that in aging mice, the lacunae were filled with viable osteocytes but paradoxically there was increased bone resorption and significant cortical porosity [33].

Fibrous cartilage often takes the place of the periosteum along grooves where tendons exert pressure against the bone. The periosteum itself is attached to bone by strong collagenous fibers called Sharpey's fibers, which extend to the outer circumferential and interstitial lamellae [34]. Pressure from muscle insertion on the fibrous membrane affects the mechanosensors almost certainly through growth factor signals from the periosteum, either locally or systemically. The periosteum also produces bone when it is stimulated appropriately [31]. Practically anything that breaks, tears, stretches, inflames, or even touches the periosteum can lead to a reactive process whereby new bone is formed. This is termed a *periosteal reaction*, also known as a *periosteitis*, which is a non-specific radiographic finding that occurs with periosteal irritation. Periosteal reactions can be broken down by pattern, but in all cases the response arises from the skeletal disease itself, not in the periosteum. With slow-growing processes, the periosteum has plenty of time to respond to the process. That is, it can produce new bone just as fast as any growing lesion. This is particularly important when considering the periosteal response to bone loss and the aging process. With aging, there is some clinical evidence that fracture healing is impaired. However there is tremendous inter-individual variation and mid-diaphyseal periosteal measures including cambium and fibrous layer thickness and cellularity do not correlate significantly with age or body mass [20]. Gender certainly plays an important role in the periosteal response to aging but the cell autonomous factors involved remain unknown [35].

The *tendon* is a tough band of fibrous connective tissue that usually connects muscle to bone and is capable of withstanding tension. It is that tension which is thought to provide the initial force on the bone that leads to signals for modeling and remodeling of the skeleton. There are no studies that report on differences in mechanical forces with aging, although certainly sarcopenia must have an impact. Tendons are similar to ligaments and fasciae; all three are made of collagen. The mechanical properties of the tendon are dependent on the collagen fiber diameter and orientation. The collagen fibrils are parallel to each other and closely packed, but show a wave-like appearance due to planar undulations, or crimps, on a scale of several micrometers. In tendons, the collagen fibers have some flexibility due to the absence of hydroxyproline and proline residues at specific locations in the amino acid sequence, which allows the formation of other conformations such as bends or internal loops in the triple helix and results in the development of crimps. The crimps in the collagen fibrils allow the tendons to have some flexibility as well as compressive stiffness. In addition, because the tendon is a multi-stranded structure made up of many partially independent fibrils and fascicles, it does not behave as a single rod, and this property also contributes to its flexibility. The uniqueness of the tendon, and its transmutation of loading from the muscle must play a role in the periosteal compensation that occurs with aging [36]. However, if that signal is dampened by sarcopenia, or reduced loading, mechanically-induced expansion of the periosteum might be impaired, leading to altered biomechanical properties and ultimately skeletal fragility.

The extent of mechanical loading (e.g. through strength training) of muscle on bone via tendons and the periosteum has been strongly associated with cortical bone mass in both cross sectional and longitudinal studies [37]. Conversely, skeletal unloading due to bed rest, zero gravity states or muscle disease results in low bone mass and skeletal fragility. However, it is unclear whether all of the effects of bone unloading are mechanically mediated or if there are soluble mediators that might be released from atrophic muscles to negatively affect skeletal remodeling. Similarly, the sarcopenia of aging is associated with falls, reduced muscle strength and fractures. Whether these age-related changes are all mechanically-mediated or are related to the intertwined factors that define aging, such as inflammatory cytokines and myokines, metabolic dysfunction, changes in innervation and/or buildup of toxic superoxides, remain to be elucidated.

3 Experimental Evidence that Periosteal Expansion Occurs During Aging

Bone loss is an inexorable feature of aging in all mammalian systems although the mechanisms are multi-factorial and species-specific. In rodents and humans, both cortical and trabecular bone are lost with advancing age although the rates differ considerably. In general, trabecular bone loss occurs first, due in part to its greater surface area compared to the cortical compartment. Inbred strains of mice lose bone from the distal femur as early as 8 weeks of age [26]. In humans, studies using quantitative CT and microCT have demonstrated trabecular thinning and slow but incessant bone loss beginning in the third decade of life in women [25]. As such, even though menopause induces an estrogen deficient state over a period of years, and this has been associated with accelerated bone loss, it is evident that the other common factors noted previously, i.e. metabolic, proteostatic, inflammatory and/or toxic, contribute to age-related bone loss. A characteristic feature of this loss, independent of gonadal steroids and in both cortical and trabecular compartments is uncoupled remodeling such that resorption accelerates beyond formation. Bone formation may increase marginally in response to the ensuing loss, but with advanced age, this response appears to be blunted. This latter feature has been variously attributed to greater reactive oxygen species, reduced stem cell pools, senescent accelerated impairment in bone formation, non-cell autonomous factors such as higher sympathetic tone and greater concentrations of toxic cytokines or ROS. Recently, Bartel et al. reported that Foxo proteins restrain age-related osteoclastogenesis by modulating H_2O_2 accumulation [38].

Mechanically, bone loss alters the skeletal microarchitecture and reduces bone strength. In the axial spine, trabecular loss leads to enhanced fragility and susceptibility to compression fractures that can have a significant impact on quality of life and morbidity [14]. In the cortical compartment, long bones undergo loss as well although it may differ by site and by the amount of trabecular bone within the appendicular skeleton. In mice, cortical remodeling is not a major feature but in rats, monkeys and humans, cortical turnover predominates in later life. The major cost to society of age-related cortical bone loss is a hip fracture. Significant architectural changes occur during the slow but inexorable process of uncoupled bone turnover. In particular, the endocortical surface of the cortex (Fig. 1) undergoes resorption, leading to an expanded marrow cavity. This loss can lead to trabecularization of the cortex and cortical porosity.

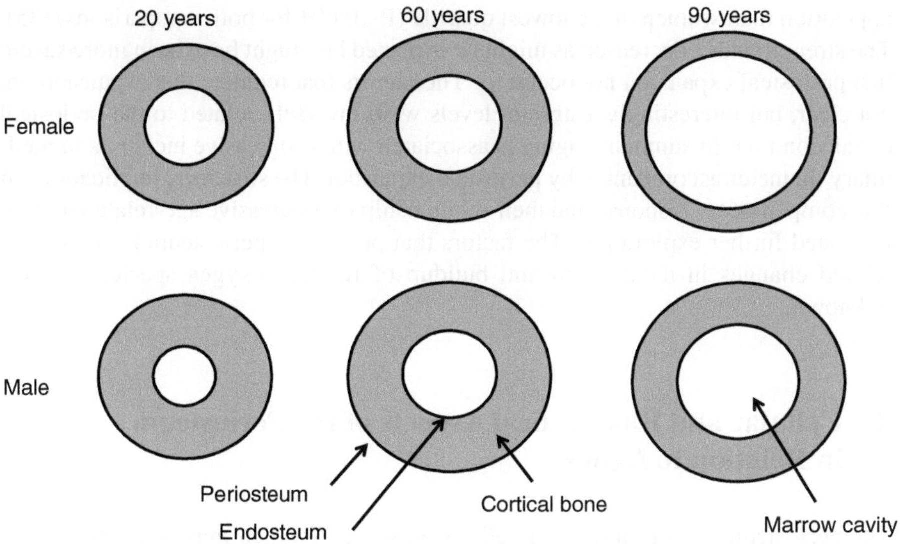

Fig. 1 Changes with aging in the endocortical and periosteal envelope by gender. A cross sectional view of a typical long bone (femur) showing different features which dynamically change with age and sex

Paradoxically, the periosteal surface of the cortex is not subject to loss and in fact expands at the same time as there is endocortical resorption [37]. Miller et al. aged genetically heterogeneous UM-HET3 mice that were produced by a cross between (C57BL/6J × BALB/cJ)F1 males and (C3H/HeJ × DBA/2J)F1 females to show that changes in bone diameter are associated with lifespan [39]. Remarkably, those mice with the greatest cortical thickness had the highest survival rate. Not surprisingly, the authors also noted that the endosteal envelope expanded by nearly 20 % due to increased bone resorption with age. Importantly, the increase in cortical area with age from 4 to 24 months was predominantly related to the increase in periosteal circumference (from 4.88 to 5.66 mm; $p < 0.01$) thereby partially maintaining cortical thickness. How periosteal expansion is related to lifespan remains to be determined in mice but may reflect the health of connective tissue or the pool of progenitors that are necessary for this compensation.

Studies in humans provide additional support for the compensation theory. Ahlborg et al. measured bone mass and the skeletal structure of the distal radius by single-photon absorptiometry every other year in 108 women, all of whom were followed from the time of menopause for a mean period of 15 years [35]. The mean (± SD) annual decrease in bone mineral density was 1.9 ± 0.7 %. The medullary bone diameter increased annually by 1.1 ± 0.9 %, and the periosteal diameter by 0.7 ± 0.3 %; while the strength index decreased by 0.7 ± 0.7 %. The expansion of the medullary diameter with a simultaneous increase in the periosteal diameter was directly correlated ($r = 0.54$, $P < 0.001$), and women in the highest quartile of medullary expansion had more loss of bone mineral density and greater periosteal apposition than women in the lowest quartile ($P < 0.001$ for both comparisons) [35]. The strength index decreased as might be expected but might have been more severe had periosteal expansion not occurred. The factors that regulate this expansion are not clear, but interestingly, estradiol levels were inversely related to the periosteal expansion rate. In summary, aging is associated with progressive increases in medullary diameter accompanied by periosteal expansion The structural implications of this compensatory response and their relationship to progressive age-related muscle loss need further exploration. The factors that permit the periosteum to resist age-related changes in metabolism and buildup of reactive oxygen species are also unknown.

4 Cellular and Biochemical Aspects of the Periosteum in Relation to Aging

The periosteum is rich in distinct cell types, some of which may be responsive to signals from the osteocyte, or from mechanical loading of the tendon. Moreover, because it is highly vascularized, the periosteum is subject to endocrine actions of hormones such as IGF-I and PTH. But the periosteum also contains significant numbers of progenitors and mesenchymal stem cells and it is the balance between mature and progenitor cells that ultimately define the function of the periosteum [37, 31]. If one of the determinants of unhealthy aging is reduced stem cell pools or reduced stem cell function, the periosteum may be at least partially protected. Alternatively it is possible that adult periosteal cells may undergo senescence and thereby be resistant to autocrine, paracrine or endocrine signals such as inflammatory cytokines. In one report immunohistochemical detection of Ki67 and p53, Nitric Oxide (NO) production and qRT- PCR of a selected gene panel for osteoblastic differentiation, bone remodeling and 'stemness' were evaluated. The authors confirmed that both Ki67 and p53 were noteworthy indicators of senescence in human periosteal precursor cells and their expression significantly correlated with cell NO production [40]. Moreover, cell age affects genes involved in bone remodeling, with a significant increase in interleukin-6 mRNA expression and receptor activator of nuclear factor kappa-B ligand (RANKL)/osteoprotegerin (OPG) ratio [40]. O'Keefe and colleagues recently reported that reduced fracture healing in aging involves decreased proliferation and differentiation of stem cells lining the bone surface. While PTH 1–34 enhances the proliferation and expansion of the periosteal stem cell population and accelerates bone formation and fracture healing, the effects are proportionately reduced during the aging process compared to young mice [41].

Another mechanism whereby periosteal expansion may be limited during aging is through the impairment of progenitor cell recruitment from muscles. Recent studies have established that muscle-derived stem cells are able to differentiate into cartilage and bone and can directly participate in fracture healing. The role of muscle-derived stem cells is particularly important in fractures associated with more severe injury to the periosteum. Muscle anabolic agents may improve function and reduce the incidence of fracture with aging as well as maintaining the muscle-bone interface [42].

5 Non-cell Autonomous Factors Regulating the Bone/Muscle Interface During Aging

Although the continued periosteal apposition that accompanies age-related bone loss is a biomechanically critical target for prophylactic treatment of bone fragility, the magnitude of periosteal expansion required to maintain strength and stiffness during aging has not been established. A new model for predicting periosteal apposition rate for men and women was developed by Jepsen et al. to better understand the complex, nonlinear interactions that exist among bone morphology, tissue-modulus, and aging [43]. Periosteal apposition rate varied up to eightfold across bone sizes, and this depended on the relationship between cortical area and total area, which varies with external size and among anatomical sites. There was a 65–145 % increase in periosteal apposition rate beyond that expected for bone loss alone. But periosteal apposition rate had to increase as much as 350 % over time to maintain stiffness for slender diaphyses, whereas robust bones required less than a 32 % increase over time. Small changes in the amount of bone accrued during growth (i.e., adult cortical area) affected periosteal apposition rate of slender bones to a much greater extent compared to robust bones. Thus bone growth places a heavy burden on the biological activity required to maintain stiffness with aging. Finally, sex-specific differences in periosteal apposition were attributable in part to differences in bone size. The results indicate that a substantial proportion of the variation in periosteal expansion required to maintain bone strength during aging can be attributed to the natural variation in adult bone width [43]. Clinical data to differentiate the biological responses that are attributable to size effects from other genetic and environmental factors are necessary.

6 Muscles, Myokines, and the Periosteum: Intertwining Factors During Aging

Muscles and bone are intimately linked beyond the biomechanical connection through bi-directional signals regulating gene expression, proliferation and differentiation. It is generally accepted that muscle cells secrete factors (myokines) that influence adjacent bone cells, but few myokines have been identified and characterized functionally. Importantly, the periosteum can serve as a barrier for locally secreted muscle factors unless the soluble substances are relatively small and can permeate the tissue. Lai et al. recently reported that PGE2, IGF-1, IL-15 and FGF-2 can penetrate the periosteal membrane and hence drive bone cell function in the cortex [44]. IGF-I is particularly important since it is synthesized by muscle and bone and has a growth promoting effect on both tissues. Low IGF-I is associated with reduced bone size and periosteal circumference as well as less muscle mass in genetically modified animal models and in humans with growth hormone deficiency (or resistance) and Type I IGF receptor resistance. Importantly, with aging, growth hormone secretion declines in mammals and circulating IGF-I decreases. IGF-I is also anti-apoptotic and as such a decline in circulatory or skeletal IGF-I could contribute to a premature decline in bone cell populations.

Other myokines have recently been studied. Sims et al. demonstrated that osteocyte-specific deletion of the co-receptor subunit utilized by IL-6 family cytokines, glycoprotein 130 (gp130), resulted in impaired bone formation in the trabecular bone, but enhanced periosteal expansion, suggesting a gp130-dependent periosteum-specific inhibition of osteoblast function, potentially induced by the local muscle fibers [45]. This consistent but inverse relationship between osteoblast induced bone formation and periosteal expansion mimics conditions such as age-related and postmenopausal osteoporosis. Similarly, Sims and colleagues reported a negative relationship between ciliary neurotrophic factor from muscle and osteoblast differentiation [45].

Myostatin is a member of the bone morphogenetic protein/transforming growth factor-β (BMP/TGFβ) super-family of secreted differentiation factors. Myostatin is a negative regulator of muscle mass as shown by increased muscle mass in myostatin deficient mice. Interestingly, these mice also exhibit increased bone mass suggesting that myostatin may also play a role in regulating bone mass. Indeed, a soluble myostatin decoy receptor (ActRIIB-Fc) increased bone and muscle mass. Bone volume fraction (BV/TV), as determined by microCT, was increased by 132 % and 27 % in the distal femur and lumbar vertebrae, respectively. Surprisingly there was no effect on cortical bone or the periosteum, suggesting that one mechanism for this myokine may be more endocrine than paracrine [46].

Exercise training benefits muscle and bone by slowing age related bone loss, but also offers protection against several of the major pillars that define an impaired healthspan. For example, metabolic disorders such as obesity and diabetes, as well as an inflammatory component can be ameliorated by exercise. Exercise has been shown to reduce mitochondrial autophagy, enhance stem cell recruitment, and slow the rise in ROS from muscle cells. Irisin, a muscle cytokine produced from surface FNDC5 proteolysis, mediates a thermogenic program in adipose tissue of mice [47]. This leads to enhanced glucose utilization in adipose tissue by increasing uncoupling protein 1 and the transcriptional co-factor Pgc1α. However, it is unclear if irisin has a direct impact on either the periosteum or cortical bone, or whether these findings are relevant in humans.

7 Sarcopenia and the Bone-Muscle Interface

Although a considerable amount of work has been done in describing sarcopenia, it was only recently that a definition was agreed upon by an international working group that incorporated functional attributes as well as the loss of muscle mass in the definition [48]. Sarcopenia is currently defined as “the age associated loss of skeletal mass and function”. The coupling of the terms “function” and “loss” is critical, as an increasing body of evidence has shown that functional impairment of aged muscle is a better correlate of frailty than the amount of muscle loss alone. In addition, other conditions where substantive muscle loss occurs such as that due to cancer (cachexia) cannot be considered as the same as the muscle loss due to aging. Remarkably, aging bone and muscle both exhibit fatty infiltration although the degree to which this feature compromises musculoskeletal strength or function is not known.

More recently, one of the initial mechanistic explanations for sarcopenia is coming under question. It has long been assumed that with age, the loss of muscle mass is due to a corresponding loss of fibers (predominantly type II), and this has been reported many times, typically using methods which compare the numbers of fibers in muscle biopsies of young individuals with those from older individuals. However, recent work in humans from the van Loon group [49] has suggested that the loss of muscle mass with age can be explained entirely by atrophy of type II fibers alone. That is, muscle mass is lost with age due to a specific atrophy of type II fibers, rather than loss of fibers per se. van Loon and colleagues conclude that the number of fibers is the same between young and old age groups, but the volume occupied by type II fibers with age is much smaller, thereby accounting for the loss in muscle mass. This directly contradicts the “fiber specific loss” with age hypothesis, as an explanation for why we lose muscle mass with age [49]. Importantly, the data in this key study was concluded in part from data collected longitudinally from the same subjects over time. Further studies are needed with methods which can longitudinally profile cell number, rather than rely on cross sectional study designs to definitively answer this long standing question.

Under normal circumstances, muscle is replenished and renewed through the action of stem cells located in the basement membrane of the myofiber. These stem cells are termed satellite cells, and have long been known to be essential for maintaining muscle mass. There has been much interest over the last few years on the role that satellite cells may play in sarcopenia. There have been reports describing both a loss of satellite cells with age, as well as a decline in the niche that allows the satellite cell to replenish the myofiber [50, 51]. Data has been presented showing that the quality of muscle can be improved through systemic administration of factors present in the blood of young animals through parabiotic pairing, in which the circulatory system of young and old animals are surgically joined [52]. However, more recently, the role of satellite cells in directly affecting muscle mass in aging was addressed in an elegant series of studies in which satellite cells were genetically ablated [53]. This allowed a direct test of the role of the satellite cell in maintaining muscle mass with advancing age. Essentially, transgenic mice were created which allowed the inactivation of satellite cells in older mice. Surprisingly, animals that had their reservoir of satellite cells genetically depleted via this genetic targeting strategy showed no difference in the rate of muscle loss compared to non-treated controls [53]. These data imply that sarcopenia is not due to either a loss of satellite cells with age in mice, or alteration of the niche preventing maturation of the satellite cell. Both of these arguments have been made as causal factors involving satellite cells as a key mechanism in sarcopenia. Similarly, there have been several studies concluding that satellite cells are not involved in driving sarcopenia due to a lack of significant difference in the numbers of satellite cells from younger versus older adults [54, 55]. In order to resolve such issues, further studies are needed, particularly addressing the challenging paradigm of longitudinal assessment of satellite cell number, fiber type, and the degree of atrophy. Perhaps the notion that sarcopenia is due to a simple exhaustion of the satellite cell pool or reservoir is overly simplistic. Notwithstanding, sarcopenia must be an important component of change in the bone-muscle interface.

8 CNS and the Bone Muscle Interface

Brain-derived neurotrophic factor (BDNF) plays important roles in neuronal differentiation/ survival, the regulation of food intake, and the pathobiology of obesity and type 2 diabetes mellitus. BDNF and its receptor are expressed in osteoblasts and chondrocytes. BDNF in vitro has a positive effect on bone cells; whether central BDNF affects bone mass in vivo is not known. Camarino et al. examined bone mass and energy use in brain-targeted BDNF conditional knockout mice (Bdnf(2lox/2lox)/93) [56]. The deletion of BDNF in the brain led to a metabolic phenotype characterized by hyperphagia, obesity, and increased abdominal white adipose tissue. Central BDNF deletion produced a marked skeletal phenotype characterized by increased femur length, elevated whole bone mineral density, and bone mineral content. Polar moment of inertia and cortical thickness were markedly increased suggesting a role for this neurotropic factor on the periosteum as well as the trabecular and cortical skeleton.

The effects of the sympathetic nervous system on bone have recently been explored and may be important during aging since several investigators have suggested there is an increase in sympathetic tone with advanced age [57]. Beta adrenergic activation of receptors on the osteoblast causes uncoupled bone remodeling such that formation is suppressed and resorption is increased within the bone marrow milieu and trabecular skeleton. The effects of adrenergic activity on the periosteum are not known, although nerve fibers are present in this highly vascular environment. However, in one model of chronically elevated SNS activity, the misty mouse, age-related changes in cortical bone were extremely pronounced. Cortical thickness was markedly reduced at 72 weeks vs wild type age-matched controls, as was trabecular bone volume [58]. Interestingly, periosteal expansion with aging did not occur in these mice leading to a much thinner bone during aging with enhanced skeletal fragility. Whether sympathetic tone prevents periosteal expansion as a compensatory mechanism during mammalian aging requires further investigation.

9 Research Directions: Musculoskeletal Aging as a Determinant of Healthspan

Aging is a physiologic process that affects the entire organism, including the musculoskeletal system, through the pillars related to healthspan. There is the direct impairment in bone formation and the acceleration in resorption that occurs over time in virtually all mammals primarily as a result of changes in the stem cell pool, as well as chronic inflammation, and greater accumulation of reactive oxygen species. There is a secondary increase in periosteal formation in response to bone loss albeit not to the degree that matches an increase in medullary expansion. There are also indirect cell non-autonomous effects in the aging animal including enhanced sympathetic tone, changes in the parathyroid/vitamin D axis, impaired renal function, and gonadal deficiency. Coincident with the aging skeleton, muscle mass is also declining and its function is reduced. As discussed above, the bone-muscle interface plays a critical role in modulating skeletal loading as well as cell signaling. Future research should start by more fully delineating how each of the pillars that compose the aging process affect bone, muscle and the interface between the two.

One major thrust should be in defining how the periosteum could be resistant to several of the determinants that impair healthy aging and its relationship to sarcopenia. Although the periosteal envelope can expand with aging, it is unclear whether the signals for that arise from the muscle, the bone matrix, from other bone cells or from an enhanced sensitivity to loading. One limitation is that studies of the periosteum have been relatively limited due to the difficulty in isolating the progenitor cells and studying them ex vivo. Even if models were developed to study the bone-muscle interface, we still do not know whether its expansion has any impact on muscle function. On the other hand, we know that by increasing periosteal surface tension, biomechanical properties improve or at least stabilize in the face of endosteal resorption. In that same vein, delineating the communication network between osteocytes (mechanical sensors) and the periosteum will be essential for defining age-related periosteal effects. A more important question is whether the periosteum is protected from several critical determinants that define aging; i.e. metabolic dysfunction, accumulation of ROS, excess mitochondrial autophagy, and cell senescence in the stem cell pool. A focus on the Foxo proteins during aging provides the first clues as to some of the protective mechanisms inherent within the cell that may be operative during aging.

Another important aspect of the bone muscle interface lies in the remarkable gender differences in the periosteal envelope across all ages. This parallels the differences in muscle mass and bone size that is observed between males and females, suggesting that there is always a factor based on size that determines the musculoskeletal mass. But it is not clear whether periosteal osteoblasts differ between males and females, and if aging has a selective effect (positive or negative) on the ability of these bone-forming cells to expand and lay down collagen.

Sarcopenia is a huge clinical problem because of the falls that result from muscle weakness. It is uncertain how progressive but modest muscle loss directly affects the skeleton and in particular the periosteum. Targeted therapy with myokine agonists or antagonists are soon to be developed for frailty, yet we know little about the mechanisms at the bone-muscle interface. Finally, the CNS plays a critical role in the maintenance of both bone and muscle. How the SNS modulates periosteal tone and muscle mass is a major question. Understanding the role of neuropeptides at the bone-muscle interface provides another targeted area for research, particularly with aging. For example, Linder et al. have shown that loss of a peptide Cthrc1, causes accelerated age-related bone loss, low energy expenditure and reduced muscle strength in C57BL6J mice (V Lindner, 2015, personal communication). Remarkably, Cthrc1 is highly expressed in the pituitary and hypothalamus and circulates in measureable quantities. Hence, CNS signals, whether they be neurotropic or hormonal, can profoundly affect the muscle-bone interface.

The new discipline of Geroscience attempts to merge the physiology of aging with an understanding of the pathophysiology of age-related diseases and the delineation of the pillars that define age-associated disorders. We can no longer afford to study major organ systems in isolation with age, and a major thrust for future studies will be in defining regulation of the bone-muscle interface and the downstream consequences that result from impairment in either tissue.

Acknowledgments AG040217 to CJR, and AR063919 awarded to SM and CJR.
Editor: John Williams, National Institute on Aging (NIA), NIH.

References

1. Manolagas SC (2000) Birth and death of bone cells: basic regulatory mechanisms and implications for the pathogenesis and treatment of osteoporosis. Endocr Rev 21(2):115–137. doi:10.1210/edrv.21.2.0395
2. Hadjiargyrou M, O'Keefe RJ (2014) The convergence of fracture repair and stem cells: interplay of genes, aging, environmental factors and disease. J Bone Miner Res 29(11):2307–2322. doi:10.1002/jbmr.2373
3. Cawthorn WP, Scheller EL, Learman BS, Parlee SD, Simon BR, Mori H, Ning X, Bree AJ, Schell B, Broome DT, Soliman SS, DelProposto JL, Lumeng CN, Mitra A, Pandit SV, Gallagher KA, Miller JD, Krishnan V, Hui SK, Bredella MA, Fazeli PK, Klibanski A, Horowitz MC, Rosen CJ, MacDougald OA (2014) Bone marrow adipose tissue is an endocrine organ that contributes to increased circulating adiponectin during caloric restriction. Cell Metab 20(2):368–375. doi:10.1016/j.cmet.2014.06.003
4. Iversen N, Krustrup P, Rasmussen HN, Rasmussen UF, Saltin B, Pilegaard H (2011) Mitochondrial biogenesis and angiogenesis in skeletal muscle of the elderly. Exp Gerontol 46(8):670–678. doi:10.1016/j.exger.2011.03.004
5. Raisz LG (1988) Local and systemic factors in the pathogenesis of osteoporosis. N Engl J Med 318(13):818–828. doi:10.1056/NEJM198803313181305
6. Colman RJ, Beasley TM, Allison DB, Weindruch R (2008) Attenuation of sarcopenia by dietary restriction in rhesus monkeys. J Gerontol A Biol Sci Med Sci 63(6):556–559
7. Evans WJ (2010) Skeletal muscle loss: cachexia, sarcopenia, and inactivity. Am J Clin Nutr 91(4):1123S–1127S. doi:10.3945/ajcn.2010.28608A

8. Freeman LM (2012) Cachexia and sarcopenia: emerging syndromes of importance in dogs and cats. J Vet Intern Med 26(1):3–17. doi:10.1111/j.1939-1676.2011.00838.x
9. Sheard PW, Anderson RD (2012) Age-related loss of muscle fibres is highly variable amongst mouse skeletal muscles. Biogerontology 13(2):157–167. doi:10.1007/s10522-011-9365-0
10. Herndon LA, Schmeissner PJ, Dudaronek JM, Brown PA, Listner KM, Sakano Y, Paupard MC, Hall DH, Driscoll M (2002) Stochastic and genetic factors influence tissue-specific decline in ageing C. elegans. Nature 419(6909):808–814. doi:10.1038/nature01135
11. Piccirillo R, Demontis F, Perrimon N, Goldberg AL (2014) Mechanisms of muscle growth and atrophy in mammals and Drosophila. Dev Dyn 243(2):201–215. doi:10.1002/dvdy.24036
12. Hepple RT (2003) Sarcopenia–a critical perspective. Sci Aging Knowledge Environ 2003(46):pe31. doi:10.1126/sageke.2003.46.pe31
13. Janssen I (2011) The epidemiology of sarcopenia. Clin Geriatr Med 27(3):355–363. doi:10.1016/j.cger.2011.03.004
14. Rosenberg IH (2011) Sarcopenia: origins and clinical relevance. Clin Geriatr Med 27(3):337–339. doi:10.1016/j.cger.2011.03.003
15. Abbatecola AM, Paolisso G, Fattoretti P, Evans WJ, Fiore V, Dicioccio L, Lattanzio F (2011) Discovering pathways of sarcopenia in older adults: a role for insulin resistance on mitochondria dysfunction. J Nutr Health Aging 15(10):890–895
16. Butikofer L, Zurlinden A, Bolliger MF, Kunz B, Sonderegger P (2011) Destabilization of the neuromuscular junction by proteolytic cleavage of agrin results in precocious sarcopenia. FASEB J 25(12):4378–4393. doi:10.1096/fj.11-191262
17. Fiser WM, Hays NP, Rogers SC, Kajkenova O, Williams AE, Evans CM, Evans WJ (2010) Energetics of walking in elderly people: factors related to gait speed. J Gerontol A Biol Sci Med Sci 65(12):1332–1337. doi:10.1093/gerona/glq137
18. Fry CS, Drummond MJ, Glynn EL, Dickinson JM, Gundermann DM, Timmerman KL, Walker DK, Dhanani S, Volpi E, Rasmussen BB (2011) Aging impairs contraction-induced human skeletal muscle mTORC1 signaling and protein synthesis. Skelet Muscle 1(1):11. doi:10.1186/2044-5040-1-11
19. Gouspillou G, Bourdel-Marchasson I, Rouland R, Calmettes G, Biran M, Deschodt-Arsac V, Miraux S, Thiaudiere E, Pasdois P, Detaille D, Franconi JM, Babot M, Trezeguet V, Arsac L, Diolez P (2014) Mitochondrial energetics is impaired in vivo in aged skeletal muscle. Aging Cell 13(1):39–48. doi:10.1111/acel.12147
20. Sanchis-Gomar F, Gomez-Cabrera MC, Vina J (2011) The loss of muscle mass and sarcopenia: non hormonal intervention. Exp Gerontol 46(12):967–969. doi:10.1016/j.exger.2011.08.012
21. Tedesco FS, Dellavalle A, Diaz-Manera J, Messina G, Cossu G (2010) Repairing skeletal muscle: regenerative potential of skeletal muscle stem cells. J Clin Invest 120(1):11–19. doi:10.1172/JCI40373
22. Walrand S, Guillet C, Salles J, Cano N, Boirie Y (2011) Physiopathological mechanism of sarcopenia. Clin Geriatr Med 27(3):365–385. doi:10.1016/j.cger.2011.03.005
23. Wang Q, Seeman E (2013) Skeletal growth and peak bone strength. In: Rosen CJ (ed) Primer on the metabolic bone diseases and disorders of mineral metabolism. Wiley-Blackwell, Ames, pp 127–134
24. Christiansen BA, Kopperdahl DL, Kiel DP, Keaveny TM, Bouxsein ML (2011) Mechanical contributions of the cortical and trabecular compartments contribute to differences in age-related changes in vertebral body strength in men and women assessed by QCT-based finite element analysis. J Bone Miner Res 26(5):974–983. doi:10.1002/jbmr.287
25. Khosla S (2013) Pathogenesis of age-related bone loss in humans. J Gerontol A Biol Sci Med Sci 68(10):1226–1235. doi:10.1093/gerona/gls163
26. Glatt V, Canalis E, Stadmeyer L, Bouxsein ML (2007) Age-related changes in trabecular architecture differ in female and male C57BL/6 J mice. J Bone Miner Res 22(8):1197–1207. doi:10.1359/jbmr.070507
27. Reeve J, Loveridge N (2014) The fragile elderly hip: mechanisms associated with age-related loss of strength and toughness. Bone 61:138–148. doi:10.1016/j.bone.2013.12.034

28. Vanderschueren D, Laurent MR, Claessens F, Gielen E, Lagerquist MK, Vandenput L, Borjesson AE, Ohlsson C (2014) Sex steroid actions in male bone. Endocr Rev 35(6):906–960. doi:10.1210/er.2014-1024
29. Seeman E (2013) Age- and menopause-related bone loss compromise cortical and trabecular microstructure. J Gerontol A Biol Sci Med Sci 68(10):1218–1225. doi:10.1093/gerona/glt071
30. Colnot C, Zhang X, Knothe Tate ML (2012) Current insights on the regenerative potential of the periosteum: molecular, cellular, and endogenous engineering approaches. J Orthop Res 30(12):1869–1878. doi:10.1002/jor.22181
31. Ferretti C, Mattioli-Belmonte M (2014) Periosteum derived stem cells for regenerative medicine proposals: boosting current knowledge. World J Stem Cells 6(3):266–277. doi:10.4252/wjsc.v6.i3.266
32. Xiong J, Piemontese M, Thostenson JD, Weinstein RS, Manolagas SC, O'Brien CA (2014) Osteocyte-derived RANKL is a critical mediator of the increased bone resorption caused by dietary calcium deficiency. Bone 66:146–154. doi:10.1016/j.bone.2014.06.006
33. Jilka RL, O'Brien CA, Roberson PK, Bonewald LF, Weinstein RS, Manolagas SC (2014) Dysapoptosis of osteoblasts and osteocytes increases cancellous bone formation but exaggerates cortical porosity with age. J Bone Miner Res 29(1):103–117. doi:10.1002/jbmr.2007
34. Aaron JE (2012) Periosteal Sharpey's fibers: a novel bone matrix regulatory system? Front Endocrinol 3:98. doi:10.3389/fendo.2012.00098
35. Ahlborg HG, Johnell O, Turner CH, Rannevik G, Karlsson MK (2003) Bone loss and bone size after menopause. N Engl J Med 349(4):327–334. doi:10.1056/NEJMoa022464
36. Moore SR, Milz S, Knothe Tate ML (2014) Periosteal thickness and cellularity in mid-diaphyseal cross-sections from human femora and tibiae of aged donors. J Anat 224(2):142–149. doi:10.1111/joa.12133
37. Carpenter RD, Carter DR (2008) The mechanobiological effects of periosteal surface loads. Biomech Model Mechanobiol 7(3):227–242. doi:10.1007/s10237-007-0087-9
38. Bartell SM, Kim HN, Ambrogini E, Han L, Iyer S, Serra Ucer S, Rabinovitch P, Jilka RL, Weinstein RS, Zhao H, O'Brien CA, Manolagas SC, Almeida M (2014) FoxO proteins restrain osteoclastogenesis and bone resorption by attenuating H2O2 accumulation. Nat Commun 5:3773. doi:10.1038/ncomms4773
39. Miller RA, Kreider J, Galecki A, Goldstein SA (2011) Preservation of femoral bone thickness in middle age predicts survival in genetically heterogeneous mice. Aging Cell 10(3):383–391. doi:10.1111/j.1474-9726.2011.00671.x
40. Ferretti C, Lucarini G, Andreoni C, Salvolini E, Bianchi N, Vozzi G, Gigante A, Mattioli-Belmonte M (2014) Human periosteal derived stem cell potential: the impact of age. Stem Cell Rev. doi:10.1007/s12015-014-9559-3
41. Yukata K, Xie C, Li TF, Takahata M, Hoak D, Kondabolu S, Zhang X, Awad HA, Schwarz EM, Beck CA, Jonason JH, O'Keefe RJ (2014) Aging periosteal progenitor cells have reduced regenerative responsiveness to bone injury and to the anabolic actions of PTH 1–34 treatment. Bone 62:79–89. doi:10.1016/j.bone.2014.02.002
42. Shah K, Majeed Z, Jonason J, O'Keefe RJ (2013) The role of muscle in bone repair: the cells, signals, and tissue responses to injury. Curr Osteoporos Rep 11(2):130–135. doi:10.1007/s11914-013-0146-3
43. Jepsen KJ, Andarawis-Puri N (2012) The amount of periosteal apposition required to maintain bone strength during aging depends on adult bone morphology and tissue-modulus degradation rate. J Bone Miner Res 27(9):1916–1926. doi:10.1002/jbmr.1643
44. Lai X, Price C, Lu XL, Wang L (2014) Imaging and quantifying solute transport across periosteum: implications for muscle-bone crosstalk. Bone 66:82–89. doi:10.1016/j.bone.2014.06.002
45. Johnson RW, White JD, Walker EC, Martin TJ, Sims NA (2014) Myokines (muscle-derived cytokines and chemokines) including ciliary neurotrophic factor (CNTF) inhibit osteoblast differentiation. Bone 64:47–56. doi:10.1016/j.bone.2014.03.053
46. Bialek P, Parkington J, Li X, Gavin D, Wallace C, Zhang J, Root A, Yan G, Warner L, Seeherman HJ, Yaworsky PJ (2014) A myostatin and activin decoy receptor enhances bone formation in mice. Bone 60:162–171. doi:10.1016/j.bone.2013.12.002

47. Rao RR, Long JZ, White JP, Svensson KJ, Lou J, Lokurkar I, Jedrychowski MP, Ruas JL, Wrann CD, Lo JC, Camera DM, Lachey J, Gygi S, Seehra J, Hawley JA, Spiegelman BM (2014) Meteorin-like is a hormone that regulates immune-adipose interactions to increase beige fat thermogenesis. Cell 157(6):1279–1291. doi:10.1016/j.cell.2014.03.065
48. Fielding RA, Vellas B, Evans WJ, Bhasin S, Morley JE, Newman AB, Abellan van Kan G, Andrieu S, Bauer J, Breuille D, Cederholm T, Chandler J, De Meynard C, Donini L, Harris T, Kannt A, Keime Guibert F, Onder G, Papanicolaou D, Rolland Y, Rooks D, Sieber C, Souhami E, Verlaan S, Zamboni M (2011) Sarcopenia: an undiagnosed condition in older adults. Current consensus definition: prevalence, etiology, and consequences. International working group on sarcopenia. J Am Med Dir Assoc 12(4):249–256. doi:10.1016/j.jamda.2011.01.003
49. Nilwik R, Snijders T, Leenders M, Groen BB, van Kranenburg J, Verdijk LB, van Loon LJ (2013) The decline in skeletal muscle mass with aging is mainly attributed to a reduction in type II muscle fiber size. Exp Gerontol 48(5):492–498. doi:10.1016/j.exger.2013.02.012
50. Alway SE, Myers MJ, Mohamed JS (2014) Regulation of satellite cell function in sarcopenia. Front Aging Neurosci 6:246. doi:10.3389/fnagi.2014.00246
51. Gopinath SD, Rando TA (2008) Stem cell review series: aging of the skeletal muscle stem cell niche. Aging Cell 7(4):590–598. doi:10.1111/j.1474-9726.2008.00399.x
52. Conboy IM, Rando TA (2012) Heterochronic parabiosis for the study of the effects of aging on stem cells and their niches. Cell Cycle 11(12):2260–2267. doi:10.4161/cc.20437
53. Fry CS, Lee JD, Mula J, Kirby TJ, Jackson JR, Liu F, Yang L, Mendias CL, Dupont-Versteegden EE, McCarthy JJ, Peterson CA (2014) Inducible depletion of satellite cells in adult, sedentary mice impairs muscle regenerative capacity without affecting sarcopenia. Nat Med. doi:10.1038/nm.3710
54. Dreyer HC, Blanco CE, Sattler FR, Schroeder ET, Wiswell RA (2006) Satellite cell numbers in young and older men 24 hours after eccentric exercise. Muscle Nerve 33(2):242–253. doi:10.1002/mus.20461
55. Roth SM, Martel GF, Ivey FM, Lemmer JT, Tracy BL, Metter EJ, Hurley BF, Rogers MA (2001) Skeletal muscle satellite cell characteristics in young and older men and women after heavy resistance strength training. J Gerontol A Biol Sci Med Sci 56(6):B240–B247
56. Camerino C, Zayzafoon M, Rymaszewski M, Heiny J, Rios M, Hauschka PV (2012) Central depletion of brain-derived neurotrophic factor in mice results in high bone mass and metabolic phenotype. Endocrinology 153(11):5394–5405. doi:10.1210/en.2012-1378
57. Quiros-Gonzalez I, Yadav VK (2014) Central genes, pathways and modules that regulate bone mass. Arch Biochem Biophys 561:130–136. doi:10.1016/j.abb.2014.06.005
58. Motyl KJ, Bishop KA, DeMambro VE, Bornstein SA, Le P, Kawai M, Lotinun S, Horowitz MC, Baron R, Bouxsein ML, Rosen CJ (2013) Altered thermogenesis and impaired bone remodeling in Misty mice. J Bone Miner Res 28(9):1885–1897. doi:10.1002/jbmr.1943

Osteoporosis and Mechanisms of Skeletal Aging

Julie Glowacki and Tamara Vokes

Contents

J. Glowacki, Ph.D. (✉)
Department of Orthopedic Surgery, Brigham and Women's Hospital,
75 Francis Street, Boston, MA, 02115, USA
e-mail: jglowacki@partners.org

T. Vokes, M.D.
Section of Endocrinology, Department of Medicine, University of Chicago, 5841 South Maryland Ave, Chicago, IL, 60637, USA
e-mail: tvokes@uchicago.edu

F. Sierra, R. Kohanski (eds.), *Advances in Geroscience*,
DOI 10.1007/978-3-319-23246-1_10

1 Clinical Aspects of Osteoporosis

Osteoporosis is a generalized skeletal disorder in which decrease in bone mass and deterioration of bone quality lead to bone fragility and increased risk of fracture. Osteoporosis is primarily a disease of the elderly, with more than 70 % of fractures being sustained by those 65 or older [1]. In fact, the risk of fractures increases exponentially with age [2] (Fig. 1). Fragility fractures, also termed osteoporotic fractures or low trauma fractures, occur when falling from a standing height during usual physical activity [3]. Fractures result from an interaction between bone strength and the mechanical force applied to it, usually during a fall. Younger individuals may experience fragility fractures when they have diseases or take medications that have harmful effects on bone. However, bone strength is influenced by bone quantity (mass) and bone quality, both of which decrease with age, thus leading to an increase in fragility fractures among the elderly. In addition, elders have an increased risk for falls, which further contributes to increased fracture incidence.

1.1 Scope and Significance of the Problem

In 2005 there were more than two million fractures in the US [1] and this number is likely to increase due to the increasing age of the population. Because the risk of osteoporotic fractures increases with age [2], this population growth will likely result in increased numbers of fractures and associated health care costs. Osteoporotic fractures result in significant morbidity, mortality, and reduced quality of life [3]. The three most common osteoporotic fractures are hip, vertebral, and wrist. [3] Hip fractures are the most devastating type of fracture in terms of both personal suffering and healthcare costs. Hip fractures are associated with increased mortality, loss of independent living, and decline in functional status [4–6]. Vertebral fractures are also associated with increased mortality as well as kyphosis, impaired breathing and digestion due to reduction in thoracic and abdominal cavities, loss of self-esteem due to change in appearance, depression, fatigue, and decreased activity [7–10]. A study of US Nationwide Inpatient Sample between 2000 and 2011 reported that among women aged 55 and over there were 4.9 million hospitalizations for osteoporotic fractures, compared with 2.9 million hospitalizations for myocardial infarctions, three million for stroke, and 700,000 for breast cancer [11]. Osteoporotic fractures accounted for nearly 50 % of hospitalizations among women 75 years and older. Although the hospitalization rates for all other diseases declined during this 11 year observation period, the rate of hospitalization for non-hip fractures actually increased [11].

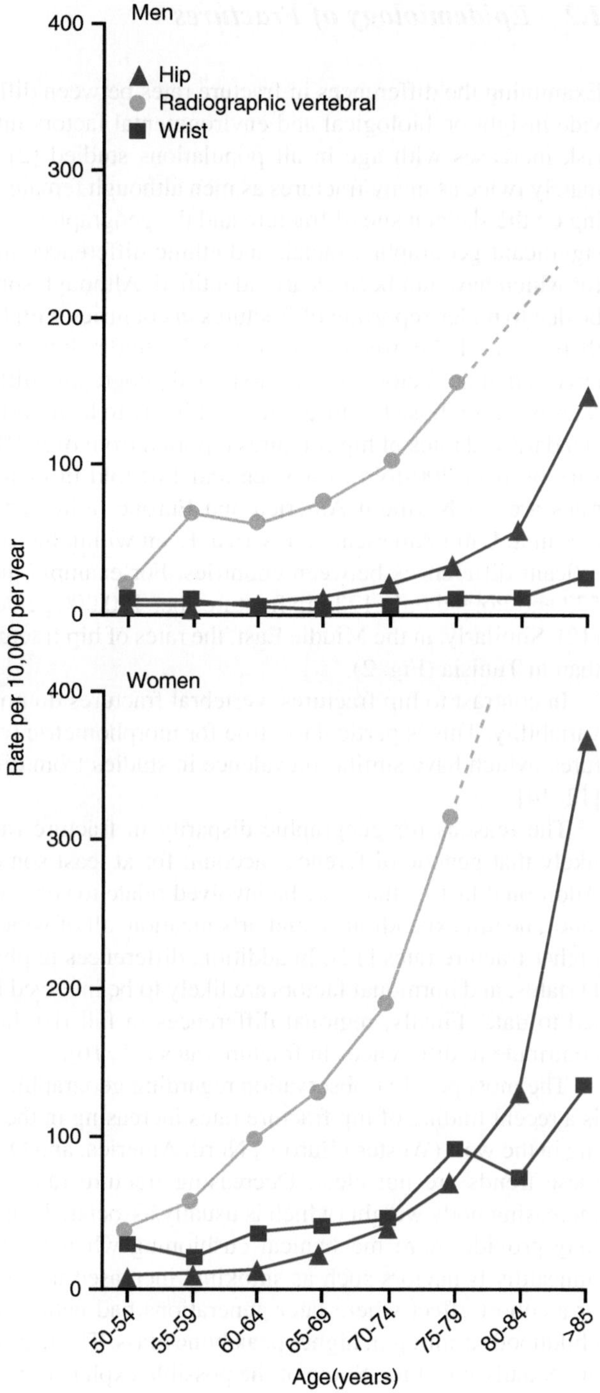

Fig. 1 Increase in fracture incidence with age for men and women. From [2]

1.2 Epidemiology of Fractures

Examining the differences in fracture rates between different populations may provide insight on biological and environmental factors influencing fragility. Fracture risk increases with age in all populations studied [2] and women have approximately twice as many fractures as men although female-to-male ratios vary depending on the skeletal site of fracture and the geographic region (Fig. 2) [12]. There are significant geographic, racial, and ethnic differences in fracture rates, the reasons for which have not been clearly identified. Although some of these differences may be due to under-reporting of fractures in countries with less developed medical care, there are probably true differences in fracture risk that are due to genetic as well as environmental factors. The best-studied geographic differences are for hip fracture rates because those fractures are most likely to be reported accurately (Fig. 2). Age-standardized rates of hip fractures reported from over 60 countries around the world vary by over 200-fold in women and 140-fold in men [12]. The highest reported rates are for Northern America and Europe, followed by Asia, the Middle East, Oceania, Latin America, and Africa. Even within the same continent there are significant differences between countries. For example, in Europe, Norway reported 532 and Poland only 173 hip fractures per 100,000 person years, a 4-fold difference [12]. Similarly, in the Middle East, the rates of hip fracture are 8 times higher in Iran than in Tunisia (Fig. 2).

In contrast to hip fractures, vertebral fractures do not show as much geographic variability. This is particularly true for morphometric (radiographic) vertebral fractures, which have similar prevalence in studies from different regions of the world [12–14].

The reasons for geographic disparity in fracture rates are not clear [12]. It is likely that genetic differences account for at least some of the observed disparity. Additional factors that may be involved relate to socio-economic status, life expectancy, health expenditures, and urbanization, all of which are higher in nations with higher fracture rates [12]. In addition, differences in physical activity, diet, vitamin D status, and hormonal factors are likely to be involved but have not been well studied to date. Finally, regional differences in fall risk have been reported and may contribute to differences in fracture rates [15, 16].

The most peculiar observation regarding geographic differences in fracture rates is a recent finding of hip fracture rates increasing in the east (China) while decreasing in the west (Western Europe, North America, and Oceania) [17]. The reasons for these trends are not clear. Decreasing fracture rates in the west may be due to increasing body weight (which is usually associated with higher bone mass and also may provide more mechanical cushioning when falling on the hip), decrease in unhealthy behaviors such as smoking, increased use of therapies for osteoporosis, or a cohort effect where later generations had better nutrition in utero and during childhood resulting in higher peak bone mass. The increasing fracture rates in China are equally puzzling. Some of the possible explanations include better reporting but also the adoption of western life-style that may lead to unfavorable changes in

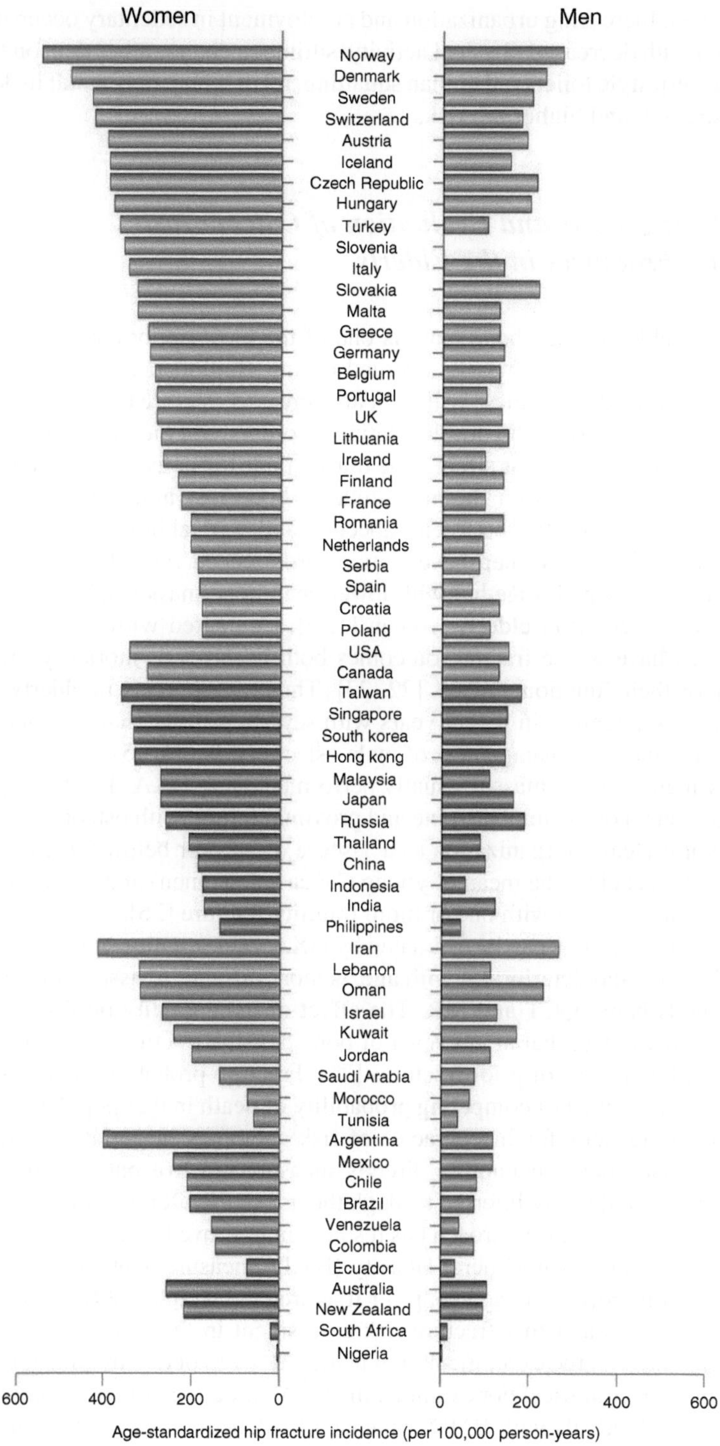

Fig. 2 Geographic differences in age-adjusted hip fracture rates. From [12]

activity level. Increasing urbanization and employment in sedentary occupations are associated with decreased physical activity, sitting on chairs rather than on the floor, use of western style toilets rather than squatting, all of which may result in decreased muscle strength and higher fall risk.

1.3 Pathogenesis and Evaluation of Osteoporosis and Fractures in the Elderly

Even in the elderly, peak bone mass is one of the most important determinants of fragility [18]. Peak bone mass is accrued during childhood and adolescence and those with low peak bone mass will be at an increased fracture risk later in life, such as is the case with those who develop eating disorders, use medications (glucocorticoids), or have diseases that affect bone during their formative years. Bone mass in the elderly also depends on the magnitude of bone loss after peak bone mass is achieved; those rates differ between trabecular and cortical bone, and between men and women [18–20]. Women have a more pronounced rate of loss during early menopause [18], which together with lower peak bone mass results in greater risk of fractures observed in elderly women [1, 6]. Compared with women, however, elderly men have worse fracture outcomes both in terms of mortality [4, 21] and recovery of their functional status [22, 23]. Thus osteoporosis in elderly men has received more attention in recent years with several professional associations providing guidelines for management of male osteoporosis [24, 25].

Assessment of bone mass is usually performed using DXA (Dual Energy X-ray Absorptiometry) of the lumbar spine and proximal femur, with osteoporosis defined by the World Health Organization as a T-score of −2.5 or below (at least 2.5 standard deviations below the mean of young Caucasian women) and severe osteoporosis as the same T-score with one or more fragility fracture [26].

Although bone mass can be estimated by DXA, bone quality, which also contributes to fragility and deteriorates with age, is more difficult to assess, the only available methods being QCT and MR. The effect of bone quality on fragility is well illustrated by the fact that at any level of bone density, fracture risk increases with age and with a history of prior fractures [27]. Fracture probability decreases in the very oldest, as a result of competing probability of death in that population, and this is one of the reasons for inaccuracies of risk estimates in geriatric studies [28]. Although bone quality cannot be directly measured in live patients, its effect on fragility can be indirectly inferred through the use of FRAX algorithm (http://www.shef.ac.uk/FRAX/). This approach has resulted in improved stratification of fracture risk by combining Bone Mineral Density (BMD) measurements with clinical risk factors (most important being age, prior fracture, and weight) [29]. Although there is universal agreement that fracture risk assessment in the elderly should include clinical risk factors, the exact method of using FRAX to derive treatment thresholds differs between countries. For example, the USA guidelines suggest pharmacologic therapy for individuals with BMD evidence of osteoporosis, or with history of hip or vertebral fractures [3]. For those with osteopenia (intermediate risk) based on BMD, therapeutic decisions rely on FRAX-derived fracture probability [3]. In contrast, the UK approach is based on FRAX-defined treatment thresholds with BMD testing only recommended if a risk estimate is borderline [30].

Prior fracture is predictive of future fracture even when controlling for age and BMD. Vertebral fractures in particular are important in this regard because they are most strongly associated with the risk of future fractures [31–33], and their association with fragility is under-estimated by FRAX [34]. Vertebral fractures, however, are clinically apparent in less than 1/3 of subjects [35], and thus require spine imaging for their detection by radiographs or Vertebral Fracture Assessment (VFA) [36]. Because vertebral fracture prevalence increases dramatically with age [6], most of the elderly evaluated for osteoporosis would be candidates for VFA or other spine imaging [3, 36–38].

Finally, the increase in fracture risk with age is related not just to increasing bone fragility, but also to increased fall risk, which is not included in the FRAX model. Falls are common in the elderly with 30–50 % of populations over 65 falling at least once per year and 15 % falling 2 or more times per year [15, 16]. Among community-dwelling individuals over 85 years of age, annual incidence of falls is over 50 % for women and around 33 % for men [39, 40]. Because falls account for 86–95 % of osteoporotic fractures [16], understanding the determinants of falls in the elderly is very important for management of osteoporosis on both individual and population levels. Falls are at least in part due to decreased muscle function which together with weight loss, self-reported fatigue, slow walking speed, and low physical activity are components of the frailty syndrome [41, 42] (as discussed in detail in Chap. 2). Frailty is associated with increased fall and fracture risk [43–45] and thus represents an important therapeutic target in geriatric medicine. It is notable that although frailty increases with age, the effect of frailty on falls and fractures is largely independent of chronological age [45] as evidenced by the fact that the association between frailty and falls or fractures was observed within each age group.

1.4 Management of Osteoporosis in the Elderly

1.4.1 Nutrition

Dietary advice should ensure adequate caloric and protein intake, low sodium (because high sodium increases urinary calcium excretion), high intake of fruits and vegetables (because this tends to produce alkalinizing effect which decreases bone loss), and adequate amounts of calcium (usually dairy products but other sources such as fortified foods are also good choices). Nutritional deficiencies are common in the elderly [46] and particularly in women with osteoporosis. Many osteoporotic women have a life-long history of eating disorders, or at least obsession with thinness, and often limit their intake of nutrients, which is particularly detrimental in advanced age. It is important, therefore, to obtain dietary history from elderly patients, both in term of quantity of food and its composition. Protein-energy malnutrition is a risk factor for sarcopenia and frailty [47, 48]. This effect is at least in part due to low IGF-I, a hormone with anabolic effects on multiple organs, including bone and muscle [49]. Nutritional assessment through dietary history or use of a validated instrument [46, 50] should form the basis for appropriate changes to the diet and consideration of protein supplementation [47, 51–53].

Even in the absence of nutritional deficiency, risk of osteoporotic fractures is higher in thin individuals – they have smaller bone size, lower BMD, often lower postmenopausal estrogen level, and possibly less "cushioning" that could prevent fractures when falling. More recently however, it has become clear that while being moderately overweight might be beneficial to bone, extreme obesity does not protect from osteoporosis and may actually have negative effects on bone, possibly through increased inflammation [54–57]. When controlled for BMD, obese subjects actually have a higher fracture risk compared to lean controls. Furthermore, the positive association between BMD and weight is due to lean body mass, whereas most reports indicate no association between BMD and fat mass [58]. All of these new findings about the relationship of bone to body weight and composition suggest that for bone strength, as well as for many other aspects of health, there may be an optimal weight that provides adequate skeletal loading but prevents the negative effects of excessive fat mass.

Calcium intake from food sources and supplements should also be assessed as this information may reveal the reasons for bone loss and fractures and also forms the basis for prescribing the amount and type of calcium supplements that are appropriate for each patient. Dairy products are recommended because they provide highly bioavailable calcium, as well as protein. If diet alone is not a sufficient source of calcium (1000–1200 mg/day), supplements should be used to make up the difference [3]. There has been a recent controversy regarding the association of increased cardiovascular risk with calcium supplementation [59, 60]. Although this debate cannot be easily resolved, it should be remembered that, from the bone health perspective, deficient calcium intake, particularly if coupled with low vitamin D is harmful as it leads to secondary hyperparathyroidism [61] and increased bone resorption, which is particularly detrimental in the elderly. Proper counseling regarding the importance of sufficient calcium intake is therefore an essential component of osteoporosis management in the geriatric clinic.

Vitamin D sufficiency is also essential in the management of osteoporosis and fall prevention because vitamin D plays an important role in both bone strength and muscle function. Because vitamin D is not generally sufficient from natural foods (other than the liver of cold water fish), most people need to get it from sun exposure, food fortification, or supplements. Sun exposure is frequently limited in the elderly, particularly those who are ill, homebound, or institutionalized [62]. Furthermore, the efficiency of cutaneous synthesis of vitamin D_3 from 7-dehydrocholesterol due to sunlight exposure declines markedly with age, particularly in those greater than 70 y, and this appears to be the primary reason for vitamin D insufficiency in the elderly. Thus, most elderly and particularly those living in northern latitudes need to use vitamin D supplements to achieve adequate vitamin D levels. Although there is a controversy regarding the target level for 25-hydroxyvitamin D in the serum, most agree that it should be between 20 and 30 ng/ml [3].

The required dose of vitamin D depends on the starting blood level and on body weight. Because vitamin D is a fat-soluble vitamin it is distributed throughout fat tissue. Consequently, individuals with larger fat mass will require higher doses and longer duration of vitamin D replacement to reach the target blood levels.

1.4.2 Physical Activity

Evaluating physical activity, including both activities of daily living and formal exercise, is essential for assessing the patients' fall risk as well as developing an individualized program of strengthening exercise and fall prevention. Frailty is strongly associated with fractures [44, 45], likely through an association with low bone mass as well as an increased fall risk due to loss of muscle strength and poor balance. Tools that have been proposed for assessment of frailty [63] are useful, particularly in research studies. It should be remembered, however, that careful observation of the patient while getting up from a chair, climbing onto an exam table, and walking in the hallway will reveal a lot of information about frailty, strength, vitality, and fall risk that can be used for developing the "activity prescription" for each patient. Additionally, examining for kyphosis is very informative because kyphosis may point to the presence of vertebral fractures which signify high bone fragility. In addition, kyphosis per se is associated with frailty, increased fall and fracture risk, and increased mortality [9, 10, 64, 65] and may be targeted by appropriate exercise programs [66].

A proper activity regimen will increase overall strength and fitness while at the same time decrease the risk of falling [67–71]. Walking, and in fit elderly individuals even jogging, are activities that provide gravity stimulus to the bone and also increase the overall fitness. A tailored exercise intervention should improve muscle strength and core strength. Muscle contraction also has an anabolic effect on the bone. Proximal muscle strength and core strength improve balance and decrease the fall risk. Strength and balance interventions should be aligned with the fitness and personal preferences of each individual and can include physical therapy, working with a personal trainer, strength training classes, yoga, Pilates, or Thai Chi. Thai Chi has been shown to improve balance and prevent falls in frail elderly [72, 73]. Multimodality approaches such as the go4life program from the NIA (www.go4life.nia.nih.gov) are particularly helpful as they combine endurance, strength and balance training. In addition to a proper exercise regimen, fall prevention should also include modification of the home environment, treatment of other medical conditions and elimination of medications that may increase fall risk.

1.4.3 Pharmacologic Therapy

According to the National Osteoporosis Foundation (NOF) guidelines, medications for osteoporosis should be considered for those who have a T-score at or below −2.5 or have sustained hip or vertebral fractures. For subjects with osteopenia, NOF recommends pharmacotherapy if their FRAX-derived 10-year probability of major osteoporotic fractures exceeds 20 % or hip fracture probability exceeds 3 %. All the available agents have shown a reduction in vertebral fractures in clinical trials but only some have documented efficacy in preventing non-vertebral fractures. Due to a lack of direct comparison trials, however, and the fact that registration trials recruited somewhat dissimilar populations, direct comparison of efficacy of different agents is not possible. As all fracture trials enrolled postmenopausal women, generally with a mean age between 68 and 75 years, pharmacotherapy seems to be efficacious among the elderly. Yet in most trials, there were relatively few subjects older than 80. Nevertheless, *post hoc* analysis of the existing trials have documented good anti-fracture efficacy of pharmacotherapy in those over 75 years of age [74–79].

In general, pharmacologic agents for osteoporosis are classified into antiresorptive and anabolic agents. In fact, however, bone resorption and formation remain coupled so that use of an antiresorptive drug (bisphosphonates, denosumab) results in suppression of both bone resorption and bone formation. Similarly, the only currently available anabolic agent, teriparatide stimulates bone formation but bone resorption appears to increase as well.

Bisphosphonates have been shown to be efficacious in older populations. Information regarding efficacy of alendronate in older women comes from *post hoc* analysis of FIT1 which showed a 38 % reduction in the risk of new vertebral fractures relative to placebo in patients over 75 years of age [74], and from a pooled analysis of FIT1 and FIT2 which showed similar risk reduction in new fractures across the ages 55–85 years [80]. A *post hoc* analysis of risedronate trials (HIP, VERN-NA and VERT-MN) showed a 44 % reduction in vertebral fractures in women 80–98 years of age [77]. Finally, a *post hoc* subgroup analysis of the two zoledronic acid trials (HORIZON and HORIZON Recurrent Fracture Trial) showed anti-fracture efficacy for women age 75 and older [75].

Other classes of therapeutic agents used for osteoporosis are also effective in the geriatric population. Subgroup analysis of women ≥75 years from the teripartide trial demonstrated a reduction in vertebral fractures (by 65 %) as well as non-vertebral fractures (by 25 %) compared with placebo [76]. *Post hoc* analysis documented fracture benefit of denosumab in women over 75 [81], and an analysis of pooled data from two strontium trials showed a significant reduction of both vertebral and non-vertebral fractures in women between 80 and 100 years [79].

Despite the evidence of its efficacy, pharmacologic therapy is underutilized in the elderly, although they have the highest fracture risk and need this therapy the most. The reasons for that are not completely clear but include poly-pharmacy, erroneous belief that fractures are a natural consequence of aging rather than disease, fear of medication side effects and perhaps, an assumption that pharmacologic agents will not have enough time to exert a benefit due to limited life expectancy in the old. However, several trials have clearly documented that fracture benefit is demonstrable in 1 year or less [77, 82–88] suggesting that even those with life expectancy of just 1–2 years would benefit from therapy. Some of the novel agents or combinations being considered may be particularly useful in geriatric populations [92].

It would seem logical that a choice between an antiresorptive and anabolic agent would be based on baseline bone turnover. In practice, however, bone turnover is not routinely assessed or used in making the therapeutic decisions. Although several biochemical markers of bone turnover have been developed and approved by the FDA, current practice guidelines do not support using turnover markers for selecting the appropriate therapy [89–91]. This is due to analytic and biological variability in the levels of these markers as well as lack of data regarding the ability of the baseline marker levels to predict the response to therapy. There is no consistent effect of aging on bone turnover markers – they increase significantly with menopause but decline thereafter. Furthermore, the increase in bone turnover markers observed in the elderly in some studies may be due to decreased renal function which increases levels of the markers that are cleared by a healthy kidney [89–91].

2 Age-Related Changes in Bone Tissue

2.1 *Age-Related Changes in Bone*

Studies of *in vitro* mechanical properties of bone show an age-related loss in yield strength and in peak strength, increasing the risk of fracture [93]. The mechanical properties of milled samples of cortical bone decrease by 7–12 % per decade in fracture toughness [94]. Bone mass as measured by DXA declines with age and contributes to mechanical instability [18]. Other factors contribute to the fragility of bone, however, independent of bone mass or volume [95]. The aging of human bone can be described at multiple hierarchical levels, from the molecular to microarchitectural to gross changes in shape and form, each of which is detrimental to fracture resistance [96]. Bone extracellular matrix is composed of approximately 35 % organic matrix, or osteoid, by dry weight and 65 % inorganic mineral, a highly substituted carbonato-calcium phosphate. As a biphasic material, bone has tensile properties attributable to the organic collagen fibers and has compressive strength and rigidity attributable to the inorganic crystals. Each component is affected differently by aging.

2.1.1 Bone Organic Matrix

Collagen is a protein that accounts for 90 % of the organic matrix of bone tissue. The self-assembly of the linear collagen molecules into fibrils provides tensile strength to bone tissue; therefore the mechanical properties of bone are influenced by collagen biochemistry. Post-translational modifications and divalent and trivalent intermolecular crosslinks (pyridinoline and deoxypyridinoline) are important aspects of collagen synthesis in bone. Abnormalities of collagen structure can arise from genetic mutations or can be induced by lathyrogenic agents [97]. In osteogenesis imperfecta, for example, mutations in collagen's amino acids can result in the formation of branched fibers that result in brittle bone. When the enzymatic formation of intermolecular crosslinks is inhibited by a lathyrogen, such as β-aminopropionitrile, found in sweet peas, bone strength and mechanical performance decrease.

It is known that aging bone is characterized by modifications in collagen by denaturation [98] or non-enzymatic glycation [99]. In contrast to the beneficial effects of enzymatic crosslinks on collagen structure and bone's material properties, the non-enzymatic crosslinking of collagen that occurs with aging and some diseases leads to bone's mechanical deterioration. The predominant source of glycation end products (AGEs) is endogenous, but there are also exogenous sources, such as foods and tobacco smoke. A study with healthy twins established that a genetic effect accounts for 74 % of normal variance in serum levels of a major AGE, *Nε-carboxymethyl* lysine (CML) [100].

AGEs in bone are of two types, with and without protein-protein crosslinking, and are formed spontaneously by glycation or oxidation. In spite of tissue turnover, there is an age-related increase in AGEs in most skeletal sites in humans [101]. Moreover, patients with osteoporotic fractures show significantly lower content of enzymatic cross-links and higher content of AGEs than non-fractured controls; the increase in AGEs occurs particularly in more highly mineralized, older regions of bone. In addition, AGEs are elevated in the serum of patients with osteoporosis [102].

2.1.2 Bone Mineral

The mineral phase of bone is best described as a highly substituted, poorly crystalline, carbonate-containing analogue of hydroxyapatite. Compared with pure mineral hydroxyapatite, chemical substitutions of its anions and cations in bone mineral result in a disarrayed lattice structure and a Ca/P ratio of less than 1.67. The most common ionic substitutions in bone mineral are carbonate, fluoride, citrate, pyrophosphate, chloride, magnesium, sodium, and potassium. Bone mineral is deposited as poorly crystalline carbonatoapatite and, with increasing age, there is an increase in its Ca/P ratio, an increase in crystal size, and a loss of substituting ions [103]. In ingenious studies, Boskey's team used Fourier transformed infrared microspectroscopy (FTIRM) to assess mineral composition and crystallinity within osteons of a diameter of approximately 150 μm in human cortical bone [104]. This approach provided powerful evidence of crystal maturation from the area of most recent mineral deposition adjacent to the Haversian canal to the oldest mineral on the periphery of the osteon. The data show a decrease in the Ca/P ratio and an increase in crystal size and order from the center to the periphery of an osteon. This conversion decreases the solubility of the mineral phase, a phenomenon that could have untoward consequences for mineral homeostasis if it were to continue unabated. Cement-like mineral is avoided under ordinary circumstances because of the normal turnover of bone's organic and mineral matrix that is achieved by the coordination of osteoclastic resorption and osteoblastic bone formation. Thus, bone remodeling can be viewed, in part, as a process of matrix rejuvenation that is central for mineral exchange and homeostasis. Bone from older individuals is more mineralized than is younger bone, attributable to the incomplete remodeling of matrix and accumulation of larger, denser crystals of mineral [105]. Thus, changes in the nature of bone mineralization with age contribute to decreased fracture toughness [106]. This means that the bone becomes more brittle and less able to absorb impact.

2.1.3 Bone Microarchitecture

Human cortical bone tissue is organized as longitudinal osteons of concentric lamellae around Haversian vascular canals. Volkmann's canals connect Haversian canals and the bone surfaces. The process of internal remodeling removes portions of the matrix and lays down new generations of osteons while maintaining structural integrity, vascularization, and cellular viability within the tissue. With advancing age, there is an imbalance between the amount of bone resorbed and deposited. Thus, porosity increases as canals grow large and coalesce [107]. Mechanical strength decreases exponentially as porosity increases [108].

The age-related loss of bone mass results in loss of strength, but microarchitectural changes are additional critical determinants of bone quality and fracture risk. These changes occur in the trabecular or cancellous interior of bones and in the dense cortical shell. The fracture resistance of bone tissue depends on matrix composition and architecture, to a large degree at the levels of mineralized collagen fibrils, interconnecting trabecular plates, and cortical porosity. Histomorphometric analyses quantify parameters of skeletal architecture, such as trabecular thickness and separation of trabecular plates in cancellous bone. Investigations of bone quality in fracture and non-fracture subjects with equivalent BMD show the relative importance of microarchitecture on fracture risk. An important study of histological specimens from subjects with or without fracture but with equivalent BMD showed significantly poorer markers of trabecular connectivity in those with fractures [109]; that study also showed age-related declines in trabecular connectivity for both fracture and non-fracture groups. Non-invasive tools such as peripheral quantitative computed tomography (pQCT) have been developed to provide three-dimensional modeling of bone *in vivo*. They show sexual dimorphism in the effects of age on trabecular microarchitecture [110]. In BMD-matched women there are significant age-related reductions in trabecular connectivity; in men, there are reductions in trabecular number, spacing, and connectivity; and in both, those reductions were correlated to increased fracture risk.

2.1.4 Microcracks

Tissue fatigue is the progressive loss of strength and stiffness that results from repeated cycles of loading. It manifests as sharp-edge microcracks in Haversian bone, approximately 30–100 μm long. Microdamage accumulates in human bone with age [111, 112]. Accumulation of even small amounts of microscopic tissue damage in human bone may have large effects on biomechanical performance [113]. There are several mechanisms that prevent microdamage from resulting in catastrophic failure; these entail crack arrest and bone turnover. The first is an advantageous feature of Haversian bone, in which crack propagation is attenuated by ultrastructural discontinuities in resorption spaces, at margins of osteons, and at lamellae. Thus, osteonal bone's microstructural features can act as barriers to arrest microcrack extension by blunting the crack tip or deflecting crack growth. The second mechanism is that bone remodeling repairs microdamage, but with aging, lower levels of turnover can retard repair and permit accumulation of microcracks [114]. Evidence indicates that microcracks in cortical bone occur in proximity to osteocyte apoptosis [115] and to sites of remodeling [116]. Although apoptosis of osteocytes is required for targeted remodeling, nearby non-apoptotic osteocytes provide RANKL to stimulate osteoclasts to initiate a resorption tunnel [117].

It is clear that linear microcracks stimulate local bone remodeling and repair by a mechanism that involves osteocytes even in rodents where cortical remodeling is typically not present. On the other hand, diffuse damage at smaller size scales, around 1 μm and less, may be repaired by a different mechanism and may not be an inevitable precursor of microcracks. With an *in vivo* rat ulnar model that introduces diffuse damage in tensile cortices without linear microcracks, Seref-Ferlenguez et al. provide convincing evidence of direct repair without remodeling [118]. This may occur by physico-chemical bridging with calcium deposition within the small gaps or with products of nearby osteocytes. The relative importance of remodeling and direct repair mechanisms in humans is uncertain in light of the fact that cortical remodeling occurs constitutively throughout the human skeleton.

2.1.5 Age-Related Changes in Bone Metabolism

Adult human bone tissue undergoes continuous renewal by a process of remodeling, in which bone-resorbing cells, the osteoclasts, degrade a quantum of mineralized matrix, after which bone-forming cells, the osteoblasts, invade and fill the voids with new organic and mineral components. This remodeling process occurs in foci and ensures the overall mechanical integrity of the skeleton while renewing the tissue, adjusting the bone architecture to mechanical forces, and repairing microdamage. These intrinsic cellular activities endow bone with the capacities of fracture healing, distraction osteogenesis (a surgical procedure to elongate bone with a device that slowly expands a healing callus), graft incorporation, implant fixation, and mineral homeostasis. By replacing mature mineralized matrix, remodeling provides new mineral that is less crystalline and more readily soluble to contribute to calcium homeostasis.

Histomorphometric evidence shows that the balance between bone resorption and formation is inadequate to conserve skeletal mass throughout the lifespan. One of the best established age-related changes in cancellous bone is the reduction in wall width [119]. The reduction is approximately one-third from young adulthood to seniority and is the result of a reduction in bone formation rate. Changes in cortical bone are less well documented. It would be anticipated that age influences bone turnover markers, but many other factors such as immobility, fractures, renal and liver impairments, medications, and circadian and seasonal variations contribute to inconsistent observations in aging studies, especially for resorption markers [89]. In addition, those confounders challenge clinical decision-making; a better biomarker for status of bone metabolism is needed.

Nevertheless, there is evidence linking histomorphometric parameters of bone turnover to serum AGEs. In one study, dynamic parameters of bone formation and static parameters of bone resorption were determined for osteoporotic and control women and men. Multiple regression analyses revealed striking correlations between serum AGEs and osteoporosis subgroups having increased bone resorption and, more specifically, with indices of osteoclast activity [102]. This is consistent with *in vitro* and animal studies showing that AGEs enhance osteoclastic bone resorption [120].

2.2 Age-related Changes in Skeletal Stem Cells

There is a growing body of information available about the effects of age on human skeletal stem cells. Distinctions can be made between those age-related changes that are caused by intrinsic cellular factors and those induced by the extrinsic somatic environment, e.g. by declines in circulating hormones or local cytokines. Marrow cells include osteoclast progenitors in the hematopoietic fraction and osteoblast progenitors called marrow stromal cells (MSCs) or mesenchymal stem cells in the adherent fraction. With marrow cells from subjects 27–82 years old, we reported age-related increases in *in vitro* osteoclast differentiation, in expression of receptors *c-fms* and *RANK* in osteoclast progenitor cells, in constitutive expression of *RANKL* with a decrease in *OPG* by MSCs, and a resulting increase in the *RANKL/OPG* ratio in elderly subjects [121]. There is also an age-related increase in pro-osteolytic IL-6 and IL-11 secretion by MSCs [122]. All of these can mediate age-related increases in bone resorption. It is notable that marrow samples from women being treated with an anti-resorptive bisphosphonate generated only 20 % the number of osteoclasts *in vitro*, as compared with marrow obtained from age-matched controls and showed a rejuvenated *RANKL/OPG* ratio [123].

Regarding osteoblast differentiation potential in human MSCs, we [124, 125] and others [126] showed an age-related decline in their differentiation to osteoblasts *in vitro*. Other important intrinsic properties of MSCs include age-related increases in senescence-associated β-galactosidase (SA β-gal), cell doubling time, apoptosis, as well as *p53* and its target genes, *p21* and *BAX* [125]. Upregulation of the *p53* pathway with age may have a critical role in mediating the reduction in both proliferation and osteoblastogenesis of MSCs. Thus, unlike pluripotent stem cells, as multipotent MSCs from bone marrow age, their intrinsic properties gradually become compromised. Effects of age on expression of WNT genes in MSCs show gender differences - findings that resolved some of the discrepancies in the literature concerning constitutive expression of WNTs: we found age-related decreases in expression of WNT5A and WNT13 for women and in expression of WNT7B and WNT14 for men [127]. Both marrow-derived and adipose-derived MSCs show age-related differences in microRNA profiles, especially those involved in cell proliferation and inflammation [128]. MicroRNA analysis of MSCs from children and adults revealed miR-196a upregulation in adults that was inversely correlated with MSC proliferation through HOXB7 targeting [129]. Rejuvenation of proliferation and differentiation was achieved by forced overexpression of HOXB7 [129]. Other striking effects of age relate to deteriorating MSC responsiveness to osteotropic hormones. The age-related decline in parathyroid hormone (PTH) receptors and PTH signaling may contribute to cellular and tissue aging and suggests receptor-based approaches to restore sensitivity to osteoanabolic PTH [130]. The age-related decline in vitamin D-1α-hydroxylase/CYP27B1 in MSCs accounts for an age-related decrease in stimulation of osteoblastogenesis by 25-hydroxyvitamin D (25OHD) [131]. Both of those adverse effects of age were corrected by treatment with PTH, which upregulated CYP27B1 [131]. In fact, in MSCs from elders, PTH upregulated the vitamin D receptor and 25(OH)D upregulated the PTH receptor [132]. The synergistic effects of PTH and 25(OH)D to rejuvenate osteoblastogenesis in MSCs from elders has been shown to entail an epigenetic mechanism [132]. These findings support the hypotheses that vitamin D metabolism in MSCs serves an autocrine/paracrine role in osteoblastogenesis and that vitamin D sufficiency is important for skeletal health throughout the lifespan. Further, they suggest that different clinical regimens of combined PTH and vitamin D for osteoporosis may be needed to optimize their synergy to stimulate bone formation in elders and in those with chronic kidney disease [133]. Similar age-related losses in response to bone-active agents, such as IGF-I, have been detected in human osteoblasts [134].

2.3 Significance of Age-Related Declines in Osteotropic Hormones

With aging and decreased ovarian, testicular, and adrenal production of estradiol (E2), testosterone (T), and dehydroepiandrosterone (DHEA), subsequent changes in anabolic mediators such as IGF-I, and osteolytic factors like Interleukin (IL-6) may contribute to changes in bone metabolism. These are examples of extrinsic factors that change with age and affect skeletal cells. The unfavorable skeletal effects of the menopause and of male hypogonadism are well known, but the effects of age-related declines in serum T on bone mass are unclear and may be related to conversion of T to E2. We reported that with age, serum DHEA sulfate and IGF-I decline and serum IL-6 increases (Fig. 3); we further found that serum DHEA sulfate and IGF-I were correlated with bone density and that serum IL-6 was inversely correlated with femoral neck bone density [135]. A unifying hypothesis on the possible mechanisms of bone loss associated with age-related declines in sex steroids (Fig. 4) holds that their declines lead to a decrease in IGF-I and other factors that

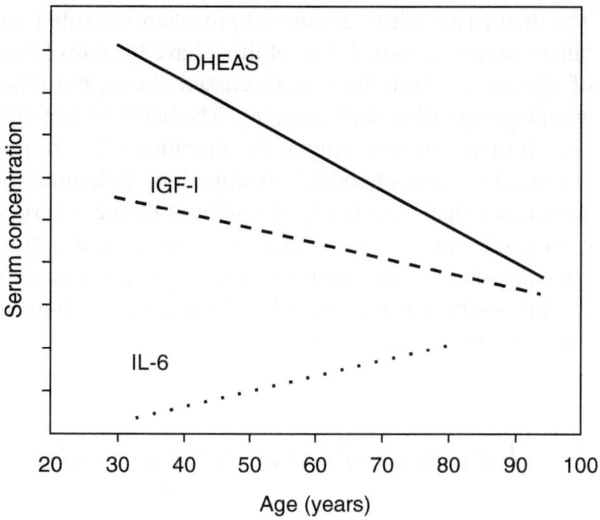

Fig. 3 Relative trajectories showing the effects of age on serum DHEAS, IGF-I, and IL-6 in healthy women (Adapted from [132])

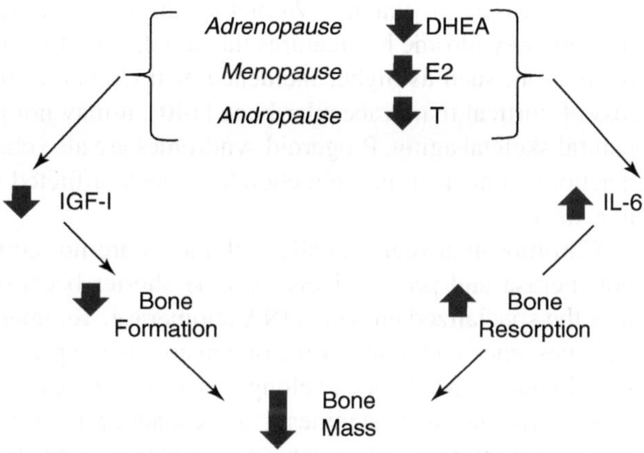

Fig. 4 An hypothesis on the mechanisms of human bone loss associated with age-related declines in sex steroids and intermediary cytokines

may contribute to decreased osteoblastic bone formation, as well as to increased cytokine mediators such as IL-6 and subsequent increased osteoclast formation and bone resorption. Each of these elements can be rejuvenated by sex steroids. For example, E2, T, and DHEA decreased secretion of IL-6 by MSCs from postmenopausal women to levels comparable to cells from young women [136].

3 Molecular and Cellular Underpinnings of Skeletal Aging – The Hallmarks of Aging

Lifespan and aging research with *C. elegans*, *Drosophila*, mice, and other species reveals major themes of interconnected aging processes that are common across species and organ systems and that help to establish potential interventions [137].

The themes also draw attention to mechanistic relationships between aging and certain chronic diseases. Some of these have been investigated in human MSCs in light of age-related bone loss, as discussed above, but others have received little or no attention regarding skeletal aging. The hallmarks of aging include genomic instability, telomere attrition, epigenetic alterations, loss of proteostasis, deregulated nutrient sensing, mitochondrial dysfunction, inflammation, cellular senescence, and stem cell exhaustion [138]. A common theme emerges from examining these processes; they may be beneficial mechanisms that optimize normal tissue homeostasis, but under chronic circumstances may become harmful to the cells or environment. Understanding the relative roles of these mechanisms in skeletal aging may point to potential therapeutic targets [139].

3.1 Hallmarks of Aging: Genetic Mechanisms

Aging is associated with accumulation of DNA damage, as are numerous premature aging diseases, for example Werner syndrome. Because the premature osteoporosis in Werner syndrome has features that are different from osteoporosis in the general population, such as higher incidence of fractures in men than women, and earlier loss of cortical than trabecular bone [140], it may not provide a relevant model for natural skeletal aging. Progeroid syndromes are also characterized by mutations and deletions in nuclear and mitochondrial DNA; afflicted individuals tend to succumb to cancer.

Chromosomal regions called telomeres are not completely replicated by DNA polymerase and progressively become shorter because mammalian somatic cells lack the specialized enzyme DNA telomerase. Telomere attrition is linked to cellular senescence and aging of the organism. Overexpression of the human telomerase gene in human MSCs led to elongation of telomeres, extension of cellular proliferation, maintenance of stemness, and enhanced bone formation [141]. It was suggested that intermittent or transient telomerase activation may be a feasible clinical intervention.

3.2 Hallmarks of Aging: Epigenetic Alterations

Epigeneic alterations include DNA methylation, histone modifications, and chromatin remodeling. Methylation generally inactivates respective promoter regions and is maintained upon replication. Highly significant differences in methylation of specific CpG sites in homeobox genes, including *DLX5* which is involved in osteoblast differentiation, were found for MSCs from young and old donors. Both hypermethylation and hypomethylation sites were detected, findings that may explain why demethylating agents do not categorically control replicative senescence [142].

Changes in histones and chromatin occur during aging and senescence of human MSCs as well, notably in acetylation of H3, H4, and a set of miRNA regions, and in decreased HMGA2, a non-histone chromosomal structural protein that regulates cell cycle arrest proteins, $p16^{INK4A}$ and $p21^{CIP1/WAF1}$ [143]. There is rising interest in ways to reverse these changes in order to use MSCs for cell-based therapies [143, 144].

3.3 *Hallmarks of Aging: Loss of Proteostasis*

Many studies have demonstrated that aging is associated with disordered protein homeostasis and impaired degradation of damaged proteins by either the autophagy-lysosomal system or the ubiquitin-proteasome system [138]. Proteasome inhibitors have been shown to increase osteogenesis in mouse models [145]. There is some information about autophagy during MSC differentiation. Undifferentiated human MSCs show an abundance of undegraded autophagic vacuoles, but autophagic turnover occurs upon induction of osteogenic differentiation [146]. The relationship between senescence and autophagy is complex, with different models showing that autophagy either protects from senescence or triggers senescence. Although there is no literature on the effects of age on autophagy in human MSCs, there is some indirect information. Low dose irradiation of cultured human MSCs reduced their proliferation and upregulated SA β-gal while impairing autophagy-driven apoptosis [147]. In those cells, low dose irradiation led to only a transient rise in apoptosis, followed by senescence of the surviving cells. Because senescent cells can secrete inflammatory cytokines and other detrimental factors (the so-called Senescence Associated Secretory Phenotype, or SASP), the accumulation of senescent cells is likely to be more damaging to the tissue than their removal by apoptosis. It is not known whether similar events occur in MSCs with aging or irradiation *in vivo*. A growing body of information about the functions of autophagy in various mouse bone cells provides a basis for future aging research [148].

3.4 *Hallmarks of Aging: Nutrient Signaling Pathways*

Research on calorie restriction to extend the lifespan of model organisms showed the pivotal roles of energy-sensing pathways that are regulated by insulin/IGF-I, sirtuins, TOR, and AMPK signaling [138]. Those mediators also interact in complex ways and have different effects at different set points, but are likely to contribute to skeletal aging [149, 150]. The growing understanding of cross-talk between bone and fat [151] and between bone and muscle [152] may provide new approaches to improve health of the aging population.

3.5 Hallmarks of Aging: Accumulation of Oxidative Stress Damage

The mitochondrial free radical theory of aging holds that progressive mitochondrial dysfunction results in increased production of reactive oxygen species (ROS), which further damage mitochondria and tissues. Recent research challenges the role of antioxidant defenses in longevity, but highlights the positive, negative, or neutral effects of ROS, depending on context. [138] Nevertheless, considerable evidence indicates a deleterious role of ROS in skeletal aging [153]. This is likely mediated by several pathways, including activation of p53, FOXOs, and NF-κB, and downstream decreases in osteoblastogenesis with increases in bone resorption [139]. Agents as diverse as estrogens, androgens, and intermittently administered PTH exert direct antioxidant effects on bone cells, shown to contribute to their bone-protective actions [154]. In contrast, common antioxidants have not been shown to have enduring beneficial effects on aging bones, but newer agents such as mitochondria-targeting compounds may have potential [155, 156].

3.6 Hallmarks of Aging: Inflammation

"Inflammaging" is the term used to describe the chronic pro-inflammatory phenotype that accompanies aging in mammals and is responsible for altered tissue activities [138]. It has been recognized that inflammation exists as an underlying feature of aging as well as many common chronic age-related diseases such as diabetes, cardiovascular disease, obesity, and osteoporosis - disorders that are generally associated with inflammatory biomarkers like serum C-reactive protein and IL-6 [157]. Persistence of the DNA damage response contributes to systemic inflammation and activation of NF-κB. It has been shown that when AGEs, which accumulate in aging bone, bind to their receptors, RAGEs, in osteoblasts, they activate NF-κB and increase production of cytokines that inhibit osteoblast proliferation and matrix synthesis [158]. Likewise, repeated stimulation of cultured human MSCs with lipopolysaccharide induces markers of senescence, reduced somewhat by silencing $p16^{INK4A}$ [159]. Thus, when present chronically, the damage response networks that normally facilitate repair and survival can compromise tissue homeostasis and lead to cellular apoptosis and senescence.

The theme of "osteoimmunology" developed to focus on the link between bone disease and immune cell activity and concerns inflammatory cytokines such as TNFα and interleukins that stimulate osteoclastogenesis through both RANKL/RANK-dependent and -independent mechanisms. [160] Tissue macrophages, T cells, and other cells in the stem cell niche are involved in these events. With aging, the relative activities of pro-inflammatory M1 macrophages and anti-inflammatory M2 macrophages favor M1 and subsequent decrease in osteoblast differentiation and increased osteoclast differentiation [160]. This information helps to explain how age-associated bone loss is linked to uncoupling of osteoblastic and osteoclastic activities in favor of bone resorption.

Cellular senescence is a beneficial response to damage and stress and prevents dysplasia and cancer by suppressing proliferation of compromised cells. With their persistence, however, damage and stress can alter the microenvironment by inducing the secretion of pro-inflammatory cytokines, a feature called the senescence-associated secretory phenotype (SASP) [157]. There is significant evidence that inflammation plays a key role in MSC senescence. Human MSCs display cell-intrinsic age-associated increases in pro-osteoclastogenic interleukins (IL-6, IL-11), p53, and phosphorylated NF-κB subunits p65 and p50 [122, 124, 125, 128]. In transwell experiments testing for paracrine interactions between human hematopoetic progenitor cells (HPCs) and MSCs, we found an age-related increase in gene expression of TNF-α in HPCs with no expression in MSCs [161]. In vitro treatment with exogenous TNF-α induced markers of senescence in MSCs from a 17-year-old donor, including increased SA-β-gal and reduced proliferation and osteoblast differentiation. Thus, these paracrine, or extrinsic effects of HPCs may contribute to skeletal aging by altering the environment for MSC differentiation to osteoblasts. There are no data about the SASP in MSCs except for a study with conditioned medium (CM) from early versus late passage MSCs. It was reported that tenth passage CM induced the full senescence response in early passage cells, that tenth passage CM contained more insulin-like growth factor binding proteins (IGFBP) 4 and 7, and that exogenous IGFBP4 and −7 each induced senescence and apoptosis in early passage MSCs [162].

4 Research Models for Human Skeletal Aging

Most of the information about mechanisms of skeletal aging summarized herein was obtained from human MSCs obtained from individuals representing a wide age range. Some of the research with those cells is motivated by their possible therapeutic use for age-related and other disorders [65]. To that end, ways to increase *ex vivo* proliferative and differentiation capacity of cells derived from the elderly become a paramount challenge. As anticipated from the Hayflick principle of replicative senescence, *in vitro* expansion of human MSCs activates markers of cellular senescence [144, 163, 164]. Furthermore, passaging *in vitro* is not the only way to induce cellular senescence; for example, treatment with TNF-α [161] or low dose gamma radiation [165] induces senescence in human MSCs. Although these are convenient ways to study cellular senescence, their findings need to be confirmed with cells from subjects across the lifespan.

Mouse models provide insights into lifespan research and the genetics of aging. There are several Senescence Accelerated Mouse (SAM) strains with shortened lifespan and acceleration of different age-associated disorders, including early development of osteopenia [166]. The klotho mouse displays a phenotype similar to human progeria and osteopenia, regulated by a gene called α-klotho [167]. There are more than 700 mouse strains available for research [168], with documentation of great differences in skeletal physiology, metabolism, gender dimorphism [169], and bone healing [156]. Specific differences in skeletal metabolism between mice and human are known, including growth, architecture, regulation and, for example, the role of bone-derived undercarboxylated osteocalcin in glucose metabolism in the former [170]. It is striking that C3H/HeJ mice are characterized by the highest values for all bone parameters, like bone mineral density, while C57BL/6 J mice have the lowest. [169] Nevertheless the C57BL/6 strain is routinely used for research on skeletal physiology and aging [170]; this raises concerns about the generalizability of findings observed in this extreme case. Other concerns about the general relevance of mouse models arise from strain differences noted in calorie restriction (CR) experiments to extend lifespan. CR does not have beneficial effects on longevity in all strains of mice and, in fact, decreases lifespan in many [171–173]. A recently recognized example of species dissimilarity is the inflammatory response of mice and human to sepsis [174]. The absence of osteonal remodeling in rodent bone may limit applicability of some findings to human pathology [175]. Nevertheless, the ovariectomized rat is an often used model to screen drugs for potential efficacy for post-menopausal osteoporosis [176].

As shown for the hallmarks of aging derived from research with diverse species, it remains an advantage to use different models to develop and test approaches for reducing the disabilities associated with human aging.

5 Future Prospects

Skeletal aging has devastating consequences due to the resulting decreased mobility and increased fracture risk. Approaches to prevent osteoporosis would have significant public health benefits. A major challenge in pharmacotherapy of osteoporosis has been the inability to uncouple bone formation and resorption. Accordingly, antiresorptive agents, which are aimed at inhibiting bone resorption, eventually also suppress bone formation while teriparatide, the only anabolic agent currently in use, stimulates not just bone formation but resorption as well. Fortunately, some of the novel agents that are currently in clinical trials hold a promise of having overcome this limitation [92], and may be particularly useful in the geriatric population. Although these advances in target-specific therapy for osteoporosis are very encouraging, it should be remembered that on a societal level the reduction in fractures in the elderly require attention to skeletal health and fitness through the entire lifetime, including improved awareness, detection, and treatment of osteoporosis in the aging population, and more wide-spread interventions for fall prevention. The growing understanding of cross-talk between bone and fat [151] and between bone and muscle [152] may provide new approaches to improve health of the aging population.

As described in this chapter, there are many voids in our knowledge about the relative roles of the hallmarks of aging as they apply to age-related loss of bone mass and increased risk of fracture in humans. It is not known how approaches designed to mitigate other chronic diseases will affect skeletal aging. It is not known how interventions designed to extend lifespan will influence skeletal aging. It is not known whether simple, inexpensive interventions like vitamin D and anti-oxidant-rich diets can diminish the rate of skeletal aging in large populations. It is not known whether or to what degree agents successful in mice are transferable to humans [141, 149, 155, 156, 177]. Several lines of evidence indicate that the decline in stem cell function during aging can involve both cell intrinsic and extrinsic mechanisms. It is not known, however, whether models of induced *in vitro* senescence or *in vivo* aging correspond with natural processes. As discoveries made from research in cell regulatory mechanisms translate to aging research, as with long noncoding RNAs for example [178], new approaches for extending the skeletal healthspan may emerge.

Research to understand the mechanistic basis for the influence of genetics on skeletal aging may yield approaches to promote healthy skeletal aging in those without the genetic advantage. Continued progress in understanding hallmarks of aging in model organisms can be tested in vertebrate species. Broader development of animal models to maximize their value for human skeletal aging research will enhance understanding and potential interventions for age-associated diseases. Use of discarded human tissue can be effective to assess clinical relevance of information gained from model species. Merging of geroscience with osteoporosis research has the potential to allow for early intervention to maximize skeletal health throughout the lifespan.

Acknowledgments The authors have no conflicts to disclose.

Editors: John Williams, National Institute on Aging (NIA) and Joan McGowan, National Institute of Arthritis and Musculoskeletal and Skin Diseases (NINDS), NIH.

References

1. Burge R, Dawson-Hughes B, Solomon DH, Wong JB, King A, Tosteson A (2007) Incidence and economic burden of osteoporosis-related fractures in the United States, 2005–2025. J Bone Miner Res 22(3):465–475
2. Sambrook P, Cooper C (2006) Osteoporosis. Lancet 367(9527):2010–2018
3. Cosman F, de Beur SJ, LeBoff MS, Lewiecki EM, Tanner B, Randall S et al (2014) Clinician's guide to prevention and treatment of osteoporosis. Osteoporos Int 25(10):2359–2381
4. Abrahamsen B, van Staa T, Ariely R, Olson M, Cooper C (2009) Excess mortality following hip fracture: a systematic epidemiological review. Osteoporos Int 20(10):1633–1650
5. Haentjens P, Magaziner J, Colon-Emeric CS, Vanderschueren D, Milisen K, Velkeniers B et al (2010) Meta-analysis: excess mortality after hip fracture among older women and men. Ann Intern Med 152(6):380–390
6. Cummings SR, Melton LJ (2002) Epidemiology and outcomes of osteoporotic fractures. Lancet 359(9319):1761–1767
7. Kado DM, Browner WS, Palermo L, Nevitt MC, Genant HK, Cummings SR (1999) Vertebral fractures and mortality in older women: a prospective study. Study of Osteoporotic Fractures Research Group. Arch Intern Med 159(11):1215–1220

8. Kado DM, Duong T, Stone KL, Ensrud KE, Nevitt MC, Greendale GA et al (2003) Incident vertebral fractures and mortality in older women: a prospective study. Osteoporos Int 14(7):589–594
9. Kado DM, Prenovost K, Crandall C (2007) Narrative review: hyperkyphosis in older persons. Ann Intern Med 147(5):330–338
10. Katzman WB, Huang MH, Lane NE, Ensrud KE, Kado DM (2013) Kyphosis and decline in physical function over 15 years in older community-dwelling women: the Study of Osteoporotic Fractures. J Gerontol A Biol Sci Med Sci 68(8):976–983
11. Singer A, Exuzides A, Spangler L, O'Malley C, Colby C, Johnston K et al (2015) Burden of illness for osteoporotic fractures compared with other serious diseases among postmenopausal women in the United States. Mayo Clin Proc 90(1):53–62
12. Cauley JA, Chalhoub D, Kassem AM, Fuleihan GH (2014) Geographic and ethnic disparities in osteoporotic fractures. Nat Rev Endocrinol 10(6):338–351
13. O'Neill TW, Felsenberg D, Varlow J, Cooper C, Kanis JA, Silman AJ (1996) The prevalence of vertebral deformity in european men and women: the European Vertebral Osteoporosis Study. J Bone Miner Res 11(7):1010–1018
14. Chen P, Krege JH, Adachi JD, Prior JC, Tenenhouse A, Brown JP et al (2009) Vertebral fracture status and the World Health Organization risk factors for predicting osteoporotic fracture risk. J Bone Miner Res 24(3):495–502
15. Karlsson MK, Ribom EL, Nilsson JA, Karlsson C, Coster M, Vonschewelov T et al (2014) International and ethnic variability of falls in older men. Scand J Public Health 42(2):194–200
16. Morrison A, Fan T, Sen SS, Weisenfluh L (2013) Epidemiology of falls and osteoporotic fractures: a systematic review. Clinicoecon Outcomes Res 5:9–18
17. Ballane G, Cauley JA, Luckey MM, Fuleihan GH (2014) Secular trends in hip fractures worldwide: opposing trends East versus West. J Bone Miner Res 29(8):1745–1755
18. Khosla S (2013) Pathogenesis of age-related bone loss in humans. J Gerontol A Biol Sci Med Sci 68(10):1226–1235
19. Seeman E (1999) The structural basis of bone fragility in men. Bone 25(1):143–147
20. Clarke BL, Khosla S (2010) Physiology of bone loss. Radiol Clin North Am 48(3):483–495
21. Kannegaard PN, van der Mark S, Eiken P, Abrahamsen B (2010) Excess mortality in men compared with women following a hip fracture. National analysis of comedications, comorbidity and survival. Age Ageing 39(2):203–209
22. Di Monaco M, Castiglioni C, Vallero F, Di Monaco R, Tappero R (2012) Sarcopenia is more prevalent in men than in women after hip fracture: a cross-sectional study of 591 inpatients. Arch Gerontol Geriatr 55(2):e48–e52
23. Di Monaco M, Castiglioni C, Vallero F, Di Monaco R, Tappero R (2012) Men recover ability to function less than women do: an observational study of 1094 subjects after hip fracture. Am J Phys Med Rehabil 91(4):309–315
24. Watts NB, Adler RA, Bilezikian JP, Drake MT, Eastell R, Orwoll ES et al (2012) Osteoporosis in men: an Endocrine Society clinical practice guideline. J Clin Endocrinol Metab 97(6):1802–1822
25. Geusens PP, van den Bergh JP (2012) Bone: new guidelines for multistep fracture prevention in men. Nat Rev Rheumatol 8(10):568–570
26. Kanis JA, Melton LJ 3rd, Christiansen C, Johnston CC, Khaltaev N (1994) The diagnosis of osteoporosis. J Bone Miner Res 9(8):1137–1141
27. Rizzoli R, Branco J, Brandi ML, Boonen S, Bruyere O, Cacoub P et al (2014) Management of osteoporosis of the oldest old. Osteoporos Int 25(11):2507–2529
28. Berry SD, Ngo L, Samelson EJ, Kiel DP (2010) Competing risk of death: an important consideration in studies of older adults. J Am Geriatr Soc 58(4):783–787
29. Kanis JA, Johnell O, Oden A, Johansson H, McCloskey E (2008) FRAX and the assessment of fracture probability in men and women from the UK. Osteoporos Int 19(4):385–397
30. Compston J, Bowring C, Cooper A, Cooper C, Davies C, Francis R et al (2013) Diagnosis and management of osteoporosis in postmenopausal women and older men in the UK: national Osteoporosis Guideline Group (NOGG) update 2013. Maturitas 75(4):392–396

31. Lindsay R, Silverman SL, Cooper C, Hanley DA, Barton I, Broy SB et al (2001) Risk of new vertebral fracture in the year following a fracture. JAMA 285(3):320–323
32. Ross PD, Genant HK, Davis JW, Miller PD, Wasnich RD (1993) Predicting vertebral fracture incidence from prevalent fractures and bone density among non-black, osteoporotic women. Osteoporos Int 3(3):120–126
33. Siris ES, Genant HK, Laster AJ, Chen P, Misurski DA, Krege JH (2007) Enhanced prediction of fracture risk combining vertebral fracture status and BMD. Osteoporos Int 18(6):761–770
34. Donaldson MG, Palermo L, Schousboe JT, Ensrud KE, Hochberg MC, Cummings SR (2009) FRAX and risk of vertebral fractures: the fracture intervention trial. J Bone Miner Res 24(11):1793–1799
35. Fink HA, Milavetz DL, Palermo L, Nevitt MC, Cauley JA, Genant HK et al (2005) What proportion of incident radiographic vertebral deformities is clinically diagnosed and vice versa? J Bone Miner Res 20(7):1216–1222
36. Lewiecki EM, Laster AJ (2006) Clinical review: clinical applications of vertebral fracture assessment by dual-energy x-ray absorptiometry. J Clin Endocrinol Metab 91(11):4215–4222
37. Ensrud KE, Schousboe JT (2011) Clinical practice. Vertebral fractures. N Engl J Med 364(17):1634–1642
38. Rosen HN, Vokes TJ, Malabanan AO, Deal CL, Alele JD, Olenginski TP et al (2013) The Official Positions of the International Society for Clinical Densitometry: vertebral fracture assessment. J Clin Densitom 16(4):482–488
39. Tinetti ME, Speechley M, Ginter SF (1988) Risk factors for falls among elderly persons living in the community. N Engl J Med 319(26):1701–1707
40. Peel NM, Kassulke DJ, McClure RJ (2002) Population based study of hospitalised fall related injuries in older people. Inj Prev 8(4):280–283
41. Rodriguez-Manas L (2014) Fried LP. Frailty in the clinical scenario, Lancet
42. Fried LP, Tangen CM, Walston J, Newman AB, Hirsch C, Gottdiener J et al (2001) Frailty in older adults: evidence for a phenotype. J Gerontol A Biol Sci Med Sci 56(3): M146–M156
43. Ensrud KE, Ewing SK, Taylor BC, Fink HA, Cawthon PM, Stone KL et al (2008) Comparison of 2 frailty indexes for prediction of falls, disability, fractures, and death in older women. Arch Intern Med 168(4):382–389
44. Ensrud KE, Ewing SK, Taylor BC, Fink HA, Stone KL, Cauley JA et al (2007) Frailty and risk of falls, fracture, and mortality in older women: the study of osteoporotic fractures. J Gerontol A Biol Sci Med Sci 62(7):744–751
45. Tom SE, Adachi JD, Anderson FA Jr, Boonen S, Chapurlat RD, Compston JE et al (2013) Frailty and fracture, disability, and falls: a multiple country study from the global longitudinal study of osteoporosis in women. J Am Geriatr Soc 61(3):327–334
46. Kaiser MJ, Bauer JM, Ramsch C, Uter W, Guigoz Y, Cederholm T et al (2010) Frequency of malnutrition in older adults: a multinational perspective using the mini nutritional assessment. J Am Geriatr Soc 58(9):1734–1738
47. Genaro Pde S, Martini LA (2010) Effect of protein intake on bone and muscle mass in the elderly. Nutr Rev 68(10):616–623
48. Waters DL, Baumgartner RN, Garry PJ, Vellas B (2010) Advantages of dietary, exercise-related, and therapeutic interventions to prevent and treat sarcopenia in adult patients: an update. Clin Interv Aging 5:259–270
49. Livingstone C (2013) Insulin-like growth factor-I (IGF-I) and clinical nutrition. Clin Sci (Lond) 125(6):265–280
50. Wijnhoven HA, Schilp J, van Bokhorst-de van der Schueren MA, de Vet HC, Kruizenga HM, Deeg DJ et al (2012) Development and validation of criteria for determining undernutrition in community-dwelling older men and women: the Short Nutritional Assessment Questionnaire 65+. Clin Nutr 31(3):351–358
51. Chevalley T, Hoffmeyer P, Bonjour JP, Rizzoli R (2010) Early serum IGF-I response to oral protein supplements in elderly women with a recent hip fracture. Clin Nutr 29(1):78–83

52. Gaffney-Stomberg E, Insogna KL, Rodriguez NR, Kerstetter JE (2009) Increasing dietary protein requirements in elderly people for optimal muscle and bone health. J Am Geriatr Soc 57(6):1073–1079
53. Mithal A, Bonjour JP, Boonen S, Burckhardt P, Degens H, El Hajj FG et al (2013) Impact of nutrition on muscle mass, strength, and performance in older adults. Osteoporos Int 24(5):1555–1566
54. Compston JE, Watts NB, Chapurlat R, Cooper C, Boonen S, Greenspan S et al (2011) Obesity is not protective against fracture in postmenopausal women: GLOW. Am J Med 124(11):1043–1050
55. Nielson CM, Srikanth P, Orwoll ES (2012) Obesity and fracture in men and women: an epidemiologic perspective. J Bone Miner Res 27(1):1–10
56. Premaor M, Parker RA, Cummings S, Ensrud K, Cauley JA, Lui LY et al (2013) Predictive value of FRAX for fracture in obese older women. J Bone Miner Res 28(1):188–195
57. Nielson CM, Marshall LM, Adams AL, LeBlanc ES, Cawthon PM, Ensrud K et al (2011) BMI and fracture risk in older men: the osteoporotic fractures in men study (MrOS). J Bone Miner Res 26(3):496–502
58. Leslie WD, Orwoll ES, Nielson CM, Morin SN, Majumdar SR, Johansson H et al (2014) Estimated lean mass and fat mass differentially affect femoral bone density and strength index but are not FRAX independent risk factors for fracture. J Bone Miner Res 29(11):2511–2519
59. Bolland MJ, Grey A, Avenell A, Gamble GD, Reid IR (2011) Calcium supplements with or without vitamin D and risk of cardiovascular events: reanalysis of the Women's Health Initiative limited access dataset and meta-analysis. BMJ 342:d2040
60. Prentice RL, Pettinger MB, Jackson RD, Wactawski-Wende J, Lacroix AZ, Anderson GL et al (2013) Health risks and benefits from calcium and vitamin D supplementation: women's Health Initiative clinical trial and cohort study. Osteoporos Int 24(2):567–580
61. Steingrimsdottir L, Gunnarsson O, Indridason OS, Franzson L, Sigurdsson G (2005) Relationship between serum parathyroid hormone levels, vitamin D sufficiency, and calcium intake. JAMA 294(18):2336–2341
62. Boucher BJ (2012) The problems of vitamin d insufficiency in older people. Aging Dis 3(4):313–329
63. Abellan van Kan G, Rolland Y, Houles M, Gillette-Guyonnet S, Soto M, Vellas B (2010) The assessment of frailty in older adults. Clin Geriatr Med 26(2):275–286
64. Kado DM, Miller-Martinez D, Lui LY, Cawthon P, Katzman WB, Hillier TA et al (2014) Hyperkyphosis, kyphosis progression, and risk of non-spine fractures in older community dwelling women: the study of osteoporotic fractures (SOF). J Bone Miner Res 29(10):2210–2216
65. Katzman WB, Harrison SL, Fink HA, Marshall LM, Orwoll E, Barrett-Connor E et al (2014) Physical function in older Men with hyperkyphosis. J Gerontol A Biol Sci Med Sci 70(5):635–40
66. Bansal S, Katzman WB, Giangregorio LM (2014) Exercise for improving age-related hyperkyphotic posture: a systematic review. Arch Phys Med Rehabil 95(1):129–140
67. Michael YL, Whitlock EP, Lin JS, Fu R, O'Connor EA, Gold R (2010) Primary care-relevant interventions to prevent falling in older adults: a systematic evidence review for the U.S. Preventive Services Task Force. Ann Intern Med 153(12):815–825
68. Howe TE, Shea B, Dawson LJ, Downie F, Murray A, Ross C et al (2011) Exercise for preventing and treating osteoporosis in postmenopausal women. Cochrane Database Syst Rev 7, CD000333
69. Robertson MC, Campbell AJ, Gardner MM, Devlin N (2002) Preventing injuries in older people by preventing falls: a meta-analysis of individual-level data. J Am Geriatr Soc 50(5):905–911
70. Sherrington C, Henschke N (2013) Why does exercise reduce falls in older people? Unrecognised contributions to motor control and cognition? Br J Sports Med 47(12):730–731

71. Giangregorio LM, Papaioannou A, Macintyre NJ, Ashe MC, Heinonen A, Shipp K et al (2014) Too Fit To Fracture: exercise recommendations for individuals with osteoporosis or osteoporotic vertebral fracture. Osteoporos Int 25(3):821–835
72. Li F, Harmer P, Fisher KJ, McAuley E, Chaumeton N, Eckstrom E et al (2005) Tai Chi and fall reductions in older adults: a randomized controlled trial. J Gerontol A Biol Sci Med Sci 60(2):187–194
73. Wayne PM, Kiel DP, Buring JE, Connors EM, Bonato P, Yeh GY et al (2012) Impact of Tai Chi exercise on multiple fracture-related risk factors in post-menopausal osteopenic women: a pilot pragmatic, randomized trial. BMC Complement Altern Med 12:7
74. Ensrud KE, Black DM, Palermo L, Bauer DC, Barrett-Connor E, Quandt SA et al (1997) Treatment with alendronate prevents fractures in women at highest risk: results from the Fracture Intervention Trial. Arch Intern Med 157(22):2617–2624
75. Boonen S, Black DM, Colon-Emeric CS, Eastell R, Magaziner JS, Eriksen EF et al (2010) Efficacy and safety of a once-yearly intravenous zoledronic acid 5 mg for fracture prevention in elderly postmenopausal women with osteoporosis aged 75 and older. J Am Geriatr Soc 58(2):292–299
76. Boonen S, Marin F, Mellstrom D, Xie L, Desaiah D, Krege JH et al (2006) Safety and efficacy of teriparatide in elderly women with established osteoporosis: bone anabolic therapy from a geriatric perspective. J Am Geriatr Soc 54(5):782–789
77. Boonen S, McClung MR, Eastell R, El-Hajj Fuleihan G, Barton IP, Delmas P (2004) Safety and efficacy of risedronate in reducing fracture risk in osteoporotic women aged 80 and older: implications for the use of antiresorptive agents in the old and oldest old. J Am Geriatr Soc 52(11):1832–1839
78. McClung MR, Geusens P, Miller PD, Zippel H, Bensen WG, Roux C et al (2001) Effect of risedronate on the risk of hip fracture in elderly women. Hip Intervention Program Study Group. N Engl J Med 344(5):333–340
79. Seeman E, Vellas B, Benhamou C, Aquino JP, Semler J, Kaufman JM et al (2006) Strontium ranelate reduces the risk of vertebral and nonvertebral fractures in women eighty years of age and older. J Bone Miner Res 21(7):1113–1120
80. Hochberg MC, Thompson DE, Black DM, Quandt SA, Cauley J, Geusens P et al (2005) Effect of alendronate on the age-specific incidence of symptomatic osteoporotic fractures. J Bone Miner Res 20(6):971–976
81. Boonen S, Adachi JD, Man Z, Cummings SR, Lippuner K, Torring O et al (2011) Treatment with denosumab reduces the incidence of new vertebral and hip fractures in postmenopausal women at high risk. J Clin Endocrinol Metab 96(6):1727–1736
82. Black DM, Delmas PD, Eastell R, Reid IR, Boonen S, Cauley JA et al (2007) Once-yearly zoledronic acid for treatment of postmenopausal osteoporosis. N Engl J Med 356(18):1809–1822
83. Black DM, Thompson DE, Bauer DC, Ensrud K, Musliner T, Hochberg MC et al (2000) Fracture risk reduction with alendronate in women with osteoporosis: the Fracture Intervention Trial. FIT Research Group. J Clin Endocrinol Metab 85(11):4118–4124
84. Boonen S, Orwoll E (2013) Fracture risk and zoledronic acid in men with osteoporosis. N Engl J Med 368(9):873
85. Cummings SR, San Martin J, McClung MR, Siris ES, Eastell R, Reid IR et al (2009) Denosumab for prevention of fractures in postmenopausal women with osteoporosis. N Engl J Med 361(8):756–765
86. Maricic M, Adachi JD, Sarkar S, Wu W, Wong M, Harper KD (2002) Early effects of raloxifene on clinical vertebral fractures at 12 months in postmenopausal women with osteoporosis. Arch Intern Med 162(10):1140–1143
87. Meunier PJ, Roux C, Seeman E, Ortolani S, Badurski JE, Spector TD et al (2004) The effects of strontium ranelate on the risk of vertebral fracture in women with postmenopausal osteoporosis. N Engl J Med 350(5):459–468
88. Roux C, Seeman E, Eastell R, Adachi J, Jackson RD, Felsenberg D et al (2004) Efficacy of risedronate on clinical vertebral fractures within six months. Curr Med Res Opin 20(4):433–439

89. Hlaing TT, Compston JE (2014) Biochemical markers of bone turnover – uses and limitations. Ann Clin Biochem 51(Pt 2):189–202
90. Vasikaran S, Eastell R, Bruyere O, Foldes AJ, Garnero P, Griesmacher A et al (2011) Markers of bone turnover for the prediction of fracture risk and monitoring of osteoporosis treatment: a need for international reference standards. Osteoporos Int 22(2):391–420
91. Naylor K, Eastell R (2012) Bone turnover markers: use in osteoporosis. Nat Rev Rheumatol 8(7):379–389
92. Watts NB, Bilezikian JP (2014) Advances in target-specific therapy for osteoporosis. J Clin Endocrinol Metab 99(4):1149–1151
93. Vashishth D (2007) The role of the collagen matrix in skeletal fragility. Curr Osteoporos Rep 5:62–66
94. Jepsen KJ (2003) The aging cortex: to crack or not to crack. Osteoporos Int 14(Suppl 5):S57–S62
95. Currey JD, Brear K, Zioupos P (1996) The effects of ageing and changes in mineral content in degrading the toughness of human femora. J Biomech 29:257–260
96. Zimmermann EA, Schaible E, Bale H et al (2011) Age-related changes in the plasticity and toughness of human cortical bone at multiple length scales. Proc Natl Acad Sci U S A 108:14416–14421
97. Paschalis EP, Tatakis DN, Robins S et al (2011) Lathyrism-induced alterations in collagen cross-links influence the mechanical properties of bone material without affecting the mineral. Bone 49:1232–1241
98. Wang X, Bank RA, TeKoppele JM, Agrawal CM (2001) The role of collagen in determining bone mechanical properties. J Orthop Res 19:1021–1026
99. Vashishth D, Gibson GJ, Khoury JI et al (2001) Influence of nonenzymatic glycation on biomechanical properties of cortical bone. Bone 28:195–201
100. Leslie RD, Beyan H, Sawtell P, Boehm BO, Spector TD, Snieder H (2003) Level of an advanced glycated end product is genetically determined: a study of normal twins. Diabetes 52:2441–2444
101. Saito M, Marumo K (2010) Collagen cross-links as a determinant of bone quality: a possible explanation for bone fragility in aging, osteoporosis, and diabetes mellitus. Osteoporos Int 21:195–214
102. Hein G, Wiegand R, Lehmann G, Stein G, Franke S (2003) Advanced glycation end-products pentosidine and N epsilon-carboxymethyllysine are elevated in serum of patients with osteoporosis. Rheumatology (Oxford) 42:1242–1246
103. Bonar LC, Roufousse AS, Sabine WK et al (1983) X-ray diffraction studies of the crystallinity of bone mineral in newly synthesized and density fractionated bone. Calcif Tissue Int 35:202–209
104. Paschalis EP, DiCarlo E, Betts F et al (1996) FTIR microspectroscopic analysis of human osteonal bone. Calcif Tissue Int 59:480–487
105. Simmons ED Jr, Pritzker KPH, Grynpas MD (1991) Age-related changes in the human femoral cortex. J Orthop Res 9:155–167
106. Burr DB, Turner CH (1999) Biomechanical measurements in age-related bone loss. In: Rosen CJ, Glowacki J, Bilezikian JP (eds) The aging skeleton. Academic, San Diego, pp 301–311
107. Kleerekoper M, Villanueva AR, Stanciu J, Rao DS, Parfitt AM (1985) The role of three-dimensional trabecular microstructure in the pathogenesis of vertebral compression fractures. Calcif Tissue Int 37:594–597
108. Gabet Y, Bab I (2011) Microarchitectural changes in the aging skeleton. Curr Osteoporos Rep 9:177–183
109. Chen H, Zhou X, Shoumura S, Emura S, Bunai Y (2010) Age- and gender-dependent changes in three-dimensional microstructure of cortical and trabecular bone at the human femoral neck. Osteoporos Int 21:627–636
110. Martin RB (1991) Determinants of the mechanical properties of bones. J Biomech ;24 Suppl 1:79–88. Review. Erratum in: J Biomech. 1992;25:1251

111. Boyde A (2002) Morphologic detail of aging bone in human vertebrae. Endocrine 17:5–14
112. Schaffler MB, Choi K, Milgrom C (1995) Aging and matrix microdamage accumulation in human compact bone. Bone 17:521–525
113. Lambers FM, Bouman AR, Rimnac CM, Hernandez CJ (2013) Microdamage caused by fatigue loading in human cancellous bone: relationship to reductions in bone biomechanical performance. PLoS One 8, e83662
114. Frost HM (1991) Some ABC's of skeletal pathophysiology. 5. Microdamage physiology. Calcif Tissue Int 49:229–231
115. Noble B (2003) Bone microdamage and cell apoptosis. Eur Cell Mater 6:46–55
116. Burr DB, Martin RB, Schaffler MB, Radin EL (1985) Bone remodeling in response to in vivo fatigue microdamage. J Biomech 18:189–200
117. Kennedy OD, Herman BC, Laudier DM et al (2012) Activation of resorption in fatigue-loaded bone involves both apoptosis and active pro-osteoclastogenic signaling by distinct osteocyte populations. Bone 50:1115–1122
118. Seref-Ferlengez Z, Basta-Pljakic J, Kennedy OD, Philemon CJ, Schaffler MB (2014) Structural and mechanical repair of diffuse damage in cortical bone in vivo. J Bone Miner Res 29:2537–2544
119. Lips P, Courpron P, Meunier PJ (1978) Mean wall thickness of trabecular bone packets in the human iliac crest: changes with age. Calcif Tissue Res 26:13–17
120. Miyata T, Notoya K, Yoshida K et al (1997) Advanced glycation end products enhance osteoclast-induced bone resorption in cultured mouse unfractionated bone cells and in rats implanted subcutaneously with devitalized bone particles. J Am Soc Nephrol 8:260–270
121. Chung PL, Zhou S, Eslami B, Shen L, LeBoff MS, Glowacki J (2014) Effect of age on regulation of human osteoclast differentiation. J Cell Biochem 115:1412–1419
122. Cheleuitte D, Mizuno S, Glowacki J (1998) In vitro secretion of cytokines by human bone marrow: effects of age and estrogen status. J Clin Endocrinol Metab 83:2043–2051
123. Eslami B, Zhou S, Van Eekeren I, LeBoff MS, Glowacki J (2011) Reduced osteoclastogenesis and RANKL expression in marrow from women taking alendronate. Calcif Tissue Int 88:272–280
124. Mueller SM, Glowacki J (2001) Age-related decline in the osteogenic potential of human bone marrow cells cultured in three-dimensional collagen sponges. J Cell Biochem 82:583–590
125. Zhou S, Greenberger JS, Epperly MW et al (2008) Age-related intrinsic changes in human marrow stromal cells and their differentiation to osteoblasts. Aging Cell 7:335–343
126. D'Ippolito G, Schiller PC, Ricordi C, Roos BA, Howard GA (1999) Age-related osteogenic potential of mesenchymal stromal stem cells from human vertebral bone marrow. J Bone Miner Res 14:1115–1122
127. Shen L, Zhou S, Glowacki J (2009) Effects of age and gender on WNT gene expression in human bone marrow stromal cells. J Cell Biochem 106:337–343
128. Pandey AC, Semon JA, Kaushal D et al (2011) MicroRNA profiling reveals age-dependent differential expression of nuclear factor κB and mitogen-activated protein kinase in adipose and bone marrow-derived human mesenchymal stem cells. Stem Cell Res Ther 2:49
129. Candini O, Spano C, Murgia A, et al. Mesenchymal progenitors aging highlights a miR-196 switch targeting HOXB7 as master regulator of proliferation and osteogenesis. Stem Cells. 2014 Nov 26. doi: 10.1002/stem.1897. [Epub ahead of print]
130. Zhou S, Bueno EM, Kim SW et al (2011) Effects of age on parathyroid hormone signaling in human marrow stromal cells. Aging Cell 10:780–788
131. Geng S, Zhou S, Glowacki J (2011) Age-related decline in osteoblastogenesis and 1α-hydroxylase/CYP27B1 in human mesenchymal stem cells: stimulation by parathyroid hormone. Aging Cell 10:962–971
132. Zhou S, Geng S, Glowacki J (2013) Histone deacetylation mediates the rejuvenation of osteoblastogenesis by the combination of 25(OH)D3 and parathyroid hormone in MSCs from elders. J Steroid Biochem Mol Biol 136:156–159

133. Geng S, Zhou S, Bi Z, Glowacki J (2013) Vitamin D metabolism in human bone marrow stromal (mesenchymal stem) cells. Metabolism 62:768–777
134. D'Avis PY, Frazier CR, Shapiro JR, Fedarko NS (1997) Age-related changes in effects of insulin-like growth factor I on human osteoblast-like cells. Biochem J 324:753–760
135. Haden ST, Glowacki J, Hurwitz S, Rosen C, LeBoff MS (2000) Effects of age on serum dehydroepiandrosterone sulfate, IGF-I, and IL-6 levels in women. Calcif Tissue Int 66:414–418
136. Gordon CM, LeBoff MS, Glowacki J (2001) Adrenal and gonadal steroids inhibit IL-6 secretion by human marrow cells. Cytokine 16:178–186
137. Kennedy BK, Berger SL, Brunet A et al (2014) Geroscience: linking aging to chronic disease. Cell 159:709–713
138. López-Otín C, Blasco MA, Partridge L, Serrano M, Kroemer G (2013) The hallmarks of aging. Cell 153:1194–1217
139. Marie PJ (2014) Bone cell senescance:Mechanisms and perspectives. J Bone Miner Res 29:1311–1321
140. Shiraki M, Aoki C, Goto M (1998) Bone and calcium metabolism in Werner's syndrome. Endocr J 45:505–512
141. Simonsen JL, Rosada C, Serakinci N et al (2002) Telomerase expression extends the proliferative life-span and maintains the osteogenic potential of human bone marrow stromal cells. Nat Biotechnol 20:592–596
142. Bork S, Pfister S, Witt H et al (2010) DNA methylation pattern changes upon long-term culture and aging of human mesenchymal stromal cells. Aging Cell 9:54–63
143. Yu KR, Kang KS (2013) Aging-related genes in mesenchymal stem cells: a mini-review. Gerontology 59:557–563
144. Zhu Y, Song X, Wang J et al (2015) Placental mesenchymal stem cells of fetal origin deposit epigenetic alterations during long-term culture under serum-free condition. Expert Opin Biol Ther 15:163–180
145. Garrett IR, Chen D, Gutierrez G et al (2003) Selective inhibitors of the osteoblast proteasome stimulate bone formation in vivo and in vitro. J Clin Invest 111:1771–1782
146. Nuschke A, Rodrigues M, Stolz DB et al (2014) Human mesenchymal stem cells/multipotent stromal cells consume accumulated autophagosomes early in differentiation. Stem Cell Res Ther 5:140e
147. Alessio N, Del Gaudio S, Capasso S et al (2015) Low dose radiation induced senescence of human mesenchymal stromal cells and impaired the autophagy process. Oncotarget 6(10):8155–8166
148. Hocking LJ, Whitehouse C, Helfrich MH (2012) Autophagy: a new player in skeletal maintenance? J Bone Miner Res 27:1439–1447
149. Mobasheri A, Shakibaei M (2013) Osteogenic effects of resveratrol in vitro: potential for the prevention and treatment of osteoporosis. Ann N Y Acad Sci 1290:59–66
150. Kneissel M, Luong-Nguyen NH, Baptist M, Cortesi R, Zumstein-Mecker S, Kossida S, O'Reilly T, Lane H, Susa M (2004) Everolimus suppresses cancellous bone loss, bone resorption, and cathepsin K expression by osteoclasts. Bone 35:1144–1156
151. Lecka-Czernik B, Stechschulte LA (2014) Bone and fat: a relationship of different shades. Arch Biochem Biophys 561:124–129
152. Novotny SA, Warren GL, Hamrick MW (2015) Aging and the muscle-bone relationship. Physiology (Bethesda) 30:8–16
153. Almeida M, O'Brien CA (2013) Basic biology of skeletal aging: role of stress response pathways. J Gerontol A Biol Sci Med Sci 68:1197–1208
154. Manolagas SC (2010) From estrogen-centric to aging and oxidative stress: a revised perspective of the pathogenesis of osteoporosis. Endocr Rev 31:266–300
155. Gokhale A, Rwigema JC, Epperly MW et al (2010) Small molecule GS-nitroxide ameliorates ionizing irradiation-induced delay in bone wound healing in a novel murine model. In Vivo 24:377–385

156. Glowacki J, Mizuno S, Kung J et al (2014) Effects of mouse genotype on bone wound healing and irradiation-induced delay of healing. In Vivo 28:189–196
157. Franceschi C, Campisi J (2014) Chronic inflammation (inflammaging) and its potential contribution to age-associated diseases. J Gerontol A Biol Sci Med Sci 69:S4–S9
158. Sanguineti R, Puddu A, Mach F, Montecucco F, Viviani GL (2014) Advanced glycation end products play adverse proinflammatory activities in osteoporosis. Mediators Inflamm 2014:975872
159. Feng X, Feng G, Xing J et al (2014) Repeated lipopolysaccharide stimulation promotes cellular senescence in human dental pulp stem cells (DPSCs). Cell Tissue Res 356:369–380
160. Abdelmagid SM, Barbe MF, Safadi FF. Role of inflammation in the aging bone. Life Sci. 2014 Dec 12. pii: S0024-3205(14)00932-1 [Epub ahead of print].
161. Glowacki J, Zhou S (2013) The role of inflammation in senescence of human MSCs. J Bone Miner Res 28:S80
162. Severino V, Alessio N, Farina A et al (2013) Insulin-like growth factor binding proteins 4 and 7 released by senescent cells promote premature senescence in mesenchymal stem cells. Cell Death Dis 4, e911
163. Jin HJ, Bae YK, Kim M et al (2013) Comparative analysis of human mesenchymal stem cells from bone marrow, adipose tissue, and umbilical cord blood as sources of cell therapy. Int J Mol Sci 14:17986–18001
164. Bellayr IH, Catalano JG, Lababidi S et al (2014) Gene markers of cellular aging in human multipotent stromal cells in culture. Stem Cell Res Ther 5:59e
165. Cmielova J, Havelek R, Soukup T et al (2012) Gamma radiation induces senescence in human adult mesenchymal stem cells from bone marrow and periodontal ligaments. Int J Radiat Biol 88:393–404
166. Takeda T (1999) Senescence-accelerated mouse (SAM): a biogerontological resource in aging research. Neurobiol Aging 20:105–110
167. Kuro-o M, Matsumura Y, Aizawa H et al (1997) Mutation of the mouse klotho gene leads to a syndrome resembling ageing. Nature 390:45–51
168. Maddatu TP, Grubb SC, Bult CJ, Bogue MA (2012) Mouse Phenome Database (MPD). Nucleic Acids Res 40(Database issue):D887–D894
169. Beamer WG, Donahue LR, Rosen CJ, Baylink DJ (1996) Genetic variability in adult bone density among inbred strains of mice. Bone 18:397–403
170. Jilka RL (2013) The relevance of mouse models for investigating age-related bone loss in humans. J Gerontol A Biol Sci Med Sci 68:1209–1217
171. Harrison DE, Archer JR (1987) Genetic differences in effects of food restriction on aging in mice. J Nutr 117:376–382
172. Forster MJ, Morris P, Sohal RS (2003) Genotype and age influence the effect of caloric intake on mortality in mice. FASEB J 17:690–692
173. Liao CY, Rikke BA, Johnson TE, Diaz V, Nelson JF (2010) Genetic variation in the murine lifespan response to dietary restriction: from life extension to life shortening. Aging Cell 9:92–95
174. Seok J, Warren HS, Cuenca AG et al (2013) Genomic responses in mouse models poorly mimic human inflammatory diseases. Proc Natl Acad Sci U S A 110:3507–3512
175. Jowsey J (1966) Studies of Haversian systems in man and some animals. J Anat 100:857–864
176. Kharode YP, Sharp MC, Bodine PV (2008) Utility of the ovariectomized rat as a model for human osteoporosis in drug discovery. Methods Mol Biol 455:111–124
177. Wahl EC, Aronson J, Liu L et al (2010) Restoration of regenerative osteoblastogenesis in aged mice: modulation of TNF. J Bone Miner Res 25:114–123
178. Grammatikakis I, Panda AC, Abdelmohsen K, Gorospe M (2014) Long noncoding RNAs(lncRNAs) and the molecular hallmarks of aging. Aging (Albany NY) 6:992–1009

[illegible]

Osteoarthritis in the Elderly

Richard F. Loeser and Martin Lotz

Contents

R.F. Loeser, MD (✉)
Division of Rheumatology, Allergy, and Immunology, Thurston Arthritis Research Center, University of North Carolina School of Medicine, Chapel Hill, NC, 27599-7280, USA
e-mail: richard_loeser@med.unc.edu

M. Lotz, MD
Department of Molecular and Experimental Medicine, The Scripps Research Institute, La Jolla, CA, 92037, USA
e-mail: mlotz@scripps.edu

F. Sierra, R. Kohanski (eds.), *Advances in Geroscience*,
DOI 10.1007/978-3-319-23246-1_11

1 Introduction

Osteoarthritis (OA) is a quintessential condition of aging. It is a slowly progressive disease of synovial joints characterized pathologically by focal destruction of the articular cartilage, a hypertrophic response in neighboring bone that results in osteophyte formation and subchondral sclerosis, variable degrees of synovial inflammation, a thickening of the joint capsule, and damage to soft tissue structures including ligaments and, in the knee, the meniscus [1]. Gross pathological changes seen in the femoral condyles of the knee joint are shown in Fig. 1. The joint tissue changes result in failure of normal joint function that is accompanied by pain and disability. OA most commonly affects the distal joints of the hands, the knees, hips, and the spine. Often called degenerative joint disease and referred to by certain practitioners and the lay public as "wear and tear arthritis", it has been considered by some to be an inevitable consequence of aging of the articular joints. However, as with many of the other chronic conditions associated with aging, this is an oversimplification of what turns out to be a multifaceted condition that cannot be explained by simple age-related degeneration of the joints.

Recent work has determined that the pathophysiology of OA is much more complex than what might be construed by the term degenerative joint disease.

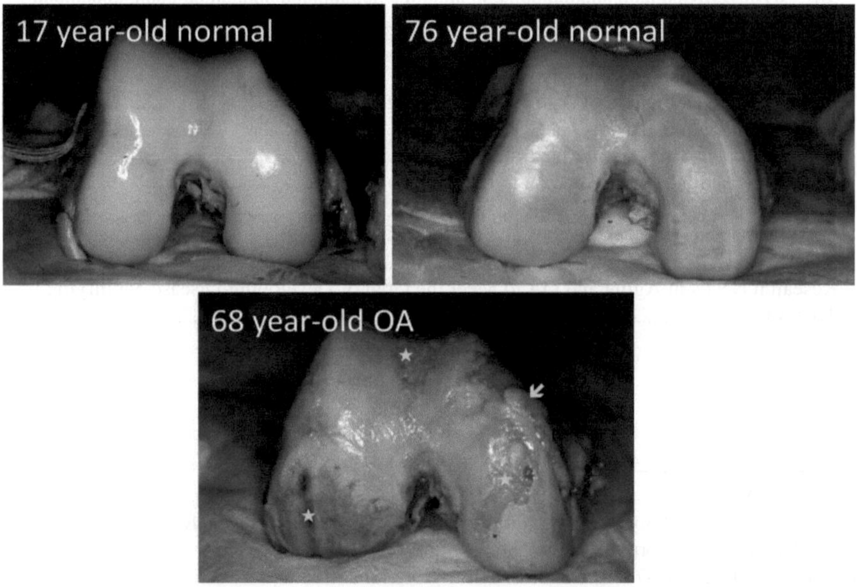

Fig. 1 OA and aging related changes in human knees. Images of the femoral condyles showing the white glass-like surface of normal cartilage in a knee from a 17 year-old. The cartilage from a macroscopically normal 76 year-old joint shows a change in cartilage color (*browning*), mainly due to advanced glycation of cartilage matrix proteins. Joint from a 68 year old shows characteristic OA-related changes, including erosions of original cartilage (*asterisks*) and formation of osteophytes at the joint margin (*arrow*)

There is evidence for the activation of a number of inflammatory pathways within the tissues affected by OA that appear to be the mediators responsible for driving joint tissue destruction [1, 2]. The finding that inflammatory factors play a major role in OA and that it is not simply a mechanical failure of the joint suggest that osteoarthr*itis*, which implies inflammation, is indeed the proper name for the condition rather than degenerative joint disease or "osteoarthr*osis*". It is also becoming clear that the multiple risk factors for OA likely follow different pathways to cause disease despite findings that end-stage OA appears to be pathologically similar in most joints.

Arthritis is a common cause of disability in older adults. Data collected from the Centers for Disease Control and Prevention indicates that arthritis and related conditions are the number one cause of disability in U.S. adults [3]. With the aging of our population, the prevalence of arthritis (of any type) in the United States is expected to rise from 47.8 million in 2005 to 67 million by 2030 with greater than 50 % of the cases of arthritis being in the 65 and older age group [4]. OA is by far the most common form of arthritis in adults affecting over 27 million Americans [5] and was recently ranked in the top ten of the diseases contributing to years lived with disability in the US [6]. OA is similarly prevalent worldwide where it is also a leading and rising cause of chronic disability [7]. Decreasing the prevalence of arthritis, including OA, was calculated to lead to a greater reduction in disability than similar reductions in coronary artery disease, stroke, cancer, diabetes, or dementia combined [8].

The prevalence of OA and its negative effects on physical function renders OA as an important co-morbidity in older adults, particularly in those who also suffer from cardiovascular disease [9]. In fact, having symptomatic OA has been associated with an increased risk of all-cause mortality with a standardized mortality ratio of 1.55 [10]. In that study, the strongest associations of OA with condition-specific mortality were with cardiovascular and dementia-associated mortality. The association with cardiovascular mortality can be attributed to the limitations in physical activity due to pain in weight bearing joints affected by OA which has been shown to contribute to reduced exercise capacity accompanied by all of the negative effects of physical inactivity [11, 12]. In addition, there is evidence that musculoskeletal pain from conditions including OA, increases the risk for falls [13] which may explain the association with dementia-associated mortality. Finally, knee OA has been associated with a greater risk for frailty in older adults [14].

These studies suggest that the development of interventions targeting common mechanisms underlying the chronic conditions of aging should include pain and disability from OA as important outcome measures. This chapter will briefly review the key risk factors and clinical features of OA, focused on the associations with age, as well as review the current management of OA, which is limited by the lack of interventions targeting the disease process. This will be followed by a more in depth discussion of the pathophysiology of OA and how it relates to the major hallmarks of aging. Future research directions and prospects for therapies will also be presented.

2 Clinical Presentation and Risk Factors for OA

The vast majority of people seeking medical care for OA do so due to joint pain. The pain most often begins insidiously and is chronic but with periods of waxing and waning. Other signs and symptoms include morning stiffness of less than an hour, impairment in function, joint crepitus with movement, and with more advanced disease, bony enlargement of the affected joints. When these signs and symptoms are present in the knee joint, close to 100 % of individuals will exhibit radiographic evidence of knee OA [15].

It is important to note that the pain associated with OA does not always correlate with the severity of disease noted by plain radiographs. Recent studies suggest that the poor correlation may be due to individual differences in central sensitization [16]. Also, standard radiographs can only detect structural changes in radiodense tissue, which is mainly the bone. The typical radiographic changes therefore include osteophytes, which are bony protuberances at the joint margins, joint space narrowing, which occurs due to cartilage loss, sclerosis, which is due to thickening of the subchondral bone, and sometimes bone cysts. Some older adults may also exhibit chondrocalcinosis which is calcification in the cartilage and, in the knee, the meniscus. Magnetic resonance imaging (MRI) is able to detect changes in soft tissues not seen by radiographs, including focal cartilage loss, synovitis, meniscal tears or degeneration, ligamentous disruption, and bone marrow lesions. Involvement of the tissues seen on MRI other than cartilage, which lacks nociceptive nerve fibers, can be an important source of pain, which also explains why standard radiographic changes do not always correlate well with symptoms.

OA occurs most commonly in the hands, feet, knees, hips or spine and may be present in a single joint (monoarticular), several joints (oligoarticular) or in multiple joints (polyarticular). Monoarticular OA is most commonly associated with a prior joint injury. Although OA of the ankle is uncommon, very severe OA can occur in ankles that have experienced significant trauma such as an ankle fracture [17]. This is also true for joints less commonly affected by OA such as the elbow and shoulder. Polyarticular OA is sometimes called generalized OA and most commonly affects the hands along with involvement of the knees and/or hips. This form of OA has been thought to have a stronger genetic component [18]. Although the genetic contribution to the overall risk of OA is estimated to be about 40 %, large genome wide association studies have found that any individual gene associated with OA has a very small relative risk in the 1.1–1.4 range [19, 20]. These studies suggest that OA may be polygenetic and that environmental factors likely interact with genetic factors to increase the risk for OA.

Clinical risk factors for OA have been well characterized (Fig. 2) with age being the strongest risk factor. Numerous studies have documented an increase in radiographic and symptomatic hand, hip, spine, and knee OA with increasing age (reviewed in [21]). The incidence of symptomatic knee OA is highest between the ages of 55–64 years where it has been calculated to range from 0.37 % per year for non-obese males to 1.02 % per year for obese females [22]. In a community-based

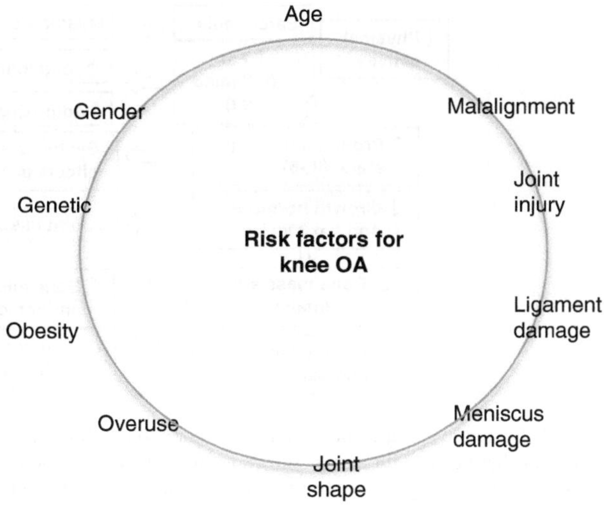

Fig. 2 Risk factors for knee OA. There are multiple predisposing factors that contribute to the development of OA either alone or in combination. These risk factors apply to other joints as well, except for meniscus damage, which is specific for OA in the knee since other joints do not contain menisci

longitudinal study in the UK, the prevalence of radiographic knee OA increased from 13.7 % at baseline, when the median age of participants was 53 years, to 47.8 % when the cohort was evaluated 14 years later [23]. Likewise, a large community based study based in the US found that about half of the participants developed symptomatic knee OA by age 85 [24]. Age is also a risk factor for a more accelerated course of OA, particularly in older adults suffering a new injury, where advancement to severe OA was seen within a year [25].

Despite the strong correlation between age and the incidence and prevalence of OA, not all older adults develop OA and not all joints are equally affected [21]. In addition to age, there are a number of other OA risk factors including obesity, prior joint injury, gender (hand and knee OA are more common in women), genetics, and anatomical factors that include joint shape and for knee OA, lower extremity alignment [26] (Fig. 2). In a meta-analysis of risk factors for knee OA in adults 50 years of age and older, the most consistent risk factors and their pooled odds ratios were obesity (OR 2.6), previous knee trauma (OR 3.86), hand OA (OR 1.49), and female gender (OR 1.84) [27]. Older age was also a risk factor but it was not possible to calculate a pooled odds ratio due to the heterogeneity in how ages were reported.

Obesity is a risk factor not only for knee OA [24, 28] but also for OA of the hips [29] and hands [30, 31]. The association of obesity with hand OA, which is not a weight bearing joint, suggests that systemic factors related to obesity and excess adipose tissue contribute to the development of OA. Adipose tissue is a source of numerous cytokines and adipokines, including IL-1, IL-6, TNFα, and leptin that have been purported to have a role in OA [32, 33]. The systemic metabolic changes associated with obesity may also contribute to OA and there is some evidence of links between OA and other obesity-related metabolic conditions including hypertension and diabetes [34–36].

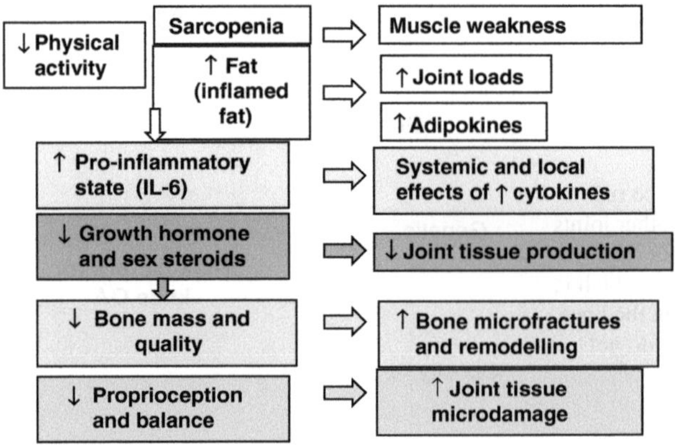

Fig. 3 Systemic aging changes that may relate to the development of OA. There are a number of changes that occur with aging, shown on the *left*, which could contribute to the development of OA. Decreased physical activity and sarcopenia resulting in muscle weakness can result in increased joint loading since muscles are important shock absorbers for the joint. The age-related increase in inflammatory fat can produce adipokines, such as leptin, and cytokines, such as IL-6, that are thought to contribute to OA. The age-related decrease in growth hormone, accompanied by a decrease in IGF-1 as well as a decrease in sex steroids, contributes to reduced anabolic activity in joint tissues. Changes in bone mass and quality can promote development of microfractures in the subchondral bone accompanied by excessive remodeling and development of bone marrow lesions seen in OA. A decrease in proprioception and balance contributes to excessive or abnormal joint loading that increases tissue microdamage (Adapted from Loeser [309])

In addition to the systemic alterations in inflammation and metabolism that result from an age-related increase in subcutaneous and visceral adipose tissue, local changes in adipose tissue could also contribute to the development of OA in older adults (Fig. 3). An age-related increase in intramuscular fat in the quadriceps muscle has been noted [37] and increased intramuscular fat in the quadriceps has been associated with knee OA [38]. This fat could contribute not only to weakness that would increase the risk of knee OA but also production of local inflammatory mediators such as IL-6 [39]. There is also evidence that the volume of the infrapatellar fat pad, which communicates with the knee joint, may increase with age [40] and could represent a local source of the adipokines, adiponectin and leptin, as well as the cytokine IL-6 [41].

Given that OA is so common in the older adult population, it is also of interest to examine factors associated with absence of OA. In a community-based study of 90 year olds living in the city of Leiden in the Netherlands, the absence of radiographic OA in the hands, hips, and knees was seen in 16 % of the 82 participants and was most strongly associated with lower BMI [42]. In addition, none of the participants who were free of OA had a family history of nodal OA of the hand, suggesting favorable genetics, and absence of knee OA was also associated with being male. Surprisingly, a history of heavy occupational work had a 7.2 odds ratio of being free from OA. The reason for this association is not clear and differs from previous studies in younger cohorts that demonstrated heavy physical labor, in particular occupations that required knee bending while lifting heavy loads, was associated with increased prevalence of knee OA [43].

3 Current Therapies

The current management of individuals with OA is mainly symptomatic and includes non-pharmacologic interventions, such as exercise and weight loss, medications to reduce pain, and, in more advanced cases, joint replacement surgery [44]. A major limitation in the management of OA is the lack of any treatment proven to slow or stop the structural progression of the disease. This is a goal of current research efforts and an area where a better understanding of the contributions of aging to OA could be of benefit, assuming the aging aspects promoting disease progression are modifiable, as proposed by the Geroscience Hypothesis.

Numerous organizations have developed OA treatment guidelines, which have recently been reviewed [45]. The guidelines stress that the management of OA needs to be individualized to each patient's needs and should utilize multiple modalities. It is generally recommended that all patients should receive non-pharmacologic modalities that include exercise, weight loss, education and self-management, and, when necessary, assistive devices such as walking aids. Pharmacologic management most often begins with the use of simple analgesics such as acetaminophen. In those that fail a trial of acetaminophen, nonsteroidal anti-inflammatory drugs (NSAIDs) can be used, although it is important to recognize individuals at risk for side-effects such as gastrointestinal bleeding. The American College of Rheumatology has recommended that since older adults are at higher risk of GI bleeding, adults 75 years and older should use topical rather than oral NSAIDs [44]. Flare-ups of joint pain can be treated with intra-articular steroids while the use of intra-articular hyaluronans has been controversial because many studies have not shown significant benefit when compared to placebo injections [44, 45]. In patients with OA of the hip or knee, joint replacement surgery (arthroplasty) is recommended when other treatments have failed and the patient continues experiencing pain and disability that interferes with their desired activities [44, 45].

Treatments that would impact the OA disease process have been referred to as disease-modifying or structure-modifying therapies. There have been several clinical trials of agents thought to have the potential for disease-modification that have failed (reviewed in [46]). Among others, these have included various inhibitors of extracellular matrix degrading enzymes, bisphosphonates used to target bone, an inhibitor of inducible nitric oxide synthase to target inflammation, and glucosamine plus chondroitin sulfate thought to target cartilage breakdown [46, 47]. A major limitation to testing potential disease-modifying drugs is that the outcome measure considered to be the "gold-standard" for efficacy is the change in joint space width on standardized radiographs as a surrogate for cartilage loss. Research is ongoing to find more sensitive outcome measures with acceptable levels of reliability, such as newer modalities of magnetic resonance imaging, which could be used to better demonstrate disease or structure-modifying effects. However, some investigators have argued that if the biomechanical aspects of OA are not addressed, such as the misalignment discussed above, then potential disease-modifying therapies will continue to fail [48]. The advanced stage of disease and heterogeneity of patient populations in regard to risk factor profile and disease mechanisms are also potential causes for failures of prior clinical trials.

4 Age-Related Alterations in Joint Tissues Relevant to OA

In established or radiographically detectable OA, all joint tissues are affected. The earliest changes appear to occur in the articular cartilage with a very close association to changes in subchondral bone. Aging-related changes in cartilage (Table 1) thus appear to be a key event in initiation of the disease process that subsequently involves the other tissues. In trying to understand mechanisms of aging that confer OA risk, cartilage has been the focus of research on joint aging.

4.1 Cartilage

Aging-associated changes in articular cartilage include reduced cartilage thickness, reduced cell density, cellular senescence with abnormal secretory profiles, impaired cellular defense mechanisms and anabolic responses [49]. The earliest changes in cartilage are enzymatic degradation of glycosaminoglycans and cartilage proteins, and loss of cartilage cells. These changes first occur in the cartilage superficial zone [50], which is exposed to shear and compressive forces during movement [51]. Cartilage cells respond to this early tissue damage with proliferation and transcriptional activation of genes involved in extracellular matrix remodeling and inflammation. Changes in extracellular matrix and cells clearly differ between aged and OA-affected cartilage. The major differences are cell depletion, reduced biosynthetic activity and senescence in aging cartilage versus cell activation and proliferation in OA cartilage [49].

Table 1 Aging-related changes in articular cartilage

Chondrocytes	Extracellular matrix
Telomere shortening	Glycation, glycoxidation
↑SA-βgal, p53, p21, p16	Carbonylation
↑Cytokine and MMP production	↓Hydration
Mitochondrial damage	GAG depletion
Oxidative stress/damage	Aggrecan size reduced
↓Growth factor response	Collagen cleavage
↓Growth factor production	Growth factor levels reduced
Autophagy defects	Calcification
Cell death	Amyloid

OA is associated with increased proteolytic activity in cartilage and synovial fluid [52], and increased degradation of collagen molecules [53, 54]. This degradation is accompanied by the loss of intrinsic cartilage fluorescence [55], especially around cells in the superficial zone. In OA, there is also a decrease in fixed charge density, due to degradation and loss of aggrecan [56].

With normal aging, there is a marked increase in the formation of advanced glycation end-products (AGEs) (Fig. 1), including pentosidine cross-links [57]. In addition to altering the biomechanical properties of cartilage, AGEs may interact with cell surface receptors including the Receptor for Advanced Glycation Endproducts (RAGE). RAGE has been detected on articular chondrocytes and there is evidence for an increase in RAGE levels in aging and OA [58]. Activation of chondrocyte RAGE has been shown to stimulate catabolic signaling pathways that result in upregulation of MMP expression and chondrocyte hypertrophy [59–61].

A highly prevalent change in aging cartilage is deposition of calcium-containing crystals, mainly calcium pyrophosphate dihydrate (CPPD) and basic calcium phosphate (BCP) [62]. In the human knee this cartilage calcification is primarily an effect of aging rather than OA and represents a precursor to increased fibrillation and OA rather than a result of OA [62]. In cartilage from patients with end-stage OA, calcification correlated with increased disease severity [63, 64]. CPPD deposition is due to increased pyrophosphate production by chondrocytes from aged cartilage [64]. The presence of calcium crystals produced by chondrocytes or released into the joint space from other tissues such as the meniscus and synovium may stimulate chondrocyte production of inflammatory mediators and extracellular matrix (ECM)-degrading enzymes and thus contribute to onset and progression of OA [65].

Amyloid deposition is a prevalent and as yet underappreciated aging-related phenomenon in cartilage even in the absence of generalized systemic amyloidosis. Almost all cartilage tissues that are removed during joint replacement surgery have Congo red positive deposits [66]. In a study of autopsy cases, there was a correlation between amyloid deposition and osteoarthritic changes [67]. The protein aggregates that are present in aging cartilage and their potential effects on cells and extracellular matrix remain to be elucidated.

Aging in human and mouse joints is also associated with a reduction in cartilage cellularity (reviewed in [68]). Diverse inducers of cell death have been proposed, including acute or chronic excessive mechanical loading, certain proinflammatory cytokines, ligands for death receptors and oxygen radicals. Consequences of cell death are immediate damage to the extracellular matrix through release of matrix degrading enzymes and inflammatory mediators. The reduced cell density may also impair the tissues' ability to maintain extracellular matrix integrity. Reduced cell density is most profound in the cartilage superficial zone, which contains the highest concentration of progenitor cells. The function of these cells in tissue homeostasis, consequences of their depletion and their role in the disease process are important topics for further research.

4.2 Bone

Articular cartilage and subchondral bone function biomechanically together as the osteochondral unit and under normal conditions transmit load pressure through the joints [69]. The morphology of the cartilage–bone interface or osteochondral junction is similar between different joints and includes the deeper non-calcified cartilage, the tidemark, calcified cartilage, the cement line, and the subchondral bone plate [70, 71]. Beneath the subchondral plate is the trabecular bone of the epiphysis, containing blood vessels, sensory nerves, endothelium and bone marrow [72].

In OA-affected joints, subchondral bone turnover is increased [73, 74] and this is related to abnormal activation and differentiation of subchondral osteoblasts [75, 76] together with increased osteoclast activity [77, 78]. The consequence of the increased subchondral bone remodeling process appears to be an increase in bone volume density. However, contrary to earlier assumptions, this altered bone is not harder or stiffer but rather, it is undermineralized [79, 80]. Changes in subchondral bone may be related to the stage of the disease process. In early-stage OA, the increased remodeling may lead to bone loss whereas in late-stage OA, remodeling decreases and subchondral bone density increases [81]. Bone from severe OA patients shows an increase of bone volume density and a reduction of its mineralization content [82, 83].

In addition to the bone remodeling in OA, there is also increased angiogenesis and vascular channels and sensory nerves invade from the subchondral bone into the calcified and non-calcified cartilage [84]. Increasing numbers of vascular channels or marrow cavities break the osteochondral interface and advance towards the tidemark in the OA joints [85]. Blood flow disturbances have also been documented in OA subchondral bone

The subchondral bone remodeling process is thought to lead to changes in signaling from bone to cartilage. Even in normal joints there is a measurable transport of solutes across the calcified cartilage, suggesting a potential cross-talk between subchondral bone and cartilage [85]. The increased subchondral bone remodeling in OA is associated with production of cytokines by bone cells and, together with changes in the barrier between bone and cartilage, supports the hypothesis that cytokines and growth factors released during subchondral bone turnover diffuse into the articular cartilage and change chondrocyte functions, contributing to disease progression [86].

While OA-associated changes in subchondral bone are well characterized, changes that are associated with aging in the absence of OA are difficult to ascertain as few studies of human OA pathology included controls without arthritis that were age matched. A comparison of knee joints from aged mice with mice that had surgically induced OA revealed that aged mice had a higher prevalence of blood vessels invading the calcified cartilage, and thinning of the subchondral bone and calcified cartilage layers, which was not seen in the joints subjected to surgical OA [85].

The temporal and mechanistic relationship of bone and cartilage change is of great interest. Changes in cartilage and subchondral bone during OA pathogenesis are closely linked [71, 87]. There is evidence at multiple scales that this is the key region in the development or progression of OA, even though the disease onset might have been triggered by an entirely different phenomenon such as by ligament or meniscus trauma or abnormal joint shape [88]. Already during early stages of OA development, morphological and functional changes in the subchondral bone occur that affect articular cartilage. It has been suggested that changes in histomorphometric parameters of subchondral bone are secondary to cartilage damage and proceed deeper into subchondral bone with increasing cartilage degeneration. However, it has also been shown that cartilage loss or further degeneration could be predicted with, or related to, increased activity within the subchondral bone [89–91]. Thus, there is substantial evidence for interactions of the two tissues in disease initiation and progression.

The relationship between OA and systemic bone mineral density (BMD), as measured at sites distant from the affected joints, has been examined in epidemiological studies which consistently demonstrated that a higher BMD is associated with a greater risk of developing subsequent radiographic OA in large joints such as knee and hip and there is also a correlation between prevalent OA and BMD [73, 92, 93]. Despite the association with incident and prevalent OA, a consistent relationship between systemic BMD and OA progression has not been found [94, 95].

The abnormal subchondral bone metabolism in OA has been targeted in animal models and clinical trials through pharmacological approaches to prevent resorption and/or to increase mineralization. Anti-resorptive drugs such osteoprotegerin, bisphosphonates, strontium ranelate, calcitonin, cathepsin K inhibitors, and estrogen were successfully tested in animal models where modulation of bone changes was associated with a reduction in cartilage damage [78, 96]. Clinical trials have been performed on a subset of these drugs including bisphosphonates, vitamin D and calcitonin and they all failed to show significant disease or structure modifying activity [78, 97]. Encouraging results from a recent clinical trial in patients with knee OA of strontium ranelate, an osteoporosis drug that inhibits subchondral bone remodeling [98] await confirmation in additional studies.

4.3 Synovium

Synovium is a membrane-like tissue that lines the inside of the joint capsule and also covers tendons and ligaments as a synovial sheath. The outer layer, also termed subintima or stroma, contains adipose and fibrous tissue that is innervated and contains blood and lymphatic vessels. The inner layer, also termed intima or synovial lining, contains a higher density of macrophages, fibroblast-like cells termed synoviocytes, as well mesenchymal progenitor or stem cells [99]. Through synovial blood vessels the joint receives oxygen and nutrients. Lymphatic vessels are responsible for fluid clearance and the transport of macromolecules to lymph nodes [100, 101]. Synovial fibroblasts are a main producer of lubricants that are secreted into the synovial fluid and mainly provide protection against shear forces [102].

Aging-related changes in the synovium that occur in the absence of diagnosed OA are poorly characterized. Amyloid deposition, which is in most cases an aging-related phenomenon, was observed in OA synovium [103] and at least some amyloid proteins can activate synoviocytes [104]. An overall increase in the thickness of the synovial membrane is typically seen in patients with diagnosed OA (reviewed in [105]). This hyperplasia is due to increased ECM formation as well as increased cell numbers. Whether this is due to proliferation of resident cells or recruitment of cells from the blood is unknown. This is also accompanied by formation of new blood vessels [70]. Predominantly perivascular leukocyte infiltrations are observed. They are usually much less intense than in rheumatoid arthritis and contain macrophages, T and B lymphocytes and plasma cells [106–108]. The significance of these synovial changes is underscored by long-term observational studies that identified the extent of synovial inflammation as a risk factor for more rapid progression of structural damage, as well as joint pain [109–111].

It is thought that these synovial changes are secondary to cartilage damage, which results in the generation of cartilage fragments, including cleavage products of extracellular matrix components such as collagen, fibronectin or cartilage oligomeric protein (COMP), which activate synoviocytes and macrophages via damage-associated molecular pattern molecule recognition receptors or Toll-like receptors [105]. Chondrocytes also produce proinflammatory cytokines and growth factors that can activate synoviocytes and recruit inflammatory cells. Calcium containing crystals, which are primarily formed in cartilage and menisci but are also produced by OA synoviocytes [112], are also potent activators of synovial fibroblasts, likely through Toll-like receptors [113].

4.4 Menisci

The menisci play an important role in both tibiofemoral compartments through load distribution and shock absorption [114–118]. Macroscopic and histopathologic analyses demonstrated a strong association between meniscus damage and OA-like changes in cartilage. Normal appearing menisci are rarely found in knees with OA [114–118].

Human meniscal degeneration in OA-affected joints has been described in detail [119–124]. However, several characteristics appear to be specific to meniscus aging in the absence of significant OA. With aging, the meniscal surface often remains intact while distinct changes in matrix stain and cellularity are observed within the meniscal substance. Tissue fibrillation and disruption is first seen at the inner rim, which spreads to the articular surfaces of the meniscus over time, and progresses to total disruption or loss of meniscus tissue, mainly in the avascular zone [125]. This is in direct contrast to degeneration in articular cartilage, which almost invariably progresses from the surface inward.

Increased Safranin O staining is observed with meniscus aging and could represent a shift from a fibroblastic to chondrocytic phenotype during early degeneration. Biochemical data [126, 127] as well as gene expression studies [128] suggest an accumulation of water-binding proteoglycans in aging and degenerating human menisci and these changes reflect an attempt at adaption or regeneration of the menisci [129, 130].

4.5 Ligaments/Tendons

Histological abnormalities in the anterior cruciate ligament (ACL) are highly prevalent in OA-affected knees and include cystic changes, disorientation of collagen fibers and mucoid degeneration [131]. Histological changes are much more severe in the ACL as compared to the posterior cruciate ligament in the same knees [132] and such age-related deterioration was not observed in the patellar tendon [133, 134]. However, histological changes in ligaments can precede cartilage histopathology [135]. Degenerated ACL were found in knees without cartilage degeneration. Also, knees with minimal cartilage degeneration can have moderate to severe ACL damage. These findings suggest that ACL degeneration can be initiated before or progresses more rapidly than cartilage degeneration, at least in a subpopulation of individuals [135].

4.6 Common Changes and Mechanisms in All Tissues

The concept that all joint tissues are affected in OA has been firmly established based on several different types of evidence [1]. There are several mechanistic changes that appear to be involved across the different tissues. Abnormal differentiation status of mesenchymal lineage cells is seen in cartilage where chondrocytes undergo hypertrophic differentiation and also show features of immature chondrocytes. In meniscus and ligaments, cells that are normally fibroblast-like express chondrogenic genes. Abnormal cell activation is also seen across joint tissues with increased expression of proinflammatory mediators, oxygen radicals and ECM degrading enzymes. There is also cell proliferation, even in cartilage, which normally has barely detectable levels of cell division. The stem cell-like populations that are present in all joint tissues also appear to be activated but instead of contributing to a successful repair response, they appear to participate in abnormal tissue remodeling and destruction. Elucidation of signaling mechanisms that mediate changes in all tissues has the potential to deliver more promising therapeutic targets.

5 Role of Major Hallmarks of Aging in OA Biology

Seven pillars or hallmarks of aging have been proposed [136], each of which has been identified to occur to some degree in joint tissues, with most work focused on the articular cartilage. These changes are summarized in Fig. 4 and will be discussed in detail below.

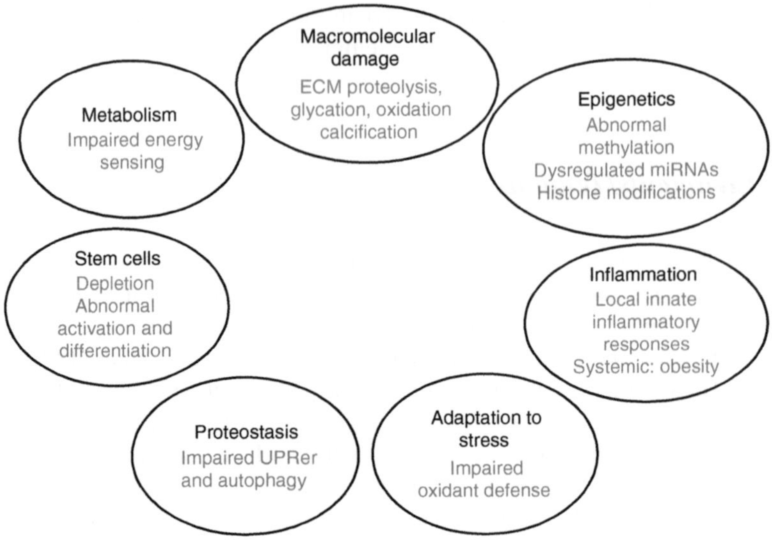

Fig. 4 'Seven pillars of aging' and the relationship to cartilage changes in osteoarthritis. Based on the recognition of conserved molecular pathways impacting aging, Kennedy at al. [136] proposed seven processes that are dysregulated in aging, emphasizing their connectedness. The figure summarizes OA and joint aging related changes in these 'seven pillars of aging'

5.1 Inflammation

A link between inflammation and OA has been made at both a systemic and a local level within joint tissues. The levels of proinflammatory cytokines, most consistently of IL-6 and TNF-α are increased in OA serum and synovial fluid and have been proposed as OA biomarkers [137, 138]. Epidemiologic studies have shown an association between serum levels of IL-6 as well as C-reactive protein (CRP) with knee OA, where higher levels correlated to risk of disease progression [139, 140]. In the Leiden cohort of 90-year-olds mentioned above, there was an association between the absence of knee OA and lower production of IL-1β and IL-6 when whole-blood samples were stimulated with lipopolysaccharide [141]. Levels of systemic markers of inflammation have also been shown to correlate with pain and function in older adults with knee OA. An increase in serum TNF-α and C-reactive protein (CRP) was associated with increased knee pain over a 5-year study [137]. Likewise, high levels of the soluble receptors for TNF-α as a surrogate for TNF activity were correlated with decreased physical ability in older adults with knee OA [142].

5.1.1 Cytokines

Cytokines are important signaling molecules that regulate cell function within and between joint tissues. Cytokines, both pro- and anti-inflammatory, anabolic and catabolic, as well as angiogenic and chemotactic have been studied for their role in

OA [143, 144]. OA-associated joint pain is at least in part mediated by cytokines such as IL-1, IL-6, TNF-α, FGF-2, NGF and certain chemokines [145, 146].

The production of cytokines by joint tissue cells is regulated by diverse extracellular stimuli, including other cytokines, enzymatic cleavage products of the extracellular matrix, and mechanical stress. Aging-related stimuli of cytokine expression in chondrocytes include advanced glycation end products [147] and amyloidogenic proteins [148]. Although some senescence markers are detectable in chondrocytes from older humans and increased expression of proinflammatory cytokines is a feature of the senescence-associated phenotype, a correlation between these phenomena in chondrocytes has not been established.

The group of cytokines that has been studied in detail for their role in OA is those that induce OA-like changes in chondrocytes. IL-1 is the prototypic catabolic and proinflammatory cytokine (reviewed [143, 144]). It affects extracellular matrix remodeling by suppressing the synthesis of cartilage-specific ECM components and by inducing the production of matrix-degrading enzymes. IL-1 also stimulates the production of other proinflammatory cytokines, prostaglandins, nitric oxide and other oxidants in chondrocytes and other joint tissue cells. Other members of the IL-1 family have similar effects, including IL-18 and members of the TNF family. Cytokines not only activate but also regulate the differentiation status of joint tissue cells. For example in chondrocytes, IL-1 suppresses Sox9, a main chondrogenic transcription factor [149].

Many proinflammatory cytokines have been implicated in causing changes in cartilage, bone and synovial tissue in spontaneous or surgically induced animal models of OA, and inhibition of these cytokines attenuated cartilage damage in animal models [144, 146]. However, attempts to modify the progression of human OA or joint pain in clinical trials with an IL-1 receptor antagonist protein (IRAP) or reduce pain with a monoclonal antibody directed against TNF-α have not been successful [150].

Anabolic cytokines, or growth factors, such as members of the bone morphogenetic proteins, transforming growth factors, insulin-like growth factors, fibroblast growth factors, platelet-derived growth factors, and connective tissue growth factor (CTGF) are produced in joint tissue cells and most are increased in OA cartilage. The function of some of these factors, such as BMP-7 or FGF-18 has been uniformly shown to promote ECM synthesis in chondrocytes and reduce the severity of OA in animal models. BMP-7 and FGF-18 have advanced to testing in clinical trials [151].

Members of the TGFβ family are important pro-chondrogenic factors. They stimulate chondrogenesis in mesenchymal stem cells, maintain the differentiation status of cultured chondrocytes and stimulate the production of ECM [152]. The chondrocyte response to TGFβ is altered in aging [153]. The abnormal TGFβ response of chondrocytes from older donors is due to changes in the ratio of TGFβ signaling components Alk1 and Alk5, which leads to a hypertrophic or pro-fibrotic phenotype and to degradation of the cartilage extracellular matrix [154, 155]. In mouse models of OA, TGFβ has been implicated in the formation of osteophytes [154].

In bone, active TGFβ1 is released during osteoclast bone resorption and recruits MSCs to form new bone at resorption sites [156]. TGFβ was elevated in OA subchondral bone in human samples and various animal models [157] and was associated with early signs of OA including bone-marrow lesions [158]. High levels of active TGFβ1 induce clustering of MSCs/osteoprogenitors in the subchondral bone marrow and formation of marrow osteoid islets. Transgenic mice with expression of active TGFβ1 in osteoblastic cells developed OA, whereas knockout of the TGFβ type II receptor in MSCs reduced OA severity in a surgical model. Neutralizing antibody to TGFβ has thus been proposed as a possible therapeutic approach in human OA [158].

5.2 *Oxidative Stress*

Reactive oxygen species (ROS) are produced by most cell types in the body, not merely as a consequence of mitochondrial respiratory activity, but rather as a physiologic mechanism for the regulation of specific cell signaling pathways [159, 160]. The modern concept of oxidative stress emphasizes that it represents an imbalance in the production and removal of ROS that results in the disruption of normal redox signaling [161]. Chondrocytes have been shown to produce the expected reactive oxygen species (ROS) made by most cells that include superoxide, hydroxyl radical, and hydrogen peroxide, as well as reactive nitrogen species, including nitric oxide [162–164]. In the growth plate, ROS serve a physiologic function to promote chondrocyte hypertrophy [165], which may be relevant to a pathologic role in OA where chondrocytes exhibiting a hypertrophic phenotype have been noted [166].

There is evidence that oxidative stress can contribute to the development of OA, although much of the evidence to date is indirect [167]. Immunohistochemical studies of human, non-human primates, and mouse articular cartilage, have shown an increase with age and with OA in the oxidative damage marker nitrotyrosine [168, 169]. Nitrotyrosine forms when nitric oxide (NO) reacts with superoxide (O_2-) to form peroxynitrite (ONOO-), which in turn reacts with protein tyrosines to form nitrotyrosine. Increased levels of intracellular ROS have also been detected in cartilage from old when compared to young adult rats using the fluorescent probe dihydrorhodamine 123 [170]. There is *in vitro* evidence that increased production of mitochondrial ROS by chondrocytes, sufficient to promote cell death, can occur in response to excessive mechanical loading [171–173]. Given that ROS are key regulators of MMP production [174], an age-related increase in the levels of ROS may contribute to the increase in MMP activity seen in aging and OA. Indeed, ROS-mediated signaling that regulates catabolic pathways that include MMP expression has been noted in response to cytokines including IL-1β and TNFα [175–178] and to stimulation of chondrocytes by fibronectin fragments [179]. Potential mechanisms by which ROS can contribute to OA are summarized in Fig. 5.

Normally, the cellular levels of ROS are controlled by the balance of ROS production and the activity of various anti-oxidants. Glutathione is a major intracellular anti-oxidant and an age-related increase in the amount of oxidized relative to reduced glutathione was detected in chondrocytes isolated from normal human

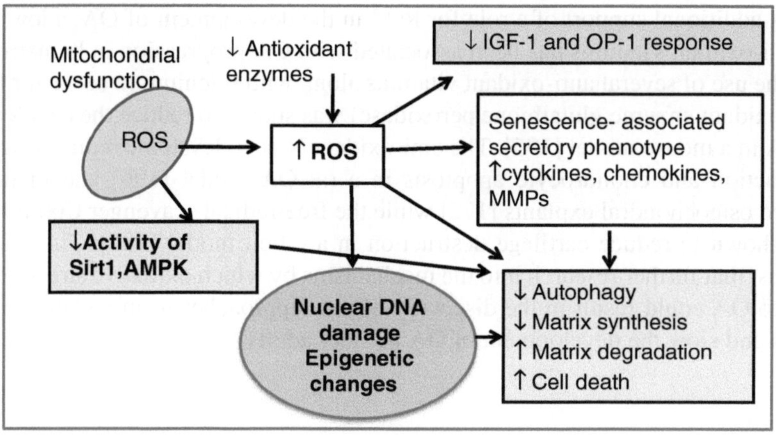

Fig. 5 Factors related to aging and oxidative stress in chondrocytes that may contribute to the development of OA. Oxidative stress is the result of an imbalance in ROS production, such as from mitochondrial dysfunction, and antioxidant enzyme function which disrupts normal cell signaling. This can contribute to reduced response to the growth factors IGF-1 and OP-1. Elevated ROS, particularly from the mitochondria, can also result in decreased Sirt1 and AMPK activity. These changes may promote the catabolic pathways associated with the senescence-associated secretory phenotype (SASP). Increased ROS can also cause nuclear DNA damage including telomere shortening as well as promote epigenetic changes. As a result of the various changes induced by elevated ROS, cells exhibit reduced autophagy and matrix synthesis and an increase in matrix degradation and cell death which all contribute to cartilage loss and OA

ankle tissue, consistent with an age-related increase in ROS [180]. The increased levels of ROS seen in aged and OA cartilage may also be due to reduced levels of anti-oxidant enzymes. Catalase and superoxide dismutase (SOD) were found to be present at lower levels with aging [170, 181] and in OA cartilage [182, 183]. Scott et al. also provided evidence for increased DNA methylation of the SOD2 promoter in human OA chondrocytes that might explain the reduced SOD2 levels [183]. SOD2 is primarily found in the mitochondria and SOD2 depletion in chondrocytes was associated with mitochondrial dysfunction [184].

Increased levels of ROS can result in DNA damage which has been noted in OA cartilage [185] including damage to mitochondrial DNA [186]. This can reduce cell viability [180] and matrix production [187] (Fig. 5). Oxidative stress can also contribute to the senescent phenotype of chondrocytes through damage to telomeres [188, 189]. Chondrocyte senescence has also been associated with increased production of oxidized low-density lipoproteins in cartilage [190]. As mentioned above, oxidative stress is associated with a disruption of normal redox signaling. Low levels of ROS are physiologic regulators of IGF-1 signaling but excessive levels may cause resistance to IGF-1, which has been noted in chondrocytes from older adults and in OA chondrocytes [191, 192]. There is growing evidence that excessive levels of ROS inhibit IGF-1 activation of the IRS-1-PI-3kinase-Akt pathway that promotes chondrocyte survival and matrix production [187, 193].

As additional support of a role for ROS in the development of OA, a low intake of anti-oxidant vitamins has been associated with OA progression in humans [194] and the use of several anti-oxidant vitamins along with selenium (a co-factor for the anti-oxidant enzyme glutathione peroxidase) was shown to reduce the development of OA in a mouse model [195]. The anti-oxidant N-acetylcysteine reduced cartilage destruction and chondrocyte apoptosis in a rat OA model [196] and in impact-loaded osteochondral explants [172] while the free-radical scavenger C60 fullerene was shown to reduce cartilage destruction in a rabbit model [197]. These studies suggest that further research into the mechanisms by which oxidative stress contributes to OA could result in the discovery of new approaches to modulate oxidative stress and slow the development of OA in older adults.

5.3 Epigenetics

Epigenetic regulation involves non-coding RNAs (ncRNAs), DNA methylation and histone modifications. As epigenetic changes are dynamic and responsive to environmental stimuli, their potential reversibility holds promise in understanding and therapeutically targeting disease mechanisms.

5.3.1 Non-coding RNAs

Among the ncRNAs, miRNAs are the most extensively studied in OA. Candidate miRs were selected based on differential expression in disease or during development for studies on cell function *in vitro* and in a limited number of cases using transgenic or knock out mice [198]. One study used a custom miR array to show that 7 of 723 miRs analyzed were differentially expressed in OA [199]. MiR-140 is the most abundant miR in cartilage [200] and its expression is reduced in OA [201]. miR-140 knock out mice showed age-related OA-like changes. Conversely, transgenic mice overexpressing miR-140 in cartilage were resistant to antigen-induced arthritis. Other miRNAs that regulate gene expression in chondrocytes are miR-125, miR-127b, miR-21, miR-148a and miR125 [202]. Upregulated miRs are potential drug targets that can be pursued by an increased availability of novel platforms to inhibit their expression or function [203].

5.3.2 DNA Methylation

OA-related methylation analyses examined a limited set of candidate genes. SOX9 expression is reduced in OA and subject to increased promoter methylation with associated repressive histone modifications [204]. A number of metalloproteinase genes including MMP3, MMP9, MMP13 and ADAMTS4 are upregulated in OA, and the promoters revealed single CpG demethylation events, possibly accounting for the observed disease-related expression changes [205–207]. Similarly, demethylation of an NF-κB-responsive enhancer facilitates transcription of inducible nitric oxide synthase (iNOS/NOS2) in chondrocytes, a gene, which is dysregulated in OA cartilage [208]. Previously, IL-1β promoter methylation was found to inversely correlate with gene expression in chondrocytes, while leptin expression was also regulated by methylation [209, 210]. GDF5, which has a SNP associated with OA, exhibits altered methylation and reduced expression [211].

Several recent studies used methylation array technology to discover differences between normal and OA cartilage [212–214]. A large variation in differentially methylated regions was observed in these various studies and this is at least in part due to comparing OA articular cartilage from different joints without including proper control groups. Large differences were observed in methylation profiles in OA cartilage from hips compared to knees [215]. In addition, the functional consequences of differentially methylated genomes were assessed by transcriptional analysis of only a small group of genes of interest, and the resulting link between DNA methylation and transcription was not further validated.

5.3.3 Histone Modifications

Posttranslational modification of nucleosomal histones, including acetylation, methylation, phosphorylation and sumoylation, play important roles in the regulation of gene transcription through remodeling of chromatin structure [216, 217]. Histone acetylation and methylation are among the best-characterized modifications. Histone methylation occurs on both lysine (K) and arginine residues. In histone H3, different lysine residues (K4, K9, K27, K36 and K79) can be methylated. Histone lysine methylation is associated with either gene activation or repression, depending on the specific residue modified [218–220]. Methylation of histone H3 lysine 4 (H3K4), H3K36 and H3K79 is generally associated with transcriptional activation, whereas methylation of H3K9 and H3K27 is associated with transcriptional repression [218–220].

Lysine methylation is controlled by the opposing activities of lysine methyltransferases (KMTs) and lysine demethylases (LSD). The level of LSD1 expression was elevated in the superficial and middle zones of OA cartilage [221]. In cultured chondrocytes, the induction of mPGES-1 expression by IL-1β was associated with decreased H3K9 methylation at the mPGES-1 promoter. Inhibition of LSD1 prevented IL-1β -induced H3K9 demethylation at the mPGES-1 promoter and mPGES-1 protein expression [221].

Histone acetylation, which is generally associated with transcriptional activation, is dynamically regulated by histone acetyl transferases (HAT) and histone deacetylases (HDAC) which are classified into four classes; class I (HDACs 1, 2, 3 & 8), class II (HDACs 4, 5, 6, 7, 9 & 10), Class III which are also referred to as sirtuins (SirT1-7), and Class IV (HDAC11) [222]. There is some evidence of altered expression of HDACs in OA cartilage. HDAC1 and HDAC2 are increased in OA chondrocytes [223]. HDAC4 mRNA was lower in cartilage from OA patients when compared with cartilage from healthy donors [224, 225]. HDAC4 was also lower in 40- to 60-year-old donors compared with specimens from 20- to 40-year-old healthy donors [225]. The reduction in HDAC4 was associated with increased Runx2 and other OA-related and chondrocyte hypertrophy-related genes in human OA cartilage, including MMP-13, Ihh and type X collagen [225]. One study found that HDAC4 expression was low in normal cartilage but increased in OA, most strongly in the chondrocyte clusters [226], which could explain the differences with the two prior studies.

In vitro studies using chondrocytes or cartilage explants generally showed protective effects of global inhibitors of class I/II HDACS such as trichostatin A (TSA) or sodium butyrate. These compounds inhibited metalloproteinase expression and protected against cartilage degradation [227]. Studies on individual HDACs showed that HDAC1 and HDAC2 overexpression suppressed transcription of cartilage anabolic genes such as ACAN and COL2A1 [223].

HDAC3 regulates chondrocyte hypertrophy and matrix content by inhibiting protein phosphatase Phlpp1 expression and promoting Akt activity. Chondrocytes lacking HDAC3 entered the hypertrophy stage sooner. Moreover, HDAC3 deficient chondrocytes had lower extracellular matrix production and smaller sizes than normal chondrocytes [228]. Inhibition of class I HDACs (HDAC-1, HDAC-2, HDAC-3) by the small molecule MS-275 or by specific siRNAs repressed cytokine-induced metalloproteinase expression in cartilage cells and cartilage explants [229].

HDAC4 has been extensively studied in OA, as it is a key regulator of chondrocyte differentiation during skeletogenesis, where it prevents chondrocyte hypertrophy by inhibiting the activity of Runx2 [230, 231]. HDAC4 overexpression in chondrocytes decreased the mRNA levels of Runx2, MMP1, MMP3, MMP-13, type X collagen, Ihh, ADAMTS-4 and -5, and increased type II collagen. Overexpression of HDAC4 also decreased IL-1β, Cox2 and iNOS and increased the expression of aggrecan, but also partially blocked the catabolic effect of IL-1β in human OA chondrocytes. Inhibition of HDAC4 by TSA had the opposite effect [225]. A challenge in using HDAC inhibitors is that their effects are qualitatively rather different as a function of duration of application. Short-term treatment of cells with HDAC inhibitors increased expression of cartilage ECM genes but prolonged treatment reduced expression of most of these genes [223, 232, 233]. Furthermore, toxicity of global HDAC inhibitors poses a major risk for use in chronic diseases such as OA. Specific inhibitors such as of class I HDACs may be a feasible OA treatment strategy.

In animal models, intra-articular injection of TSA in rabbits with experimental OA alleviated the extent of cartilage erosion, concomitant with reduced expression of IL-1 and matrix-degrading enzymes [234]. Systemically administered TSA also protected cartilage in the destabilized medial meniscus (DMM) model of surgically-induced OA [229].

Among the Class III HDACs, SIRT1 has been linked to aging and caloric intake, and also to various age-associated diseases such as type II diabetes, Alzheimer's disease and osteoporosis. SIRT1 activity or levels were altered in OA cartilage [235]. SIRT1 expression was clearly detected in the non-OA cartilage while MMP-13 and ADAMTS-5 were undetectable. In contrast, in the OA cartilage, SIRT1 expression was decreased while MMP-13 and ADAMTS-5 were increased [236]. In human chondrocytes, SIRT1 plays a role in cartilage ECM synthesis and promotes cell survival, even under proinflammatory stress. Overexpression of SIRT1 significantly inhibited the up-regulation of those genes caused by IL-1β, while the inhibition of SIRT1 further increased them [237].

SIRT1 KO mice up to 3 weeks of age exhibited low levels of type 2 collagen, aggrecan, and glycosaminoglycan while protein levels of MMP-13 were elevated, leading to increased cartilage breakdown, that was evident even in heterozygous mice [236]. Additional results showed elevated chondrocyte apoptosis in SIRT1 KO mice [238]. The histological OA score was significantly higher in 1-year-old SIRT1-KO mice compared to control mice. In the surgical OA model, SIRT1-KO mice showed accelerated OA progression compared with control mice and this was associated with increases in type X collagen, MMP-13, ADAMTS-5, apoptotic markers, and acetylated NF-κB p65 {Matsuzaki, 2014 #244}. These observations indicate that SIRT1 is involved in cartilage biology and sirtuin activation could potentially serve as a drug target in treating OA even at its early stages [235]. Resveratrol, which among other effects increases the activity of sirtuins, was tested by intraarticular injection in a rabbit knee OA model. It reduced cartilage destruction, the apoptosis rate of chondrocytes and the level of NO in synovial fluid [239].

5.4 *Autophagy*

Autophagy is a cellular response to various types of stress and a central homeostasis mechanism to eliminate damaged organelles, long-lived or aberrant proteins, protein aggregates and superfluous portions of the cytoplasm [240]. Substrates are enclosed in a double membrane, the autophagosome, which fuses with lysosomes, allowing enzymatic substrate degradation. Cleavage products are recycled for use in biosynthesis or as energy sources [240]. Autophagy is required for lifespan extension in various organisms, and many autophagy-related proteins are directly regulated by longevity pathways [241].

Conceptually, autophagy in normal adult articular cartilage is an important mechanism for cellular homeostasis, in particular as chondrocytes in normal cartilage are undergoing very low levels of proliferation. Thus, cells in the superficial zone display a robust expression of autophagy proteins BECN1, ATG5, and MAP1LC3 [242]. When MAP1LC3 is tagged with GFP, the highest GFP signal is observed in cells present in the superficial and middle zones of the knee articular cartilage [242]. Few cells in the deep cartilage zone exhibit detectable levels of GFP-MAP1LC3 signal [242]. As with other tissues, starvation increases the number of autophagosomes in chondrocytes [243].

Cartilage aging in humans and mice is associated with reduced expression of autophagy regulators ULK1 (unc-51 like autophagy activating kinase 1), MAP1LC3, and BECN1, which is most profound in the superficial zone in the weight bearing areas of cartilage [242]. Cartilage that is deficient in autophagy has reduced cellularity and extracellular matrix damage [242]. In the GFP-LC3 autophagy reporter mouse, basal autophagy activity was detected in young (6 months) liver and knee articular cartilage, with higher levels in cartilage than in liver in the same animals. In mice aged 28 months there was a reduction in the total number of autophagic vesicles. With increasing age, the expression of Atg5 and LC3 decreased, followed by a reduction in cartilage cellularity and an increase in the apoptosis marker PARP p85. Cartilage structural damage progressed in an age-dependent manner, subsequent to autophagy changes [244]. In contrast to the reduction in autophagic proteins in nonproliferating chondrocytes, the cell clusters in OA cartilage express high levels of these proteins [245], consistent with observations that certain chondrocytes in OA cartilage display numerous autophagic MAP1LC3 puncta [246, 247].

The reduction in autophagy protein levels and activity lends strong support to the hypothesis that basal autophagic activity decreases with age, thus contributing to the accumulation of damaged organelles and macromolecules and susceptibility to OA as an aging-related disease. Indeed, prior studies demonstrated mitochondrial dysfunction in various animal models and in human OA [248]. In addition, mitochondrial DNA mutations increase in OA chondrocytes [249]. Damaged mitochondria, producing high levels of ROS, promote pro-inflammatory signaling as they initiate formation of inflammasomes and activation of other inflammatory pathways [250]. In knee chondrocytes, IL1- or nitric oxide-dependent increase in expression of MAP1LC3 and BECN1 activates autophagy [251]. Furthermore, autophagy activation prevents IL1-mediated suppression of cartilage matrix degradation while reducing the levels of MMP13, ADAMTS5, and ROS. Given that one of the cytoprotective functions of autophagy is removal of damaged mitochondria, the IL1–induced OA-like gene expression changes might possibly occur through reduction in the intracellular ROS level via elimination of damaged mitochondria. Inhibition of autophagy caused OA-like gene expression changes, while the induction of autophagy by rapamycin reduced the MMP13 and ADAMTS5 expression induced by IL-1β [252].

These observations on autophagy defects in aging and OA-affected cartilage led to studies to test autophagy activators in OA models. A major focus of these studies was the protein kinase mammalian target of rapamycin (mTOR), which as part of the complex mTORC1, a key regulator and suppressor of autophagy. Excess mTOR activation has been linked to aging on the basis of results from genetic and pharmacological studies [253]. Major effects of mTOR are the inhibition of autophagy and the stimulation of protein synthesis. Chronic mTOR activation thus can potentially lead to increased accumulation of aggregation-prone proteins [254]. Senescent cells are enlarged or hypertrophic and these phenotypes as well as the abnormal gene expression can be reversed by the mTOR inhibitor rapamycin [255]. Abnormal mTOR signaling has been associated with autophagy to promote the secretory phenotype of senescent cells and the release of factors known to contribute to defective renewal and dysfunction of aging tissue [256].

The expression of mTOR mRNA and protein was increased in OA-affected human and mouse articular cartilage [257] (Fig. 6). The cartilage-specific mTOR

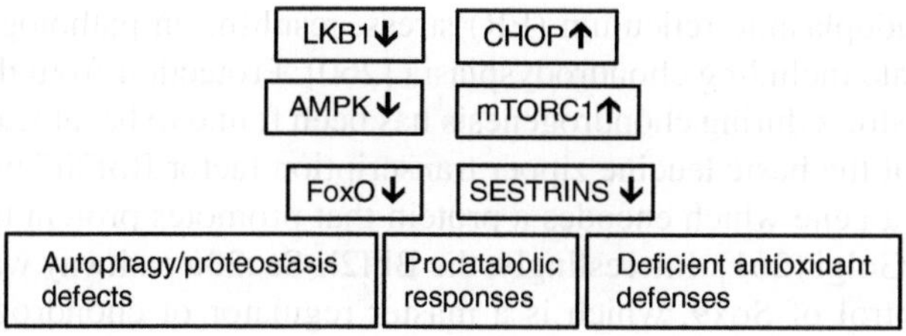

Fig. 6 Impaired nutrient and energy sensing in OA cartilage. Several recent findings on abnormal signaling mechanisms in OA appear to converge on AMPK and mTOR (mTORC1), which are central regulators of the cellular response to changes in energy and nutrient supply. Key abnormalities in OA are reduced AMPK activation and increased mTOR activation. Potential mechanisms involved in mTOR hyperactivation include deficient sestrin expression. Decreased AMPK expression and activation is related to reduced LKB1 expression and increased CHOP expression as a feature of the unfolded protein response (UPR) to endoplasmic reticulum (ER) stress. Dysregulation of these signaling mechanisms contributes to key changes in OA-affected cartilage, including defective autophagy and proteostasis, pro-catabolic (ECM-degrading and proinflammatory) responses and deficient anti-oxidant defenses

knock-out mouse showed reduced severity of OA induced by destabilization of the medial meniscus. This was associated with increased autophagy and decreased chondrocyte death [257]. Pharmacologic inhibition of mTORC-1 by rapamycin activates autophagy in chondrocytes and has protective effects against mechanical injury-induced ECM damage and cell death in cartilage explants [258]. In experimental OA in mice, rapamycin also reduced the severity of cartilage degradation and this was associated with a preservation of cartilage cellularity and a reduction in inflammatory mediators [259]. Intraarticular injection of rapamycin also reduced the severity of experimental OA in mice [236].

5.5 *Damage to Macromolecules and Proteostasis*

As detailed above, a hallmark of OA is degradation of extracellular matrix proteins that results in impaired joint tissue function. The most common age-related change in cartilage matrix proteins that may contribute to the development of OA is the accumulation of advanced glycation end products (AGEs) that are produced by the spontaneous nonenzymatic glycation of long-lived proteins such as collagen [57]. The role of AGEs in OA was discussed above in Sect. 4.1 on cartilage aging changes in OA.

Because articular chondrocytes are the only cell type present in cartilage and are therefore responsible for production and maintenance of the articular cartilage, they are required to synthesize large amounts of extracellular matrix proteins such as the collagens, proteoglycans, and cartilage oligomeric protein that may make chondrocytes susceptible to disruptions in proteostasis. For example, chondrocyte expression of mutant type X collagen was shown to induce the unfolded protein response

(UPR) and endoplasmic reticulum (ER) stress, resulting in pathologic changes in the growth plate including chondrodysplasia [260]. Protection from the deleterious effects of ER stress during chondrogenesis has been found to be, at least in part, due to activation of the basic leucine zipper transcription factor Bbf2h7 which up regulates Sec23a, a gene which encodes a protein that promotes protein transport from the ER to the Golgi [261]. Interestingly, the Bbf2h-Sec23a pathway was found to be under the control of Sox9, which is a master regulator of chondrogenesis [262]. Whether this pathway also may regulate ER stress responses in articular cartilage relevant to OA has not been studied.

There is growing evidence that disrupted bioenergy sensing and proteostasis may contribute to OA [263] (Fig. 6). The serine/threonine protein kinase AMP activated protein kinase (AMPK) is a master regulator of energy homeostasis. AMPK activity is constitutively present in normal articular chondrocytes, but is decreased in human knee OA chondrocytes [264]. Both IL-1β and TNFα induce a marked loss of AMPK activity in normal articular chondrocytes. Conversely, AMPK pharmacological activators attenuate cartilage explant and monolayer cultured chondrocyte procatabolic responses to IL-1β and TNFα [264]. These compounds also increased the NAD^{+}/NADH ratio and the expression of SIRT1, SOD2 and catalase. Peroxisome proliferator-activated receptor γ coactivator 1α (PGC-1α) and FoxO3A, major AMPK downstream targets, mediate the chondroprotective effect of AMPK activation [265]. Decreased AMPK activity in articular chondrocytes has the potential to disrupt cartilage homeostasis by promoting matrix catabolism, thereby contributing to progression of OA.

The presence of ER stress markers in association with altered production of type VI collagen has been noted in human OA cartilage [266] and the ER stress-induced protein C/EBP homologous protein (CHOP) has been shown to mediate chondrocyte apoptosis in a mouse OA model [267]. Increased levels of CHOP and other ER stress markers were noted in human OA cartilage [268, 269] and CHOP expression was associated with increased chondrocyte catabolic activity [269]. Chondrocyte death in response to ER stress induced by thapsigargin was found to be inhibited by the pro-survival protein Akt1 [270]. Excess CHOP activity is limited by AMPK activity in chondrocytes [269]. An inhibitor of Akt1 that is induced by ER stress, tribbles homolog 3 (TRB3), was found to be increased in OA cartilage, while overexpression of TRB3 in chondrocytes reduced matrix synthesis and promoted cell death through inhibition of chondrocyte Akt phosphorylation [271]. Thus further studies on mechanisms relevant to proteostasis in cartilage may discover important links to OA.

5.6 Stem Cells

Stem cell depletion is recognized as a hallmark of aging [272] and it has been suggested that the regenerative activity of bone marrow-derived stem cells (BMSC) declines after the age of 30 [273]. A large number of studies examined whether aging or OA is associated with changes in BMSC [274]. The frequency of BMSC in the mononuclear cell fraction of bone marrow aspirates was reported not to correlate with the diagnosis of OA or osteoporosis [275–280] or to decline in older donors [281, 282]. The proliferation rate of BMSC during culture expansion decreased with age in some studies [277, 282–285] but not in others [286, 287]. There is also a substantial variability in proliferation rates of BMSCs in older donors, which does not appear to be related to the presence of diabetes or OA [288].

Similar to most other tissues, joint tissues in adult humans are thought to contain cell populations with certain phenotypic and functional properties of mesenchymal stem cells (MSC) or progenitor cells. Articular cartilage, synovium, menisci, ligaments, tendons [289], infrapatellar fat pad, synovial fluid [290] and periosteum [291] contain cells that express MSC surface markers and upon isolation and culture under specific conditions, they have multilineage differentiation potential towards chondrocytes, osteoblasts and adipocytes. In human articular cartilage, the highest concentration of cells expressing progenitor cell markers, including Stro-1, Notch-1, CD105, CD166 is in the superficial zone and this is reduced with donor age [292, 293].

The function of these cells in the maintenance of articular cartilage and other joint tissues under normal conditions is currently unclear. Interestingly, there is a high level of expression of these stem cell markers in the cell clusters in OA cartilage [292], a histological hallmark of OA [294]. Cells in these clusters produce a large number of mediators involved in joint inflammation and tissue remodeling. The abnormal activation and differentiation pattern of cluster cells in OA cartilage has been interpreted as chondrocyte de-dedifferentiation [295] where the differentiated articular chondrocytes change gene expression patterns in response to the different extracellular signaling environment. An alternative hypothesis is that cluster formation is the result of progenitor cell proliferation. A systematic analysis of the fate of cells expressing progenitor markers during the development of OA is required to address this hypothesis.

Surgical injury to articular cartilage is also associated with proliferation of progenitor cells that produce new extracellular matrix [293]. In addition, migratory chondrogenic progenitor cells, apparently originating from subchondral bone or bone marrow were also described in areas of OA joints where the original cartilage had been degraded [296]. It thus appears that traumatic or OA-related injury to articular cartilage activates proliferation of these mesenchymal progenitor cells.

Osteophytes are new bone tissues that are formed in OA affected joints and represent radiographic hallmarks of OA. While osteophytes most commonly form at the joint margins and originate from the periosteum, a tissue rich in stem cells, similar structures can also develop in areas of exposed subchondral bone, in ligaments and tendons [297]. The formation of osteophytes appears to be initiated by MSCs that undergo condensation, chondrogenic differentiation, and proliferation [298]. The chondrocytes then undergo hypertrophic differentiation, promoting the formation of blood vessels that allow recruitment of osteoblasts and osteoclasts that remodel the cartilaginous tissue into bone in a process similar to endochondral ossification [299].

Bone marrow-derived mesenchymal stem cells (BMSCs) have been studied extensively as a cell source for engineering of cartilage and other tissues [273]. Direct injection of MSCs into the joint can slow the degradation of articular cartilage or even regenerate it in OA animal models [300, 301]. A proof of concept study in humans who received injections of autologous adipose tissue-derived MSCs into OA-affected knees showed encouraging results [302].

For differentiation potential towards single lineages, an unaltered differentiation capacity with age was reported for adipogenesis [276, 277, 282, 303] and chondrogenesis [278, 282, 303]. For osteogenesis, reports on maintained differentiation capacity [275–277, 285, 303, 304] contrasted others showing a decline in osteogenesis with age [281, 282, 284].

Since there is no specific marker to identify MSC in bone marrow [274] and the cells are analyzed after culture, the question whether there is a change in MSC numbers and/or function with age is, therefore, an important question which still remains open. Despite these aging-related changes that were observed in some studies, bone marrow stem cells from older donors still appear to be a suitable cell source for tissue engineering, as proliferating and differentiating MSC were isolated even from very old donors, so that the ability to isolate MSC is per se not a limitation for therapeutic use [288].

6 Research Needs in Aging and OA to Meet the Goal of Developing Disease-Modifying Therapies

A major long-term goal of research in OA is to delay onset and develop disease modifying therapies that would slow or stop structural progression while also reducing pain and improving physical function. We are getting closer to that goal. When compared to the previous century of work, rapid progress in the understanding of the biology of OA has been made over the past two decades with the recognition that OA is not simply due to "wear and tear" of the articular cartilage. As detailed in the sections above, the biology of OA and the contribution of aging are much more complex than ever imagined. It includes pathological changes in all of the tissues that make up the affected joint(s) driven not only by abnormal joint mechanics that result in excessive or abnormal loading of the joint but also by the activity of a host of inflammatory mediators as well as by aging changes that promote catabolic over anabolic activity and reduced cell survival. This improved understanding of OA within the context of aging provides hope that new disease-modifying therapies can be developed but there is still much work to be done first.

A summary of the research needs in OA most relevant to aging includes:

- Determining the key pathways which lead to OA downstream from specific risk factors
- Understanding the role of systemic vs. local inflammation and determine which of the many inflammatory factors are key drivers of OA

- Determining if targeting a specific hallmark of aging will decrease the progression of OA
- Developing better age appropriate pre-clinical models of OA
- Developing clinical, biochemical, and imaging biomarkers that can accurately phenotype individuals with various forms of OA to predict those at greatest risk of progression and target therapy to a specific phenotype

A major challenge moving forward is determining which of the many pathways recently found to contribute to OA are critical to promoting disease progression. It is likely that the pathway to OA will vary for each individual based on specific risk factors in conjunction with age. For example, preventing the development of post-traumatic knee OA in a 20 year-old with an acute ACL tear will require a different strategy than for slowing the progression of knee OA in an obese 60 year-old. The former would likely require targeting an inflammatory pathway initiated at the time of ACL tear while the latter would require targeting a key metabolic aspect of obesity that contributes to OA. It is not known which inflammatory mediators are key drivers of OA and the relative role of local versus systemic inflammation associated with aging or "inflamm-aging" in the OA process is also not clear. Likewise more work is needed to determine if targeting one of the major hallmarks of aging, such as oxidative stress, epigenetic alterations, metabolism and autophagy, or ER stress will be of benefit in reducing the progression of OA. Finally, there is a need to know if protecting chondrocytes from dying and/or inducing endogenous stem cells to promote repair is feasible.

Another need to advance the field is the development of pre-clinical models of OA that better predict efficacy in humans. A number of the treatments mentioned above, such as bisphosphonates, calcitonin, and iNOS inhibitors, worked great in the pre-clinical models in which they were tested but failed in large (and expensive) clinical trials in humans [46]. A major issue is that pre-clinical studies are routinely performed in young animals in which OA is induced surgically. Using microarrays to evaluate changes in gene expression after the induction of OA in young and older adult mice, we noted significant age-related differences that suggest the genes and pathways activated as OA develops in older animals may be very different from those in younger animals [305]. Surgical induction of OA in young animals might only be a good model system for predicting efficacy in preventing post-traumatic OA in young humans, while predicting efficacy in older adult humans will require use of animals that more closely match the age of the human population for which an intervention is being developed.

There is a movement to think of OA as a condition with multiple phenotypes that might be segregated by clinical findings, including risk factor assessment, along with biochemical and imaging biomarkers to evaluate specific joint tissue involvement and disease activity [306–308]. However, we currently lack biochemical markers sensitive and specific enough to phenotype patients and although advances are being made quite rapidly in imaging, there is a lack of agreement on the most useful modalities. Clinical, imaging and biochemical biomarker data from the Osteoarthritis Initiative (OAI), an ongoing longitudinal multicenter study of close to

5000 men and women aged 45–79 years with knee OA or at risk for knee OA, is expected to provide results that can be used to better phenotype people with knee OA and predict who is at greatest risk of progression. This work and the many other ongoing studies in aging and OA provide confidence that progress in the field will continue to be made that should lead to new therapies for OA.

Acknowledgments Work in the Loeser laboratory was supported by National Institutes of Health grants AG044034 and AR049003. Work in the Lotz laboratory was supported by National Institutes of Health grant AG007996.

Editor: John Williams, National Institute on Aging (NIA), NIH.

References

1. Loeser RF, Goldring SR, Scanzello CR, Goldring MB (2012) Osteoarthritis: a disease of the joint as an organ. Arthritis Rheum 64:1697–1707. doi:10.1002/art.34453
2. Goldring MB, Goldring SR (2007) Osteoarthritis. J Cell Physiol 213(3):626–634
3. Centers for Disease Control and Prevention (CDC) (2006) Prevalence of doctor-diagnosed arthritis and arthritis-attributable activity limitation – United States, 2003–2005. MMWR Morb Mortal Wkly Rep 55(40):1089–1092
4. Hootman JM, Helmick CG (2006) Projections of US prevalence of arthritis and associated activity limitations. Arthritis Rheum 54(1):226–229
5. Lawrence RC, Felson DT, Helmick CG, Arnold LM, Choi H, Deyo RA et al (2008) Estimates of the prevalence of arthritis and other rheumatic conditions in the United States: part II. Arthritis Rheum 58(1):26–35
6. Collaborators, U. S. B. o. D. (2013) The state of US health, 1990–2010: burden of diseases, injuries, and risk factors. JAMA 310(6):591–608. doi:10.1001/jama.2013.13805
7. Vos T, Flaxman AD, Naghavi M, Lozano R, Michaud C, Ezzati M et al (2012) Years lived with disability (YLDs) for 1160 sequelae of 289 diseases and injuries 1990–2010: a systematic analysis for the Global Burden of Disease Study 2010. Lancet 380(9859):2163–2196. doi:10.1016/S0140-6736(12)61729-2
8. Boult C, Altmann M, Gilbertson D, Yu C, Kane RL (1996) Decreasing disability in the 21st century: the future effects of controlling six fatal and nonfatal conditions. Am J Public Health 86(10):1388–1393
9. Ong KL, Wu BJ, Cheung BM, Barter PJ, Rye KA (2013) Arthritis: its prevalence, risk factors, and association with cardiovascular diseases in the United States, 1999 to 2008. Ann Epidemiol 23(2):80–86. doi:10.1016/j.annepidem.2012.11.008
10. Nuesch E, Dieppe P, Reichenbach S, Williams S, Iff S, Juni P (2011) All cause and disease specific mortality in patients with knee or hip osteoarthritis: population based cohort study. BMJ 342:d1165. doi:10.1136/bmj.d1165
11. Philbin EF, Groff GD, Ries MD, Miller TE (1995) Cardiovascular fitness and health in patients with end-stage osteoarthritis. Arthritis Rheum 38(6):799–805
12. Dunlop DD, Semanik P, Song J, Manheim LM, Shih V, Chang RW (2005) Risk factors for functional decline in older adults with arthritis. Arthritis Rheum 52(4):1274–1282
13. Leveille SG, Bean J, Bandeen-Roche K, Jones R, Hochberg M, Guralnik JM (2002) Musculoskeletal pain and risk for falls in older disabled women living in the community. J Am Geriatr Soc 50(4):671–678
14. Misra D, Felson DT, Silliman RA, Nevitt M, Lewis CE, Torner J et al (2014) Knee osteoarthritis and frailty: findings from the multicenter osteoarthritis study and osteoarthritis initiative. J Gerontol A Biol Sci Med Sci. doi:10.1093/gerona/glu102

15. Zhang W, Doherty M, Leeb BF, Alekseeva L, Arden NK, Bijlsma JW et al (2009) EULAR evidence-based recommendations for the diagnosis of hand osteoarthritis: report of a task force of ESCISIT. Ann Rheum Dis 68(1):8–17. doi:10.1136/ard.2007.084772. [Practice Guideline Research Support, Non-U.S. Gov't Review]
16. Finan PH, Buenaver LF, Bounds SC, Hussain S, Park RJ, Haque UJ et al (2013) Discordance between pain and radiographic severity in knee osteoarthritis: findings from quantitative sensory testing of central sensitization. Arthritis Rheum 65(2):363–372. doi:10.1002/art.34646
17. Thomas RH, Daniels TR (2003) Ankle arthritis. J Bone Joint Surg Am 85-A(5):923–936
18. Valdes AM, Spector TD (2011) Genetic epidemiology of hip and knee osteoarthritis. Nat Rev Rheumatol 7(1):23–32. doi:10.1038/nrrheum.2010.191
19. arc, O. C., arc, O. C., Zeggini E, Panoutsopoulou K, Southam L, Rayner NW (2012) Identification of new susceptibility loci for osteoarthritis (arcOGEN): a genome-wide association study. Lancet 380(9844):815–823. doi:10.1016/S0140-6736(12)60681-3
20. Rodriguez-Fontenla C, Calaza M, Evangelou E, Valdes AM, Arden N, Blanco FJ et al (2014) Assessment of osteoarthritis candidate genes in a meta-analysis of nine genome-wide association studies. Arthritis Rheumatol 66(4):940–949. doi:10.1002/art.38300
21. Anderson SA, Loeser RF (2010) Why is osteoarthritis an age-related disease? Best Pract Res Clin Rheumatol 24(1):15–26. doi:10.1016/j.berh.2009.08.006, S1521-6942(09)00084-9 [pii]
22. Losina E, Weinstein AM, Reichmann WM, Burbine SA, Solomon DH, Daigle ME et al (2013) Lifetime risk and age at diagnosis of symptomatic knee osteoarthritis in the US. Arthritis Care Res (Hoboken) 65(5):703–711. doi:10.1002/acr.21898. [Research Support, N.I.H., Extramural Research Support, U.S. Gov't, P.H.S.]
23. Leyland KM, Hart DJ, Javaid MK, Judge A, Kiran A, Soni A et al (2012) The natural history of radiographic knee osteoarthritis: a fourteen-year population-based cohort study. Arthritis Rheum 64(7):2243–2251. doi:10.1002/art.34415
24. Murphy L, Schwartz TA, Helmick CG, Renner JB, Tudor G, Koch G et al (2008) Lifetime risk of symptomatic knee osteoarthritis. Arthritis Rheum 59(9):1207–1213
25. Driban JB, Eaton CB, Lo GH, Ward RJ, Lu B, McAlindon TE (2014) Association of knee injuries with accelerated knee osteoarthritis progression: data from the osteoarthritis initiative. Arthritis Care Res (Hoboken) 66(11):1673–1679. doi:10.1002/acr.22359
26. Johnson VL, Hunter DJ (2014) The epidemiology of osteoarthritis. Best Pract Res Clin Rheumatol 28(1):5–15. doi:10.1016/j.berh.2014.01.004
27. Blagojevic M, Jinks C, Jeffery A, Jordan KP (2010) Risk factors for onset of osteoarthritis of the knee in older adults: a systematic review and meta-analysis. Osteoarthritis Cartilage 18(1):24–33. doi:10.1016/j.joca.2009.08.010, S1063-4584(09)00225-8 [pii]
28. Felson DT, Anderson JJ, Naimark A, Walker AM, Meenan RF (1988) Obesity and knee osteoarthritis. The Framingham Study. Ann Intern Med 109(1):18–24
29. Karlson EW, Mandl LA, Aweh GN, Sangha O, Liang MH, Grodstein F (2003) Total hip replacement due to osteoarthritis: the importance of age, obesity, and other modifiable risk factors. Am J Med 114(2):93–98
30. Oliveria SA, Felson DT, Cirillo PA, Reed JI, Walker AM (1999) Body weight, body mass index, and incident symptomatic osteoarthritis of the hand, hip, and knee. Epidemiology 10(2):161–166
31. Visser AW, Ioan-Facsinay A, de Mutsert R, Widya RL, Loef M, de Roos A et al (2014) Adiposity and hand osteoarthritis: the Netherlands Epidemiology of Obesity study. Arthritis Res Ther 16(1):R19. doi:10.1186/ar4447
32. Issa RI, Griffin TM (2012) Pathobiology of obesity and osteoarthritis: integrating biomechanics and inflammation. Pathobiol Aging Age Relat Dis 2(2012). doi:10.3402/pba.v2i0.17470
33. Hui W, Litherland GJ, Elias MS, Kitson GI, Cawston TE, Rowan AD et al (2012) Leptin produced by joint white adipose tissue induces cartilage degradation via upregulation and activation of matrix metalloproteinases. Ann Rheum Dis 71(3):455–462. doi:10.1136/annrheumdis-2011-200372. [Research Support, Non-U.S. Gov't]
34. Zhuo Q, Yang W, Chen J, Wang Y (2012) Metabolic syndrome meets osteoarthritis. Nat Rev Rheumatol 8(12):729–737. doi:10.1038/nrrheum.2012.135

35. Sellam J, Berenbaum F (2013) Is osteoarthritis a metabolic disease? Joint Bone Spine 80(6):568–573. doi:10.1016/j.jbspin.2013.09.007
36. Jungmann PM, Kraus MS, Alizai H, Nardo L, Baum T, Nevitt MC et al (2013) Association of metabolic risk factors with cartilage degradation assessed by T2 relaxation time at the knee: data from the osteoarthritis initiative. Arthritis Care Res (Hoboken) 65(12):1942–1950. doi:10.1002/acr.22093
37. Borkan GA, Hults DE, Gerzof SG, Robbins AH, Silbert CK (1983) Age changes in body composition revealed by computed tomography. J Gerontol 38(6):673–677
38. Kumar D, Karampinos DC, MacLeod TD, Lin W, Nardo L, Li X et al (2014) Quadriceps intramuscular fat fraction rather than muscle size is associated with knee osteoarthritis. Osteoarthritis Cartilage 22(2):226–234. doi:10.1016/j.joca.2013.12.005
39. Addison O, Drummond MJ, LaStayo PC, Dibble LE, Wende AR, McClain DA et al (2014) Intramuscular fat and inflammation differ in older adults: the impact of frailty and inactivity. J Nutr Health Aging 18(5):532–538. doi:10.1007/s12603-014-0019-1
40. Chuckpaiwong B, Charles HC, Kraus VB, Guilak F, Nunley JA (2010) Age-associated increases in the size of the infrapatellar fat pad in knee osteoarthritis as measured by 3T MRI. J Orthop Res 28(9):1149–1154. doi:10.1002/jor.21125
41. Distel E, Cadoudal T, Durant S, Poignard A, Chevalier X, Benelli C (2009) The infrapatellar fat pad in knee osteoarthritis: an important source of interleukin-6 and its soluble receptor. Arthritis Rheum 60(11):3374–3377. doi:10.1002/art.24881
42. Goekoop RJ, Kloppenburg M, Kroon HM, Dirkse LE, Huizinga TW, Westendorp RG et al (2011) Determinants of absence of osteoarthritis in old age. Scand J Rheumatol 40(1):68–73. doi:10.3109/03009742.2010.500618
43. Felson DT (2004) Risk factors for osteoarthritis: understanding joint vulnerability. Clin Orthop Relat Res (427 Suppl):S16–21. doi:00003086-200410001-00005 [pii]
44. Hochberg MC, Altman RD, April KT, Benkhalti M, Guyatt G, McGowan J et al (2012) American College of Rheumatology 2012 recommendations for the use of nonpharmacologic and pharmacologic therapies in osteoarthritis of the hand, hip, and knee. Arthritis Care Res (Hoboken) 64(4):455–474. [Consensus Development Conference Practice Guideline Review]
45. Nelson AE, Allen KD, Golightly YM, Goode AP, Jordan JM (2014) A systematic review of recommendations and guidelines for the management of osteoarthritis: the Chronic Osteoarthritis Management Initiative of the U.S. Bone and Joint Initiative. Semin Arthritis Rheum 43(6):701–712. doi:10.1016/j.semarthrit.2013.11.012
46. Tonge DP, Pearson MJ, Jones SW (2014) The hallmarks of osteoarthritis and the potential to develop personalised disease-modifying pharmacological therapeutics. Osteoarthritis Cartilage 22(5):609–621. doi:10.1016/j.joca.2014.03.004
47. Yang S, Eaton CB, McAlindon TE, Lapane KL (2014) Effects of glucosamine and chondroitin on treating knee osteoarthritis: an analysis with marginal structural models. Arthritis Rheumatol. doi:10.1002/art.38932
48. Guermazi A, Roemer FW, Felson DT, Brandt KD (2013) Motion for debate: osteoarthritis clinical trials have not identified efficacious therapies because traditional imaging outcome measures are inadequate. Arthritis Rheum 65(11):2748–2758. doi:10.1002/art.38086
49. Lotz M, Loeser RF (2012) Effects of aging on articular cartilage homeostasis. Bone 51(2):241–248. doi:10.1016/j.bone.2012.03.023
50. Hollander AP, Pidoux I, Reiner A, Rorabeck C, Bourne R, Poole AR (1995) Damage to type II collagen in aging and osteoarthritis starts at the articular surface, originates around chondrocytes, and extends into the cartilage with progressive degeneration. J Clin Invest 96(6):2859–2869
51. Nguyen QT, Wong BL, Chun J, Yoon YC, Talke FE, Sah RL (2010) Macroscopic assessment of cartilage shear: effects of counter-surface roughness, synovial fluid lubricant, and compression offset. J Biomech 43(9):1787–1793. doi:10.1016/j.jbiomech.2010.02.014, S0021-9290(10)00097-7 [pii]
52. Dean DD, Martel-Pelletier J, Pelletier JP, Howell DS, Woessner JF Jr (1989) Evidence for metalloproteinase and metalloproteinase inhibitor imbalance in human osteoarthritic cartilage. J Clin Invest 84(2):678–685

53. Dodge GR, Poole AR (1989) Immunohistochemical detection and immunochemical analysis of type II collagen degradation in human normal, rheumatoid, and osteoarthritic articular cartilages and in explants of bovine articular cartilage cultured with interleukin 1. J Clin Invest 83(2):647–661. doi:10.1172/JCI113929
54. Bank RA, Krikken M, Beekman B, Stoop R, Maroudas A, Lafeber FP et al (1997) A simplified measurement of degraded collagen in tissues: application in healthy, fibrillated and osteoarthritic cartilage. Matrix Biol 16(5):233–243
55. Gibson GJ, Verner JJ, Nelson FR, Lin DL (2001) Degradation of the cartilage collagen matrix associated with changes in chondrocytes in osteoarthrosis. Assessment by loss of background fluorescence and immunodetection of matrix components. J Orthop Res 19(1):33–42. doi:10.1016/S0736-0266(00)00008-5, S0736-0266(00)00008-5 [pii]
56. Maroudas A (1979) Physico-chemical properties of articular cartilage. In: Freeman MAR (ed) Adult articular cartilage. Pitman Medical, Turnbridge Wells, pp 215–290
57. Verzijl N, Bank RA, TeKoppele JM, DeGroot J (2003) AGEing and osteoarthritis: a different perspective. Curr Opin Rheumatol 15(5):616–622
58. Loeser RF, Yammani RR, Carlson CS, Chen H, Cole A, Im HJ et al (2005) Articular chondrocytes express the receptor for advanced glycation end products: potential role in osteoarthritis. Arthritis Rheum 52(8):2376–2385
59. Yammani RR, Carlson CS, Bresnick AR, Loeser RF (2006) Increase in production of matrix metalloproteinase 13 by human articular chondrocytes due to stimulation with S100A4: role of the receptor for advanced glycation end products. Arthritis Rheum 54(9):2901–2911
60. Steenvoorden MM, Huizinga TW, Verzijl N, Bank RA, Ronday HK, Luning HA et al (2006) Activation of receptor for advanced glycation end products in osteoarthritis leads to increased stimulation of chondrocytes and synoviocytes. Arthritis Rheum 54(1):253–263
61. Cecil DL, Johnson K, Rediske J, Lotz M, Schmidt AM, Terkeltaub R (2005) Inflammation-induced chondrocyte hypertrophy is driven by receptor for advanced glycation end products. J Immunol 175(12):8296–8302
62. Mitsuyama H, Healey RM, Terkeltaub RA, Coutts RD, Amiel D (2007) Calcification of human articular knee cartilage is primarily an effect of aging rather than osteoarthritis. Osteoarthritis Cartilage 15(5):559–565. doi:10.1016/j.joca.2006.10.017, S1063-4584(06)00319-0 [pii]
63. Fuerst M, Bertrand J, Lammers L, Dreier R, Echtermeyer F, Nitschke Y et al (2009) Calcification of articular cartilage in human osteoarthritis. Arthritis Rheum 60(9):2694–2703. doi:10.1002/art.24774
64. Rosen F, McCabe G, Quach J, Solan J, Terkeltaub R, Seegmiller JE et al (1997) Differential effects of aging on human chondrocyte responses to transforming growth factor beta: increased pyrophosphate production and decreased cell proliferation. Arthritis Rheum 40(7):1275–1281
65. Terkeltaub RA (2002) What does cartilage calcification tell us about osteoarthritis? J Rheumatol 29(3):411–415
66. Niggemeyer O, Steinhagen J, Deuretzbacher G, Zustin J, Ruther W (2011) Amyloid deposition in osteoarthritis of the hip. Arch Orthop Trauma Surg 131(5):637–643. doi:10.1007/s00402-010-1187-z
67. Ladefoged C (1986) Amyloid deposits in the knee joint at autopsy. Ann Rheum Dis 45(8):668–672
68. Grogan SP, D'Lima DD (2010) Joint aging and chondrocyte cell death. Int J Clin Rheumatol 5(2):199–214. doi:10.2217/ijr.10.3
69. Radin EL, Rose RM (1986) Role of subchondral bone in the initiation and progression of cartilage damage. Clin Orthop Relat Res (213):34–40
70. Walsh DA, Bonnet CS, Turner EL, Wilson D, Situ M, McWilliams DF (2007) Angiogenesis in the synovium and at the osteochondral junction in osteoarthritis. Osteoarthritis Cartilage 15(7):743–751. doi:10.1016/j.joca.2007.01.020, S1063-4584(07)00058-1 [pii]
71. Lories RJ, Luyten FP (2011) The bone-cartilage unit in osteoarthritis. Nat Rev Rheumatol 7(1):43–49. doi:10.1038/nrrheum.2010.197, nrrheum.2010.197 [pii]

72. Lyons TJ, McClure SF, Stoddart RW, McClure J (2006) The normal human chondro-osseous junctional region: evidence for contact of uncalcified cartilage with subchondral bone and marrow spaces. BMC Musculoskelet Disord 7:52. doi:10.1186/1471-2474-7-52, 1471-2474-7-52 [pii]
73. Hart DJ, Cronin C, Daniels M, Worthy T, Doyle DV, Spector TD (2002) The relationship of bone density and fracture to incident and progressive radiographic osteoarthritis of the knee: the Chingford Study. Arthritis Rheum 46(1):92–99. doi:10.1002/1529-0131(200201)46: 1<92::AID-ART10057>3.0.CO;2-#
74. Hunter DJ, Hart D, Snieder H, Bettica P, Swaminathan R, Spector TD (2003) Evidence of altered bone turnover, vitamin D and calcium regulation with knee osteoarthritis in female twins. Rheumatology (Oxford) 42(11):1311–1316. doi:10.1093/rheumatology/keg373keg373 [pii]
75. Hilal G, Martel-Pelletier J, Pelletier JP, Duval N, Lajeunesse D (1999) Abnormal regulation of urokinase plasminogen activator by insulin-like growth factor 1 in human osteoarthritic subchondral osteoblasts. Arthritis Rheum 42(10):2112–2122. doi:10.1002/1529-0131(199910)42:10<2112::AID-ANR11>3.0.CO;2-N
76. Hilal G, Martel-Pelletier J, Pelletier JP, Ranger P, Lajeunesse D (1998) Osteoblast-like cells from human subchondral osteoarthritic bone demonstrate an altered phenotype in vitro: possible role in subchondral bone sclerosis. Arthritis Rheum 41(5):891–899
77. Bettica P, Cline G, Hart DJ, Meyer J, Spector TD (2002) Evidence for increased bone resorption in patients with progressive knee osteoarthritis: longitudinal results from the Chingford study. Arthritis Rheum 46(12):3178–3184. doi:10.1002/art.10630
78. Kwan Tat S, Lajeunesse D, Pelletier JP, Martel-Pelletier J (2010) Targeting subchondral bone for treating osteoarthritis: what is the evidence? Best Pract Res Clin Rheumatol 24(1):51–70. doi:10.1016/j.berh.2009.08.004, S1521-6942(09)00087-4 [pii]
79. Li B, Aspden RM (1997) Composition and mechanical properties of cancellous bone from the femoral head of patients with osteoporosis or osteoarthritis. J Bone Miner Res 12(4):641–651. doi:10.1359/jbmr.1997.12.4.641
80. Couchourel D, Aubry I, Delalandre A, Lavigne M, Martel-Pelletier J, Pelletier JP et al (2009) Altered mineralization of human osteoarthritic osteoblasts is attributable to abnormal type I collagen production. Arthritis Rheum 60(5):1438–1450. doi:10.1002/art.24489
81. Burr DB, Gallant MA (2012) Bone remodelling in osteoarthritis. Nat Rev Rheumatol 8(11):665–673. doi:10.1038/nrrheum.2012.130
82. Fazzalari NL, Parkinson IH (1997) Fractal properties of subchondral cancellous bone in severe osteoarthritis of the hip. J Bone Miner Res 12(4):632–640. doi:10.1359/jbmr.1997.12.4.632
83. Grynpas MD, Alpert B, Katz I, Lieberman I, Pritzker KP (1991) Subchondral bone in osteoarthritis. Calcif Tissue Int 49(1):20–26
84. Suri S, Walsh DA (2012) Osteochondral alterations in osteoarthritis. Bone 51(2):204–211. doi:10.1016/j.bone.2011.10.010, S8756-3282(11)01290-7 [pii]
85. Pan J, Wang B, Li W, Zhou X, Scherr T, Yang Y et al (2012) Elevated cross-talk between subchondral bone and cartilage in osteoarthritic joints. Bone 51(2):212–217. doi:10.1016/j.bone.2011.11.030. [Research Support, N.I.H., Extramural Research Support, Non-U.S. Gov't]
86. Lajeunesse D (2004) The role of bone in the treatment of osteoarthritis. Osteoarthritis Cartilage 12 Suppl A:S34–S38, doi:S1063458403002504 [pii]
87. Goldring SR, Goldring MB (2010) Bone and cartilage in osteoarthritis: is what's best for one good or bad for the other? Arthritis Res Ther 12(5):143. doi:10.1186/ar3135, ar3135 [pii]
88. Weinans H, Siebelt M, Agricola R, Botter SM, Piscaer TM, Waarsing JH (2012) Pathophysiology of peri-articular bone changes in osteoarthritis. Bone 51(2):190–196. doi:10.1016/j.bone.2012.02.002, S8756-3282(12)00057-9 [pii]
89. Dieppe P, Cushnaghan J, Young P, Kirwan J (1993) Prediction of the progression of joint space narrowing in osteoarthritis of the knee by bone scintigraphy. Ann Rheum Dis 52(8):557–563
90. Matsui H, Shimizu M, Tsuji H (1997) Cartilage and subchondral bone interaction in osteoarthrosis of human knee joint: a histological and histomorphometric study. Microsc Res Tech

37(4):333–342. doi:10.1002/(SICI)1097-0029(19970515)37:4<333::AID-JEMT8>3.0.CO;2-L, [pii]10.1002/(SICI)1097-0029(19970515)37:4<333::AID-JEMT8>3.0.CO;2-L
91. Carlson CS, Loeser RF, Purser CB, Gardin JF, Jerome CP (1996) Osteoarthritis in cynomolgus macaques. III: effects of age, gender, and subchondral bone thickness on the severity of disease. J Bone Miner Res 11(9):1209–1217
92. Nevitt MC, Zhang Y, Javaid MK, Neogi T, Curtis JR, Niu J et al (2010) High systemic bone mineral density increases the risk of incident knee OA and joint space narrowing, but not radiographic progression of existing knee OA: the MOST study. Ann Rheum Dis 69(1):163–168. doi:10.1136/ard.2008.099531, ard.2008.099531 [pii]
93. Bergink AP, Uitterlinden AG, Van Leeuwen JP, Hofman A, Verhaar JA, Pols HA (2005) Bone mineral density and vertebral fracture history are associated with incident and progressive radiographic knee osteoarthritis in elderly men and women: the Rotterdam Study. Bone 37(4):446–456. doi:10.1016/j.bone.2005.05.001., S8756-3282(05)00220-6 [pii]
94. Hochberg MC, Lethbridge-Cejku M, Tobin JD (2004) Bone mineral density and osteoarthritis: data from the Baltimore Longitudinal Study of Aging. Osteoarthritis Cartilage 12 Suppl A:S45–S48, doi:S1063458403002462 [pii]
95. Zhang Y, Hannan MT, Chaisson CE, McAlindon TE, Evans SR, Aliabadi P et al (2000) Bone mineral density and risk of incident and progressive radiographic knee osteoarthritis in women: the Framingham Study. J Rheumatol 27(4):1032–1037
96. Kadri A, Ea HK, Bazille C, Hannouche D, Liote F, Cohen-Solal ME (2008) Osteoprotegerin inhibits cartilage degradation through an effect on trabecular bone in murine experimental osteoarthritis. Arthritis Rheum 58(8):2379–2386. doi:10.1002/art.23638
97. Karsdal MA, Bay-Jensen AC, Lories RJ, Abramson S, Spector T, Pastoureau P et al (2014) The coupling of bone and cartilage turnover in osteoarthritis: opportunities for bone antiresorptives and anabolics as potential treatments? Ann Rheum Dis 73(2):336–348. doi:10.1136/annrheumdis-2013-204111, annrheumdis-2013-204111 [pii]
98. Reginster JY (2014) Efficacy and safety of strontium ranelate in the treatment of knee osteoarthritis: results of a double-blind randomised, placebo-controlled trial. Ann Rheum Dis 73(2):e8. doi:10.1136/annrheumdis-2013-204194, annrheumdis-2013-204194 [pii]
99. Sakaguchi Y, Sekiya I, Yagishita K, Muneta T (2005) Comparison of human stem cells derived from various mesenchymal tissues: superiority of synovium as a cell source. Arthritis Rheum 52(8):2521–2529. doi:10.1002/art.21212
100. Wilkinson LS, Edwards JC (1991) Demonstration of lymphatics in human synovial tissue. Rheumatol Int 11(4–5):151–155
101. Walsh DA, Verghese P, Cook GJ, McWilliams DF, Mapp PI, Ashraf S et al (2012) Lymphatic vessels in osteoarthritic human knees. Osteoarthritis Cartilage 20(5):405–412. doi:10.1016/j.joca.2012.01.012, S1063-4584(12)00036-2 [pii]
102. Hui AY, McCarty WJ, Masuda K, Firestein GS, Sah RL (2012) A systems biology approach to synovial joint lubrication in health, injury, and disease. Wiley Interdiscip Rev Syst Biol Med 4(1):15–37. doi:10.1002/wsbm.157
103. Mitrovic DR, Stankovic A, Quintero M, Ryckewaert A (1985) Amyloid deposits in human knee and hip joints. Rheumatol Int 5(2):83–89
104. Satomura K, Torigoshi T, Koga T, Maeda Y, Izumi Y, Jiuchi Y et al (2013) Serum amyloid A (SAA) induces pentraxin 3 (PTX3) production in rheumatoid synoviocytes. Mod Rheumatol 23(1):28–35. doi:10.1007/s10165-012-0630-0
105. Scanzello CR, Goldring SR (2012) The role of synovitis in osteoarthritis pathogenesis. Bone 51(2):249–257. doi:10.1016/j.bone.2012.02.012
106. Lindblad S, Hedfors E (1987) Arthroscopic and immunohistologic characterization of knee joint synovitis in osteoarthritis. Arthritis Rheum 30(10):1081–1088
107. Revell PA, Mayston V, Lalor P, Mapp P (1988) The synovial membrane in osteoarthritis: a histological study including the characterisation of the cellular infiltrate present in inflammatory osteoarthritis using monoclonal antibodies. Ann Rheum Dis 47(4):300–307
108. Krenn V, Morawietz L, Burmester GR, Kinne RW, Mueller-Ladner U, Muller B et al (2006) Synovitis score: discrimination between chronic low-grade and high-grade synovitis. Histopathology 49(4):358–364. doi:10.1111/j.1365-2559.2006.02508.x, HIS2508 [pii]

109. Ayral X, Pickering EH, Woodworth TG, Mackillop N, Dougados M (2005) Synovitis: a potential predictive factor of structural progression of medial tibiofemoral knee osteoarthritis – results of a 1 year longitudinal arthroscopic study in 422 patients. Osteoarthritis Cartilage 13(5):361–367
110. Hill CL, Hunter DJ, Niu J, Clancy M, Guermazi A, Genant H et al (2007) Synovitis detected on magnetic resonance imaging and its relation to pain and cartilage loss in knee osteoarthritis. Ann Rheum Dis 66(12):1599–1603. doi:10.1136/ard.2006.067470, ard.2006.067470 [pii]
111. Baker K, Grainger A, Niu J, Clancy M, Guermazi A, Crema M et al (2010) Relation of synovitis to knee pain using contrast-enhanced MRIs. Ann Rheum Dis 69(10):1779–1783. doi:10.1136/ard.2009.121426, ard.2009.121426 [pii]
112. Sun Y, Mauerhan DR, Franklin AM, Zinchenko N, Norton HJ, Hanley EN Jr et al (2014) Fibroblast-like synoviocytes induce calcium mineral formation and deposition. Arthritis 2014:812678. doi:10.1155/2014/812678
113. McCarthy GM, Macius AM, Christopherson PA, Ryan LM, Pourmotabbed T (1998) Basic calcium phosphate crystals induce synthesis and secretion of 92 kDa gelatinase (gelatinase B/matrix metalloprotease 9) in human fibroblasts. Ann Rheum Dis 57(1):56–60
114. Englund M, Guermazi A, Lohmander LS (2009) The meniscus in knee osteoarthritis. Rheum Dis Clin North Am 35(3):579–590. doi:10.1016/j.rdc.2009.08.004, S0889-857X(09)00062-3 [pii]
115. Englund M (2009) Meniscal tear – a common finding with often troublesome consequences. J Rheumatol 36(7):1362–1364. doi:10.3899/jrheum.090335, 36/7/1362 [pii]
116. Englund M, Guermazi A, Roemer FW, Aliabadi P, Yang M, Lewis CE et al (2009) Meniscal tear in knees without surgery and the development of radiographic osteoarthritis among middle-aged and elderly persons: the Multicenter Osteoarthritis Study. Arthritis Rheum 60(3):831–839
117. Englund M (2009) The role of the meniscus in osteoarthritis genesis. Med Clin North Am 93(1):37–43. doi:10.1016/j.mcna.2008.08.005, S0025-7125(08)00126-0 [pii]
118. Ding C, Martel-Pelletier J, Pelletier JP, Abram F, Raynauld JP, Cicuttini F et al (2007) Meniscal tear as an osteoarthritis risk factor in a largely non-osteoarthritic cohort: a cross-sectional study. J Rheumatol 34(4):776–784, doi:07/13/0313 [pii]
119. Rodeo SA, Seneviratne A, Suzuki K, Felker K, Wickiewicz TL, Warren RF (2000) Histological analysis of human meniscal allografts. A preliminary report. J Bone Joint Surg Am 82-A(8):1071–1082
120. Krenn V, Kurz B, Krukemeyer MG, Knoess P, Jakobs M, Poremba C et al (2010) Histopathological degeneration score of fibrous cartilage. Low- and high-grade meniscal degeneration. Z Rheumatol 69(7):644–652. doi:10.1007/s00393-010-0609-1
121. Mesiha M, Zurakowski D, Soriano J, Nielson JH, Zarins B, Murray MM (2007) Pathologic characteristics of the torn human meniscus. Am J Sports Med 35(1):103–112. doi:10.1177/0363546506293700, 0363546506293700 [pii]
122. Meister K, Indelicato PA, Spanier S, Franklin J, Batts J (2004) Histology of the torn meniscus: a comparison of histologic differences in meniscal tissue between tears in anterior cruciate ligament-intact and anterior cruciate ligament-deficient knees. Am J Sports Med 32(6):1479–1483. doi:10.1177/036354650326218232/6/1479 [pii]
123. Fisseler-Eckhoff A, Muller KM (2009) Histopathological meniscus diagnostic. Orthopade 38(6):539–545. doi:10.1007/s00132-008-1401-7
124. Ghadially FN, Lalonde JM, Wedge JH (1983) Ultrastructure of normal and torn menisci of the human knee joint. J Anat 136(Pt 4):773–791
125. Pauli C, Grogan SP, Patil S, Otsuki S, Hasegawa A, Koziol J et al (2011) Macroscopic and histopathologic analysis of human knee menisci in aging and osteoarthritis. Osteoarthritis Cartilage 19(9):1132–1141. doi:10.1016/j.joca.2011.05.008
126. Adams ME, Billingham ME, Muir H (1983) The glycosaminoglycans in menisci in experimental and natural osteoarthritis. Arthritis Rheum 26(1):69–76
127. Herwig J, Egner E, Buddecke E (1984) Chemical changes of human knee joint menisci in various stages of degeneration. Ann Rheum Dis 43(4):635–640

128. Katsuragawa Y, Saitoh K, Tanaka N, Wake M, Ikeda Y, Furukawa H et al (2010) Changes of human menisci in osteoarthritic knee joints. Osteoarthritis Cartilage 18(9):1133–1143. doi:10.1016/j.joca.2010.05.017. [Research Support, Non-U.S. Gov't]
129. Nakano T, Dodd CM, Scott PG (1997) Glycosaminoglycans and proteoglycans from different zones of the porcine knee meniscus. J Orthop Res 15(2):213–220. doi:10.1002/jor.1100150209
130. Cheung HS (1987) Distribution of type I, II, III and V in the pepsin solubilized collagens in bovine menisci. Connect Tissue Res 16(4):343–356
131. Mullaji AB, Marawar SV, Simha M, Jindal G (2008) Cruciate ligaments in arthritic knees: a histologic study with radiologic correlation. J Arthroplasty 23(4):567–572. doi:10.1016/j.arth.2007.05.024
132. Levy YD, Hasegawa A, Patil S, Koziol JA, Lotz MK, D'Lima DD (2013) Histopathological changes in the human posterior cruciate ligament during aging and osteoarthritis: correlations with anterior cruciate ligament and cartilage changes. Ann Rheum Dis 72(2):271–277. doi:10.1136/annrheumdis-2012-201730
133. Blevins FT, Hecker AT, Bigler GT, Boland AL, Hayes WC (1994) The effects of donor age and strain rate on the biomechanical properties of bone-patellar tendon-bone allografts. Am J Sports Med 22(3):328–333
134. Flahiff CM, Brooks AT, Hollis JM, Vander Schilden JL, Nicholas RW (1995) Biomechanical analysis of patellar tendon allografts as a function of donor age. Am J Sports Med 23(3):354–358
135. Hasegawa A, Otsuki S, Pauli C, Miyaki S, Patil S, Steklov N et al (2012) Anterior cruciate ligament changes in the human knee joint in aging and osteoarthritis. Arthritis Rheum 64(3):696–704. doi:10.1002/art.33417. [Research Support, N.I.H., Extramural]
136. Kennedy BK, Berger SL, Brunet A, Campisi J, Cuervo AM, Epel ES et al (2014) Geroscience: linking aging to chronic disease. Cell 159(4):709–713. doi:10.1016/j.cell.2014.10.039
137. Stannus OP, Jones G, Blizzard L, Cicuttini FM, Ding C (2013) Associations between serum levels of inflammatory markers and change in knee pain over 5 years in older adults: a prospective cohort study. Ann Rheum Dis 72(4):535–540. doi:10.1136/annrheumdis-2011-201047, annrheumdis-2011-201047 [pii]
138. Lotz M, Martel-Pelletier J, Christiansen C, Brandi ML, Bruyere O, Chapurlat R et al (2013) Value of biomarkers in osteoarthritis: current status and perspectives. Ann Rheum Dis 72(11):1756–1763. doi:10.1136/annrheumdis-2013-203726, annrheumdis-2013-203726 [pii]
139. Spector TD, Hart DJ, Nandra D, Doyle DV, Mackillop N, Gallimore JR et al (1997) Low-level increases in serum C-reactive protein are present in early osteoarthritis of the knee and predict progressive disease. Arthritis Rheum 40(4):723–727
140. Livshits G, Zhai G, Hart DJ, Kato BS, Wang H, Williams FM et al (2009) Interleukin-6 is a significant predictor of radiographic knee osteoarthritis: the Chingford study. Arthritis Rheum 60(7):2037–2045
141. Goekoop RJ, Kloppenburg M, Kroon HM, Frolich M, Huizinga TW, Westendorp RG et al (2010) Low innate production of interleukin-1beta and interleukin-6 is associated with the absence of osteoarthritis in old age. Osteoarthritis Cartilage 18(7):942–947. doi:10.1016/j.joca.2010.03.016, S1063-4584(10)00116-0 [pii]
142. Penninx BW, Abbas H, Ambrosius W, Nicklas BJ, Davis C, Messier SP et al (2004) Inflammatory markers and physical function among older adults with knee osteoarthritis. J Rheumatol 31(10):2027–2031
143. Malemud CJ (2010) Anticytokine therapy for osteoarthritis: evidence to date. Drugs Aging 27(2):95–115. doi:10.2165/11319950-000000000-00000
144. Kapoor M, Martel-Pelletier J, Lajeunesse D, Pelletier JP, Fahmi H (2011) Role of proinflammatory cytokines in the pathophysiology of osteoarthritis. Nat Rev Rheumatol 7(1):33–42. doi:10.1038/nrrheum.2010.196
145. Lee AS, Ellman MB, Yan D, Kroin JS, Cole BJ, van Wijnen AJ et al (2013) A current review of molecular mechanisms regarding osteoarthritis and pain. Gene 527(2):440–447. doi:10.1016/j.gene.2013.05.069

146. Miller RE, Miller RJ, Malfait AM (2014) Osteoarthritis joint pain: the cytokine connection. Cytokine 70(2):185–193. doi:10.1016/j.cyto.2014.06.019, S1043-4666(14)00206-3 [pii]
147. Chen YJ, Sheu ML, Tsai KS, Yang RS, Liu SH (2013) Advanced glycation end products induce peroxisome proliferator-activated receptor gamma down-regulation-related inflammatory signals in human chondrocytes via Toll-like receptor-4 and receptor for advanced glycation end products. PLoS One 8(6):e66611. doi:10.1371/journal.pone.0066611PONE-D-13-01610 [pii]
148. Akasaki Y, Reixach N, Matsuzaki T, Alvarez-Garcia O, Olmer M, Iwamoto Y et al (2015) Transthyretin deposition in articular cartilage: a novel mechanism in the pathogenesis of osteoarthritis. Arthritis Rheumatol 67:2097–2107
149. Kolettas E, Muir HI, Barrett JC, Hardingham TE (2001) Chondrocyte phenotype and cell survival are regulated by culture conditions and by specific cytokines through the expression of Sox-9 transcription factor. Rheumatology (Oxford) 40(10):1146–1156
150. Chevalier X, Eymard F, Richette P (2013) Biologic agents in osteoarthritis: hopes and disappointments. Nat Rev Rheumatol 9(7):400–410. doi:10.1038/nrrheum.2013.44, nrrheum.2013.44 [pii]
151. Anz AW, Hackel JG, Nilssen EC, Andrews JR (2014) Application of biologics in the treatment of the rotator cuff, meniscus, cartilage, and osteoarthritis. J Am Acad Orthop Surg 22(2):68–79. doi:10.5435/JAAOS-22-02-68, 22/2/68 [pii]
152. Wang W, Rigueur D, Lyons KM (2014) TGFbeta signaling in cartilage development and maintenance. Birth Defects Res C Embryo Today 102(1):37–51. doi:10.1002/bdrc.21058
153. Loeser RF (2013) Aging processes and the development of osteoarthritis. Curr Opin Rheumatol 25(1):108–113. doi:10.1097/BOR.0b013e32835a9428. [Research Support, N.I.H., Extramural Research Support, Non-U.S. Gov't Review]
154. Blaney Davidson EN, van der Kraan PM, van den Berg WB (2007) TGF-beta and osteoarthritis. Osteoarthritis Cartilage 15(6):597–604. doi:10.1016/j.joca.2007.02.005, S1063-4584(07)00065-9 [pii]
155. Plaas A, Velasco J, Gorski DJ, Li J, Cole A, Christopherson K et al (2011) The relationship between fibrogenic TGFbeta1 signaling in the joint and cartilage degradation in post-injury osteoarthritis. Osteoarthritis Cartilage 19(9):1081–1090. doi:10.1016/j.joca.2011.05.003
156. Tang Y, Wu X, Lei W, Pang L, Wan C, Shi Z et al (2009) TGF-beta1-induced migration of bone mesenchymal stem cells couples bone resorption with formation. Nat Med 15(7):757–765. doi:10.1038/nm.1979, nm.1979 [pii]
157. Zhen G, Cao X (2014) Targeting TGFbeta signaling in subchondral bone and articular cartilage homeostasis. Trends Pharmacol Sci 35(5):227–236. doi:10.1016/j.tips.2014.03.005, S0165-6147(14)00039-X [pii]
158. Zhen G, Wen C, Jia X, Li Y, Crane JL, Mears SC et al (2013) Inhibition of TGF-beta signaling in mesenchymal stem cells of subchondral bone attenuates osteoarthritis. Nat Med 19(6):704–712. doi:10.1038/nm.3143
159. Janssen-Heininger YM, Mossman BT, Heintz NH, Forman HJ, Kalyanaraman B, Finkel T et al (2008) Redox-based regulation of signal transduction: principles, pitfalls, and promises. Free Radic Biol Med 45(1):1–17, [Research Support, Non-U.S. Gov't Review]. ACS Chem Biol 5(1):47–62. doi:10.1021/cb900258z
160. Paulsen CE, Carroll KS (2010) Orchestrating redox signaling networks through regulatory cysteine switches 5(1):47–62
161. Jones DP (2006) Redefining oxidative stress. Antioxid Redox Signal 8(9–10):1865–1879. doi:10.1089/ars.2006.8.1865
162. Studer R, Jaffurs D, Stefanovic-Racic M, Robbins PD, Evans CH (1999) Nitric oxide in osteoarthritis. Osteoarthritis Cartilage 7(4):377–379
163. Hiran TS, Moulton PJ, Hancock JT (1997) Detection of superoxide and NADPH oxidase in porcine articular chondrocytes. Free Radic Biol Med 23(5):736–743
164. Tiku ML, Shah R, Allison GT (2000) Evidence linking chondrocyte lipid peroxidation to cartilage matrix protein degradation: possible role in cartilage aging and the pathogenesis of osteoarthritis. J Biol Chem 275:20069–20076

165. Morita K, Miyamoto T, Fujita N, Kubota Y, Ito K, Takubo K et al (2007) Reactive oxygen species induce chondrocyte hypertrophy in endochondral ossification. J Exp Med 204(7):1613–1623
166. Fosang AJ, Beier F (2011) Emerging frontiers in cartilage and chondrocyte biology. Best Pract Res Clin Rheumatol 25(6):751–766. doi:10.1016/j.berh.2011.11.010
167. Henrotin YE, Bruckner P, Pujol JP (2003) The role of reactive oxygen species in homeostasis and degradation of cartilage. Osteoarthritis Cartilage 11(10):747–755
168. Loeser RF, Carlson CS, Carlo MD, Cole A (2002) Detection of nitrotyrosine in aging and osteoarthritic cartilage: correlation of oxidative damage with the presence of interleukin-1beta and with chondrocyte resistance to insulin-like growth factor 1. Arthritis Rheum 46(9):2349–2357
169. Hui W, Young DA, Rowan AD, Xu X, Cawston TE, Proctor CJ (2014) Oxidative changes and signalling pathways are pivotal in initiating age-related changes in articular cartilage. Ann Rheum Dis. doi:10.1136/annrheumdis-2014-206295
170. Jallali N, Ridha H, Thrasivoulou C, Underwood C, Butler PE, Cowen T (2005) Vulnerability to ROS-induced cell death in ageing articular cartilage: the role of antioxidant enzyme activity. Osteoarthritis Cartilage 13(7):614–622
171. Kurz B, Lemke A, Kehn M, Domm C, Patwari P, Frank EH et al (2004) Influence of tissue maturation and antioxidants on the apoptotic response of articular cartilage after injurious compression. Arthritis Rheum 50(1):123–130
172. Martin JA, McCabe D, Walter M, Buckwalter JA, McKinley TO (2009) N-acetylcysteine inhibits post-impact chondrocyte death in osteochondral explants. J Bone Joint Surg Am 91(8):1890–1897. doi:10.2106/JBJS.H.00545, 91/8/1890 [pii]
173. Goodwin W, McCabe D, Sauter E, Reese E, Walter M, Buckwalter JA et al (2010) Rotenone prevents impact-induced chondrocyte death. J Orthop Res 28(8):1057–1063. doi:10.1002/jor.21091
174. Nelson KK, Melendez JA (2004) Mitochondrial redox control of matrix metalloproteinases. Free Radic Biol Med 37(6):768–784
175. Lo YY, Cruz TF (1995) Involvement of reactive oxygen species in cytokine and growth factor induction of c-fos expression in chondrocytes. J Biol Chem 270(20):11727–11730
176. Lo YY, Wong JM, Cruz TF (1996) Reactive oxygen species mediate cytokine activation of c-Jun NH2- terminal kinases. J Biol Chem 271(26):15703–15707
177. Mendes AF, Caramona MM, Carvalho AP, Lopes MC (2003) Differential roles of hydrogen peroxide and superoxide in mediating IL-1-induced NF-kappaB activation and iNOS expression in bovine articular chondrocytes. J Cell Biochem 88(4):783–793
178. Ahmad R, Sylvester J, Ahmad M, Zafarullah M (2011) Involvement of H-Ras and reactive oxygen species in proinflammatory cytokine-induced matrix metalloproteinase-13 expression in human articular chondrocytes. Arch Biochem Biophys 507(2):350–355. doi:10.1016/j.abb.2010.12.032. [Research Support, Non-U.S. Gov't]
179. Del Carlo M, Schwartz D, Erickson EA, Loeser RF (2007) Endogenous production of reactive oxygen species is required for stimulation of human articular chondrocyte matrix metalloproteinase production by fibronectin fragments. Free Radic Biol Med 42(9):1350–1358
180. Jr Del Carlo M, Loeser RF (2003) Increased oxidative stress with aging reduces chondrocyte survival: correlation with intracellular glutathione levels. Arthritis Rheum 48(12):3419–3430
181. Ruiz-Romero C, Calamia V, Mateos J, Carreira V, Martinez-Gomariz M, Fernandez M et al (2008) Mitochondrial dysregulation of osteoarthritic human articular chondrocytes analyzed by proteomics: a decrease in mitochondrial superoxide dismutase points to a redox imbalance. Mol Cell Proteomics 8(1):179–189
182. Aigner T, Fundel K, Saas J, Gebhard PM, Haag J, Weiss T et al (2006) Large-scale gene expression profiling reveals major pathogenetic pathways of cartilage degeneration in osteoarthritis. Arthritis Rheum 54(11):3533–3544
183. Scott JL, Gabrielides C, Davidson RK, Swingler TE, Clark IM, Wallis GA et al (2010) Superoxide dismutase downregulation in osteoarthritis progression and end-stage disease. Ann Rheum Dis 69(8):1502–1510. doi:10.1136/ard.2009.119966, ard.2009.119966 [pii]

184. Gavriilidis C, Miwa S, von Zglinicki T, Taylor RW, Young DA (2013) Mitochondrial dysfunction in osteoarthritis is associated with down-regulation of superoxide dismutase 2. Arthritis Rheum 65(2):378–387. doi:10.1002/art.37782
185. Davies CM, Guilak F, Weinberg JB, Fermor B (2008) Reactive nitrogen and oxygen species in interleukin-1-mediated DNA damage associated with osteoarthritis. Osteoarthritis Cartilage 16(5):624–630
186. Grishko VI, Ho R, Wilson GL, Pearsall AW 4th (2009) Diminished mitochondrial DNA integrity and repair capacity in OA chondrocytes. Osteoarthritis Cartilage 17(1):107–113. doi:10.1016/j.joca.2008.05.009, S1063-4584(08)00167-2 [pii]
187. Loeser RF, Gandhi U, Long DL, Yin W, Chubinskaya S (2014) Aging and oxidative stress reduce the response of human articular chondrocytes to insulin-like growth factor 1 and osteogenic protein 1. Arthritis Rheumatol 66(8):2201–2209. doi:10.1002/art.38641
188. Yudoh K, Nguyen T, Nakamura H, Hongo-Masuko K, Kato T, Nishioka K (2005) Potential involvement of oxidative stress in cartilage senescence and development of osteoarthritis: oxidative stress induces chondrocyte telomere instability and downregulation of chondrocyte function. Arthritis Res Ther 7(2):R380–R391
189. Brandl A, Hartmann A, Bechmann V, Graf B, Nerlich M, Angele P (2011) Oxidative stress induces senescence in chondrocytes. J Orthop Res 29:1114–1120. doi:10.1002/jor.21348
190. Zushi S, Akagi M, Kishimoto H, Teramura T, Sawamura T, Hamanishi C (2009) Induction of bovine articular chondrocyte senescence with oxidized low-density lipoprotein through lectin-like oxidized low-density lipoprotein receptor 1. Arthritis Rheum 60(10):3007–3016. doi:10.1002/art.24816
191. Loeser RF, Shanker G, Carlson CS, Gardin JF, Shelton BJ, Sonntag WE (2000) Reduction in the chondrocyte response to insulin-like growth factor 1 in aging and osteoarthritis: studies in a non-human primate model of naturally occurring disease. Arthritis Rheum 43(9):2110–2120
192. Loeser RF (2009) Aging and osteoarthritis: the role of chondrocyte senescence and aging changes in the cartilage matrix. Osteoarthritis Cartilage 17(8):971–979. doi:10.1016/j.joca.2009.03.002. [Research Support, N.I.H., Extramural Research Support, Non-U.S. Gov't Review]
193. Yin W, Park JI, Loeser RF (2009) Oxidative stress inhibits insulin-like growth factor-I induction of chondrocyte proteoglycan synthesis through differential regulation of phosphatidylinositol 3-Kinase-Akt and MEK-ERK MAPK signaling pathways. J Biol Chem 284(46):31972–31981. doi:10.1074/jbc.M109.056838
194. McAlindon TE, Jacques P, Zhang Y, Hannan MT, Aliabadi P, Weissman B et al (1996) Do antioxidant micronutrients protect against the development and progression of knee osteoarthritis? Arthritis Rheum 39(4):648–656
195. Kurz B, Jost B, Schunke M (2002) Dietary vitamins and selenium diminish the development of mechanically induced osteoarthritis and increase the expression of antioxidative enzymes in the knee joint of STR/1N mice. Osteoarthritis Cartilage 10(2):119–126
196. Nakagawa S, Arai Y, Mazda O, Kishida T, Takahashi KA, Sakao K et al (2010) N-acetylcysteine prevents nitric oxide-induced chondrocyte apoptosis and cartilage degeneration in an experimental model of osteoarthritis. J Orthop Res 28(2):156–163. doi:10.1002/jor.20976
197. Yudoh K, Shishido K, Murayama H, Yano M, Matsubayashi K, Takada H et al (2007) Water-soluble C60 fullerene prevents degeneration of articular cartilage in osteoarthritis via down-regulation of chondrocyte catabolic activity and inhibition of cartilage degeneration during disease development. Arthritis Rheum 56(10):3307–3318
198. Dudek KA, Lafont JE, Martinez-Sanchez A, Murphy CL (2010) Type II collagen expression is regulated by tissue-specific miR-675 in human articular chondrocytes. J Biol Chem 285(32):24381–24387. doi:10.1074/jbc.M110.111328, M110.111328 [pii]
199. Matsukawa T, Sakai T, Yonezawa T, Hiraiwa H, Hamada T, Nakashima M et al (2013) MicroRNA-125b regulates the expression of aggrecanase-1 (ADAMTS-4) in human osteoarthritic chondrocytes. Arthritis Res Ther 15(1):R28. doi:10.1186/ar4164, ar4164 [pii]
200. Miyaki S, Asahara H (2012) Macro view of microRNA function in osteoarthritis. Nat Rev Rheumatol 8(9):543–552. doi:10.1038/nrrheum.2012.128, nrrheum.2012.128 [pii]

201. Miyaki S, Nakasa T, Otsuki S, Grogan SP, Higashiyama R, Inoue A et al (2009) MicroRNA-140 is expressed in differentiated human articular chondrocytes and modulates interleukin-1 responses. Arthritis Rheum 60(9):2723–2730. doi:10.1002/art.24745
202. Miyaki S, Sato T, Inoue A, Otsuki S, Ito Y, Yokoyama S et al (2010) MicroRNA-140 plays dual roles in both cartilage development and homeostasis. Genes Dev 24(11):1173–1185. doi:10.1101/gad.1915510. [Research Support, N.I.H., Extramural Research Support, Non-U.S. Gov't]
203. Li Z, Rana TM (2014) Therapeutic targeting of microRNAs: current status and future challenges. Nat Rev Drug Discov 13(8):622–638. doi:10.1038/nrd4359, nrd4359 [pii]
204. Bhutani N, Burns DM, Blau HM (2011) DNA demethylation dynamics. Cell 146(6):866–872. doi:10.1016/j.cell.2011.08.042, S0092-8674(11)01012-9 [pii]
205. Kim KI, Park YS, Im GI (2013) Changes in the epigenetic status of the SOX-9 promoter in human osteoarthritic cartilage. J Bone Miner Res 28(5):1050–1060. doi:10.1002/jbmr.1843
206. Roach HI, Aigner T (2007) DNA methylation in osteoarthritic chondrocytes: a new molecular target. Osteoarthritis Cartilage 15(2):128–137
207. Roach HI, Yamada N, Cheung KS, Tilley S, Clarke NM, Oreffo RO et al (2005) Association between the abnormal expression of matrix-degrading enzymes by human osteoarthritic chondrocytes and demethylation of specific CpG sites in the promoter regions. Arthritis Rheum 52(10):3110–3124
208. Hashimoto K, Otero M, Imagawa K, de Andres MC, Coico JM, Roach HI et al (2013) Regulated transcription of human matrix metalloproteinase 13 (MMP13) and interleukin-1beta (IL1B) genes in chondrocytes depends on methylation of specific proximal promoter CpG sites. J Biol Chem 288(14):10061–10072. doi:10.1074/jbc.M112.421156
209. de Andres MC, Imagawa K, Hashimoto K, Gonzalez A, Roach HI, Goldring MB et al (2013) Loss of methylation in CpG sites in the NF-kappaB enhancer elements of inducible nitric oxide synthase is responsible for gene induction in human articular chondrocytes. Arthritis Rheum 65(3):732–742. doi:10.1002/art.37806
210. Hashimoto K, Oreffo RO, Gibson MB, Goldring MB, Roach HI (2009) DNA demethylation at specific CpG sites in the IL1B promoter in response to inflammatory cytokines in human articular chondrocytes. Arthritis Rheum 60(11):3303–3313. doi:10.1002/art.24882. [Research Support, N.I.H., Extramural Research Support, Non-U.S. Gov't]
211. Iliopoulos D, Malizos KN, Tsezou A (2007) Epigenetic regulation of leptin affects MMP-13 expression in osteoarthritic chondrocytes: possible molecular target for osteoarthritis therapeutic intervention. Ann Rheum Dis 66(12):1616–1621. doi:10.1136/ard.2007.069377, ard.2007.069377 [pii]
212. Reynard LN, Bui C, Canty-Laird EG, Young DA, Loughlin J (2011) Expression of the osteoarthritis-associated gene GDF5 is modulated epigenetically by DNA methylation. Hum Mol Genet 20(17):3450–3460. doi:10.1093/hmg/ddr253, ddr253 [pii]
213. Delgado-Calle J, Fernandez AF, Sainz J, Zarrabeitia MT, Sanudo C, Garcia-Renedo R et al (2013) Genome-wide profiling of bone reveals differentially methylated regions in osteoporosis and osteoarthritis. Arthritis Rheum 65(1):197–205. doi:10.1002/art.37753
214. Jeffries MA, Donica M, Baker LW, Stevenson ME, Annan AC, Humphrey MB et al (2014) Genome-wide DNA methylation study identifies significant epigenomic changes in osteoarthritic cartilage. Arthritis Rheumatol 66(10):2804–2815. doi:10.1002/art.38762
215. den Hollander W, Ramos YF, Bos SD, Bomer N, van der Breggen R, Lakenberg N et al (2014) Knee and hip articular cartilage have distinct epigenomic landscapes: implications for future cartilage regeneration approaches. Ann Rheum Dis 73(12):2208–2212. doi:10.1136/annrheumdis-2014-205980
216. Bannister AJ, Kouzarides T (2011) Regulation of chromatin by histone modifications. Cell Res 21(3):381–395. doi:10.1038/cr.2011.22, cr201122 [pii]
217. Zentner GE, Henikoff S (2013) Regulation of nucleosome dynamics by histone modifications. Nat Struct Mol Biol 20(3):259–266. doi:10.1038/nsmb.2470, nsmb.2470 [pii]
218. Black JC, Van Rechem C, Whetstine JR (2012) Histone lysine methylation dynamics: establishment, regulation, and biological impact. Mol Cell 48(4):491–507. doi:10.1016/j.molcel.2012.11.006, S1097-2765(12)00937-9 [pii]

219. Fischle W (2012) One, two, three: how histone methylation is read. Epigenomics 4(6):641–653. doi:10.2217/epi.12.56
220. Verrier L, Vandromme M, Trouche D (2011) Histone demethylases in chromatin cross-talks. Biol Cell 103(8):381–401. doi:10.1042/BC20110028, BC20110028 [pii]
221. El Mansouri FE, Nebbaki SS, Kapoor M, Afif H, Martel-Pelletier J, Pelletier JP et al (2014) Lysine-specific demethylase 1-mediated demethylation of histone H3 lysine 9 contributes to interleukin 1beta-induced microsomal prostaglandin E synthase 1 expression in human osteoarthritic chondrocytes. Arthritis Res Ther 16(3):R113. doi:10.1186/ar4564, ar4564 [pii]
222. de Ruijter AJ, van Gennip AH, Caron HN, Kemp S, van Kuilenburg AB (2003) Histone deacetylases (HDACs): characterization of the classical HDAC family. Biochem J 370(Pt 3):737–749. doi:10.1042/BJ20021321BJ20021321 [pii]
223. Hong S, Derfoul A, Pereira-Mouries L, Hall DJ (2009) A novel domain in histone deacetylase 1 and 2 mediates repression of cartilage-specific genes in human chondrocytes. FASEB J 23(10):3539–3552. doi:10.1096/fj.09-133215, fj.09-133215 [pii]
224. Higashiyama R, Miyaki S, Yamashita S, Yoshitaka T, Lindman G, Ito Y et al (2010) Correlation between MMP-13 and HDAC7 expression in human knee osteoarthritis. Mod Rheumatol 20(1):11–17. doi:10.1007/s10165-009-0224-7
225. Cao K, Wei L, Zhang Z, Guo L, Zhang C, Li Y et al (2014) Decreased histone deacetylase 4 is associated with human osteoarthritis cartilage degeneration by releasing histone deacetylase 4 inhibition of runt-related transcription factor-2 and increasing osteoarthritis-related genes: a novel mechanism of human osteoarthritis cartilage degeneration. Arthritis Res Ther 16(6):491. doi:10.1186/s13075-014-0491-3, s13075-014-0491-3 [pii]
226. Lu J, Sun Y, Ge Q, Teng H, Jiang Q (2014) Histone deacetylase 4 alters cartilage homeostasis in human osteoarthritis. BMC Musculoskelet Disord 15:438. doi:10.1186/1471-2474-15-438, 1471-2474-15-438 [pii]
227. Young DA, Lakey RL, Pennington CJ, Jones D, Kevorkian L, Edwards DR et al (2005) Histone deacetylase inhibitors modulate metalloproteinase gene expression in chondrocytes and block cartilage resorption. Arthritis Res Ther 7(3):R503–R512. doi:10.1186/ar1702
228. Bradley EW, Carpio LR, Westendorf JJ (2013) Histone deacetylase 3 suppression increases PH domain and leucine-rich repeat phosphatase (Phlpp)1 expression in chondrocytes to suppress Akt signaling and matrix secretion. J Biol Chem 288(14):9572–9582. doi:10.1074/jbc.M112.423723, M112.423723 [pii]
229. Culley KL, Hui W, Barter MJ, Davidson RK, Swingler TE, Destrument AP et al (2013) Class I histone deacetylase inhibition modulates metalloproteinase expression and blocks cytokine-induced cartilage degradation. Arthritis Rheum 65(7):1822–1830. doi:10.1002/art.37965
230. Vega RB, Matsuda K, Oh J, Barbosa AC, Yang X, Meadows E et al (2004) Histone deacetylase 4 controls chondrocyte hypertrophy during skeletogenesis. Cell 119(4):555–566. doi:10.1016/j.cell.2004.10.024, S0092867404010347 [pii]
231. Shimizu E, Selvamurugan N, Westendorf JJ, Olson EN, Partridge NC (2010) HDAC4 represses matrix metalloproteinase-13 transcription in osteoblastic cells, and parathyroid hormone controls this repression. J Biol Chem 285(13):9616–9626. doi:10.1074/jbc.M109.094862, M109.094862 [pii]
232. Furumatsu T, Tsuda M, Yoshida K, Taniguchi N, Ito T, Hashimoto M et al (2005) Sox9 and p300 cooperatively regulate chromatin-mediated transcription. J Biol Chem 280(42):35203–35208. doi:10.1074/jbc.M502409200, M502409200 [pii]
233. Huh YH, Ryu JH, Chun JS (2007) Regulation of type II collagen expression by histone deacetylase in articular chondrocytes. J Biol Chem 282(23):17123–17131. doi:10.1074/jbc.M700599200, M700599200 [pii]
234. Chen WP, Bao JP, Hu PF, Feng J, Wu LD (2010) Alleviation of osteoarthritis by Trichostatin A, a histone deacetylase inhibitor, in experimental osteoarthritis. Mol Biol Rep 37(8):3967–3972. doi:10.1007/s11033-010-0055-9
235. Dvir-Ginzberg M, Steinmeyer J (2013) Towards elucidating the role of SirT1 in osteoarthritis. Front Biosci (Landmark Ed) 18:343–355, doi:4105 [pii]

236. Matsuzaki T, Matsushita T, Takayama K, Matsumoto T, Nishida K, Kuroda R et al (2014) Disruption of Sirt1 in chondrocytes causes accelerated progression of osteoarthritis under mechanical stress and during ageing in mice. Ann Rheum Dis 73(7):1397–1404. doi:10.1136/annrheumdis-2012-202620, annrheumdis-2012-202620 [pii]
237. Matsushita T, Sasaki H, Takayama K, Ishida K, Matsumoto T, Kubo S et al (2013) The overexpression of SIRT1 inhibited osteoarthritic gene expression changes induced by interleukin-1beta in human chondrocytes. J Orthop Res 31(4):531–537. doi:10.1002/jor.22268
238. Gabay O, Zaal KJ, Sanchez C, Dvir-Ginzberg M, Gagarina V, Song Y et al (2013) Sirt1-deficient mice exhibit an altered cartilage phenotype. Joint Bone Spine 80(6):613–620. doi:10.1016/j.jbspin.2013.01.001, S1297-319X(13)00021-3 [pii]
239. Wang J, Gao JS, Chen JW, Li F, Tian J (2012) Effect of resveratrol on cartilage protection and apoptosis inhibition in experimental osteoarthritis of rabbit. Rheumatol Int 32(6):1541–1548. doi:10.1007/s00296-010-1720-y
240. Mizushima N (2009) Physiological functions of autophagy. Curr Top Microbiol Immunol 335:71–84. doi:10.1007/978-3-642-00302-8_3
241. Levine B, Kroemer G (2008) Autophagy in the pathogenesis of disease. Cell 132(1):27–42. doi:10.1016/j.cell.2007.12.018, S0092-8674(07)01685-6 [pii]
242. Carames B, Taniguchi N, Otsuki S, Blanco FJ, Lotz M (2010) Autophagy is a protective mechanism in normal cartilage, and its aging-related loss is linked with cell death and osteoarthritis. Arthritis Rheum 62(3):791–801. doi:10.1002/art.27305
243. Carames B, Kiosses WB, Akasaki Y, Brinson DC, Eap W, Koziol J et al (2013) Glucosamine activates autophagy in vitro and in vivo. Arthritis Rheum 65(7):1843–1852. doi:10.1002/art.37977
244. Carames B, Olmer M, Kiosses W, Lotz M (2015) The relationship of autophagy defects and cartilage damage during joint aging in a mouse model. Arthritis Rheumatol 67:1568–1576
245. Almonte-Becerril M, Navarro-Garcia F, Gonzalez-Robles A, Vega-Lopez MA, Lavalle C, Kouri JB (2010) Cell death of chondrocytes is a combination between apoptosis and autophagy during the pathogenesis of Osteoarthritis within an experimental model. Apoptosis 15(5):631–638. doi:10.1007/s10495-010-0458-z
246. Bohensky J, Terkhorn SP, Freeman TA, Adams CS, Garcia JA, Shapiro IM et al (2009) Regulation of autophagy in human and murine cartilage: hypoxia-inducible factor 2 suppresses chondrocyte autophagy. Arthritis Rheum 60(5):1406–1415. doi:10.1002/art.24444
247. Chang J, Wang W, Zhang H, Hu Y, Wang M, Yin Z (2013) The dual role of autophagy in chondrocyte responses in the pathogenesis of articular cartilage degeneration in osteoarthritis. Int J Mol Med 32(6):1311–1318. doi:10.3892/ijmm.2013.1520
248. Blanco FJ, Rego I, Ruiz-Romero C (2011) The role of mitochondria in osteoarthritis. Nat Rev Rheumatol 7(3):161–169. doi:10.1038/nrrheum.2010.213, nrrheum.2010.213 [pii]
249. Kim J, Xu M, Xo R, Mates A, Wilson GL, Pearsall AW 4th et al (2010) Mitochondrial DNA damage is involved in apoptosis caused by pro-inflammatory cytokines in human OA chondrocytes. Osteoarthritis Cartilage 18(3):424–432. doi:10.1016/j.joca.2009.09.008, S1063-4584(09)00250-7 [pii]
250. Tschopp J (2011) Mitochondria: sovereign of inflammation? Eur J Immunol 41(5):1196–1202. doi:10.1002/eji.201141436
251. Sanchez-Adams J, Athanasiou A (2009) The knee meniscus: a complex tissue of diverse cells. Cell Mol Bioeng 2(3):332–340
252. Sasaki H, Takayama K, Matsushita T, Ishida K, Kubo S, Matsumoto T et al (2012) Autophagy modulates osteoarthritis-related gene expression in human chondrocytes. Arthritis Rheum 64(6):1920–1928. doi:10.1002/art.34323
253. Rubinsztein DC, Marino G, Kroemer G (2011) Autophagy and aging. Cell 146(5):682–695. doi:10.1016/j.cell.2011.07.030, S0092-8674(11)00828-2 [pii]
254. Ravikumar B, Vacher C, Berger Z, Davies JE, Luo S, Oroz LG et al (2004) Inhibition of mTOR induces autophagy and reduces toxicity of polyglutamine expansions in fly and mouse models of Huntington disease. Nat Genet 36(6):585–595. doi:10.1038/ng1362ng1362 [pii]

255. Demidenko ZN, Zubova SG, Bukreeva EI, Pospelov VA, Pospelova TV, Blagosklonny MV (2009) Rapamycin decelerates cellular senescence. Cell Cycle 8(12):1888–1895, doi:8606 [pii]
256. Narita M, Young AR, Arakawa S, Samarajiwa SA, Nakashima T, Yoshida S et al (2011) Spatial coupling of mTOR and autophagy augments secretory phenotypes. Science 332(6032):966–970. doi:10.1126/science.1205407, science.1205407 [pii]
257. Zhang Y, Vasheghani F, Li YH, Blati M, Simeone K, Fahmi H et al (2014) Cartilage-specific deletion of mTOR upregulates autophagy and protects mice from osteoarthritis. Ann Rheum Dis. doi:10.1136/annrheumdis-2013-204599, annrheumdis-2013-204599 [pii]
258. Carames B, Taniguchi N, Seino D, Blanco FJ, D'Lima D, Lotz M (2012) Mechanical injury suppresses autophagy regulators and pharmacologic activation of autophagy results in chondroprotection. Arthritis Rheum 64(4):1182–1192. doi:10.1002/art.33444. [Research Support, N.I.H., Extramural]
259. Carames B, Hasegawa A, Taniguchi N, Miyaki S, Blanco FJ, Lotz M (2012) Autophagy activation by rapamycin reduces severity of experimental osteoarthritis. Ann Rheum Dis 71(4):575–581. doi:10.1136/annrheumdis-2011-200557. [Research Support, N.I.H., Extramural Research Support, Non-U.S. Gov't]
260. Rajpar MH, McDermott B, Kung L, Eardley R, Knowles L, Heeran M et al (2009) Targeted induction of endoplasmic reticulum stress induces cartilage pathology. PLoS Genet 5(10):e1000691. doi:10.1371/journal.pgen.1000691
261. Saito A, Hino S, Murakami T, Kanemoto S, Kondo S, Saitoh M et al (2009) Regulation of endoplasmic reticulum stress response by a BBF2H7-mediated Sec23a pathway is essential for chondrogenesis. Nat Cell Biol 11(10):1197–1204. doi:10.1038/ncb1962
262. Hino K, Saito A, Kido M, Kanemoto S, Asada R, Takai T et al (2014) Master regulator for chondrogenesis, Sox9, regulates transcriptional activation of the endoplasmic reticulum stress transducer BBF2H7/CREB3L2 in chondrocytes. J Biol Chem 289(20):13810–13820. doi:10.1074/jbc.M113.543322
263. Liu-Bryan R, Terkeltaub R (2014) Emerging regulators of the inflammatory process in osteoarthritis. Nat Rev Rheumatol. doi:10.1038/nrrheum.2014.162
264. Terkeltaub R, Yang B, Lotz M, Liu-Bryan R (2011) Chondrocyte AMP-activated protein kinase activity suppresses matrix degradation responses to proinflammatory cytokines interleukin-1beta and tumor necrosis factor alpha. Arthritis Rheum 63(7):1928–1937. doi:10.1002/art.30333. [Research Support, N.I.H., Extramural Research Support, U.S. Gov't, Non-P.H.S.]
265. Zhao X, Petursson F, Viollet B, Lotz M, Terkeltaub R, Liu-Bryan R (2014) PGC-1alpha and FOXO3a mediate chondroprotection by AMP-activated Protein Kinase. Arthritis Rheumatol. doi:10.1002/art.38791
266. Nugent AE, Speicher DM, Gradisar I, McBurney DL, Baraga A, Doane KJ et al (2009) Advanced osteoarthritis in humans is associated with altered collagen VI expression and upregulation of ER-stress markers Grp78 and bag-1. J Histochem Cytochem 57(10):923–931. doi:10.1369/jhc.2009.953893
267. Uehara Y, Hirose J, Yamabe S, Okamoto N, Okada T, Oyadomari S et al (2014) Endoplasmic reticulum stress-induced apoptosis contributes to articular cartilage degeneration via C/EBP homologous protein. Osteoarthritis Cartilage 22(7):1007–1017. doi:10.1016/j.joca.2014.04.025
268. Takada K, Hirose J, Senba K, Yamabe S, Oike Y, Gotoh T et al (2011) Enhanced apoptotic and reduced protective response in chondrocytes following endoplasmic reticulum stress in osteoarthritic cartilage. Int J Exp Pathol 92(4):232–242. doi:10.1111/j.1365-2613.2010.00758.x
269. Husa M, Petursson F, Lotz M, Terkeltaub R, Liu-Bryan R (2013) C/EBP homologous protein drives pro-catabolic responses in chondrocytes. Arthritis Res Ther 15(6):R218. doi:10.1186/ar4415
270. Price J, Zaidi AK, Bohensky J, Srinivas V, Shapiro IM, Ali H (2010) Akt-1 mediates survival of chondrocytes from endoplasmic reticulum-induced stress. J Cell Physiol 222(3):502–508. doi:10.1002/jcp.22001

271. Cravero JD, Carlson CS, Im HJ, Yammani RR, Long D, Loeser RF (2009) Increased expression of the Akt/PKB inhibitor TRB3 in osteoarthritic chondrocytes inhibits insulin-like growth factor 1-mediated cell survival and proteoglycan synthesis. Arthritis Rheum 60(2):492–500
272. Lopez-Otin C, Blasco MA, Partridge L, Serrano M, Kroemer G (2013) The hallmarks of aging. Cell 153(6):1194–1217. doi:10.1016/j.cell.2013.05.039
273. Caplan AI (2005) Review: mesenchymal stem cells: cell-based reconstructive therapy in orthopedics. Tissue Eng 11(7–8):1198–1211. doi:10.1089/ten.2005.11.1198
274. Baker N, Boyette LB, Tuan RS (2015) Characterization of bone marrow-derived mesenchymal stem cells in aging. Bone 70C:37–47. doi:10.1016/j.bone.2014.10.014, S8756-3282(14)00387-1 [pii]
275. Oreffo RO, Bennett A, Carr AJ, Triffitt JT (1998) Patients with primary osteoarthritis show no change with ageing in the number of osteogenic precursors. Scand J Rheumatol 27(6):415–424
276. Justesen J, Stenderup K, Eriksen EF, Kassem M (2002) Maintenance of osteoblastic and adipocytic differentiation potential with age and osteoporosis in human marrow stromal cell cultures. Calcif Tissue Int 71(1):36–44. doi:10.1007/s00223-001-2059-x
277. Stenderup K, Justesen J, Clausen C, Kassem M (2003) Aging is associated with decreased maximal life span and accelerated senescence of bone marrow stromal cells. Bone 33(6):919–926, doi:S8756328203002679 [pii]
278. Scharstuhl A, Schewe B, Benz K, Gaissmaier C, Buhring HJ, Stoop R (2007) Chondrogenic potential of human adult mesenchymal stem cells is independent of age or osteoarthritis etiology. Stem Cells 25(12):3244–3251. doi:10.1634/stemcells.2007-0300, 2007–0300 [pii]
279. Stenderup K, Justesen J, Eriksen EF, Rattan SI, Kassem M (2001) Number and proliferative capacity of osteogenic stem cells are maintained during aging and in patients with osteoporosis. J Bone Miner Res 16(6):1120–1129. doi:10.1359/jbmr.2001.16.6.1120
280. Payne KA, Didiano DM, Chu CR (2010) Donor sex and age influence the chondrogenic potential of human femoral bone marrow stem cells. Osteoarthritis Cartilage 18(5):705–713. doi:10.1016/j.joca.2010.01.011, S1063-4584(10)00041-5 [pii]
281. D'Ippolito G, Schiller PC, Ricordi C, Roos BA, Howard GA (1999) Age-related osteogenic potential of mesenchymal stromal stem cells from human vertebral bone marrow. J Bone Miner Res 14(7):1115–1122. doi:10.1359/jbmr.1999.14.7.1115, jbm375 [pii]
282. Stolzing A, Jones E, McGonagle D, Scutt A (2008) Age-related changes in human bone marrow-derived mesenchymal stem cells: consequences for cell therapies. Mech Ageing Dev 129(3):163–173. doi:10.1016/j.mad.2007.12.002, S0047-6374(07)00179-0 [pii]
283. Mendes SC, Tibbe JM, Veenhof M, Bakker K, Both S, Platenburg PP et al (2002) Bone tissue-engineered implants using human bone marrow stromal cells: effect of culture conditions and donor age. Tissue Eng 8(6):911–920. doi:10.1089/107632702320934010
284. Zhou S, Greenberger JS, Epperly MW, Goff JP, Adler C, Leboff MS et al (2008) Age-related intrinsic changes in human bone-marrow-derived mesenchymal stem cells and their differentiation to osteoblasts. Aging Cell 7(3):335–343. doi:10.1111/j.1474-9726.2008.00377.x, ACE377 [pii]
285. Fickert S, Schroter-Bobsin U, Gross AF, Hempel U, Wojciechowski C, Rentsch C et al (2011) Human mesenchymal stem cell proliferation and osteogenic differentiation during long-term ex vivo cultivation is not age dependent. J Bone Miner Metab 29(2):224–235. doi:10.1007/s00774-010-0215-y
286. Phinney DG, Kopen G, Righter W, Webster S, Tremain N, Prockop DJ (1999) Donor variation in the growth properties and osteogenic potential of human marrow stromal cells. J Cell Biochem 75(3):424–436. doi:10.1002/(SICI)1097-4644(19991201)75:3<424::AID-JCB8>3.0.CO;2[--]8 [pii]
287. Suva D, Garavaglia G, Menetrey J, Chapuis B, Hoffmeyer P, Bernheim L et al (2004) Non-hematopoietic human bone marrow contains long-lasting, pluripotential mesenchymal stem cells. J Cell Physiol 198(1):110–118. doi:10.1002/jcp.10396

288. Dexheimer V, Mueller S, Braatz F, Richter W (2011) Reduced reactivation from dormancy but maintained lineage choice of human mesenchymal stem cells with donor age. PLoS One 6(8):e22980. doi:10.1371/journal.pone.0022980PONE-D-11-04667 [pii]
289. Segawa Y, Muneta T, Makino H, Nimura A, Mochizuki T, Ju YJ et al (2009) Mesenchymal stem cells derived from synovium, meniscus, anterior cruciate ligament, and articular chondrocytes share similar gene expression profiles. J Orthop Res 27(4):435–441. doi:10.1002/jor.20786. [Research Support, Non-U.S. Gov't]
290. Sekiya I, Ojima M, Suzuki S, Yamaga M, Horie M, Koga H et al (2012) Human mesenchymal stem cells in synovial fluid increase in the knee with degenerated cartilage and osteoarthritis. J Orthop Res 30(6):943–949. doi:10.1002/jor.22029
291. Ferretti C, Lucarini G, Andreoni C, Salvolini E, Bianchi N, Vozzi G et al (2014) Human periosteal derived stem cell potential: the impact of age. Stem Cell Rev. doi:10.1007/s12015-014-9559-3
292. Grogan SP, Miyaki S, Asahara H, D'Lima DD, Lotz MK (2009) Mesenchymal progenitor cell markers in human articular cartilage: normal distribution and changes in osteoarthritis. Arthritis Res Ther 11(3):R85. doi:10.1186/ar2719, ar2719 [pii]
293. Otsuki S, Grogan SP, Miyaki S, Kinoshita M, Asahara H, Lotz MK (2010) Tissue neogenesis and STRO-1 expression in immature and mature articular cartilage. J Orthop Res 28(1):96–102. doi:10.1002/jor.20944
294. Lotz MK, Otsuki S, Grogan SP, Sah R, Terkeltaub R, D'Lima D (2010) Cartilage cell clusters. Arthritis Rheum 62(8):2206–2218. doi:10.1002/art.27528
295. Sandell LJ, Aigner T (2001) Articular cartilage and changes in arthritis. An introduction: cell biology of osteoarthritis. Arthritis Res 3(2):107–113
296. Koelling S, Kruegel J, Irmer M, Path JR, Sadowski B, Miro X et al (2009) Migratory chondrogenic progenitor cells from repair tissue during the later stages of human osteoarthritis. Cell Stem Cell 4(4):324–335. doi:10.1016/j.stem.2009.01.015, S1934-5909(09)00018-6 [pii]
297. van der Kraan PM, van den Berg WB (2007) Osteophytes: relevance and biology. Osteoarthritis Cartilage 15(3):237–244. doi:10.1016/j.joca.2006.11.006, S1063-4584(06)00327-X [pii]
298. Singh S, Jones BJ, Crawford R, Xiao Y (2008) Characterization of a mesenchymal-like stem cell population from osteophyte tissue. Stem Cells Dev 17(2):245–254. doi:10.1089/scd.2007.0146
299. Hashimoto S, Creighton-Achermann L, Takahashi K, Amiel D, Coutts RD, Lotz M (2002) Development and regulation of osteophyte formation during experimental osteoarthritis. Osteoarthritis Cartilage 10(3):180–187
300. Patrascu JM, Kruger JP, Boss HG, Ketzmar AK, Freymann U, Sittinger M et al (2013) Polyglycolic acid-hyaluronan scaffolds loaded with bone marrow-derived mesenchymal stem cells show chondrogenic differentiation in vitro and cartilage repair in the rabbit model. J Biomed Mater Res B Appl Biomater 101(7):1310–1320. doi:10.1002/jbm.b.32944
301. Al Faqeh H, Nor Hamdan BM, Chen HC, Aminuddin BS, Ruszymah BH (2012) The potential of intra-articular injection of chondrogenic-induced bone marrow stem cells to retard the progression of osteoarthritis in a sheep model. Exp Gerontol 47(6):458–464. doi:10.1016/j.exger.2012.03.018, S0531-5565(12)00073-3 [pii]
302. Jo CH, Lee YG, Shin WH, Kim H, Chai JW, Jeong EC et al (2014) Intra-articular injection of mesenchymal stem cells for the treatment of osteoarthritis of the knee: a proof-of-concept clinical trial. Stem Cells 32(5):1254–1266. doi:10.1002/stem.1634
303. Murphy JM, Dixon K, Beck S, Fabian D, Feldman A, Barry F (2002) Reduced chondrogenic and adipogenic activity of mesenchymal stem cells from patients with advanced osteoarthritis. Arthritis Rheum 46(3):704–713. doi:10.1002/art.10118, 10.1002/art.10118 [pii]
304. Siddappa R, Licht R, van Blitterswijk C, de Boer J (2007) Donor variation and loss of multipotency during in vitro expansion of human mesenchymal stem cells for bone tissue engineering. J Orthop Res 25(8):1029–1041. doi:10.1002/jor.20402

305. Loeser RF, Olex A, McNulty MA, Carlson CS, Callahan M, Ferguson C et al (2012) Microarray analysis reveals age-related differences in gene expression during the development of osteoarthritis in mice. Arthritis Rheum 64(3):705–717. doi:10.1002/art.33388
306. McGonagle D, Tan AL, Carey J, Benjamin M (2010) The anatomical basis for a novel classification of osteoarthritis and allied disorders. J Anat 216(3):279–291. doi:10.1111/j.1469-7580.2009.01186.x
307. Karsdal MA, Christiansen C, Ladel C, Henriksen K, Kraus VB, Bay-Jensen AC (2014) Osteoarthritis – a case for personalized health care? Osteoarthritis Cartilage 22(1):7–16. doi:10.1016/j.joca.2013.10.018
308. van der Esch M, Knoop J, van der Leeden M, Roorda LD, Lems WF, Knol DL et al (2015) Clinical phenotypes in patients with knee osteoarthritis: a study in the Amsterdam osteoarthritis cohort. Osteoarthritis Cartilage. doi:10.1016/j.joca.2015.01.006
309. Loeser RF (2012) The effects of aging on the development of osteoarthritis. HSS J 8:18–19

Diabetes and Aging

Nicolas Musi and Andrzej Bartke

Contents

N. Musi, M.D. (✉)
Department of Medicine, Sam and Ann Barshop Institute for Longevity and Aging Studies, San Antonio Geriatric Research, Education and Clinical Center, University of Texas Health Science Center at San Antonio, San Antonio, TX, USA
e-mail: musi@uthscsa.edu

A. Bartke, Ph.D.
Department of Medicine, Geriatrics Research, Southern Illinois University School of Medicine, Springfield, IL, USA
e-mail: abartke@siumed.edu

F. Sierra, R. Kohanski (eds.), *Advances in Geroscience*,
DOI 10.1007/978-3-319-23246-1_12

1 Introduction

Age is a major risk factor for type 2 diabetes mellitus (T2DM), similarly to what is known about the etiology of most chronic diseases. Indeed, before the recent "epidemic" of childhood and adolescent obesity, T2DM was called "adult onset diabetes." The importance of chronological age as a risk factor for T2DM is only one of many links between diabetes and aging. The genetic and environmental control of the biological process of aging is intertwined with the development and consequences of T2DM in a highly complex network of interactions that are probably unique to diabetes.

Symptoms and diagnosis of T2DM reflect failure of insulin-secreting pancreatic β cells on the background of progressive increase in insulin resistance in target tissues. Insulin resistance commonly accompanies aging, although it is not clear whether insulin resistance is primarily a cause or a consequence of aging and whether it represents a protective/compensatory response to hyperinsulinemia [9]. While controversy persists about the role of aging in insulin resistance, and vice versa, it is widely accepted that maintaining insulin sensitivity through physical activity or pharmacological agents can prevent/delay the development of T2DM in high-risk young and older adults [84].

There is considerable evidence that insulin and homologous hormones in invertebrate species are intricately involved in the control of aging and lifespan [62, 124]. In experimental organisms ranging from yeast to mice, reducing (but not eliminating) insulin/insulin-like growth factor (IGF) signaling by genetic or dietary means can lead to slower aging, reduced susceptibility to age-related disease and significant, often remarkably large, extension of longevity. The increase in longevity can be quite impressive – more than twofold increase in *C. elegans* and more than 1.5-fold increase in mice. Importantly, a role of insulin/IGF signaling in the control of aging and longevity also applies to primates, including humans [106, 133]. The beneficial effects of reduced insulin/IGF signaling on longevity could be viewed as the first paradox in the relationship between insulin and aging. A severe reduction of the capacity to produce insulin leads to diabetes, a serious, life-threatening disease. In contrast, chronic hypoinsulinemia induced by calorie restriction or mutations related to growth hormone (GH) signaling are associated, most likely causally, with delayed aging and long, healthy life.

The complex interplay between insulin action, diabetes and aging does not end here. Progression of T2DM, with its practically unavoidable deterioration of glucose homeostasis, is associated with the emergence of functional deficits and pathological alterations that typically accompany aging [35]. Indeed, many diabetologists view accelerated aging as a result of T2DM (as well as type 1 diabetes) and believe that patients with T2DM are physiologically older than their chronological age. This difference may be as great as 10 years or, for new patients, approximately 1 year for each year since the diagnosis of their disease. This relationship between T2DM and aging at times has been emphasized by the simple but powerful statement that "diabetes is accelerated aging" [DeFronzo, RA, personal communication]. Although a disease (diabetes) cannot be equated with a physiological process (aging), it is indeed striking that some of their consequences are very similar. Regarding accelerated aging as one of the consequences of diabetes is consistent with many clinical and epidemiological findings. It is also important to remember that diabetes disturbs insulin secretion and its intracellular actions, which are mediated by signaling pathways known to be involved in the control of aging.

In terms of broader relationships between aging and disease as embodied in the concept of geroscience and explored throughout this volume, the suggestion that diabetes may lead to accelerated aging is extremely important. It provides a mechanistic explanation of why diabetes, similarly to chronological age, increases the risk for cardiovascular disease, cancer, frailty and dementia. Uncovering cause:effect relationships between insulin signaling, aging and diabetes is greatly complicated by the fact that circulating insulin concentration and insulin sensitivity are mutually dependent and both are altered in pre-diabetic conditions such as glucose intolerance and the metabolic syndrome, and in patients with T2DM. Although chronic insulin exposure promotes insulin resistance by well-known mechanisms, insulin resistance coexists with insulin deficiency in T2DM; meanwhile long-lived mice with GH-related mutations and offspring of exceptionally long-lived people exhibit reductions in both insulin levels and insulin resistance [8, 133]. These relationships are further complicated by differential alterations in insulin sensitivity and various steps of insulin signaling in different organs of the same individual [19], as well as by the emerging evidence that insulin resistance can be either detrimental or protective [10]. Is this another paradox related to diabetes?

Discussion of the interplay between aging and diabetes would not be complete without a reference to obesity. Aging is associated with progressive changes in the distribution and secretory activity of adipose tissue, as well as adipogenesis and adipocyte senescence, and most often also with a gradual, often very striking increase in adiposity [125]. Obesity is one of the important features of metabolic syndrome and a key risk factor for T2DM. It is also independently associated with an increased risk of cancer and cardiovascular disease, thus resembling the effects of both aging and diabetes. In turn, regulation of lipid metabolism, food intake and adiposity are disturbed in T2DM.

2 Studies in Experimental Animals Link Glucose Homeostasis and Insulin Signaling with Healthy Aging and Longevity

2.1 Animals with Reduced Longevity

Much of the evidence for causal links between glucose homeostasis and aging is derived from mice with genetic or dietary interventions that alter insulin signaling. In this species, obesity induced either by mutations or high-fat diet (HFD) leads to a reduced lifespan, as well as insulin resistance, diabetes and functional deficits resembling those that normally occur during aging [95, 96, 121]. Mice with morbid obesity due to hereditary deficiency of leptin or leptin receptors provide a particularly striking example of these associations [67, 89, 102]. Animals heterozygous for the lethal yellow (A^Y) mutation at the agouti locus (often referred to as "agouti mice") are obese, hyperinsulinemic and hyperglycemic and more likely to develop cancer, an aging-related disease [134].

Interestingly, some genetic interventions allow dissociating abnormal glucose homeostasis and accelerated aging from obesity. Transgenic mice overexpressing growth hormone (GH) have reduced adiposity (percent of body fat) during most of their adult life [103], are insulin resistant and hyperinsulinemic, live much shorter than their genetically normal siblings, and exhibit numerous characteristics resembling aging that develop at an inappropriately early chronological age [6]. Pertinent to the subject of this chapter, these "giant mice" are more susceptible to age-related diseases including kidney inflammatory disease, glomerulosclerosis and cancer [129]. Moreover, blood pressure is elevated in these insulin-resistant transgenic mice [78].

2.2 *Animals with Extended Longevity*

Causal relationships between insulin signaling, glucose homeostasis, age-associated disease and longevity are strongly supported by studies in mice in which somatotropic signaling is suppressed by spontaneous mutations or targeted gene disruptions. Remarkable extension of average and maximal longevity of mice of both sexes lacking GH or GH receptors is associated with enhanced insulin sensitivity, reduced or "low-normal" levels of blood glucose, and resistance to the detrimental impact of high-fat diet on insulin signaling [8, 22, 90]. These characteristics, together with reduced blood pressure [57], could be described as a phenotype opposite to metabolic syndrome or "prediabetes." Importantly, these animals exhibit numerous features of delayed aging, including improved maintenance of cognitive, immune and neuromuscular function, collagen properties and glucose homeostasis at the age when these parameters exhibit decline in their normal (wild type) siblings [8, 22]. Incidence of cancer and various pathological changes associated with aging are delayed and/or reduced in these insulin-sensitive, long-lived mutants [73, 74].

Our hypothesis that improved insulin signaling/action is one of the key mechanisms responsible for extension of longevity in GH-related mutants was supported by experiments exposing these animals to calorie restriction (CR) throughout most of their post-natal lives. In most strains of mice, CR increases insulin sensitivity, slows aging and extends longevity. In Ames dwarf mice, which lack GH, CR led to a further increase in insulin sensitivity and further extension of their already remarkably long lifespan [7]. In contrast, in GH-resistant GHRKO (a.k.a. "Laron dwarf") mice, CR did not further increase insulin action (Fig. 1), had no effect on longevity in males and caused a small increase in maximal (but not average) longevity in

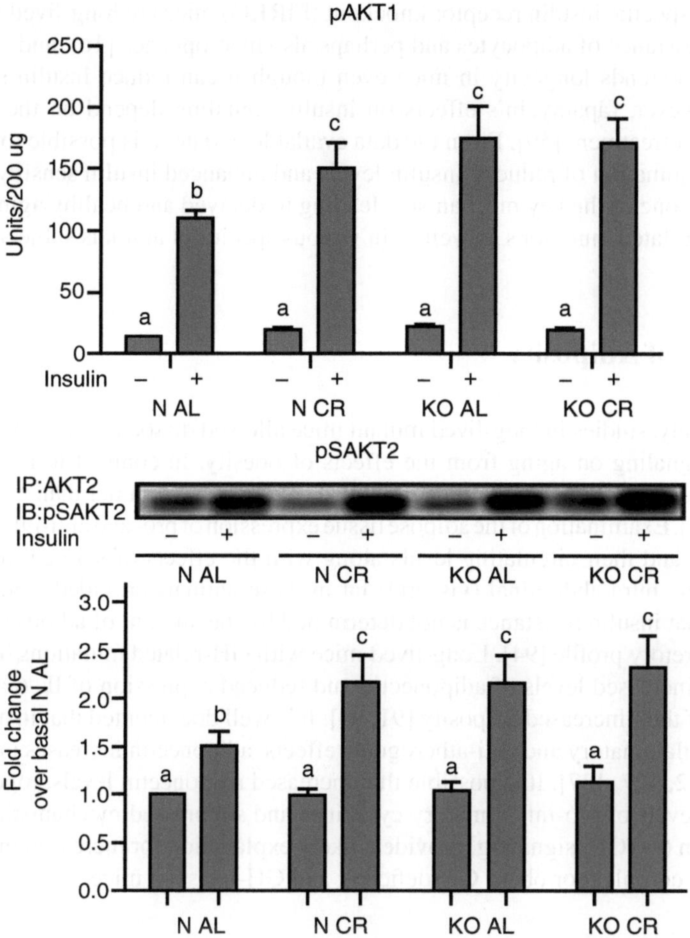

Fig. 1 Phosphorylation of Akt1 and Akt2 in response to acute insulin stimulation in the skeletal muscle of long-lived growth hormone receptor deleted (GHRKO; KO) and normal (*N*) male mice fed ad libitum (*AL*) or subjected to chronic 30 % calorie restriction (*CR*). Results indicate that GHRKO mice are more insulin sensitive than N (control animals). CR produces the expected increase in insulin sensitivity of N mice but no further increase in the GHRKO mutants. Importantly, this CR regimen extends longevity in N but not in GHRKO males [19]

females [18]. Additional evidence for a cause:effect relationship between insulin signaling and aging was obtained in a recent study in which experimental suppression of insulin sensitivity in long-lived GHRKO mice attenuated phenotypic markers of delayed aging [4]. In contrast to these observations, mice with deletion of insulin receptor substrate 1 (IRS1 –/–) and transgenic mice overexpressing klotho are long-lived in spite of enhanced rather than reduced insulin resistance [86, 116]. It was suggested that insulin resistance may act to reduce the strength of the insulin signals [86] or protect target organs from excessive insulin exposure [9]. Similarly, fat tissue-specific insulin receptor knockout (FIRKO) mice are long-lived in spite of insulin resistance of adipocytes and perhaps also macrophages [16], and rapamycin treatment extends longevity in mice even though it can reduce insulin sensitivity [68]. However, rapamycin's effects on insulin signaling depend on the dose and duration of treatment [49]. From the data available to date, it is possible to conclude that a combination of reduced insulin levels and enhanced insulin sensitivity might emerge as one of the key mechanisms leading to delayed and healthy aging in mice with GH-related mutations as well as in various species of animals subjected to CR.

3 Role of Adiposity

Interestingly, studies in long-lived mutant mice allowed dissociation of the effects of insulin signaling on aging from the effects of obesity. In contrast to animals subjected to CR, long-lived GH-related mutants have increased rather than reduced adiposity [13]. Examination of the adipose tissue expression of pro- and anti-inflammatory cytokines and their circulating levels along with the effects of surgical removal of most of the intra-abdominal (visceral) fat in these animals provided evidence suggesting that insulin resistance is not determined by the amount of adipose tissue but by its secretory profile [94]. Long-lived mice with GH-related mutations have unexpectedly increased levels of adiponectin and reduced expression of IL-6 and TNFα in spite of their increased adiposity [91, 94]. It is well documented that in addition to its anti-inflammatory and anti-atherogenic effects, adiponectin increases insulin sensitivity [12, 127, 137]. It is possible that increased adiponectin levels, together with reduced levels of pro-inflammatory cytokines and suppressed mechanistic target of rapamycin (mTOR) signaling, provide a likely explanation for increased insulin sensitivity in corpulent or obese GH-deficient and GH-resistant mice.

4 Pathophysiology and Natural History of Type 2 Diabetes

In young and middle-aged individuals, T2DM occurs as a consequence of two pathophysiologic alterations, insulin resistance and β-cell failure. Both genetic and environmental factors (obesity, physical inactivity) contribute to the development of insulin resistance. Resistance to the actions of insulin in skeletal muscle, liver, and adipose tissue appears early in the natural history of the disease. In muscle, insulin resistance is manifested as decreased insulin-mediated glucose disposal; in the liver it is manifested as impaired suppression of hepatic glucose output; and in adipose tissue insulin resistance manifests as increased lipolysis rates, resulting in increased plasma free fatty acid concentration that further impairs insulin action in muscle and liver (i.e. lipotoxicity). During the early stages of insulin resistance, β-cells can compensate by augmenting insulin secretion to maintain normal glucose tolerance. However, in subjects destined to develop diabetes (~10–20 % of all insulin-resistant individuals), the β-cells eventually will fail, leading to the onset of overt diabetes. The resultant hyperglycemia causes a further decline in insulin sensitivity (i.e. glucotoxicity), but it is the progressive β-cell failure that determines the rate of disease progression [44].

5 Aging as a Risk Factor for Type 2 Diabetes

Substantial evidence has demonstrated that increasing age is associated with impaired glucose homeostasis [2, 43, 107]. The Baltimore Longitudinal Study of Aging showed a progressive decline in glucose tolerance from the third through the ninth decade of life [118]. During an oral glucose tolerance test, the mean fasting plasma glucose increased ~1 mg/dl per decade, and the 2 h glucose increased ~5 mg/dl per decade. This decline in glucose tolerance was also evident in the National Health and Nutrition Examination Survey (NHANES) III, which showed that the percentage of physician-diagnosed diabetes (fasting glucose ≥126 mg/dl) is 3.9 % in subjects aged 40–49 years, whereas prevalence increases to 13.2 % in subjects ≥75 years of age [66]. The percentage of subjects with undiagnosed diabetes also increases from 7.1 to 14.1 % within these age groups. Approximately 50–60 % of subjects aged ≥65 have diabetes or impaired glucose tolerance (IGT), and ~25–50 % (depending upon the population) of subjects with IGT will ultimately convert to type 2 diabetes [47], Other than chronological age, the factors responsible for such high prevalence of IGT and T2DM in the aging population are not clear. However, age-dependent decreases in (i) insulin sensitivity and (ii) β cell function are thought to play important roles in the deterioration of glucose homeostasis that occurs with advancing age.

5.1 Molecular Basis for Peripheral Insulin Resistance

As mentioned above, insulin resistance is characteristic of peripheral tissues (i.e. muscle, adipose) from obese and T2DM subjects [43]. The first step in the insulin signaling transduction pathway is binding of insulin to the α subunits of the insulin receptor in the cell surface. The activated insulin receptor then tyrosine phosphorylates and activates downstream insulin receptor substrate (IRS) proteins, such as IRS-1 (IRS-1). Tyrosine phosphorylation of IRS-1 leads to its association with the p85 subunit of phosphatidylinositol 3-kinase (PI-3 kinase) [132]. Activation of PI-3 kinase leads to the phosphorylation/activation of a series of enzymes and proteins, such as phosphoinositide-dependent kinase (PDK)-1, protein kinase C (PKC) λ/ζ, Akt, and the RabGAP protein AS160 [50, 142, 143]. The phosphorylation/activation of these signaling intermediaries results in the translocation of GLUT4 glucose transporters to the cell membrane and the uptake of glucose [50, 142, 143]. A wide array of abnormalities distinguish insulin-resistant muscle from normal muscle, including decreased insulin receptor tyrosine kinase activity, decreased IRS-1-associated tyrosine phosphorylation, and decreased insulin-stimulated PI 3-kinase activation [28, 41, 145]. Other defects reported in insulin-resistant muscle include decreases in insulin-stimulated PKC activity [50], and AS160 phosphorylation [80]. Insulin-mediated GLUT4 translocation also is reduced in insulin resistant subjects [54], due in part to impairments in insulin signaling described above. These molecular abnormalities are strongly correlated with decreased insulin-stimulated glucose disposal in muscle [41, 145]. Some of these cellular processes will be further discussed in Sect. 5.2

Insulin-stimulated glucose transport in muscles from nondiabetic rodents decreases with age [58, 61]. The majority of reports that have examined the effect of age on insulin sensitivity in humans also have demonstrated reduced insulin sensitivity [48]. Some studies have reported that decreased insulin sensitivity with aging is not apparent when results are expressed by lean body mass. However, when glucose disposal is measured using the hyperinsulinemic euglycemic clamp technique over a range of insulin doses, the plasma insulin concentration required to achieve half-maximal glucose disposal is considerably lower in younger compared with older subjects (shift to the right) [112]. This decrease in insulin-stimulated glucose disposal is evident whether glucose disposal rates are plotted per kg of whole body weight or lean body mass. Other dose-response studies using the euglycemic clamp also have shown impaired insulin sensitivity in older subjects [51, 105]. Petersen et al. demonstrated that, in response to physiologic hyperinsulinemia (20 mU/kg.min insulin clamp), older subjects have a ~40 % reduction in peripheral glucose disposal compared to younger subjects who were matched for body mass index (BMI) and lean body mass [105]. Importantly, the impairment in insulin sensitivity in older subjects was still evident when glucose disposal was expressed per lean mass [77]. In addition, peripheral insulin resistance in aging has been demonstrated using other techniques, such as forearm glucose uptake and the frequently sampled intravenous glucose tolerance test (minimal model) [32]. Studies in rodents and human subjects from various groups also have shown that skeletal muscle from aging animals has defects in the insulin transduction pathway as described above [55, 75, 85, 87].

The underlying pathogenic mechanism responsible for the reduction in insulin action that occurs with aging is unclear. Probable factors contributing to age-associated insulin resistance include adiposity/lipotoxicity, inflammation, and mitochondrial dysfunction. These are discussed below and illustrated in Fig. 2.

5.2 Role of Adiposity/Lipotoxicity on Insulin Resistance

Approximately 30–40 % of older U.S. adults are obese (CDC/NHANES). The high prevalence of obesity in this population is multifactorial, including decreased physical activity, lower oxidative capacity, and muscle wasting. Obesity is associated with impaired glucose metabolism, although the mechanism by which excess adipose tissue alters glucose homeostasis is unclear. Plasma concentration of free fatty

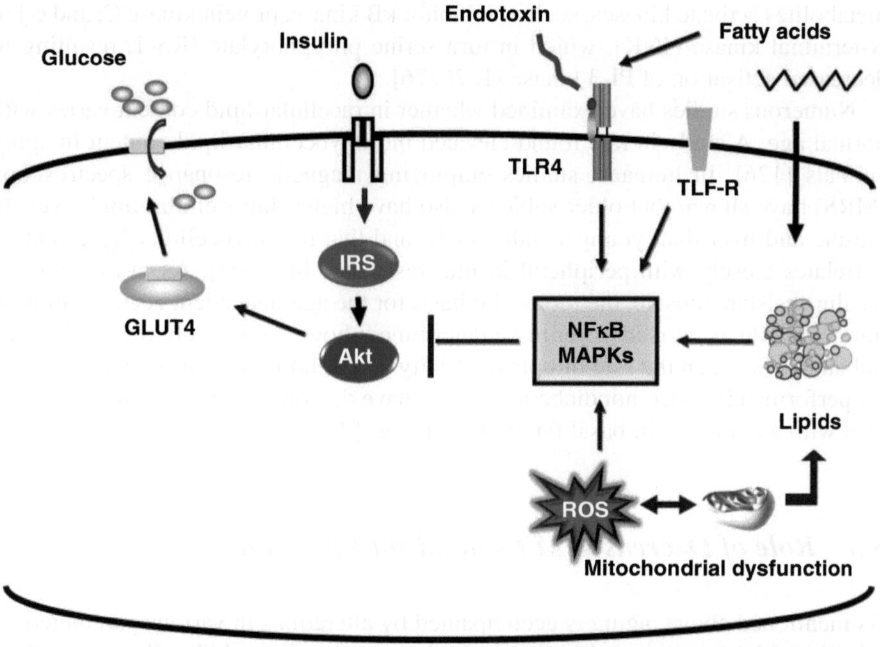

Fig. 2 Model of pathogenesis of muscle insulin resistance with aging. During aging, there is a pro-inflammatory state evidenced by increased expression and activity of mediators such as NFκB and MAP-kinases, caused by elevated levels of inflammatory stimuli, including plasma free fatty acids and endotoxin (which signal via TLR4), intracellular lipids (e.g. ceramides, diacylglycerol), cytokines (e.g. TNFα) and ROS. Decreases in mitochondrial function also contribute by promoting accumulation of intracellular lipids and increased ROS production. The activation of NFκB and MAP-kinases impair insulin action at the Akt and IRS levels, which eventually result in decreased insulin-mediated glucose disposal

acids (FFA) is commonly elevated in older subjects, and several lines of evidence implicate a deleterious effect of elevated plasma FFA level on muscle insulin sensitivity. For example, an experimental elevation of FFA induces muscle insulin resistance in normal glucose-tolerant subjects [17], whereas a reduction in plasma FFA concentrations rapidly ameliorates insulin resistance in insulin-resistant individuals [113]. In addition to the circulation, a variety of approaches has confirmed the existence of increased lipid content in insulin-resistant skeletal muscle [5, 60, 128]. Triglycerides account for most intramyocellular lipids. While triglycerides probably do not impair insulin action per se, metabolites of triglycerides/FFA, particularly diacylglycerol and ceramides, have been shown to have a deleterious effect on insulin action [63, 141]. The cause for the accumulation of intracellular lipids in insulin-resistant muscle is unknown, although one possibility is a reduction in mitochondrial oxidative capacity [105], which appears as a major hallmark of aging (see below). This elevation in intramyocellular lipids is thought to initiate a reverberating negative feedback cycle by decreasing insulin signaling and aggravating the insulin resistance that is already present. Specifically, these intracellular lipid metabolites activate kinases, such as inhibitor κB kinase, protein kinase C, and c-jun N-terminal kinase (JNK), which in turn serine phosphorylate IRS-1, resulting in decreased activation of PI-3 kinase [1, 26, 76].

Numerous studies have examined whether intracellular lipid content varies with normal age. A study in rats found elevated intramyocellular lipid content in aging animals [126]. In humans, studies employing magnetic resonance spectroscopy (MRS) have shown that older subjects also have higher intracellular lipid levels in muscle and liver than younger individuals, and that intramyocellular lipid content correlates closely with peripheral insulin resistance [40, 105]. As in the case for insulin-resistant muscle, the molecular basis for the age-dependent accumulation of intramyocellular fat is also yet to be determined; however, it likely results from an imbalance between the rate of uptake of fatty acids and fat oxidation. Indeed, studies performed in older, nondiabetic subjects have demonstrated that aging is associated with a reduction in basal fat oxidation rates [27, 122].

5.3 *Role of Decreased Mitochondrial Function*

As mentioned above, aging is accompanied by alterations in various parameters of mitochondrial function and structure. In the context of metabolic diseases such as obesity and type 2 diabetes, particularly relevant are mitochondrial alterations described in skeletal muscle, which is a key tissue responsible for substrate (glucose, FFA) uptake and oxidation. Mitochondrial alterations described in aging muscle include reductions in mitochondrial number, ATP production, and respiration, abnormal structure, and, in some cases, elevated reactive oxygen species generation [reviewed in [79].

The cause for the reduction in mitochondrial function observed with aging is not clear. According to the free radical and mitochondrial theories of aging the decreases in mitochondrial function are the result of cumulative oxidative damage to mitochondrial molecules (mtDNA, proteins, and lipids) [65, 97]. Consistent with this theory, studies performed in human muscle have shown that aging is associated with oxidative damage to mtDNA and proteins [reviewed in 79]. Some [130], albeit not all [55], studies also have found increased lipid peroxidation in muscle from older subjects. Concerning the age-related reductions in oxidative capacity, various studies [88, 100, 110] have reported that aging is accompanied by a reduction in the activity of the energy-sensing enzyme AMP-activated protein kinase (AMPK). AMPK works as a fuel gauge, being activated robustly by energy-consuming stimuli such as muscle contraction, hypoxia, and ischemia [70, 135]. Upon stimulation, AMPK functions to restore cellular ATP by modifying diverse metabolic and cellular pathways, including increased fat oxidation and glucose transport. AMPK promotes fat oxidation in tissues by phosphorylating and inactivating acetyl CoA carboxylase (ACC), resulting in decreased synthesis of malonyl-CoA, an inhibitor of carnitine palmitoyltransferase I (CPT-1). The reduction in malonyl CoA relieves the inhibition of CPT-1 and promotes CPT-1 mediated transport of fatty acids into the mitochondria for oxidation. Because the end result of AMPK activation is an increase in fat oxidation, decreases in AMPK activity, as seen with aging, could lead to an excessive accumulation of intramyocelular lipids, which in turn would impair insulin action/sensitivity. Other key regulators of mitochondrial biogenesis and oxidative capacity reported to be altered with aging are the transcriptional coactivator peroxisome proliferator-activated receptor γ coactivator (PGC)-1α and the NAD-dependent deacetylase sirtuin-1. PGC-1α controls mitochondrial biogenesis and cellular metabolism by regulating the expression of numerous proteins involved in the Krebs cycle, oxidative phosphorylation, and mtDNA replication/transcription [64]. AMPK increases PGC-1α gene expression, and sirtuin-1 enhances PGC-1α activity through deacetylation on lysine residues. Similar to AMPK, the expression of both PGC-1α [55] and sirtuin-1 [31] are reduced in aging tissues, suggesting that a coordinated downregulation of the AMPK-PGC-1α-sirtuin 1 axis may play a role in the reductions of mitochondrial number, oxidative capacity, and cellular metabolic functions that occur with aging.

While, in general there is agreement with the notion that aging is associated with alterations in mitochondrial number, structure, and function, it is debated whether primary mitochondrial abnormalities (independent of aging) are sufficient to induce insulin resistance [59, 72]. Conflicting results about the role of mitochondrial dysfunction on the pathogenesis of insulin resistance, obesity, and type diabetes are due in part to differences in animal species, tissues, and human populations studied, and in the conditions and methods employed to assess mitochondrial function [79, 104]. Despite the ongoing debate, it is probable that increasing mitochondrial oxidative capacity via lifestyle interventions (i.e. physical activity) or pharmacological agents would result in improved metabolic outcomes during aging.

5.4 Role of Inflammation

Aging may be considered a state of low-grade "sterile" inflammation that could play a role in the high prevalence of glucose metabolism abnormalities in older subjects [24, 30, 37]. Increased low-grade inflammatory activity in older subjects could either cause age-related diseases or be a marker of diseases that occur with aging. Bruunsgaard et al. determined that low-grade increases in the levels of pro-inflammatory cytokines in older subjects were independent of the presence of medical disorders [23], although medical disorders can exacerbate this phenomenon.

The source for the elevated cytokine levels with aging is not clear. Cytokine levels in the circulation reflect production from many tissues including inflammatory cells (monocytes/macrophages, T cells, etc.), senescent cells and adipose tissue. In the young, approximately 25 % of IL-6 is derived from fat tissue [98], and adipocytes can secrete TNFα in addition to IL-6. Aging is associated with increases in abdominal fat mass and visceral obesity is associated with increased circulating levels of these cytokines [42, 52, 81]. There is some evidence suggesting that omental fat produces more cytokines than subcutaneous fat tissue [52], and this may explain why visceral obesity has a greater detrimental effect on insulin sensitivity [53]. The production of cytokines by adipose tissue can also be modulated by the interaction between adipocytes and macrophages within the adipose tissue [82]. The number of macrophages in adipose tissue directly correlates with adiposity [131, 136] and insulin resistance [46], and adipose tissue expansion correlates with the accumulation of macrophages and the proinflammatory phenotype [20, 92]. Another potential source of inflammation with aging is the accumulation of senescent preadipocytes [125]. Senescent cells typically have a pro-inflammatory secretory profile, termed the senescence-associated secretory phenotype (SASP), that may propagate the inflammatory adipose tissue microenvironment as well as promote inflammation throughout the whole body.

In addition to cytokines produced by adipose tissue and inflammatory cells, another potential source of inflammation during aging is the microbiome and its products. Studies in flies [109], rodents [21] and humans [56] have shown that aging is accompanied by alterations in intestinal microbiota composition and intestinal barrier integrity. In line with these findings, our group recently showed that older subjects have increased plasma concentration of endotoxin (a marker of altered barrier integrity) in association with insulin resistance, sarcopenia, and increased inflammatory signaling (toll like-receptor 4, NFκB, MAPK) in muscle [56]. Thus, it is possible that endotoxin, and other yet unidentified microbial products, could be involved in the inflammatory state and consequent metabolic alterations of aging.

A potential mechanism linking aging, inflammation, and metabolic disease is immune sensing through the NLRP3 inflammasome [140]. The NLRP3 inflammasome is a multiprotein cytoplasmic complex composed of NLRP3, the adaptor molecule ASC, and the cysteine protease caspase-1. Stimulation of the inflammasome by pathogen-associated molecular patterns (PAMPs) leads to the activation of caspase-1, which cleaves the pro-forms of the cytokines IL-1β, IL-18 and IL-33 to their active and secreted forms. A role of the inflammasome in aging-related inflammation and associated pathologies is suggested by findings that NLRP3 ablation protects against glucose intolerance, bone loss, and thymic involution in aged mice [140].

6 Aging, Diabetes and Insulin Signaling in the Brain

The well-documented association of diabetes with chronic age-related disease and geriatric conditions [35] includes increased risk of cognitive impairment, dementia, brain atrophy and Alzheimer's disease [35, 38, 99]. Potential mechanisms of these associations include several well-recognized hallmarks of aging, including increased accumulation of advanced glycation end-products (AGEs), oxidative stress, inflammation, defective proteostasis and metabolic abnormalities, as well as altered insulin signaling within the brain [38, 83].

Insulin receptors, as well as various proteins involved in the intra-cellular transmission of insulin signals, are expressed in various brain regions [83]. However, the role of insulin in the control of glucose metabolism in the central nervous system has been questioned. The controversy surrounding this issue is likely related to the fact that, in healthy subjects, glucose uptake by the brain is already maximally stimulated by normal insulin levels and therefore does not respond to further insulin stimulation [71]. In the context of Alzheimer's disease, some investigators refer to the brain insulin resistance as "type III diabetes" [34, 123]. Recent studies in obese patients suggest that iron overload in the brain may be caused by local insulin resistance and could represent yet another potential mechanism of the detrimental influence of diabetes on cognitive performance [15] and risk for Alzheimer's and Parkinson's disease [115].

Complex relationships between brain function and obesity, insulin resistance, diabetes and its complications also involve the role of hypothalamic function in the control of peripheral metabolism and aging. High-fat diet and obesity promote hypothalamic inflammation and insulin resistance [96, 111], and the hypothalamus controls multiple facets of peripheral metabolism [11, 14, 101]. Insulin signaling within the hypothalamus influences hepatic gluconeogenesis [25], lipogenesis and lipolysis in the adipose tissue [117], as well as circulating levels of branch chain amino acids, which are known to be elevated in T2DM [119]. Recent elegant studies in the Cai laboratory [144] linked hypothalamic inflammation with the control of aging. Interestingly, ongoing studies in our laboratory [69, Bartke unpublished] indicate that expression of IL-1β and other pro-inflammatory cytokines is reduced in the hypothalamus of long-lived mutant mice. Further studies will be necessary to elucidate the role of hypothalamic inflammation and insulin resistance in the development of whole-body metabolic abnormalities that lead to diabetes.

More work will also be needed to identify mechanisms responsible for the increased risk of cognitive decline and Alzheimer's disease in patients with diabetes. Epidemiological studies provide evidence that the risk of dementia in diabetic patients is reduced by treatment with metformin [33]. Metformin and related drugs have also been shown to reduce the risk of cancer and to extend longevity of experimental animals [3, 93]. These findings imply that drugs of this class can slow down and/or delay the aging process. The apparent "anti-aging" action of metformin could have contributed to its beneficial effects on cognition in diabetes patients. However, regardless of the mechanisms involved, evidence for cognitive benefits of diabetes treatment strengthens the suggestion for etiological links between diabetes and dementia. These findings also generate interest in the exciting possibility that diabetes drugs could be useful for prevention and/or treatment of Alzheimer's disease [139]. Consistent with this notion, animal studies and early clinical trials suggest that intranasal administration of insulin in order to overcome insulin resistance and enhanced brain metabolism leads to reductions in β amyoloid and tauopathy, as well as improvements in brain function and cognition in Alzheimer's disease [36, 108, 138].

7 Prevention of Diabetes Versus Anti-aging Interventions

The prevalence of pre-diabetes (IGT) among older adults is increasing [29]. Because life expectancy is also increasing, the number of older individuals with diabetes and/or at risk of developing its complications (blindness, kidney failure, amputations, neuropathic pain, cardiovascular disease, etc.) will be substantial. Therefore, strategies for diabetes prevention are urgently needed. Since physical inactivity and obesity are common in older subjects, lifestyle interventions are a logical diabetes preventative strategy. In addition, exercise improves mitochondrial and vascular function, which are reduced/impaired with aging [55, 114, 120]. Physical activity is effective in improving insulin action in older subjects [55]. In line with these findings, the Diabetes Prevention Program (DPP) clinical trial, conducted in pre-diabetic (IGT) subjects showed that lifestyle intervention proved exceptionally effective in preventing diabetes in older individuals [39].

Pharmacological interventions with insulin-sensitizing agents also have been evaluated for diabetes prevention in older subjects. In parallel to lifestyle changes, the DPP also tested the effect of metformin on diabetes prevention. In contrast to lifestyle modification, metformin seemed to be less effective in preventing conversion to diabetes in older subjects versus middle-aged individuals [39]. Our group conducted a multi-center diabetes prevention trial in pre-diabetic subjects using pioglitazone [45], a potent insulin-sensitizer. Pioglitazone was highly effective in preventing diabetes conversion (72 % overall reduction), and it was as effective in older (mean age = 66 years) as in middle-aged (mean age = 46 years) individuals in improving insulin sensitivity and in preventing diabetes (Espinoza S, Tripathy D, Defronzo RA, Musi N, unpublished, 2015).

Since the glucose metabolism alterations seen in older subjects may be caused by "primary" aging-mediated cellular changes (mitochondrial dysfunction, oxidative damage, cellular senescence, inflammation), another strategy for diabetes prevention is to target the aging process instead of "secondary" metabolic/endocrine perturbations (β cell dysfunction, insulin resistance). This approach would have the added benefit of potentially preventing other aging-related diseases such as cardiovascular disease, cancer, neurodegeneration and arthritis at the same time. This is in fact the central tenet of the Geroscience Hypothesis, which is awaiting experimental testing. The apparent beneficial effects of metformin and physical activity on many of these diseases, exemplifies the possibility of preventing/treating them through modifying basic mechanisms of aging. As research in aging biology advances and novel molecular targets are identified, trials using agents that modify these targets should be conducted for the testing of interventions to prevent diabetes and other diseases of aging in the elderly.

8 Closing Remarks

Aging is accompanied by various changes in metabolic processes at the cellular, tissue and whole body levels, including decreases in oxidative capacity, intracellular lipid accumulation, insulin resistance, and β cell dysfunction. These metabolic changes contribute to the higher prevalence of obesity and T2DM that are important causes of disability and death in older people. A better understanding of the molecular basis for the age-induced metabolic alterations will help design strategies to preserve metabolic homeostasis and prevent these diseases that affect millions of people around the world.

Acknowledgments N.M. has been supported by grants from the American Diabetes Association, the National Institutes of Health (DK-80157, DK-089229 and AG-030979), and by the San Antonio Nathan Shock Center (AG-013319). A.B has been supported by grants from the National Institute on Aging (AG-019899 and AG-031736), and by the SIU Geriatrics Research Initiative.

Editor: Aaron Pawlik, National Institute of Diabetes and Digestive and Kidney Diseases (NIDDK), NIH.

References

1. Aguirre V, Uchida T, Yenush L, Davis R, White MF (2000) The c-Jun NH(2)-terminal kinase promotes insulin resistance during association with insulin receptor substrate-1 and phosphorylation of Ser(307). J Biol Chem 275:9047–9054
2. Andres R (1971) Aging and diabetes. Med Clin North Am 55:835–846
3. Anisimov VN, Bartke A (2013) The key role of growth hormone-insulin-IGF-1 signaling in aging and cancer. Crit Rev Oncol Hematol 87(3):201–223
4. Arum O, Boparai RK, Saleh JK, Wang F, Dirks AL, Turner JG, Kopchick JJ, Liu JL, Khardori RK, Bartke A (2014) Specific suppression of insulin sensitivity in growth hormone receptor gene-disrupted (GHR-KO) mice attenuates phenotypic features of slow aging. Aging Cell 13:981–1000
5. Bachmann OP, Dahl DB, Brechtel K, Machann J, Haap M, Maier T, Loviscach M, Stumvoll M, Claussen CD, Schick F, Haring HU, Jacob S (2001) Effects of intravenous and dietary lipid challenge on intramyocellular lipid content and the relation with insulin sensitivity in humans. Diabetes 50:2579–2584
6. Bartke A (2003) Can growth hormone (GH) accelerate aging? Evidence from GH-transgenic mice. Neuroendocrinology 78:210–216
7. Bartke A, Wright JC, Mattison JA, Ingram DK, Miller RA, Roth GS (2001) Extending the lifespan of long-lived mice. Nature 414:412
8. Bartke A, Sun LY, Longo V (2013) Somatotropic signaling: trade-offs between growth, reproductive development, and longevity. Physiol Rev 93:571–598
9. Barzilai N, Ferrucci L (2012) Insulin resistance and aging: a cause or a protective response? J Gerontol A Biol Sci Med Sci 67:1329–1331
10. Barzilai N, Huffman DM, Muzumdar RH, Bartke A (2012) The critical role of metabolic pathways in aging. Diabetes 61(6):1315–1322. http://doi.org/10.2337/db11-1300
11. Belgardt BF, Mauer J, Wunderlich FT, Ernst MB, Pal M, Spohn G, Brönneke HS, Brodesser S, Hampel B, Schauss AC, Brüning JC (2010) Hypothalamic and pituitary c-Jun N-terminal kinase 1 signaling coordinately regulates glucose metabolism. Proc Natl Acad Sci U S A 107(13):6028–6033
12. Berg AH, Combs TP, Du X, Brownlee M, Scherer PE (2001) The adipocyte-secreted protein Acrp30 enhances hepatic insulin action. Nat Med 7:947–953
13. Berryman DE, List EO, Palmer AJ, Chung MY, Wright-Piekarski J, Lubbers E, O'Connor P, Okada S, Kopchick JJ (2010) Two-year body composition analyses of long-lived GHR null mice. J Gerontol A Biol Sci Med Sci 65:31–40
14. Martínez B, de Morentin P, González CR, Saha AK, Martins L, Diéguez C, Vidal-Puig A, Tena-Sempere M, López M (2011) Hypothalamic AMP-activated protein kinase as a mediator of whole body energy balance. Rev Endocr Metab Disord 12(3):127–140
15. Blasco G, Puig J, Daunis-i-Estradella J, Molina X, Xifra G, Fernádez-Aranda F, Pedraza S, Wifredo R, Portero-Otín M, Fernádez-Real JM (2014) Brain iron overload, insulin resistance

and cognitive performance in obese subjects: a preliminary MRI case-control study. Diabetes Care 37:3076–3083
16. Blüher M, Kahn B, Kahn CR (2003) Extended longevity in mice lacking the insulin receptor in adipose tissue. Science 299:572–574
17. Boden G, Chen X (1995) Effects of fat on glucose uptake and utilization in patients with non-insulin-dependent diabetes. J Clin Invest 96:1261–1268
18. Bonkowski MS, Rocha JS, Masternak MM, Al Regaiey KA, Bartke A (2006) Targeted disruption of growth hormone receptor interferes with the beneficial actions of calorie restriction. Proc Natl Acad Sci U S A 103:7901–7905
19. Bonkowski MS, Dominici FP, Arum O, Rocha JS, Al Regaiey KA, Westbrook R, Spong A, Panici J, Masternak MM, Kopchick JJ, Bartke A (2009) Disruption of growth hormone receptor prevents calorie restriction from improving insulin action and longevity. PLoS One 4:e4567
20. Bourlier V, Zakaroff-Girard A, Miranville A, De Barros S, Maumus M, Sengenes C, Galitzky J, Lafontan M, Karpe F, Frayn KN, Bouloumie A (2008) Remodeling phenotype of human subcutaneous adipose tissue macrophages. Circulation 117:806–815
21. Brasili E, Mengheri E, Tomassini A, Capuani G, Roselli M, Finamore A, Sciubba F, Marini F, Miccheli A (2013) Lactobacillus acidophilus La5 and Bifidobacterium lactis Bb12 induce different age-related metabolic profiles revealed by 1H-NMR spectroscopy in urine and feces of mice. J Nutr 143:1549–1557
22. Brown-Borg HM, Bartke A (2012) GH and IGF1: roles in energy metabolism of long-living GH mutant mice. J Gerontol A Biol Sci Med Sci 67:652–660
23. Bruunsgaard H, Ladelund S, Pedersen AN, Schroll M, Jorgensen T, Pedersen BK (2003) Predicting death from tumour necrosis factor-alpha and interleukin-6 in 80-year-old people. Clin Exp Immunol 132:24–31
24. Bruunsgaard H, Pedersen M, Pedersen BK (2001) Aging and proinflammatory cytokines. Curr Opin Hematol 8:131–136
25. Buettner C, Camacho RC (2008) Hypothalamic control of hepatic glucose production and its potential role in insulin resistance. Endocrinol Metab Clin North Am 37(4):825–840
26. Cai D, Yuan M, Frantz DF, Melendez PA, Hansen L, Lee J, Shoelson SE (2005) Local and systemic insulin resistance resulting from hepatic activation of IKK-beta and NF-kappaB. Nat Med 11:183–190
27. Calles-Escandon J, Poehlman ET (1997) Aging, fat oxidation and exercise. Aging (Milano) 9:57–63
28. Caro JF, Sinha MK, Raju SM, Ittoop O, Pories WJ, Flickinger EG, Meelheim D, Dohm GL (1987) Insulin receptor kinase in human skeletal muscle from obese subjects with and without noninsulin dependent diabetes. J Clin Invest 79:1330–1337
29. Caspersen CJ, Thomas GD, Beckles GL, Bullard KM (2015) Secular changes in prediabetes indicators among older-adult Americans, 1999–2010. Am J Prev Med 48:253–263
30. Catania A, Airaghi L, Motta P, Manfredi MG, Annoni G, Pettenati C, Brambilla F, Lipton JM (1997) Cytokine antagonists in aged subjects and their relation with cellular immunity. J Gerontol A Biol Sci Med Sci 52:B93–B97
31. Chang HC, Guarente L (2013) SIRT1 mediates central circadian control in the SCN by a mechanism that decays with aging. Cell 153:1448–1460
32. Chen M, Bergman RN, Pacini G, Porte D Jr (1985) Pathogenesis of age-related glucose intolerance in man: Insulin resistance and decreased beta-cell function. J Clin Endocrinol Metab 60:13–20
33. Cheng C, Lin CH, Tsai YW, Tsai CJ, Chou PH, Lan TH (2014) Type 2 diabetes and antidiabetic medications in relations to dementia and diagnosis. J Gerontol A Biol Sci Med Sci 69(10):1299–1305
34. Cholerton B, Baker LD, Craft S (2011) Insulin resistance and pathological brain ageing. Diabet Med 28:1463–1475
35. Cigolle CT, Lee PG, Langa KM, Lee YY, Tian Z, Blaum CS (2011) Geriatric conditions develop in middle-aged adults with diabetes. J Gen Intern Med 26:272–279

36. Claxton A, Baker LD, Hanson A, Trittschuh EH, Cholerton B, Morgan A, Callaghan M, Arbuckle M, Behl C, Craft S (2015) Long-acting intranasal insulin detemir improves cognition for adults with mild cognitive impairment or early-stage Alzheimer's disease dementia. J Alzheimers Dis 44:897–906
37. Cohen HJ, Pieper CF, Harris T, Rao KM, Currie MS (1997) The association of plasma IL-6 levels with functional disability in community-dwelling elderly. J Gerontol A Biol Sci Med Sci 52:M201–M208
38. Craft S (2009) The role of metabolic disorders in Alzheimer's disease and vascular dementia: two roads converged. Arch Neurol 66(3):300–305
39. Crandall J, Schade D, Ma Y, Fujimoto WY, Barrett-Connor E, Fowler S, Dagogo-Jack S, Andres R (2006) The influence of age on the effects of lifestyle modification and metformin in prevention of diabetes. J Gerontol A Biol Sci Med Sci 61:1075–1081
40. Cree MG, Newcomer BR, Katsanos CS, Sheffield-Moore M, Chinkes D, Aarsland A, Urban R, Wolfe RR (2004) Intramuscular and liver triglycerides are increased in the elderly. J Clin Endocrinol Metab 89:3864–3871
41. Cusi K, Maezono K, Osman A, Pendergrass M, Patti ME, Pratipanawatr T, DeFronzo RA, Kahn CR, Mandarino LJ (2000) Insulin resistance differentially affects the PI 3-kinase- and MAP kinase-mediated signaling in human muscle. J Clin Invest 105:311–320
42. Dandona P, Weinstock R, Thusu K, Abdel-Rahman E, Aljada A, Wadden T (1998) Tumor necrosis factor-alpha in sera of obese patients: fall with weight loss. J Clin Endocrinol Metab 83:2907–2910
43. DeFronzo RA (1981) Glucose intolerance and aging. Diabetes Care 4:493–501
44. DeFronzo RA (2009) Banting lecture. From the triumvirate to the ominous octet: a new paradigm for the treatment of type 2 diabetes mellitus. Diabetes 58:773–795
45. DeFronzo RA, Tripathy D, Schwenke DC, Banerji M, Bray GA, Buchanan TA, Clement SC, Henry RR, Hodis HN, Kitabchi AE, Mack WJ, Mudaliar S, Ratner RE, Williams K, Stentz FB, Musi N, Reaven PD (2011) Pioglitazone for diabetes prevention in impaired glucose tolerance. N Engl J Med 364:1104–1115
46. Di Gregorio GB, Yao-Borengasser A, Rasouli N, Varma V, Lu T, Miles LM, Ranganathan G, Peterson CA, McGehee RE, Kern P (2005) Expression of CD68 and macrophage chemoattractant protein-1 genes in human adipose and muscle tissues: association with cytokine expression, insulin resistance, and reduction by pioglitazone. Diabetes 54: 2305–2313
47. Edelstein SL, Knowler WC, Bain RP, Andres R, Barrett-Connor EL, Dowse GK, Haffner SM, Pettitt DJ, Sorkin JD, Muller DC, Collins VR, Hamman RF (1997) Predictors of progression from impaired glucose tolerance to NIDDM: an analysis of six prospective studies. Diabetes 46(4):701–710
48. Elahi D, Muller DC (2000) Carbohydrate metabolism in the elderly. Eur J Clin Nutr 54(Suppl 3):S112–S120
49. Fang Y, Westbrook R, Hill C, Boparai RK, Arum O, Spong A, Wang F, Javors MA, Chen J, Sun LY, Bartke A (2013) Duration of rapamycin treatment has differential effects on metabolism in mice. Cell Metab 17:456–462
50. Farese RV, Sajan MP, Standaert ML (2005) Insulin-sensitive protein kinases (atypical protein kinase C and protein kinase B/Akt): actions and defects in obesity and type II diabetes. Exp Biol Med (Maywood) 230:593–605
51. Fink RI, Kolterman OG, Griffin J, Olefsky JM (1983) Mechanisms of insulin resistance in aging. J Clin Invest 71:1523–1535
52. Fried SK, Bunkin DA, Greenberg AS (1998) Omental and subcutaneous adipose tissues of obese subjects release interleukin-6: depot difference and regulation by glucocorticoid. J Clin Endocrinol Metab 83:847–850
53. Gastaldelli A, Miyazaki Y, Pettiti M, Matsuda M, Mahankali S, Santini E, DeFronzo RA, Ferrannini E (2002) Metabolic effects of visceral fat accumulation in type 2 diabetes. J Clin Endocrinol Metab 87:5098–5103

54. Garvey WT, Maianu L, Zhu JH, Brechtel-Hook G, Wallace P, Baron AD (1998) Evidence for defects in the trafficking and translocation of GLUT4 glucose transporters in skeletal muscle as a cause of human insulin resistance. J Clin Invest 101:2377–2386
55. Ghosh S, Lertwattanarak R, Lefort N, Molina-Carrion M, Joya-Galeana J, Bowen BP, Garduno-Garcia Jde J, Abdul-Ghani M, Richardson A, DeFronzo RA, Mandarino L, Van Remmen H, Musi N (2011) Reduction in reactive oxygen species production by mitochondria from elderly subjects with normal and impaired glucose tolerance. Diabetes 60:2051–2060
56. Ghosh S, Lertwattanarak R, Garduno Jde J, Galeana JJ, Li J, Zamarripa F, Lancaster JL, Mohan S, Hussey S, Musi N (2015) Elevated muscle TLR4 expression and metabolic endotoxemia in human aging. J Gerontol A Biol Sci Med Sci 70:232–246
57. Giani JF, Miquet JG, Munoz MC et al (2012) Upregulation of the angiotensin-converting enzyme 2/angiotensin-(1–7)/mas receptor axis in the heart and the kidney of growth hormone receptor knock-out mice. Growth Horm IGF Res 22:224–233
58. Goodman MN, Dluz SM, McElaney MA, Belur E, Ruderman NB (1983) Glucose uptake and insulin sensitivity in rat muscle: changes during 3–96 weeks of age. Am J Physiol 244:E93–E100
59. Goodpaster BH (2013) Mitochondrial deficiency is associated with insulin resistance. Diabetes 62(4):1032–1035
60. Goodpaster BH, Theriault R, Watkins SC, Kelley DE (2000) Intramuscular lipid content is increased in obesity and decreased by weight loss. Metabolism 49:467–472
61. Guarente L, Kenyon C (2000) Genetic pathways that regulate ageing in model organisms. Nature 408(6809):255–262
62. Grundleger ML, Godbole VY, Thenen SW (1980) Age-dependent development of insulin resistance of soleus muscle in genetically obese (ob/ob) mice. Am J Physiol 239:E363–E371
63. Hajduch E, Balendran A, Batty IH, Litherland GJ, Blair AS, Downes CP, Hundal HS (2001) Ceramide impairs the insulin-dependent membrane recruitment of protein kinase B leading to a loss in downstream signalling in L6 skeletal muscle cells. Diabetologia 44:173–183
64. Handschin C, Speigelman BM (2006) Peroxisome proliferator-activated receptor gamma coactivator 1 coactivators, energy homeostasis, and metabolism. Endocr Rev 27(7):728–735
65. Harman D (1972) Free radical theory of aging: dietary implications. Am J Clin Nutr 25(8):839–843
66. Harris MI, Flegal KM, Cowie CC, Eberhardt MS, Goldstein DE, Little RR, Wiedmeyer HM, Byrd-Holt DD (1998) Prevalence of diabetes, impaired fasting glucose, and impaired glucose tolerance in U.S. adults. The third national health and nutrition examination survey, 1988–1994. Diabetes Care 21:518–524
67. Harrison DE, Archer JR, Astle CM (1984) Effects of food restriction on aging: separation of food intake and adiposity. Proc Natl Acad Sci U S A 81:1835–1838
68. Harrison DE, Strong R, Sharp ZD, Nelson JF, Astle CM, Flurkey K, Nadon NL, Wilkinson JE, Frenkel K, Carter CS, Pahor M, Javors MA, Fernandez E, Miller RA (2009) Rapamycin fed late in life extends lifespan in genetically heterogeneous mice. Nature 460:392–395
69. Hascup ER, Wang F, Kopchick JJ, Bartke A (2015) Inflammatory and glutamatergic homeostasis are involved in successful aging. J Gerontol A Biol Sci Med Sci Pii: doi: 10.1093/gerona/glv010. PubMed PMID: 25711529
70. Hayashi T, Hirshman MF, Kurth EJ, Winder WW, Goodyear LJ (1998) Evidence for 5' amp-activated protein kinase mediation of the effect of muscle contraction on glucose transport. Diabetes 47:1369–1373
71. Hirvonen J, Virtanen KA, Nummenmaa L, Hannukainen JC, Honka MJ, Bucci M, Nesterov SV, Parkkola R, Rinne J, Iozzo P, Nuutila P (2011) Effects of insulin on brain glucose metabolism in impaired glucose tolerance. Diabetes 60:443–447
72. Holloszy JO (2013) "Deficiency" of mitochondria in muscle does not cause insulin resistance. Diabetes 62(4):1036–1040
73. Ikeno Y, Hubbard GB, Lee S et al (2009) Reduced incidence and delayed occurrence of fatal neoplastic diseases in growth hormone receptor/binding protein knockout mice. J Gerontol A Biol Sci Med Sci 64:522–529

74. Ikeno Y, Hubbard GB, Lee S et al (2013) Do Ames dwarf and calorie-restricted mice share common effects on age-related pathology? Pathobiol Aging Age Relat Dis 3. http://www.pathobiologyofaging.net/index.php/pba/article/view/20833
75. Iossa S, Mollica MP, Lionetti L, Crescenzo R, Tasso R, Liverini G (2004) A possible link between skeletal muscle mitochondrial efficiency and age-induced insulin resistance. Diabetes 53:2861–2866
76. Itani SI, Zhou Q, Pories WJ, MacDonald KG, Dohm GL (2000) Involvement of protein kinase C in human skeletal muscle insulin resistance and obesity. Diabetes 49:1353–1358
77. Jackson RA, Blix PM, Matthews JA, Hamling JB, Din BM, Brown DC, Belin J, Rubenstein AH, Nabarro JD (1982) Influence of ageing on glucose homeostasis. J Clin Endocrinol Metab 55:840–848
78. Jara A, Benner CM, Sim D, Liu X, List EO, Householder LA, Berryman DE, Kopchick JJ (2014) Elevated systolic blood pressure in male GH transgenic mice is age dependent. Endocrinology 155(3):975–986. doi:10.1210/en.2013-1899
79. Johnson ML, Robinson MM, Nair KS (2013) Skeletal muscle aging and the mitochondrion. Trends Endocrinol Metab 24(5):247–256
80. Karlsson HK, Zierath JR, Kane S, Krook A, Lienhard GE, Wallberg-Henriksson H (2005) Insulin-stimulated phosphorylation of the Akt substrate AS160 is impaired in skeletal muscle of type 2 diabetic subjects. Diabetes 54:1692–1697
81. Kern PA, Ranganathan S, Li C, Wood L, Ranganathan G (2001) Adipose tissue tumor necrosis factor and interleukin-6 expression in human obesity and insulin resistance. Am J Physiol Endocrinol Metab 280:E745–E751
82. Klaus S, Keijer J (2004) Gene expression profiling of adipose tissue: Individual, depot-dependent, and sex-dependent variabilities. Nutrition 20:115–120
83. Kleinridders A, Ferris Ham Cai W, Kahn CR (2014) Insulin action on brain regulates systemic metabolism and brain function. Diabetes 63(7):2232–2243
84. Knowler WC, Barrett-Connor E, Fowler SE, Hamman RF, Lachin JM, Walker EA, Nathan DM (2002) Reduction in the incidence of type 2 diabetes with lifestyle intervention or metformin. N Engl J Med 346:393–403
85. Korach-Andre M, Gounarides J, Deacon R, Beil M, Sun D, Gao J, Laurent D (2005) Age and muscle-type modulated role of intramyocellular lipids in the progression of insulin resistance in nondiabetic Zucker rats. Metabolism 54:522–528
86. Kurosu H, Yamamoto M, Clark JD, Pastor JV, Nandi A, Gurnani P, McGuinness OP, Chikuda H, Yamaguchi M, Kawaguchi H, Shimomura I, Takayama Y, Herz J, Kahn CR, Rosenblatt KP, Kuro-o M (2005) Suppression of aging in mice by the hormone Klotho. Science 309:1829–1833
87. Larkin LM, Reynolds TH, Supiano MA, Kahn BB, Halter JB (2001) Effect of aging and obesity on insulin responsiveness and GLUT-4 glucose transporter content in skeletal muscle of Fischer 344 x brown Norway rats. J Gerontol A Biol Sci Med Sci 56:B486–B492
88. Li M, Verdijk LB, Sakamoto K, Ely B, van Loon LJ, Musi N (2012) Reduced AMPK-ACC and mTOR signaling in muscle from older men, and effect of resistance exercise. Mech Ageing Dev 133:655–664
89. Lindstrom P (2007) The physiology of obese-hyperglycemic mice [ob/ob mice]. ScientificWorldJournal 7:666–685
90. List EO, Sackmann-Sala L, Berryman DE, Funk K, Kelder B, Gosney ES, Okada S, Ding J, Cruz-Topete D, Kopchick JJ (2011) Endocrine parameters and phenotypes of the growth hormone receptor gene disrupted (GHR–/–) mouse. Endocr Rev 32:356–386
91. Lubbers ER, List EO, Jara A, Sackman-Sala L et al (2013) Adiponectin in mice with altered GH action: links to insulin sensitivity and longevity? J Endocrinol 216:363–374
92. Lumeng CN, DelProposto JB, Westcott DJ, Saltiel AR (2008) Phenotypic switching of adipose tissue macrophages with obesity is generated by spatiotemporal differences in macrophage subtypes. Diabetes 57:3239–3246
93. Martin-Montalvo A, Mercken EM, Mitchell SJ, Palacios HH, Mote PL, Scheibye-Knudsen M, Gomes AP, Ward TM, Minor RK, Blouin MJ, Schwab M, Pollak M, Zhang Y, Yu Y,

Becker KG, Bohr VA, Ingram DK, Sinclair DA, Wolf NS, Spindler SR, Bernier M, de Cabo R (2013) Metformin improves healthspan and lifespan in mice. Nat Commun 4:2192
94. Masternak MM, Bartke A, Wang F, Spong A, Gesing A, Fang Y, Salmon AB, Hughes LF, Liberati T, Boparai R, Kopchick JJ, Westbrook R (2012) Metabolic effects of intra-abdominal fat in GHRKO mice. Aging Cell 11:73–81
95. McArdle MA, Finucane OM, Connaughton RM, McMorrow AM, Roche HM (2013) Mechanisms of obesity-induced inflammation and insulin resistance: insights into the emerging role of nutritional strategies. Front Endocrinol 4:52
96. Milanski M, Arruda AP, Coope A, Ignacio-Souza LM, Nunez CE, Roman EA, Romanatto T, Pascoal LB, Caricilli AM, Torsoni MA, Prada PO, Saad MJ, Velloso LA (2012) Inhibition of hypothalamic inflammation reverses diet-induced insulin resistance in the liver. Diabetes 61:1455–1462
97. Miquel J, Economos AC, Flemming J, Johnson JE Jr (1980) Mitochondrial role in cell aging. Exp Gerontol 15(6):575–591
98. Mohamed-Ali V, Goodrick S, Rawesh A, Katz DR, Miles JM, Yudkin JS, Klein S, Coppack SW (1997) Subcutaneous adipose tissue releases interleukin-6, but not tumor necrosis factor-alpha, in vivo. J Clin Endocrinol Metab 82:4196–4200
99. Moran C, Phan TG, Chen J, Blizzard L, Beare R, Venn A, Münch G, Wood AG, Forbes J, Greenaway TM, Pearson S, Srikanth V (2013) Brain atrophy in type 2 diabetes: regional distribution and influence on cognition. Diabetes Care 36(12):4036–4042
100. Mortensen B, Poulsen P, Wegner L, Stender-Petersen KL, Ribel-Madsen R, Friedrichsen M, Birk JB, Vaag A, Wojtaszewski JF (2009) Genetic and metabolic effects on skeletal muscle AMPK in young and older twins. Am J Physiol Endocrinol Metab 297: E956–E964
101. Obici S, Feng Z, Morgan K, Stein D, Karkanias G, Rossetti L (2002) Central administration of oleic acid inhibits glucose production and food intake. Diabetes 51(2):271–275
102. Okada-Iwabu M, Yamauchi T, Iwabu M, Honma T, Hamagami K, Matsuda K, Yamaguchi M, Tanabe H, Kimura-Someya T, Shirouzu M, Ogata H, Tokuyama K, Ueki K, Nagano T, Tanaka A, Yokoyama S, Kadowaki T (2013) A small-molecule AdipoR agonist for type 2 diabetes and short life in obesity. Nature 503:493–499
103. Palmer AJ, Chung MY, List EO, Walker J, Okada S, Kopchick JJ, Berryman DE (2009) Age-related changes in body composition of bovine growth hormone transgenic mice. Endocrinology 150:1353–1360
104. Perry CG, Kane DA, Lanza IR, Neufer PD (2013) Methods for assessing mitochondrial function in diabetes. Diabetes 62(4):1041–1053
105. Petersen KF, Befroy D, Dufour S, Dziura J, Ariyan C, Rothman DL, DiPietro L, Cline GW, Shulman GI (2003) Mitochondrial dysfunction in the elderly: possible role in insulin resistance. Science 300:1140–1142
106. Paolisso G, Fambardella A, Ammendola S, D'Amore A, Balbi V, Varricchio M, M'Onofrio F (1996) Glucose tolerance and insulin action in healthy centenarians. Am J Physiol 270(5 Pt 1):E890–E894
107. Reaven GM, Chen N, Hollenbeck C, Chen YD (1989) Effect of age on glucose tolerance and glucose uptake in healthy individuals. J Am Geriatr Soc 37:735–740
108. Reger MA, Watson GS, Green PS, Wilkinson CW, Baker LD, Cholerton B, Fishel MA, Plymate SR, Breitner JC, DeGroodt W, Mehta P, Craft S (2008) Intranasal insulin improves cognition and modulates beta-amyloid in early AD. Neurology 70:440–448
109. Rera M, Clark RI, Walker DW (2012) Intestinal barrier dysfunction links metabolic and inflammatory markers of aging to death in Drosophila. Proc Natl Acad Sci U S A 109:21528–21533
110. Reznick RM, Zong H, Li J, Morino K, Moore IK, Yu HJ, Liu ZX, Dong J, Mustard KJ, Hawley SA, Befroy D, Pypaert M, Hardie DG, Young LH, Shulman GI (2007) Aging-associated reductions in AMP-activated protein kinase activity and mitochondrial biogenesis. Cell Metab 5:151–156

111. Rother E, Kuschewski R, Alcazar MA, Oberthuer A, Bae-Gart I, Vohlen C, Roth B, Dötsch J (2012) Hypothalamic JNK1 and IKKβ activation and impaired early postnatal glucose metabolism after maternal perinatal high-fat feeding. Endocrinology 153(2):770–781
112. Rowe JW, Minaker KL, Pallotta JA, Flier JS (1983) Characterization of the insulin resistance of aging. J Clin Invest 71:1581–1587
113. Santomauro AT, Boden G, Silva ME, Rocha DM, Santos RF, Ursich MJ, Strassmann PG, Wajchenberg BL (1999) Overnight lowering of free fatty acids with Acipimox improves insulin resistance and glucose tolerance in obese diabetic and nondiabetic subjects. Diabetes 48(9):1836–1841
114. Santos-Parker JR, LaRocca TJ, Seals DR (2014) Aerobic exercise and other healthy lifestyle factors that influence vascular aging. Adv Physiol Educ 38:296–307
115. Schipper HM, Song W (2015) A heme oxygenase-1 transducer model of degenerative and developmental brain disorders. Int J Mol Sci 16(3):5400–5419
116. Selman C, Lingard S, Choudhury AI et al (2008) Evidence for lifespan extension and delayed age-related biomarkers in insulin receptor substrate 1 null mice. FASEB J 22:807–818
117. Scherer T, O'Hare J, Diggs-Andrews K, Schweiger M, Cheng B, Lindtner C, Zielinski E, Vempati P, Su K, Dighe S, Milsom T, Puchowicz M, Scheja L, Zechner R, Fisher SJ, Previs SF, Buettner C (2011) Brain insulin controls adipose tissue lipolysis and lipogenesis. Cell Metab 13:183–194
118. Shimokata H, Muller DC, Fleg JL, Sorkin J, Ziemba AW, Andres R (1991) Age as independent determinant of glucose tolerance. Diabetes 40:44–51
119. Shin AC, Fasshauer M, Filatova N, Grundell LA, Zielinski E, Zhou JY, Scherer T, Lindtner C, White PJ, Lapworth AL, Ilkayeva O, Knippschild U, Wolf AM, Scheja L, Grove KL, Smith RD, Qian WJ, Lynch CJ, Newgard CB, Buettner C (2014) Brain insulin lowers circulating bcaa levels by inducing hepatic bcaa catabolism. Cell Metab 20:898–909
120. Short KR, Vittone JL, Bigelow ML, Proctor DN, Rizza RA, Coenen-Schimke JM, Nair KS (2003) Impact of aerobic exercise training on age-related changes in insulin sensitivity and muscle oxidative capacity. Diabetes 52:1888–1896
121. Smith BA, Edwards MS, Ballachey BE, Cramer DA, Sutherland TM (1991) Body weight and longevity in genetically obese and non-obese mice fed fat-modified diets. Growth Dev Aging 55:81–89
122. Solomon TP, Marchetti CM, Krishnan RK, Gonzalez F, Kirwan JP (2008) Effects of aging on basal fat oxidation in obese humans. Metabolism 57:1141–1147
123. Talbot K, Han L, Schneider JA, Wilson RS, Bennett DA, Arnold SE (2006) Expression of pIRS-1 (S312 and S616) is elevated in MCI and AD and correlates with cognitive impairment and neurofibrillary pathology. Alzheimers Dement 2:S54
124. Tartar M, Barke A, Antebi A (2003) The endocrine regulation of aging by insulin-like signals. Science 299(5611):1346–1351
125. Tchkonia T, Morbeck DE, Von Zglinicki T, Van Deursen J, Lustgarten J, Scrable H, Khosla S, Jensen MD, Kirkland JL (2010) Fat tissue, aging, and cellular senescence. Aging Cell 9:667–684
126. Tucker MZ, Turcotte LP (2003) Aging is associated with elevated muscle triglyceride content and increased insulin-stimulated fatty acid uptake. Am J Physiol Endocrinol Metab 285:E827–E835
127. Villarreal-Molina MT, Antuna-Puente B (2012) Adiponectin: anti-inflammatory and cardioprotective effects. Biochimie 94:2143–2149
128. Virkamaki A, Korsheninnikova E, Seppala-Lindroos A, Vehkavaara S, Goto T, Halavaara J, Hakkinen AM, Yki-Jarvinen H (2001) Intramyocellular lipid is associated with resistance to in vivo insulin actions on glucose uptake, antilipolysis, and early insulin signaling pathways in human skeletal muscle. Diabetes 50:2337–2343
129. Wanke R, Wolf E, Hermanns W, Folger S, Buchmüller T, Brem G (1992) The GH-transgenic mouse as an experimental model for growth research: clinical and pathological studies. Horm Res 37:74–87

130. Wei YH, Kao SH, Lee HC (1996) Simultaneous increase of mitochondrial DNA deletions and lipid peroxidation in human aging. Ann N Y Acad Sci 786:24–43
131. Weisberg SP, McCann D, Desai M, Rosenbaum M, Leibel RL, Ferrante AW Jr (2003) Obesity is associated with macrophage accumulation in adipose tissue. J Clin Invest 112:1796–1808
132. White MF (1998) The IRS-signalling system: a network of docking proteins that mediate insulin action. Mol Cell Biochem 182:3–11
133. Wijsman CA, Rozing MP, Streefland TC, le Cessie S, Mooijaart SP, Slagboom PE, Westendorp RG, Pijl H, van Heemst D, Leiden Logevity Study group (2011) Familial longevity is marked by enhanced insulin sensitivity. Aging Cell 10(1):114–121
134. Wolff GL, Roberts DW, Mountjoy KG (1999) Physiological consequences of ectopic agouti gene expression: the yellow obese mouse syndrome. Physiol Genomics 1:151–163
135. Xing Y, Musi N, Fujii N, Zou L, Luptak I, Hirshman MF, Goodyear LJ, Tian R (2003) Glucose metabolism and energy homeostasis in mouse hearts overexpressing dominant negative alpha2 subunit of AMP-activated protein kinase. J Biol Chem 278:28372–28377
136. Xu H, Barnes GT, Yang Q, Tan G, Yang D, Chou CJ, Sole J, Nichols A, Ross JS, Tartaglia LA, Chen H (2003) Chronic inflammation in fat plays a crucial role in the development of obesity-related insulin resistance. J Clin Invest 112(12):1821–1830. 1821-1830. doi:10.1172/JCI19451
137. Yamauchi T, Kamon J, Waki H, Terauchi Y, Kubota N, Hara K, Mori Y, Ide T, Murakami K, Tsuboyama-Kasaoka N, Ezaki O, Akanuma Y, Gavrilova O, Vinson C, Reitman ML, Kagechika H, Shudo K, Yoda M, Nakano Y, Tobe K, Nagai R, Kimura S, Tomita M, Froguel P, Kadowaki T (2001) The fat-derived hormone adiponectin reverses insulin resistance associated with both lipoatrophy and obesity. Nat Med 7:941–946
138. Yang Y, Ma D, Wang Y, Jiang T, Hu S, Zhang M, Yu X, Gong CX (2013) Intranasal insulin ameliorates tau hyperphosphorylation in a rat model of type 2 diabetes. J Alzheimers Dis 33:329–338
139. Yarchoan M, Arnold SE (2014) Repurposing diabetes drugs for brain insulin resistance in alzheimer disease. Diabetes 63:2253–2261
140. Youm YH, Grant RW, McCabe LR, Albarado DC, Nguyen KY, Ravussin A, Pistell P, Newman S, Carter R, Laque A, Munzberg H, Rosen CJ, Ingram DK, Salbaum JM, Dixit VD (2013) Canonical Nlrp3 inflammasome links systemic low-grade inflammation to functional decline in aging. Cell Metab 18:519–532
141. Yu C, Chen Y, Cline GW, Zhang D, Zong H, Wang Y, Bergeron R, Kim JK, Cushman SW, Cooney GJ, Atcheson B, White MF, Kraegen EW, Shulman GI (2002) Mechanism by which fatty acids inhibit insulin activation of insulin receptor substrate-1 (irs-1)-associated phosphatidylinositol 3-kinase activity in muscle. J Biol Chem 277:50230–50236
142. Zaid H, Antonescu CN, Randhawa VK, Klip A (2008) Insulin action on glucose transporters through molecular switches, tracks and tethers. Biochem J 413(2):201–215
143. Zeigerer A, McBrayer MK, McGraw TE (2004) Insulin stimulation of GLUT4 exocytosis, but not its inhibition of endocytosis, is dependent on RabGAP AS160. Mol Biol Cell 15:4406–4415
144. Zhang G, Li J, Purkayastha S, Tang Y, Zhang H, Yin Y, Li B, Liu G, Cai D (2013) Hypothalamic programming of systemic ageing involving IKKβ, NF-kB and GnRH. Nature 497(7448):211–216
145. Zierath JR, Krook A, Wallberg-Henriksson H (1998) Insulin action in skeletal muscle from patients with NIDDM. Mol Cell Biochem 182:153–160

Renal Aging and Transplantation

John P. Higgins and Stuart K. Kim

Contents

1 Introduction

The kidney plays an important role in human aging because it shows stereotyped changes in morphology and physiology beginning around age 50. Kidneys show clear changes in structure and morphology with age. Starting at about age 50, the weight and volume of the kidney shrinks by about one third [1]. The glomerulus is a network of capillaries that is located at the beginning of the nephron that filters

J.P. Higgins, Ph.D.
Department Pathology, Stanford University Medical Center, Stanford, CA 94305, USA
e-mail: john.higgins@stanford.edu

S.K. Kim, Ph.D. (✉)
Department Developmental Biology and Genetics, Stanford University Medical Center, Stanford, CA 94305, USA
e-mail: stuartkm@stanford.edu

F. Sierra, R. Kohanski (eds.), *Advances in Geroscience*,
DOI 10.1007/978-3-319-23246-1_13

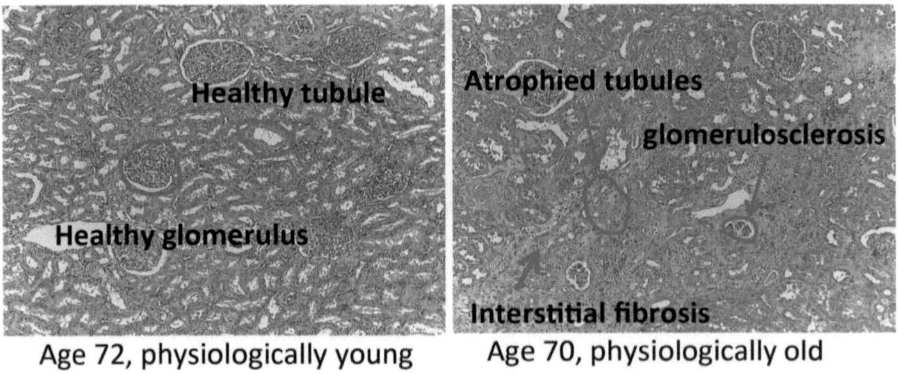

Fig. 1 Physiological kidney aging. Shown are two kidneys from elderly donors with a similar chronological age. The left kidney is physiologically younger than the right kidney. The kidney on the right shows classic signs of structural and morphological changes associated with renal aging, decreased kidney function, and poor renal transplant outcome. The glomeruli show signs of scarring (glomerulosclerosis). The cells in the interstitial space have thickened extracellular membranes indicative of fibrosis (interstitial fibrosis). The tubules are smaller, have thickened walls and have atrophied (tubular atrophy). Genetic and molecular biomarkers for physiological aging could be used to distinguish kidney donors based on physiological age. This could rescue renal organs such as the one on the left from exclusion, possibly making them eligible for renal transplant. The net effect would be to expand the pool of renal donors available for patients with end stage renal disease

blood to form urine. The number of glomeruli declines by one third to one half in old age through obsolescence or glomerulosclerosis. The tubules associated with the sclerosed glomeruli cease to function and the filtration capacity of each kidney declines. As the tubules atrophy, the tubular epithelium shrinks, the tubules contract and the basement membranes of the tubules thicken. Interstitial fibrosis increases with age, and refers to an increase in connective tissue in the space between the tubules. With age, the walls of arterioles become thick, caused by a deposition of hyaline. Hyaline is composed of plasma protein (for example C3 and IgM) that has leaked across the endothelium and accumulated in the wall of the arterioles. In Fig. 1, the kidney section on the right is from a 70 yo and shows many of the morphological hallmarks of aging: atrophied tubules, glomerulosclerosis and interstitial fibrosis.

The rate of filtration of the blood through the glomerulus (i.e. the glomerular filtration rate; GFR) is one of the primary indicators of renal function. On average, the glomerular filtration rate begins to decline at age 40, although the rate of decline is different in different individuals [1–3]. The loss of renal function due to advancing age may become clinically significant over a normal human life span. In the elderly, glomerular filtration rate often reaches levels low enough to indicate chronic kidney disease. By age 70, 35 % of people have moderate chronic kidney disease (stage 3) according to the National Health and Nutrition Examination Survey [4]. A healthy GFR is $\geq$90 ml/min for an adult, but when the GFR falls to <15 ml/min the patient is considered to have end stage renal disease. Patients with end stage renal disease require dialysis in order to survive as the blood no longer receives adequate renal filtration, but simply going on dialysis doubles the 5-year risk for mortality. Renal transplantation is preferable to dialysis for end stage renal disease because the donated kidney can function at a relatively normal level and restore glomerular filtration rate. Both quality of life and survival are greatly improved by transplantation compared to dialysis [6]. The decline in glomerular filtration rate is likely caused by structural changes to the glomerulus, the interstitium and the arterioles [3, 7].

Understanding the genetic and molecular mechanisms that contribute to kidney aging will advance our basic understanding of the aging process in humans. Furthermore, aging research on the kidney could have important clinical applications. In the long run, a better understanding of renal aging could lead to strategies or treatments to delay the aging process. This could delay or prevent chronic kidney disease and reduce the number of people suffering from end stage renal disease.

In the short run, one promising opportunity is to use knowledge of aging to develop biomarkers in order to measure physiological age, as opposed to chronological age. For instance, from a cohort of elderly, it would be desirable to be able to identify those that have physiologically young kidneys. Figure 1 illustrates two kidneys from donors of similar chronological age of about 70 years. The kidney on the left retains a youthful morphological appearance, equivalent to the appearance of kidneys from middle-aged donors, suggesting that this kidney is physiologically young. The kidney on the right shows classic signs of aging. Figure 1 illustrates the concept that adding information from histopathological and molecular biomarkers to chronological age can improve our knowledge of the true age of an organ better than chronological age alone

Individuals with kidneys that are physiologically younger are likely to show a lower incidence of renal disease as they grow older. Furthermore, donor age is the major criterion for success of a kidney in renal transplantation [8, 9], which means that individuals with kidneys that are physiologically young are likely to be better renal transplantation donors than individuals with kidneys that are physiologically old irrespective of their chronological age. Instead of categorically discarding all of the organs from donors above a certain age, it may be possible to select a subset of organs that are physiologically young and suitable for transplantation (Fig. 2). Renal transplant outcome declines gradually with age, and the difference between youthful and elderly kidney donors is relative but not absolute. With elderly renal donors, the fraction of renal transplants that are successful (as measured by graft survival after 1 and 5 years) is lower than the fraction of successful transplants from youthful donors. Still, some of the transplants from elderly donors are successful. Figure 2 illustrates the concept of using physiological age to help increase the pool of renal transplant donors. Exclusion criteria based on chronological age alone become increasingly strong as the donor ages (gradient arrow). Aging biomarkers could be used to provide information about the physiological age of the tissue, which might permit certain prospective donors (dots shown in red) by expanding the criteria to include physiological in addition to chronological age. This strategy would expand the pool of kidney organs suitable for transplantation, and thereby

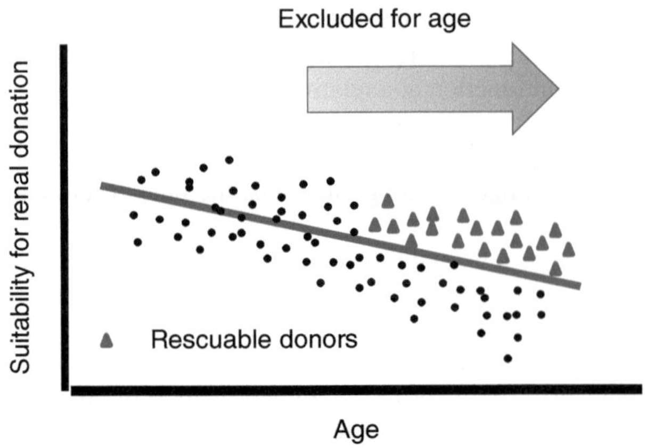

Fig. 2 Kidney aging and renal transplantation. Figure depicts how some kidneys from elderly donors may be suitable for renal transplantation. With increasing age of the donor, there is a steady decline in the percent of renal transplants that survive 1 and 5 years after transplantation. Nevertheless, there are many renal transplants from elderly donors that last for a suitable length of time. In principle, aging biomarkers could be used to identify kidneys that are physiologically young, and perhaps could be used to rescue organs that are currently discarded due to old age. The gradient *arrow* indicates how donor age becomes a stronger criterion for exclusion with increasing chronological age. The *red triangles* indicate kidney donors that may still be suitable for renal transplantation, even though their chronological age may have exceeded an exclusion criteria cut-off that is currently used. The *black dots* indicate individual donor kidneys

allow patients with end stage renal disease to receive a transplant and end their time on dialysis treatment.

2 Renal Transplantation

Renal transplantation is the best option for patients with end stage renal disease. The survival rates for kidneys in recipients following renal transplantation are 80 % after 1 year and about 60 % after 5 years [6, 10, 11]. Renal transplantation can possibly extend the lives of patients by 10–15 years compared to dialysis [6].

However, there are many more patients with end stage renal disease than there are renal transplantation donors. In 2014, there were 101,513 people in the United States on the waiting list for kidney transplantations. At the same time, there were only about 13,125 donor kidneys available [12]. Some patients with end stage renal disease receive a kidney from a living donor. In the donation process, there is a large number of volunteers that offer to donate their kidney. Most of the volunteered kidneys are excluded from becoming a kidney donor for medical reasons, including old age. In a recent study of kidney donors at Stanford University from 2007 to 2009, it was found that 92 % of potential donor kidneys were excluded from consideration, exacerbating the shortage of kidneys available for transplantation [13]. As a result of the shortage of donor kidneys, many patients with end stage renal disease do not have the opportunity to receive a donor kidney for renal transplantation, an operation that would extend their lives. In principle, improvements in the criteria for exclusion might allow one to rescue potential donor kidneys that might be suitable for renal transplantation even though they are currently excluded from renal transplantation.

2.1 Predictors of Renal Transplant Outcome

Currently, there are three main criteria affecting the success of renal transplantation: ABO blood type and HLA histocompatibility matching, kidney preservation time and donor age. Donor kidneys that do not match the recipient for ABO blood type and HLA histocompatibility are at risk for graft rejection. Many kidneys are stored cold while awaiting the transplantation procedure, especially kidneys from deceased donors.

The third factor, age, is important for renal transplantation success as greater age of the donor diminishes the chance of success of the renal transplant. On the average, kidneys from older donors have a shorter graft survival time than those from younger donors. The short-term difference is relatively minor, but is amplified with the passage of time: About 95 % of kidneys have a graft survival greater than 1 year when the donor was younger than age 50, and about 85 % of kidneys have a 1 year graft survival rate when the donor was over age 65 [14]. At 5 years after the transplantation, there is about a 25 % increase in renal survival in kidneys from younger donors compared to those from elder donors [15]. Thus, while on average there is a drop off in graft survival from elder donors, there is also a significant number of exceptions where a kidney from an elder donor has a long graft survival time [16–21].

If we could better understand why old kidney age affects graft survival, it might be possible to identify kidneys from the elderly population that are still fit for renal transplantation. One way to do this is to develop a set of biomarkers for physiological age that could predict renal transplant outcome better than chronological age, or at least that could be used to improve transplant outcome in combination with chronological age. That is, among elderly donors of the same age, the kidney aging biomarker should be able to identify donors with a higher chance for long term graft survival. This might be one way to expand the pool of donor kidneys available for renal transplantation. Several molecular assays are being developed as biomarkers for renal graft survival.

Recent studies have begun to identify biomarkers of aging that can be used to help predict how well a kidney will perform in renal transplantation. Aging is a complex process dependent on many different mechanisms and pathways. One of the cellular pathways that may contribute to aging of the kidney is cell senescence [22–26]. Cell senescence could impact renal function if it prevented cell division necessary to replace lost or damaged cells. However, cell turnover in the kidney is normally low compared to other tissues with high rates of cell turnover, such as the hematopoietic system or the lining of the gut [27]. Thus, it is unclear how strongly cell senescence would impact renal function under normal circumstances. However, disease or injury could result in loss of renal cells by apoptosis, and in this situation cell proliferation would be required in order to replace dead cells.

As kidneys grow old, there is an increased frequency of senescent cells. Senescence could contribute to renal aging in at least three ways. First, cell senescence may prevent new cells from replenishing cells that are lost from disease or damage. Second, cell senescence could lead to an increase of macromolecular damage. When a cell divides, there is a burst of new synthesis of all of the macromolecules needed to form the new cells (DNA, RNA, protein etc.). Conversely, in any post-mitotic cell such as a senescent cell, there is no net gain of RNA and protein, so new RNA and proteins are synthesized via transcription and translation at a much lower rate. Assuming that RNA and protein levels in senescent cells are at steady state, then the levels of transcription and translation are set to merely replace RNA as it is lost via RNA degradation and protein as it is lost by degradation via proteolysis machinery such as autophagosomes or the proteasome. As a result, macromolecules in a non-dividing cell have a longer molecular half-life than those in a dividing cell. The increased molecular half-life exposes all of the molecules to increased susceptibility for damage accumulation; for instance, macromolecules in a non-dividing cell would be expected to have higher levels of damage from reactive oxygen species. Third, senescent cells secrete a variety of signaling molecules and cytokines, a phenomenon referred to as the senescence-associated secretory phenotype [28]. The senescence-associated cytokines include factors such as interleukin-6 and interleukin-1β that can activate inflammatory signaling pathways. One possibility is increased abundance of senescent cells during aging contributes to chronic inflammation.

One marker of senescent cells is expression of the cell cycle regulator CDKN2A/p16. CDKN2A/p16 plays an important role in cell cycle regulation by decelerating progression from the G1 to the S phase [29]. In the normal cell cycle, CDKN2A/p16 acts to inhibit cell division by binding CDK4/6, which ultimately inhibits the activity of transcription factors such as E2F1 and arrests cell proliferation [30]. High levels of CDKN2A/p16 expression prevent cell division and are a hallmark of cell senescence [31]. Expression of CDKN2A/p16 increases with age in the kidney [32, 33]. Several studies have shown that expression levels of CDKN2A/p16 can be used as a biomarker of aging in order to predict renal transplant outcome. At the time of the renal transplant, a kidney biopsy was obtained and CDKN2A/p16 levels were measured [22, 34, 35]. Age, CDKN2A/p16 levels, and a combination of Age/CDKN2A/p16 levels were evaluated as predictors for renal transplant success. CDKN2A expression (biological age) was found to be better than donor age (chronological age) in predicting organ function [35]. However, CDKN2A/p16 levels combined with chronological age was found to be the most powerful predictor for renal function following transplantation [22]. The key concept is that CDKN2A/p16 levels may be measuring biological age, and that biological age may be better than simple chronological age as a predictor of future renal function following transplantation.

Another strategy to develop biomarkers for renal transplantation is to identify gene expression signatures that can predict renal graft survival. In this approach, gene expression from the entire genome is measured from kidney biopsies at the time of transplantation using DNA chips. The renal transplants are then separated into two groups based on success of the graft, and the gene expression data are analyzed to identify differences in expression between kidneys that were or were not successful in the renal transplant. In one study, 31 renal allografts were separated into low and high GFR after 1 year following transplantation. Then, expression profiles taken at the time of transplantation were analyzed, resulting in the identification of 52 genes that showed significantly different expression profiles between the high- and low-functioning kidneys [36]. The gene expression profile of these 52 genes at the time of transplantation was able to predict the success of the renal transplantation over a medium term.

In a second study, 92 renal allografts were separated based on whether or not patients required dialysis during the first week (delayed graft function), and then gene expression data were analyzed to identify 206 genes whose expression showed a significant difference. This study suggests that preimplant gene expression profiles may be able to identify kidneys of poor quality that perform poorly in transplantation. This information may eventually improve organ allocation [37].

3 Hallmarks of Renal Aging

To better understand renal aging, it is useful to consider mechanisms that are involved in aging in other tissues, and even other species such as mice, flies, worms and yeast. The human aging process can be thought of as a clock that spans about 80 years. Aging mechanisms guide the rate at which this clock proceeds, and the most central pathways are part of the clock mechanism itself. As we grow old, aging affects many of the underlying networks in the kidney. There is accumulation of damage of diverse types to cellular components such as DNA, proteins and lipids. There are changes in gene expression and epigenetic networks with age. Cells lose their ability to divide and undergo senescence. There is a steady increase in the thickness of the extracellular matrix that is a major determinant of fibrosis. It is possible that changes in each of these networks serves as part of a molecular aging clock, that changes over time and dictates the rate of functional decline of the kidney.

A hallmark of aging not only changes as we grow old, but it also plays an important functional role in the physiological decline of the kidney with old age. Genetic and pharmacological experiments that reset the clock in old cells or organs to the young state should have a beneficial effect. By contrast, experiments in which the aging pathway has been reset in young cells to the old state should cause rapid aging to ensue.

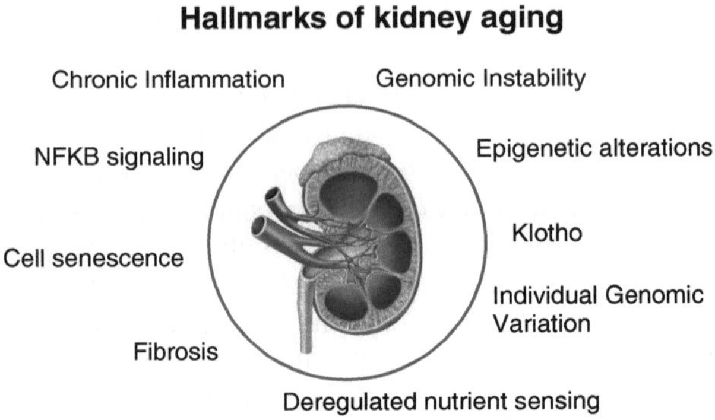

Fig. 3 Hallmarks of kidney aging. Shown are nine pathways that play important roles in renal aging

These hallmarks of kidney aging might collectively form a type of aging clock that dictate the functional and physiological state of the kidney over a lifetime. Although these pathways progress during aging in everyone, the rate of progression can vary between people. For people of the same chronological age, the aging clock might be slightly more advanced in one person than the other. With additional studies, hallmarks of aging could one day become very important because they could be used as biomarkers to report the true physiological age of a person or tissue, rather than mere chronological age. Not only would the aging biomarkers associate with the current functional state of the kidney better than chronological age, but the aging biomarkers would be better than chronological age at predicting the future trajectory of renal decay.

A recent review describes nine hallmarks of aging that form the conceptual pillars to understand changes as one grows old [38]. These hallmarks are: genomic instability, telomere attrition, epigenetic alterations, loss of proteostasis, deregulated nutrient-sensing, mitochondrial dysfunction, cellular senescence, stem cell exhaustion, and altered intercellular communication. In addition to the general hallmarks, kidney aging is known to involve increased Klotho expression, chronic inflammation, and fibrosis (Fig. 3). Below, we consider each of these hallmarks and summarize what is known about how that hallmark may contribute to human renal aging.

One of the most important hallmarks of kidney aging is chronic inflammation [39]. Chronic inflammation occurs when there is an increased abundance of immune cells – B cells, T cells, neutrophils and macrophages. Low levels of activity of the immune cells lead to a low grade inflammatory response that contributes to fibrosis and tissue damage with age. One of the causes of chronic inflammation in the kidney may be increased systemic levels of inflammatory cytokines, such as IL1, IL6, and TNFα, in old age [40, 41]. The inflammatory response leads to increased production and accumulation of fibrinogen and C-reactive protein by the liver, leading to increased systemic levels of inflammatory biomarkers. The level of chronic inflammation is higher in patients with chronic kidney disease compared to healthy age-matched controls, suggesting that chronic inflammation may play a role in the etiology of this disease [39]. However, the relationship between cause and effect between chronic inflammation and renal aging is unclear; specifically, it is unclear whether chronic inflammation causes renal damage to accumulate with age, whether age-related renal damage causes chronic inflammation, or whether both act together in a complicated feedback loop with increasingly dire consequences.

The inflammatory cytokines that are responsible for chronic inflammation could arise from several sources. One source is from immune cells (B cells, T cells and macrophages) that become dysregulated in old age [42, 43]. A second source is from adipocytes, which are known to produce many inflammatory cytokines, including TNFα and IL-6 [44]. Cytokines secreted from adipose tissue would enter the blood system and increase chronic inflammation throughout the body, including the kidney. A third source is from senescent cells, which secrete many inflammatory cytokines as part of the senescence phenotype [45]. Senescent cells increase in number in old age, leading to increased production of the inflammatory cytokines, a phenomenon termed the senescent-associated secretory phenotype [28]. A fourth source is from cells within the kidney itself. DNA microarray analysis showed that aged human kidneys have increased expression of certain inflammatory cytokines and chemokines [46]. One of the results of chronic inflammation is the recruitment of immune cells that secrete inflammatory cytokines, which may further increase chronic inflammation in the kidney as part of a positive feedback loop.

The inflammatory cytokine TNFα activates initiates a signaling cascade leading to activation of the transcription factor NFκB. The TNFα signaling pathway is induced when TNFα binds the TNF receptor (TNFR1), which leads to the degradation of IKB, a protein that normally keeps the NFκB complex inactive in the cytoplasm. Once IKB is degraded, the NFκB complex is no longer tethered in the cytoplasm and enters the nucleus where it activates expression of its target genes, which include many inflammatory cytokines and chemokines. Activity of NFκB in the kidney has been found to increase with age in the rat [47]. One of the major functions of NFKB is to mediate inflammatory and innate immune responses. Besides TNFα, the inflammatory cytokine IL-1, lipopoysacharide, and reactive oxygen species can activate NFκB.

Another hallmark of renal aging is fibrosis of the interstitium or scarring of the glomeruli [48]. The glomeruli are the functional unit of the kidney responsible for filtering the blood. The tubules and interstitium constitute 90 % of the volume of the kidney. Interstitial fibrosis and glomerulosclerosis increase with age, characterized by an increased thickening of the extracellular matrix. Extracellular matrix is composed primarily of collagen. Matrix metalloproteases are zinc-dependent endopeptidases responsible for degrading collagen and proteoglycans, and may function to help remodel the extracellular matrix. Excess production of collagen or altered expression of matrix metalloproteases could play a role in thickening of the extracellular matrix in old age. Mesenchymal cells (i.e. fibroblasts and myofibroblasts) produce extracellular matrix. Increased numbers or altered functions of these cells in old age could be responsible for increased collagen deposition and fibrosis. Activated interstitial mesenchymal cells are thought to contribute directly to renal fibrosis by secreting fibrotic factors and extracellular matrix proteins that accumulate in the interstitial space and disrupt normal epithelial architecture [48].

Changes in DNA methylation and DNA repair play a role in fibrosis. Bechtel et al. found that increased DNA methylation plays a key role in renal fibrosis. They performed a genome-wide screen to look for DNA methylation differences between fibroblasts from fibrotic and non-fibrotic kidneys, and found 12 genes that showed increased levels of DNA methylation in the fibrotic kidneys [49]. They then inhibited DNA methylation using 5-azacytidine and found that fibrosis was lessened [49]. These results indicate that increased DNA methylation plays a causative role in causing renal fibrosis.

Another clue about the mechanisms responsible for renal fibrosis was obtained by studying a heriditary form of chronic kidney disease. Karyomegalic interstitial nephritis is a rare hereditary form of chronic kidney disease with only 12 known families worldwide [50]. Kidneys in patients with karyomegalic interstitial nephritis have enlarged hyperchromatic nuclei and develop chronic kidney disease in their third decade. The specific diagnostic traits associated with chronic kidney disease in these patients include interstitial fibrosis, tubular atrophy and microcyst formation. These symptoms are seen in normal elder subjects as part of kidney aging, suggesting that karyomegalic interstitial nephritis involves an accelerated rate of fibrosis in the kidney.

The genetic cause for karyomegalic interstitial nephritis has recently been found to be due to mutations in the FAN1 gene. FAN1 plays a role in DNA repair. Specifically, FAN1 is required to repair interstrand DNA crosslinks, so that covalently cross-linked DNA strands cannot separate during S-phase in patients with a non-functional form of FAN1 [51, 52]. In renal tissue from patients with karyomegalic interstitial nephritis, there was an increased level of double strand breaks. These results show that DNA damage caused by reduced activity of FAN1 can potentiate renal fibrosis.

Nutrient sensing is a general hallmark of aging in all tissues and most species. Caloric restriction can extend lifespan of nearly every species tested, including yeast, worms, flies, mice and primates [53]. Recent work has begun to unravel the molecular mechanisms underlying lifespan extension due to caloric restriction [54–56]. These molecular mechanisms involve key modulators of aging such as the sirtuins and mTOR.

Sirtuins are a family of protein deacetylases that are involved in a diverse array of cellular processes such as life span regulation, fat mobilization in human cells, insulin secretion and caloric restriction [57]. In mammals, there are seven sirtuin genes (SIRT1 to SIRT7). Activation of sirtuins has been shown to extend lifespan in diverse organisms including yeast, worms, flies and mice [58]. In the kidney, SIRT1 is abundantly expressed in renal medullary interstitial cells, where it is cytoprotective and participates in the regulation of blood pressure and sodium balance [59, 60]. Sirtuins require nicotinamide adenine dinucleotide (NAD+) as a cofactor, and hence could be responsive to changes in metabolic state of the cell. Levels of nicotinamide adenine dinucleotide show a marked decline in the kidney in old age, which may decrease activity of sirtuins thereby contributing to cellular dysfunction in old age [59, 61]. Regulation of sodium balance by SIRT1 involves repression of expression of the α-subunit of the epithelial sodium channel, ENaC [59].

Genetic experiments by He et al., 2010 have shown that Sirt1 activity plays an important role in kidney function [60]. On the one hand, lowering Sirt1 activity impairs renal function. For example, reduced Sirt1 expression in mouse renal cells *in vitro* leads to reduced resistance to oxidative stress. Genetically mutant mice with reduced Sirt1 activity ($Sirt1^{+/-}$) show lower levels of cyclooxygenase-2 (COX2) expression and impaired kidney function under several conditions. $Sirt1^{+/-}$ mice show more renal apoptosis and fibrosis in mice that have suffered kidney injury by unilateral ureteral obstruction. On the other hand, increasing Sirt1 activity improves renal function. Sirt1 activity can be pharmacologically increased either by resveratrol or by the drug SRT2183. Pharmacologic activation of Sirt1 improves cell survival in response to oxidative stress, attenuates renal apoptosis and fibrosis, and increases expression of COX2 in the renal medulla.

mTOR is a conserved kinase that acts in a nutrient sensing pathway to regulate cell growth and longevity. In yeast, worms and flies, mutants with reduced TOR activity have longer lifespans [56]. In mice, inhibition of mTOR activity with the drug rapamycin results in extended lifespan [56]. During normal kidney aging, mTOR expression shows a marked increase in glomerular mesangial cells in old age [62]. This result suggests that increased levels of mTOR in the kidney may contribute to poor cell regulation and cell senescence in old age.

Klotho plays an important role in aging of the kidney [63]. Klotho is a transmembrane protein primarily expressed in the distal tubule cells of the kidney and the brain choroid plexus. Loss-of-function mutations in the Klotho gene in mice are associated with symptoms of premature aging, including hyperphosphatemia and a shorter lifespan [64]. Conversely, overexpression of Klotho extends lifespan [64]. In humans, serum levels of Klotho decrease with age after age 40 years and there are low levels of Klotho in patients with Chronic Kidney Disease [65–67]. One of the main functions of Klotho is to act as a co-receptor for Fibroblast Growth Factor 23 [68]. However, there is also evidence that Klotho can also regulate the Insulin-like Growth Factor signaling pathway, can participate in Ca^{2+} homeostasis, phosphate homeostasis and can relieve oxidative stress [69].

Changes in gene expression occur with age, and can be used to predict the physiological age of kidneys. DNA microarrays have been used to define the changes in gene expression that accompany the renal aging process. One study analyzed RNA from 74 patients ranging in age from 27 to 92 years, and found 985 genes to change expression with age [46]. Among these 74 individuals, there was a good correlation between the gene expression signature and the biological age of the kidney. For example, for some kidney samples, the gene expression signatures did not resemble kidneys of their chronological age but rather kidneys from individuals that were younger. Subsequent histological examination of these renal samples showed that they had lower levels of interstitial fibrosis, arterial hyalinosis and glomerulosclerosis than typically found in kidneys of that chronological age. That is, the gene expression signatures were able to accurately predict that these renal samples had a biological age that was younger than their chronological age. Similarly, the gene expression signatures were also able to predict renal samples with a biological age that is older than their chronological age. A second study analyzed RNA isolated from 20 kidney samples and identified about 500 genes that changed expression with age [70]. The sets of kidney aging genes showed a large degree of overlap in the two DNA microarray studies.

What are the upstream transcription factors responsible for causing the changes in expression of the age-regulated kidney genes? In order to identify the upstream regulators, ChIP-seq data from the ENCODE consortium were examined. The ENCODE consortium has defined the in vivo binding sites for a large number of transcription factors (about 160) [71]. For each transcription factor, they used ChIP seq to first immunoprecipitate the transcription factor from tissue culture cells, and then sequenced the bound DNA in the immunoprecipitate. Each experiment resulted in a list of target genes bound by that transcription factor in specific cell lines. To find transcription factors that bind to the age-related kidney genes, a bioinformatics screen was performed to search for transcription factors that had ChIP seq datasets showing a large degree of overlap with the kidney age-related genes [72]. The top three transcription factors that showed enrichment to binding the aging-regulated genes were STAT1, STAT3 and NFκB. These three transcription factors are all known to be involved in mediating the inflammatory response.

STAT1, STAT3 and NFκB appear to mediate transcriptional changes during the kidney aging process *in vivo* [72]. All three transcription factors show higher levels of activation in old age. When the transcription factors are activated by inflammatory cytokines in human renal epithelial cells, the resulting changes in gene expression recapitulate the gene expression changes that occur during kidney aging to a large extent. These data indicate that activation of these three transcription factors during kidney aging may contribute to a large fraction of the aging transcriptional program. The finding that NFκB binds to and regulates genes during aging confirms and extends previous work showing that increased NFκB activity contributes to aging phenotypes in many tissues [73–75].

4 Genetic Differences May Explain Some of the Individual Variation in Kidney Aging

There is a great deal of variation in the rate of loss of kidney function with age between different individuals. In old age, many people have lost a considerable fraction of kidney function, leading to chronic kidney disease or end stage renal disease. Other people have relatively mild loss of kidney function. Genetic variation accounts for some of the individual variation in kidney aging. Genome-wide association studies (GWAS) have begun to define DNA polymorphisms that are associated with either reduced glomerular filtration rate or chronic kidney disease. In principle, genetic algorithms may one day be used in the young to help predict which individuals are at risk for a rapid loss of renal function as they grow old, and which individuals are more likely to have a slower rate of loss of kidney function with age. These genetic risk scores would be able to look-up the information from each of the individual polymorphisms and create one combined score to be used to predict future renal rates of aging.

Genome wide association studies have recently been successful at identifying single nucleotide polymorphisms (SNPs) associated with a variety of renal phenotypes associated with age. An example is glomerular filtration rate, which is an estimate of renal function based on levels of creatinine in the urine. The studies are controlled for age, so DNA variants associated with GFR in these studies would identify loci that predispose one to having higher or lower GFR for a given age. There are at least two ways in which a SNP could be associated with variation in GFR. First, the SNP could be associated with a different baseline of renal function (higher or lower GFR) but not with a difference in the rate of renal aging. In this case, GFR would decline similarly in individuals that vary at the SNP, but the GFR would have a different baseline so that those with a higher baseline would be less likely to have renal disease as they grow old. Second, the SNP could be associated with a different rate of renal aging. In this case, allelic variation at this SNP may have little effect on young adults as all might have a similar starting point for renal function. However, individuals with one allele could show more rapid rates of renal decline leading to larger and larger differences in GFR as one ages.

Another powerful phenotype used in kidney GWA studies is chronic kidney disease. Chronic kidney disease is diagnosed when an individual has glomerular filtration rate below 60 ml/min/1.73 m^2 for 3 or more months, or if the individual has other signs of kidney damage. The gene association studies using chronic kidney disease as a phenotype are similar to the GWAS using glomerular filtration rate as a trait because one of the main criteria for diagnosing chronic kidney disease is glomerular filtration rate. Chronic kidney disease splits up individuals based on GFR levels above or below the 60 ml/min/1.73 m^2 cutoff and GFR assigns a continuous variable to each individual.

Kidney GWA studies have also been performed for albuminuria and diabetic nephropathy. Albuminuria refers to albumin in the urine, which is another indication of poor kidney function. Diabetic nephropathy is characterized by three related conditions: albuminuria, low glomerular filtration rate, and high blood pressure.

A total of 67,073 individuals have been examined to search for SNPs associated with GFR, chronic kidney disease, albuminuria and diabetic nephropathy [76, 77]. These studies have found 27 SNPs associated with these parameters of kidney function. These SNPs may be useful in helping to identify individuals with high and low renal function. This information could be useful to individuals when they are alive by providing information useful for precision medicine. In addition, the genetic information could be useful to help identify which kidneys from elderly donors may still be viable for use in renal transplantations.

In order to predict renal function, one would not evaluate each of the kidney SNPs one at a time, but rather one would develop an algorithm that could evaluate all of the SNPs together and provide a score that summarizes the information from all of the SNPs combined. A preliminary study was performed combining information from 16 kidney SNPs to create a genetic risk score for chronic kidney disease [76]. The genotype score was evaluated in 2129 test subject and was partially successful in predicting chronic kidney disease risk. Carrying a high number of risk alleles was partially able to predict those at increased risk for chronic kidney disease. However, the effect of the genotype was small, and not necessarily an improvement over clinical factors such as lifestyle, blood pressure and the presence of Type 2 Diabetes. Beyond renal disease, genetic algorithms have been developed for Type 2 Diabetes mellitus, incident myocardial infarction, coronary heart disease, myocardial infarction and stroke in women [78–81]. In each case, the genetic algorithms had only a mild effect in predicting disease risk beyond current clinical tests, indicating that improved methods or more complete data will be required for these algorithms to become widely used. Future studies may be able to expand the number of SNPs known to be associated with renal function, leading to improved genetic algorithms with greater predictive power.

One study has specifically searched for SNPs that are associated with different levels of kidney function, independent of age [82]. To do this study, longitudinal data was used from the Baltimore Longitudinal Study of Aging and from the InCHIANTI cohort. Both of these cohorts have followed individuals over a number of years, so it was possible to follow the loss of renal function with age for each individual. However, the total number of people in each cohort was relatively small (1–3 thousand), precluding a straightforward GWA study for association with different rates of loss of GFR.

Instead of searching through every SNP in the genome, the study used a genomic convergence approach to systematically select SNPs that are most likely to be involved in kidney aging. The first assumption was that genes whose expression changes with age would be enriched for those with functional effect on the rate of renal aging or renal function (630 aging-related genes). The next step was to search for SNPs that are associated with differential expression of these 630 genes (expression quantitative trait loci; eQTLs) because higher or lower expression levels may affect the age at which the expression of a gene dips below a functional threshold (110 SNPs associated with differential expression in kidney aging related genes). The last step was to determine whether any of the 110 SNPs was associated with different levels of renal function independent of age in the Baltimore Longitudinal Study of Aging or the InCHIANTI cohort. Two linked SNPs (rs1711437 and rs1784418) located within the matrix metalloproteinase 20 gene (MMP20) were found to associate with loss of GFR independent of age at a significant level following correction for multiple hypothesis testing. For an individual who carries the A allele at rs1711437, his or her creatinine clearance is approximately that of someone 4–5 years younger who does not carry the A allele. In the BLSA population, the genotype of rs1711437 explains 2.1 % of the variation in creatinine clearance and in the InCHIANTI population, the genotype explains 0.9 % of the variation.

Future work will help find the missing heritability in DNA variants associated with kidney function, disease and aging. These studies may use full genome sequence rather than DNA chip analysis, enabling one to analyze rare variants. Larger cohorts may become available that will increase statistical power. Genetic algorithms may become available that are more powerful and able to better predict renal function. These algorithms may gain effectiveness by capitalizing on the complex genetic interactions among the SNPs associated with renal aging. These interactions are termed epistatic interactions, and refer to all of the possible ways in which two SNPs may be associated with renal aging that are not merely the additive sum of each individual SNP. For instance, the two SNPs may act redundantly, such that a protective allele in one SNP is as effective as a protective allele in both SNPs. Alternatively, the two SNPs may act synergistically, such that protective alleles in both SNPs are far more effective than would be predicted by adding up the effects of each allele individually. Future work may lead to ways to use DNA information to predict current and future renal function. For the purposes of renal transplantation, this could be key as a criterion to be used to help select which potential donor kidneys are most likely to retain adequate function following renal transplantation.

5 Conclusion

As we enter the era of precision medicine, individual genetic and biomarker data can be used to provide information about the physiological age of a person, rather than simple chronological age. Recent work has begun to unravel some of the mechanisms underlying kidney aging (Fig. 3). These mechanisms include increasing levels of cell senescence, chronic inflammation, fibrosis, and transcriptional regulation of the aging gene network. A better and more complete understanding of the molecular underpinnings of aging may one day enable the development of biomarkers that are able to report true biological age, as opposed to chronological age. These aging biomarkers could be used to more accurately ascertain which kidneys are most suitable for renal transplantation. This precision medicine approach of using personalized aging biomarkers may enable one to expand the pool of available kidneys for transplantation without diminishing the length of graft survival or the quality of the transplant.

Acknowledgments We would like to thank Zachary O'Brown for helpful comments. This work was funded by NIA RO1 AG025941.

Editor: John Williams, National Institute on Aging (NIA), NIH.

References

1. Lindeman RD (1990) Overview: renal physiology and pathophysiology of aging. Am J Kidney Dis 16(4):275–282. doi:S0272638690001184 [pii]
2. Davies DF, Shock NW (1950) Age changes in glomerular filtration rate, effective renal plasma flow, and tubular excretory capacity in adult males. J Clin Invest 29(5):496–507. doi:10.1172/JCI102286
3. Hoang K, Tan JC, Derby G, Blouch KL, Masek M, Ma I, Lemley KV, Myers BD (2003) Determinants of glomerular hypofiltration in aging humans. Kidney Int 64(4):1417–1424. doi:kid207 [pii] 10.1046/j.1523-1755.2003.00207.x

4. Levey AS, Stevens LA, Schmid CH, Zhang YL, Castro AF 3rd, Feldman HI, Kusek JW, Eggers P, Van Lente F, Greene T, Coresh J (2009) A new equation to estimate glomerular filtration rate. Ann Intern Med 150(9):604–612
5. K/DOQI (2002) K/DOQI clinical practice guidelines for chronic kidney disease: evaluation, classification, and stratification. Am J Kidney Dis 39(2 Suppl 1):S1–S266
6. Wolfe RA, Ashby VB, Milford EL, Ojo AO, Ettenger RE, Agodoa LY, Held PJ, Port FK (1999) Comparison of mortality in all patients on dialysis, patients on dialysis awaiting transplantation, and recipients of a first cadaveric transplant. N Engl J Med 341(23):1725–1730. doi:10.1056/NEJM199912023412303
7. Nyengaard JR, Bendtsen TF (1992) Glomerular number and size in relation to age, kidney weight, and body surface in normal man. Anat Rec 232(2):194–201. doi:10.1002/ar.1092320205
8. Tan JC, Busque S, Workeneh B, Ho B, Derby G, Blouch KL, Sommer FG, Edwards B, Myers BD (2010) Effects of aging on glomerular function and number in living kidney donors. Kidney Int 78(7):686–692. doi:10.1038/ki.2010.128ki2010128 [pii]
9. Anderson S, Halter JB, Hazzard WR, Himmelfarb J, Horne FM, Kaysen GA, Kusek JW, Nayfield SG, Schmader K, Tian Y, Ashworth JR, Clayton CP, Parker RP, Tarver ED, Woolard NF, High KP (2009) Prediction, progression, and outcomes of chronic kidney disease in older adults. J Am Soc Nephrol 20(6):1199–1209. doi:10.1681/ASN.2008080860 ASN.2008080860 [pii]
10. Hariharan S, Johnson CP, Bresnahan BA, Taranto SE, McIntosh MJ, Stablein D (2000) Improved graft survival after renal transplantation in the United States, 1988 to 1996. N Engl J Med 342(9):605–612. doi:MJBA-420901 [pii] 10.1056/NEJM200003023420901
11. Gjertson DW (1996) A multi-factor analysis of kidney graft outcomes at one and five years posttransplantation: 1996 UNOS update. Clin Transpl 10:343–360
12. United Network for Organ Sharing (UNOS) (2015) http://www.unos.org/donation/index.php?topic=data. Accessed 22 Feb 2015
13. Lapasia JB, Kong SY, Busque S, Scandling JD, Chertow GM, Tan JC (2011) Living donor evaluation and exclusion: the Stanford experience. Clin Transplant 25(5):697–704. doi:10.1111/j.1399-0012.2010.01336.x
14. de Fijter JW (2005) The impact of age on rejection in kidney transplantation. Drugs Aging 22(5):433–449, doi:2257 [pii]
15. Terasaki PI, Gjertson DW, Cecka JM, Takemoto S, Cho YW (1997) Significance of the donor age effect on kidney transplants. Clin Transplant 11(5 Pt 1):366–372
16. Lee CM, Scandling JD, Shen GK, Salvatierra O, Dafoe DC, Alfrey EJ (1996) The kidneys that nobody wanted: support for the utilization of expanded criteria donors. Transplantation 62(12):1832–1841
17. Prommool S, Jhangri GS, Cockfield SM, Halloran PF (2000) Time dependency of factors affecting renal allograft survival. J Am Soc Nephrol 11(3):565–573
18. Smits JM, Persijn GG, van Houwelingen HC, Claas FH, Frei U (2002) Evaluation of the Eurotransplant Senior Program. The results of the first year. Am J Transplant 2(7):664–670
19. Oppenheimer F, Aljama P, Asensio Peinado C, Bustamante Bustamante J, Crespo Albiach JF, Guirado Perich L (2004) The impact of donor age on the results of renal transplantation. Nephrol Dial Transplant (19 Suppl 3):iii11–iii15. doi:10.1093/ndt/gfh1008 19/suppl_3/iii11 [pii]
20. Cohen B, Smits JM, Haase B, Persijn G, Vanrenterghem Y, Frei U (2005) Expanding the donor pool to increase renal transplantation. Nephrol Dial Transplant 20(1):34–41. doi:gfh506 [pii] 10.1093/ndt/gfh506
21. Frei U, Noeldeke J, Machold-Fabrizii V, Arbogast H, Margreiter R, Fricke L, Voiculescu A, Kliem V, Ebel H, Albert U, Lopau K, Schnuelle P, Nonnast-Daniel B, Pietruck F, Offermann R, Persijn G, Bernasconi C (2008) Prospective age-matching in elderly kidney transplant recipients–a 5-year analysis of the Eurotransplant Senior Program. Am J Transplant 8(1):50–57. doi:10.1111/j.1600-6143.2007.02014.x AJT2014 [pii]

22. McGlynn LM, Stevenson K, Lamb K, Zino S, Brown M, Prina A, Kingsmore D, Shiels PG (2009) Cellular senescence in pretransplant renal biopsies predicts postoperative organ function. Aging Cell 8(1):45–51. doi:ACE447 [pii] 10.1111/j.1474-9726.2008.00447.x
23. Yang H, Fogo AB (2010) Cell senescence in the aging kidney. J Am Soc Nephrol 21(9):1436–1439. doi:10.1681/ASN.2010020205 ASN.2010020205 [pii]
24. Melk A, Halloran PF (2001) Cell senescence and its implications for nephrology. J Am Soc Nephrol 12(2):385–393
25. Wills LP, Schnellmann RG (2011) Telomeres and telomerase in renal health. J Am Soc Nephrol 22(1):39–41. doi:10.1681/ASN.2010060662 22/1/39 [pii]
26. Naesens M (2011) Replicative senescence in kidney aging, renal disease, and renal transplantation. Discov Med 11(56):65–75
27. Fontana L, Partridge L (2015) Promoting health and longevity through diet: from model organisms to humans. Cell 161(1):106–118. doi:10.1016/j.cell.2015.02.020
28. Campisi J, Robert L (2014) Cell senescence: role in aging and age-related diseases. Interdiscip Top Gerontol 39:45–61. doi:000358899 [pii] 10.1159/000358899 [doi]
29. Kamb A (1995) Cell-cycle regulators and cancer. Trends Genet 11(4):136–140. doi:S0168-9525(00)89027-7 [pii]
30. Vidal A, Koff A, Kamb A (2000) Cell-cycle inhibitors: three families united by a common cause Cell-cycle regulators and cancer. Gene 247(1–2):1–15. doi:S0378-1119(00)00092-5 [pii] S0168-9525(00)89027-7 [pii]
31. Chandler H, Peters G (2013) Stressing the cell cycle in senescence and aging. Curr Opin Cell Biol 25(6):765–771. doi:10.1016/j.ceb.2013.07.005
32. Melk A, Schmidt BM, Takeuchi O, Sawitzki B, Rayner DC, Halloran PF (2004) Expression of p16INK4a and other cell cycle regulator and senescence associated genes in aging human kidney. Kidney Int 65(2):510–520. doi:10.1111/j.1523-1755.2004.00438.x
33. Melk A, Schmidt BM, Vongwiwatana A, Rayner DC, Halloran PF (2005) Increased expression of senescence-associated cell cycle inhibitor p16INK4a in deteriorating renal transplants and diseased native kidney. Am J Transplant 5(6):1375–1382. doi:AJT [pii] 10.1111/j.1600-6143.2005.00846.x
34. Koppelstaetter C, Schratzberger G, Perco P, Hofer J, Mark W, Ollinger R, Oberbauer R, Schwarz C, Mitterbauer C, Kainz A, Karkoszka H, Wiecek A, Mayer B, Mayer G (2008) Markers of cellular senescence in zero hour biopsies predict outcome in renal transplantation. Aging Cell 7(4):491–497. doi:ACE398 [pii] 10.1111/j.1474-9726.2008.00398.x
35. Gingell-Littlejohn M, McGuinness D, McGlynn LM, Kingsmore D, Stevenson KS, Koppelstaetter C, Clancy MJ, Shiels PG (2013) Pre-transplant CDKN2A expression in kidney biopsies predicts renal function and is a future component of donor scoring criteria. PLoS One 8(7):e68133. doi:10.1371/journal.pone.0068133, PONE-D-12-38457 [pii]
36. Kainz A, Perco P, Mayer B, Soleiman A, Steininger R, Mayer G, Mitterbauer C, Schwarz C, Meyer TW, Oberbauer R (2007) Gene-expression profiles and age of donor kidney biopsies obtained before transplantation distinguish medium term graft function. Transplantation 83(8):1048–1054. doi:10.1097/01.tp.0000259960.56786.ec 00007890-200704270-00011 [pii]
37. Mas VR, Scian MJ, Archer KJ, Suh JL, David KG, Ren Q, Gehr TWB, King AL, Posner MP, Mueller TF, Maluf DG (2011) Pretransplant transcriptome profiles identify among kidneys with delayed graft function those with poorer quality and outcome. Mol Med 17(11–12):1311–1322. doi:10.2119/molmed.2011.00159
38. Lopez-Otin C, Blasco MA, Partridge L, Serrano M, Kroemer G (2013) The hallmarks of aging. Cell 153(6):1194–1217. doi:10.1016/j.cell.2013.05.039
39. Mei C, Zheng F (2009) Chronic inflammation potentiates kidney aging. Semin Nephrol 29(6):555–568. doi:10.1016/j.semnephrol.2009.07.002 S0270-9295(09)00144-2 [pii]
40. Cesari M, Penninx BW, Pahor M, Lauretani F, Corsi AM, Rhys Williams G, Guralnik JM, Ferrucci L (2004) Inflammatory markers and physical performance in older persons: the InCHIANTI study. J Gerontol A Biol Sci Med Sci 59(3):242–248

41. Krabbe KS, Pedersen M, Bruunsgaard H (2004) Inflammatory mediators in the elderly. Exp Gerontol 39(5):687–699. doi:10.1016/j.exger.2004.01.009 S0531556504000531 [pii]
42. Dorshkind K, Montecino-Rodriguez E, Signer RA (2009) The ageing immune system: is it ever too old to become young again? Nat Rev Immunol 9(1):57–62. doi:10.1038/nri2471, nri2471 [pii]
43. Wu D, Meydani SN (2008) Age-associated changes in immune and inflammatory responses: impact of vitamin E intervention. J Leukoc Biol 84(4):900–914. doi:10.1189/jlb.0108023 jlb.0108023 [pii]
44. Dixit VD (2008) Adipose-immune interactions during obesity and caloric restriction: reciprocal mechanisms regulating immunity and health span. J Leukoc Biol 84(4):882–892. doi:10.1189/jlb.0108028 jlb.0108028 [pii]
45. Coppe JP, Desprez PY, Krtolica A, Campisi J (2010) The senescence-associated secretory phenotype: the dark side of tumor suppression. Annu Rev Pathol 5:99–118. doi:10.1146/annurev-pathol-121808-102144
46. Rodwell GE, Sonu R, Zahn JM, Lund J, Wilhelmy J, Wang L, Xiao W, Mindrinos M, Crane E, Segal E, Myers BD, Brooks JD, Davis RW, Higgins J, Owen AB, Kim SK (2004) A transcriptional profile of aging in the human kidney. PLoS Biol 2(12):e427. doi:10.1371/journal.pbio.0020427
47. Jung KJ, Go EK, Kim JY, Yu BP, Chung HY (2009) Suppression of age-related renal changes in NF-kappaB and its target gene expression by dietary ferulate. J Nutr Biochem 20(5):378–388. doi:10.1016/j.jnutbio.2008.04.008 S0955-2863(08)00099-5 [pii]
48. Hewitson TD (2009) Renal tubulointerstitial fibrosis: common but never simple. Am J Physiol Renal Physiol 296(6):F1239–F1244. doi:10.1152/ajprenal.90521.2008 90521.2008 [pii]
49. Bechtel W, McGoohan S, Zeisberg EM, Muller GA, Kalbacher H, Salant DJ, Muller CA, Kalluri R, Zeisberg M (2010) Methylation determines fibroblast activation and fibrogenesis in the kidney. Nat Med 16(5):544–550. doi:10.1038/nm.2135 nm.2135 [pii]
50. Zhou W, Otto EA, Cluckey A, Airik R, Hurd TW, Chaki M, Diaz K, Lach FP, Bennett GR, Gee HY, Ghosh AK, Natarajan S, Thongthip S, Veturi U, Allen SJ, Janssen S, Ramaswami G, Dixon J, Burkhalter F, Spoendlin M, Moch H, Mihatsch MJ, Verine J, Reade R, Soliman H, Godin M, Kiss D, Monga G, Mazzucco G, Amann K, Artunc F, Newland RC, Wiech T, Zschiedrich S, Huber TB, Friedl A, Slaats GG, Joles JA, Goldschmeding R, Washburn J, Giles RH, Levy S, Smogorzewska A, Hildebrandt F (2012) FAN1 mutations cause karyomegalic interstitial nephritis, linking chronic kidney failure to defective DNA damage repair. Nat Genet 44(8):910–915. doi:ng.2347 [pii] 10.1038/ng.2347
51. Deans AJ, West SC (2011) DNA interstrand crosslink repair and cancer. Nat Rev Cancer 11(7):467–480. doi:nrc3088 [pii] 10.1038/nrc3088
52. Smogorzewska A, Desetty R, Saito TT, Schlabach M, Lach FP, Sowa ME, Clark AB, Kunkel TA, Harper JW, Colaiacovo MP, Elledge SJ (2010) A genetic screen identifies FAN1, a Fanconi anemia-associated nuclease necessary for DNA interstrand crosslink repair. Mol Cell 39(1):36–47. doi:S1097-2765(10)00464-8 [pii] 10.1016/j.molcel.2010.06.023
53. Fontana L, Partridge L, Longo VD (2010) Extending healthy life span–from yeast to humans. Science 328(5976):321–326. doi:328/5976/321 [pii] 10.1126/science.1172539 [doi]
54. Guarente L (2007) Sirtuins in aging and disease. Cold Spring Harb Symp Quant Biol 72:483–488. doi:10.1101/sqb.2007.72.024
55. Imai S, Guarente L (2014) NAD+ and sirtuins in aging and disease. Trends Cell Biol 24(8):464–471. doi:S0962-8924(14)00063-4 [pii] 10.1016/j.tcb.2014.04.002 [doi]
56. Kaeberlein M (2013) mTOR inhibition: from aging to autism and beyond. Scientifica (Cairo) 2013:849186. doi:10.1155/2013/849186
57. Sauve AA, Wolberger C, Schramm VL, Boeke JD (2006) The biochemistry of sirtuins. Annu Rev Biochem 75:435–465. doi:10.1146/annurev.biochem.74.082803.133500
58. Guarente L (2013) Calorie restriction and sirtuins revisited. Genes Dev 27(19):2072–2085. doi:27/19/2072 [pii] 10.1101/gad.227439.113

59. Hao CM, Haase VH (2010) Sirtuins and their relevance to the kidney. J Am Soc Nephrol 21(10):1620–1627. doi:ASN.2010010046 [pii] 10.1681/ASN.2010010046 [doi]
60. He W, Wang Y, Zhang M-Z, You L, Davis LS, Fan H, Yang H-C, Fogo AB, Zent R, Harris RC, Breyer MD, Hao C-M (2010) Sirt1 activation protects the mouse renal medulla from oxidative injury. J Clin Invest 120(4):1056–1068. doi:10.1172/jci41563
61. Braidy N, Guillemin GJ, Mansour H, Chan-Ling T, Poljak A, Grant R (2011) Age related changes in NAD+ metabolism oxidative stress and Sirt1 activity in wistar rats. PLoS One 6(4):e19194. doi:10.1371/journal.pone.0019194
62. Zhuo L, Cai G, Liu F, Fu B, Liu W, Hong Q, Ma Q, Peng Y, Wang J, Chen X (2009) Expression and mechanism of mammalian target of rapamycin in age-related renal cell senescence and organ aging. Mech Ageing Dev 130(10):700–708. doi:S0047-6374(09)00112-2 [pii] 10.1016/j.mad.2009.08.005 [doi]
63. Kuro-o M (2011) Klotho and the aging process. Korean J Intern Med 26(2):113–122. doi:10.3904/kjim.2011.26.2.113
64. Kuro-o M, Matsumura Y, Aizawa H, Kawaguchi H, Suga T, Utsugi T, Ohyama Y, Kurabayashi M, Kaname T, Kume E, Iwasaki H, Iida A, Shiraki-Iida T, Nishikawa S, Nagai R, Nabeshima YI (1997) Mutation of the mouse klotho gene leads to a syndrome resembling ageing. Nature 390(6655):45–51. doi:10.1038/36285
65. Koh N, Fujimori T, Nishiguchi S, Tamori A, Shiomi S, Nakatani T, Sugimura K, Kishimoto T, Kinoshita S, Kuroki T, Nabeshima Y (2001) Severely reduced production of klotho in human chronic renal failure kidney. Biochem Biophys Res Commun 280(4):1015–1020. doi:10.1006/bbrc.2000.4226 [doi] S0006291X00942268 [pii]
66. Pedersen L, Pedersen SM, Brasen CL, Rasmussen LM (2013) Soluble serum Klotho levels in healthy subjects. Comparison of two different immunoassays. Clin Biochem 46(12):1079–1083. doi:S0009-9120(13)00242-7 [pii] 10.1016/j.clinbiochem.2013.05.046 [doi]
67. Yamazaki Y, Imura A, Urakawa I, Shimada T, Murakami J, Aono Y, Hasegawa H, Yamashita T, Nakatani K, Saito Y, Okamoto N, Kurumatani N, Namba N, Kitaoka T, Ozono K, Sakai T, Hataya H, Ichikawa S, Imel EA, Econs MJ, Nabeshima Y (2010) Establishment of sandwich ELISA for soluble alpha-Klotho measurement: age-dependent change of soluble alpha-Klotho levels in healthy subjects. Biochem Biophys Res Commun 398(3):513–518. doi:S0006-291X(10)01250-7 [pii] 10.1016/j.bbrc.2010.06.110 [doi]
68. Olauson H, Vervloet MG, Cozzolino M, Massy ZA, Urena Torres P, Larsson TE (2014) New insights into the FGF23-Klotho axis. Semin Nephrol 34(6):586–597. doi:S0270-9295(14)00145-4 [pii] 10.1016/j.semnephrol.2014.09.005 [doi]
69. Xu Y, Sun Z (2015) Molecular basis of klotho: from gene to function in aging. Endocr Rev 36(2):174–193. doi:10.1210/er.2013-1079
70. Melk A, Mansfield ES, Hsieh SC, Hernandez-Boussard T, Grimm P, Rayner DC, Halloran PF, Sarwal MM (2005) Transcriptional analysis of the molecular basis of human kidney aging using cDNA microarray profiling. Kidney Int 68(6):2667–2679. doi:KID738 [pii] 10.1111/j.1523-1755.2005.00738.x
71. Gerstein MB, Kundaje A, Hariharan M, Landt SG, Yan KK, Cheng C, Mu XJ, Khurana E, Rozowsky J, Alexander R, Min R, Alves P, Abyzov A, Addleman N, Bhardwaj N, Boyle AP, Cayting P, Charos A, Chen DZ, Cheng Y, Clarke D, Eastman C, Euskirchen G, Frietze S, Fu Y, Gertz J, Grubert F, Harmanci A, Jain P, Kasowski M, Lacroute P, Leng J, Lian J, Monahan H, O'Geen H, Ouyang Z, Partridge EC, Patacsil D, Pauli F, Raha D, Ramirez L, Reddy TE, Reed B, Shi M, Slifer T, Wang J, Wu L, Yang X, Yip KY, Zilberman-Schapira G, Batzoglou S, Sidow A, Farnham PJ, Myers RM, Weissman SM, Snyder M (2012) Architecture of the human regulatory network derived from ENCODE data. Nature 489(7414):91–100. doi:nature11245 [pii] 10.1038/nature11245 [doi]
72. O'Brown ZK, Van Nostrand EL, Higgins JP, Kim SK (2015) The inflammatory transcription factors NFκB, STAT1 and STAT3 drive aging of the human kidney (submitted)
73. Adler AS, Kawahara TL, Segal E, Chang HY (2008) Reversal of aging by NFkappaB blockade. Cell Cycle 7(5):556–559, doi:5490 [pii]

74. Adler AS, Sinha S, Kawahara TL, Zhang JY, Segal E, Chang HY (2007) Motif module map reveals enforcement of aging by continual NF-kappaB activity. Genes Dev 21(24):3244–3257. doi:gad.1588507 [pii] 10.1101/gad.1588507 [doi]
75. Southworth LK, Owen AB, Kim SK (2009) Aging mice show a decreasing correlation of gene expression within genetic modules. PLoS Genet 5(12):e1000776. doi:10.1371/journal.pgen.1000776
76. Kottgen A, Pattaro C, Boger CA, Fuchsberger C, Olden M, Glazer NL, Parsa A, Gao X, Yang Q, Smith AV, O'Connell JR, Li M, Schmidt H, Tanaka T, Isaacs A, Ketkar S, Hwang SJ, Johnson AD, Dehghan A, Teumer A, Pare G, Atkinson EJ, Zeller T, Lohman K, Cornelis MC, Probst-Hensch NM, Kronenberg F, Tonjes A, Hayward C, Aspelund T, Eiriksdottir G, Launer LJ, Harris TB, Rampersaud E, Mitchell BD, Arking DE, Boerwinkle E, Struchalin M, Cavalieri M, Singleton A, Giallauria F, Metter J, de Boer IH, Haritunians T, Lumley T, Siscovick D, Psaty BM, Zillikens MC, Oostra BA, Feitosa M, Province M, de Andrade M, Turner ST, Schillert A, Ziegler A, Wild PS, Schnabel RB, Wilde S, Munzel TF, Leak TS, Illig T, Klopp N, Meisinger C, Wichmann HE, Koenig W, Zgaga L, Zemunik T, Kolcic I, Minelli C, Hu FB, Johansson A, Igl W, Zaboli G, Wild SH, Wright AF, Campbell H, Ellinghaus D, Schreiber S, Aulchenko YS, Felix JF, Rivadeneira F, Uitterlinden AG, Hofman A, Imboden M, Nitsch D, Brandstatter A, Kollerits B, Kedenko L, Magi R, Stumvoll M, Kovacs P, Boban M, Campbell S, Endlich K, Volzke H, Kroemer HK, Nauck M, Volker U, Polasek O, Vitart V, Badola S, Parker AN, Ridker PM, Kardia SL, Blankenberg S, Liu Y, Curhan GC, Franke A, Rochat T, Paulweber B, Prokopenko I, Wang W, Gudnason V, Shuldiner AR, Coresh J, Schmidt R, Ferrucci L, Shlipak MG, van Duijn CM, Borecki I, Kramer BK, Rudan I, Gyllensten U, Wilson JF, Witteman JC, Pramstaller PP, Rettig R, Hastie N, Chasman DI, Kao WH, Heid IM, Fox CS (2010) New loci associated with kidney function and chronic kidney disease. Nat Genet 42(5):376–384. doi:ng.568 [pii] 10.1038/ng.568
77. O'Seaghdha CM, Fox CS (2012) Genome-wide association studies of chronic kidney disease: what have we learned? Nat Rev Nephrol 8(2):89–99. doi:nrneph.2011.189 [pii] 10.1038/nrneph.2011.189 [doi]
78. Meigs JB, Shrader P, Sullivan LM, McAteer JB, Fox CS, Dupuis J, Manning AK, Florez JC, Wilson PW, D'Agostino RB Sr, Cupples LA (2008) Genotype score in addition to common risk factors for prediction of type 2 diabetes. N Engl J Med 359(21):2208–2219. doi:359/21/2208 [pii] 10.1056/NEJMoa0804742
79. Talmud PJ, Hingorani AD, Cooper JA, Marmot MG, Brunner EJ, Kumari M, Kivimaki M, Humphries SE (2010) Utility of genetic and non-genetic risk factors in prediction of type 2 diabetes: Whitehall II prospective cohort study. BMJ 340:b4838
80. Ripatti S, Tikkanen E, Orho-Melander M, Havulinna AS, Silander K, Sharma A, Guiducci C, Perola M, Jula A, Sinisalo J, Lokki ML, Nieminen MS, Melander O, Salomaa V, Peltonen L, Kathiresan S (2010) A multilocus genetic risk score for coronary heart disease: case-control and prospective cohort analyses. Lancet 376(9750):1393–1400. doi:S0140-6736(10)61267-6 [pii] 10.1016/S0140-6736(10)61267-6
81. Paynter NP, Chasman DI, Pare G, Buring JE, Cook NR, Miletich JP, Ridker PM (2010) Association between a literature-based genetic risk score and cardiovascular events in women. JAMA 303(7):631–637. doi:303/7/631 [pii] 10.1001/jama.2010.119
82. Wheeler HE, Metter EJ, Tanaka T, Absher D, Higgins J, Zahn JM, Wilhelmy J, Davis RW, Singleton A, Myers RM, Ferrucci L, Kim SK (2009) Sequential use of transcriptional profiling, expression quantitative trait mapping, and gene association implicates MMP20 in human kidney aging. PLoS Genet 5(10):e1000685. doi:10.1371/journal.pgen.1000685

Asthma and Aging

Nicola A. Hanania and Paula Busse

Contents

N.A. Hanania, M.D., M.S. (✉)
Department of Medicine, Section of Pulmonaryand Critical Care Medicine,
Baylor College of Medicine, Houston, TX, USA
e-mail: Hanania@bcm.edu

P. Busse, M.D.
Division of Clinical Immunology, Department of Medicine,
Icahn School of Medicine at Mount Sinai, New York, NY, USA
e-mail: Paula.busse@mssm.edu

F. Sierra, R. Kohanski (eds.), *Advances in Geroscience*,
DOI 10.1007/978-3-319-23246-1_14

1 Introduction

Asthma is a chronic airway disease, driven by complex interactions of inflammatory cells including eosinophils, lymphocytes, neutrophils and mast cells, and their mediators. Increased airway inflammation in asthma produces its characteristic features; airway hyperresponsiveness (AHR) to stimuli such as aeroallergens, histamine or methacholine and reversible airflow obstruction [1, 2]. However, in some patients, in particular those with a history of long-standing asthma, airflow obstruction may become only partially reversible. In susceptible individuals, clinical features of asthma are recurrent episodes of wheezing, breathlessness, chest tightness, and cough, particularly at night and/or in the early morning. These symptoms are usually associated with widespread but variable airflow limitation that is at least partially reversible either spontaneously or with treatment. While it is not clear whether airway hyperresponsiveness is acquired or is genetically determined to appear with the appropriate stimulus, it is thought that airway inflammation is the main stimulus as it may be induced by a number of inciting events including viral respiratory infections, allergen exposure in sensitized individuals, cold air, and exposure to noxious agents such as ozone and sulfur dioxide. Allergic or atopic reactions in the upper (nose, sinuses) and lower airways are both important in the pathogenesis of asthma in childhood and young adulthood. However, their role in the elderly is less clear. Atopy is defined by the presence of detectable IgE antibodies to environmental antigens and can be manifested as asthma, eczema and/or seasonal and perennial allergic rhinitis. In elderly patients with or without asthma, an elevated level of IgE may be an important risk factor for the development of chronic airflow obstruction [3].

Similar to other chronic diseases in this age group, asthma in the elderly population has a major impact on the patient's well-being and significantly impairs health status. Patients may consequently suffer from poor general health, symptoms of depression, and significant limitations of daily activity [4–9]. The exact prevalence of asthma in the aging population is not entirely clear as many studies do not clearly distinguish asthma from other obstructive lung diseases, but it appears to be similar to younger adults. According to the Center for Disease Control and Prevention (CDC), the overall prevalence of asthma in the US population >18 years of age is 7 % (www.cdc.gov/asthma/most_recent_data.htm) [10]. Large cross-sectional data base surveys including the National Health and Nutrition Examination Survey (NHANES) 2005–2006 and the National Health Interview Survey (NHIS) have reported that between 6.9–8 % of the U.S. population over the age of 60 years has a current diagnosis of asthma based upon a physician diagnosis [11, 12]. This translates to more than 2.7 million adults >65 age in the US with a diagnosis of current asthma. (www.cdc.gov/asthma/most_recent_data.htm) Asthma in the elderly is associated with a significant numbers of emergency department visits and hospitalizations, leading to substantial healthcare costs [13, 14]. Elderly patients with asthma are >5 times more likely to die from their disease than younger individuals and while mortality rates in some age groups have decreased, this is not true of the elderly [13–15]. According to the CDC, asthma deaths in the elderly account for more than 50 % of asthma fatalities annually with an approximately 5.8 asthma deaths per 100,000 in this group reported in the years 2001–2003 [10, 16]. Although the majority of elderly patients with asthma have long-standing asthma that may have developed early in life, some develop asthma late in life. LOA may occur at any age, even in the eighth or ninth decades of life and when it does, moderate to severe symptoms are more likely [17]. Despite the frequent occurrence of asthma in the elderly, it is a diagnosis that has been frequently overlooked and even when discovered it is often under treated [5, 18–22]. There are a number of important reasons that may explain the under diagnosis and under treatment of asthma in the elderly and these will be discussed in this chapter.

2 Pathophysiology and Risk Factors

2.1 Airway Inflammation and Aging

The immune system is complex and comprised of an innate arm, providing an initial defense against pathogens, and the adaptive arm, a subsequent antigen-specific response. Non cellular components including C-reactive protein, mannose-binding protein and complement, and a cellular component including phagocytic cells (neutrophils, macrophages/monocytes), epithelial cells and natural killer cells comprise the innate side. Cells of the innate arm use pattern recognition receptors (PRR), such as Toll-like receptors (TLR), to identify conserved pathogen associated molecular profiles (PAMPs) expressed on the surface of pathogens. The actions of the innate response are not long-lived, but are an important initial event, triggering activation of antigen-specific responses of the adaptive immune response which include humoral immune defenses (mediated through B cells) and cellular responses (mediated by T cells).

With increasing age, there are alterations in both the innate and adaptive immune responses. One phenomenon is termed "immunosenescence" in which the adaptive arm of the immune system response is "blunted" after a pathogenic threat or tissue injury. Cellular senescence is due to an irreversible loss of cellular replication and eventually results in impaired tissue repair. Senescence either develops via a continued stepwise accumulation of DNA damage occurring during replication (replicative senescence), or abruptly by activation of oncogenes (typically associated with premature aging) [23, 24]. However, despite an inability to proliferate, senescent cells remain alive, but function at a diminished or altered capacity. This may include an increased low-grade basal systemic inflammation (characterized by increased IL1-β, IL-6 and TNF-α) in the absence of an overt infection, referred to as "inflamm-aging." [25]

Immunosescence research has focused primarily upon cancer and autoimmunity, but not asthma. The underlying mechanisms of immunosenescence and inflamm-aging are complex, and a consequence of several processes, including both "random" (e.g., environmental exposures, accumulation of reactive oxygen species

(ROS) from metabolic activity, mutagenesis) and "regulated" (e.g., genetic) aspects. Alteration and loss of mitochondrial function plays a key role in cellular changes with aging. Decreased mitochondrial function increases generation of ROS resulting in accumulation of oxidative products [26]. Furthermore, damage of mitochondria leads to release of mitochondrial damage-associated molecular patterns which are similar in structure to many of the conserved patterns on bacterial surfaces (PAMPs), which in turn, activate the innate immune response [27]. A loss of mitochondrial function alters protein synthesis and protein folding, necessary for proteostasis. Additionally, accumulation of damaged cellular and organelle components and macromolecules may induce ongoing low-grade systemic inflammation. These products can be subsequently recognized as "danger" signals, initiating ongoing inflammation [25]. Shortening of telomeres (necessary to protect the chromosomal ends) may signal cell cycle arrest or apoptosis [28, 29] or replicative senescence, which in turn induces the release of pro-inflammatory proteins [26]. Shortening of telomere length in peripheral blood mononuclear cells (PBMCs) has recently been demonstrated to correlate with increased asthma severity in older patients with asthma [30]. Mechanistically, microRNAs (miRs, regulatory elements which can either suppress or activate gene expression, usually at the post-translational level) have been implicated in immunosenescence and inflamm-aging, increasing the expression of pro-inflammatory proteins [31], and decreasing cell proliferation and function [31, 32].

Older individuals with fewer features of immunosenescence may have a prolonged lifespan [33]. Conversely, specific features of immunosenescence are associated with increased morbidity and mortality [34], and low-grade systemic inflammation with more clinically "frail" individuals [35, 36]. However, how the effects of immunosenescence translate to airway inflammation and its regulation in older patients with asthma is not well established. Additionally, whether asthma is a distinct inflammatory phenotype in older patients is unknown, important and unclear, yet it may alter treatment of the disease. The following section will address what is known about the effect of increased age on the innate and adaptive immune responses and how these changes may alter airway inflammation of asthma in older adults.

2.2 *Aging-Related Changes in Cellular Components of the Innate Immune Response*

2.2.1 Epithelial Cells

The respiratory epithelium is an important component of the innate immune system, and is composed of ciliated and secretory cells. Ciliated cells propel inhaled antigens and irritants trapped in mucus produced by goblet cells, proximally up the tracheobronchial tree via mucociliary clearance. Additionally, airway epithelial cells produce nitric oxide, secrete cytokines [e.g. type I and III interferons (IFNs)], growth factors (e.g. granulocyte/macrophage colony-stimulating factor), IgA and antimicrobial proteins, which act as an initial immune clearance mechanism. Adjacent epithelial airway cells are connected by tight and adherent junctions, which form a physical barrier against entrance of microbes and antigens. In younger patients with asthma, airway epithelial cells have disrupted tight cellular junctions, an increased susceptibility to apoptosis and an impaired production of interferons [37]. Although changes in the airway epithelial cells with aging in individuals *without* airway disease have not been investigated in detail, there is evidence that there are alterations. In non-smoking healthy individuals, ciliary beat frequency and clearance decrease with age [38, 39]. Additionally, aged airway epithelial cells have a decreased barrier function [40]. In aged mice, repeated inhalation of toxicants decreased airway epithelial cell repair and exacerbated p38 MAPK-dependent pro-inflammatory cytokine expression [41]. These age-associated changes may enhance alterations in airway epithelial cells characteristic of some younger patients with asthma or increase the risks of respiratory tract infections, thus exacerbating underlying asthma. Additionally, it is possible that individuals with asthma are at increased risk of "accelerated" cellular aging, as has been suggested with chronic obstructive pulmonary disease (COPD) [42] (see chapter "Aging in COPD and idiopathic pulmonary fibrosis" by C. Sanchez in this volume).

2.2.2 Neutrophils/Polymorphonuclear Cells

Neutrophils or polymorphonuclear cells (PMNs) have several important functions in the initial clearance of pathogens including phagocytosis, recruitment and maturation of dendritic cells, transport of antigens to lymph nodes and secretion of neutrophil extracellular traps (NETs), which immobilize bacteria. With increasing age, neutrophil recruitment is altered, preventing chemotaxis of PMNs to the site of inflammation [43]. Additionally, neutrophils from aged mice produce fewer NETs after either infection or non-specific cellular activation [44]. Aged neutrophils also have a decreased capacity for phagocytosis [45, 46]. Potential mechanisms include a decreased surface expression of CD16 [47] (necessary for Fc-mediated phagocytosis) and decreased downstream signaling of TLRs [48, 49]. Normally, PMNs produce superoxide anions, which are converted to reactive oxygen species (ROS), for intracellular killing after phagocytosis. With advanced age, there is a higher constitutive ROS expression by neutrophils which may be due to increased cellular activation (suggested by increased extracellular CD11b and HLA-DR expression) and altered regulation of cGMP and cAMP for ROS production [50–53]. Increased production of free radicals is a theory of aging and damages local tissues [54]. Despite increased basal production of ROS, induction of ROS in response to a bacterial infection is suppressed in older compared to younger patients [50, 55]. This may occur secondary to a decreased glucose accumulation (required for ROS production) in aged PMNs [56].

Despite diminished anti-pathogen activity, the number of neutrophils in the airway of aging individuals without airway disease increases [57, 58]. Whether this is secondary to decreased neutrophil apoptosis [59] or to increased systemic inflammation with aging is not clearly established. Additionally, the impact this has on asthma in aging is not known. Emerging data suggests that older patients with asthma may have increased airway neutrophils (measured from induced sputum) compared to younger patients [60, 61]. Furthermore, increased airway neutrophilia in older asthma subjects corresponds to increased levels of sputum neutrophil mediators including MMP-9, neutrophil elastase and IL-8. This resembles changes seen in a phenotype of severe asthma noted in some younger adults [61] . Determining underlying airway inflammation in older adults with asthma is important as neutrophilic asthma is often less responsive to corticosteroid treatment [62, 63], therefore alternative therapies may be indicated for older patients with asthma.

2.2.3 Eosinophils

There is a limited understanding of the age-associated changes in eosinophil number and function in older patients with asthma. In younger patients with asthma, in particular those with an "allergic phenotype," the airway eosinophil plays an important role in airway inflammation and AHR [64] . However, aged eosinophils may have decreased effector functions and be less important for AHR. In a study examining age-related changes in eosinophil function, peripheral eosinophils from subjects with asthma (55–80 years of age) exhibited decreased degranulation in response to IL-5 stimulation and a trend for decreased superoxide production, when compared to cells from patients 20–40 years of age [60]. However, study of another in vitro eosinophil effector function, leukotriene C4 (LTC_4) production, revealed no difference between older and younger subjects with asthma [65]. The role of the eosinophil in AHR in older patients with asthma is not well established. In a mouse model of asthma, antigen sensitized and airway challenged aged mice developed greater bronchoalveolar fluid (BALF) eosinophilia in comparison to younger mice, however AHR was lower in the former, suggesting that increased airway eosinophilia was not correlated to AHR [66]. However, the development of AHR later in life was associated with elevated peripheral blood eosinophil counts in males (mean age 60 years) enrolled in the Normative Aging Study [67]. Thus, eosinophils may be less functional in the elderly; their role in AHR, however, requires further investigation.

2.2.4 Dendritic Cells

Dendritic cells (DCs) lie at the interface between the innate and adaptive immune responses and play key roles in antigen presentation. There are 2 main subpopulations of DCs, plasmacytoid (pDC) and myeloid (mDC). Both DC populations express TLRs and produce type I and type III interferons (IFN), necessary for anti-viral responses, although pDCs produce greater amounts and more rapidly. The impact of age-related changes in DC in asthma is not clear. However, there are several alterations of DC function with aging which may alter asthma in older patients.

The increased production of IL-6 and TNF-α by DCs during aging might contribute to the observed inflammaging [68–70]. Additionally, there are several changes associated with DCs during aging which may increase susceptibility to respiratory infections and asthma exacerbation, including decreased IFN production [71], phagocytosis [68], migration [72, 73] and TLR function, as well as a poor antibody response to influenza immunization [70]. Although there are likely several mechanisms underlying alteration of DC function with aging, epigenetic modifications of DNA producing hypomethylation may contribute [74].

2.3 *Aging-Related Changes in the Adaptive Immune Response*

In contrast to changes in immune function of the innate system with aging, changes in adaptive immunity are better defined. With aging, the ability to mount an adaptive immune response declines [75]. These changes produce several clinical outcomes including a reduced ability to produce specific and long-lasting antibodies to vaccinations [76, 77] and a lack of a rapid immunologic response when encountering pathogens. Similar to the innate immune response, the effects of age on the adaptive response and asthma in older patients is not well characterized.

2.3.1 T Lymphocytes

T cells play a key role in the adaptive arm, in particular, the cellular immune response, necessary for eradication of viruses and intracellular pathogens and cytokine production. T cells are identified by surface expression of CD3 and are further categorized into CD4 or CD8 T cells which have distinct functions. CD8+ T cells recognize class I MHC-associated peptides on infected cells, releasing granular contents to promote their cytolysis. In contrast, naïve CD4+ T cells recognize MHC II peptides on antigen presentation cells, and after activation, differentiate into subsets including: Th1 (characterized by secretion of interleukin (IL)-2 and interferon (IFN)-γ), that function to increase macrophage phagocytosis and CD8+ T-cell proliferation; Th-2, which produce IL-4, −5, −13, stimulating eosinophil activation and production of IgE antibodies by B cells; and Th-17 (characterized by secretion of IL-17a, IL-21 and IL-22 which recruit and active neutrophils).

With increased age, the number of circulating naive T lymphocytes significantly decreases. Reduction of naïve T cells has been largely attributed to thymus involution with aging and replacement of tissue with adipose, which produces additional pro-inflammatory mediators affecting thymopoiesis [78]. However, recent work suggests that a decrease in naïve T cells with aging may be instead due to alterations of peripheral cell division [79], secondary to lower systemic levels of IL-7, which normally protects against telomere loss and apoptosis [80], thereby shortening the survival of naïve T cells [81–83]. Although the number of naïve T cells decreases with aging, the total number of circulating T cells remain relatively constant due to the survival and increased resistance to apoptosis of memory T cells [84, 85]. The increased survival of memory CD4+ was reported to be due to changes in the aged mouse microenvironment [84], and survival of memory CD8+ T cells to chronic viral stimulation, in particular Cytomegalovirus (CMV) [86].

With aging, there are several alterations which affect the T cell receptor (TCR). Aging T cells in humans lose surface CD28, necessary for TCR signaling. Loss of CD28 expression may occur from repeated antigen stimulation over time, producing a cellular stage of replicative senescence as demonstrated by shortened telomers and reduced proliferation [87–89]. Furthermore, type I interferons and TNF-α can decrease CD28 expression [90, 91]. The consequences of loss of CD28 include decreased IL-2 secretion and cellular proliferation, and increased resistance to apoptosis [92–94]. In addition, TCR diversity decreases with age [95, 96], which likely decreases viral clearance [97].

2.3.2 Th17 Cells

Th17 cells are a subgroup of T cells which secrete IL-17A, IL-17F and IL-22, which in turn stimulates IL-23 production. Th17 cells are differentiated from naïve T cells when stimulated by IL-6 (increased expression associated with aging) and TGF-β, which upregulate Th17 transcription factors retinoid acid-related orphan nuclear hormone receptor-γt (ROR-γt) and STAT3 [98]. The frequency of Th17 cells most likely increases with aging [99, 100], which clinically may translate to decreased viral clearance, increased neutrophilic inflammation increased transplant rejection [101], increased risk of autoimmune diseases and cancer [102] and mortality during a viral infection [103].

Recent data in aged mouse models of asthma also demonstrate an increased expression of airway Th17 expression [104, 105]. Murine models of allergic asthma suggest that IL-17A and IL-17F are important in airway inflammation, particularly neutrophilia [106, 107], enhanced Th2 associated eosinophila [108], increased AHR and airway mucus gene expression [109, 110]. In humans, IL-17A and IL-17F expression is increased in the airways of patients with asthma [111, 112]. Increased IL-17 in older patients with asthma likely contributes to increased airway neutrophilia and may be an additional target for therapy of asthma in this group of patients with difficult-to-treat asthma and increased morbidity and mortality from their disease.

2.3.3 Regulatory T Cells

Regulatory T cells (Tregs) suppress several effector functions of Th1, Th2 and Th17 cells. Tregs may protect against autoimmune diseases such as multiple sclerosis and Type I diabetes, but excessive numbers and activity may lead to increased susceptibility to infection and cancer. Treg cells found in the lung are in the CD4+ CD25^{hi} population. Natural Tregs (nTregs) are derived from the thymus, express the transcription factor Foxp3, and mediate suppression primarily via cell contact. Peripheral antigen-induced adaptive Tregs (iTregs) are either Foxp3+ or Foxp3-, and their major suppressive action is mediated via IL-10 and TGF-β.

The effect of age on the numbers and function of Tregs is less clear. Numbers of cord blood CD4+ $CD25^{hi}$ are elevated at 2.3–9.5 % of the total CD4+ T-cells (suggested to support fetal development of the immune system); they decline within the first 36 months of life, and then remain stable in young and middle aged healthy adults [113]. Some investigators have shown an increase in CD4+ CD25+ and CD4+ Foxp3+ T cells [114, 115] in healthy elderly compared to younger subjects, whereas other groups have found no significant differences in numbers of CD4+ Foxp3+ T cells [116]. The proportion and numbers of CD4+ Foxp3+ cells have been shown to be increased in lymphoid organs, but not in lungs or blood of aged (>20 month) C57BL/6 mice compared to younger mice [114, 117]. Earlier studies suggested that the suppressive activity of Treg cells declines with increasing age [118], but more recent studies have suggested that it is preserved or even enhanced [115–117].

The precise role of nTregs and iTregs in asthma is not clearly established, but most studies in murine models and in humans with asthma suggest a suppression of airway inflammation and AHR [119–121]. Peripheral numbers of Treg cells are lower in younger patients with asthma compared to age-matched subjects without airway disease [122]. Antigen tolerance in murine models of asthma, induced by inhalation or low-dose oral feeding of allergen to young mice suppressed several features of allergic asthma and increased Treg cell numbers [123–125]. On the other hand, aged mice fed low-dose antigen prior to antigen sensitization and challenge developed decreased AHR, antigen-specific serum IgE, BALF eosinophilia and mucus hypersecretion compared to age-matched non-tolerant mice. Additionally, the percent of lung tissue containing CD4+ CD25+ Foxp3+ cells was increased, but not to the same extent as in young mice, in antigen-tolerant aged mice suggesting that with increased age, Treg cells may continue to suppress AHR [126]. Similar to younger patients with asthma, one study reported that older patients with asthma had decreased peripheral Treg cell numbers compared to age-matched normal controls [127]. However, the importance of Treg cells in airway inflammation in older patients with asthma has not been widely investigated.

2.3.4 B Cells

B cells produce antigen specific antibodies. With increased age, the potential for the bone marrow to generate new naïve B cells from hematopoietic stem cells (HSC) diminishes [128, 129]. Although the ability of B cells to produce antibodies remains intact with aging [130], the quality of antibodies produced decreases, as exemplified by lower antigen affinity and avidity [131]. This is likely due to several alterations during aging, including deficient somatic hypermutation, responsible for the enhancement of antibody specificity for antigens [132], altered cytokine production by aged T cells and decreased expression of CD154 (CD40L), required for class switching of antibody production [133]. Therefore, the age-related decline in diversity and quality of antibodies, particularly in the setting of new antigen exposure, may impact the ability of the elderly to clear pathogens and provide protection against a repeat exposure to the same pathogen.

2.3.5 Cytokine Secretion

Cytokine secretion by the adaptive immune system is altered with aging, although study results are varied. Some studies suggest increased IFN-γ secretion by activated peripheral CD4+ and CD8+ cells [134] due to increased methylation of the IFN-γ promoter [135], whereas several others have shown a shift towards Type 2 cytokine production [136]. In a study of healthy subjects in three different age groups (between 21–30 years, 80–81 years, and 100–103 years), flow cytometry determined CD4+ and CD8+ T cell populations with respect to intracellular IL-4 (Th2) and IFN-γ (Th1) expression upon PMA (phorbol myristic acid) stimulation [137]. When the ratio of IFN-γ/IL-4 producing cells was examined, there were no age-related differences in the CD4+ T cells; however, the IFN-γ/IL-4 ratio in the CD8+ T cells was lower in both the older age groups, suggesting an age-related shift from Th1 to Th2 cytokine profiles. Additionally, aging has been associated with a decreased CD8+ IFN-γ response to virus and cytolytic activity [138]. The underlying effects of cytokine secretion changes with aging and asthma are not clear at this point; however, allergic disorders are usually secondary to increased Th2 cell expression.

2.4 Role of Allergic Sensitization and Aging

Atopy is defined as the predisposition towards developing an IgE-mediated hypersensitivity reaction. It is documented by the presence of at least one detectable antigen-specific IgE to a common environmental (e.g. pollens, dust mites, animals with fur, cockroaches) or food antigen [139]. Antigen-specific IgE is produced by B cells and binds to the high-affinity IgE receptors on mast cells and basophils. Once re-exposed, the specific antigen binds to the IgE molecules, causing these cells to cross-link. This process initiates cellular degranulation with the release of preformed mediators including histamine, leukotrienes and the synthesis of cytokines which contribute to allergic symptoms and airway bronchospasm. Detection of antigen-specific IgE is done either through skin prick testing (SPTs) or measurement in serum. Although SPTs are easily performed and offer results quickly, atrophy of the skin and declining skin mast cells with aging may decrease their sensitivity in older patients [140].

Total IgE declines with increasing age in the general population, including individuals with and without atopy [141–143]. Cross-sectional studies including the Tucson Epidemiological Study and the National Health and Nutrition Examination Survey (NHANES 2005–2006) demonstrate that total serum IgE peaks by 20 years of age and is lowest after 70 years [11, 141]. However, some studies have not supported a trend of decreased total IgE with increased age [144–147]. Allergen-specific IgE also decreases with aging. Studies including healthy subjects *with and without asthma and other atopic diseases*, suggest that approximately 40–50 % of subjects under the age of 50 years and approximately 35 % of subjects older than 50 are IgE sensitized to at least one allergen [143, 145, 147–149]. The increase in total and allergen-specific IgE production with age is most likely due to multiple factors such as alterations in cytokine profiles, changes in B cell antibody production, and DC and T lymphocyte function with aging, as described previously.

For many years, asthma in older patients was characterized as non-atopic or intrinsic [150, 151]. However, over the past two decades, it has been demonstrated that atopy is not uncommon in older patients with asthma. The reported percentage of older patients with atopic versus non-atopic asthma varies and may depend upon the characteristics of the population investigated. Studies examining the presence of IgE-sensitization in older patients with asthma in the United States describe sensitization rates between 23.9 % to as high as 74 % in one study [152–155]. Multicenter studies from the Netherlands [156] and France [157] reported that approximately 35 % of older patients with asthma were sensitized to at least one allergen. The most common aeroallergen to which older patients are sensitized is not consistent among studies, but typically includes cats [154], dust mites [155, 158], and cockroaches [159].

Atopy in children with asthma increases disease morbidity [160, 161]. Whether this relationship occurs in older patients with asthma is not known. Studies including older subjects with asthma, suggest that an atopic history may increase disease severity [162, 163]. In a study of 45 patients > 65 years of age recruited from an asthma clinic in New York City, cockroach sensitization was associated with more severe asthma as measured by airflow limitation and hyperinflation [159]. Data from NHANES 2005–2006 suggested that the relationship between antigen sensitization and asthma outcomes (e.g. hospitalizations, health care use) were not different in subjects 20–40 years old vs. ≥55 years old [164]. However, no studies have been conducted to date with antigen challenge of older patients with asthma and measurement of subsequent airway function.

Some studies suggest that antigen sensitization plays a role in late onset asthma (LOA) [153, 165]. In the Normative Aging Study, men developing airway hyperresponsiveness after 49 years of age were more likely to be sensitized to cat (23.9 % versus 4.4 %) compared to age-matched controls [153]. Nearly 50 % of the 40 patients in the Tucson Epidemiologic study of obstructive lung diseases who developed asthma after the age of 60 years were skin prick positive to at least one antigen, compared with 26 % of the age-matched control population without asthma [152]. In a study of 21 patients with asthma onset after 65 years of age, 81 % demonstrated a positive skin prick test to at least one allergen compared to a group of 14 patients developing asthma at <65 years, and in whom 57 % were allergen sensitized [165]. However, some studies suggest that IgE Ag sensitization may play a less significant role in LOA. For example, a French study recruited 1,485 patients (mean age 73 years) with a diagnosis of asthma from a total of 379 lung specialists to examine disease characteristics. Only 14.7 % of those developing asthma after 65 years of age were sensitized to at least one antigen by skin prick testing, whereas 60.1 % of those developing asthma prior to 21 years of age were antigen sensitized [157].

Although there are several unanswered questions regarding the roles of allergen sensitization in older patients with asthma including its role in disease inception, disease progression, trigger and management, allergen sensitization evaluation should be a component of asthma management and evaluation in this group, in particular patients with persistent asthma. Additionally, some older patients with asthma have enjoyed clinical improvement in their asthma after treatment with anti-IgE therapy [166, 167], emphasizing that allergen sensitization should be evaluated in older patients with asthma.

2.5 Role of Cigarette Smoke and Irritants

Exposure to cigarette smoke increases the risk of developing asthma in children [168]. Although first and second-hand cigarette smoke exposures are traditionally associated with the development of COPD in older adults, it also increases the risk of developing asthma later in life. Between 1/3 to ½ of older adults with asthma report a current smoking history [5, 19, 169]. In the Normative Aging Study, a history of current smoking and presence of atopy was associated with increased airway hyperresponsiveness to methacholine challenge [170]. Additionally, in 48 patients over the age of 70 years with late onset asthma (mean onset 58 years), those with a smoking history had increased frequency of antigen-specific IgE to common aeroallergens and increased generation of LTB4 by peripheral leukocytes suggesting that cigarette smoking may enhance IgE production [171]. Furthermore, cigarette smoke is a common trigger of exacerbation in adults with asthma and exacerbates the rate of lung function decline in both young and old patients with asthma [172, 173].

Exposure to biomass fuels, cleaning products, food preparation and chemicals increase the risk of developing asthma in younger adults [174]. Many of these exposures occur in the work-space, which may not be applicable to older patients if they have retired. Although the effect of irritants on the development of asthma in older patients is less well characterized than in younger patients, exposure to dust particles, art supplies, and cleaning products has been reported to induce asthma in older patients [175]. Additionally, outdoor air pollution, especially increased ozone, NO_2 and SO_2 may exacerbate and induce asthma in older patients; however, this needs further study [176]. Diesel exhaust is another potential environmental issue and it increases pulmonary neutrophilia in aged compared to younger mice with a prolonged pulmonary inflammation [177].

2.6 Role of Infections

Immunosenescence increases susceptibility to infections in older patients. Respiratory infections in children and young adults (e.g. rhinovirus, Respiratory Syncytial Virus (RSV), influenza, parvovirus) can affect asthma at several points, including disease inception (particularly in the presence of allergen sensitization), exacerbation, and worsening asthma control [178–186]. However, the effect of infection in older patients with asthma on these stages is less defined. Observational studies have reported that nearly 50 % of subjects with asthma onset after the age of 60 years reported a prior respiratory infection [19]. To address age-related differences in the effect of a particular respiratory infection, influenza A, on antigen sensitization and its effect on subsequent allergen-induced AHR and inflammation, aged and young mice were sensitized to Ag after an acute viral infection. There were significant age-associated differences in the response to infection and sensitization with greater antigen-specific IgE production in aged infected mice, but no alteration of AHR [104]. Detection of respiratory infections is more difficult in older patients as viral culture and rapid antigen testing, which are used frequently in the pediatric population are less sensitive in the older population. In a prospective cohort study among healthy older adults and older adults at high-risk (those with chronic heart or lung disease), RSV infection was seen annually in 3–7 % of health patients and 4–10 % in high-risk subjects [187]. Additionally, RSV was noted by RT-PCR detection in 7.2 % of older adults hospitalized with asthma in this study. Although viral infections have been traditionally investigated as an asthma trigger, the role of *Chlamydia pneumonaie* and other atypical infections has recently come into focus, in particular with adult-onset asthma [188].

Vaccination is an appropriate measure to attempt to alter the immune response and decrease the risk and progression of infectious diseases in older patients. Although vaccination of older patients decreases the progression of many infectious diseases, the immunologic protective response is decreased in the aged [77, 189]. In children and young adults, both influenza and pneumococcal vaccines reduce respiratory infections and rates of asthma exacerbation in patients with asthma [190, 191]. The effect of influenza vaccination or pneumococcal vaccination has not been evaluated in reduction of asthma symptoms in older patients. Recently, high doses of trivalent inactivated influenza vaccine have been shown to increase the antibody response and increase protection against influenza in older patients [192], which may offer some protection against asthma exacerbations secondary to viral infections in older patients.

3 Diagnosis and Clinical Assessment

3.1 Clinical Phenotypes

Data on the clinical phenotypes of asthma in the elderly have been derived from both longitudinal community surveys and case studies [5, 19, 150, 152, 193–195]. Two clinical phenotypes have been reported based on the onset and duration of the disease [150, 193, 196]. Patients with late-onset asthma (LOA) first develop asthma symptoms after the age of 65 years, although particular studies define LOA as first asthma diagnosis after the age of 40 years. Some studies of elderly asthmatics have shown that as a group, as many as 40 % will have their first attack after the age of 40 years [150, 152, 197]. Patients with LOA tend to be less atopic, have a higher baseline FEV_1, and a more pronounced bronchodilator response than those with long-standing asthma (LSA). Patients with LSA develop asthma symptoms early in life, typically before the age of 12 years of age, and tend to have higher incidence of atopic diseases, more severe and irreversible or partially reversible airway obstruction, and greater airway hyperinflation. The duration of the disease in this group is an important determinant of severity and of development of irreversible airflow obstruction [198].

3.2 Clinical Presentation

3.2.1 Symptoms

The onset of wheezing, shortness of breath and cough in an elderly patient is likely to cause concern. Symptoms of asthma that are common in younger patients, including episodic wheezing, shortness of breath, and chest tightness, are also characteristic of asthma in the elderly. Symptoms are generally worse at night and with exertion. Dyspnea is a common symptom of many other chronic disorders in older patients such as cardiac or other lung diseases, therefore asthma as an etiology of these symptoms may be overlooked. Many elderly patients limit their activity to avoid getting dyspneic, and others assume that their dyspnea is resulting from their aging process and, thus, neglect seeking medical attention early in their disease process. However, aging per se does not cause dyspnea, and an etiology needs to be always pursued in assessing an elderly patient who complains of breathlessness. Cough is a prominent symptom of asthma and may occasionally be the only presenting symptom [199]. Wheezing, on the other hand, may not be as prominent, and its presence is not very specific and does not correlate with severity of obstruction.

3.2.2 Triggers

Asthma symptoms are often precipitated by an upper respiratory tract infection. However symptoms of asthma in the elderly are non-specific and may be caused by a variety of other conditions. Symptoms can also often be triggered by medications, such as aspirin, non-steroidal antiinflammatory agents, or beta-blockers, commonly used by this population. History of atopy is a strong predictor of asthma in this age group, and allergic rhinitis, sinusitis, and nasal polyps are not uncommon.

3.2.3 Signs

Physical examination in elderly patients with asthma is usually nonspecific and may misguide the diagnosis. Physical examination should focus on ruling other causes for respiratory symptoms such as cardiovascular disease and identifying comorbid conditions such as rhinitis/sinusitis and atopic dermatitis.

3.2.4 Physiologic Measurements

The diagnosis of asthma in the elderly should be confirmed by objective measures which are often also helpful in staging the severity of the disease. Lung function testing is especially important in this age group since there is an age-related reduction in the perception of dyspnea seen in the elderly [200]. Unfortunately, objective measures of lung function such as spirometry and peak flow measurements are generally underutilized in elderly patients and this also contributes to the delay in diagnosis [19, 201]. Furthermore, spirometry may be difficult to perform in some situations, because of physical and poor cognitive impairment. In addition, a major problem is the difficulty in defining the lower limits of predicted normal values in this age group, which may vary in different patients. Elderly patients with asthma may also demonstrate an impaired acute bronchodilator response, which can lead to a misdiagnosis of COPD. This poor response may result from the decreased number of β-adrenergic receptors on smooth airway muscles that has been described with the aging process. Airway obstruction may be absent at the time of testing in approximately 8 % of elderly and further testing which may include methacholine challenge testing or even cardiopulmonary exercise stress testing may be needed to facilitate the diagnosis. Measurement of airway hyperresponsiveness to methacholine may not be an accurate test in the elderly [202], although a few studies suggest a higher prevalence of airway hyperresponsiveness in elderly asthmatics than younger populations. Peak expiratory flow variability may also be helpful in the diagnosis and follow-up of patients with obstructive airway diseases, but poor coordination and muscle weakness in some patients may lead to an inaccurate reading [203, 204]. However, a prospective study failed to demonstrate any advantage of peak flow monitoring over symptom monitoring as an asthma management strategy for older adults with moderate-severe asthma when used in a comprehensive asthma management program [205]. Other tests such as measuring the carbon monoxide diffusing capacity of the lung (DLCO) may help distinguish between asthma and COPD (emphysema).

3.2.5 Diagnostic Challenges

Studies have consistently shown that symptoms caused by asthma are frequently overlooked by elderly patients and their physicians. Several factors contribute to the under diagnosis and misdiagnosis of asthma in this age group. A major factor is that symptoms of asthma are also common to other diseases seen in this age group. The hallmark symptoms of asthma, including shortness of breath, wheeze and cough, are non-specific in the elderly and are mimicked by and often confused with such diseases as congestive heart failure, emphysema and chronic bronchitis (COPD), chronic aspiration, gastroesophageal reflux (GERD), and tracheobronchial tumors. Since smoking is an important risk factor for asthma-like symptoms of wheeze, cough and sputum production, asthma is frequently confused with COPD. Lung function is generally lower in those who smoke compared with those who do not smoke due to concomitant COPD [152]. Distinguishing between COPD and asthma in older patients may be very challenging and in some patients asthma cannot be distinguished from COPD with our current diagnostic tests [206–208]. This group of patients has recently been classified as "Asthma-COPD overlap syndrome" or (ACOS). Symptoms of ACOS include variability of respiratory symptoms, and airflow limitation which is not fully reversible. Although there is not a distinct definition for ACOS, such patients tend to have a poorer quality of life, frequent exacerbations and a more rapid lung decline than patients with asthma or COPD alone [207–209].

It has been known for more than a century that early morning wheezing is a prominent symptom of congestive heart failure; it has been called cardiac asthma as it can mimic the clinical picture of asthma. Typical symptoms of gastroesophageal reflux in the elderly, such as vomiting and heartburn, may be absent. In a study of elderly patients with esophageal reflux proven by intraesophageal pH monitoring, chronic cough, hoarseness and wheezing were present in 57 % of patients [210, 211]. In addition to causing asthma-like symptoms, there is also evidence that GERD may be a cause of worsening asthma.

Unlike younger adults with asthma, a family or personal history of atopy is usually absent in older patients. Blood and sputum eosinophilia are common, but not universal. Since large community studies have shown that most patients first develop asthma in childhood or adolescence, many physicians have had the misconception that asthma is a childhood disease.

In addition, elderly patients have been shown to have a reduced perception of bronchoconstriction that further delays medical intervention. In fact, many elderly patients are fearful of having an illness and dying and are therefore reluctant to admit they are having symptoms. Even when they do so, they may underestimate them or consider them a result of normal aging. Underreporting of symptoms in the elderly may have many causes including depression, cognitive impairment, social isolation, denial, and confusing symptoms with those of other comorbid illnesses.

3.2.6 Assessing Asthma Severity and Control

Evidence-based guidelines for asthma advocate that assessing severity and control of the disease should be based on assessing current impairment and future risk. Assessing impairment includes questions about day symptoms, night symptoms, activity limitation, use of rescue medication and lung function measures. In addition the use of asthma questionnaires such as the Asthma Control Test (ACT) [212] or the Asthma Control Questionnaire (ACQ) is recommended in assessing asthma control. In patients who are newly diagnosed and who have not yet been started on controller medication, assessing severity of the disease is recommended at first encounter. Asthma severity is classified into either *intermittent* (no interference of daily activity, and symptoms occur less than twice per week during the daytime or less than two nights per week) or *persistent* (limitation of normal activity and more frequent symptoms). Persistent asthmatics are further classified into mild, moderate or severe based upon increasing impairment [213]. In patients who are already receiving controller medication, assessing of asthma control is recommended on a periodic basis. Asthma control level may be well controlled to not well controlled or very poorly controlled based on the above criteria. Assessment of asthma control is essential to define management strategies described below.

4 Management

The goal of asthma therapy is to reduce symptoms and impairments in function imposed by the disease and to improve the quality of life. There are two sets of published guidelines for the management of asthma, one by the National Asthma Education and Prevention Program (NAEPP) [213] and the other by the Global Initiative for Asthma (GINA) [214]. However, these asthma guidelines are based upon studies done in patients younger than 60 years of age, and extrapolated to patients over the age of 60 years. As asthma in a 12 year old is likely to be very different from a 65 year old patient, future guidelines should include studies done specifically on older patients and may need specific age-appropriate modifications.

4.1 Non-pharmacologic Management

Non-pharmacologic components of management include asthma-education and control of environmental factors. Educating patients with asthma about their disease and how to assess and manage exacerbations reduces urgent care visits, asthma-related health care costs, and improves health status and quality of life and adherence to medication regimens medication in both younger and older patients [214–217]. Instructions, or "action plans" for routine asthma care should be easy to read and understand for the patient.

4.2 Pharmacologic Management: Challenges

Treatment of asthma in patients of all ages should address the severity of the patient's disease and how well it is controlled. Treating patients to control their disease and allow for increased quality of life, while minimizing potential medication side effects are a major goal, particularly in older patients who often receive multiple medications and therefore are at a potentially greater risk for side effects. Older patients frequently have a poor inhaler technique, due to decreased cognitive function [218, 219], or a physical impairment, and while some breath-activated medication devices may be easier for elderly patients to self administer [220], some still may have difficulties with its administration [221]. For these patients, a spacer can be attached to an HFA pressured metered dose (MDI) or some medications may be delivered by nebulizer (including corticosteroids for daily controller use, although this is not currently an FDA indicationfor older patients with asthma). There are several medications used to treat congestive heart failure and glaucoma, such as non-selective β-blockers [222], aspirin and NSAIDs which can worsen asthma in some patients.

4.2.1 Bronchodilating Medications

All patients with asthma should have a prescription for a "rescue medication" to treat acute symptoms. Rescue bronchodilators are either short-acting beta 2-receptor agonists (SABAs) or anticholinergic agents. SABAs are relatively safe in the elderly if used on an as-needed basis to treat exacerbations, though mild systemic absorption can produce tachycardia and tremor. The combination of diuretics that do not spare potassium (e.g. thiazides) and overuse of β2-agonists may produce significant hypokalemia and hypomagnesemia, increasing the risk of cardiac arrhythmias [223]. One report suggested that ipatrotropium in elderly asthmatics was associated with a slight increase in mortality, which the authors concluded was secondary to these patients having more severe asthma than those patients not receiving ipatrotropium [224]. However, anticholinergics, due to their atropine-like effects, may produce adverse side effects in the elderly including a dry mouth, urinary hesitancy, constipation and exacerbation of glaucoma. Long-acting β2-agonists should only be used as an add-on therapy in patients who have properly used inhaled corticosteroids (ICS) without relief, as they are not effective monotherapy for asthma [225].

4.2.2 Anti-Inflammatory Medications

Patients with persistent asthma should receive a daily anti-inflammatory medication, such as an ICS to suppress airway inflammation [226, 227]. Elderly patients with asthma are often under-prescribed ICS [5, 22, 228, 229]. ICS may produce local side effects including hoarseness or oral candidiasis, which can be prevented by using a spacer or by rinsing the mouth after use. A concern when prescribing ICS to older patients is increasing the risk of osteoporosis and bone fractures, especially for women and if the dose of ICS is >1000mcg/day budeosonide equivalent [230–233]. Therefore, patients should be given an ICS with the lowest dose to control their disease,[234] and one with a lower oral bioavailability. To decrease the effects of corticosteroids on bone resorption, patients should be encouraged to exercise, avoid excess alcohol intake, and use daily supplemental calcium with vitamin D. Observational studies in the elderly have suggested that ICS have a small, but significant risk of subcapsular and nuclear cataracts [235–237]. Observational studies have also suggested that elderly patients treated with ICS may have a small risk for developing glaucoma; however, further studies are needed [238–240]. Therefore, patients receiving ICS should have yearly ophthalmologic exams.

Corticosteroids improve asthma control and symptoms in some, but not all older patients. A large database review of patients 65 years or older hospitalized at least once for asthma and followed 12 months after discharge, demonstrated that those given ICS had a 29 % reduction in asthma readmission and a 39 % reduction in all-cause mortality in the subsequent 12 months [241]. However, some older patients with asthma may have a component of fixed airway obstruction [150]. The NAEPP has recommended for some older patients, that a 2 week trial of oral corticosteroids (0.3–0.5 mg/kg) be administered with repeat lung function measurement after the course to assess for possible reversal of obstruction and clinical benefit [242].

Leukotriene modifiers are a class of anti-inflammatory agents that inhibit the effects of leukotrienes, which are potent bronchoconstrictors, recruit inflammatory cells to the airways and induce mucus hypersecretion. When compared with ICS, leukotriene receptor antagonists generally do not have as favorable an outcome in improving FEV_1, symptom scores and other outcome measures [213]. Two studies have investigated the role of leukotriene modifiers in patients of different ages with asthma, and have concluded that their effectiveness may be limited in the elderly patients compared with younger counterparts, but continue to improve asthma symptoms without reducing the need for rescue therapy [243, 244].

4.2.3 Methylxanthines

Methylxanthines (e.g. Theophylline) increase intracellular cyclic adenosine monophosphate which bronchodilates the airways, and in lower doses have anti-inflammatory properties [245]. Its use in asthma, especially in an older group, is limited by its relatively weak bronchodilator properties and many side effects and drug interactions [246].

5 Future Research Needs

In 2008, the National Institute on Aging and the National Heart, Lung, and Blood Institute co-sponsored a conference on asthma in older individuals [247]. This program explored data from animal models and human studies of asthma in older patients and concluded that airway inflammation in asthma and its clinical response to therapy in older patients likely differs from younger patients. Since this conference, there are still several remaining unanswered questions regarding asthma in older patients. For example, how do age-related changes in the innate and adaptive immune responses impact airway inflammation in older patients with asthma and does it differ from younger patients with asthma? Understanding the pathophysiology and underlying airway inflammation in older adults with asthma and the different phenotypes and endotypes of asthma in this population is a major unmet need as this group of patients has high rates of morbidity and mortality. Furthermore, with the expected increase in the elderly populations, including elderly asthmatic patients in clinical trials is essential, and particular attention should be paid to also address how differences in inflammatory mechanisms affect responses to therapy.

6 Summary

Asthma is a major public health problem which is frequently overlooked in the geriatric population. While much has been uncovered about the pathogenesis, course and outcomes of asthma in children and young adults, studies in the aging population have been scarce or non-existent. Asthma in the elderly has at least two distinct phenotypes based on the onset of the disease. While the characteristics of long-standing asthma that starts early in life may be similar to the general asthma population, more studies are needed to uncover details about asthma that develops late in life which can have distinct clinical features and may have different course of response to therapy. Our knowledge about management of asthma in this population is based on extrapolation from studies in the younger population. Although future studies are needed to investigate the response to existing and novel interventions in the elderly, current guidelines recommend that management of asthma in this population should not differ from that of younger patients. Careful monitoring of compliance with therapy and of adverse events to medication is essential in this population. Despite severe symptoms and physiologic impairment, most elderly patients with asthma improve with therapy and can lead active productive lives.

References

1. Bousquet J, Jeffery PK, Busse WW, Johnson M, Vignola AM (2000) Asthma. From bronchoconstriction to airways inflammation and remodeling. Am J Respir Crit Care Med 161:1720–1745
2. Busse WW (2010) The relationship of airway hyperresponsiveness and airway inflammation: airway hyperresponsiveness in asthma: its measurement and clinical significance. Chest 138:4S–10S
3. Dow L, Coggon D, Campbell MJ, Osmond C, Holgate ST (1992) The interaction between immunoglobulin E and smoking in airflow obstruction in the elderly. Am Rev Respir Dis 146:402–407
4. Slavin RG (2004) The elderly asthmatic patient. Allergy Asthma Proc 25:371–373
5. Enright PL, McClelland RL, Newman AB, Gottlieb DJ, Lebowitz MD (1999) Underdiagnosis and undertreatment of asthma in the elderly. Cardiovascular Health Study Research Group. Chest 116:603–613
6. Dyer CA, Hill SL, Stockley RA, Sinclair AJ (1999) Quality of life in elderly subjects with a diagnostic label of asthma from general practice registers. Eur Respir J 14:39–45
7. Nejjari C, Tessier JF, Barberger-Gateau P, Jacqmin H, Dartigues JF, Salamon R (1994) Functional status of elderly people treated for asthma-related symptoms: a population based case–control study. Eur Respir J 7:1077–1083
8. Krauskopf KA, Sofianou A, Goel MS, Wolf MS, Wilson EA, Martynenko ME, Halm EA, Leventhal H, Feldman JM, Federman AD, Wisnivesky JP (2013) Depressive symptoms, low adherence, and poor asthma outcomes in the elderly. J Asthma 50:260–266
9. Ross JA, Yang Y, Song PX, Clark NM, Baptist AP (2013) Quality of life, health care utilization, and poor asthma outcomes in older adults with asthma. J Allergy Clin Immunol Pract 1:157–162
10. Mannino DM, Homa DM, Akinbami LJ, Moorman JE, Gwynn C, Redd SC (2002) Surveillance for asthma – United States, 1980–1999. MMWR Surveill Summ 51:1–13

11. Gergen PJ, Arbes SJ Jr, Calatroni A, Mitchell HE, Zeldin DC (2009) Total IgE levels and asthma prevalence in the US population: results from the National Health and Nutrition Examination Survey 2005–2006. J Allergy Clin Immunol 124:447–453
12. Oraka E, Kim HJ, King ME, Callahan DB (2012) Asthma prevalence among US elderly by age groups: age still matters. J Asthma 49:593–599
13. Tsai CL, Lee WY, Hanania NA, Camargo CA Jr (2012) Age-related differences in clinical outcomes for acute asthma in the United States, 2006–2008. J Allergy Clin Immunol 129:1252–1258
14. Tsai CL, Delclos GL, Huang JS, Hanania NA, Camargo CA (2013) Age-related differences in asthma outcomes in the United States, 1998–2006. Ann Allergy Asthma Immunol (in press)
15. Moorman JE, Mannino DM (2001) Increasing U.S. asthma mortality rates: who is really dying? J Asthma 38:65–71
16. Mannino DM, Gagnon RC, Petty TL, Lydick E (2000) Obstructive lung disease and low lung function in adults in the United States: data from the National Health and Nutrition Examination Survey, 1988–1994. Arch Intern Med 160:1683–1689
17. Savage-Brown A, Mannino DM, Redd SC (2005) Lung disease and asthma severity in adults with asthma: data from the Third National Health and Nutrition Examination. J Asthma 42:519–523
18. Banerjee DK, Lee GS, Malik SK, Daly S (1987) Underdiagnosis of asthma in the elderly. Br J Dis Chest 81:23–29
19. Bauer BA, Reed CE, Yunginger JW, Wollan PC, Silverstein MD (1997) Incidence and outcomes of asthma in the elderly. A population-based study in Rochester, Minnesota. Chest 111:303–310
20. Boezen HM, Rijcken B, Schouten JP, Postma DS (1998) Breathlessness in elderly individuals is related to low lung function and reversibility of airway obstruction. Eur Respir J 12:805–810
21. Dow L, Fowler L, Phelps L, Waters K, Coggon D, Kinmonth AL, Holgate ST (2001) Prevalence of untreated asthma in a population sample of 6000 older adults in Bristol, UK. Thorax 56:472–476
22. Parameswaran K, Hildreth AJ, Chadha D, Keaney NP, Taylor IK, Bansal SK (1998) Asthma in the elderly: underperceived, underdiagnosed and undertreated; a community survey. Respir Med 92:573–577
23. Schmid G, Kramer MP, Maurer M, Wandl S, Wesierska-Gadek J (2007) Cellular and organismal ageing: role of the p53 tumor suppressor protein in the induction of transient and terminal senescence. J Cell Biochem 101:1355–1369
24. Din S, Konstandin MH, Johnson B, Emathinger J, Volkers M, Toko H, Collins B, Ormachea L, Samse K, Kubli DA, De La Torre A, Kraft AS, Gustafsson AB, Kelly DP, Sussman MA (2014) Metabolic dysfunction consistent with premature aging results from deletion of Pim kinases. Circ Res 115:376–387
25. Franceschi C, Campisi J (2014) Chronic inflammation (inflammaging) and its potential contribution to age-associated diseases. J Gerontol A Biol Sci Med Sci 69(Suppl 1):S4–S9
26. Passos JF, Simillion C, Hallinan J, Wipat A, von Zglinicki T (2009) Cellular senescence: unravelling complexity. Age (Dordr) 31:353–363
27. Zhang Q, Raoof M, Chen Y, Sumi Y, Sursal T, Junger W, Brohi K, Itagaki K, Hauser CJ (2010) Circulating mitochondrial DAMPs cause inflammatory responses to injury. Nature 464:104–107
28. di d'Adda FF, Reaper PM, Clay-Farrace L, Fiegler H, Carr P, von Zglinicki T, Saretzki G, Carter NP, Jackson SP (2003) A DNA damage checkpoint response in telomere-initiated senescence. Nature 426:194–198
29. Gilson E, Geli V (2007) How telomeres are replicated. Nat Rev Mol Cell Biol 8:825–838
30. Kyoh S, Venkatesan N, Poon AH, Nishioka M, Lin TY, Baglole CJ, Eidelman DH, Hamid Q (2013) Are leukocytes in asthmatic patients aging faster? A study of telomere length and disease severity. J Allergy Clin Immunol 132:480–482

31. Olivieri F, Rippo MR, Procopio AD, Fazioli F (2013) Circulating inflamma-miRs in aging and age-related diseases. Front Genet 4:121
32. Olivieri F, Lazzarini R, Recchioni R, Marcheselli F, Rippo MR, Di NS, Albertini MC, Graciotti L, Babini L, Mariotti S, Spada G, Abbatecola AM, Antonicelli R, Franceschi C, Procopio AD (2013) MiR-146a as marker of senescence-associated pro-inflammatory status in cells involved in vascular remodelling. Age (Dordr) 35:1157–1172
33. Franceschi C, Monti D, Sansoni P, Cossarizza A (1995) The immunology of exceptional individuals: the lesson of centenarians. Immunol Today 16:12–16
34. Ferguson FG, Wikby A, Maxson P, Olsson J, Johansson B (1995) Immune parameters in a longitudinal study of a very old population of Swedish people: a comparison between survivors and nonsurvivors. J Gerontol A Biol Sci Med Sci 50:B378–B382
35. Franceschi C, Capri M, Monti D, Giunta S, Olivieri F, Sevini F, Panourgia MP, Invidia L, Celani L, Scurti M, Cevenini E, Castellani GC, Salvioli S (2007) Inflammaging and anti-inflammaging: a systemic perspective on aging and longevity emerged from studies in humans. Mech Ageing Dev 128:92–105
36. Fulop T, Larbi A, Kotb R, de Angelis F, Pawelec G (2011) Aging, immunity, and cancer. Discov Med 11:537–550
37. Holgate ST, Roberts G, Arshad HS, Howarth PH, Davies DE (2009) The role of the airway epithelium and its interaction with environmental factors in asthma pathogenesis. Proc Am Thorac Soc 6:655–659
38. Ho JC, Chan KN, Hu WH, Lam WK, Zheng L, Tipoe GL, Sun J, Leung R, Tsang KW (2001) The effect of aging on nasal mucociliary clearance, beat frequency, and ultrastructure of respiratory cilia. Am J Respir Crit Care Med 163:983–988
39. Svartengren M, Falk R, Philipson K (2005) Long-term clearance from small airways decreases with age. Eur Respir J 26:609–615
40. Ghadially R, Brown BE, Sequeira-Martin SM, Feingold KR, Elias PM (1995) The aged epidermal permeability barrier. Structural, functional, and lipid biochemical abnormalities in humans and a senescent murine model. J Clin Invest 95:2281–2290
41. Zhou F, Onizawa S, Nagai A, Aoshiba K (2011) Epithelial cell senescence impairs repair process and exacerbates inflammation after airway injury. Respir Res 12:78
42. Shaykhiev R, Crystal RG (2013) Innate immunity and chronic obstructive pulmonary disease: a mini-review. Gerontology 59:481–489
43. Sapey E, Greenwood H, Walton G, Mann E, Love A, Aaronson N, Insall RH, Stockley RA, Lord JM (2014) Phosphoinositide 3-kinase inhibition restores neutrophil accuracy in the elderly: toward targeted treatments for immunosenescence. Blood 123:239–248
44. Tseng CW, Kyme PA, Arruda A, Ramanujan VK, Tawackoli W, Liu GY (2012) Innate immune dysfunctions in aged mice facilitate the systemic dissemination of methicillin-resistant S aureus. PLoS One 7, e41454
45. Fortin CF, Larbi A, Lesur O, Douziech N, Fulop T Jr (2006) Impairment of SHP-1 down-regulation in the lipid rafts of human neutrophils under GM-CSF stimulation contributes to their age-related, altered functions. J Leukoc Biol 79:1061–1072
46. Plackett TP, Boehmer ED, Faunce DE, Kovacs EJ (2004) Aging and innate immune cells. J Leukoc Biol 76:291–299
47. Butcher SK, Chahal H, Nayak L, Sinclair A, Henriquez NV, Sapey E, O'Mahony D, Lord JM (2001) Senescence in innate immune responses: reduced neutrophil phagocytic capacity and CD16 expression in elderly humans. J Leukoc Biol 70:881–886
48. Kong KF, Delroux K, Wang X, Qian F, Arjona A, Malawista SE, Fikrig E, Montgomery RR (2008) Dysregulation of TLR3 impairs the innate immune response to West Nile virus in the elderly. J Virol 82:7613–7623
49. van Duin D, Mohanty S, Thomas V, Ginter S, Montgomery RR, Fikrig E, Allore HG, Medzhitov R, Shaw AC (2007) Age-associated defect in human TLR-1/2 function. J Immunol 178:970–975
50. Verschoor CP, Loukov D, Naidoo A, Puchta A, Johnstone J, Millar J, Lelic A, Novakowski KE, Dorrington MG, Loeb M, Bramson JL, Bowdish DM (2015) Circulating TNF and mitochondrial DNA are major determinants of neutrophil phenotype in the advanced-age, frail elderly. Mol Immunol 65:148–156

51. Chaves MM, Costa DC, de Oliveira BF, Rocha MI, Nogueira-Machado JA (2009) Role PKA and p38 MAPK on ROS production in neutrophil age-related: lack of IL-10 effect in older subjects. Mech Ageing Dev 130:588–591
52. Baptista G, Dupuy AM, Jaussent A, Durant R, Ventura E, Sauguet P, Picot MC, Jeandel C, Cristol JP (2012) Low-grade chronic inflammation and superoxide anion production by NADPH oxidase are the main determinants of physical frailty in older adults. Free Radic Res 46:1108–1114
53. Coelho HB, Steinberg PC, Caldeira CD, Nogueira-Machado JA, Martins CM (2005) Aging: functional metabolic balance among cAMP, cGMP and reactive oxygen intermediate generation by human granulocytes. Gerontology 51:363–368
54. Oliveira BF, Nogueira-Machado JA, Chaves MM (2010) The role of oxidative stress in the aging process. ScientificWorldJournal 10:1121–1128
55. Ogawa K, Suzuki K, Okutsu M, Yamazaki K, Shinkai S (2008) The association of elevated reactive oxygen species levels from neutrophils with low-grade inflammation in the elderly. Immun Ageing 5:13
56. Qian F, Guo X, Wang X, Yuan X, Chen S, Malawista SE, Bockenstedt LK, Allore HG, Montgomery RR (2014) Reduced bioenergetics and toll-like receptor 1 function in human polymorphonuclear leukocytes in aging. Aging 6:131–139
57. Meyer KC, Rosenthal NS, Soergel P, Peterson K (1998) Neutrophils and low-grade inflammation in the seemingly normal aging human lung. Mech Ageing Dev 104(2):169–181
58. Meyer KC, Soergel P (1999) Variation of bronchoalveolar lymphocyte phenotypes with age in the physiologically normal human lung. Thorax 54(8):697–700
59. Fulop T Jr, Fouquet C, Allaire P, Perrin N, Lacombe G, Stankova J, Rola-Pleszczynski M, Gagne D, Wagner JR, Khalil A, Dupuis G (1997) Changes in apoptosis of human polymorphonuclear granulocytes with aging. Mech Ageing Dev 96:15–34
60. Mathur SK, Schwantes EA, Jarjour NN, Busse WW (2008) Age-related changes in eosinophil function in human subjects. Chest 133(2):412–419
61. Nyenhuis SM, Schwantes EA, Evans MD, Mathur SK (2010) Airway neutrophil inflammatory phenotype in older subjects with asthma. J Allergy Clin Immunol 125:1163–1165
62. Nair P, Aziz-Ur-Rehman A, Radford K (2015) Therapeutic implications of neutrophilic asthma. Curr Opin Pulm Med 21:33–38, Ref Type: Generic
63. Kane B, Fowler SJ, Niven R (2015) Refractory asthma – beyond step 5, the role of new and emerging adjuvant therapies. Chron Respir Dis 12:69–77
64. Brannan JD (2010) Bronchial hyperresponsiveness in the assessment of asthma control: airway hyperresponsiveness in asthma: its measurement and clinical significance. Chest 138:11S–17S
65. Nyenhuis SM, Schwantes EA, Mathur SK (2010) Characterization of leukotrienes in a pilot study of older asthma subjects. Immun Ageing 7:8
66. Busse PJ, Zhang TF, Srivastava K, Schofield B, Li XM (2007) Effect of ageing on pulmonary inflammation, airway hyperresponsiveness and T and B cell responses in antigen-sensitized and -challenged mice. Clin Exp Allergy 37:1392–1403
67. Annema JT, Sparrow D, O'Connor GT, Rijcken B, Koeter GH, Postma DS, Weiss ST (1995) Chronic respiratory symptoms and airway responsiveness to methacholine are associated with eosinophilia in older men: the Normative Aging Study. Eur Respir J 8:62–69
68. Agrawal A, Agrawal S, Cao JN, Su H, Osann K, Gupta S (2007) Altered innate immune functioning of dendritic cells in elderly humans: a role of phosphoinositide 3-kinase-signaling pathway. J Immunol 178:6912–6922
69. Agrawal A, Tay J, Ton S, Agrawal S, Gupta S (2009) Increased reactivity of dendritic cells from aged subjects to self-antigen, the human DNA. J Immunol 182:1138–1145
70. Panda A, Qian F, Mohanty S, van Duin D, Newman FK, Zhang L, Chen S, Towle V, Belshe RB, Fikrig E, Allore HG, Montgomery RR, Shaw AC (2010) Age-associated decrease in TLR function in primary human dendritic cells predicts influenza vaccine response. J Immunol 184:2518–2527
71. Canaday DH, Amponsah NA, Jones L, Tisch DJ, Hornick TR, Ramachandra L (2010) Influenza-induced production of interferon-alpha is defective in geriatric individuals. J Clin Immunol 30:373–383

72. Grolleau-Julius A, Harning EK, Abernathy LM, Yung RL (2008) Impaired dendritic cell function in aging leads to defective antitumor immunity. Cancer Res 68:6341–6349
73. Zhao J, Zhao J, Legge K, Perlman S (2011) Age-related increases in PGD(2) expression impair respiratory DC migration, resulting in diminished T cell responses upon respiratory virus infection in mice. J Clin Invest 121:4921–4930
74. Agrawal A, Tay J, Yang GE, Agrawal S, Gupta S (2010) Age-associated epigenetic modifications in human DNA increase its immunogenicity. Aging 2:93–100
75. Weng NP (2006) Aging of the immune system: how much can the adaptive immune system adapt? Immunity 24:495–499
76. Gagliardi AM, Gomes Silva BN, Torloni MR, Soares BG (2012) Vaccines for preventing herpes zoster in older adults. Cochrane Database Syst Rev 10, CD008858
77. Jefferson T, Rivetti D, Rivetti A, Rudin M, Di PC, Demicheli V (2005) Efficacy and effectiveness of influenza vaccines in elderly people: a systematic review. Lancet 366:1165–1174
78. Flores KG, Li J, Sempowski GD, Haynes BF, Hale LP (1999) Analysis of the human thymic perivascular space during aging. J Clin Invest 104:1031–1039
79. den Braber I, Mugwagwa T, Vrisekoop N, Westera L, Mogling R, de Boer AB, Willems N, Schrijver EH, Spierenburg G, Gaiser K, Mul E, Otto SA, Ruiter AF, Ackermans MT, Miedema F, Borghans JA, de Boer RJ, Tesselaar K (2012) Maintenance of peripheral naive T cells is sustained by thymus output in mice but not humans. Immunity 36:288–297
80. Yang Y, An J, Weng NP (2008) Telomerase is involved in IL-7-mediated differential survival of naive and memory CD4+ T cells. J Immunol 180:3775–3781
81. Zediak VP, Maillard I, Bhandoola A (2007) Multiple prethymic defects underlie age-related loss of T progenitor competence. Blood 110:1161–1167
82. Tsuboi I, Morimoto K, Hirabayashi Y, Li GX, Aizawa S, Mori KJ, Kanno J, Inoue T (2004) Senescent B lymphopoiesis is balanced in suppressive homeostasis: decrease in interleukin-7 and transforming growth factor-beta levels in stromal cells of senescence-accelerated mice. Exp Biol Med (Maywood) 229:494–502
83. Cho RH, Sieburg HB, Muller-Sieburg CE (2008) A new mechanism for the aging of hematopoietic stem cells: aging changes the clonal composition of the stem cell compartment but not individual stem cells. Blood 111:5553–5561
84. Timm JA, Thoman ML (1999) Maturation of CD4+ lymphocytes in the aged microenvironment results in a memory-enriched population. J Immunol 162:711–717
85. Tsukamoto H, Clise-Dwyer K, Huston GE, Duso DK, Buck AL, Johnson LL, Haynes L, Swain SL (2009) Age-associated increase in lifespan of naive CD4 T cells contributes to T-cell homeostasis but facilitates development of functional defects. Proc Natl Acad Sci U S A 106:18333–18338
86. Khan N, Shariff N, Cobbold M, Bruton R, Ainsworth JA, Sinclair AJ, Nayak L, Moss PA (2002) Cytomegalovirus seropositivity drives the CD8 T cell repertoire toward greater clonality in healthy elderly individuals. J Immunol 169:1984–1992
87. Nociari MM, Telford W, Russo C (1999) Postthymic development of CD28-CD8+ T cell subset: age-associated expansion and shift from memory to naive phenotype. J Immunol 162:3327–3335
88. Globerson A, Effros RB (2000) Ageing of lymphocytes and lymphocytes in the aged. Immunol Today 21:515–521
89. Valenzuela HF, Effros RB (2002) Divergent telomerase and CD28 expression patterns in human CD4 and CD8 T cells following repeated encounters with the same antigenic stimulus. Clin Immunol 105:117–125
90. Fletcher JM, Vukmanovic-Stejic M, Dunne PJ, Birch KE, Cook JE, Jackson SE, Salmon M, Rustin MH, Akbar AN (2005) Cytomegalovirus-specific CD4+ T cells in healthy carriers are continuously driven to replicative exhaustion. J Immunol 175:8218–8225
91. Bryl E, Vallejo AN, Weyand CM, Goronzy JJ (2001) Down-regulation of CD28 expression by TNF-alpha. J Immunol 167:3231–3238
92. Topp MS, Riddell SR, Akatsuka Y, Jensen MC, Blattman JN, Greenberg PD (2003) Restoration of CD28 expression in CD28-CD8+ memory effector T cells reconstitutes antigen-induced IL-2 production. J Exp Med 198:947–955

93. Chiu WK, Fann M, Weng NP (2006) Generation and growth of CD28nullCD8+ memory T cells mediated by IL-15 and its induced cytokines. J Immunol 177:7802–7810
94. Alonso-Arias R, Moro-Garcia MA, Lopez-Vazquez A, Rodrigo L, Baltar J, Garcia FM, Jaurrieta JJ, Lopez-Larrea C (2011) NKG2D expression in CD4+ T lymphocytes as a marker of senescence in the aged immune system. Age (Dordr) 33:591–605
95. Naylor K, Li G, Vallejo AN, Lee WW, Koetz K, Bryl E, Witkowski J, Fulbright J, Weyand CM, Goronzy JJ (2005) The influence of age on T cell generation and TCR diversity. J Immunol 174:7446–7452
96. Britanova OV, Putintseva EV, Shugay M, Merzlyak EM, Turchaninova MA, Staroverov DB, Bolotin DA, Lukyanov S, Bogdanova EA, Mamedov IZ, Lebedev YB, Chudakov DM (2014) Age-related decrease in TCR repertoire diversity measured with deep and normalized sequence profiling. J Immunol 192:2689–2698
97. Yager EJ, Ahmed M, Lanzer K, Randall TD, Woodland DL, Blackman MA (2008) Age-associated decline in T cell repertoire diversity leads to holes in the repertoire and impaired immunity to influenza virus. J Exp Med 205:711–723
98. Korn T, Oukka M, Kuchroo V, Bettelli E (2007) Th17 cells: effector T cells with inflammatory properties. Semin Immunol 19:362–371
99. Ouyang X, Yang Z, Zhang R, Arnaboldi P, Lu G, Li Q, Wang W, Zhang B, Cui M, Zhang H, Liang-Chen J, Qin L, Zheng F, Huang B, Xiong H (2011) Potentiation of Th17 cytokines in aging process contributes to the development of colitis. Cell Immunol 266:208–217
100. Schmitt V, Rink L, Uciechowski P (2013) The Th17/Treg balance is disturbed during aging. Exp Gerontol 48:1379–1386
101. Tesar BM, Du W, Shirali AC, Walker WE, Shen H, Goldstein DR (2009) Aging augments IL-17 T-cell alloimmune responses. Am J Transplant 9:54–63
102. Malaguarnera M, Cristaldi E, Romano G, Malaguarnera L (2012) Autoimmunity in the elderly: implications for cancer. J Cancer Res Ther 8:520–527
103. Stout-Delgado HW, Du W, Shirali AC, Booth CJ, Goldstein DR (2009) Aging promotes neutrophil-induced mortality by augmenting IL-17 production during viral infection. Cell Host Microbe 6:446–456
104. Birmingham JM, Gillespie VL, Srivastava K, Li XM, Busse PJ (2014) Influenza A infection enhances antigen-induced airway inflammation and hyperresponsiveness in young but not aged mice. Clin Exp Allergy 44:1188–1199
105. Brandenberger C, Li N, Jackson-Humbles DN, Rockwell CE, Wagner JG, Harkema JR (2014) Enhanced allergic airway disease in old mice is associated with a Th17 response. Clin Exp Allergy 44:1282–1292
106. Hellings PW, Kasran A, Liu Z, Vandekerckhove P, Wuyts A, Overbergh L, Mathieu C, Ceuppens JL (2003) Interleukin-17 orchestrates the granulocyte influx into airways after allergen inhalation in a mouse model of allergic asthma. Am J Respir Cell Mol Biol 28:42–50
107. Oda N, Canelos PB, Essayan DM, Plunkett BA, Myers AC, Huang SK (2005) Interleukin-17F induces pulmonary neutrophilia and amplifies antigen-induced allergic response. Am J Respir Crit Care Med 171:12–18
108. Wakashin H, Hirose K, Maezawa Y, Kagami S, Suto A, Watanabe N, Saito Y, Hatano M, Tokuhisa T, Iwakura Y, Puccetti P, Iwamoto I, Nakajima H (2008) IL-23 and Th17 cells enhance Th2-cell-mediated eosinophilic airway inflammation in mice. Am J Respir Crit Care Med 178:1023–1032
109. Chen Y, Thai P, Zhao YH, Ho YS, DeSouza MM, Wu R (2003) Stimulation of airway mucin gene expression by interleukin (IL)-17 through IL-6 paracrine/autocrine loop. J Biol Chem 278:17036–17043
110. Finkelman FD, Hogan SP, Hershey GK, Rothenberg ME, Wills-Karp M (2010) Importance of cytokines in murine allergic airway disease and human asthma. J Immunol 184:1663–1674
111. Molet S, Hamid Q, Davoine F, Nutku E, Taha R, Page N, Olivenstein R, Elias J, Chakir J (2001) IL-17 is increased in asthmatic airways and induces human bronchial fibroblasts to produce cytokines. J Allergy Clin Immunol 108:430–438

112. Kawaguchi M, Onuchic LF, Li XD, Essayan DM, Schroeder J, Xiao HQ, Liu MC, Krishnaswamy G, Germino G, Huang SK (2001) Identification of a novel cytokine, ML-1, and its expression in subjects with asthma. J Immunol 167:4430–4435
113. Dejaco C, Duftner C, Schirmer M (2006) Are regulatory T-cells linked with aging? Exp Gerontol 41:339–345
114. Lages CS, Suffia I, Velilla PA, Huang B, Warshaw G, Hildeman DA, Belkaid Y, Chougnet C (2008) Functional regulatory T cells accumulate in aged hosts and promote chronic infectious disease reactivation. J Immunol 181:1835–1848
115. Gregg R, Smith CM, Clark FJ, Dunnion D, Khan N, Chakraverty R, Nayak L, Moss PA (2005) The number of human peripheral blood CD4+ CD25 high regulatory T cells increases with age. Clin Exp Immunol 140:540–546
116. Hwang KA, Kim HR, Kang I (2009) Aging and human CD4(+) regulatory T cells. Mech Ageing Dev 130:509–517
117. Chiu BC, Stolberg VR, Zhang H, Chensue SW (2007) Increased Foxp3(+) Treg cell activity reduces dendritic cell co-stimulatory molecule expression in aged mice. Mech Ageing Dev 128:618–627
118. Tsaknaridis L, Spencer L, Culbertson N, Hicks K, Latocha D, Chou YK, Whitham RH, Bakke A, Jones RE, Offner H, Bourdette DN, Vandenbark AA (2003) Functional assay for human CD4+ CD25+ Treg cells reveals an age-dependent loss of suppressive activity. J Neurosci Res 74:296–308
119. Hartl D, Koller B, Mehlhorn AT, Reinhardt D, Nicolai T, Schendel DJ, Griese M, Krauss-Etschmann S (2007) Quantitative and functional impairment of pulmonary CD4+ CD25hi regulatory T cells in pediatric asthma. J Allergy Clin Immunol 119:1258–1266
120. Kearley J, Barker JE, Robinson DS, Lloyd CM (2005) Resolution of airway inflammation and hyperreactivity after in vivo transfer of CD4+ CD25+ regulatory T cells is interleukin 10 dependent. J Exp Med 202:1539–1547
121. Kearley J, Robinson DS, Lloyd CM (2008) CD4+ CD25+ regulatory T cells reverse established allergic airway inflammation and prevent airway remodeling. J Allergy Clin Immunol 122:617–624
122. Ling EM, Smith T, Nguyen XD, Pridgeon C, Dallman M, Arbery J, Carr VA, Robinson DS (2004) Relation of CD4+ CD25+ regulatory T-cell suppression of allergen-driven T-cell activation to atopic status and expression of allergic disease. Lancet 363:608–615
123. Mucida D, Kutchukhidze N, Erazo A, Russo M, Lafaille JJ, Curotto de Lafaille MA (2005) Oral tolerance in the absence of naturally occurring Tregs. J Clin Invest 115:1923–1933
124. Ostroukhova M, Seguin-Devaux C, Oriss TB, Dixon-McCarthy B, Yang L, Ameredes BT, Corcoran TE, Ray A (2004) Tolerance induced by inhaled antigen involves CD4(+) T cells expressing membrane-bound TGF-beta and FOXP3. J Clin Invest 114:28–38
125. Russo M, Nahori MA, Lefort J, Gomes E, de Castro KA, Rodriguez D, Ribeiro OG, Adriouch S, Gallois V, de Faria AM, Vargaftig BB (2001) Suppression of asthma-like responses in different mouse strains by oral tolerance. Am J Respir Cell Mol Biol 24:518–526
126. Birmingham JM, Patil S, Li XM, Busse PJ (2013) The effect of oral tolerance on the allergic airway response in younger and aged mice. J Asthma 50:122–132
127. Vale-Pereira S, Todo-Bom A, Geraldes L, Schmidt-Weber C, Akdis CA, Mota-Pinto A (2011) FoxP3, GATA-3 and T-bet expression in elderly asthma. Clin Exp Allergy 41:490–496
128. McKenna RW, Asplund SL, Kroft SH (2004) Immunophenotypic analysis of hematogones (B-lymphocyte precursors) and neoplastic lymphoblasts by 4-color flow cytometry. Leuk Lymphoma 45:277–285
129. Kogut I, Scholz JL, Cancro MP, Cambier JC (2012) B cell maintenance and function in aging. Semin Immunol 24:342–349
130. Dailey RW, Eun SY, Russell CE, Vogel LA (2001) B cells of aged mice show decreased expansion in response to antigen, but are normal in effector function. Cell Immunol 214:99–109
131. Burns EA, Goodwin JS (1997) Immunodeficiency of aging. Drugs Aging 11:374–397
132. Miller C, Kelsoe G (1995) Ig VH hypermutation is absent in the germinal centers of aged mice. J Immunol 155:3377–3384

133. Eaton SM, Burns EM, Kusser K, Randall TD, Haynes L (2004) Age-related defects in CD4 T cell cognate helper function lead to reductions in humoral responses. J Exp Med 200:1613–1622
134. Yen CJ, Lin SL, Huang KT, Lin RH (2000) Age-associated changes in interferon-gamma and interleukin-4 secretion by purified human CD4+ and CD8+ T cells. J Biomed Sci 7:317–321
135. Yano S, Ghosh P, Kusaba H, Buchholz M, Longo DL (2003) Effect of promoter methylation on the regulation of IFN-gamma gene during in vitro differentiation of human peripheral blood T cells into a Th2 population. J Immunol 171:2510–2516
136. Jing Y, Gravenstein S, Chaganty NR, Chen N, Lyerly KH, Joyce S, Deng Y (2007) Aging is associated with a rapid decline in frequency, alterations in subset composition, and enhanced Th2 response in CD1d-restricted NKT cells from human peripheral blood. Exp Gerontol 42:719–732
137. Sandmand M, Bruunsgaard H, Kemp K, Andersen-Ranberg K, Pedersen AN, Skinhoj P, Pedersen BK (2002) Is ageing associated with a shift in the balance between Type 1 and Type 2 cytokines in humans? Clin Exp Immunol 127:107–114
138. Jiang J, Bennett AJ, Fisher E, Williams-Bey Y, Shen H, Murasko DM (2009) Limited expansion of virus-specific CD8 T cells in the aged environment. Mech Ageing Dev 130:713–721
139. Salo PM, Calatroni A, Gergen PJ, Hoppin JA, Sever ML, Jaramillo R, Arbes SJ Jr, Zeldin DC (2011) Allergy-related outcomes in relation to serum IgE: results from the National Health and Nutrition Examination Survey 2005–2006. J Allergy Clin Immunol 127:1226–1235
140. King MJ, Lockey RF (2003) Allergen prick-puncture skin testing in the elderly. Drugs Aging 20:1011–1017
141. Barbee RA, Halonen M, Lebowitz M, Burrows B (1981) Distribution of IgE in a community population sample: correlations with age, sex, and allergen skin test reactivity. J Allergy Clin Immunol 68:106–111
142. Sapigni T, Biavati P, Simoni M, Viegi G, Baldacci S, Carrozzi L, Modena P, Pedreschi M, Vellutini M, Paoletti P (1998) The Po River Delta Respiratory Epidemiological Survey: an analysis of factors related to level of total serum IgE. Eur Respir J 11:278–283
143. Stoy PJ, Roitman-Johnson B, Walsh G, Gleich GJ, Mendell N, Yunis E, Blumenthal MN (1981) Aging and serum immunoglobulin E levels, immediate skin tests, RAST. J Allergy Clin Immunol 68(6):421–426
144. Jarvis D, Luczynska C, Chinn S, Burney P (1995) The association of age, gender and smoking with total IgE and specific IgE. Clin Exp Allergy 25(11):1083–1091
145. Kerkhof M, Droste JH, de Monchy JG, Schouten JP, Rijcken B (1996) Distribution of total serum IgE and specific IgE to common aeroallergens by sex and age, and their relationship to each other in a random sample of the Dutch general population aged 20–70 years. Dutch ECRHS Group, European Community Respiratory Health Study. Allergy 51:770–776
146. Kerkhof M, Dubois AE, Postma DS, Schouten JP, de Monchy JG (2003) Role and interpretation of total serum IgE measurements in the diagnosis of allergic airway disease in adults. Allergy 58:905–911
147. Omenaas E, Bakke P, Elsayed S, Hanoa R, Gulsvik A (1994) Total and specific serum IgE levels in adults: relationship to sex, age and environmental factors. Clin Exp Allergy 24(6):530–539
148. Wuthrich B, Schindler C, Medici TC, Zellweger JP, Leuenberger P (1996) IgE levels, atopy markers and hay fever in relation to age, sex and smoking status in a normal adult Swiss population. SAPALDIA (Swiss Study on Air Pollution and Lung Diseases in Adults) Team. Int Arch Allergy Immunol 111:396–402
149. Nakazawa T, Houjyo S, Dobashi K, Sato K (1994) Influence of aging and sex on specific IgE antibody production. Intern Med 33:396–401
150. Braman SS, Kaemmerlen JT, Davis SM (1991) Asthma in the elderly. A comparison between patients with recently acquired and long-standing disease. Am Rev Respir Dis 143:336–340
151. Reed CE (2006) The natural history of asthma. J Allergy Clin Immunol 118:543–548
152. Burrows B, Barbee RA, Cline MG, Knudson RJ, Lebowitz MD (1991) Characteristics of asthma among elderly adults in a sample of the general population. Chest 100:935–942

153. Litonjua AA, Sparrow D, Weiss ST, O'Connor GT, Long AA, Ohman JL Jr (1997) Sensitization to cat allergen is associated with asthma in older men and predicts new-onset airway hyperresponsiveness. The Normative Aging Study. Am J Respir Crit Care Med 156:23–27
154. Huss K et al (2001) Asthma severity, atopic status, allergen exposure, and quality of life in elderly persons. Ann Allergy Asthma Immunol 86:524–530
155. Busse PJ, Lurslurchachai L, Sampson HA, Halm EA, Wisnivesky J (2010) Perennial allergen-specific immunoglobulin E levels among inner-city elderly asthmatics. J Asthma 47:781–785
156. Niemeijer NR, de Monchy JG (1992) Age-dependency of sensitization to aero-allergens in asthmatics. Allergy 47:431–435
157. Zureik M, Orehek J (2002) Diagnosis and severity of asthma in the elderly: results of a large survey in 1,485 asthmatics recruited by lung specialists. Respiration 69:223–228
158. King MJ, Bukantz SC, Phillips S, Mohapatra SS, Tamulis T, Lockey RF (2004) Serum total IgE and specific IgE to Dermatophagoides pteronyssinus, but not eosinophil cationic protein, are more likely to be elevated in elderly asthmatic patients. Allergy Asthma Proc 25(5):321–325
159. Rogers L, Cassino C, Berger KI, Goldring RM, Norman RG, Klugh T, Reibman J (2002) Asthma in the elderly: cockroach sensitization and severity of airway obstruction in elderly nonsmokers. Chest 122:1580–1586
160. Vargas PA, Simpson PM, Gary WJ, Goel R, Feild CR, Tilford JM, Jones SM (2004) Characteristics of children with asthma who are enrolled in a Head Start program. J Allergy Clin Immunol 114:499–504
161. Bacharier LB, Dawson C, Bloomberg GR, Bender B, Wilson L, Strunk RC (2003) Hospitalization for asthma: atopic, pulmonary function, and psychological correlates among participants in the Childhood Asthma Management Program. Pediatrics 112:e85–e92
162. Parameswaran K, Hildreth AJ, Taylor IK, Keaney NP, Bansal SK (1999) Predictors of asthma severity in the elderly: results of a community survey in Northeast England. J Asthma 36(7):613–618
163. Scichilone N, Messina M, Battaglia S, Catalano F, Bellia V (2005) Airway hyperresponsiveness in the elderly: prevalence and clinical implications. Eur Respir J 25:364–375
164. Busse PJ, Cohn RD, Salo PM, Zeldin DC (2013) Characteristics of allergic sensitization among asthmatic adults older than 55 years: results from the National Health and Nutrition Examination Survey, 2005–2006. Ann Allergy Asthma Immunol 110:247–252
165. Ariano R, Panzani RC, Augeri G (1998) Late onset asthma clinical and immunological data: importance of allergy. J Investig Allergol Clin Immunol 8:35–41
166. Korn S, Schumann C, Kropf C, Stoiber K, Thielen A, Taube C, Buhl R (2010) Effectiveness of omalizumab in patients 50 years and older with severe persistent allergic asthma. Ann Allergy Asthma Immunol 105:313–319
167. Maykut RJ, Kianifard F, Geba GP (2008) Response of older patients with IgE-mediated asthma to omalizumab: a pooled analysis. J Asthma 45:173–181
168. Weitzman M, Gortmaker S, Walker DK, Sobol A (1990) Maternal smoking and childhood asthma. Pediatrics 85:505–511
169. Park J, Kim TB, Joo H, Lee JS, Lee SD, Oh YM (2013) Diseases concomitant with asthma in middle-aged and elderly subjects in Korea: a population-based study. Allergy Asthma Immunol Res 5:16–25
170. O'Connor GT, Sparrow D, Segal MR, Weiss ST (1989) Smoking, atopy, and methacholine airway responsiveness among middle-aged and elderly men. The Normative Aging Study. Am Rev Respir Dis 140:1520–1526
171. Mitsunobu F, Ashida K, Hosaki Y, Tsugeno H, Okamoto M, Nishida N, Nagata T, Tanizaki Y, Tanimoto M (2004) Influence of long-term cigarette smoking on immunoglobulin E-mediated allergy, pulmonary function, and high-resolution computed tomography lung densitometry in elderly patients with asthma. Clin Exp Allergy 34:59–64
172. James AL, Palmer LJ, Kicic E, Maxwell PS, Lagan SE, Ryan GF, Musk AW (2005) Decline in lung function in the Busselton Health Study: the effects of asthma and cigarette smoking. Am J Respir Crit Care Med 171:109–114

173. Kim YK, Kim SH, Tak YJ, Jee YK, Lee BJ, Kim SH, Park HW, Jung JW, Bahn JW, Chang YS, Choi DC, Chang SI, Min KU, Kim YY, Cho SH (2002) High prevalence of current asthma and active smoking effect among the elderly. Clin Exp Allergy 32:1706–1712
174. Karjalainen A, Kurppa K, Martikainen R, Klaukka T, Karjalainen J (2001) Work is related to a substantial portion of adult-onset asthma incidence in the Finnish population. Am J Respir Crit Care Med 164:565–568
175. Forastiere F, Balmes J, Scarinci M, Tager IB (1998) Occupation, asthma, and chronic respiratory symptoms in a community sample of older women. Am J Respir Crit Care Med 157:1864–1870
176. McDonnell WF, Abbey DE, Nishino N, Lebowitz MD (1999) Long-term ambient ozone concentration and the incidence of asthma in nonsmoking adults: the AHSMOG Study. Environ Res 80:110–121
177. Sunil VR, Patel KJ, Mainelis G, Turpin BJ, Ridgely S, Laumbach RJ, Kipen HM, Nazarenko Y, Veleeparambil M, Gow AJ, Laskin JD, Laskin DL (2009) Pulmonary effects of inhaled diesel exhaust in aged mice. Toxicol Appl Pharmacol 241:283–293
178. Nafstad P, Brunekreef B, Skrondal A, Nystad W (2005) Early respiratory infections, asthma, and allergy: 10-year follow-up of the Oslo Birth Cohort. Pediatrics 116:e255–e262
179. Sigurs N, Gustafsson PM, Bjarnason R, Lundberg F, Schmidt S, Sigurbergsson F, Kjellman B (2005) Severe respiratory syncytial virus bronchiolitis in infancy and asthma and allergy at age 13. Am J Respir Crit Care Med 171:137–141
180. Stein RT, Sherrill D, Morgan WJ, Holberg CJ, Halonen M, Taussig LM, Wright AL, Martinez FD (1999) Respiratory syncytial virus in early life and risk of wheeze and allergy by age 13 years. Lancet 354:541–545
181. Hollams EM, Deverell M, Serralha M, Suriyaarachchi D, Parsons F, Zhang G, de Klerk N, Holt BJ, Ladyman C, Sadowska A, Rowe J, Loh R, Sly PD, Holt PG (2009) Elucidation of asthma phenotypes in atopic teenagers through parallel immunophenotypic and clinical profiling. J Allergy Clin Immunol 124:463–470, 470
182. Lemanske RF Jr, Jackson DJ, Gangnon RE, Evans MD, Li Z, Shult PA, Kirk CJ, Reisdorf E, Roberg KA, Anderson EL, Carlson-Dakes KT, Adler KJ, Gilbertson-White S, Pappas TE, DaSilva DF, Tisler CJ, Gern JE (2005) Rhinovirus illnesses during infancy predict subsequent childhood wheezing. J Allergy Clin Immunol 116:571–577
183. Kusel MM, de Klerk NH, Holt PG, Kebadze T, Johnston SL, Sly PD (2006) Role of respiratory viruses in acute upper and lower respiratory tract illness in the first year of life: a birth cohort study. Pediatr Infect Dis J 25:680–686
184. Kusel MM, de Klerk NH, Kebadze T, Vohma V, Holt PG, Johnston SL, Sly PD (2007) Early-life respiratory viral infections, atopic sensitization, and risk of subsequent development of persistent asthma. J Allergy Clin Immunol 119:1105–1110
185. Holt PG, Strickland DH, Sly PD (2012) Virus infection and allergy in the development of asthma: what is the connection? Curr Opin Allergy Clin Immunol 12:151–157
186. Beasley R, Coleman ED, Hermon Y, Holst PE, O'Donnell TV, Tobias M (1988) Viral respiratory tract infection and exacerbations of asthma in adult patients. Thorax 43:679–683
187. Falsey AR, Hennessey PA, Formica MA, Cox C, Walsh EE (2005) Respiratory syncytial virus infection in elderly and high-risk adults. N Engl J Med 352:1749–1759
188. ten Brinke A, van Dissel JT, Sterk PJ, Zwinderman AH, Rabe KF, Bel EH (2001) Persistent airflow limitation in adult-onset nonatopic asthma is associated with serologic evidence of Chlamydia pneumoniae infection. J Allergy Clin Immunol 107:449–454
189. Koivula I, Sten M, Leinonen M, Makela PH (1997) Clinical efficacy of pneumococcal vaccine in the elderly: a randomized, single-blind population-based trial. Am J Med 103:281–290
190. Tata LJ, West J, Harrison T, Farrington P, Smith C, Hubbard R (2003) Does influenza vaccination increase consultations, corticosteroid prescriptions, or exacerbations in subjects with asthma or chronic obstructive pulmonary disease? Thorax 58:835–839
191. Ansaldi F, Turello V, Lai P, Bastone G, De LS, Rosselli R, Durando P, Sticchi L, Gasparini R, Delfino E, Icardi G (2005) Effectiveness of a 23-valent polysaccharide vaccine in preventing pneumonia and non-invasive pneumococcal infection in elderly people: a large-scale retrospective cohort study. J Int Med Res 33:490–500

192. DiazGranados CA, Dunning AJ, Kimmel M, Kirby D, Treanor J, Collins A, Pollak R, Christoff J, Earl J, Landolfi V, Martin E, Gurunathan S, Nathan R, Greenberg DP, Tornieporth NG, Decker MD, Talbot HK (2014) Efficacy of high-dose versus standard-dose influenza vaccine in older adults. N Engl J Med 371:635–645
193. Weiner P, Magadle R, Waizman J, Weiner M, Rabner M, Zamir D (1998) Characteristics of asthma in the elderly. Eur Respir J 12:564–568
194. Busse PJ (2007) Allergic respiratory disease in the elderly. Am J Med 120:498–502
195. Park HW, Song WJ, Kim SH, Park HK, Kim SH, Kwon YE, Kwon HS, Kim TB, Chang YS, Cho YS, Lee BJ, Jee YK, Jang AS, Nahm DH, Park JW, Yoon HJ, Cho YJ, Choi BW, Moon HB, Cho SH (2015) Classification and implementation of asthma phenotypes in elderly patients. Ann Allergy Asthma Immunol 114:18–22
196. Cassino C, Berger KI, Goldring RM, Norman RG, Kammerman S, Ciotoli C, Reibman J (2000) Duration of asthma and physiologic outcomes in elderly nonsmokers. Am J Respir Crit Care Med 162:1423–1428
197. Braman SS (2005) Asthma in the elderly. Exp Lung Res 31 Suppl 1:6–7
198. Quadrelli SA, Roncoroni A (2001) Features of asthma in the elderly. J Asthma 38:377–389
199. National Institute of Health (2007) NAEPP working group: consideration for diagnosing and managing asthma in the elderly. Publication # 96–3662. Ref Type: Generic
200. Connolly MJ, Crowley JJ, Charan NB, Nielson CP, Vestal RE (1992) Reduced subjective awareness of bronchoconstriction provoked by methacholine in elderly asthmatic and normal subjects as measured on a simple awareness scale. Thorax 47:410–413
201. Bellia V, Pistelli R, Catalano F, Antonelli-Inc, Grassi V, Melillo G, Olivieri D, Rengo F (2000) Quality control of spirometry in the elderly. The SA.R.A. study. SAlute Respiration nell'Anziano = Respiratory Health in the Elderly. Am J Respir Crit Care Med 161:1094–1100
202. Cuttitta G, Cibella F, Bellia V, Grassi V, Cossi S, Bucchieri S, Bonsignore G (2001) Changes in FVC during methacholine-induced bronchoconstriction in elderly patients with asthma: bronchial hyperresponsiveness and aging. Chest 119:1685–1690
203. Enright PL, Burchette RJ, Peters JA, Lebowitz MD, McDonnell WF, Abbey DE (1997) Peak flow lability: association with asthma and spirometry in an older cohort. Chest 112:895–901
204. Enright PL, McClelland RL, Buist AS, Lebowitz MD (2001) Correlates of peak expiratory flow lability in elderly persons. Chest 120:1861–1868
205. Buist AS, Vollmer WM, Wilson SR, Frazier EA, Hayward AD (2006) A randomized clinical trial of peak flow versus symptom monitoring in older adults with asthma. Am J Respir Crit Care Med 174:1077–1087
206. Abramson MJ, Schattner RL, Sulaiman ND, Del Colle EA, Aroni R, Thien F (2012) Accuracy of asthma and COPD diagnosis in Australian general practice: a mixed methods study. Prim Care Respir J 21:167–173
207. Gibson PG, Simpson JL (2009) The overlap syndrome of asthma and COPD: what are its features and how important is it? Thorax 64:728–735
208. Bujarski S, Parulekar AD, Sharafkhaneh A, Hanania NA (2015) The asthma COPD overlap syndrome (ACOS). Curr Allergy Asthma Rep 15:509
209. Kauppi P, Kupiainen H, Lindqvist A, Tammilehto L, Kilpelainen M, Kinnula VL, Haahtela T, Laitinen T (2011) Overlap syndrome of asthma and COPD predicts low quality of life. J Asthma 48:279–285
210. Raiha I, Impivaara O, Seppala M, Knuts LR, Sourander L (1993) Determinants of symptoms suggestive of gastroesophageal reflux disease in the elderly. Scand J Gastroenterol 28:1011–1014
211. Raiha I, Hietanen E, Sourander L (1991) Symptoms of gastro-oesophageal reflux disease in elderly people. Age Ageing 20:365–370
212. Nathan RA, Sorkness CA, Kosinski M, Schatz M, Li JT, Marcus P, Murray JJ, Pendergraft TB (2004) Development of the asthma control test: a survey for assessing asthma control. J Allergy Clin Immunol 113:59–65
213. National Asthma Education and Prevention Program, Third Expert Panel on the Diagnosis and Management of Asthma, Expert Panel Report 3: Guidelines for the Diagnosis and

Management of Asthma (2007). Bethesda (MD): National Heart, Lung and Blood Institute (US)
214. Bateman ED, Hurd SS, Barnes PJ, Bousquet J, Drazen JM, FitzGerald M, Gibson P, Ohta K, O'Byrne P, Pedersen SE, Pizzichini E, Sullivan SD, Wenzel SE, Zar HJ (2008) Global strategy for asthma management and prevention: GINA executive summary. Eur Respir J 31:143–178
215. Janson SL, Fahy JV, Covington JK, Paul SM, Gold WM, Boushey HA (2003) Effects of individual self-management education on clinical, biological, and adherence outcomes in asthma. Am J Med 115:620–626
216. Baptist AP, Talreja N, Clark NM (2011) Asthma education for older adults: results from the National Asthma Survey. J Asthma 48:133–138
217. Baptist AP, Ross JA, Yang Y, Song PX, Clark NM (2013) A randomized controlled trial of a self-regulation intervention for older adults with asthma. J Am Geriatr Soc 61:747–753
218. Allen SC, Jain M, Ragab S, Malik N (2003) Acquisition and short-term retention of inhaler techniques require intact executive function in elderly subjects. Age Ageing 32:299–302
219. Armitage JM, Williams SJ (1988) Inhaler technique in the elderly. Age Ageing 17:275–278
220. Chapman KR, Love L, Brubaker H (1993) A comparison of breath-actuated and conventional metered-dose inhaler inhalation techniques in elderly subjects. Chest 104:1332–1337
221. Pounsford JC (1997) Nebulisers for the elderly. Thorax 52(Suppl 2):S53–S55
222. Kendall MJ (1997) Clinical relevance of pharmacokinetic differences between beta blockers. Am J Cardiol 80:15J–19J
223. Holland OB, Nixon JV, Kuhnert L (1981) Diuretic-induced ventricular ectopic activity. Am J Med 70:762–768
224. Sin DD, Tu JV (2000) Lack of association between ipratropium bromide and mortality in elderly patients with chronic obstructive airway disease. Thorax 55:194–197
225. Burgel PR, Lazarus SC, Tam DC, Ueki IF, Atabai K, Birch M, Nadel JA (2001) Human eosinophils induce mucin production in airway epithelial cells via epidermal growth factor receptor activation. J Immunol 167:5948–5954
226. Holgate ST, Baldwin CJ, Tattersfield AE (1977) beta-adrenergic agonist resistance in normal human airways. Lancet 2:375–377
227. Djukanovic R, Wilson JW, Britten KM, Wilson SJ, Walls AF, Roche WR, Howarth PH, Holgate ST (1992) Effect of an inhaled corticosteroid on airway inflammation and symptoms in asthma. Am Rev Respir Dis 145:669–674
228. Kline JN, Waldschmidt TJ, Businga TR, Lemish JE, Weinstock JV, Thorne PS, Krieg AM (1998) Modulation of airway inflammation by CpG oligodeoxynucleotides in a murine model of asthma. J Immunol 160:2555–2559
229. Sin DD, Tu JV (2001) Underuse of inhaled steroid therapy in elderly patients with asthma. Chest 119:720–725
230. Marystone JF, Barrett-Connor EL, Morton DJ (1995) Inhaled and oral corticosteroids: their effects on bone mineral density in older adults. Am J Public Health 85:1693–1695
231. de Vries F (2013) Severity of obstructive airway disease and risk of fracture. Bone 52:711
232. Braun-Fahrlander C, Riedler J, Herz U, Eder W, Waser M, Grize L, Maisch S, Carr D, Gerlach F, Bufe A, Lauener RP, Schierl R, Renz H, Nowak D, von Mutius E (2002) Environmental exposure to endotoxin and its relation to asthma in school-age children. N Engl J Med 347:869–877
233. Suissa S, Baltzan M, Kremer R, Ernst P (2004) Inhaled and nasal corticosteroid use and the risk of fracture. Am J Respir Crit Care Med 169:83–88
234. Mortimer KJ, Harrison TW, Tattersfield AE (2005) Effects of inhaled corticosteroids on bone. Ann Allergy Asthma Immunol 94:15–21
235. Cumming RG, Mitchell P, Leeder SR (1997) Use of inhaled corticosteroids and the risk of cataracts. N Engl J Med 337:8–14
236. Garbe E, Suissa S, LeLorier J (1998) Association of inhaled corticosteroid use with cataract extraction in elderly patients. JAMA 280:539–543
237. Ernst P, Baltzan M, Deschenes J, Suissa S (2006) Low-dose inhaled and nasal corticosteroid use and the risk of cataracts. Eur Respir J 27:1168–1174

238. Leone FT, Fish JE, Szefler SJ, West SL (2003) Systematic review of the evidence regarding potential complications of inhaled corticosteroid use in asthma: collaboration of American College of Chest Physicians, American Academy of Allergy, Asthma, and Immunology, and American College of Allergy, Asthma, and Immunology. Chest 124:2329–2340
239. Mitchell P, Cumming RG, Mackey DA (1999) Inhaled corticosteroids, family history, and risk of glaucoma. Ophthalmology 106:2301–2306
240. Garbe E, LeLorier J, Boivin JF, Suissa S (1997) Inhaled and nasal glucocorticoids and the risks of ocular hypertension or open-angle glaucoma. JAMA 277:722–727
241. Sin DD, Tu JV (2001) Inhaled corticosteroid therapy reduces the risk of rehospitalization and all-cause mortality in elderly asthmatics. Eur Respir J 17:380–385
242. National Asthma Education and Prevention Program (National Heart, Lung, and Blood Institute). Working Group on Asthma in the Elderly. NAEPP Working Group report : considerations for diagnosing and managing asthma in the elderly (1996). Bethesda (MD): National Institutes of Health, National Heart, Lung, and Blood Institute (US)
243. Korenblat PE, Kemp JP, Scherger JE, Minkwitz MC, Mezzanotte W (2000) Effect of age on response to zafirlukast in patients with asthma in the Accolate Clinical Experience and Pharmacoepidemiology Trial (ACCEPT). Ann Allergy Asthma Immunol 84:217–225
244. Creticos P, Knobil K, Edwards LD, Rickard KA, Dorinsky P (2002) Loss of response to treatment with leukotriene receptor antagonists but not inhaled corticosteroids in patients over 50 years of age. Ann Allergy Asthma Immunol 88:401–409
245. Barnes PJ (2003) Theophylline: new perspectives for an old drug. Am J Respir Crit Care Med 167:813–818
246. Shannon M, Lovejoy FH Jr (1990) The influence of age vs peak serum concentration on life-threatening events after chronic theophylline intoxication. Arch Intern Med 150:2045–2048
247. Hanania NA, King MJ, Braman SS, Saltoun C, Wise RA, Enright P, Falsey AR, Mathur SK, Ramsdell JW, Rogers L, Stempel DA, Lima JJ, Fish JE, Wilson SR, Boyd C, Patel KV, Irvin CG, Yawn BP, Halm EA, Wasserman SI, Sands MF, Ershler WB, Ledford DK (2011) Asthma in the elderly: current understanding and future research needs – a report of a National Institute on Aging (NIA) workshop. J Allergy Clin Immunol 128:S4–S24

Aging in COPD and Idiopathic Pulmonary Fibrosis

Cecilia G. Sanchez

Contents

C.G. Sanchez, Ph.D.
Section of Pulmonary Diseases, Critical Care and Environmental Medicine, Tulane University Medical School, SL-9, 1430 Tulane Avenue, New Orleans, LA 70112, USA
e-mail: csanche3@tulane.edu

F. Sierra, R. Kohanski (eds.), *Advances in Geroscience*,
DOI 10.1007/978-3-319-23246-1_15

1 Clinical Aspects and Pathophysiology of COPD

The term Chronic Obstructive Pulmonary Fibrosis (COPD) refers to two distinct disease entities previously considered separate: pulmonary emphysema and chronic (obstructive) bronchitis, both of them may be present in the same individual to different degrees. The Global Initiative for Chronic Obstructive Lung Disease (GOLD) defines chronic obstructive pulmonary disease (COPD) as "a preventable and treatable disease, characterized by a persistent airflow limitation that is progressive and associated with an abnormal inflammatory response of the lungs [1]. COPD is the fourth leading cause of death in the world and will rise to the third leading cause of death by 2030 [2, 3]. COPD itself is a risk factor for other age-related disorders including cardiovascular disease, type II diabetes and other functional deteriorations [4–6]. Importantly, COPD is reported to be two to three times more prevalent in persons over 60 years of age [7].

The histopathological characterization of COPD includes the loss of lung elasticity due to emphysema and small airway obstruction due to inflammatory narrowing and fibrosis. The clinical manifestations include dyspnea, chronic cough, low exercise capacity, wheezing, and frequent or longer-lasting bronchial infections. The airway obstruction progresses with time and exacerbations of the disease tend to arise about once per year [8].

Smoking is the greatest risk factor for COPD [1, 7]. But COPD can occur in nonsmokers, in which other risk factors include asthma, advanced age, low educational level, occupational and domestic (from cooking and heating) exposure to toxins, a history of airway infections in childhood and genetic predisposition [9–11]. The understanding of COPD pathogenesis and the factors that influence its heterogeneity is a complex and evolving area of research.

1.1 Pathogenesis

It is well accepted that impairment in lung function is the result of chronic inflammatory processes, many of which are initiated years before the initial symptoms are apparent. At the molecular level, the pathogenesis of COPD includes a proteinase-antiproteinase imbalance, immunological mechanisms including systemic inflammation, oxidant-antioxidant imbalance, increased apoptosis and ineffective repair. At the cellular level, COPD represents a complex interplay among lung epithelial cells, endothelial cells, neutrophils, macrophages, and multiple subpopulations of both $CD8^+$ and $CD4^+$ T cells [12–15] (Fig. 1).

The combination of tissue damage, release of inflammatory mediators, cytokines, and chemokines leads to the activation of epithelial cells and endothelial cells. These cells are a source of chemoattractant cytokines, including CXCL1, CXCL8, CXCL10, CCL2, and CCL5 [16]. The expression of chemokines and adhesion molecules by activated epithelial and endothelial cells drives the accumulation of inflammatory infiltrates consisting of neutrophils, macrophages, and $CD8^+$ T

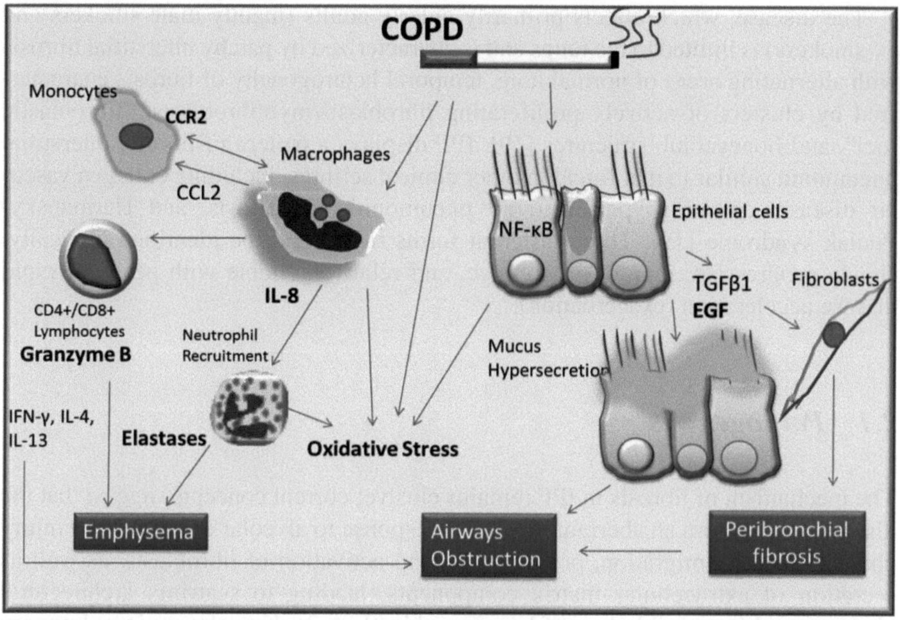

Fig. 1 Schematic representation of the main contributing factors for the development and progression of COPD. Cigarette smoking causes airway inflammation and remodeling. The inflammatory infiltrate is composed of macrophages, CD8+lymphocytes, and neutrophils. In addition, goblet cell hyperplasia, squamous metaplasia, peribronchial fibrosis, and smooth muscle hypertrophy is observed

cells [16–18]. In addition, the disease is correlated with accelerated apoptosis of alveolar and pulmonary vascular endothelial cells [18] (Fig. 1).

Evidence for a prominent role of aging in COPD progression is growing and cell senescence is one of the possible molecular pathways for development of COPD [2]. In spite of these advances, there is still a fundamental lack of knowledge about the cellular, molecular, epigenetic and genetic causes of COPD, as well as the role of aging.

2 Clinical Aspects and Pathophysiology of Idiopathic Pulmonary Fibrosis

Idiopathic pulmonary fibrosis (IPF) is a chronic, progressive and irreversible scarring of the lungs of unknown origin that causes tissues' inability to transport oxygen to the blood. An estimated 100,000 people in the U.S. have IPF and the mean survival is approximately 3–5 years from the time of diagnosis with a very heterogeneous rate of progression [19–23]. In 2030, there will be more than 70 million people in the U.S. aged 65 and older, and because IPFis strongly associated with advanced age, they will be at risk for pulmonary fibrosis.

The disease, which affects primarily elderly adults (mainly male smokers and ex-smokers) is limited to the lungs and is characterized by patchy interstitial fibrosis with alternating areas of normal lung, temporal heterogeneity of fibrosis characterized by clusters of actively proliferating fibroblasts/myofibroblasts, "fibroblastic foci", and honeycomb structures [19]. IPF displays a pattern of fibrotic interstitial pneumonia similar to that found in other clinical settings, including collagen vascular disease, chronic hypersensitivity pneumonitis, asbestosis, and Hermansky-Pudlak syndrome [19]. Three different forms of IPF can be identified clinically: slowly progressive, rapidly progressive, and relatively stable with peaks of rapid disease acceleration (exacerbation).

2.1 Pathogenesis

The mechanism of fibrosis in IPF remains elusive; current concepts suggest that the disease results from an aberrant reparative response to alveolar epithelial cell injury characterized by migration, proliferation, and activation of fibroblasts, as well as secretion of extracellular matrix components, leading to scarring, architectural changes, and irreversible loss of lung function (Fig. 2). Key players in pulmonary fibrosis include:

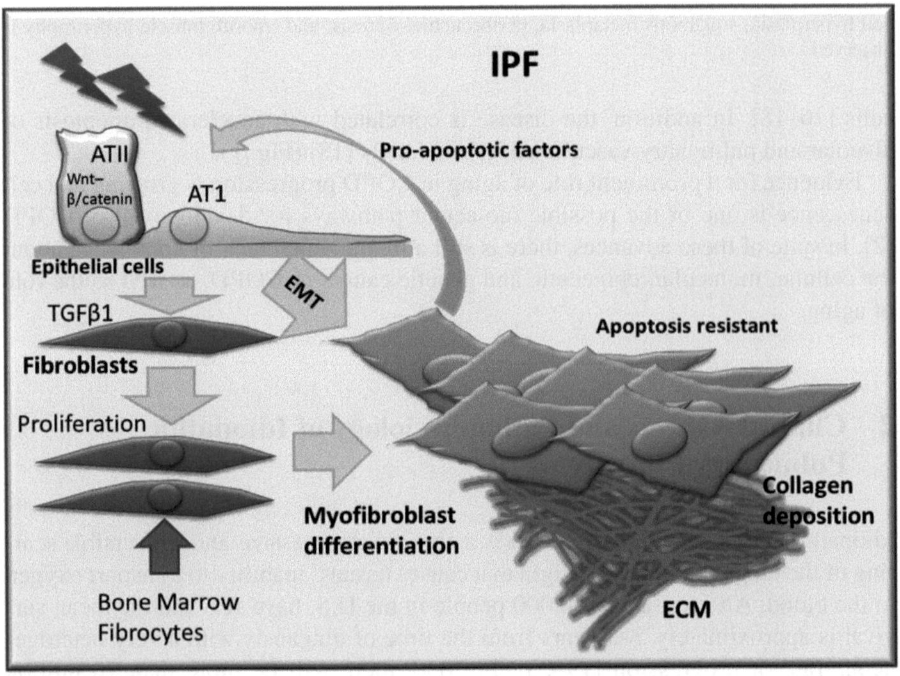

Fig. 2 Schematic representation of key players involved in the development and progression of Idiopathic pulmonary fibrosis

Epithelial Injury It is proposed that microinjuries to alveolar epithelial cells induce fibrogenesis. Abnormal or injured epithelial cells secrete growth factors that favor the recruitment of resident fibroblasts and fibrocytes that differentiate into myofibroblasts [24]. Those epithelial cells also release inflammatory mediators that initiate an anti-fibrinolytic coagulation cascade and trigger platelet activation and blood clot formation. This process is followed by activation of leukocytes at the site of tissue injury.

IPF Fibroblasts and Myofibroblast Differentiation IPF fibroblasts are characterized by excessive proliferation possibly linked to decreased PTEN expression and aberrant activation of PI3 kinase. Fibroblasts can trans-differentiate to a myofibroblast phenotype, which are major producers of excessive extracellular matrix. Continuous accumulation of myofibroblasts prevents proper re-epithelialization. Myofibroblasts express features of both fibroblasts and smooth muscle cells, and they can be recruited to the lungs or they can differentiate from resident fibroblasts [25]. Myofibroblasts may also be derived from epithelial cells undergoing Epithelial Mesenchymal Transition (EMT), a process triggered in part by the Wnt–β-catenin signaling. This pathway is constitutively active in epithelial cells from IPF patients (and in mice treated with bleomycin as a model of pulmonary fibrosis).

Transforming Growth Factor Beta 1 (TGF-β1) Signaling Pathway TGF-β1 is an important profibrotic cytokine that triggers fibroblast proliferation and activation. TGF-β1 further exacerbates the inflammatory response by stimulating the differentiation of Th17 cells. TGF-β1 induces a number of growth factors and cytokines that participate in fibrosis, including, fibroblast growth factor (FGF-2), connective tissue growth factor (CTGF), insulin-like growth factor (IGF), platelet derived growth factor (PDGF) and interleukins (ILs).

Oxidative Stress and Mitochondrial Dysfunction PTEN-induced putative kinase 1 (PINK1), is a regulator of mitophagy and maintains mitochondrial integrity by regulating diverse aspects of mitochondrial function, including membrane potential, respiration, calcium homeostasis, structure, and mitochondrial DNA integrity. Recently, it had been shown that PINK1 is deregulated in the aging lung and in pulmonary fibrosis. Deficiency of PINK1 expression promotes fibrogenesis, possibly by inducing the production of mitochondrial reactive oxygen species (ROS), which perpetuate profibrotic inflammatory responses [26–29]. The activation of the Nalp3 inflammasome and IL-1β secretion is largely driven by ROS [30–32].

Recent genome-wide studies have identified several genetic variants critical for epithelial integrity as a risk factor for IPF, including genes for cell-cell adhesion and migration [33, 34]. Furthermore, polymorphisms in common variants in TERT and TERC and oligonucleotide/oligosaccharide-binding fold containing1 (OBFC1), required for telomere length, indicates telomere attrition in IPF [35, 36]. Toll interacting protein (TOLLIP), is another genetic variant, important for innate immunity and epithelial mesenchymal transition [37]. None of these associations have been investigated mechanistically, so their role in IPF remains to be studied.

The animal model most extensively used to study pulmonary fibrosis is the bleomycin model, which recapitulates several of the features seen in the IPF patients [38]. Importantly, age-related differences are noted in this model of pulmonary fibrosis. Older mice do not resolve fibrosis as effectively as do younger mice. Twenty-four-month-old mice exhibit increased fibrogenesis, collagen deposition, and activation of TGF-β signaling compared to 3-month-old mice [39].

An interesting question is the heterogeneity of outcomes: why does a patient in the 60s with a history of smoking, shorter telomeres, alveolar epithelial senescence, oxidative stress and mitochondrial dysfunction will develop pulmonary fibrosis and not COPD, or why in some cases, emphysema and pulmonary fibrosis can occur in the same patient (Fig. 3) [40–42]. One major difference is that COPD is viewed as an inflammatory disorder while IPF is considered an abnormal wound healing disorder [43].

2.2 *Current Treatment*

Patients with IPF are currently referred early for lung transplantation, due to the unpredictable nature and the high mortality rate of the disease. Recently, nintedanib and pirfenidone, two compounds with pleiotropic anti-fibrotic properties, have been proven effective in reducing lung function decline and disease progression in IPF [44]. Nintedanib is a tyrosine kinase inhibitor that targets receptors thought to be involved in the pathogenesis of IPF such as receptors for the platelet-derived growth factor, vascular endothelial growth factor and fibroblast growth factor [45]. Pirfenidone has anti-fibrotic and anti-inflammatory properties [46–48]. Unfortunately, neither nintedanib nor pirfenidone is a cure for IPF [44].

3 Molecular Hallmarks of Aging in the Development and Progression of COPD and Pulmonary Fibrosis

Biological aging, usually linked to chronological age, contributes to the deterioration of pulmonary function. However, features of biological aging can also occur earlier in life ("accelerated aging") as a result of failure to maintain cellular homeostasis (among other factors), including deficiencies in cellular maintenance or repair, DNA damage, epigenetic alterations, and loss of proteostasis. Some of these are induced by environmental factors such as cigarette smoke, viruses, particles, etc., but are also likely to interact with the aging-dependent changes in these processes. Irrespective of the proximal cause (aging or environment), loss of cellular homeostasis promotes tissue injury, involving tissue remodeling, airspace enlargement, inflammation, and/or lung fibrogenesis (Fig. 2).

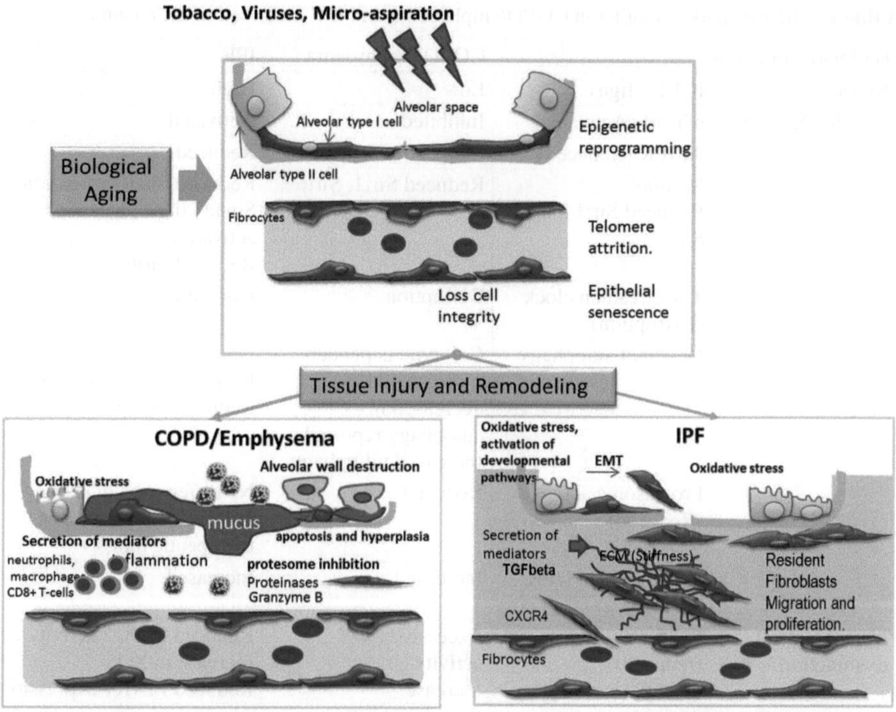

Fig. 3 Schematic representation of the key players involved in the development and/or progression of COPD/Emphysema or IPF in the elderly. **IPF** (*bottom right*)**.** In a genetically predisposed and/or pro-fibrotic lung, disrupted alveolar epithelium and basement membrane promote the release of proinflammatory cytokines and chemokines. These soluble mediators may activate residential and/or circulating cells including fibrocytes. Profibrogenic molecules such as PDGF and TGF-β1 are secreted by inflammatory, epithelial, and endothelial cells. Fibroblasts respond to these changes by proliferation and differentiation into myofibroblasts, promoting abnormal collagen deposition. The aggregates of mesenchymal cells and collagen are called "fibrotic foci". The resulting fibrotic lung has no potential for regeneration and repair. **COPD** (*bottom left*). In susceptible subjects (e.g., with severe α-1 antitrypsin deficiency), aging, cigarette smoke and other environmental factors promote the activation of the inflammatory and structural cells of the lower airways of the lungs with the release of mediators causing inflammation, remodeling and emphysema. Cigarette smoking enhances the age-related accumulation of endogenous ROS and DNA damage, NF-κB activation, and oxidative stress responses. Further amplification of inflammation, accumulation of damaged proteins, apoptosis and hyperplasia are also the result of epigenetic changes and antioxidant/nitrosant activity

There is no doubt that the pathogenesis of emphysema in COPD, as well as idiopathic pulmonary fibrosis are characterized by of the appearance of accelerated aging of the lung, including inflammatory pathways, antioxidant responses, epigenetic modifiers and proteolytic pathways [49–51]. The processes of initiation and perpetuation of COPD/emphysema and pulmonary fibrosis are not well understood, and considering the main pillars of aging might provide important new insights. Table 1 summarizes the aging hallmarks associated with COPD and IPF.

Table 1 The hallmarks of aging in COPD/emphysema and IPF described in the chapter

Hallmarks of aging		COPD/emphysema	IPF
Nutrient Sensing Systems	IGF1 (High)	Low	High
	mTOR (Activated)	Inhibited	Activated
	AMPK (Reduced)	Reduced	Reduced
	Sirtuins Reduced Sirt1 and Sirt6.	Reduced Sirt1, Sirt6	Reduced Sirt1 expression Smad3 dependent Sirt1 activation. Reduced Sirt6.
	The circadian clock (Disruption)	Disruption	Disruption
Proteostasis	Deficient autophagy	Cell type dependent Increase and decreased in autophagy reported. Increased mitophagy	Disruption Reduced macro-autophagy Reduced mitophagy
	Proteasome activity (Reduced)	Reduced	No differences in activity. Protesome inhibition as a therapeutic tool.
	ER stress and UPR response	Increased	Increased
Mitochondrial dysfunction	COX activity (reduced) mtROS (increased) Altered mitochondrial dynamics Reduced PIK1 expression	Excessive COX-2 activity Increase mitochondrial fragmentation	Reduced COX-2 activity Increase mtROS Reduced PINK1 expression
Genomic instability Microsatellite instability	Increase DNA breaks and Deficient reparative mechanisms. Excision repair (reduced) Loss of heterozygosity (LOH)	Double-strand breaks and activation of DNA damage response. LOH	Double-strand breaks and activation of DNA damage response. LOH
Telomere attrition	Short telomere length	Short telomere length	Telomere related disorder (familial cases)-TERT, TERC, OBFC1polymorphism. Short telomere length.
Cellular Senescence and SAPS	Increase	Increase	Increase

(continued)

Table 1 (continued)

Hallmarks of aging		COPD/emphysema	IPF
Epigenetic alterations	DNA methylation Increase global hypomethylation	Increase global hypomethylation DNA Hypomethylation and DNA Hypermethylation of specific genes.	No LINE1-hypomethylation Hypomethylation and hypermethylation for specific genes.
	MicroRNAs Lack of alteration in miRNA expression in the aging mouse lung.	Upregulated miRs include: miR223, 1274a, 144, 374a, 664, 17–92, 576-3p,513a-p5, 25, 99b, 125b-1, 24. Downregulated miRs include: miR21, mIr923, mIR937, miR422a.	Upregulation of miRNAs 21, 34a,145, 154, 155, 199a-5p Downregulation of miR Let7d, 17–92, 26a, 29, 200 family, and 326.
	Splicing (altered)	Splicing variants	Differential in IPF Lung Tissue.
	Histone modifications Increase (H4K16) Variations in methylation of histones H3 and H4	Reduced HDACs 2, 5, and 8, HDAC10 expression. Sirt1 downregulation. H3 and H4 panacetylation	Altered protein-levels of Class-I- (HDAC1, 2, 3 and 8) and Class-II-HDACs (HDAC4, 5, 7, 9, 10), and of the Class-III-HDAC Sirtuin-1 is significantly elevated in IPF lungs.
Intercellular communications and inflammaging	Cell-cell contacts (altered) Immunoregulation (altered) Changes in ECM	Loss of cell-cell contacts Changes in ECM Elastases Granzyme B Increase extracellular ATP levels High PDE4 Low cAMP Inflammaging/SAPS	Reduction in gap-junctions and changes intercellular communication. Changes in ECM composition. Increase elastic fibers, stiffness Increase active TGFbeta Changes in MMPs and TIMPs Inflammaging/SASP
	The microbiome (altered)		
	Hormones (disruption) 17β-estradiol/estrogen receptor alpha signaling pathway;melatonin	Low testosterone levels Reduced melatonin levels Decrease in DHEA	Decrease in DHEA
Stem cell exhaustion	Reduced tissue regeneration, healing. Changes in stem cells	Reduced tissue regeneration, healing.	Reduced tissue regeneration, healing.

3.1 Defects in Stress Recognition and Nutrient Sensing Systems

3.1.1 Insulin-Like Growth Factor-1

Circulating IGF-1 and IGF-1 signaling is reduced in aging and reduced IGF-1 signaling increases longevity in organisms from worms to mice. Low levels of circulating IGF-I can negatively influence the progression of disease, and are common in patients with COPD/emphysema [52]. For example, exogenous IGF1 reduces diaphragm fiber atrophy in animal models of emphysema [53]. In apparent contrast in the IPF-lung, IGF-I is highly expressed by lung fibroblasts, interstitial macrophages, alveolar epithelial cells, and ciliated columnar epithelial cell [54–56]. However, it is possible that IGF-1 signaling is actually impaired due to increased expression of IGBP3 and IGFBP5, members of a family of IGF-binding proteins that bind to IGFs with high affinity and restrict access of IGF1 to the IGF1 receptor. In keeping with this, overexpression of IGFBP5 promotes fibrogenesis in lung tissues [57, 58].

3.1.2 mTOR Regulation

Reduced activity of the mechanistic target of rapamycin (mTOR)/AKT pathway increases longevity, as demonstrated both by genetic manipulations in flies and worms, and by pharmacological inhibition with rapamycin in mice. As in other tissues, activity of this pathway has been shown to be increased in the aged mouse lung, with detrimental consequences [59]. However, Rtp801-null mice exhibit increased mTOR signaling and are substantially protected against pulmonary injury from smoke exposure [60]. Rtp801, also known as Redd1, is a suppressor of mTOR signaling which is overexpressed in human emphysematous lungs and in lungs exposed to cigarette smoke. It appears therefore that the age-related increase in mTOR might be protective of the lungs against some types of injury [61].

Conversely, the mTOR pathway has been shown to be activated in pulmonary fibrosis. mTOR is mainly expressed in hyperplastic alveolar epithelial cells and in some mesenchymal cells. mTOR expression in pulmonary fibrosis patients significantly correlates with the fibrosis score and decline in lung function [62], indicating that this age-related hallmark may be associated with the prognosis of pulmonary fibrosis. Treatment with TGFβ and bleomycin also result in the activation of the AKT/mTOR pathway and the consequent phosphorylation of p70S6 Kinase, ribosomal S6 protein, and 4E-BP1. The effects of TGFβ and bleomycin on extracellular matrix deposition are reduced by pre-treatment with rapamycin, an mTOR inhibitor [63]. Taken together, it appears that the age-related increase in mTOR activity might be protective against injuries leading to COPD, but deleterious because of promoting the development of fibrosis.

Inhibition of autophagy is a downstream effect of mTOR activity, and this appears to also play a role in myofibroblast differentiation. Enhanced mTOR activity is observed in myofibroblasts within fibroblastic foci from IPF patients [64, 65].

The Akt/mTOR pathway desensitizes IPF fibroblasts from polymerized collagen-driven stress by suppressing autophagic activity, which produces a viable IPF fibroblast phenotype in collagen [66]. It is also possible that mTOR activation and the consequent blockage in autophagy promote the initial cell fate change to myofibroblasts [65, 67]. The time-dependent regulation of mTOR-autophagy during the initiation and progression of pulmonary fibrosis needs to be further evaluated [29]. It is safe to conclude that mTOR plays complex and potentially counteracting roles in lung disease.

3.1.3 Adenosine Monophosphate-Activated Protein Kinase (AMPK)

Independent of effects on the mTOR pathway, AMPK can also modulate inflammatory responses and oxidative stress. AMPK attenuates inflammatory lung injury by phosphorylating its downstream targets, including sirtuin1 (SIRT1), peroxisome proliferator-activated receptor gamma (PPARγ) coactivator-1 alpha (PGC-1α), p53, and FOXO3a. AMPK mRNA and protein are significantly reduced in skeletal muscles from rat modes of COPD, in comparison to control rats [68]. AMPKα1-deficient mice show increased fibrosis after bleomycin exposure, compared to control mice. Furthermore, metformin, an AMPK activator, decreases the expression levels of fibronectin and collagen in cultured fibroblasts and suppresses NOX4 activation [69].

3.1.4 Sirtuins

Activation of sirtuins has been shown to extend lifespan and increase healthspan in models ranging from yeast to mice. The sirtuin family of proteins contributes to interactions among autophagy, metabolism and aging (for reviews, see [70, 71]). Substrates of SIRT1 have significant roles in the formation of autophagosomes, fatty acid oxidation, glucose homeostasis, circadian rhythm and life span [72–78]. SIRT1 expression in the lung is reduced during aging and COPD [79, 80], as well as emphysema [81]. SIRT1 negatively regulates expression of the proteinase MMP9, which is increased in chronic inflammatory diseases [82]. Moreover, genetic or pharmacological activation of SIRT1 protects mice from elastase and cigarette smoke-induced emphysema in addition to attenuating stress-induced premature cellular senescence [83].

The specific role of SIRT1 in the process of lung fibrogenesis needs further studies. Recently SIRT1 has being shown to be a crucial regulator of TGFβ/Smad signaling in systemic scleroderma, whereby mice with fibroblast-specific knockdown of SIRT1 are in fact less susceptible to bleomycin- or TBRIact-induced fibrosis [84]. Another sirtuin, SIRT6, is also being studied in COPD and IPF. Reduced SIRT6 levels in COPD and IPF have been associated with cellular senescence [85–87]. SIRT6 negatively regulates cigarette and TGFβ-induced senescence [87–89], so it appears that SIRT6 may have a protective effect against COPD and IPF.

3.1.5 The Clock

Circadian rhythms are altered during aging, both at the level of the central nervous system in the suprachiasmatic nucleus, and at the individual cellular level in peripheral tissues. At the molecular level, the circadian rhythm consists of interlocking transcriptional/translational feedback loops of core clock genes and oscillatory metabolic products. The clock modulates stress responses and physiological processes unique to each organ [90–94]. Importantly, some respiratory pathologies, such as pulmonary edema, asthma, and allergic attacks, peak at certain times during the circadian cycle [95]. Animal models of pulmonary fibrosis have revealed the effects of day/night cycling in the fibrotic response, with a 'clock-gated' pulmonary response to oxidative injury. Furthermore, lungs from mice carrying a Clock gene mutation are characterized by an increased oxidative burden and increased collagen deposition around the bronchioles, even in the absence of bleomycin challenge [96]. Basal autophagy and other metabolic pathways are rhythmically activated in a clock-dependent manner [97], supporting the significance of the circadian clock as a bioenergetic regulator of human physiology and pathophysiology [98, 99].

3.1.6 Therapeutic Approaches Targeting Bioenergetics Sensors

Bioenergetic sensors have been proposed as potential targets for interventions to modulate autophagy and metabolism and to slow the progression of age-related diseases like COPD/emphysema and pulmonary fibrosis. For example, rapamycin (an mTOR inhibitor) reduced fibrosis in a mouse model of COPD [100], although it did not prevent fibrosis in the murine model using bleomycin [101]. Activating AMPK with 5-aminoimidazole-4-carboxamide-1-b-4-ribofuranoside or metformin leads to protection from airway inflammation and remodeling in a murine asthma model, at least partially through decreased oxidative stress [102].

The polyphenol resveratrol is known to activate SIRT1, inhibit phosphoinositide-3-kinase (PI3K), a pro-aging kinase, and regulate FOXO3A and mTOR signaling [103–106]. Resveratrol also promotes autophagy, and inhibits inflammation. Resveratrol reduces bleomycin-induced pulmonary fibrosis and ventricular remodeling in old rats with COPD [107–109]. In fact, resveratrol reduces the release of inflammatory mediators from human airway smooth muscle cells and relieves alveolar epithelial cells from endoplasmic reticulum (ER) stress and apoptosis in animal models of COPD [110, 111].

Melatonin, N-acetyl-5-methoxytryptamine, is a naturally occurring compound found in animals, plants, and microbes. Melatonin is produced by the pineal gland in vertebrates and it is involved in circadian rhythms through the activation of melatonin receptors. Melatonin is also known to exert a powerful antioxidant activity. Melatonin has been suggested as a 'geroprotector', as an agent to treat age-associated inflammatory diseases and to increase quality of life in elderly patients [112, 113]. Clinical studies demonstrated a decrease in oxidative stress in patients treated with melatonin, detected by 8-isoprostane levels in exhaled breath after 1, 2, and 3 months of melatonin treatment, with significant improvement of dyspnea, although there were no significant changes in lung function or exercise capacity [112]. At the cellular level, melatonin has been shown to inhibit mucin 5 AC (MUC5AC) production via suppression of MAPK signaling in human airway epithelial cells [114]. Furthermore, melatonin attenuates neutrophil inflammation and mucus secretion in cigarette smoke-induced chronic obstructive pulmonary diseases via the suppression of Erk-Sp1 signaling [115].

Finally, it is expected that more geroprotectors and "chronotherapeutic strategies" for intervention against human chronic lung diseases will be proposed in the near future, based in part on studies of stress recognition and nutrient sensing systems [112].

3.2 Loss of Proteostasis: The Role of Autophagy and Proteasomal Digestion

Aberrant proteostasis contributes to COPD, severe emphysema, and pulmonary fibrosis [116]. Quality control is maintained by the autophagy-lysosomal system and the ubiquitin-proteasomal system; both these proteolytic systems decline with age [117, 118]. In a recently published proteomics analysis of IPF, 89 differentially expressed proteins were identified, of which 51 were upregulated and 38 downregulated. Increased expression was observed for proteins involved in unfolded protein response (UPR), heat-shock proteins, and DNA damage stress [119]. In COPD, recent studies demonstrated that cigarette smoke promotes the accumulation of ubiquitinated protein aggregates in the insoluble protein fraction of cigarette smoke extract (CSE)-treated human bronchial epithelium cells. Interestingly, these aggregates disappear after induction of autophagy [120]. We review recent advances in our understanding of the contribution of proteolytic system failure to COPD and IPF.

3.2.1 Autophagy

The recycling and clearance of proteins and mitochondria by autophagy has beneficial effects on aging, including increased energy production and decreased ROS production. Nevertheless, the role of autophagy in the pathogenesis of COPD seems to be complex and cell type specific. In addition to its role in recycling proteins and mitochondria, autophagy is implicated in processes such as cilliar homeostasis and response to hypoxia.

In the lungs of patients with COPD, LC3B-II and autophagosome formation are increased [121], and increased mitophagy in alveolar epithelial cells was found to contribute to the pathogenesis of COPD [122]. Furthermore, increased autophagy also results in decreased resistance to emphysema in animals exposed to cigarette smoke [121, 123]. On the other hand, autophagic flux is impaired in alveolar macrophages, which could contribute to the deficient xenophagy seen in COPD patients as well as accumulation of ubiquitinated aggregates in bronchial epithelial cells exposed to cigarette smoke [120, 124].

In contrast to COPD, autophagy has been shown to be reduced in IPF [64, 65], and autophagy inhibition is sufficient to induce the acceleration of epithelial cell senescence and fibroblasts-myofibroblast differentiation [64, 65]. Deficient autophagy is also associated with the anti-apoptotic features of persistent myofibroblasts and the progression of IPF [64, 65, 125]. A recent report using bleomycin animal models indicates that deletion of the essential autophagy gene ATG7 in endothelial cells leads to marked changes in the architecture of endothelial cells, and increased susceptibility to pulmonary fibrosis [126]. These findings underscore autophagic deficiency as a contributing factor in the development and/or establishment of pulmonary fibrosis. Interestingly, both upregulation and downregulation of autophagy have been associated with fibrosis in various organs, highlighting the diverse nature of the roles that autophagy may play in the various phases of the response to stress and repair in different tissues [127–129].

Mechanistically, it has been proposed that cigarette smoke induces mitophagy by stabilizing PINK1 in pulmonary epithelial cells. Genetic deficiency in PINK1 protects lung epithelial cells from cigarette smoke-induced cell death and mitochondrial dysfunction [122]. In contrast to COPD, Pink1−/− deficient mice are more susceptible to pulmonary fibrosis than are wild-type mice, PINK1 expression is reduced in animal models of pulmonary fibrosis and in biopsies of IPF patients, suggesting a key role for mitochondrial homeostasis in the pathogenesis and the progression of lung fibrosis [26]. Recent studies focused on the hSP-C^{I73T} mutation associated with interstitial lung diseases, like IPF, indicate that a disruption of autophagy-dependent proteostasis in hSP-C^{I73T} is accompanied by an increase in mitochondrial biomass, and a decrease in mitochondrial membrane potential [130]. This agrees with recent findings by our group that describe an age-dependent decline in the autophagic response to bleomycin, a decrease in PINK1 expression with aging, and deficient PINK1 recruitment to the mitochondria in a TGFβ-dependent manner, favor the profibrotic phenotype of the aging lung [29].

Recent studies also demonstrate that autophagy regulates cilia length through ciliophagy and ciliogenesis, which control the sensitivity of the cell to stressors such as cigarette smoke [131–133]. Autophagy-deficient mice are protected from cigarette-smoke-associated ciliary dysfunction [134]. Elevated expression of cilium genes is associated with more extensive microscopic honeycombing and higher expression levels of both the airway mucin gene MUC5B and the metalloproteinase MMP7, a gene recently implicated in attenuating ciliated cell differentiation during wound repair [135]. Interestingly, new findings indicate that a functional hedgehog pathway machinery is required for the effects of TGF-β1 on normal and IPF fibroblasts during myofibroblastic differentiation [136]. However, a causal role of autophagy in ciliogenesis an IPF pathogenesis remains to be determined.

In vivo studies addressing the changes in autophagy during aging and the temporal relationship between autophagy, cell fate determination and fibrogenesis are missing, in part due to the difficulties in studying autophagic flux in vivo at different time points in the fibrotic process as well as due to cell type differences. An improved understanding of the specific mechanisms by which dysfunctional autophagy and mitophagy can promote cell type-specific features characteristic of emphysema and/or pulmonary fibrosis may lead to an understanding of this dynamic and complex process and the identification of new targets for both diagnostic and therapeutic approaches [123].

3.2.2 Proteasome Regulation

The proteasome regulatory network, a system in charge of the degradation of 90 % of cellular proteins, has been shown to be dysregulated during aging. Proteasomes can be directly inhibited by oxidative stress [137, 138] and lipofuscin, both of which accumulate with aging [139, 140]. Inhibition of proteasome activity increases senescence in fibroblasts [141, 142], and proteasome activity decreases significantly with age in the lung [143]. Protein degradation by the ubiquitin-proteasomal system plays a positive role in modulating TGF-β1 expression and signaling, a key player in pulmonary fibrosis. In fact, proteasomal inhibition is one of the approaches used in various animal models of tissue fibrosis to regulate TGF-β1 signaling and consequently fibrogenesis. This has been reviewed elsewhere [144]. However, while the IPF lung is characterized by a higher content of proteasomes, no significant differences in proteasome peptidase activity in IPF lungs compared to control lungs have been found [145]. This would suggest that if age-dependent dysregulation of proteasome activity plays a role in the etiology of IPF, it would be permissive rather than causal.

In contrast several lines of evidence support a role of proteasome dysfunction in the pathogenesis of emphysema, including the accumulation of ubiquitinated proteins and the deubiquitinating enzyme UCHL-1 [146]. Altered expression of genes involved in protein ubiquitination has been found in COPD patients [147]. For example, a high expression of the valosin-containing protein retrograde translocation complex (VCP-Rma1-gp78) correlates with the severity of emphysema in COPD and the overexpression of inflammatory, ER stress, and apoptotic mediators like NFκB, GADD-153/CHOP, and p-eIF2α in lung tissues [146]. VCP-Rma1-gp78 plays a key role in both protein extraction from the ER and ubiquitin-proteasome mediated protein degradation by ERAD [148]. Those studies propose that VCP mediates the proteasomal degradation of HDAC2 and Nrf2, as a potential mechanism for corticosteroid resistance and increased oxidative stress observed in COPD subjects with emphysema.

3.2.3 ER Stress and the UPR Response

Injury, viral infections, and defects in protein folding can promote ER stress and the consequent UPR in epithelial cells. This mechanism can promote homeostasis and cellular survival. Prolonged stress, by contrast, can contribute to apoptosis and the initiation of fibrotic remodeling [149]. UPR and ER stress are detected in alveolar epithelial cells in lungs of patients with emphysema, COPD, and IPF. The activation of these pathways may result from altered surfactant protein processing [119, 149, 150]. Activation of the UPR system in epithelial cells may induce secretion of the profibrotic mediator, TGF-β1 [151]. Furthermore, UPR can be induced by TGF-β1, through ROS generation, promoting myofibroblast differentiation in human lung fibroblasts [152, 153].

3.3 Mitochondrial Dysfunction

Mitochondrial metabolism is known to play a central role in mediating longevity via nutrient-sensing pathways and dietary restriction [154]. Mitochondrial homeostasis controls ROS production from respiration. In consequence, mitochondrial dysfunction is often accompanied by increased ROS levels that can contribute to cellular dysfunction and disease etiology. Mitochondrial ROS can directly damage proteins, RNA, nuclear and mitochondrial DNA, and promote senescence and/or apoptosis [155]. Mitochondrial ROS is also derived from the NADH oxidase NOX4, which is important for the transformation of lung fibroblasts to myofibroblasts and consequently, collagen deposition [30, 156].

The antioxidant transcription factor nuclear factor erythroid 2-related factor 2 (NRF2) is a "master regulator" that promotes cell survival through the coordinated induction of phase II and antioxidant defense enzymes to counteract oxidative stress and modulate redox signaling [157, 158]. Deficient NRF2 activity has been associated with the pathogenesis of chronic lung diseases such as asthma, COPD, and IPF, thus contributing to excessive oxidative stress in the lung [159, 160]. Therefore, pharmacological targeting of NRF2 is a potential therapeutic strategy under current study [161, 162].

Another potential regulator is Prohibitin-1 (PHB1), a mitochondrial protein that interacts with the NADH dehydrogenase protein complex. It is known to play a crucial role in preserving normal mitochondrial function, morphology and mitophagy, and also is linked to aging. PHB1 is significantly down-regulated in bronchial epithelial cells from patients with COPD. Deficient PHB1 is probably a contributor to oxidative stress in the COPD lung [163].

Several defects in mitochondrial dynamics, including fission, fusion, biogenesis and mitophagy have been reported in various lung diseases. Reduced expression of PINK1 is observed in the lungs of aging mice and IPF patients [26], and it is associated with pulmonary fibrosis [29]. In the absence of PINK1, cells often develop fragmented mitochondria, due to imbalanced fission/fusion dynamics. Whereas the coordinated induction of fission and autophagy is believed to facilitate the removal of damaged mitochondria through mitophagy, excessive fission may cause apoptosis [164]. Disruption of mitochondrial dynamics may also be relevant to the pathogenesis of COPD, as indicated by the exacerbation of mitochondrial fragmentation observed in Cigarette Smoke Extract-exposed bronchial epithelial cells [165]. Long-term CSE exposure significantly increased the expression of oxidative phosphorylation proteins as well as specific markers of mitochondrial fission/fusion and oxidative stress [166]. Mitochondrial fragmentation can also be induced in human airway smooth muscle cells by CSE exposure, via the increased expression of the mitochondrial fission protein dynamin-related protein 1 (Drp1) and decreased expression of the fusion protein mitofusin2 (Mfn2). Importantly, the inhibition of Drp1 prevents the effects of CSE on mitochondrial networks and ROS generation, whereas blocking Mfn2 has the opposite effect [167]. In consequence, mitochondrial-targeted interventions are currently developed as the studies to elucidate the role of mitochondrial metabolism and recycling in lung aging, COPD, and pulmonary fibrosis.

3.4 Genomic Instability and Telomere Attrition

Smoking is reported to be one of the most important environmental factors that cause DNA damage [168]. Alveolar type I and type II cells and endothelial cells in patients with COPD showed higher levels of double-strand breaks and DNA damage response than those in asymptomatic smokers and non-smokers [169]. DNA damage and the activation of the DNA damage response are also seen in IPF [119, 170].

Telomere length has been proposed as a marker for physiological aging. Telomeres protect the DNA at the end of chromosomes from degradation, remodeling, and gene-gene fusion, and constitute a marker for aging [171]. The replicative cycle, ROS and inflammation promote telomere shortening [172]. Conversely, telomere erosion versus maintenance and repair can determine cell fate. Telomere length abnormalities and the expression of senescence markers such as p16 and p21 in alveolar cells have been demonstrated in emphysema, COPD/emphysema, and IPF [173–175]. In the case of familial IPF, 10 % of the patients have a mutation on the reverse transcriptase component of telomerase, TERT, and/or the RNA template component TERC [176, 177]. It is also known that 20 % of the patients with telomerase mutations that cause congenital dyskeratosis also develop pulmonary fibrosis [178, 179]. Interestingly, models of injury and fibrosis in mice with telomerase deficiency reveal that TERT induction is associated with the increased survival of lung fibroblasts, which favors the development of fibrosis instead of injury resolution, whereas telomerase deficiency does not result in a predisposition to enhanced fibrosis in bleomycin-treated lungs i [180, 181].

Telomere-deficient mice do not develop spontaneous lung emphysema; however, short telomere length is a susceptibility factor in COPD/emphysema; short telomeres lower the threshold of cigarette smoke-induced damage [182]. Peripheral leukocytes, alveolar epithelial cells, endothelial cells, and fibroblasts from patients with COPD contain shorter telomeres compared to cells from healthy lungs [175, 183, 184]. Interestingly, shorter telomere length in peripheral leukocytes has been associated with increased comorbidity, as well as total and cancer mortality in patients with COPD [183]. In contrast, parenchymal fibroblasts from emphysema patients do not have altered telomere length despite showing markers of senescence such as β-galactosidase, suggesting that cell senescence in this case might be driven by stress, rather than exhaustive proliferation [185].

3.5 Cellular Senescence

Accelerated cellular senescence is considered a possible molecular pathway by which COPD occurs [2]. Pulmonary vascular endothelial cells, alveolar epithelial cells, and fibroblasts from COPD/emphysema patients show signs of premature senescence compared to controls, as determined by decreased telomerase activity, reduced telomere length, and increased expression of p16 and p21 markers. Furthermore, senescence can be induced in vitro by the exposure of fibroblasts and alveolar epithelial cells to cigarette smoke [186, 187]. It has been proposed that after cigarette smoke exposure, senescent cells may contribute to the pathogenesis of emphysema, while fibroblasts that resist cigarette smoke-induced cellular senescence may contribute to the pathogenesis of pulmonary fibrosis and possibly to fibrotic lesions through a TGF-β1-mediated pathway [188]. Nevertheless, small airways, alveolar epithelia and resident fibroblasts are characterised by different renewal strategies, and telomere dysfunction and cellular senescence could be expected to act differently in these compartments and to be associated to different pathologic outcomes.

3.6 Epigenetic Alterations

3.6.1 DNA Methylation

Epigenetic drift refers to alterations in the genomic landscape of DNA methylation as a function of age. Both DNA hypomethylation of some loci and hypermethylation of others have been shown to be differentially present in sputum and small airway epithelial cells derived from COPD patients compared to healthy individuals [189, 190].

In the context of age-related lung diseases, Alu and LINE-1 hypomethylation in circulating leukocytes was found to be associated with increased age and lung function decline [191]. Methylation changes in promoter regions drive the differential expression of specific genes important for the response against inflammation and oxidative stress in COPD, such as the transcriptional downregulation of antioxidant NRF2 and PTEN, and the transcriptional upregulation of HDAC6 in COPD patients [131, 190, 192, 193].

DNA hypermethylation and hypomethylation in IPF also correlate with the transcriptional regulation of key genes, such as the hypermethylation of COX2 and prostaglandin E receptor 2 PTGER2 [194, 195]. The hypermethylation of Thy1, a fibrosis suppressor, has been associated with the pathogenesis of IPF [196, 197]. In IPF, hypermethylation of the promoter for the miR-17–92 cluster, which is critical for lung development and lung epithelial cell homeostasis, correlates with the reduced expression of miR-17–92 and increased expression of DNMT-1, a target of miRNAs in this cluster [198].

3.6.2 Histone Modifications

In addition to DNA methylation, another major aspect of epigenetic regulation is represented by post-translational modification of core histones. Many changes occur as a function of age; for example, histone acetylation on lysine 16 of histone H4 (H4K16) increases gradually, due at least in part to a reduction of sirtuin 1 (SIRT1) deacetylase protein level [199–201]. Variations in methylation of histones H3 and H4, depending on residues, have also been also reported. The tri-methylated lysines 9 and 36 of histone H3 (H3K9me3 and H3K36me3 respectively), and the mono-methylated lysine 20 of histone H4 (H4K20me) can also change as a function of age [202–204]. In theory, the promotion of "healthy aging" could be pursued by developing epigenetics drugs able to cope with the "aged epigenome" [205].

There is evidence that the imbalance between histone acetylases (HAC) and deacetylases (HDAC) contribute to specific alterations in histone modifications and gene expression that are characteristic of COPD/emphysema and pulmonary fibrosis [206]. Indeed, cigarette smoke can modulate histone post-translational modifications through reductions in HDAC activity and expression [206, 207]. One example is the downregulation of HDAC2 by cigarette smoke in human macrophages and lung epithelial cells in vitro and in mouse lung in vivo [207]. HDAC2 activity normally delays cellular senescence by negatively regulating pro-senescent genes, such as p21 and p16 [208].

Reductions in mRNA abundance for HDACs 5, and 8, and a decrease in overall HDAC activity are also observed in COPD lung tissue and macrophages [51]. Decreased expression of HDACs correlates with increased H4 acetylation at the IL8 promoter and consequently, elevated expression, promoting inflammation [51]. In concordance, elevated H4 pan-acetylation, a marker of "permissive chromatin", is observed in lung tissue and airway smooth muscle cells from patients with COPD [206, 209]. Elevated H4 pan-acetylation in the vascular endothelial growth factor (VEGF) promoter correlates with reduced VEGF expression in airway smooth muscle from patients with COPD [209].

Sirtuins (see Sect. 3.1.4) are histone deacetylases and reduced SIRT1 activity due to cigarette smoke exposure correlates with increased H4 pan-acetylation and MMP9 expression in COPD [82, 210]. Cigarette smoke-driven reduction of SIRT1 expression in lung epithelial cells promotes FOXO3 and p53 acetylation, regulators of cell proliferation and cellular senescence (see Section 3.5) [210, 211].

Protein levels of Class-I HDACs (1, 2, 3 and 8), Class-II HDACs (4, 5, 7, 9, 10), and of the Class-III HDAC Sirtuin-1 were found to be significantly elevated in IPF lungs compared to healthy counterparts [212]. It has also been demonstrated that the process of myofibroblast differentiation is HDAC4 dependent and requires the phosphorylation of Akt [213]. Histone modifications are involved in the activation of some pro-fibrotic and repression of anti-fibrotic genes, while modifications at the Fas promoter are responsible for decreased Fas expression and apoptosis resistance in fibrotic lung fibroblasts [214]. COX-2-deficient mice are more susceptible to pulmonary fibrogenesis than are wild-type mice [215, 216], and a deficiency in COX2 expression is associated with deficient histone H3 and H4 acetylation, with a consequent increase in the recruitment of histone corepressor complexes to the COX-2 promoter [197]. Furthermore, recent studies demonstrated a marked increase in histone modifying enzymes and their respective binding proteins at the COX-2 promoter in lung fibroblasts from IPF patients, compared to those from nonfibrotic lungs [195]. Importantly, treatment with histone deacetylase inhibitors promotes fibroblast apoptosis and ameliorates pulmonary fibrosis in mice, inhibits the expression of fibrotic markers in IPF lung fibroblasts and restores cytokine-induced COX-2 mRNA and protein expression [217, 218]. These studies suggest that histone deacetylase inhibitors may offer a new therapeutic strategy in IPF.

3.6.3 MicroRNAs

While alterations in micro-RNA expression could contribute to the age-associated impairment of lung function, a lack of alterations in miRNA expression in the "normal aging lung" was initially described [219]. The total level of miRNA expression is reduced in smokers compared to non-smokers, probably due to the reduced activity of the endonuclease Dicer following cigarette exposure [220]. Reduced miRNA expression has been detected in whole lung tissues, airway epithelia, and alveolar macrophages of smokers. Furthermore, differential expression of miRNAs are detected in whole lung tissue, lung fibroblasts, cells from bronchoalveolar lavage, sputum, serum, and plasma samples from patients with COPD [220–224]. miRNA and mRNA expression profiles enriched for biological pathways that may be relevant to the pathogenesis of COPD including TGF β, Wnt and focal adhesion pathways have been described. For instance, differentially expression of miR15b in COPD regulates Smad7 in bronchial epithelial cells [224]. COPD patients have an abnormal repression of miR-199a-5p compared to unaffected smokers, probably contributing to the adaptive immune balance favoring a Th1 and Th17 profile [222]. The increase in miR-101 in COPD correlates with reduced CFTR expression, which may contribute to mucus accumulation, chronic infection and inflammation [225]. Importantly, recent studies demonstrated that the expression of specific microRNAs such as miR-638 correlates with emphysema severity and specific gene expression networks related to the oxidative stress response in aging emphysematous tissue as well as lung fibroblasts [226].

Fibroblasts from IPF patients with a highly progressive disease exhibit reduced expression of Dicer1 and Argonaute compared to patients with a slowly progressive disease. As Dicer and Argonaute are involved in miRNA biogenesis and silencing of gene expression, it is expected that miRNA biogenesis may contribute to the progression of IPF [227]. Forty-three microRNAs were found deregulated in samples from IPF patients [228]. Several specific miRNAs have been implicated in fibrogenic processes [198, 229–232]. The correlation between the findings in IPF patient samples and in the bleomycin-treated animal models indicate a miRNA profibrotic signature that includes the upregulation of miR 21, 145, 155, and 199a and the downregulation of miR Let7d, 17–92, 26a, 29, 200 family, and 326 (for a review see [233]); many of these altered miRNAs regulate TGF-β signaling, inflammation, and tissue remodeling [231, 232, 234].

An important connection between microRNAs and the pathogenesis of IPF is centered in the abnormal activation of epithelial cells and fibroblasts by reactivation of developmental programs such as Wnt/B-catenin and Sonic hedgehog signaling pathways. For example, miR-154 causes activation of the WNT pathway in normal human lung fibroblasts, regulating cellular migration and proliferation, and its levels are increased in IPF fibroblasts and lung fibroblasts treated with TGF-β. Upregulation of the Wnt/B-catenin signaling has also been associated with a decrease in miR-375 and miR487b [228, 235].

Contrary to the situation of IPF, a decrease in Wnt/B-catenin signaling contributes to parenchymal tissue destruction and impaired repair capacity in emphysema. In fact, activation of the Wnt/B-catenin pathway attenuates experimental emphysema [236].

3.6.4 Long Non-coding RNAs

Multiple functions are attributed to lncRNAs, such as regulation of transcription, mRNA splicing, mRNA decay, and gene neighborhood localization [237–239]. The expression profile of lncRNAs is significantly altered in fibrotic lung tissue, as demonstrated in bleomycin animal models of pulmonary fibrosis [240]. Until now, no specific lncRNAs have been associated with the pathogenesis of COPD/emphysema or pulmonary fibrosis. In consequence, the pathobiological relevance to lung remains to be established.

3.6.5 Differential Splicing

As a major source of protein diversity, alternative splicing plays critical roles in differentiation, development and disease. RNA-Seq technology allows us to understand how alternative splicing might affect the structure of the final protein products [241]. Splicing becomes less tightly controlled in aged individuals, based on data from the InChianti study [242]. Some splicing variants are known to occur in conjunction with fibrosis. For example, the inclusion of at least one of two extra exons, termed Extra Type III Domain A (EDA) and Extra Type III Domain B (EDB), is a feature of cellular fibronectin in IPF [243, 244]. While this is a nascent field, further RNA-seq experiments might identify additional splicing variants associated with lung disease and prognosis [241, 245].

3.7 *Inflammation*

3.7.1 Immunosenescence and Inflammaging

The immune system undergoes profound transformations with age, and strong similarities in inflammation are evident among aging, COPD and IPF, including lymphocyte senescence, neutrophil accumulation, NF-κB activation, and an increase in IL-6/IL-8/TNFα levels [246]. Disruption in the balance between inflammation and immune activation after local and systemic insults contributes to increased morbidity and mortality in the elderly [247–249], and older lungs respond to insults in a different manner from that of younger counterparts. For instance, the expression of cytokines interleukin IL17A, IL6, and CXCL is induced in older lungs to a greater degree than in young lungs after house dust exposure [250] and sputum from older asthmatics reveals higher numbers of neutrophils, IL-8, and neutrophil elastase [251] [252]. However, the source of inflammatory factors involved in the pathogenesis of either COPD or pulmonary fibrosis is unknown. Alveolar macrophages, lymphocytes, senescent epithelial cells, and mesenchymal cells constitute potential candidates.

Chronic inflammation in the elderly can exacerbate responses to lung injury, promoting alveolar destruction, tissue remodeling and the development of chronic inflammatory lung diseases such as COPD, interstitial pneumonia, persistent lower respiratory tract infection, and pulmonary fibrosis [85, 253–257]. Notably, chronic lung injury, inhaled gases, and particles from cigarette smoke mimic the effects of inflammaging [258–264] (Fig. 4).

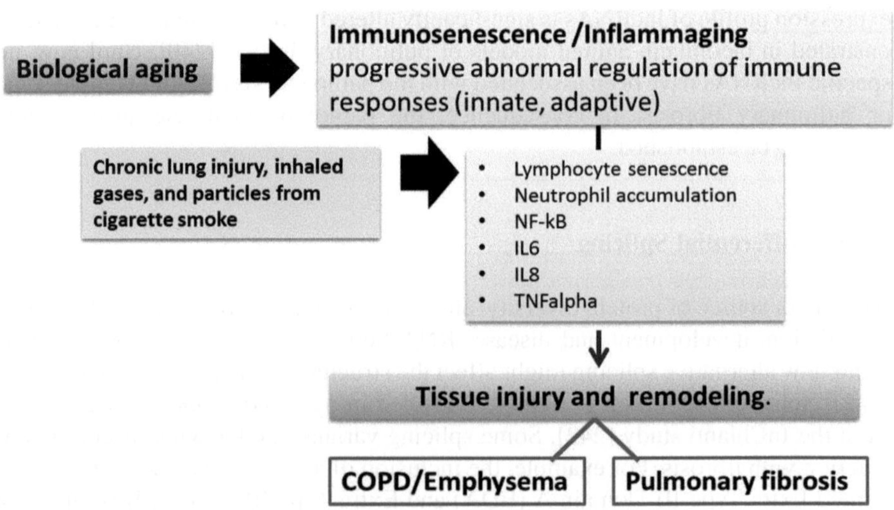

Fig. 4 Schematic representation of the relationship between immunosenescence induced by biological aging and environmental exposures in the development of pulmonary fibrosis and/or COPD and emphysema

Systemic immune-related defects described in COPD/emphysema and IPF patients include the following: ineffective phagocytic removal of apoptosed cells bodies [265–267], defects in innate immunity toll-like receptor sensors and a marked downregulation of CD28 in circulating CD4 T cells [268–270]. The downregulation of CD28 is a marker of CD8 T cell replicative senescence.

In spite of the observations cited above, the role of inflammation in the progression of pulmonary fibrosis is challenged by the lack of response to anti-inflammatory treatments and immunosuppressants. In addition, pulmonary fibrosis can be developed in mice by the overexpression of the profibrotic mediator, TGF-β1, without a significant inflammatory component [271]. Nevertheless, a pathogenic role for inflammation cannot be excluded in the initiation of the disease. In fact, the inability to identify the initiating inflammatory process in human IPF may explain the failure of the anti-inflammatory therapies.

3.7.2 Macrophages

Macrophages are essential components of innate immunity, contribute to acute and chronic inflammatory responses by releasing both pro- and anti-inflammatory mediators, as well as angiogenic, mitogenic, and under pathological conditions, profibrotic proteins [272–274]. M2-activated macrophages secrete cytokines that stimulate collagen production in lung fibroblasts. Moreover, monomeric collagen type I favors the shifting of alveolar macrophages to the profibrotic M2 type, which may perpetuate fibrosis [274–277]. The M2 phenotype is predominant in the lungs of IPF and COPD patients, with higher levels in patients who were smokers [278–280].

3.7.3 Lymphocytes

CD8+ T cells are frequently found in the airways of patients with COPD and are considered relevant for the pathogenesis of emphysema [281, 282]. Indeed, overexpression of T cell cytokines such as IFN-γ and IL-13 induce emphysematous changes in mice [283–285] and CD8+ T cells are required for cigarette smoke-induced emphysema. It is therefore suggested that CD8+ T cells induce the production of macrophage elastase, which degrades elastin, directly causing lung destruction [286].

Different types of CD4+ T cells also appear to be associated with COPD/emphysema [287]. IFN-γ–producing Th1 cells are increased in the airways and parenchymal tissues of patients with COPD [288]. IL17-producing Th17 cells accumulate in the bronchial mucosa and submucosa of patients with COPD [289, 290]. Furthermore, the expansion of CD28− CD4+ T cells, seen during aging [291], is found in patients with COPD, and correlates with impaired lung function. Moreover, these cells exhibit increased expression of perforin, and granzyme B (GzmB) [287, 292, 293]. GzmB released from cytotoxic lymphocytes protects against viral infection and it also contributes to extracellular matrix degradation and remodeling and, consequently, the emphysematous phenotype [294, 295].

3.7.4 Airway, Alveolar Epithelial and Mesenchymal Cells

These cells may also be an important source of inflammatory mediators by several mechanisms (for a review, see [296]). Clara cells are found in the epithelium of bronchioles and secrete surfactants; they may also have functions in maintenance of the bronchiolar epithelium. Loss of Clara cells in the small airways of smokers leads to decreased production of anti-inflammatory Clara cell protein or secretoglobin 1A1 [297], while squamous cells in the airway epithelium of individuals with COPD produce IL-1 alpha and IL-1 beta [298]. In COPD, airway fibroblast express high levels of IL-6, IL-8 [296, 299–301] and integrin alphavB8, which activates TGFβ, in turn stimulating CCL1 and CCL20 production in airway fibroblasts in an autocrine manner [302]. In pulmonary fibrosis, epithelial cells release chemokines that recruit inflammatory monocytes and neutrophils. Furthermore, a subpopulation of fibroblasts promotes inflammation, survival, and fibrosis through the constitutive activation of signal transducer and activator of transcription (STAT) 3 [303]. STAT3 mediates fibroblast chemotaxis through the induction of oncostatin M [304, 305].

3.7.5 TGF-β Signaling

Sequestration of latent TGF-β1 in the extracellular matrix is crucial for proper mobilization of this cytokine and its subsequent activation. TGF-β expression, activation and signaling have been shown to increase in several tissues, including the lung, as a function of aging [39, 87, 306–308]. Older mice (24 months old) are more susceptible to bleomycin exposure and develop extensive fibrosis, showing high levels of Smad3 phosphorylation (a consequence of canonical TGF-β signaling) [39]. In concordance, Smad3 null mice, deficient in TGF-β signal transmission, are resistant to bleomycin- and TGF-β-mediated fibrosis. However, the role of TGF-β1 seems to differ in emphysema and pulmonary fibrosis as Smad3 null mice develop spontaneous age-related airspace enlargement, consistent with emphysema. Furthermore, $\alpha_v\beta_6$ integrin null mouse, which is compromised in the ability to activate latent TGF-β, develops an emphysema-like response [41, 81, 309, 310]. Taken together, TGF-β and Smad3 seem to play a key role in the transition from inflammation to chronic fibrosis and/or from inflammation to emphysema. This can explain why both pathologies can appear within the same lung specimen [81].

3.8 Altered Intercellular Communications

Age-related alterations in the communication among resident cells, immune cells, and the lung microbiome promote susceptibility to age-related diseases [311, 312]. Changes in intercellular communication can occur due loss of cell-cell contact, changes in the extracellular matrix, hormones, electrical and chemical signals. Changes at many of these levels have been described in lung pathology. For example, age-related lung fibrosis worsens after lung injury in mice deficient in CD151, a regulator of cell-cell adhesion [313].

3.8.1 The Extracellular Matrix

Alterations in the extracellular matrix (ECM) protein profile and architecture are detected in several organs and tissues as a function of physiological aging. Initial studies in the lung from aged rats demonstrated elevation in TGF-β1 protein level, increase in the expression of some metalloproteinases and decreased expression of TIMPS [314].

Interestingly, emphysematous and fibrotic changes in the ECM, induced by elastase and bleomycin respectively, can be preserved in decellularized lungs and used to evaluate the behavior of other cells, such as engraftment of epithelial cells. It appears, therefore, that decellularized emphysematous scaffolds lack the necessary extracellular matrix architecture to support sustained cell growth [315]. Furthermore, IPF scaffolds support fibroblast survival, proliferation, and more importantly, differentiation to a myofibroblast phenotype, characteristic of fibrotic tissue [316].

Metalloproteinases The increased expression of matrix metalloproteinases (MMPs) MMP2/9 and the decreased mRNA expression of tissue inhibitors of metalloproteinases (TIMPS) are characteristic of the aging lung [314]. These proteases and anti-proteases secreted into the extracellular milieu are characteristic features of the pathogenesis of COPD and IPF [39, 317–320]. The relevance of protease imbalance for COPD pathogenesis is reflected by hereditary deficiency of α_1-antitrypsin or α_1-antichymotrypsin, which drive emphysematous ECM remodeling in patients [321]. In IPF, the mechanism involves changes in expression, turnover and/or deposition of ECM components and promotion of tissue remodeling, apoptosis, migration, proliferation, and angiogenesis. For a review, see [322].

3.8.2 Hormones

Hormones can also define intercellular communications. Disruption of the 17β-estradiol/estrogen receptor alpha signaling pathway is observed in aging. Animal models for acute lung injury and inflammation treated with 17β-estradiol exhibit reduced lung inflammation in a gender-independent, age-dependent manner [323]. Testosterone levels have been found to be low in COPD patients, with a prevalence of hypogonadism in men with COPD between 22 and 69 % [324]. However, the therapeutic efficacy of testosterone replacement therapy in COPD patients remains controversial [325–327]. The most abundant steroid in humans is dehydroepiandrosterone (DHEA), and patients with moderately severe COPD have lower concentrations of DHEA than smokers with chronic bronchitis or mild COPD [328]. Both DHEA and its sulfated form (DHEA-S) have been previously linked decreased function of the immune system observed with aging and this decline was also observed in IPF patients [329]. DHEA decreases lung fibroblast proliferation, increases apoptosis, reduces the fibroblast to myofibroblast differentiation and collagen production mediated by TGF-β1 or PDGF. In consequence, DHEA represents a putative therapeutic option and a strong support of the concept that IPF may be a disease of accelerated aging [329].

As discussed in Sect. 3.1.5, circadian rhythms play an important role in COPD and fibrosis. Melatonin is a key regulator of circadian rhythm homeostasis, and in consequence, melatonin was found to be significantly reduced during the exacerbation period in patients with COPD [330]. A low daily dose of melatonin has been shown to protect lungs from histopathological changes in rabbits exposed to smoke [331]. In animal models of pulmonary fibrosis, melatonin significantly attenuates bleomycin-induced myofibroblast differentiation, and alleviates ER stress and the ER stress-mediated epithelial-mesenchymal transition [332].

3.9 Stem Cell Exhaustion

Age-related defects in epithelial precursors act in concert with environmental toxic exposure to promote the breakdown of epithelial regeneration, leading to the chronic and irreversible alveolar loss characteristic of chronic lung injury, emphysema, and pulmonary fibrosis. Pathogenic models of COPD and IPF propose that premature cellular senescence leading to stem cell exhaustion likely affects distinct progenitor cells, such as mesenchymal stem cells in COPD, and alveolar epithelial precursors in IPF [179]. Unfortunately, our knowledge regarding lung resident stem cells is still emerging, and the changes in these cellular populations during aging and/or age-related diseases of the lung remain to be elucidated. Nevertheless, several studies provide evidence for the existence of human lung epithelial stem/progenitor cells [333–336]. Furthermore, convincing evidence for the existence of resident stem cells comes from a case study that reported compensatory lung growth with an increase in alveolar number in a 33-year-old woman, 15 years after a right-sided pneumonectomy for the treatment of lung adenocarcinoma [337].

The regenerative potential of pulmonary and extra-pulmonary stem and progenitor cells raises the hope for successful treatment options against pulmonary fibrosis, as shown by studies using human amniotic epithelial cells and bone marrow-derived epithelial progenitors cells [338, 339]. Bone marrow mesenchymal stem cells (B-MSCs) protect against the progression of emphysema and pulmonary fibrosis by increasing epithelial cell regeneration and reducing alveolar apoptosis [340, 341]. Furthermore, B-MSCs suppress the inflammatory response [342]. Currently, several studies are listed on the www.clinicaltrial.gov website using different MSC preparations and registered MSC products in patients with COPD and IPF [343, 344].

Nevertheless, it is important to consider that in addition to their reparative properties, MSCs can be a critical factor in the development of dysfunctional lung remodeling [345–349]. Furthermore, resident tissue-specific mesenchymal progenitor cells can eventually contribute to fibrogenesis in human lung allografts [349].

4 Conclusions and Future Directions

The lung is a unique organ that is directly exposed to high levels of oxygen and other reactive compounds. It will be important to further understand the interaction between the hallmarks of aging and the environment in the development of age-related lung diseases. Two generalizations can be taken from the findings reviewed in this chapter. First, the diverse pillars of aging appear to be more permissive than causal in the development of COPD and IPF. If borne out by future research, this would identify an important intersection between the biology of aging and environmental risk-factors for the prevalence of lung diseases among the elderly. Second, some aspects of aging biology may be protective in one disease but permissive in the other, as illustrated by the relatively different prevalence of fibrosis in IPF versus COPD (e.g., cellular senescence as described in Sect. 3.5)

Here we described major similarities as well as differences in the aging features in two age-related lung diseases, COPD/emphysema and IPF. Several preclinical studies using modulators of proteostasis, selective epigenetic modifiers as well as hormetic compounds to promote restoration of some of the aging hallmarks were presented. Modulation of endogenous stem cells may also help restore normal regenerative processes and correct the cellular and structural architecture of the lung and provide immunomodulation and trophic support for epithelial regeneration. It is expected that future studies will provide additional interventions to promote healthy lung aging and prevention of the onset of age-related lung diseases such as COPD and IPF.

We expect that future studies on molecular hallmarks of aging in young, middle age, and old age will provide a better understanding of the progressive decline of lung function in "normal aging" and to separate normal compensatory mechanisms occurring during aging from pathologic changes. Studies that also consider the diversity of cell types present in the lung may lead to an improved understanding of the dynamic and complex process of aging and the identification of new targets for early diagnostics, interventions and therapeutic approaches against age-related lung diseases.

The discoveries of gene variants and changes in gene expression in COPD and IPF that reliably predict outcomes have the potential to revolutionize the prognostic perspective and impact on therapeutic approaches. Emerging approaches to studying genetic/epigenetic/environment interactions, which impact disease pathogenesis are promising leads for novel biomarkers. By adding the perspective of the major pillars of aging, geroscience approaches will add a physiological layer to the efforts in personalized medicine currently focused on gene × environment interactions. This should incorporate aging as an essential parameter to match subjects with optimal therapeutic regimens while minimizing side effects.

Acknowledgments Dr. Sanchez work was supported by the National Institute of General Medical Sciences of the National Institutes of Health under Award Number P20GM103629-04 and the American Thoracic Society and Scleroderma Foundation award 552114G1.

Editor: Anthony Punturieri, National Heart, Lung and Blood Institute (NHLBI), NIH.

References

1. Vestbo J et al (2013) Global strategy for the diagnosis, management, and prevention of chronic obstructive pulmonary disease: GOLD executive summary. Am J Respir Crit Care Med 187(4):347–365
2. Lee J et al (2011) Is the aging process accelerated in chronic obstructive pulmonary disease? Curr Opin Pulm Med 17(2):90–97
3. Cornwell WD et al (2010) Pathogenesis of inflammation and repair in advanced COPD. Semin Respir Crit Care Med 31(3):257–266
4. Huiart L, Ernst P, Suissa S (2005) Cardiovascular morbidity and mortality in COPD. Chest 128(4):2640–2646
5. Rana JS et al (2004) Chronic obstructive pulmonary disease, asthma, and risk of type 2 diabetes in women. Diabetes Care 27(10):2478–2484
6. Ozge C, Ozge A, Unal O (2006) Cognitive and functional deterioration in patients with severe COPD. Behav Neurol 17(2):121–130
7. Mannino DM (2003) Chronic obstructive pulmonary disease: definition and epidemiology. Respir Care 48(12):1185–1191; discussion 1191–3
8. Tashkin DP (2013) Variations in FEV(1) decline over time in chronic obstructive pulmonary disease and its implications. Curr Opin Pulm Med 19(2):116–124
9. Lamprecht B et al (2011) COPD in never smokers: results from the population-based burden of obstructive lung disease study. Chest 139(4):752–763
10. Thomsen M et al (2013) Characteristics and outcomes of chronic obstructive pulmonary disease in never smokers in Denmark: a prospective population study. Lancet Respir Med 1(7):543–550
11. Salvi SS, Barnes PJ (2009) Chronic obstructive pulmonary disease in non-smokers. Lancet 374(9691):733–743
12. Bosken CH et al (1992) Characterization of the inflammatory reaction in the peripheral airways of cigarette smokers using immunocytochemistry. Am Rev Respir Dis 145(4 Pt 1):911–917
13. Lams BE et al (1998) Immunopathology of the small-airway submucosa in smokers with and without chronic obstructive pulmonary disease. Am J Respir Crit Care Med 158(5 Pt 1):1518–1523
14. Saetta M et al (1999) CD8+ve cells in the lungs of smokers with chronic obstructive pulmonary disease. Am J Respir Crit Care Med 160(2):711–717
15. Amin K et al (2003) Relationship between inflammatory cells and structural changes in the lungs of asymptomatic and never smokers: a biopsy study. Thorax 58(2):135–142
16. Larsson K (2007) Aspects on pathophysiological mechanisms in COPD. J Intern Med 262(3):311–340
17. Barnes PJ (2008) Immunology of asthma and chronic obstructive pulmonary disease. Nat Rev Immunol 8(3):183–192
18. Henson PM, Cosgrove GP, Vandivier RW (2006) State of the art. Apoptosis and cell homeostasis in chronic obstructive pulmonary disease. Proc Am Thorac Soc 3(6):512–516
19. Raghu G et al (2011) An official ATS/ERS/JRS/ALAT statement: idiopathic pulmonary fibrosis: evidence-based guidelines for diagnosis and management. Am J Respir Crit Care Med 183(6):788–824
20. Bjoraker JA et al (1998) Prognostic significance of histopathologic subsets in idiopathic pulmonary fibrosis. Am J Respir Crit Care Med 157(1):199–203
21. Flaherty KR et al (2002) Clinical significance of histological classification of idiopathic interstitial pneumonia. Eur Respir J 19(2):275–283
22. Nicholson AG et al (2000) The prognostic significance of the histologic pattern of interstitial pneumonia in patients presenting with the clinical entity of cryptogenic fibrosing alveolitis. Am J Respir Crit Care Med 162(6):2213–2217

23. American Thoracic, S. and S. European Respiratory (2002) American Thoracic Society/European Respiratory Society International Multidisciplinary Consensus Classification of the Idiopathic Interstitial Pneumonias. This joint statement of the American Thoracic Society (ATS), and the European Respiratory Society (ERS) was adopted by the ATS board of directors, June 2001 and by the ERS Executive Committee, June 2001. Am J Respir Crit Care Med 165(2):277–304
24. Seibold MA et al (2011) A common MUC5B promoter polymorphism and pulmonary fibrosis. N Engl J Med 364(16):1503–1512
25. Selman M et al (2001) Idiopathic pulmonary fibrosis: prevailing and evolving hypotheses about its pathogenesis and implications for therapy. Ann Intern Med 134(2):136–151
26. Bueno M et al (2015) PINK1 deficiency impairs mitochondrial homeostasis and promotes lung fibrosis. J Clin Invest 25(2):521–538
27. Patel AS et al (2015) Epithelial cell mitochondrial dysfunction and PINK1 are induced by transforming growth factor- beta1 in pulmonary fibrosis. PLoS One 10(3):e0121246
28. Jain M et al (2013) Mitochondrial reactive oxygen species regulate transforming growth factor-beta signaling. J Biol Chem 288(2):770–777
29. Sosulski ML et al (2015) Deregulation of selective autophagy during aging and pulmonary fibrosis: the role of TGFbeta1. Aging Cell [Epub ahead of print]
30. Hecker L et al (2009) NADPH oxidase-4 mediates myofibroblast activation and fibrogenic responses to lung injury. Nat Med 15(9):1077–1081
31. Griffith B et al (2009) NOX enzymes and pulmonary disease. Antioxid Redox Signal 11(10):2505–2516
32. Cheresh P et al (2013) Oxidative stress and pulmonary fibrosis. Biochim Biophys Acta 1832(7):1028–1040
33. Kumar A et al (2014) Genome sequencing of idiopathic pulmonary fibrosis in conjunction with a medical school human anatomy course. PLoS One 9(9):e106744
34. Peljto AL et al (2015) The MUC5B promoter polymorphism is associated with idiopathic pulmonary fibrosis in a Mexican cohort but is rare among Asian ancestries. Chest 147(2):460–464
35. Coghlan MA et al (2014) Sequencing of idiopathic pulmonary fibrosis-related genes reveals independent single gene associations. BMJ Open Respir Res 1(1):e000057
36. Fingerlin TE et al (2013) Genome-wide association study identifies multiple susceptibility loci for pulmonary fibrosis. Nat Genet 45(6):613–620
37. Noth I et al (2013) Genetic variants associated with idiopathic pulmonary fibrosis susceptibility and mortality: a genome-wide association study. Lancet Respir Med 1(4):309–317
38. Peng R et al (2013) Bleomycin induces molecular changes directly relevant to idiopathic pulmonary fibrosis: a model for "active" disease. PLoS One 8(4):e59348
39. Sueblinvong V et al (2012) Predisposition for disrepair in the aged lung. Am J Med Sci 344(1):41–51
40. Selman M, Pardo A (2014) Revealing the pathogenic and aging-related mechanisms of the enigmatic idiopathic pulmonary fibrosis. an integral model. Am J Respir Crit Care Med 189(10):1161–1172
41. Gauldie J et al (2006) Smad3 signaling involved in pulmonary fibrosis and emphysema. Proc Am Thorac Soc 3(8):696–702
42. Jankowich MD et al (2008) Heterogeneity in combined pulmonary fibrosis and emphysema. Respiration 75(4):411–417
43. Mahavadi P et al (2010) Epithelial stress and apoptosis underlie Hermansky-Pudlak syndrome-associated interstitial pneumonia. Am J Respir Crit Care Med 182(2):207–219
44. Tzouvelekis A, Bonella F, Spagnolo P (2015) Update on therapeutic management of idiopathic pulmonary fibrosis. Ther Clin Risk Manag 11:359–370
45. Chaudhary NI et al (2007) Inhibition of PDGF, VEGF and FGF signalling attenuates fibrosis. Eur Respir J 29(5):976–985

46. Iyer SN, Gurujeyalakshmi G, Giri SN (1999) Effects of pirfenidone on transforming growth factor-beta gene expression at the transcriptional level in bleomycin hamster model of lung fibrosis. J Pharmacol Exp Ther 291(1):367–373
47. Gurujeyalakshmi G, Hollinger MA, Giri SN (1999) Pirfenidone inhibits PDGF isoforms in bleomycin hamster model of lung fibrosis at the translational level. Am J Physiol 276(2 Pt 1):L311–L318
48. Iyer SN et al (1998) Lung fibrosis is ameliorated by pirfenidone fed in diet after the second dose in a three-dose bleomycin-hamster model. Exp Lung Res 24(1):119–132
49. Joos L, Pare PD, Sandford AJ (2002) Genetic risk factors of chronic obstructive pulmonary disease. Swiss Med Wkly 132(3–4):27–37
50. Demeo DL et al (2006) The SERPINE2 gene is associated with chronic obstructive pulmonary disease. Am J Hum Genet 78(2):253–264
51. Ito K et al (2005) Decreased histone deacetylase activity in chronic obstructive pulmonary disease. N Engl J Med 352(19):1967–1976
52. Kythreotis P et al (2009) Plasma leptin and insulin-like growth factor I levels during acute exacerbations of chronic obstructive pulmonary disease. BMC Pulm Med 9:11
53. Fournier M et al (2003) Insulin-like growth factor I prevents corticosteroid-induced diaphragm muscle atrophy in emphysematous hamsters. Am J Physiol Regul Integr Comp Physiol 285(1):R34–R43
54. Uh ST et al (1998) Morphometric analysis of insulin-like growth factor-I localization in lung tissues of patients with idiopathic pulmonary fibrosis. Am J Respir Crit Care Med 158(5 Pt 1):1626–1635
55. Ruan W, Ying K (2010) Abnormal expression of IGF-binding proteins, an initiating event in idiopathic pulmonary fibrosis? Pathol Res Pract 206(8):537–543
56. Honeyman L et al (2013) MicroRNA profiling implicates the insulin-like growth factor pathway in bleomycin-induced pulmonary fibrosis in mice. Fibrogenesis Tissue Repair 6(1):16
57. Pilewski JM et al (2005) Insulin-like growth factor binding proteins 3 and 5 are overexpressed in idiopathic pulmonary fibrosis and contribute to extracellular matrix deposition. Am J Pathol 166(2):399–407
58. Yasuoka H, Yamaguchi Y, Feghali-Bostwick CA (2009) The pro-fibrotic factor IGFBP-5 induces lung fibroblast and mononuclear cell migration. Am J Respir Cell Mol Biol 41(2):179–188
59. Calhoun C et al (2015) Senescent cells contribute to the physiological remodeling of aged lungs. J Gerontol A Biol Sci Med Sci [Epub ahead of print]
60. Yoshida T et al (2010) Rtp801, a suppressor of mTOR signaling, is an essential mediator of cigarette smoke-induced pulmonary injury and emphysema. Nat Med 16(7):767–773
61. Nadon AM et al (2014) Rtp801 suppression of epithelial mTORC1 augments endotoxin-induced lung inflammation. Am J Pathol 184(9):2382–2389
62. Park JS et al (2014) Clinical significance of mTOR, ZEB1, ROCK1 expression in lung tissues of pulmonary fibrosis patients. BMC Pulm Med 14:168
63. Goc A et al (2011) TGFbeta- and bleomycin-induced extracellular matrix synthesis is mediated through Akt and mammalian target of rapamycin (mTOR). J Cell Physiol 226(11):3004–3013
64. Patel AS et al (2012) Autophagy in idiopathic pulmonary fibrosis. PLoS One 7(7):e41394
65. Araya J et al (2013) Insufficient autophagy in idiopathic pulmonary fibrosis. Am J Physiol Lung Cell Mol Physiol 304(1):L56–69
66. Nho RS, Hergert P (2014) IPF fibroblasts are desensitized to type I collagen matrix-induced cell death by suppressing low autophagy via aberrant Akt/mTOR kinases. PLoS One 9(4):e94616
67. Richeldi L et al (2013) Mapping the future for pulmonary fibrosis: report from the 17th International Colloquium on Lung and Airway Fibrosis. Eur Respir J 42(1):230–238
68. Qi Y et al (2014) Inhibition of AMPK expression in skeletal muscle by systemic inflammation in COPD rats. Respir Res 15(1):156

69. Park CS et al (2012) Metformin reduces airway inflammation and remodeling via activation of AMP-activated protein kinase. Biochem Pharmacol 84(12):1660–1670
70. Park S, Mori R, Shimokawa I (2013) Do sirtuins promote mammalian longevity? A critical review on its relevance to the longevity effect induced by calorie restriction. Mol Cells 35(6):474–480
71. Verdin E (2014) The many faces of sirtuins: coupling of NAD metabolism, sirtuins and lifespan. Nat Med 20(1):25–27
72. Purushotham A et al (2009) Hepatocyte-specific deletion of SIRT1 alters fatty acid metabolism and results in hepatic steatosis and inflammation. Cell Metab 9(4):327–338
73. He Q et al (2014) Shorter men live longer: association of height with longevity and FOXO3 genotype in American men of Japanese ancestry. PLoS One 9(5):e94385
74. Eijkelenboom A et al (2013) FOXO3 selectively amplifies enhancer activity to establish target gene regulation. Cell Rep 5(6):1664–1678
75. Flachsbart F et al (2009) Association of FOXO3A variation with human longevity confirmed in German centenarians. Proc Natl Acad Sci U S A 106(8):2700–2705
76. Marinkovic D et al (2007) Foxo3 is required for the regulation of oxidative stress in erythropoiesis. J Clin Invest 117(8):2133–2144
77. Giannakou ME, Partridge L (2004) The interaction between FOXO and SIRT1: tipping the balance towards survival. Trends Cell Biol 14(8):408–412
78. Rodgers JT et al (2005) Nutrient control of glucose homeostasis through a complex of PGC-1alpha and SIRT1. Nature 434(7029):113–118
79. Luo C et al (2013) A cigarette component acrolein induces accelerated senescence in human diploid fibroblast IMR-90 cells. Biogerontology 14(5):503–511
80. Sundar IK, Yao H, Rahman I (2013) Oxidative stress and chromatin remodeling in chronic obstructive pulmonary disease and smoking-related diseases. Antioxid Redox Signal 18(15):1956–1971
81. Warburton D, Shi W, Xu B (2013) TGF-beta-Smad3 signaling in emphysema and pulmonary fibrosis: an epigenetic aberration of normal development? Am J Physiol Lung Cell Mol Physiol 304(2):L83–L85
82. Nakamaru Y et al (2009) A protein deacetylase SIRT1 is a negative regulator of metalloproteinase-9. FASEB J 23(9):2810–2819
83. Yao H, Rahman I (2012) Perspectives on translational and therapeutic aspects of SIRT1 in inflammaging and senescence. Biochem Pharmacol 84(10):1332–1339
84. Zerr P et al (2014) Sirt1 regulates canonical TGF-beta signalling to control fibroblast activation and tissue fibrosis. Ann Rheum Dis 205740. Published Online 1 September
85. Ito K, Barnes PJ (2009) COPD as a disease of accelerated lung aging. Chest 135(1):173–180
86. Chun P (2015) Role of sirtuins in chronic obstructive pulmonary disease. Arch Pharm Res 38(1):1–10
87. Minagawa S et al (2011) Accelerated epithelial cell senescence in IPF and the inhibitory role of SIRT6 in TGF-beta-induced senescence of human bronchial epithelial cells. Am J Physiol Lung Cell Mol Physiol 300(3):L391–L401
88. Takasaka N et al (2014) Autophagy induction by SIRT6 through attenuation of insulin-like growth factor signaling is involved in the regulation of human bronchial epithelial cell senescence. J Immunol 192(3):958–968
89. Liu R et al (2014) Oxidative stress induces endothelial cell senescence via downregulation of Sirt6. Biomed Res Int 2014:902842
90. Ko CH, Takahashi JS (2006) Molecular components of the mammalian circadian clock. Hum Mol Genet 15(Spec No 2):R271–7
91. Panda S, Hogenesch JB, Kay SA (2002) Circadian rhythms from flies to human. Nature 417(6886):329–335
92. Stangherlin A, Reddy AB (2013) Regulation of circadian clocks by redox homeostasis. J Biol Chem 288(37):26505–26511

93. Gossan N et al (2013) The circadian clock in murine chondrocytes regulates genes controlling key aspects of cartilage homeostasis. Arthritis Rheum 65(9):2334–2345
94. Patel SA, Velingkaar NS, Kondratov RV (2014) Transcriptional control of antioxidant defense by the circadian clock. Antioxid Redox Signal 20(18):2997–3006
95. Froy O (2010) Metabolism and circadian rhythms – implications for obesity. Endocr Rev 31(1):1–24
96. Pekovic-Vaughan V et al (2014) The circadian clock regulates rhythmic activation of the NRF2/glutathione-mediated antioxidant defense pathway to modulate pulmonary fibrosis. Genes Dev 28(6):548–560
97. Ma D, Lin JD (2012) Circadian regulation of autophagy rhythm through transcription factor C/EBPbeta. Autophagy 8(1):124–125
98. Eckel-Mahan KL et al (2013) Reprogramming of the circadian clock by nutritional challenge. Cell 155(7):1464–1478
99. Masri S et al (2013) Circadian acetylome reveals regulation of mitochondrial metabolic pathways. Proc Natl Acad Sci U S A 110(9):3339–3344
100. Korfhagen TR et al (2009) Rapamycin prevents transforming growth factor-alpha-induced pulmonary fibrosis. Am J Respir Cell Mol Biol 41(5):562–572
101. Madala SK et al (2011) Rapamycin regulates bleomycin-induced lung damage in SP-C-deficient mice. Pulm Med 2011:653524
102. Park SJ et al (2012) L-2-oxothiazolidine-4-carboxylic acid or alpha-lipoic acid attenuates airway remodeling: involvement of nuclear factor-kappaB (NF-kappaB), nuclear factor erythroid 2p45-related factor-2 (Nrf2), and hypoxia-inducible factor (HIF). Int J Mol Sci 13(7):7915–7937
103. Lin CH et al (2014) Resveratrol enhanced FOXO3 phosphorylation via synergetic activation of SIRT1 and PI3K/Akt signaling to improve the effects of exercise in elderly rat hearts. Age (Dordr) 36(5):9705
104. Eo SH, Cho HS, Kim SJ (2014) Resveratrol regulates type II collagen and COX-2 expression via the ERK, p38 and Akt signaling pathways in rabbit articular chondrocytes. Exp Ther Med 7(3):640–648
105. Liu M, Liu F (2011) Resveratrol inhibits mTOR signaling by targeting DEPTOR. Commun Integr Biol 4(4):382–384
106. Harikumar KB, Aggarwal BB (2008) Resveratrol: a multitargeted agent for age-associated chronic diseases. Cell Cycle 7(8):1020–1035
107. Akgedik R et al (2012) Effect of resveratrol on treatment of bleomycin-induced pulmonary fibrosis in rats. Inflammation 35(5):1732–1741
108. Sener G et al (2007) Resveratrol alleviates bleomycin-induced lung injury in rats. Pulm Pharmacol Ther 20(6):642–649
109. Hu YX et al (2013) Resveratrol attenuates left ventricular remodeling in old rats with COPD induced by cigarette smoke exposure and LPS instillation. Can J Physiol Pharmacol 91(12):1044–1054
110. Li Y et al (2014) Resveratrol attenuates endoplasmic reticulum stress and alveolar epithelial apoptosis in a rat model of chronic obstructive pulmonary disease. Zhonghua Jie He He Hu Xi Za Zhi 37(1):30–35
111. Knobloch J et al (2010) Resveratrol impairs the release of steroid-resistant inflammatory cytokines from human airway smooth muscle cells in chronic obstructive pulmonary disease. J Pharmacol Exp Ther 335(3):788–798
112. Ito K, Colley T, Mercado N (2012) Geroprotectors as a novel therapeutic strategy for COPD, an accelerating aging disease. Int J Chron Obstruct Pulmon Dis 7:641–652
113. Benloucif S, Masana MI, Dubocovich ML (1997) Responsiveness to melatonin and its receptor expression in the aging circadian clock of mice. Am J Physiol 273(6 Pt 2):R1855–R1860
114. Shin IS et al (2014) Melatonin inhibits MUC5AC production via suppression of MAPK signaling in human airway epithelial cells. J Pineal Res 56(4):398–407

115. Shin IS et al (2015) Melatonin attenuates neutrophil inflammation and mucus secretion in cigarette smoke-induced chronic obstructive pulmonary diseases via the suppression of Erk-Sp1 signaling. J Pineal Res 58(1):50–60
116. Bouchecareilh M, Balch WE (2011) Proteostasis: a new therapeutic paradigm for pulmonary disease. Proc Am Thorac Soc 8(2):189–195
117. Menzies FM, Moreau K, Rubinsztein DC (2011) Protein misfolding disorders and macroautophagy. Curr Opin Cell Biol 23(2):190–197
118. Koga H, Kaushik S, Cuervo AM (2011) Protein homeostasis and aging: the importance of exquisite quality control. Ageing Res Rev 10(2):205–215
119. Korfei M et al (2011) Comparative proteomic analysis of lung tissue from patients with idiopathic pulmonary fibrosis (IPF) and lung transplant donor lungs. J Proteome Res 10(5):2185–2205
120. Tran I et al (2014) Role of cigarette smoke-induced aggresome-formation in COPD-emphysema pathogenesis. Am J Respir Cell Mol Biol 53(2):159–173
121. Chen ZH et al (2010) Autophagy protein microtubule-associated protein 1 light chain-3B (LC3B) activates extrinsic apoptosis during cigarette smoke-induced emphysema. Proc Natl Acad Sci U S A 107(44):18880–18885
122. Mizumura K et al (2014) Mitophagy-dependent necroptosis contributes to the pathogenesis of COPD. J Clin Invest 124(9):3987–4003
123. Ryter SW et al (2012) Autophagy in pulmonary diseases. Annu Rev Physiol 74:377–401
124. Monick MM et al (2010) Identification of an autophagy defect in smokers' alveolar macrophages. J Immunol 185(9):5425–5435
125. Ricci A et al (2013) Decreased expression of autophagic beclin 1 protein in idiopathic pulmonary fibrosis fibroblasts. J Cell Physiol 228(7):1516–1524
126. Singh KK et al (2015) The essential autophagy gene ATG7 modulates organ fibrosis via regulation of endothelial-to-mesenchymal transition. J Biol Chem 290(5):2547–59
127. Bernard M et al (2014) Autophagy fosters myofibroblast differentiation through MTORC2 activation and downstream upregulation of CTGF. Autophagy 10(12):2193–207
128. Hernandez-Gea V, Friedman SL (2012) Autophagy fuels tissue fibrogenesis. Autophagy 8(5):849–850
129. Del Principe D et al (2013) Fibroblast autophagy in fibrotic disorders. J Pathol 229(2):208–220
130. Hawkins A et al (2015) A non-BRICHOS SFTPC mutant (SP-CI73T) linked to interstitial lung disease promotes a late block in macroautophagy disrupting cellular proteostasis and mitophagy. Am J Physiol Lung Cell Mol Physiol 308(1):L33–L47
131. Lam HC et al (2013) Histone deacetylase 6-mediated selective autophagy regulates COPD-associated cilia dysfunction. J Clin Invest 123(12):5212–5230
132. Tang Z, Zhu M, Zhong Q (2014) Self-eating to remove cilia roadblock. Autophagy 10(2):379–381
133. Tang Z et al (2013) Autophagy promotes primary ciliogenesis by removing OFD1 from centriolar satellites. Nature 502(7470):254–257
134. Cloonan SM et al (2014) "Ciliophagy": the consumption of cilia components by autophagy. Autophagy 10(3):532–534
135. Yang IV et al (2013) Expression of cilium-associated genes defines novel molecular subtypes of idiopathic pulmonary fibrosis. Thorax 68(12):1114–1121
136. Moshai EF et al (2014) Targeting the hedgehog-glioma-associated oncogene homolog pathway inhibits bleomycin-induced lung fibrosis in mice. Am J Respir Cell Mol Biol 51(1):11–25
137. Farout L, Friguet B (2006) Proteasome function in aging and oxidative stress: implications in protein maintenance failure. Antioxid Redox Signal 8(1–2):205–216
138. Jung T et al (2009) Age-related differences in oxidative protein-damage in young and senescent fibroblasts. Arch Biochem Biophys 483(1):127–135

139. Sitte N et al (2000) Proteasome inhibition by lipofuscin/ceroid during postmitotic aging of fibroblasts. FASEB J 14(11):1490–1498
140. Rock KL et al (1994) Inhibitors of the proteasome block the degradation of most cell proteins and the generation of peptides presented on MHC class I molecules. Cell 78(5):761–771
141. Torres C, Lewis L, Cristofalo VJ (2006) Proteasome inhibitors shorten replicative life span and induce a senescent-like phenotype of human fibroblasts. J Cell Physiol 207(3):845–853
142. Chondrogianni N, Gonos ES (2004) Proteasome inhibition induces a senescence-like phenotype in primary human fibroblasts cultures. Biogerontology 5(1):55–61
143. Breusing N et al (2009) Inverse correlation of protein oxidation and proteasome activity in liver and lung. Mech Ageing Dev 130(11–12):748–753
144. Weiss CH et al (2010) Proteasomal regulation of pulmonary fibrosis. Proc Am Thorac Soc 7(1):77–83
145. Baker TA et al (2014) Proteasomes in lungs from organ donors and patients with end-stage pulmonary diseases. Physiol Res 63(3):311–319
146. Min T et al (2011) Critical role of proteostasis-imbalance in pathogenesis of COPD and severe emphysema. J Mol Med (Berl) 89(6):577–593
147. Stepaniants S et al (2014) Genes related to emphysema are enriched for ubiquitination pathways. BMC Pulm Med 14:187
148. Meusser B et al (2005) ERAD: the long road to destruction. Nat Cell Biol 7(8):766–772
149. Korfei M et al (2008) Epithelial endoplasmic reticulum stress and apoptosis in sporadic idiopathic pulmonary fibrosis. Am J Respir Crit Care Med 178(8):838–846
150. Lawson WE et al (2008) Endoplasmic reticulum stress in alveolar epithelial cells is prominent in IPF: association with altered surfactant protein processing and herpesvirus infection. Am J Physiol Lung Cell Mol Physiol 294(6):L1119–L1126
151. Roberson EC et al (2012) Influenza induces endoplasmic reticulum stress, caspase-12-dependent apoptosis, and c-Jun N-terminal kinase-mediated transforming growth factor-beta release in lung epithelial cells. Am J Respir Cell Mol Biol 46(5):573–581
152. Lenna S, Trojanowska M (2012) The role of endoplasmic reticulum stress and the unfolded protein response in fibrosis. Curr Opin Rheumatol 24(6):663–668
153. Baek HA et al (2012) Involvement of endoplasmic reticulum stress in myofibroblastic differentiation of lung fibroblasts. Am J Respir Cell Mol Biol 46(6):731–739
154. Bratic A, Larsson NG (2013) The role of mitochondria in aging. J Clin Invest 123(3):951–957
155. Ott M et al (2007) Mitochondria, oxidative stress and cell death. Apoptosis 12(5):913–922
156. Bondi CD et al (2010) NAD(P)H oxidase mediates TGF-beta1-induced activation of kidney myofibroblasts. J Am Soc Nephrol 21(1):93–102
157. Niture SK et al (2010) Nrf2 signaling and cell survival. Toxicol Appl Pharmacol 244(1):37–42
158. Hybertson BM et al (2011) Oxidative stress in health and disease: the therapeutic potential of Nrf2 activation. Mol Aspects Med 32(4–6):234–246
159. Cho HY, Kleeberger SR (2007) Genetic mechanisms of susceptibility to oxidative lung injury in mice. Free Radic Biol Med 42(4):433–445
160. Cho HY, Reddy SP, Kleeberger SR (2006) Nrf2 defends the lung from oxidative stress. Antioxid Redox Signal 8(1–2):76–87
161. Malhotra D et al (2011) Denitrosylation of HDAC2 by targeting Nrf2 restores glucocorticosteroid sensitivity in macrophages from COPD patients. J Clin Invest 121(11):4289–4302
162. Kumar V et al (2011) Novel chalcone derivatives as potent Nrf2 activators in mice and human lung epithelial cells. J Med Chem 54(12):4147–4159
163. Soulitzis N et al (2012) Downregulation of lung mitochondrial prohibitin in COPD. Respir Med 106(7):954–961
164. Bueler H (2010) Mitochondrial dynamics, cell death and the pathogenesis of Parkinson's disease. Apoptosis 15(11):1336–1353

165. Hara H et al (2013) Mitochondrial fragmentation in cigarette smoke-induced bronchial epithelial cell senescence. Am J Physiol Lung Cell Mol Physiol 305(10):L737–L746
166. Hoffmann RF et al (2013) Prolonged cigarette smoke exposure alters mitochondrial structure and function in airway epithelial cells. Respir Res 14:97
167. Aravamudan B et al (2014) Cigarette smoke-induced mitochondrial fragmentation and dysfunction in human airway smooth muscle. Am J Physiol Lung Cell Mol Physiol 306(9):L840–L854
168. Wistuba II, Mao L, Gazdar AF (2002) Smoking molecular damage in bronchial epithelium. Oncogene 21(48):7298–7306
169. Aoshiba K et al (2012) DNA damage as a molecular link in the pathogenesis of COPD in smokers. Eur Respir J 39(6):1368–1376
170. Kuwano K et al (1996) P21Waf1/Cip1/Sdi1 and p53 expression in association with DNA strand breaks in idiopathic pulmonary fibrosis. Am J Respir Crit Care Med 154(2 Pt 1):477–483
171. Moyzis RK et al (1988) A highly conserved repetitive DNA sequence, (TTAGGG)n, present at the telomeres of human chromosomes. Proc Natl Acad Sci U S A 85(18):6622–6626
172. Serra V et al (2003) Extracellular superoxide dismutase is a major antioxidant in human fibroblasts and slows telomere shortening. J Biol Chem 278(9):6824–6830
173. Tsuji T, Aoshiba K, Nagai A (2006) Alveolar cell senescence in patients with pulmonary emphysema. Am J Respir Crit Care Med 174(8):886–893
174. Kashima K et al (2006) Decrease of telomeres and increase of interstitial telomeric sites in chromosomes of short-term cultured gastric carcinoma cells detected by fluorescence in situ hybridization. Anticancer Res 26(4B):2849–2855
175. Houben JM et al (2009) Telomere shortening in chronic obstructive pulmonary disease. Respir Med 103(2):230–236
176. Armanios MY et al (2007) Telomerase mutations in families with idiopathic pulmonary fibrosis. N Engl J Med 356(13):1317–1326
177. Tsakiri KD et al (2007) Adult-onset pulmonary fibrosis caused by mutations in telomerase. Proc Natl Acad Sci U S A 104(18):7552–7557
178. Utz JP et al (2005) Usual interstitial pneumonia complicating dyskeratosis congenita. Mayo Clin Proc 80(6):817–821
179. Chilosi M, Poletti V, Rossi A (2012) The pathogenesis of COPD and IPF: distinct horns of the same devil? Respir Res 13:3
180. Liu T et al (2007) Telomerase activity is required for bleomycin-induced pulmonary fibrosis in mice. J Clin Invest 117(12):3800–3809
181. Degryse AL et al (2012) Telomerase deficiency does not alter bleomycin-induced fibrosis in mice. Exp Lung Res 38(3):124–134
182. Alder JK et al (2011) Telomere length is a determinant of emphysema susceptibility. Am J Respir Crit Care Med 184(8):904–912
183. Lee J et al (2012) The relationship between telomere length and mortality in chronic obstructive pulmonary disease (COPD). PLoS One 7(4):e35567
184. Savale L et al (2009) Shortened telomeres in circulating leukocytes of patients with chronic obstructive pulmonary disease. Am J Respir Crit Care Med 179(7):566–571
185. Muller KC et al (2006) Lung fibroblasts from patients with emphysema show markers of senescence in vitro. Respir Res 7:32
186. Nyunoya T et al (2006) Cigarette smoke induces cellular senescence. Am J Respir Cell Mol Biol 35(6):681–688
187. Tsuji T, Aoshiba K, Nagai A (2004) Cigarette smoke induces senescence in alveolar epithelial cells. Am J Respir Cell Mol Biol 31(6):643–649
188. Kanaji N et al (2014) Fibroblasts that resist cigarette smoke-induced senescence acquire profibrotic phenotypes. Am J Physiol Lung Cell Mol Physiol 307(5):L364–L373

189. Guzman L et al (2012) Analysis of aberrant methylation on promoter sequences of tumor suppressor genes and total DNA in sputum samples: a promising tool for early detection of COPD and lung cancer in smokers. Diagn Pathol 7:87
190. Vucic EA et al (2014) DNA methylation is globally disrupted and associated with expression changes in chronic obstructive pulmonary disease small airways. Am J Respir Cell Mol Biol 50(5):912–922
191. Lange NE et al (2012) Alu and LINE-1 methylation and lung function in the normative ageing study. BMJ Open 2(5)
192. Boutten A et al (2011) NRF2 targeting: a promising therapeutic strategy in chronic obstructive pulmonary disease. Trends Mol Med 17(7):363–371
193. Bozinovski S et al (2006) Akt in the pathogenesis of COPD. Int J Chron Obstruct Pulmon Dis 1(1):31–38
194. Huang SK et al (2010) Hypermethylation of PTGER2 confers prostaglandin E2 resistance in fibrotic fibroblasts from humans and mice. Am J Pathol 177(5):2245–2255
195. Coward WR et al (2014) A central role for G9a and EZH2 in the epigenetic silencing of cyclooxygenase-2 in idiopathic pulmonary fibrosis. FASEB J 28(7):3183–3196
196. Sanders YY et al (2008) Thy-1 promoter hypermethylation: a novel epigenetic pathogenic mechanism in pulmonary fibrosis. Am J Respir Cell Mol Biol 39(5):610–618
197. Robinson CM et al (2012) Hypoxia-induced DNA hypermethylation in human pulmonary fibroblasts is associated with Thy-1 promoter methylation and the development of a pro-fibrotic phenotype. Respir Res 13:74
198. Dakhlallah D et al (2013) Epigenetic regulation of miR-17 ~ 92 contributes to the pathogenesis of pulmonary fibrosis. Am J Respir Crit Care Med 187(4):397–405
199. Gonzalo S (2010) Epigenetic alterations in aging. J Appl Physiol (1985) 109(2):586–97
200. Pruitt K et al (2006) Inhibition of SIRT1 reactivates silenced cancer genes without loss of promoter DNA hypermethylation. PLoS Genet 2(3):e40
201. Vaquero A et al (2007) SIRT1 regulates the histone methyl-transferase SUV39H1 during heterochromatin formation. Nature 450(7168):440–444
202. Fraga MF, Esteller M (2007) Epigenetics and aging: the targets and the marks. Trends Genet 23(8):413–418
203. Sarg B et al (2002) Postsynthetic trimethylation of histone H4 at lysine 20 in mammalian tissues is associated with aging. J Biol Chem 277(42):39195–39201
204. Jung JW et al (2010) Histone deacetylase controls adult stem cell aging by balancing the expression of polycomb genes and jumonji domain containing 3. Cell Mol Life Sci 67(7):1165–1176
205. Cencioni C et al (2013) Oxidative stress and epigenetic regulation in ageing and age-related diseases. Int J Mol Sci 14(9):17643–17663
206. Szulakowski P et al (2006) The effect of smoking on the transcriptional regulation of lung inflammation in patients with chronic obstructive pulmonary disease. Am J Respir Crit Care Med 174(1):41–50
207. Adenuga D et al (2009) Histone deacetylase 2 is phosphorylated, ubiquitinated, and degraded by cigarette smoke. Am J Respir Cell Mol Biol 40(4):464–473
208. Yao H, Rahman I (2012) Role of histone deacetylase 2 in epigenetics and cellular senescence: implications in lung inflammaging and COPD. Am J Physiol Lung Cell Mol Physiol 303(7):L557–L566
209. Raidl M et al (2007) Impaired TNFalpha-induced VEGF expression in human airway smooth muscle cells from smokers with COPD: role of MAPkinases and histone acetylation – effect of dexamethasone. Cell Biochem Biophys 49(2):98–110
210. Rajendrasozhan S et al (2008) Deacetylases and NF-kappaB in redox regulation of cigarette smoke-induced lung inflammation: epigenetics in pathogenesis of COPD. Antioxid Redox Signal 10(4):799–811
211. Brunet A et al (2004) Stress-dependent regulation of FOXO transcription factors by the SIRT1 deacetylase. Science 303(5666):2011–2015

212. Mahavadi P et al (2014) Altered surfactant homeostasis and alveolar epithelial cell stress in amiodarone-induced lung fibrosis. Toxicol Sci 142(1):285–297
213. Guo W et al (2009) Abrogation of TGF-beta1-induced fibroblast-myofibroblast differentiation by histone deacetylase inhibition. Am J Physiol Lung Cell Mol Physiol 297(5):L864–L870
214. Huang SK et al (2013) Histone modifications are responsible for decreased Fas expression and apoptosis resistance in fibrotic lung fibroblasts. Cell Death Dis 4:e621
215. Bonner JC et al (2002) Susceptibility of cyclooxygenase-2-deficient mice to pulmonary fibrogenesis. Am J Pathol 161(2):459–470
216. Keerthisingam CB et al (2001) Cyclooxygenase-2 deficiency results in a loss of the anti-proliferative response to transforming growth factor-beta in human fibrotic lung fibroblasts and promotes bleomycin-induced pulmonary fibrosis in mice. Am J Pathol 158(4):1411–1422
217. Coward WR et al (2009) Defective histone acetylation is responsible for the diminished expression of cyclooxygenase 2 in idiopathic pulmonary fibrosis. Mol Cell Biol 29(15):4325–4339
218. Sanders YY et al (2014) Histone deacetylase inhibition promotes fibroblast apoptosis and ameliorates pulmonary fibrosis in mice. Eur Respir J 43(5):1448–1458
219. Williams AE et al (2007) microRNA expression in the aging mouse lung. BMC Genomics 8:172
220. Gross TJ et al (2014) A microRNA processing defect in smokers' macrophages is linked to SUMOylation of the endonuclease DICER. J Biol Chem 289(18):12823–12834
221. Molina-Pinelo S et al (2014) MicroRNA clusters: dysregulation in lung adenocarcinoma and COPD. Eur Respir J 43(6):1740–1749
222. Chatila WM et al (2014) Blunted expression of miR-199a-5p in regulatory T cells of patients with chronic obstructive pulmonary disease compared to unaffected smokers. Clin Exp Immunol 177(1):341–352
223. Soeda S et al (2013) Clinical relevance of plasma miR-106b levels in patients with chronic obstructive pulmonary disease. Int J Mol Med 31(3):533–539
224. Ezzie ME et al (2012) Gene expression networks in COPD: microRNA and mRNA regulation. Thorax 67(2):122–131
225. Hassan F et al (2012) MiR-101 and miR-144 regulate the expression of the CFTR chloride channel in the lung. PLoS One 7(11):e50837
226. Christenson SA et al (2013) miR-638 regulates gene expression networks associated with emphysematous lung destruction. Genome Med 5(12):114
227. Oak SR et al (2011) A micro RNA processing defect in rapidly progressing idiopathic pulmonary fibrosis. PLoS One 6(6):e21253
228. Milosevic J et al (2012) Profibrotic role of miR-154 in pulmonary fibrosis. Am J Respir Cell Mol Biol 47(6):879–887
229. Liu G et al (2010) miR-21 mediates fibrogenic activation of pulmonary fibroblasts and lung fibrosis. J Exp Med 207(8):1589–1597
230. Das S et al (2014) MicroRNA-326 regulates profibrotic functions of transforming growth factor-beta in pulmonary fibrosis. Am J Respir Cell Mol Biol 50(5):882–892
231. Lino Cardenas CL et al (2013) miR-199a-5p Is upregulated during fibrogenic response to tissue injury and mediates TGFbeta-induced lung fibroblast activation by targeting caveolin-1. PLoS Genet 9(2):e1003291
232. Pandit KV, Milosevic J, Kaminski N (2011) MicroRNAs in idiopathic pulmonary fibrosis. Transl Res 157(4):191–199
233. Comer BS et al (2015) Epigenetic targets for novel therapies of lung diseases. Pharmacol Ther 147:91–110
234. Yang S et al (2012) miR-21 regulates chronic hypoxia-induced pulmonary vascular remodeling. Am J Physiol Lung Cell Mol Physiol 302(6):L521–L529

235. Wang Y et al (2013) miR-375 regulates rat alveolar epithelial cell trans-differentiation by inhibiting Wnt/beta-catenin pathway. Nucleic Acids Res 41(6):3833–3844
236. Kneidinger N et al (2011) Activation of the WNT/beta-catenin pathway attenuates experimental emphysema. Am J Respir Crit Care Med 183(6):723–733
237. Hacisuleyman E et al (2014) Topological organization of multichromosomal regions by the long intergenic noncoding RNA Firre. Nat Struct Mol Biol 21(2):198–206
238. Gong C, Maquat LE (2011) "Alu" strious long ncRNAs and their role in shortening mRNA half-lives. Cell Cycle 10(12):1882–1883
239. Natoli G, Andrau JC (2012) Noncoding transcription at enhancers: general principles and functional models. Annu Rev Genet 46:1–19
240. Cao G et al (2013) Differential expression of long non-coding RNAs in bleomycin-induced lung fibrosis. Int J Mol Med 32(2):355–364
241. Deng N et al (2013) Detecting splicing variants in idiopathic pulmonary fibrosis from non-differentially expressed genes. PLoS One 8(7):e68352
242. Pilling LC et al (2012) Genomics and successful aging: grounds for renewed optimism? J Gerontol A Biol Sci Med Sci 67(5):511–519
243. White ES et al (2010) Control of fibroblast fibronectin expression and alternative splicing via the PI3K/Akt/mTOR pathway. Exp Cell Res 316(16):2644–2653
244. Muro AF et al (2008) An essential role for fibronectin extra type III domain A in pulmonary fibrosis. Am J Respir Crit Care Med 177(6):638–645
245. Nance T et al (2014) Transcriptome analysis reveals differential splicing events in IPF lung tissue. PLoS One 9(5):e97550
246. Ito K, Mercado N (2014) STOP accelerating lung aging for the treatment of COPD. Exp Gerontol 59:21–27
247. Bruunsgaard H et al (2003) Elevated levels of tumor necrosis factor alpha and mortality in centenarians. Am J Med 115(4):278–283
248. Bruunsgaard H, Pedersen BK (2003) Age-related inflammatory cytokines and disease. Immunol Allergy Clin North Am 23(1):15–39
249. Gomez CR et al (2008) Innate immunity and aging. Exp Gerontol 43(8):718–728
250. Brandenberger C et al (2014) Enhanced allergic airway disease in old mice is associated with a Th17 response. Clin Exp Allergy 44(10):1282–1292
251. Nyenhuis SM et al (2010) Airway neutrophil inflammatory phenotype in older subjects with asthma. J Allergy Clin Immunol 125(5):1163–1165
252. Wilson JW (2005) Inflammation and remodelling in the ageing airway. Med J Aust 183(1 Suppl):S33–S34
253. Soler Artigas M et al (2011) Genome-wide association and large-scale follow up identifies 16 new loci influencing lung function. Nat Genet 43(11):1082–1090
254. Chung HY et al (2009) Molecular inflammation: underpinnings of aging and age-related diseases. Ageing Res Rev 8(1):18–30
255. Kovacs EJ et al (2009) Aging and innate immunity in the mouse: impact of intrinsic and extrinsic factors. Trends Immunol 30(7):319–324
256. Franceschi C et al (2000) Inflamm-aging. An evolutionary perspective on immunosenescence. Ann N Y Acad Sci 908:244–254
257. Park GY, Christman JW (2006) Involvement of cyclooxygenase-2 and prostaglandins in the molecular pathogenesis of inflammatory lung diseases. Am J Physiol Lung Cell Mol Physiol 290(5):L797–L805
258. Canan CH et al (2014) Characterization of lung inflammation and its impact on macrophage function in aging. J Leukoc Biol 96(3):473–480
259. Ren X et al (2014) Age-related activation of MKK/p38/NF-kappaB signaling pathway in lung: from mouse to human. Exp Gerontol 57:29–40
260. Chilosi M et al (2013) Premature lung aging and cellular senescence in the pathogenesis of idiopathic pulmonary fibrosis and COPD/emphysema. Transl Res 162(3):156–173

261. MacNee W (2009) Accelerated lung aging: a novel pathogenic mechanism of chronic obstructive pulmonary disease (COPD). Biochem Soc Trans 37(Pt 4):819–823
262. Tuder RM, Yun JH, Graham BB (2008) Cigarette smoke triggers code red: p21CIP1/WAF1/SDI1 switches on danger responses in the lung. Am J Respir Cell Mol Biol 39(1):1–6
263. Provinciali M, Cardelli M, Marchegiani F (2011) Inflammation, chronic obstructive pulmonary disease and aging. Curr Opin Pulm Med 17(Suppl 1):S3–S10
264. Weiskopf D, Weinberger B, Grubeck-Loebenstein B (2009) The aging of the immune system. Transpl Int 22(11):1041–1050
265. Vandivier RW, Henson PM, Douglas IS (2006) Burying the dead: the impact of failed apoptotic cell removal (efferocytosis) on chronic inflammatory lung disease. Chest 129(6):1673–1682
266. Pisetsky DS, Fairhurst AM (2007) The origin of extracellular DNA during the clearance of dead and dying cells. Autoimmunity 40(4):281–284
267. Mahoney JA, Rosen A (2005) Apoptosis and autoimmunity. Curr Opin Immunol 17(6):583–588
268. Hogaboam CM, Trujillo G, Martinez FJ (2012) Aberrant innate immune sensing leads to the rapid progression of idiopathic pulmonary fibrosis. Fibrogenesis Tissue Repair 5(Suppl 1 (Proceedings of fibroproliferative disorders: from biochemical analysis to targeted therapies Petro E. Petrides and David Brenner)):S3
269. Trujillo G et al (2010) TLR9 differentiates rapidly from slowly progressing forms of idiopathic pulmonary fibrosis. Sci Transl Med 2(57):57ra82
270. Gilani SR et al (2010) CD28 down-regulation on circulating CD4 T-cells is associated with poor prognoses of patients with idiopathic pulmonary fibrosis. PLoS One 5(1):e8959
271. Sime PJ et al (1997) Adenovector-mediated gene transfer of active transforming growth factor-beta1 induces prolonged severe fibrosis in rat lung. J Clin Invest 100(4):768–776
272. Mosser DM, Edwards JP (2008) Exploring the full spectrum of macrophage activation. Nat Rev Immunol 8(12):958–969
273. Baran CP et al (2007) Important roles for macrophage colony-stimulating factor, CC chemokine ligand 2, and mononuclear phagocytes in the pathogenesis of pulmonary fibrosis. Am J Respir Crit Care Med 176(1):78–89
274. Prasse A et al (2006) A vicious circle of alveolar macrophages and fibroblasts perpetuates pulmonary fibrosis via CCL18. Am J Respir Crit Care Med 173(7):781–792
275. Stahl M et al (2013) Lung collagens perpetuate pulmonary fibrosis via CD204 and M2 macrophage activation. PLoS One 8(11), e81382
276. Murray LA et al (2011) TGF-beta driven lung fibrosis is macrophage dependent and blocked by Serum amyloid P. Int J Biochem Cell Biol 43(1):154–162
277. Kiemle-Kallee J et al (1991) Alveolar macrophages in idiopathic pulmonary fibrosis display a more monocyte-like immunophenotype and an increased release of free oxygen radicals. Eur Respir J 4(4):400–406
278. Prasse A et al (2009) Serum CC-chemokine ligand 18 concentration predicts outcome in idiopathic pulmonary fibrosis. Am J Respir Crit Care Med 179(8):717–723
279. Tiev KP et al (2011) Serum CC chemokine ligand-18 predicts lung disease worsening in systemic sclerosis. Eur Respir J 38(6):1355–1360
280. Xiao L et al (2003) Evaluation of interleukin-13 in the serum and bronchoalveolar lavage fluid of patients with idiopathic pulmonary fibrosis. Zhonghua Jie He He Hu Xi Za Zhi 26(11):686–688
281. Sharma G, Hanania NA, Shim YM (2009) The aging immune system and its relationship to the development of chronic obstructive pulmonary disease. Proc Am Thorac Soc 6(7):573–580
282. Gorska MM et al (2010) Uncoordinated 119 preferentially induces Th2 differentiation and promotes the development of asthma. J Immunol 184(8):4488–4496
283. Zheng T et al (2000) Inducible targeting of IL-13 to the adult lung causes matrix metalloproteinase- and cathepsin-dependent emphysema. J Clin Invest 106(9):1081–1093

284. Wang Z et al (2000) Interferon gamma induction of pulmonary emphysema in the adult murine lung. J Exp Med 192(11):1587–1600
285. Hoshino T et al (2007) Pulmonary inflammation and emphysema: role of the cytokines IL-18 and IL-13. Am J Respir Crit Care Med 176(1):49–62
286. Maeno T et al (2007) CD8+ T cells are required for inflammation and destruction in cigarette smoke-induced emphysema in mice. J Immunol 178(12):8090–8096
287. Lambers C et al (2009) T cell senescence and contraction of T cell repertoire diversity in patients with chronic obstructive pulmonary disease. Clin Exp Immunol 155(3):466–475
288. Grumelli S et al (2004) An immune basis for lung parenchymal destruction in chronic obstructive pulmonary disease and emphysema. PLoS Med 1(1):e8
289. Di Stefano A et al (2009) T helper type 17-related cytokine expression is increased in the bronchial mucosa of stable chronic obstructive pulmonary disease patients. Clin Exp Immunol 157(2):316–324
290. Doe C et al (2010) Expression of the T helper 17-associated cytokines IL-17A and IL-17 F in asthma and COPD. Chest 138(5):1140–1147
291. Weyand CM et al (1998) Functional properties of CD4+ CD28- T cells in the aging immune system. Mech Ageing Dev 102(2–3):131–147
292. Zhang C et al (2014) Hypomethylation of perforin regulatory elements in CD4+ T cells from rat spleens contributes to the development of autoimmune emphysema. Respirology 19(3):376–381
293. Sullivan AK et al (2006) Activated oligoclonal CD4+ T cells in the lungs of patients with severe emphysema. Proc Am Thorac Soc 3(6):486
294. Romero V, Andrade F (2008) Non-apoptotic functions of granzymes. Tissue Antigens 71(5):409–416
295. Ngan DA et al (2009) The possible role of granzyme B in the pathogenesis of chronic obstructive pulmonary disease. Ther Adv Respir Dis 3(3):113–129
296. Alkhouri H et al (2014) Regulation of pulmonary inflammation by mesenchymal cells. Pulm Pharmacol Ther 29(2):156–165
297. Lumsden AB, McLean A, Lamb D (1984) Goblet and Clara cells of human distal airways: evidence for smoking induced changes in their numbers. Thorax 39(11):844–849
298. Araya J et al (2007) Squamous metaplasia amplifies pathologic epithelial-mesenchymal interactions in COPD patients. J Clin Invest 117(11):3551–3562
299. Gosselink JV et al (2010) Differential expression of tissue repair genes in the pathogenesis of chronic obstructive pulmonary disease. Am J Respir Crit Care Med 181(12):1329–1335
300. Zhang J et al (2012) Pro-inflammatory phenotype of COPD fibroblasts not compatible with repair in COPD lung. J Cell Mol Med 16(7):1522–1532
301. Larsson-Callerfelt AK et al (2013) Defective alterations in the collagen network to prostacyclin in COPD lung fibroblasts. Respir Res 14:21
302. Kitamura H et al (2011) Mouse and human lung fibroblasts regulate dendritic cell trafficking, airway inflammation, and fibrosis through integrin alphavbeta8-mediated activation of TGF-beta. J Clin Invest 121(7):2863–2875
303. Lindahl GE et al (2013) Microarray profiling reveals suppressed interferon stimulated gene program in fibroblasts from scleroderma-associated interstitial lung disease. Respir Res 14:80
304. Nagahama KY et al (2013) Oncostatin M modulates fibroblast function via signal transducers and activators of transcription proteins-3. Am J Respir Cell Mol Biol 49(4):582–591
305. Botelho FM et al (2013) Pulmonary expression of oncostatin M (OSM) promotes inducible BALT formation independently of IL-6, despite a role for IL-6 in OSM-driven pulmonary inflammation. J Immunol 191(3):1453–1464
306. van der Kraan PM et al (2012) Age-dependent alteration of TGF-beta signalling in osteoarthritis. Cell Tissue Res 347(1):257–265
307. Doyle KP et al (2010) TGFbeta signaling in the brain increases with aging and signals to astrocytes and innate immune cells in the weeks after stroke. J Neuroinflammation 7:62

308. Biernacka A, Frangogiannis NG (2011) Aging and cardiac fibrosis. Aging Dis 2(2):158–173
309. Morty RE, Konigshoff M, Eickelberg O (2009) Transforming growth factor-beta signaling across ages: from distorted lung development to chronic obstructive pulmonary disease. Proc Am Thorac Soc 6(7):607–613
310. Bonniaud P et al (2004) Smad3 null mice develop airspace enlargement and are resistant to TGF-beta-mediated pulmonary fibrosis. J Immunol 173(3):2099–2108
311. Lopez-Otin C et al (2013) The hallmarks of aging. Cell 153(6):1194–1217
312. Heintz C, Mair W (2014) You are what you host: microbiome modulation of the aging process. Cell 156(3):408–411
313. Tsujino K et al (2012) Tetraspanin CD151 protects against pulmonary fibrosis by maintaining epithelial integrity. Am J Respir Crit Care Med 186(2):170–180
314. Zhao ZJ, Liu XM, Zhou GP (2008) Changes of matrix metalloproteinase 2,9 and tissue inhibitor of metalloproteinase 1, 2, 3 expression level in aged rat lung. Beijing Da Xue Xue Bao 40(1):101–104
315. Wagner DE et al (2014) Comparative decellularization and recellularization of normal versus emphysematous human lungs. Biomaterials 35(10):3281–3297
316. Booth AJ et al (2012) Acellular normal and fibrotic human lung matrices as a culture system for in vitro investigation. Am J Respir Crit Care Med 186(9):866–876
317. McAloon CJ et al (2009) Matrix metalloprotease polymorphisms are associated with gas transfer in alpha 1 antitrypsin deficiency. Ther Adv Respir Dis 3(1):23–30
318. Elkington PT, Friedland JS (2006) Matrix metalloproteinases in destructive pulmonary pathology. Thorax 61(3):259–266
319. Dancer RC, Wood AM, Thickett DR (2011) Metalloproteinases in idiopathic pulmonary fibrosis. Eur Respir J 38(6):1461–1467
320. Suga M et al (2000) Characteristic elevation of matrix metalloproteinase activity in idiopathic interstitial pneumonias. Am J Respir Crit Care Med 162(5):1949–1956
321. Gooptu B, Ekeowa UI, Lomas DA (2009) Mechanisms of emphysema in alpha1-antitrypsin deficiency: molecular and cellular insights. Eur Respir J 34(2):475–488
322. Pardo A, Selman M (2012) Role of matrix metaloproteases in idiopathic pulmonary fibrosis. Fibrogenesis Tissue Repair 5(Suppl 1 (Proceedings of fibroproliferative disorders: from biochemical analysis to targeted therapies Petro E Petrides and David Brenner)):S9
323. Vegeto E et al (2010) Estrogen receptor-alpha as a drug target candidate for preventing lung inflammation. Endocrinology 151(1):174–184
324. Corbo GM et al (2014) Serum level of testosterone, dihydrotestosterone and IGF-1 during an acute exacerbation of COPD and their relationships with inflammatory and prognostic indices: a pilot study. Minerva Med 105(4):289–294
325. Debigare R et al (2003) Catabolic/anabolic balance and muscle wasting in patients with COPD. Chest 124(1):83–89
326. Balasubramanian V, Naing S (2012) Hypogonadism in chronic obstructive pulmonary disease: incidence and effects. Curr Opin Pulm Med 18(2):112–117
327. Pan L et al (2014) Effects of anabolic steroids on chronic obstructive pulmonary disease: a meta-analysis of randomised controlled trials. PLoS One 9(1):e84855
328. Podzolkov VI et al (2012) Comparative characteristic of the hormonal profile in men with stable obstructive pulmonary disease and smokers. Klin Med (Mosk) 90(6):26–30
329. Mendoza-Milla C et al (2013) Dehydroepiandrosterone has strong antifibrotic effects and is decreased in idiopathic pulmonary fibrosis. Eur Respir J 42(5):1309–1321
330. Gumral N et al (2009) Antioxidant enzymes and melatonin levels in patients with bronchial asthma and chronic obstructive pulmonary disease during stable and exacerbation periods. Cell Biochem Funct 27(5):276–283
331. Unlu M et al (2006) Effects of melatonin on the oxidant/antioxidant status and lung histopathology in rabbits exposed to cigarette smoke. Respirology 11(4):422–428

332. Zhao H et al (2014) Melatonin inhibits endoplasmic reticulum stress and epithelial-mesenchymal transition during bleomycin-induced pulmonary fibrosis in mice. PLoS One 9(5):e97266
333. Kajstura J et al (2011) Evidence for human lung stem cells. N Engl J Med 364(19):1795–1806
334. Gong X et al (2014) Isolation and characterization of lung resident mesenchymal stem cells capable of differentiating into alveolar epithelial type II cells. Cell Biol Int 38(4):405–411
335. Lama VN et al (2007) Evidence for tissue-resident mesenchymal stem cells in human adult lung from studies of transplanted allografts. J Clin Invest 117(4):989–996
336. Majka SM et al (2005) Identification of novel resident pulmonary stem cells: form and function of the lung side population. Stem Cells 23(8):1073–1081
337. Butler JP et al (2012) Evidence for adult lung growth in humans. N Engl J Med 367(3):244–247
338. Blyszczuk P et al (2011) Profibrotic potential of prominin-1+ epithelial progenitor cells in pulmonary fibrosis. Respir Res 12:126
339. Murphy S et al (2011) Human amnion epithelial cells prevent bleomycin-induced lung injury and preserve lung function. Cell Transplant 20(6):909–923
340. Zhen G et al (2008) Mesenchymal stem cells transplantation protects against rat pulmonary emphysema. Front Biosci 13:3415–3422
341. Ortiz LA et al (2003) Mesenchymal stem cell engraftment in lung is enhanced in response to bleomycin exposure and ameliorates its fibrotic effects. Proc Natl Acad Sci U S A 100(14):8407–8411
342. Guan XJ et al (2013) Mesenchymal stem cells protect cigarette smoke-damaged lung and pulmonary function partly via VEGF-VEGF receptors. J Cell Biochem 114(2):323–335
343. McQualter JL et al (2014) Harnessing the potential of lung stem cells for regenerative medicine. Int J Biochem Cell Biol 56:82–91
344. Ruiz EJ, Oeztuerk-Winder F, Ventura JJ (2014) A paracrine network regulates the cross-talk between human lung stem cells and the stroma. Nat Commun 5:3175
345. Bonner JC (2010) Mesenchymal cell survival in airway and interstitial pulmonary fibrosis. Fibrogenesis Tissue Repair 3:15
346. Pierro M, Thebaud B (2010) Mesenchymal stem cells in chronic lung disease: culprit or savior? Am J Physiol Lung Cell Mol Physiol 298(6):L732–L734
347. Yan X et al (2007) Injured microenvironment directly guides the differentiation of engrafted Flk-1(+) mesenchymal stem cell in lung. Exp Hematol 35(9):1466–1475
348. Walker N et al (2011) Resident tissue-specific mesenchymal progenitor cells contribute to fibrogenesis in human lung allografts. Am J Pathol 178(6):2461–2469
349. Todd JL, Palmer SM (2011) Bronchiolitis obliterans syndrome: the final frontier for lung transplantation. Chest 140(2):502–508

Age-Related Macular Degeneration and Vision Impairment

Charles Wright and Jayakrishna Ambati

Contents

C. Wright • J. Ambati (✉)
Department of Ophthalmology and Visual Sciences, University of KentucKY
740 S. Limestone Street, Lexington, KY 40536-0284, USA
e-mail: wright.charles@uky.edu; jayakrishna.ambati@uky.edu

F. Sierra, R. Kohanski (eds.), *Advances in Geroscience*,
DOI 10.1007/978-3-319-23246-1_16

1 Introduction

The eye is the organ that allows for vision, the ability to see the world. For a person to be able to see, light must enter through the transparent cornea in the front of the eye, be focused by the lens, and detected by the light-sensitive retina in the back of the interior of the eye. Photoreceptors, cells in the outer retina that contain the light-sensitive opsin proteins, transmit the visual signal through the inner retina, where the signal is amplified, parsed, processed, and finally sent through the optic nerve to the brain for higher visual processing and image interpretation. In humans, photoreceptors can be divided into two primary cell types: rods and cones. Rod photoreceptors, which greatly outnumber cone photoreceptors, are incredibly sensitive to light and are primarily responsible for vision in dim light conditions. Cone photoreceptors, on the other hand, operate primarily under bright light conditions and are responsible for providing color-rich and detailed vision. Rods and cones are not evenly distributed throughout the eye; rods are found throughout the entire eye, with the exception of the very center of the retina. In humans, cone photoreceptors are predominantly found in a structure called the macula, which is located in the center of the retina and where light is most focused from the lens. For healthy vision, each component of this pathway must work in concert, and degeneration or injury to any one of these anatomical structures can lead to visual impairment or blindness. The retinal pigmented epithelium (RPE) is a cell monolayer underlying the retina that is responsible for several important functions, one of the most important being the maintenance of photoreceptor health. As individuals age, they can sometimes experience a progressive loss of RPE cells in the macula, and this loss of RPE cells compromises the overlying photoreceptors and can result in severe visual impairment or blindness. This age-associated degeneration of cells in the macula is known as age-related macular degeneration (AMD).

2 Clinical Aspects

2.1 *Clinical Presentation*

The initial form of AMD (known as early AMD) is marked by the presence of yellowish or whitish punctate extracellular deposits (drusen) that are found between the RPE and the underlying basement membrane known as Bruch's membrane, and that can be seen during a fundus examination. The presence of drusen alone does not necessarily indicate AMD, other diseases can also present with drusen distinct from those noted in AMD patients [1–3]. For drusen to be considered pathognomic of AMD, they must be >63 μm in diameter and be "soft" in nature (i.e., have poorly-defined edges and cluster together in the fundus) [4]. By commonly-accepted clinical diagnostic criteria [4], individuals who present with these drusen (along with Bruch's membrane thickening) without associated pigmentary disturbances are classified as having early AMD. At these early stages of the disease, patients commonly have no reported visual deficiencies. Focal hypo- and hyperpigmentation of the RPE are also found in the fundus with drusen [5] in patients with intermediate AMD [4]. Typically, visual deficits are not noted until more advanced stages of AMD, which can be divided into two separate categories: (1) the non-exudative (or "dry") form of the disease known as geographic atrophy (GA) that presents with large regions of pigmentary abnormalities in the fundus and cell death in the RPE, without blood vessel infiltration, and (2) the exudative (or "wet") form of the disease known as neovascular AMD (nvAMD) that presents with infiltration of blood vessels into the retina that have a high likelihood of leakage [4]. The severity of AMD can be graded on a classification system [6] developed by the Age-Related Eye Disease Study (AREDS) group, which is based on the previously published Wisconsin age-related maculopathy grading system [7]. The advent of spectral domain optical coherence tomography (SD-OCT) imaging has further aided in the monitoring of AMD [8–10].

2.2 *Visual Impairment*

Unlike the earlier stages of AMD, in which visual impairment is not detectable or not noticeable by the patient, later stages of AMD are associated with profound visual deficits. Progression from early AMD to one or both of the advanced forms of AMD may take 10 years or more. For patients with AMD, the amount of visual acuity loss correlates with the percentage of the fovea (a particularly cone photoreceptor-dense location in the macula) affected [11]. Once patients develop GA, individuals with <10 % foveal involvement and 20/20 visual acuity (by Snellen chart) may take as little as 5 years to have >80 % foveal involvement, at which point visual acuity may be worse than 20/200 [11]. Likewise, patients with untreated nvAMD have a poor prognosis with respect to visual function: 3 years after diagnosis with nvAMD, >75 % of patients who receive no treatment will have 20/200 or worse [12]. Despite the devastating visual outcomes for many late-stage AMD patients, it should be noted that patients with untreated nvAMD typically have a worse visual prognosis than their GA counterparts, so although nvAMD patients account for only 10–15 % of all AMD cases [13], 75 % of all blindness resulting from AMD can be attributed to the neovascular form of the disease [14].

2.3 *Known Risk Factors*

Although the exact etiology of AMD is not known, many studies have been performed to identify factors that correlate with increased incidence of the disease in the population. The greatest risk factor for AMD is age [15]; individuals 65 years or older are more likely to be diagnosed with AMD, and individuals 85 years or older are more likely to have advanced forms of the disease [16]. Race also appears to be a risk factor. Individuals who identify as non-Hispanic Caucasian are more likely to develop AMD than non-Hispanic African-American or individuals of Hispanic descent [17]. Also, individuals who have a family history of AMD are more likely to develop the disease [18]. In addition to these risk factors (i.e., age, race, and family history), other environmental risk factors have been shown to increase risk for AMD development and progression. After age, smoking is widely considered to be the next strongest risk factor for developing the disease. Individuals who have smoked at one point or who currently smoke are more likely to develop the disease and to have advanced forms of the disease as compared to individuals who never smoked [16]. Finally, dietary fat intake and obesity are also known to associate with AMD, and some reports indicate that individuals with healthy diets and who regularly exercise may experience lower incidence of the disease [18].

2.4 Prevalence

Because age is the risk factor that is reported to best correlate with AMD development and progression, it follows that AMD prevalence increases in older populations. Several studies have been done to estimate the prevalence of AMD in the general population, such as the Beaver Dam Eye Study [19], Rotterdam Eye Study [20], and Blue Mountains Eye Study [21]. In individuals 18 years of age or older, AMD patients only comprise approximately 1 % of the population, but in adults 65–74 years old, the prevalence of AMD approaches 9 % [22]. Of the patients afflicted with more advanced forms of AMD, nvAMD appears to be more common than GA [16]. Depending on the age group and population being surveyed, nvAMD can be twice as common as GA [16]. Because a relatively large percentage of the aging population suffers from AMD as compared to other age-related diseases, millions of people worldwide are affected. The number of affected individuals is expected to rise as the aging also comprise the fastest growing segment of the general population.

2.5 Current Treatment Options—Dry AMD

Currently, the only treatment for nonexudative AMD is the AREDS vitamin supplement formulation. Based on reports published beginning in the late 1980s [23–26], it was hypothesized that dietary supplementation with a mixture of antioxidants, vitamins, and zinc could aid in the prevention or slowing of AMD progression. A large-scale clinical trial was conducted at the turn of the century with the goal of determining whether AMD patients who took a formulation of high-dose vitamin C and E, beta carotene, and zinc would benefit from a slowing of AMD progression and visual acuity loss [27]; this study became known as the AREDS1 study [28].

The AREDS group concluded that this combined vitamin supplement (i.e., the AREDS formulation) slowed the progression of dry AMD in patients. Later, another large-scale clinical trial was performed in the AREDS2 study, which added lutein, zeaxanthin, and omega-3 fatty acids to the AREDS1 formulation to test whether these additional supplements could further slow visual acuity loss and AMD progression [29]. It was found that the AREDS2 supplement formulation did not significantly improve upon the efficacy of the AREDS1 formulation [29]. Furthermore, the AREDS2 formulation did not significantly improve upon the AREDS1 formulation in patients harboring mutations reported to confer AMD risk when patients that did not harbor these mutations were excluded from statistical analysis [30].

This is not to say that the AREDS1 formulation could be considered a success in preventing AMD progression and vision loss, however. Almost immediately after the findings of the AREDS1 formulation were published [27], the AREDS1 study's conclusion was challenged. Careful examination of the AREDS1 dataset revealed that there was no statistically significant effect of the AREDS1 formulation on visual acuity, and even whether there was any effect on the progression of early and intermediate AMD to late-stage dry AMD was questioned [31]. Other independent large-scale studies found that patients taking AREDS1 formulation were at higher risk for developing choroidal neovascularization (CATT & CAPT) [32, 33]. Meta-analysis of other small-scale studies also indicate no significant therapeutic benefit from the AREDS vitamin formulation [34]. A 10-year follow-up of patients included in the original AREDS1 study found that the progressive visual loss of AMD patients continued unabated [35], casting further doubt on the efficacy of the AREDS vitamin formulation in treating the disease. In light of these results, many AMD physicians and researchers now believe it should not be recommended that AMD patients take the AREDS supplement [34, 36]. In addition to being of questionable efficacy, it was argued that the AREDS vitamin supplement could also carry the risk of being harmful to AMD patients' health, because of the association of some of these vitamins with exacerbated lung cancer progression [37, 38]. The fact that smoking is a risk factor for AMD patients [16] has led many physicians to feel that not only are the AREDS vitamins ineffective at treating dry AMD, they may also predispose patients to a greater risk of potentially life-threatening off-target effects. In summary, there is no safe and effective treatment approved for the treatment of dry AMD.

2.6 Current Treatment Options—Wet AMD

Unlike dry AMD, there has been considerably more success with respect to treatment development for nvAMD. In decades past, destructive treatments such as thermal laser photocoagulation and photodynamic therapy with verteporfin were employed for nvAMD [39–41]. Over the past decade, the U.S. Federal Drug Administration (FDA) and regulatory bodies in other countries have approved several molecular therapies targeting vascular endothelial growth factor A (VEGF-A)

[42], a potent pro-angiogenic factor that has been implicated in the formation of the unstable, leaky blood vessels that invade the retina in nvAMD [43]. There are currently four anti-VEGF-A treatments used to treat nvAMD patients in the United States: bevacizumab (Avastin) [44], ranibizumab (Lucentis) [44], pegaptanib (Macugen) [44], and aflibercept (Eylea) [42]. Bevacizumab and ranibizumab are both derived from antibodies raised against human VEGF-A; bevacizumab, originally approved as a therapy for colorectal cancer, is the full-length anti-VEGF antibody that has been used in an "off-label" fashion to treat nvAMD, while ranibizumab was specifically approved for use in nvAMD patients and only consists of the Fab portion (the region that recognizes the human VEGF-A epitope) of the antibody [45]. Pegaptanib sodium is a pegylated aptamer (i.e., a nucleic acid designed to bind to a specific biological target) that specifically binds VEGF-A_{165}, one of the many isoforms of VEGF-A [44]. Aflibercept is a fusion protein that consists of the VEGF binding region found on VEGF receptors-1 and –2 fused to the Fc portion of human IgG1 immunoglobulin [42]. These drugs are injected into the vitreous humor of nvAMD patients on a monthly basis or as needed [45]. Extensive off-label use of bevacizumab instead of ranibizumab because of lower cost prompted the National Institutes of Health (NIH) to fund the Comparison of Age-Related Macular Degeneration Treatment Trials (CATT) Study to test whether bevacizumab was comparable to ranibizumab at preserving or improving visual acuity in nvAMD patients [45]. CATT, and subsequently similar studies in other countries, reported that bevacizumab and ranibizumab have clinically similar effects on visual acuity in nvAMD patients [45]. Although these studies indicate bevacizumab has safety and efficacy profiles similar to ranibizumab, bevacizumab is not currently approved for the treatment of nvAMD. Despite the success of treating a subset of nvAMD patients with these anti-VEGF-A therapies, many nvAMD patients do not experience significant reductions in blood vessel penetration or visual improvement [46]. Treatment options for anti-VEGF-resistant nvAMD are currently limited. For patients that do respond well to treatment, repeated use of these drugs is associated with the development of GA [47–50], which is currently untreatable.

3 Cell Types Implicated in AMD Pathology

3.1 Retinal Pigmented Epithelium (RPE)

Both late stage forms of AMD, GA and nvAMD, are ultimately diseases of RPE dysfunction and degeneration. The RPE serves multiple roles in the maintenance of healthy vision, including maintenance of photoreceptor cell health (e.g., phagocytosis of shed rod and cone outer segments [51], transport of nutrients from the choriocapillaris to the metabolically-demanding photoreceptors [52]) and the formation of the outer blood-retinal barrier separating the intraocular tissues from the systemic blood circulation that aids in maintaining the eye as an "immune privileged" site [53]. In the case of GA, the advanced form of dry AMD is marked by large, confluent regions of RPE atrophy [13]; this suggests that one of the primary RPE functions, maintenance of photoreceptor cell health, is compromised. Likewise, another primary function of the RPE layer, formation of the blood-retinal barrier, is compromised in nvAMD [50]. Because patients can simultaneously present with both GA and nvAMD [16], the loss of both of these critical functions should not be considered to be mutually exclusive from one another. Compromised RPE cell function ultimately associates with the deterioration of other tissues on either side of the RPE, such as the retina and choriocapillaris [13].

3.2 Immune Cell Involvement

To date, there has been little evidence in the literature suggesting immune cell (e.g., macrophage) infiltration in individuals with early or intermediate AMD or in GA patients [50, 54]. For individuals with nvAMD, though, immune cell infiltration is particularly important with respect to angiogenesis. Of particular interest are macrophages, which have been implicated in multiple studies of nvAMD [55–59]. Macrophages, involved in the clearance of cellular detritus and foreign matter, have been proposed to play a role in AMD pathogenesis for quite some time—but whether their presence is pro- or anti-angiogenic is still a matter for debate. Evidence for both pro- and anti-angiogenic roles of macrophages abound in the literature. Macrophage depletion by genetic ablation of the chemoattractant *Ccl2* or *Ccr2*, necessary for macrophage recruitment to the retina, has been shown to promote angiogenesis in mouse models [58]. On the other hand, genetic ablation of the cytokine IL-10 has also been shown to decrease macrophage activity and this inhibits angiogenesis instead [60]. Other data indicate that macrophages may also have a potent pro-angiogenic role in nvAMD, as shown in numerous studies of human neovascular membranes and in mouse choroidal neovascularization (CNV) models [55–57, 59]. Because of the potential dual-role of macrophages in nvAMD, many researchers hypothesize that macrophage polarization may be important in modulating angiogenesis in nvAMD. Macrophages can interchangeably adopt either one of two polarization states, M1 or M2, which determine their activity in tissue. M1 macrophages are typically understood to assume pro-inflammatory roles in tissue, while M2 macrophages are involved in wound repair activities [50]. With respect to AMD, pro-inflammatory M1 macrophages are thought to be anti-angiogenic, and M2 macrophages are thought to be pro-angiogenic. Direct evidence for macrophage polarization modulating angiogenesis in nvAMD is sparse at the moment. There is some evidence to suggest that IL-10, which is known to be involved in switching macrophages from M1 to M2 polarization, may be upregulated in mice with laser-induced CNV [61]. Future studies will hopefully resolve current questions concerning macrophage polarization with respect to nvAMD.

The retina also contains resident macrophages called microglia that are involved in spontaneous CNV formation in mice. Genetic ablation of the macrophage-associated *CX3C chemokine receptor 1* (*CX3CR1*) results in accumulation of retinal microglia in the subretinal space, and these mice develop spontaneous CNV [62]. *Ccl2*- and *Ccr2*-deficient mice also develop spontaneous retinal lesions in addition to spontaneous CNV formation [63]. Despite the associations of these genes with microglia involvement in spontaneous CNV formation in mice, there is still no clear evidence to date of microglia accumulation in the subretinal space of human neovascular membranes [64]. At this time, it is unclear what role microglia play in nvAMD progression, but future work may be able to explore microglia in further detail.

Work in animal models of nvAMD implicated neutrophil involvement in CNV formation, but there is also compelling evidence to indicate that neutrophil infiltration of CNVs may be the result of experimental artifact. Studies in mouse nvAMD models have reported that neutrophils may be involved in disruption of Bruch's membrane integrity following laser-induced CNV [65] and that neutrophils may directly promote CNV in laser-induced mouse models of nvAMD [66]. Other studies, on the other hand, indicate that neutrophils may only be peripherally involved in CNV formation and that even if neutrophils are associated with CNV formation in the mouse, this finding is not relevant with respect to human nvAMD cases. For example, laser-induced CNV in mice with impaired macrophage infiltration exhibit minimal neovascularization [67]; this suggests neutrophil involvement in CNV formation is minimal. Furthermore, histological analysis of human neovascular tissue indicated minimal neutrophil intrusion into the retina [68] as compared to mouse laser-induced CNV tissue. Because of the known role neutrophils play with respect to injury response, it is possible that the reported neutrophil involvement in nvAMD mouse models is a result of the injury incurred by laser rather than angiogenesis.

4 Aging Processes Contributing to AMD

4.1 Inflammation

Evidence now indicates that AMD has an important inflammatory component to its pathology, as do many aging-related diseases. The RPE expresses the components of many different pro-inflammatory pathways, including Toll-like receptors (TLRs), inflammasome-associated proteins, complement cascade proteins, and other pro-inflammatory cytokines. These pro-inflammatory components are currently an active topic of research, and although the relative contributions of each pro-inflammatory pathway on AMD progression in patients is still unclear, many lines of evidence indicate that age-related inflammatory dysregulation is involved in disease development and progression. These pro-inflammatory mediators are thus attractive targets for future therapeutic approaches. TLR, NLRP3, and complement cascade pathways are currently targets of active investigation.

TLRs are a class of transmembrane receptor proteins that are a primary component of the innate immune system. These TLR proteins are classified as pattern recognition receptors (PRRs), where each TLR is capable of recognizing a type of pathogen-associated molecular pattern (PAMP) and inducing a signal transduction pathway to initiate an immune response [69]. TLR signal transduction pathways have been well-characterized and have also been implicated in retinal degeneration. One such TLR, TLR3, is best known for its role in recognizing double-stranded RNA (dsRNA) [70]. Unlike other TLR proteins, which rely upon Myeloid differentiation primary response gene 88 (MyD88)-mediated intracellular signaling, TLR3 relies upon TIR-domain-containing adapter-inducing interferon-ß (TRIF) for downstream intracellular signaling [69]. dsRNA-induced TLR3 signal transduction ultimately results in activation of the transcription factor Interferon response factor 3 (IRF3) for expression of type I interferon (IFN) and other pro-inflammatory cytokines [69].

In addition to TLR signaling, recent evidence of the involvement of the NOD-like receptor family, pyrin containing domain protein 3 (NLRP3; also known as the NACHT, LRR, and PYD domains-containing protein 3, or NALP3) inflammasome gives further credence to the role of age-related inflammation in AMD progression. Some TLRs, such as TLR4, which senses extracellular lipopolysaccharide (LPS) found on Gram-negative bacteria, lead to downstream inflammasome activity. The NLRP3 inflammasome, a protein complex that contains NLRP3 protein, Apoptosis-associated speck-like protein containing a CARD (ASC; encoded by the *PYCARD* gene), and cleaved Caspase-1, acts as a platform for interleukin (IL)-18 and IL-1ß cytokine maturation [71]. These cytokines can then be secreted; extracellular IL-18 is known to then bind and activate IL-18 receptor (IL-18R) to induce either pyroptotic or apoptotic cell death pathways [72]. Inflammasome activity is the result of two distinct phases: the first step is priming, in which inflammasome-associated gene products (e.g., NLRP3, Pro-Caspase-1, Pro-IL-18, etc.) are upregulated, and the second step is activation, in which the inflammasome protein complex assembles and begins producing mature, cleaved proteins capable of producing a biological response (e.g., cleaved Caspase-1, cleaved IL-18, etc.) [71]. Despite the potential overlap between TLR signal transduction pathways and the NLRP3 inflammasome, recent evidence suggests the RPE degeneration can proceed via the NLRP3 inflammasome independently of TLR receptors. Also, unlike TLR signal transduction, which typically requires a specific ligand to initiate downstream inflammasome or interferon secretion pathways, NLRP3 inflammasome activators are varied. To date, mitochondrial toxins, reactive oxygen species (ROS), various organic molecules (e.g., nigericin, LPS, etc.), and ions (e.g., potassium) have all been demonstrated to induce NLRP3 inflammasome activation [73].

The complement cascade is a central component of the innate immune response in humans; although the complement cascade can be initiated by multiple different pathogenic insults, the end result is the creation of the membrane attack complex (MAC) which is used to lyse invading pathogens [74]. The MAC complex can be assembled via activation of either the classical, alternative, or lectin pathways; all pathways converge upon C3 activation for assembly of the C5b-9 protein complex (i.e., MAC) [15]. The classical pathway requires antibody binding to antigen for activation by the C1 protein complex [15], and this same set of complement proteins (C2, C4, etc.) is activated through recognition of mannose residues on pathogen surfaces by Mannose-binding lectin-associated serine proteases (MASP) 1 and MASP2 [75]. Unlike the classical and lectin pathways, the alternative pathway makes use of a different set of complement activating proteins (e.g., CFB, CFD) for C3 activation [15].

4.1.1 TLR Signaling

TLR3 signal transduction triggered by dsRNA has been shown to be activated in a sequence-independent manner by dsRNAs that are 21-nt in length or longer [76]. dsRNA is found in large quantities in the drusen of patients with GA, and consistent with this observation, RPE degeneration occurs in mice following subretinal injection of 21-nt [77]. Furthermore, TLR3 SNPs have been associated with a protective effect against GA in human patients [78], suggesting that TLR3 signaling may play a role in AMD progression, although these findings are not universal among studies. Paradoxically, TLR3 signaling that exacerbates one form of advanced AMD (GA) may also inhibit angiogenesis in nvAMD, the other form of advanced AMD. TLR3 signaling has been shown to inhibit both angiogenic and lymphangiogenic events [79, 80], but despite the potential utility of the TLR3 signaling pathway with respect to angiogenic inhibition, the parallel apoptotic events associated with TLR3 signal transduction have largely precluded this pathway as a viable therapeutic target. At least one TLR3 polymorphism may be protective in patients [81], but its veracity is still controversial; this is discussed later in this chapter. In spite of unclear genetic data, however, multiple labs have independently shown that TLR3 activation is capable of inducing RPE cell death [82–84].

4.1.2 Inflammasome

The NLRP3 inflammasome has been implicated in AMD under the effect of various stressors. Carboxyethylpyrrole (CEP)-adducted proteins, the result of free radical-catalyzed oxidation of docosahexaenoate (DHA) that causes damaging covalent modifications to proteins [85], have been shown to prime the NLRP3 inflammasome [86]; these CEP adducts have also been shown to accumulate with age in the retina [85]. In addition to CEP protein adducts, certain drusen components are known to be capable of inducing NLRP3 inflammasome activation. Unfortunately, aside from the fact that drusen correlates with AMD, it is not known whether drusen are directly capable of inducing pro-inflammatory events. The complement cascade intermediate C1q has been shown to activate the NLRP3 inflammasome [86]; other complement cascade proteins have also been implicated in AMD and will be discussed in further detail below. Recently, it has also been shown that DICER1, a protein normally associated with microRNA (miRNA) biogenesis, may also play a role in GA via NLRP3 inflammasome activation because of its ability to degrade cytotoxic *Alu* RNA transcripts [77]. *Alu* elements are retrotransposons that are found interspersed throughout the genome (over 1 million copies are present) [87] and were until recently thought to be mostly transcriptionally inactive. GA patients have reduced DICER1 levels in the retina and RPE, and the resulting accumulation of cytotoxic *Alu* RNA transcripts results in inflammasome activation in an NF-κB-dependent [88], TLR-independent manner [89]. Mature IL-18 production by *Alu* RNA-mediated insult then causes MyD88- and Caspase-8-dependent RPE apoptosis [90]. Interestingly, elevated circulating levels of IL-18 have been reported in patients with dry AMD [91, 92]. Most likely, given the diverse range of substances capable of inducing the NLRP3 inflammasome, more inflammasome agonists will be found to associate with AMD progression.

4.1.3 Complement Pathway Activation

In addition to the potential involvement of C1q in inflammasome activity, the complement cascade may play a role in AMD. Of the three complement activation pathways (classical, alternative, and lectin), the alternative pathway may be associated with AMD progression. Recognition of the complement system as playing a potential role of AMD came in 1992 with the finding that subretinal membranes surgically removed from AMD patients contained complement proteins C1q, C3c, and C3d [93]. It was also found later that C5 and C9 complement components could be found in both hard and soft drusen in AMD patients [94] and that immunohistological sections of AMD patient samples were positively labeled for C3, C5, and C5b-9 [95]. The potential role of the complement system, particularly the alternative pathway, was further highlighted by a genome-wide association study (GWAS) that found roughly 50 % of the heritability for AMD could be associated with a polymorphism in the complement factor H (CFH) gene (Y402H); individuals with the CFH^{Y402H} variant were found to be significantly more likely to develop AMD [96–98]. In the alternative complement activation pathway, CFH plays an inhibitory role by regulating C3 convertase (C3bBb) activity [15]. Putative pathogenic CFH variants have been found to have reduced binding affinity for proteins found in the retina [99–101], including C reactive protein (CRP), a protein that has increased expression in response to inflammation and that is known to activate the complement system via C1q [102]. Other polymorphisms in CFH and other complement proteins will be discussed later in the chapter.

Despite reports of complement proteins appearing in drusen and statistical associations of complement gene polymorphisms with AMD in human patients, a precise mechanism for complement-mediated cell stress and cell death remains elusive. For example, C1q is capable of inducing NLRP3 inflammasome activation in cell culture, but the precise mechanism for this phenomenon remains unknown [86]. Furthermore, although complement proteins that appear in drusen may be capable of inducing an inflammasome response and causing cell death, these drusen components are insoluble; after all, drusen are deposits of insoluble cell debris. Because of this, it is still not entirely clear how insoluble materials could be capable of producing a cell response in vivo. Experiments reporting the cytotoxicity of these complement proteins are performed with soluble complement proteins and depletion of endogenous negative complement regulators; they do not recapitulate the conditions that are present within the aging eye.

In addition to uncertainties with respect to mechanism from cell culture data, mouse models have been unable to reproduce AMD pathology when the *Cfh* gene is ablated. Despite a strong statistical association between CFH polymorphisms and AMD development in human patients, *Cfh* null mice exhibit an extremely weak phenotype. *Cfh* mutant mice exhibit no appreciable photoreceptor degeneration, even at 2 years of age [103]. Also, AMD patients have regions of RPE and retinal atrophy; there seems to be very little effect on the RPE of *Cfh* mutant mice despite the fact that they accumulate more C3 in the eye [103]; this suggests that although CFH may inhibit C3 deposition, C3 itself may not be sufficient to induce AMD phenotypes. This finding has been reproduced in humans; despite the fact that C3 deposition occurs in the eye, many humans with C3 or C5b-9 deposition never develop AMD [104, 105]. Perhaps most concerning is the fact that none of the clinical trials for inhibition of complement factors have yet met with any success in human patients. This raises concerns about the utility of the *Cfh* mutant mouse in developing treatments given the fact that the disease phenotype in these mice is very weak and therefore does not seem to accurately recapitulate the human condition. Many of these approaches have passed Phase I clinical trials without any appreciable safety concerns, but they have not shown any functional benefit with respect to inhibiting disease progression and visual function loss in patients [50]. Thus, although the *CFH* gene has been associated with AMD susceptibility in multiple studies, these genetic findings have not as of yet translated into a comprehensive molecular pathological mechanism. Much more work remains to be done to elucidate the role of the complement cascade in AMD beyond the current genetic associations [106]

4.1.4 Cytokines and Chemotactic Signaling

Increases in the expression of numerous pro-inflammatory cytokines are associated with AMD, further highlighting the importance of inflammation in disease formation and progression. In addition to specific suspected pro-inflammatory pathways involved in AMD progression that have already been discussed in this chapter, there are multiple other reports of cytokines involved in pro-inflammatory events and cytokines and chemotactic signaling factors involved in angiogenic events. Pro-inflammatory cytokines IL-17 and IL-22, for example, are recently identified cytokines that may be upregulated in patients with non-exudative AMD [107, 108]. With respect to nvAMD, multiple anti-inflammatory pro-angiogenic cytokines and chemotactic signaling proteins have been implicated. IL-10 is of particular interest as it has been shown to promote M1 to M2 macrophage polarization [108] and has been reported to be increased in the serum of AMD patients [109]. Other groups report that IL-6 may be a good marker of nvAMD progression [110]. The eotaxins CC-chemokine ligand (CCL) 11, CCL24, CCL26, CXC-chemokine ligand (CXCL) 10 and CC-chemokine receptor (CCR) 3 have been shown to be upregulated prior to blood vessel invasion in the retina [59, 111]. The increases in CCL11 and CCL24 expression do not appear to be restricted to the retina, as other studies have shown their serum levels increase in AMD patients [111, 112]. Future work will hopefully identify some as potential therapeutic targets.

4.2 Angiogenesis

Angiogenesis, the process by which new blood vessels are formed, is necessary for development [113], wound repair [114], and a number of other vital processes in healthy individuals. Given the fundamental role that angiogenesis plays in complex organisms, dysregulation of angiogenic signaling pathways can have far-reaching consequences. Inside the eye, dysregulation of angiogenic signaling events are associated not just with nvAMD but also corneal neovascularization [115] and diabetic retinopathy [116], both of which are vision-threatening. Outside the eye, dysregulation of angiogenic processes can have life-threatening consequences, such as preeclampsia [117], diabetic nephropathy [118], and cancer [119]. Like inflammation, regulation of angiogenesis is impaired with age [120]. In the context of AMD, angiogenesis plays a crucial role in nvAMD progression through the formation of new blood vessels during both CNV and retinal angiomatous proliferation (RAP) [121, 122]. Both CNV and RAP result in the formation of aberrant blood vessels in the retina, but the two terms refer to the differing origin of the blood vessels; CNV refers to blood vessels originating from the choroid that invade the retina, while RAP refers to blood vessels originating from the retina that migrate externally toward the choroid [123]. Fortunately, despite differences in the origin of these aberrant blood vessels, both RAP and CNV may respond favorably to anti-VEGF therapies [121, 122]. In light of the importance that anti-VEGF therapies have with respect to current nvAMD treatments, understanding the VEGF signaling pathway is critical to understanding nvAMD.

VEGF-A is considered the most important regulator of angiogenesis in many tissues [124], and it is expressed in a variety of isoforms [125]; with respect to angiogenesis in nvAMD, VEGF-A_{165} is considered the most biologically relevant [124]. Increased VEGF-A expression has been associated with nvAMD in human tissue samples [126]. VEGF-A is capable of extracellular binding to Vascular endothelial growth factor receptors (VEGFRs; part of a class of tyrosine kinase receptors) to induce intracellular signaling events that ultimately result in endothelial cell migration and blood vessel growth [127]. There are two VEGF-A receptors that have been shown to mediate angiogenesis in the eye: VEGFR1 (also known as fms-like tyrosine kinase 1), which binds VEGF-A with high affinity but has weak tyrosine phosphorylation upon binding, and VEGFR2 (also known as fetal liver kinase 1, or FLK1, and as kinase insert domain receptor, or KDR), which has a weaker affinity for VEGF-A but a stronger tyrosine phosphorylation in response to binding [127]. It is commonly thought that VEGFR2 is predominantly responsible for angiogenic events in a variety of tissues [127], including the retina [124]. Nevertheless, VEGFR1 is also thought to play an important role in regulating angiogenesis through the action of one of its splice variants, soluble VEGFR1 (sVEGFR1; also known as soluble

FLT-1, or sFLT-1). sFLT-1, which has been shown to have reduced expression in both the eye (where it is primarily expressed in the RPE) [128] and in the serum in nvAMD patients [129], acts as a soluble decoy receptor for VEGF-A to prevent its binding to VEGFR2 for angiogenesis [130]. In addition to VEGF-A overexpression in nvAMD patient's eyes, there are other anti-angiogenic factors with decreased expression that have been reported. One such protein is pigment epithelium-derived factor (PEDF), a potent inhibitor of angiogenesis in the eye, which is highly expressed when oxygen is abundant in the microenvironment [131]; in nvAMD, PEDF levels were reported to be decreased [132]. PEDF acts by antagonizing VEGF-A signaling through cleavage of the transmembrane domain of membrane-bound VEGFR1 [133].

Chemokine and chemokine receptors are also implicated in angiogenesis because of the roles that macrophages and microglia have in nvAMD pathology (see Sect. 3.2). CCL2, CCR2, CX3CR1 and CCR3 have all been implicated in nvAMD, and they all regulate macrophage or microglia recruitment to Bruch's membrane [58, 59, 62]. Because macrophages can have either pro- or anti-angiogenic effects with respect to AMD, the role of macrophage involvement appears to be more nuanced than in the case with VEGF-A (where increased VEGF-A expression is more easily correlated with angiogenesis). Future research is required for a more complete picture to be formed for the involvement of macrophages and VEGF-A signaling pathways in nvAMD.

The relative contributions of CNV and RAP to nvAMD are still unclear. It was originally suspected that nvAMD patients presenting with RAP account for only 12–15 % of all cases [121], but more recent work suggests that RAP may be present in as much as 1 out of every 3 nvAMD cases [134]. Given the fact that most clinical trials for anti-VEGF therapies have excluded nvAMD patients presenting with RAP and that three times as many nvAMD patients may present with this form of aberrant blood vessel growth than previously thought, more attention to the molecular underpinnings of RAP development may be warranted. Fortunately, the VEGF signaling pathway still appears to be an effective therapeutic target for a subset of nvAMD patients presenting with either RAP, CNV, or both. It is thought that VEGF is overexpressed in the retina in patients presenting with RAP much in the same way as nvAMD patients with CNV [135], thus making both sets of patients responsive to treatment. To date, multiple independent studies have indicated that different anti-VEGF therapies (bevacizumab, ranibizumab, and pegaptanib) are all effective at treating RAP lesions [136–138], although it should be noted that the reported therapeutic benefit was observed in relatively small sample sizes. Other therapies for anti-VEGF-resistant nvAMD patients are limited at this time.

4.3 Protein Homeostasis

The RPE is an extremely metabolically active cell type. Its many functions include, but are not limited to, visual cycle metabolism (the recycling of vitamin A isomers that are required for visual function) [139] and outer segment phagocytosis [140];

these processes are notoriously metabolically demanding and can lead to an overabundance of reactive oxygen species (ROS) [141]. The ROS generated by the RPE can result in significant oxidative damage to proteins and other macromolecules within the cell, and to cope with this constant onslaught of ROS damage, the cell relies on autophagy. Autophagy is the process by which cells can clear damaged proteins and organelles via lysosomal degradation [142]. Because of the constant insult from ROS on the RPE over the course of the lifetime, AMD researchers hypothesize that impaired clearance of these damaged cellular components because of impaired autophagy could lead to drusen formation [28, 143].

There is some evidence in the literature to support this hypothesis; aging patients are known to accumulate protein adducts known as advanced glycation end products (AGEs) that have been associated with some age-related diseases [144]. AGEs form under oxidative conditions through Maillard reactions, which nonenzymatically covalently glycate proteins [144]. These AGEs are capable of binding receptors for advanced glycation end products (RAGEs), and it is possible that RAGE signaling may lead to either VEGF-A upregulation that could promote angiogenesis [145] or innate immune activation (e.g., NLRP3 inflammasome) via NF-κB [146]. It is important to note, however, that these findings came primarily from cell culture data; despite a reported increase in AGEs in the serum of AMD patients [147], it is unclear if the reported observations of AGE in endothelial cell culture studies will be replicated in human tissue samples. A proteome analysis of drusen taken from human tissue samples also identified CEP-adducted proteins (formed from the oxidation of docosahexaenoic acid), which have been interpreted as being potentially relevant to AMD [148]. Furthermore, anti-CEP autoantibodies were identified in the sera of AMD patients in higher amounts than in age-matched controls [85]. Despite the potentially clinical relevance of the finding on anti-CEP autoantibodies in the sera of AMD patients, there has yet to be a defined molecular pathway leading from CEP protein adduct formation to antibody production [108]. Future research will hopefully provide the missing links to this phenomenon.

Amyloid ß (Aß) is most commonly known for its suspected role in causing Alzheimer's disease (AD), but it has also been identified as a component of drusen in human AMD patients [149]. Aß is formed from cleavage of Amyloid ß precursor protein (APP) by ß and γ secretases and accumulates in extracellular, insoluble amyloid plaques in patients with AD; the function of both Aß and the APP protein precursor are still a matter of debate [150]. Interestingly, Aß was found in drusen also positive for activated complement proteins [149], suggesting a functional link between the two with respect to AMD pathogenesis. Later work indicated that Aß could cause upregulation of complement factor B (CFB), a protein involved in the alternative complement activation pathway [151]. Further highlighting the fact that many aging processes are interrelated and can contribute to disease formation is the fact that Aß could also activate the NLRP3 inflammasome through lysosomal destabilization, in addition to being potentially linked to the alternative complement pathway [152]. Because the lysosome is central to autophagic processes, this finding further emphasizes the importance of protein homeostasis to age-related disease processes.

4.4 Metabolism

As mentioned in Sect. 4.3, one of the primary functions of the RPE is photoreceptor outer segment phagocytosis [140]. Photoreceptor outer segments are outgrowths of the plasma membrane on photoreceptors that are densely packed with the visual pigment molecules rhodopsin (in rods) and opsins (in cones) [153]. Each day, based on a circadian rhythm cycle [154], the outermost plasma membrane discs containing visual pigments (which are closest to the apical processes of the RPE) are shed so they can be phagocytized and broken down by lysosomes in the RPE [51]. The phagocytosis of shed outer segments by the RPE is compensated by the continuous renewal of new outer segment discs by the photoreceptors and allows for the removal of older visual pigment proteins that become damaged as the result of oxidative stress [51]. One of the metabolic byproducts of visual pigment activation by photons is the retinoid all-*trans*-retinal, which is typically recycled back into 11-*cis*-retinal by other visual cycle proteins for reformation of functional visual pigment in outer segments still attached to their respective photoreceptors [155]. Under oxidative conditions, two molecules of all-trans-retinal can react with ethanolamine to form N-retinyl-N-retinylidene ethanolamine (known more commonly as A2E) [156]. When A2E in outer segments is phagocytized by the RPE, it accumulates because the cell has no method for metabolizing it back into all-*trans*-retinal. A2E forms one of the primary components of lipofuscin, an autofluorescent accumulation that collects in the RPE with age [157]. Because early studies with high-performance liquid chromatography (HPLC) found that patients with AMD may have increased amounts of A2E [158], it was long thought that the age-related accumulation of lipofuscin may be a contributing factor to AMD development [157]. A proposed mechanism of action for A2E cytotoxicity was inhibition of efficient lysosomal activity that could then lead to apoptosis [156]. As more research was performed examining the A2E in AMD, though, the proposed causative role of A2E in AMD development was challenged. Initial measurements of A2E accumulation in AMD patients were performed with whole eyecups; separate HPLC measurements of macula and periphery indicated that A2E accumulated in the periphery of the eye, not in the macula [158]. Other work indicated that lipofuscin only poorly predicted the spread of atrophic regions in GA [159], and perhaps most damaging to the hypothesis that A2E contributed to lipofuscin accumulation and RPE cell death was the finding that lipofuscin and A2E distribution in the RPE did not even correlate with one another [160].

That is not to say there is no suspected role for metabolism in AMD, however. Disruptions in metal homeostasis in the aging eye may contribute to AMD [161]. Although one study has implicated reduced copper and zinc in the eyes of AMD patients as compared to age-matched controls [162], iron accumulation in particular may have a contributing role in AMD [163]. In healthy eyes, iron is an important cofactor necessary for many proteins, including proteins that perform necessary visual functions [164]. Because of the toxic nature of iron and the fact that it is required for many proteins, its transport and metabolism must be carefully regulated by the cell. Iron-induced retinal degeneration has been noted in patients and in animal models of hereditary iron overload, aceruloplasminemia, hereditary hemochromatosis, and pantothenate kinase associated neurodegeneration (PKAN) [163]. The levels of free iron (i.e., iron ions that are not bound to proteins or other macromolecules) are also known to increase with age, possibly because of reduced ability of the aging body to clear the ions from tissues [165]. This increase in free iron concentration occurs throughout the entire body, but is particularly relevant to AMD because patients were reported to have an increase of free iron in the affected macula [166] and were reported to also have iron as a drusen component [167]. Furthermore, AMD patients were found to have increased mRNA expression of the iron transporter gene transferrin (responsible for binding iron for transport into the cell), which is highly expressed in the RPE [168]. Also, mice deficient in both the iron-binding protein ceruloplasmin (Cp; also binds free iron) and hephaestin (Heph; necessary for cellular excretion of excess iron) appear to develop retinal pathologies similar to those seen in AMD patients, such as sub-RPE deposits and neovascularization [169]. Collectively, these data suggest that metabolic events favoring intracellular accumulation of iron may contribute to AMD pathogenesis. The toxicity of iron in biological systems is largely attributed to the result of Fenton chemistry, the process by which free divalent cations can catalyze the formation of hydrogen peroxide from water [170]. The resulting oxidative stress is then thought to contribute to cell death via apoptosis, but it is also possible that free divalent cations like iron could directly induce an inflammatory response. After all, it is already well known that ROS intermediates can cause NLRP3 inflammasome activation, which has been implicated in AMD by multiple different causative factors (please refer to Sect. 4.1.2). Indeed, recent evidence also suggests that iron also activates the NLRP3 inflammasome via sequestration of Poly(rC) binding protein 2 (PCBP2), which is required for efficient DICER1-mediated clearance of *Alu* RNAs, and that this mechanism is independent of ROS generation [171].

4.5 Genetic Associations

As is most likely the case with many complex diseases like AMD, a single mutation alone may not be sufficient to cause disease; genetic variants may predispose individuals to developing AMD, but other environmental risk factors are most likely necessary for interactions with those genetic variants [5]. Some of these genetic variants confer more risk for AMD development than other genetic variants. One such polymorphism, CFH^{Y402H}, has already been discussed earlier in this chapter (please refer to Sect. 4.1.3). The CFH^{Y402H} variant is thought to contribute to as much as 50 % heritability to the disease [96–98], but the exact mechanistic pathway for CFH involvement in AMD is still unclear. The relative contributions of other, less common or rare genetic variants are still a matter for further investigation. Another potential confound with respect to interpreting the genetic data in the literature is the fact that some variants found to be significant in one study may not be found to be significant in another. Reasons for discrepancies between studies include differences in risk conferred by each polymorphism in specific populations, insufficient statistical power in the study, or the presence of false positives in the dataset. It is also worth noting that individuals of Caucasian ancestry are more likely to develop clinical features of AMD than African Americans even though the rates of late AMD do not differ significantly between the two groups, although the exact reason for this is currently unknown [172]. In this section of the chapter, known or suspected polymorphisms will be discussed with respect to the known functional roles these genes have in healthy cells, but it is important to keep in mind that many of these polymorphisms have been reported without subsequent analysis of their molecular consequences. For many reported polymorphisms, further work is needed to ascertain their functional importance in disease processes and to understand why genetic mutations present from birth do not manifest as being pathogenic until patients are older.

4.5.1 Inflammatory Pathway Genes

To date, CFH variants are the most strongly associated genetic variants with respect to AMD. The finding that the CFH^{Y402H} polymorphism could contribute to AMD [96–98] was significant because it was the first report of a strong genetic association with a complex disease [50]. Furthermore, a later GWAS study indicated that although the CFH^{Y402H} variant did strongly associate with AMD, 20 other CFH variants also showed association with AMD [173], suggesting CFH variants account for a large amount of the heritability reported in AMD patients. In addition to CFH polymorphisms, other complement pathway factors have also been associated with AMD, even though the relative contributions of these other polymorphisms is still debated [15]. $C2^{E318D}$, $C3^{R102G}$, and CFB^{R32Q} polymorphisms were each found to independently associate with AMD risk in 2011 [174]. Another CFB variant, CFB^{L9H}, was identified later [175]. The association of $C3^{R102G}$ with AMD has also been replicated in another independent study [176]. Other rare complement factor variants, such as $C3^{K155Q}$ [177, 178], CFH^{R1210C} [178], CFH^{R53C}, CFH^{D90G} [179], CFI^{G119R} [180], and $C9^{P167S}$ [181] have all been identified as being highly penetrant in their association with AMD. The exact functional consequences of these variants with respect to protein function are still being studied.

In Sect. 4.1.1, the contribution of TLR signaling to AMD was discussed; TLR3 signaling has been reported to both inhibit neovascularization [79] and induce retinal degeneration [76], but the involvement of genetic polymorphisms in TLR signaling pathway genes is less clear. Initial findings of a potential protective effect against GA conferred by a $TLR3^{L412F}$ polymorphism [78], a potential susceptibility to AMD conferred by a TLR^{D299G} polymorphism [182], and a potential association of AMD with a TLR7 polymorphism were not replicated in later studies [183, 184]. One group later reported that the $TLR3^{L412F}$ variant may have a protective effect against GA because of a reduced ability of the receptor to bind dsRNA [185], but this finding has not been independently replicated by other laboratories. Because of the previous independent genetic studies that concluded this variant was unlikely to be significant, the contributions of TLR polymorphisms to AMD is at best uncertain.

4.5.2 Angiogenic Signaling Pathway Genes

With the recent interest in pharmacogenetics (i.e., the influence genetic polymorphisms have in the response to drug treatment), researchers have focused on the potential influence of VEGF-A polymorphisms on the nvAMD patient's response to anti-VEGF therapy. Despite the occasional report of VEGF-A polymorphism association with anti-VEGF therapy treatment outcome [186], rigorous analysis indicates that VEGF-A or VEGF-A receptor polymorphisms have no effect on treatment response [179]. Another study has implicated a CCR3 polymorphism (rs3091250) as a potential risk factor for AMD, but this result has not yet been replicated in other laboratories (it should be noted that another CCR3 variant, rs3091312, was not found to significantly associate with AMD) [187]. Stronger statistical associations of genetic variants in the *VEGFR1* gene that are predicted by computer modeling to alter splice sites and RNA secondary structure have also been reported [188] but have still yet to be examined at the protein level.

4.5.3 Genes with Other Functions

Genes with known functions in retinoid clearance from photoreceptors, lipid metabolism, and extracellular matrix maintenance have also been implicated in AMD [5, 189]. High-density lipoprotein metabolism pathways have been implicated in AMD because of an association of the hepatic lipase (*LIPC*) gene [190], and this gene association was supported by findings from another study [191]. APOE, a gene involved in low-density lipoprotein (LDL) metabolism, has also had polymorphisms associated with AMD [192]; this finding highlights the possible role that lipid dysmetabolism may have in disease development and progression. TIMP3, an inhibitor of matrix metalloproteinases [193], has also been associated with AMD along with *COL8A1*, a gene involved in the *FRK/COL10A1* extracellular collagen matrix pathway [194]. It is possible, but not yet clear, that these polymorphisms may affect the maintenance of Bruch's membrane with age.

4.5.4 Chromosome 10q26 Locus

Along with CFH, a genomic region with one of the strongest genetic associations with AMD is the chromosome 10q26 locus containing the two open reading frames (ORFs), *HTRA1* and *ARMS2* [195]. Because of the great degree of linkage disequilibrium in this region, genetic approaches alone have been unsuccessful in identifying the gene responsible for the strong association with AMD [195]. Studies of these two genes are further confounded by the fact that the *ARMS2* gene has a largely unknown function because it is only present in primates and thus precludes the use of a mouse model for studies of gene function [196]. Attempts to define the cellular and tissue localization of ARMS2 protein have produced inconsistent findings [197, 198]. Because of this, ARMS2 has not been definitively implicated in AMD pathology despite the strong associations of individual gene variants with the disease. HTRA1, on the other hand, is a serine protease with a well-defined function in extracellular matrix degradation, suggesting it may have an involvement in degradation of Bruch's membrane to promote RPE atrophy. HTRA1 is also an attractive candidate for AMD pathological association because of its presence within drusen [199]. Unfortunately, HTRA1 has produced inconsistent results based on hypotheses of the proposed mechanism of action. It would be expected that HTRA1 overexpression would be connected to AMD pathology in patients because of its ability to degrade extracellular matrix, but this has not been found in humans [200]. Because of this, the HTRA1 protein remains an attractive candidate for AMD association, but like ARMS2, ultimately has an uncertain role in disease progression based on current data. Future research into the functions of HTRA1 and ARMS2 in AMD patients may resolve these questions. Again, it is worth noting that despite these strong genetic associations, it is still unclear why these genetic influences do not become relevant until old age.

4.6 Epigenetics

Currently, there are few published studies examining epigenetics and AMD risk [201, 202]. There is some evidence to suggest that DNA hypermethylation of the *GSTM5* promoter, which encodes for the glutathione *S*-transferases mu1 and mu5, may result in reduced expression of these proteins in AMD patients [203]. Future work may further explore this potential link and other possible avenues of study.

5 Future Therapeutic Prospects

Unlike nvAMD, which has multiple current therapeutic interventions approved for use in patients, dry AMD has no current FDA-approved therapies with the exception of the AREDS formulation which, as discussed, does not appear to prevent the development of geographic atrophy [35]. Multiple clinical trials with the goal of inhibiting the complement pathway have had little success in preventing or restoring vision loss [50, 204]. It is not currently clear why these complement inhibition therapies have not had any clinical success with respect to AMD, but successful interventions seen in animal models that do not translate to human patients suggests current models are not reflective of human disease pathology and that there are gaps in our understanding of the role of complement in AMD. Also, many of these therapies have been developed in younger mice, suggesting that perhaps studies in older mice would better reflect the conditions found in aging AMD patients. The creation of better animal models will ultimately allow for better treatment validation so the field can make progress with the goals set forth in the NEI Audacious Goals Initiative [205, 206]. Nonetheless, other complement inhibition therapies, e.g., anti-complement factor D antibody (lampalizumab), are currently in clinical trials, and it is possible these new methods will be more successful than other approaches taken in the past. Because of the potential role that autophagy may play in AMD progression (please refer to Sect. 4.3), this pathway is currently a focus of interest for therapeutic intervention. Recent work indicated that RPE cell stress induced through oxidative stress may activate the Protein kinase B (AKT)/Mammalian target of rapamycin (mTOR) pathway, which could be inhibited with rapamcyin administration in a mouse model [207]. It is interesting to note that rapamycin is not only of interest in AMD research, but is also of interest in many other age-related diseases discussed in this book, because of observations that long-term rapamycin administration may promote longevity [208].

Given the advent of induced pluripotent stem (iPS) cells, cell replacement therapy is another current avenue of research that may show promise for the treatment of AMD [28]. Stem cells derived from mouse bone marrow were capable of establishing a monolayer in subretinal regions damaged by sodium iodate after subretinal injection [209]. Other work with adult hematopoietic stem cells (HSCs) indicated that these cells could assume RPE-like characteristics after transgenic expression of RPE65 [210], a visual cycle protein specific to the RPE in the eye [155]. Mice with RPE damage from sodium iodate exhibited restored visual function and prevented retinal degeneration upon subretinal injection of these HSCs [210]. Cell replacement therapy therefore offers the potential for benefit in AMD patients, but significant hurdles remain with respect to assessing its efficacy in the clinic. Despite the success of this technique in mouse models, it is unclear whether this method could rescue vision in an eye in which AMD has already caused significant RPE and photoreceptor cell atrophy. After all, restoration of the cell layer responsible for photoreceptor cell health is ultimately futile if the photoreceptors have already been lost by the time of cell replacement therapy. Furthermore, many of these proof-of-concept studies for stem cell therapies were performed in younger mice. Given the senescence of stem cells in older patients, it is currently unclear whether these therapies could be used with stem cells from aging individuals.

Recent work has also raised the possibility that nucleoside reverse transcriptase inhibitors (NRTIs) may have potential therapeutic against AMD pathology [211]. This finding is based on previous work that indicated that DICER1 deficiency could induce NLRP3 inflammasome activation and RPE apoptosis via *Alu* RNA accumulation [77, 89, 90, 212]. *Alu* RNA-induced RPE cytotoxicity requires the purinoreceptor P2X7 for cell death [88], and evidence suggests that NRTI's mechanism of action may involve blockade of the receptor [211]. NRTI treatment of AMD could offer several advantages over other therapeutic approaches currently under investigation, such as the fact that many NRTIs are already approved by the FDA for treatment of other diseases, like human immunodeficiency virus (HIV). Because AMD is a chronic disease that patients could have for several decades, drugs with long-term safety are particularly needed. Many HIV patients take NRTIs for decades and these drugs are thus promising candidates for AMD treatment.

With respect to nvAMD, IL-18 has been the subject of recent debate as it has been proposed as a potential treatment for nvAMD, based on reports that recombinant IL-18 administration could inhibit CNV [213] and that neutralization of endogenous IL-18 by antibody administration results in increased CNV [86]. However, five independent laboratories were unable to reproduce the original findings [214]. Hence, IL-18 is most likely not therapeutic for AMD and, based on findings of IL-18 with respect to RPE degeneration induced by *Alu* RNA accumulation, will most likely prove harmful to patients already experiencing cell atrophy because of AMD.

6 Future Research Needs

Despite the advances made in the understanding of AMD pathology, biology, and treatment, much work remains because of the complexity of the disease. Treatments for nvAMD have allowed patients who would otherwise have debilitating vision loss to maintain quality of life. That being said, nvAMD treatments have much room for improvement. Even though the risk of infection with each injection of anti-VEGF-A drug is low, repeated injections of drug for several years increases the possibility of an eventual eye infection. Also, nvAMD patients who receive anti-VEGF-A therapy also have a high likelihood to develop GA, for which there is no current treatment. Because of these reasons, there is a need for future treatments that will allow sustained therapeutic benefit and that do not exacerbate the progression of the dry form of the disease. Furthermore, the pitfalls of current nvAMD therapies highlight another important need in the field today: the need for treatments for the early and intermediate forms of AMD and the advanced dry form, GA. The therapeutic benefit of the AREDS formulation, which is currently the only option available for non-nvAMD patients, is still debated and may be limited in its efficacy. It appears that the formulation may even cause more harm than good in some patients. The reasons for this dearth of options for all AMD patients are manifold but ultimately reflect the lack of basic understanding of the biology underlying AMD pathology, including the impact of aging on development of the disease. The vast majority of the data in the literature concerning AMD development and progression are correlative observations (e.g., drusen formation, complement deposition, etc.) and do not offer molecular mechanistic explanations for their actual role in the disease. For example, CFH variants associate with AMD risk but as of yet, there is little data on how these associations translate into an actual disease mechanism that could be targeted by therapy. Because of this, there is no clear demarcation between normal aging in the eye and AMD.

The first CFH variant identified was found to be highly significant, and other significant associations between individual variants and AMD have been made since then, but these variants have not allowed for the creation of an animal model that can fully recapitulate the patient phenotype. Mouse models will always have confounding factors because of the fact that mice do not have a macula, but mice mutant for genes identified in GWAS studies have consistently failed to accurately reflect observations seen in the clinic with respect to RPE atrophy and visual function loss. Moreover, despite not having a macula, mouse models of nvAMD have been useful in their ability to advance clinical therapies for macular nvAMD in humans. This begs the question about whether the reason there is no effective treatment for early, intermediate, or dry AMD is because of the fact that current commonly accepted models do not adequately reflect the disease and the impact of aging on development of the disease has not been thoroughly tested. The number of failed clinical trials supports this hypothesis. The biggest challenge facing AMD research today is the fact that the field needs to find a way to tie clinical observations (including, but not limited to, genetic associations) to experimental models that accurately reflect the disease to allow for a comprehensive understanding of molecular disease processes so new, effective treatments can be created.

Acknowledgments Dr. Ambati is funded through an NIH Director's Pioneer Award, NEI, NIGMS, Ellison Medical Foundation, Harrington Discovery Institute, Foundation Fighting Blindness.

Disclosures: J.A. is a co-founder of iVeena (Holdings, Pharma, Delivery) LLCs, and has received honoraria and travel support from Allergan, Inc.

Editor: Grace Shen, National Eye Institute (NEI), NIH.

References

1. Marmorstein L (2004) Association of EFEMP1 with malattia leventinese and age-related macular degeneration: a mini-review. Ophthalmic Genet 25(3):219–226. doi:10.1080/13816810490498305
2. Gerber DM, Munier FL, Niemeyer G (2003) Cross-sectional study of visual acuity and electroretinogram in two types of dominant drusen. Invest Ophthalmol Vis Sci 44(2):493–496
3. Boon CJ, van de Kar NC, Klevering BJ, Keunen JE, Cremers FP, Klaver CC, Hoyng CB, Daha MR, den Hollander AI (2009) The spectrum of phenotypes caused by variants in the CFH gene. Mol Immunol 46(8–9):1573–1594. doi:10.1016/j.molimm.2009.02.013
4. Ferris FL 3rd, Wilkinson CP, Bird A, Chakravarthy U, Chew E, Csaky K, Sadda SR, Beckman Initiative for Macular Research Classification Committee (2013) Clinical classification of age-related macular degeneration. Ophthalmology 120(4):844–851. doi:10.1016/j.ophtha.2012.10.036
5. Ratnapriya R, Chew EY (2013) Age-related macular degeneration-clinical review and genetics update. Clin Genet 84(2):160–166. doi:10.1111/cge.12206
6. Davis MD, Gangnon RE, Lee LY, Hubbard LD, Klein BE, Klein R, Ferris FL, Bressler SB, Milton RC, Age-Related Eye Disease Study Group (2005) The Age-Related Eye Disease Study severity scale for age-related macular degeneration: AREDS Report No. 17. Arch Ophthalmol 123(11):1484–1498. doi:10.1001/archopht.123.11.1484
7. Klein R, Davis MD, Magli YL, Segal P, Klein BE, Hubbard L (1991) The Wisconsin age-related maculopathy grading system. Ophthalmology 98(7):1128–1134
8. de Sisternes L, Simon N, Tibshirani R, Leng T, Rubin DL (2014) Quantitative SD-OCT imaging biomarkers as indicators of age-related macular degeneration progression. Invest Ophthalmol Vis Sci. doi:10.1167/iovs.14-14918

9. Yehoshua Z, Rosenfeld PJ, Gregori G, Feuer WJ, Falcao M, Lujan BJ, Puliafito C (2011) Progression of geographic atrophy in age-related macular degeneration imaged with spectral domain optical coherence tomography. Ophthalmology 118(4):679–686. doi:10.1016/j.ophtha.2010.08.018
10. de Bruin DM, Burnes DL, Loewenstein J, Chen Y, Chang S, Chen TC, Esmaili DD, de Boer JF (2008) In vivo three-dimensional imaging of neovascular age-related macular degeneration using optical frequency domain imaging at 1050 nm. Invest Ophthalmol Vis Sci 49(10):4545–4552. doi:10.1167/iovs.07-1553
11. Sarks JP, Sarks SH, Killingsworth MC (1988) Evolution of geographic atrophy of the retinal pigment epithelium. Eye 2(Pt 5):552–577. doi:10.1038/eye.1988.106
12. Wong TY, Chakravarthy U, Klein R, Mitchell P, Zlateva G, Buggage R, Fahrbach K, Probst C, Sledge I (2008) The natural history and prognosis of neovascular age-related macular degeneration: a systematic review of the literature and meta-analysis. Ophthalmology 115(1):116–126. doi:10.1016/j.ophtha.2007.03.008
13. Ambati J, Fowler BJ (2012) Mechanisms of age-related macular degeneration. Neuron 75(1):26–39. doi:10.1016/j.neuron.2012.06.018
14. Nivison-Smith L, Milston R, Madigan M, Kalloniatis M (2014) Age-related macular degeneration: linking clinical presentation to pathology. Optom Vis Sci 91(8):832–848. doi:10.1097/OPX.0000000000000281
15. Bora NS, Matta B, Lyzogubov VV, Bora PS (2014) Relationship between the complement system, risk factors and prediction models in age-related macular degeneration. Mol Immunol. doi:10.1016/j.molimm.2014.07.012
16. Smith W, Assink J, Klein R, Mitchell P, Klaver CC, Klein BE, Hofman A, Jensen S, Wang JJ, de Jong PT (2001) Risk factors for age-related macular degeneration: Pooled findings from three continents. Ophthalmology 108(4):697–704
17. Zhang X, Cotch MF, Ryskulova A, Primo SA, Nair P, Chou CF, Geiss LS, Barker LE, Elliott AF, Crews JE, Saaddine JB (2012) Vision health disparities in the United States by race/ethnicity, education, and economic status: findings from two nationally representative surveys. Am J Ophthalmol 154(6 Suppl):S53–S62.e51. doi:10.1016/j.ajo.2011.08.045
18. Seddon JM (2013) Genetic and environmental underpinnings to age-related ocular diseases. Invest Ophthalmol Vis Sci 54(14):ORSF28–ORSF30. doi:10.1167/iovs.13-13234
19. Klein R, Klein BE, Franke T (1993) The relationship of cardiovascular disease and its risk factors to age-related maculopathy. The Beaver Dam Eye Study. Ophthalmology 100(3):406–414
20. Vingerling JR, Dielemans I, Hofman A, Grobbee DE, Hijmering M, Kramer CF, de Jong PT (1995) The prevalence of age-related maculopathy in the Rotterdam Study. Ophthalmology 102(2):205–210
21. Mitchell P, Smith W, Attebo K, Wang JJ (1995) Prevalence of age-related maculopathy in Australia. The Blue Mountains Eye Study. Ophthalmology 102(10):1450–1460
22. Klein R, Klein BE (2013) The prevalence of age-related eye diseases and visual impairment in aging: current estimates. Invest Ophthalmol Vis Sci 54(14):ORSF5–ORSF13. doi:10.1167/iovs.13-12789
23. Antioxidant status and neovascular age-related macular degeneration. Eye Disease Case-control Study Group (1993) Arch Ophthalmol 111(1):104–109
24. Goldberg J, Flowerdew G, Smith E, Brody JA, Tso MO (1988) Factors associated with age-related macular degeneration. An analysis of data from the first National Health and Nutrition Examination Survey. Am J Epidemiol 128(4):700–710
25. Seddon JM, Ajani UA, Sperduto RD, Hiller R, Blair N, Burton TC, Farber MD, Gragoudas ES, Haller J, Miller DT et al (1994) Dietary carotenoids, vitamins A, C, and E, and advanced age-related macular degeneration. Eye Disease Case-control Study Group. JAMA 272(18):1413–1420
26. VandenLangenberg GM, Mares-Perlman JA, Klein R, Klein BE, Brady WE, Palta M (1998) Associations between antioxidant and zinc intake and the 5-year incidence of early age-related maculopathy in the Beaver Dam Eye Study. Am J Epidemiol 148(2):204–214

27. Age-Related Eye Disease Study Research Group (2001) A randomized, placebo-controlled, clinical trial of high-dose supplementation with vitamins C and E, beta carotene, and zinc for age-related macular degeneration and vision loss: AREDS report no. 8. Arch Ophthalmol 119(10):1417–1436
28. Bowes Rickman C, Farsiu S, Toth CA, Klingeborn M (2013) Dry age-related macular degeneration: mechanisms, therapeutic targets, and imaging. Invest Ophthalmol Vis Sci 54(14):ORSF68–ORSF80. doi:10.1167/iovs.13-12757
29. Age-Related Eye Disease Study 2 Research Group (2013) Lutein+zeaxanthin and omega-3 fatty acids for age-related macular degeneration: the Age-Related Eye Disease Study 2 (AREDS2) randomized clinical trial. JAMA 309(19):2005–2015. doi:10.1001/jama.2013.4997
30. Age-Related Eye Disease Study 2 Research Group, Chew EY, Clemons TE, Sangiovanni JP, Danis RP, Ferris FL 3rd, Elman MJ, Antoszyk AN, Ruby AJ, Orth D, Bressler SB, Fish GE, Hubbard GB, Klein ML, Chandra SR, Blodi BA, Domalpally A, Friberg T, Wong WT, Rosenfeld PJ, Agron E, Toth CA, Bernstein PS, Sperduto RD (2014) Secondary analyses of the effects of lutein/zeaxanthin on age-related macular degeneration progression: AREDS2 report No. 3. JAMA Ophthalmol 132(2):142–149. doi:10.1001/jamaophthalmol.2013.7376
31. Ambati J, Ambati BK (2002) Age-related eye disease study caveats. Arch Ophthalmol 120(7):997; author reply 997–999
32. Complications of Age-Related Macular Degeneration Prevention Trial Research Group (2006) Laser treatment in patients with bilateral large drusen: the complications of age-related macular degeneration prevention trial. Ophthalmology 113(11):1974–1986. doi:10.1016/j.ophtha.2006.08.015
33. Maguire MG, Daniel E, Shah AR, Grunwald JE, Hagstrom SA, Avery RL, Huang J, Martin RW, Roth DB, Castellarin AA, Bakri SJ, Fine SL, Martin DF, Comparison of Age-Related Macular Degeneration Treatments Trials (2013) Incidence of choroidal neovascularization in the fellow eye in the comparison of age-related macular degeneration treatments trials. Ophthalmology 120(10):2035–2041. doi:10.1016/j.ophtha.2013.03.017
34. Evans J (2008) Antioxidant supplements to prevent or slow down the progression of AMD: a systematic review and meta-analysis. Eye 22(6):751–760. doi:10.1038/eye.2008.100
35. Chew EY, Clemons TE, Agron E, Sperduto RD, Sangiovanni JP, Davis MD, Ferris FL 3rd, Age-Related Eye Disease Study Research Group (2014) Ten-year follow-up of age-related macular degeneration in the age-related eye disease study: AREDS report no. 36. JAMA Ophthalmol 132(3):272–277. doi:10.1001/jamaophthalmol.2013.6636
36. Krishnadev N, Meleth AD, Chew EY (2010) Nutritional supplements for age-related macular degeneration. Curr Opin Ophthalmol 21(3):184–189. doi:10.1097/ICU.0b013e32833866ee
37. The alpha-tocopherol, beta-carotene lung cancer prevention study: design, methods, participant characteristics, and compliance. The ATBC Cancer Prevention Study Group (1994) Ann Epidemiol 4(1):1–10
38. Omenn GS, Goodman GE, Thornquist MD, Balmes J, Cullen MR, Glass A, Keogh JP, Meyskens FL Jr, Valanis B, Williams JH Jr, Barnhart S, Cherniack MG, Brodkin CA, Hammar S (1996) Risk factors for lung cancer and for intervention effects in CARET, the Beta-Carotene and Retinol Efficacy Trial. J Natl Cancer Inst 88(21):1550–1559
39. Donati G, Kapetanios AD, Pournaras CJ (1999) Principles of treatment of choroidal neovascularization with photodynamic therapy in age-related macular degeneration. Semin Ophthalmol 14(1):2–10. doi:10.153/SOPH01400002
40. Margherio RR, Margherio AR, DeSantis ME (2000) Laser treatments with verteporfin therapy and its potential impact on retinal practices. Retina 20(4):325–330
41. Bressler NM, Bressler SB (2000) Photodynamic therapy with verteporfin (Visudyne): impact on ophthalmology and visual sciences. Invest Ophthalmol Vis Sci 41(3):624–628
42. Scott AW, Bressler SB (2013) Long-term follow-up of vascular endothelial growth factor inhibitor therapy for neovascular age-related macular degeneration. Curr Opin Ophthalmol 24(3):190–196. doi:10.1097/ICU.0b013e32835fefee

43. Witmer AN, Vrensen GF, Van Noorden CJ, Schlingemann RO (2003) Vascular endothelial growth factors and angiogenesis in eye disease. Prog Retin Eye Res 22(1):1–29
44. Brown DM, Regillo CD (2007) Anti-VEGF agents in the treatment of neovascular age-related macular degeneration: applying clinical trial results to the treatment of everyday patients. Am J Ophthalmol 144(4):627–637. doi:10.1016/j.ajo.2007.06.039
45. Group TCR (2011) Ranibizumab and Bevacizumab for neovascular age-related macular degeneration. N Engl J Med. doi:10.1056/NEJMoa1102673
46. Tranos P, Vacalis A, Asteriadis S, Koukoula S, Vachtsevanos A, Perganta G, Georgalas I (2013) Resistance to antivascular endothelial growth factor treatment in age-related macular degeneration. Drug Des Devel Ther 7:485–490. doi:10.2147/DDDT.S43470
47. Sayanagi K, Sharma S, Kaiser PK (2009) Photoreceptor status after antivascular endothelial growth factor therapy in exudative age-related macular degeneration. Br J Ophthalmol 93(5):622–626. doi:10.1136/bjo.2008.151977
48. Saint-Geniez M, Kurihara T, Sekiyama E, Maldonado AE, D'Amore PA (2009) An essential role for RPE-derived soluble VEGF in the maintenance of the choriocapillaris. Proc Natl Acad Sci U S A 106(44):18751–18756. doi:10.1073/pnas.0905010106
49. Comparison of Age-related Macular Degeneration Treatments Trials Research Group, Martin DF, Maguire MG, Fine SL, Ying GS, Jaffe GJ, Grunwald JE, Toth C, Redford M, Ferris FL 3rd (2012) Ranibizumab and bevacizumab for treatment of neovascular age-related macular degeneration: two-year results. Ophthalmology 119(7):1388–1398. doi:10.1016/j.ophtha.2012.03.053
50. Ambati J, Atkinson JP, Gelfand BD (2013) Immunology of age-related macular degeneration. Nat Rev Immunol 13(6):438–451. doi:10.1038/nri3459
51. Mazzoni F, Safa H, Finnemann SC (2014) Understanding photoreceptor outer segment phagocytosis: use and utility of RPE cells in culture. Exp Eye Res 126:51–60. doi:10.1016/j.exer.2014.01.010
52. Lehmann GL, Benedicto I, Philp NJ, Rodriguez-Boulan E (2014) Plasma membrane protein polarity and trafficking in RPE cells: past, present and future. Exp Eye Res 126:5–15. doi:10.1016/j.exer.2014.04.021
53. Rizzolo LJ (2014) Barrier properties of cultured retinal pigment epithelium. Exp Eye Res 126:16–26. doi:10.1016/j.exer.2013.12.018
54. Cherepanoff S, McMenamin P, Gillies MC, Kettle E, Sarks SH (2010) Bruch's membrane and choroidal macrophages in early and advanced age-related macular degeneration. Br J Ophthalmol 94(7):918–925. doi:10.1136/bjo.2009.165563
55. Oh H, Takagi H, Takagi C, Suzuma K, Otani A, Ishida K, Matsumura M, Ogura Y, Honda Y (1999) The potential angiogenic role of macrophages in the formation of choroidal neovascular membranes. Invest Ophthalmol Vis Sci 40(9):1891–1898
56. Killingsworth MC, Sarks JP, Sarks SH (1990) Macrophages related to Bruch's membrane in age-related macular degeneration. Eye 4(Pt 4):613–621. doi:10.1038/eye.1990.86
57. Espinosa-Heidmann DG, Suner IJ, Hernandez EP, Monroy D, Csaky KG, Cousins SW (2003) Macrophage depletion diminishes lesion size and severity in experimental choroidal neovascularization. Invest Ophthalmol Vis Sci 44(8):3586–3592
58. Ambati J, Anand A, Fernandez S, Sakurai E, Lynn BC, Kuziel WA, Rollins BJ, Ambati BK (2003) An animal model of age-related macular degeneration in senescent Ccl-2- or Ccr-2-deficient mice. Nat Med 9(11):1390–1397. doi:10.1038/nm950
59. Takeda A, Baffi JZ, Kleinman ME, Cho WG, Nozaki M, Yamada K, Kaneko H, Albuquerque RJ, Dridi S, Saito K, Raisler BJ, Budd SJ, Geisen P, Munitz A, Ambati BK, Green MG, Ishibashi T, Wright JD, Humbles AA, Gerard CJ, Ogura Y, Pan Y, Smith JR, Grisanti S, Hartnett ME, Rothenberg ME, Ambati J (2009) CCR3 is a target for age-related macular degeneration diagnosis and therapy. Nature 460(7252):225–230. doi:10.1038/nature08151
60. Apte RS, Richter J, Herndon J, Ferguson TA (2006) Macrophages inhibit neovascularization in a murine model of age-related macular degeneration. PLoS Med 3(8), e310. doi:10.1371/journal.pmed.0030310
61. Kelly J, Ali Khan A, Yin J, Ferguson TA, Apte RS (2007) Senescence regulates macrophage activation and angiogenic fate at sites of tissue injury in mice. J Clin Invest 117(11):3421–3426. doi:10.1172/JCI32430

62. Combadiere C, Feumi C, Raoul W, Keller N, Rodero M, Pezard A, Lavalette S, Houssier M, Jonet L, Picard E, Debre P, Sirinyan M, Deterre P, Ferroukhi T, Cohen SY, Chauvaud D, Jeanny JC, Chemtob S, Behar-Cohen F, Sennlaub F (2007) CX3CR1-dependent subretinal microglia cell accumulation is associated with cardinal features of age-related macular degeneration. J Clin Invest 117(10):2920–2928. doi:10.1172/JCI31692
63. Tuo J, Bojanowski CM, Zhou M, Shen D, Ross RJ, Rosenberg KI, Cameron DJ, Yin C, Kowalak JA, Zhuang Z, Zhang K, Chan CC (2007) Murine ccl2/cx3cr1 deficiency results in retinal lesions mimicking human age-related macular degeneration. Invest Ophthalmol Vis Sci 48(8):3827–3836. doi:10.1167/iovs.07-0051
64. Penfold PL, Liew SC, Madigan MC, Provis JM (1997) Modulation of major histocompatibility complex class II expression in retinas with age-related macular degeneration. Invest Ophthalmol Vis Sci 38(10):2125–2133
65. Zhou J, He S, Zhang N, Spee C, Zhou P, Ryan SJ, Kannan R, Hinton DR (2010) Neutrophils compromise retinal pigment epithelial barrier integrity. J Biomed Biotechnol 2010:289360. doi:10.1155/2010/289360
66. Zhou J, Pham L, Zhang N, He S, Gamulescu MA, Spee C, Ryan SJ, Hinton DR (2005) Neutrophils promote experimental choroidal neovascularization. Mol Vis 11:414–424
67. Tsutsumi-Miyahara C, Sonoda KH, Egashira K, Ishibashi M, Qiao H, Oshima T, Murata T, Miyazaki M, Charo IF, Hamano S, Ishibashi T (2004) The relative contributions of each subset of ocular infiltrated cells in experimental choroidal neovascularisation. Br J Ophthalmol 88(9):1217–1222. doi:10.1136/bjo.2003.036392
68. Penfold PL, Provis JM, Billson FA (1987) Age-related macular degeneration: ultrastructural studies of the relationship of leucocytes to angiogenesis. Graefes Arch Clin Exp Ophthalmol 225(1):70–76
69. Kawasaki T, Kawai T (2014) Toll-like receptor signaling pathways. Front Immunol 5:461. doi:10.3389/fimmu.2014.00461
70. Tatematsu M, Seya T, Matsumoto M (2014) Beyond dsRNA: toll-like receptor 3 signalling in RNA-induced immune responses. Biochem J 458(2):195–201. doi:10.1042/BJ20131492
71. Gross O, Thomas CJ, Guarda G, Tschopp J (2011) The inflammasome: an integrated view. Immunol Rev 243(1):136–151. doi:10.1111/j.1600-065X.2011.01046.x
72. Sagulenko V, Thygesen SJ, Sester DP, Idris A, Cridland JA, Vajjhala PR, Roberts TL, Schroder K, Vince JE, Hill JM, Silke J, Stacey KJ (2013) AIM2 and NLRP3 inflammasomes activate both apoptotic and pyroptotic death pathways via ASC. Cell Death Differ 20(9):1149–1160. doi:10.1038/cdd.2013.37
73. Munoz-Planillo R, Kuffa P, Martinez-Colon G, Smith BL, Rajendiran TM, Nunez G (2013) K(+) efflux is the common trigger of NLRP3 inflammasome activation by bacterial toxins and particulate matter. Immunity 38(6):1142–1153. doi:10.1016/j.immuni.2013.05.016
74. Ricklin D, Lambris JD (2013) Complement in immune and inflammatory disorders: therapeutic interventions. J Immunol 190(8):3839–3847. doi:10.4049/jimmunol.1203200
75. Hajela K, Kojima M, Ambrus G, Wong KH, Moffatt BE, Ferluga J, Hajela S, Gal P, Sim RB (2002) The biological functions of MBL-associated serine proteases (MASPs). Immunobiology 205(4–5):467–475
76. Kleinman ME, Kaneko H, Cho WG, Dridi S, Fowler BJ, Blandford AD, Albuquerque RJ, Hirano Y, Terasaki H, Kondo M, Fujita T, Ambati BK, Tarallo V, Gelfand BD, Bogdanovich S, Baffi JZ, Ambati J (2012) Short-interfering RNAs induce retinal degeneration via TLR3 and IRF3. Mol Ther 20(1):101–108. doi:10.1038/mt.2011.212
77. Kaneko H, Dridi S, Tarallo V, Gelfand BD, Fowler BJ, Cho WG, Kleinman ME, Ponicsan SL, Hauswirth WW, Chiodo VA, Karikó K, Yoo JW, Lee D-K, Hadziahmetovic M, Song Y, Misra S, Chaudhuri G, Buaas FW, Braun RE, Hinton DR, Zhang Q, Grossniklaus HE, Provis JM, Madigan MC, Milam AH, Justice NL, Albuquerque RJC, Blandford AD, Bogdanovich S, Hirano Y, Witta J, Fuchs E, Littman DR, Ambati BK, Rudin CM, Chong MMW, Provost P, Kugel JF, Goodrich JA, Dunaief JL, Baffi JZ, Ambati J (2011) DICER1 deficit induces Alu RNA toxicity in age-related macular degeneration. Nature. doi:10.1038/nature09830
78. Yang Z, Stratton C, Francis PJ, Kleinman ME, Tan PL, Gibbs D, Tong Z, Chen H, Constantine R, Yang X, Chen Y, Zeng J, Davey L, Ma X, Hau VS, Wang C, Harmon J, Buehler J, Pearson

E, Patel S, Kaminoh Y, Watkins S, Luo L, Zabriskie NA, Bernstein PS, Cho W, Schwager A, Hinton DR, Klein ML, Hamon SC, Simmons E, Yu B, Campochiaro B, Sunness JS, Campochiaro P, Jorde L, Parmigiani G, Zack DJ, Katsanis N, Ambati J, Zhang K (2008) Toll-like receptor 3 and geographic atrophy in age-related macular degeneration. N Engl J Med 359(14):1456–1463. doi:10.1056/NEJMoa0802437
79. Kleinman ME, Yamada K, Takeda A, Chandrasekaran V, Nozaki M, Baffi JZ, Albuquerque RJC, Yamasaki S, Itaya M, Pan Y, Appukuttan B, Gibbs D, Yang Z, Karikó K, Ambati BK, Wilgus TA, DiPietro LA, Sakurai E, Zhang K, Smith JR, Taylor EW, Ambati J (2008) Sequence- and target-independent angiogenesis suppression by siRNA via TLR3. Nature 452(7187):591–597. doi:10.1038/nature06765
80. Cho WG, Albuquerque RJC, Kleinman ME, Tarallo V, Greco A, Nozaki M, Green MG, Baffi JZ, Ambati BK, De Falco M, Alexander JS, Brunetti A, De Falco S, Ambati J (2009) Small interfering RNA-induced TLR3 activation inhibits blood and lymphatic vessel growth. Proc Natl Acad Sci U S A 106(17):7137–7142. doi:10.1073/pnas.0812317106
81. Gorin MB (2012) Genetic insights into age-related macular degeneration: controversies addressing risk, causality, and therapeutics. Mol Aspects Med 33(4):467–486. doi:10.1016/j.mam.2012.04.004
82. Shiose S, Chen Y, Okano K, Roy S, Kohno H, Tang J, Pearlman E, Maeda T, Palczewski K, Maeda A (2011) Toll-like receptor 3 is required for development of retinopathy caused by impaired all-trans-retinal clearance in mice. J Biol Chem 286(17):15543–15555. doi:10.1074/jbc.M111.228551
83. Murakami Y, Matsumoto H, Roh M, Giani A, Kataoka K, Morizane Y, Kayama M, Thanos A, Nakatake S, Notomi S, Hisatomi T, Ikeda Y, Ishibashi T, Connor KM, Miller JW, Vavvas DG (2014) Programmed necrosis, not apoptosis, is a key mediator of cell loss and DAMP-mediated inflammation in dsRNA-induced retinal degeneration. Cell Death Differ 21(2):270–277. doi:10.1038/cdd.2013.109
84. Klettner A, Koinzer S, Meyer T, Roider J (2013) Toll-like receptor 3 activation in retinal pigment epithelium cells – Mitogen-activated protein kinase pathways of cell death and vascular endothelial growth factor secretion. Acta Ophthalmol 91(3):E211–E218. doi:10.1111/Aos.12031
85. Gu X, Meer SG, Miyagi M, Rayborn ME, Hollyfield JG, Crabb JW, Salomon RG (2003) Carboxyethylpyrrole protein adducts and autoantibodies, biomarkers for age-related macular degeneration. J Biol Chem 278(43):42027–42035. doi:10.1074/jbc.M305460200
86. Doyle SL, Campbell M, Ozaki E, Salomon RG, Mori A, Kenna PF, Farrar GJ, Kiang AS, Humphries MM, Lavelle EC, O'Neill LA, Hollyfield JG, Humphries P (2012) NLRP3 has a protective role in age-related macular degeneration through the induction of IL-18 by drusen components. Nat Med 18(5):791–798. doi:10.1038/nm.2717
87. Burns KH, Boeke JD (2012) Human transposon tectonics. Cell 149(4):740–752. doi:10.1016/j.cell.2012.04.019
88. Kerur N, Hirano Y, Tarallo V, Fowler BJ, Bastos-Carvalho A, Yasuma T, Yasuma R, Kim Y, Hinton DR, Kirschning CJ, Gelfand BD, Ambati J (2013) TLR-independent and P2X7-dependent signaling mediate Alu RNA-induced NLRP3 inflammasome activation in geographic atrophy. Invest Ophthalmol Vis Sci 54(12):7395–7401. doi:10.1167/iovs.13-12500
89. Tarallo V, Hirano Y, Gelfand BD, Dridi S, Kerur N, Kim Y, Cho WG, Kaneko H, Fowler BJ, Bogdanovich S, Albuquerque RJ, Hauswirth WW, Chiodo VA, Kugel JF, Goodrich JA, Ponicsan SL, Chaudhuri G, Murphy MP, Dunaief JL, Ambati BK, Ogura Y, Yoo JW, Lee DK, Provost P, Hinton DR, Nunez G, Baffi JZ, Kleinman ME, Ambati J (2012) DICER1 loss and Alu RNA induce age-related macular degeneration via the NLRP3 inflammasome and MyD88. Cell 149(4):847–859. doi:10.1016/j.cell.2012.03.036
90. Kim Y, Tarallo V, Kerur N, Yasuma T, Gelfand BD, Bastos-Carvalho A, Hirano Y, Yasuma R, Mizutani T, Fowler BJ, Li S, Kaneko H, Bogdanovich S, Ambati BK, Hinton DR, Hauswirth WW, Hakem R, Wright C, Ambati J (2014) DICER1/Alu RNA dysmetabolism induces Caspase-8-mediated cell death in age-related macular degeneration. Proc Natl Acad Sci U S A 111(45):16082–16087. doi:10.1073/pnas.1403814111

91. Cao S, Ko A, Partanen M, Pakzad-Vaezi K, Merkur AB, Albiani DA, Kirker AW, Wang A, Cui JZ, Forooghian F, Matsubara JA (2013) Relationship between systemic cytokines and complement factor H Y402H polymorphism in patients with dry age-related macular degeneration. Am J Ophthalmol 156(6):1176–1183. doi:10.1016/j.ajo.2013.08.003
92. Ijima R, Kaneko H, Ye F, Nagasaka Y, Takayama K, Kataoka K, Kachi S, Iwase T, Terasaki H (2014) Interleukin-18 induces retinal pigment epithelium degeneration in mice. Invest Ophthalmol Vis Sci 55(10):6673–6678. doi:10.1167/iovs.14-15367
93. Baudouin C, Peyman GA, Fredj-Reygrobellet D, Gordon WC, Lapalus P, Gastaud P, Bazan NG (1992) Immunohistological study of subretinal membranes in age-related macular degeneration. Jpn J Ophthalmol 36(4):443–451
94. Hageman GS, Luthert PJ, Victor Chong NH, Johnson LV, Anderson DH, Mullins RF (2001) An integrated hypothesis that considers drusen as biomarkers of immune-mediated processes at the RPE-Bruch's membrane interface in aging and age-related macular degeneration. Prog Retin Eye Res 20(6):705–732
95. Anderson DH, Mullins RF, Hageman GS, Johnson LV (2002) A role for local inflammation in the formation of drusen in the aging eye. Am J Ophthalmol 134(3):411–431
96. Haines JL, Hauser MA, Schmidt S, Scott WK, Olson LM, Gallins P, Spencer KL, Kwan SY, Noureddine M, Gilbert JR, Schnetz-Boutaud N, Agarwal A, Postel EA, Pericak-Vance MA (2005) Complement factor H variant increases the risk of age-related macular degeneration. Science (New York, NY) 308(5720):419–421. doi:10.1126/science.1110359
97. Edwards AO, Ritter R, Abel KJ, Manning A, Panhuysen C, Farrer LA (2005) Complement factor H polymorphism and age-related macular degeneration. Science (New York, NY) 308(5720):421–424. doi:10.1126/science.1110189
98. Klein RJ, Zeiss C, Chew EY, Tsai J-Y, Sackler RS, Haynes C, Henning AK, SanGiovanni JP, Mane SM, Mayne ST, Bracken MB, Ferris FL, Ott J, Barnstable C, Hoh J (2005) Complement factor H polymorphism in age-related macular degeneration. Science (New York, NY) 308(5720):385–389. doi:10.1126/science.1109557
99. Clark SJ, Perveen R, Hakobyan S, Morgan BP, Sim RB, Bishop PN, Day AJ (2010) Impaired binding of the age-related macular degeneration-associated complement factor H 402H allotype to Bruch's membrane in human retina. J Biol Chem 285(39):30192–30202. doi:10.1074/jbc.M110.103986
100. Sjoberg AP, Trouw LA, Clark SJ, Sjolander J, Heinegard D, Sim RB, Day AJ, Blom AM (2007) The factor H variant associated with age-related macular degeneration (His-384) and the non-disease-associated form bind differentially to C-reactive protein, fibromodulin, DNA, and necrotic cells. J Biol Chem 282(15):10894–10900. doi:10.1074/jbc.M610256200
101. Laine M, Jarva H, Seitsonen S, Haapasalo K, Lehtinen MJ, Lindeman N, Anderson DH, Johnson PT, Jarvela I, Jokiranta TS, Hageman GS, Immonen I, Meri S (2007) Y402H polymorphism of complement factor H affects binding affinity to C-reactive protein. J Immunol 178(6):3831–3836
102. Black S, Kushner I, Samols D (2004) C-reactive protein. J Biol Chem 279(47):48487–48490. doi:10.1074/jbc.R400025200
103. Coffey PJ, Gias C, McDermott CJ, Lundh P, Pickering MC, Sethi C, Bird A, Fitzke FW, Maass A, Chen LL, Holder GE, Luthert PJ, Salt TE, Moss SE, Greenwood J (2007) Complement factor H deficiency in aged mice causes retinal abnormalities and visual dysfunction. Proc Natl Acad Sci U S A 104(42):16651–16656. doi:10.1073/pnas.0705079104
104. Anderson DH, Radeke MJ, Gallo NB, Chapin EA, Johnson PT, Curletti CR, Hancox LS, Hu J, Ebright JN, Malek G, Hauser MA, Rickman CB, Bok D, Hageman GS, Johnson LV (2010) The pivotal role of the complement system in aging and age-related macular degeneration: hypothesis re-visited. Prog Retin Eye Res 29(2):95–112. doi:10.1016/j.preteyeres.2009.11.003
105. Mullins RF, Schoo DP, Sohn EH, Flamme-Wiese MJ, Workamelahu G, Johnston RM, Wang K, Tucker BA, Stone EM (2014) The membrane attack complex in aging human choriocapillaris: relationship to macular degeneration and choroidal thinning. Am J Pathol 184(11):3142–3153. doi:10.1016/j.ajpath.2014.07.017

106. Chakravarti A, Clark AG, Mootha VK (2013) Distilling pathophysiology from complex disease genetics. Cell 155(1):21–26. doi:10.1016/j.cell.2013.09.001
107. Liu B, Wei L, Meyerle C, Tuo J, Sen HN, Li Z, Chakrabarty S, Agron E, Chan CC, Klein ML, Chew E, Ferris F, Nussenblatt RB (2011) Complement component C5a promotes expression of IL-22 and IL-17 from human T cells and its implication in age-related macular degeneration. J Transl Med 9:1–12. doi:10.1186/1479-5876-9-111
108. Ardeljan D, Chan CC (2013) Aging is not a disease: distinguishing age-related macular degeneration from aging. Prog Retin Eye Res 37:68–89. doi:10.1016/j.preteyeres.2013.07.003
109. Nassar K, Grisanti S, Elfar E, Luke J, Luke M, Grisanti S (2014) Serum cytokines as biomarkers for age-related macular degeneration. Graefes Arch Clin Exp Ophthalmol. doi:10.1007/s00417-014-2738-8
110. Seddon JM, George S, Rosner B, Rifai N (2005) Progression of age-related macular degeneration: prospective assessment of C-reactive protein, interleukin 6, and other cardiovascular biomarkers. Arch Ophthalmol 123(6):774–782. doi:10.1001/archopht.123.6.774
111. Mo FM, Proia AD, Johnson WH, Cyr D, Lashkari K (2010) Interferon gamma-inducible protein-10 (IP-10) and eotaxin as biomarkers in age-related macular degeneration. Invest Ophthalmol Vis Sci 51(8):4226–4236. doi:10.1167/iovs.09-3910
112. Sharma NK, Prabhakar S, Gupta A, Singh R, Gupta PK, Gupta PK, Anand A (2012) New biomarker for neovascular age-related macular degeneration: eotaxin-2. DNA Cell Biol 31(11):1618–1627. doi:10.1089/dna.2012.1786
113. Plate KH, Breier G, Risau W (1994) Molecular mechanisms of developmental and tumor angiogenesis. Brain Pathol 4(3):207–218
114. Arnold F, West DC (1991) Angiogenesis in wound healing. Pharmacol Ther 52(3):407–422
115. Chang JH, Garg NK, Lunde E, Han KY, Jain S, Azar DT (2012) Corneal neovascularization: an anti-VEGF therapy review. Surv Ophthalmol 57(5):415–429. doi:10.1016/j.survophthal.2012.01.007
116. Salam A, Mathew R, Sivaprasad S (2011) Treatment of proliferative diabetic retinopathy with anti-VEGF agents. Acta Ophthalmol 89(5):405–411. doi:10.1111/j.1755-3768.2010.02079.x
117. Kita N, Mitsushita J (2008) A possible placental factor for preeclampsia: sFlt-1. Curr Med Chem 15(7):711–715
118. Zent R, Pozzi A (2007) Angiogenesis in diabetic nephropathy. Semin Nephrol 27(2):161–171. doi:10.1016/j.semnephrol.2007.01.007
119. Cao Y (2014) VEGF-targeted cancer therapeutics-paradoxical effects in endocrine organs. Nat Rev Endocrinol 10(9):530–539. doi:10.1038/nrendo.2014.114
120. Lahteenvuo J, Rosenzweig A (2012) Effects of aging on angiogenesis. Circ Res 110(9):1252–1264. doi:10.1161/CIRCRESAHA.111.246116
121. Gupta B, Jyothi S, Sivaprasad S (2010) Current treatment options for retinal angiomatous proliferans (RAP). Br J Ophthalmol 94(6):672–677. doi:10.1136/bjo.2009.166975
122. Scott AW, Bressler SB (2010) Retinal angiomatous proliferation or retinal anastomosis to the lesion. Eye 24(3):491–496. doi:10.1038/eye.2009.311
123. Grossniklaus HE, Kang SJ, Berglin L (2010) Animal models of choroidal and retinal neovascularization. Prog Retin Eye Res 29(6):500–519. doi:10.1016/j.preteyeres.2010.05.003
124. de Oliveira Dias JR, Rodrigues EB, Maia M, Magalhaes O Jr, Penha FM, Farah ME (2011) Cytokines in neovascular age-related macular degeneration: fundamentals of targeted combination therapy. Br J Ophthalmol 95(12):1631–1637. doi:10.1136/bjo.2010.186361
125. Arcondeguy T, Lacazette E, Millevoi S, Prats H, Touriol C (2013) VEGF-A mRNA processing, stability and translation: a paradigm for intricate regulation of gene expression at the post-transcriptional level. Nucleic Acids Res 41(17):7997–8010. doi:10.1093/nar/gkt539
126. Kliffen M, Sharma HS, Mooy CM, Kerkvliet S, de Jong PT (1997) Increased expression of angiogenic growth factors in age-related maculopathy. Br J Ophthalmol 81(2):154–162
127. Goel HL, Mercurio AM (2013) VEGF targets the tumour cell. Nat Rev Cancer 13(12):871–882. doi:10.1038/nrc3627
128. Luo L, Uehara H, Zhang X, Das SK, Olsen T, Holt D, Simonis JM, Jackman K, Singh N, Miya TR, Huang W, Ahmed F, Bastos-Carvalho A, Le YZ, Mamalis C, Chiodo VA, Hauswirth WW, Baffi J, Lacal PM, Orecchia A, Ferrara N, Gao G, Young-Hee K, Fu Y, Owen L,

Albuquerque R, Baehr W, Thomas K, Li DY, Chalam KV, Shibuya M, Grisanti S, Wilson DJ, Ambati J, Ambati BK (2013) Photoreceptor avascular privilege is shielded by soluble VEGF receptor-1. Elife 2, e00324. doi:10.7554/eLife.00324

129. Uehara H, Mamalis C, McFadden M, Taggart M, Stagg B, Passi S, Earle P, Chakravarthy U, Hogg RE, Ambati BK (2014) The reduction of serum soluble Flt-1 in patients with neovascular age-related macular degeneration. Am J Ophthalmol. doi:10.1016/j.ajo.2014.09.036
130. Ambati BK, Nozaki M, Singh N, Takeda A, Jani PD, Suthar T, Albuquerque RJ, Richter E, Sakurai E, Newcomb MT, Kleinman ME, Caldwell RB, Lin Q, Ogura Y, Orecchia A, Samuelson DA, Agnew DW, St Leger J, Green WR, Mahasreshti PJ, Curiel DT, Kwan D, Marsh H, Ikeda S, Leiper LJ, Collinson JM, Bogdanovich S, Khurana TS, Shibuya M, Baldwin ME, Ferrara N, Gerber HP, De Falco S, Witta J, Baffi JZ, Raisler BJ, Ambati J (2006) Corneal avascularity is due to soluble VEGF receptor-1. Nature 443(7114):993–997. doi:10.1038/nature05249
131. Dawson DW, Volpert OV, Gillis P, Crawford SE, Xu H, Benedict W, Bouck NP (1999) Pigment epithelium-derived factor: a potent inhibitor of angiogenesis. Science (New York, NY) 285(5425):245–248
132. Holekamp NM, Bouck N, Volpert O (2002) Pigment epithelium-derived factor is deficient in the vitreous of patients with choroidal neovascularization due to age-related macular degeneration. Am J Ophthalmol 134(2):220–227
133. Cai J, Jiang WG, Grant MB, Boulton M (2006) Pigment epithelium-derived factor inhibits angiogenesis via regulated intracellular proteolysis of vascular endothelial growth factor receptor 1. J Biol Chem 281(6):3604–3613. doi:10.1074/jbc.M507401200
134. Jung JJ, Chen CY, Mrejen S, Gallego-Pinazo R, Xu L, Marsiglia M, Boddu S, Freund KB (2014) The incidence of neovascular subtypes in newly diagnosed neovascular age-related macular degeneration. Am J Ophthalmol 158(4):769–779 e762. doi:10.1016/j.ajo.2014.07.006
135. Yannuzzi LA, Negrao S, Iida T, Carvalho C, Rodriguez-Coleman H, Slakter J, Freund KB, Sorenson J, Orlock D, Borodoker N (2001) Retinal angiomatous proliferation in age-related macular degeneration. Retina 21(5):416–434
136. Spaide RF, Laud K, Fine HF, Klancnik JM Jr, Meyerle CB, Yannuzzi LA, Sorenson J, Slakter J, Fisher YL, Cooney MJ (2006) Intravitreal bevacizumab treatment of choroidal neovascularization secondary to age-related macular degeneration. Retina 26(4):383–390. doi:10.1097/01.iae.0000238561.99283.0e
137. Rosenfeld PJ, Brown DM, Heier JS, Boyer DS, Kaiser PK, Chung CY, Kim RY, MARINA Study Group (2006) Ranibizumab for neovascular age-related macular degeneration. N Engl J Med 355(14):1419–1431. doi:10.1056/NEJMoa054481
138. Gragoudas ES, Adamis AP, Cunningham ET Jr, Feinsod M, Guyer DR, Group VISiONCT (2004) Pegaptanib for neovascular age-related macular degeneration. N Engl J Med 351(27):2805–2816. doi:10.1056/NEJMoa042760
139. Saari JC (2012) Vitamin A metabolism in rod and cone visual cycles. Annu Rev Nutr 32:125–145
140. Li W (2013) Phagocyte dysfunction, tissue aging and degeneration. Ageing Res Rev 12(4):1005–1012. doi:10.1016/j.arr.2013.05.006
141. Jarrett SG, Boulton ME (2012) Consequences of oxidative stress in age-related macular degeneration. Mol Aspects Med 33(4):399–417. doi:10.1016/j.mam.2012.03.009
142. Eskelinen EL (2008) New insights into the mechanisms of macroautophagy in mammalian cells. Int Rev Cell Mol Biol 266:207–247. doi:10.1016/S1937-6448(07)66005-5
143. Buschini E, Piras A, Nuzzi R, Vercelli A (2011) Age related macular degeneration and drusen: neuroinflammation in the retina. Prog Neurobiol 95(1):14–25. doi:10.1016/j.pneurobio.2011.05.011
144. Baynes JW (2001) The role of AGEs in aging: causation or correlation. Exp Gerontol 36(9):1527–1537
145. Ma W, Lee SE, Guo J, Qu W, Hudson BI, Schmidt AM, Barile GR (2007) RAGE ligand upregulation of VEGF secretion in ARPE-19 cells. Invest Ophthalmol Vis Sci 48(3):1355–1361. doi:10.1167/iovs.06-0738

146. Schmidt AM, Hori O, Chen JX, Li JF, Crandall J, Zhang J, Cao R, Yan SD, Brett J, Stern D (1995) Advanced glycation endproducts interacting with their endothelial receptor induce expression of vascular cell adhesion molecule-1 (VCAM-1) in cultured human endothelial cells and in mice. A potential mechanism for the accelerated vasculopathy of diabetes. J Clin Invest 96(3):1395–1403. doi:10.1172/JCI118175
147. Ni J, Yuan X, Gu J, Yue X, Gu X, Nagaraj RH, Crabb JW, Group CGaPAS (2009) Plasma protein pentosidine and carboxymethyllysine, biomarkers for age-related macular degeneration. Mol Cell Proteomics 8(8):1921–1933. doi:10.1074/mcp.M900127-MCP200
148. Crabb JW, Miyagi M, Gu X, Shadrach K, West KA, Sakaguchi H, Kamei M, Hasan A, Yan L, Rayborn ME, Salomon RG, Hollyfield JG (2002) Drusen proteome analysis: an approach to the etiology of age-related macular degeneration. Proc Natl Acad Sci U S A 99(23):14682–14687. doi:10.1073/pnas.222551899
149. Johnson LV, Leitner WP, Rivest AJ, Staples MK, Radeke MJ, Anderson DH (2002) The Alzheimer's A beta -peptide is deposited at sites of complement activation in pathologic deposits associated with aging and age-related macular degeneration. Proc Natl Acad Sci U S A 99(18):11830–11835. doi:10.1073/pnas.192203399
150. Nicolas M, Hassan BA (2014) Amyloid precursor protein and neural development. Development 141(13):2543–2548. doi:10.1242/dev.108712
151. Wang J, Ohno-Matsui K, Yoshida T, Shimada N, Ichinose S, Sato T, Mochizuki M, Morita I (2009) Amyloid-beta up-regulates complement factor B in retinal pigment epithelial cells through cytokines released from recruited macrophages/microglia: another mechanism of complement activation in age-related macular degeneration. J Cell Physiol 220(1):119–128. doi:10.1002/jcp.21742
152. Halle A, Hornung V, Petzold GC, Stewart CR, Monks BG, Reinheckel T, Fitzgerald KA, Latz E, Moore KJ, Golenbock DT (2008) The NALP3 inflammasome is involved in the innate immune response to amyloid-beta. Nat Immunol 9(8):857–865. doi:10.1038/ni.1636
153. Besharse JC, Pfenninger KH (1980) Membrane assembly in retinal photoreceptors I. Freeze-fracture analysis of cytoplasmic vesicles in relationship to disc assembly. J Cell Biol 87(2 Pt 1):451–463
154. LaVail MM (1980) Circadian nature of rod outer segment disc shedding in the rat. Invest Ophthalmol Vis Sci 19(4):407–411
155. Redmond TM, Yu S, Lee E, Bok D, Hamasaki D, Chen N, Goletz P, Ma JX, Crouch RK, Pfeifer K (1998) Rpe65 is necessary for production of 11-cis-vitamin A in the retinal visual cycle. Nat Genet 20(4):344–351
156. Suter M, Reme C, Grimm C, Wenzel A, Jaattela M, Esser P, Kociok N, Leist M, Richter C (2000) Age-related macular degeneration. The lipofusion component N-retinyl-N-retinylidene ethanolamine detaches proapoptotic proteins from mitochondria and induces apoptosis in mammalian retinal pigment epithelial cells. J Biol Chem 275(50):39625–39630. doi:10.1074/jbc.M007049200
157. Eldred GE (1995) Lipofuscin fluorophore inhibits lysosomal protein degradation and may cause early stages of macular degeneration. Gerontology 41(Suppl 2):15–28
158. Bhosale P, Serban B, Bernstein PS (2009) Retinal carotenoids can attenuate formation of A2E in the retinal pigment epithelium. Arch Biochem Biophys 483(2):175–181. doi:10.1016/j.abb.2008.09.012
159. Hwang JC, Chan JW, Chang S, Smith RT (2006) Predictive value of fundus autofluorescence for development of geographic atrophy in age-related macular degeneration. Invest Ophthalmol Vis Sci 47(6):2655–2661. doi:10.1167/iovs.05-1027
160. Ablonczy Z, Higbee D, Anderson DM, Dahrouj M, Grey AC, Gutierrez D, Koutalos Y, Schey KL, Hanneken A, Crouch RK (2013) Lack of correlation between the spatial distribution of A2E and lipofuscin fluorescence in the human retinal pigment epithelium. Invest Ophthalmol Vis Sci 54(8):5535–5542. doi:10.1167/iovs.13-12250
161. Ugarte M, Osborne NN, Brown LA, Bishop PN (2013) Iron, zinc, and copper in retinal physiology and disease. Surv Ophthalmol 58(6):585–609. doi:10.1016/j.survophthal.2012.12.002

162. Erie JC, Good JA, Butz JA, Pulido JS (2009) Reduced zinc and copper in the retinal pigment epithelium and choroid in age-related macular degeneration. Am J Ophthalmol 147(2):276–282.e271. doi:10.1016/j.ajo.2008.08.014
163. Wong RW, Richa DC, Hahn P, Green WR, Dunaief JL (2007) Iron toxicity as a potential factor in AMD. Retina 27(8):997–1003. doi:10.1097/IAE.0b013e318074c290
164. Kiser PD, Farquhar ER, Shi W, Sui X, Chance MR, Palczewski K (2012) Structure of RPE65 isomerase in a lipidic matrix reveals roles for phospholipids and iron in catalysis. Proc Natl Acad Sci U S A 109(41):E2747–E2756
165. Dunaief JL (2006) Iron induced oxidative damage as a potential factor in age-related macular degeneration: the Cogan Lecture. Invest Ophthalmol Vis Sci 47(11):4660–4664. doi:10.1167/iovs.06-0568
166. Hahn P, Ying GS, Beard J, Dunaief JL (2006) Iron levels in human retina: sex difference and increase with age. Neuroreport 17(17):1803–1806. doi:10.1097/WNR.0b013e3280107776
167. Hahn P, Milam AH, Dunaief JL (2003) Maculas affected by age-related macular degeneration contain increased chelatable iron in the retinal pigment epithelium and Bruch's membrane. Arch Ophthalmol 121(8):1099–1105. doi:10.1001/archopht.121.8.1099
168. Chowers I, Wong R, Dentchev T, Farkas RH, Iacovelli J, Gunatilaka TL, Medeiros NE, Presley JB, Campochiaro PA, Curcio CA, Dunaief JL, Zack DJ (2006) The iron carrier transferrin is upregulated in retinas from patients with age-related macular degeneration. Invest Ophthalmol Vis Sci 47(5):2135–2140. doi:10.1167/iovs.05-1135
169. Hahn P, Qian Y, Dentchev T, Chen L, Beard J, Harris ZL, Dunaief JL (2004) Disruption of ceruloplasmin and hephaestin in mice causes retinal iron overload and retinal degeneration with features of age-related macular degeneration. Proc Natl Acad Sci U S A 101(38):13850–13855. doi:10.1073/pnas.0405146101
170. Jomova K, Vondrakova D, Lawson M, Valko M (2010) Metals, oxidative stress and neurodegenerative disorders. Mol Cell Biochem 345(1–2):91–104. doi:10.1007/s11010-010-0563-x
171. Gelfand BD, Wright CB, Kim Y, Yasuma T, Yasuma R, Li S, Fowler BJ, Bastos-Carvalho A, Kerur N, Uittenbogaard A, Han YS, Lou D, Kleinman ME, McDonald WH, Nunez G, Georgel P, Dunaief JL, Ambati J (2015) Iron toxicity in the retina requires alu RNA and the NLRP3 inflammasome. Cell Rep 11(11):1686–1693. doi:10.1016/j.celrep.2015.05.023
172. Chang MA, Bressler SB, Munoz B, West SK (2008) Racial differences and other risk factors for incidence and progression of age-related macular degeneration: Salisbury Eye Evaluation (SEE) Project. Invest Ophthalmol Vis Sci 49(6):2395–2402. doi:10.1167/iovs.07-1584
173. Li M, Atmaca-Sonmez P, Othman M, Branham KE, Khanna R, Wade MS, Li Y, Liang L, Zareparsi S, Swaroop A, Abecasis GR (2006) CFH haplotypes without the Y402H coding variant show strong association with susceptibility to age-related macular degeneration. Nat Genet 38(9):1049–1054. doi:10.1038/ng1871
174. Seddon JM, Reynolds R, Yu Y, Daly MJ, Rosner B (2011) Risk models for progression to advanced age-related macular degeneration using demographic, environmental, genetic, and ocular factors. Ophthalmology 118(11):2203–2211. doi:10.1016/j.ophtha.2011.04.029
175. Gold B, Merriam JE, Zernant J, Hancox LS, Taiber AJ, Gehrs K, Cramer K, Neel J, Bergeron J, Barile GR, Smith RT, Group AMDGCS, Hageman GS, Dean M, Allikmets R (2006) Variation in factor B (BF) and complement component 2 (C2) genes is associated with age-related macular degeneration. Nat Genet 38(4):458–462. doi:10.1038/ng1750
176. Maller JB, Fagerness JA, Reynolds RC, Neale BM, Daly MJ, Seddon JM (2007) Variation in complement factor 3 is associated with risk of age-related macular degeneration. Nat Genet 39(10):1200–1201. doi:10.1038/ng2131
177. Helgason H, Sulem P, Duvvari MR, Luo H, Thorleifsson G, Stefansson H, Jonsdottir I, Masson G, Gudbjartsson DF, Walters GB, Magnusson OT, Kong A, Rafnar T, Kiemeney LA, Schoenmaker-Koller FE, Zhao L, Boon CJ, Song Y, Fauser S, Pei M, Ristau T, Patel S, Liakopoulos S, van de Ven JP, Hoyng CB, Ferreyra H, Duan Y, Bernstein PS, Geirsdottir A, Helgadottir G, Stefansson E, den Hollander AI, Zhang K, Jonasson F, Sigurdsson H, Thorsteinsdottir U, Stefansson K (2013) A rare nonsynonymous sequence variant in C3 is

associated with high risk of age-related macular degeneration. Nat Genet 45(11):1371–1374. doi:10.1038/ng.2740

178. Zhan X, Larson DE, Wang C, Koboldt DC, Sergeev YV, Fulton RS, Fulton LL, Fronick CC, Branham KE, Bragg-Gresham J, Jun G, Hu Y, Kang HM, Liu D, Othman M, Brooks M, Ratnapriya R, Boleda A, Grassmann F, von Strachwitz C, Olson LM, Buitendijk GH, Hofman A, van Duijn CM, Cipriani V, Moore AT, Shahid H, Jiang Y, Conley YP, Morgan DJ, Kim IK, Johnson MP, Cantsilieris S, Richardson AJ, Guymer RH, Luo H, Ouyang H, Licht C, Pluthero FG, Zhang MM, Zhang K, Baird PN, Blangero J, Klein ML, Farrer LA, DeAngelis MM, Weeks DE, Gorin MB, Yates JR, Klaver CC, Pericak-Vance MA, Haines JL, Weber BH, Wilson RK, Heckenlively JR, Chew EY, Stambolian D, Mardis ER, Swaroop A, Abecasis GR (2013) Identification of a rare coding variant in complement 3 associated with age-related macular degeneration. Nat Genet 45(11):1375–1379. doi:10.1038/ng.2758
179. Hagstrom SA, Ying GS, Pauer GJ, Sturgill-Short GM, Huang J, Maguire MG, Martin DF, Comparison of Age-Related Macular Degeneration Treatments Trials Research Group (2014) VEGFA and VEGFR2 gene polymorphisms and response to anti-vascular endothelial growth factor therapy: comparison of age-related macular degeneration treatments trials (CATT). JAMA Ophthalmol 132(5):521–527. doi:10.1001/jamaophthalmol.2014.109
180. van de Ven JP, Nilsson SC, Tan PL, Buitendijk GH, Ristau T, Mohlin FC, Nabuurs SB, Schoenmaker-Koller FE, Smailhodzic D, Campochiaro PA, Zack DJ, Duvvari MR, Bakker B, Paun CC, Boon CJ, Uitterlinden AG, Liakopoulos S, Klevering BJ, Fauser S, Daha MR, Katsanis N, Klaver CC, Blom AM, Hoyng CB, den Hollander AI (2013) A functional variant in the CFI gene confers a high risk of age-related macular degeneration. Nat Genet 45(7):813–817. doi:10.1038/ng.2640
181. Seddon JM, Yu Y, Miller EC, Reynolds R, Tan PL, Gowrisankar S, Goldstein JI, Triebwasser M, Anderson HE, Zerbib J, Kavanagh D, Souied E, Katsanis N, Daly MJ, Atkinson JP, Raychaudhuri S (2013) Rare variants in CFI, C3 and C9 are associated with high risk of advanced age-related macular degeneration. Nat Genet 45(11):1366–1370. doi:10.1038/ng.2741
182. Zareparsi S, Buraczynska M, Branham KE, Shah S, Eng D, Li M, Pawar H, Yashar BM, Moroi SE, Lichter PR, Petty HR, Richards JE, Abecasis GR, Elner VM, Swaroop A (2005) Toll-like receptor 4 variant D299G is associated with susceptibility to age-related macular degeneration. Hum Mol Genet 14(11):1449–1455. doi:10.1093/hmg/ddi154
183. Edwards AO, Chen D, Fridley BL, James KM, Wu Y, Abecasis G, Swaroop A, Othman M, Branham K, Iyengar SK, Sivakumaran TA, Klein R, Klein BE, Tosakulwong N (2008) Toll-like receptor polymorphisms and age-related macular degeneration. Invest Ophthalmol Vis Sci 49(4):1652–1659. doi:10.1167/iovs.07-1378
184. Cho Y, Wang JJ, Chew EY, Ferris FL 3rd, Mitchell P, Chan CC, Tuo J (2009) Toll-like receptor polymorphisms and age-related macular degeneration: replication in three case-control samples. Invest Ophthalmol Vis Sci 50(12):5614–5618. doi:10.1167/iovs.09-3688
185. Zhou P, Fan L, Yu KD, Zhao MW, Li XX (2011) Toll-like receptor 3 C1234T may protect against geographic atrophy through decreased dsRNA binding capacity. FASEB J 25(10):3489–3495. doi:10.1096/fj.11-189258
186. Zhao L, Grob S, Avery R, Kimura A, Pieramici D, Lee J, Rabena M, Ortiz S, Quach J, Cao G, Luo H, Zhang M, Pei M, Song Y, Tornambe P, Goldbaum M, Ferreyra H, Kozak I, Zhang K (2013) Common variant in VEGFA and response to anti-VEGF therapy for neovascular age-related macular degeneration. Curr Mol Med 13(6):929–934
187. Sharma NK, Gupta A, Prabhakar S, Singh R, Bhatt AK, Anand A (2013) CC chemokine receptor-3 as new target for age-related macular degeneration. Gene 523(1):106–111. doi:10.1016/j.gene.2013.03.052
188. Owen LA, Morrison MA, Ahn J, Woo SJ, Sato H, Robinson R, Morgan DJ, Zacharaki F, Simeonova M, Uehara H, Chakravarthy U, Hogg RE, Ambati BK, Kotoula M, Baehr W, Haider NB, Silvestri G, Miller JW, Tsironi EE, Farrer LA, Kim IK, Park KH, DeAngelis MM (2014) FLT1 genetic variation predisposes to neovascular AMD in ethnically diverse popula-

tions and alters systemic FLT1 expression. Invest Ophthalmol Vis Sci 55(6):3543–3554. doi:10.1167/iovs.14-14047
189. Lim LS, Mitchell P, Seddon JM, Holz FG, Wong TY (2012) Age-related macular degeneration. Lancet 379(9827):1728–1738. doi:10.1016/S0140-6736(12)60282-7
190. Reynolds R, Rosner B, Seddon JM (2010) Serum lipid biomarkers and hepatic lipase gene associations with age-related macular degeneration. Ophthalmology 117(10):1989–1995. doi:10.1016/j.ophtha.2010.07.009
191. Neale BM, Fagerness J, Reynolds R, Sobrin L, Parker M, Raychaudhuri S, Tan PL, Oh EC, Merriam JE, Souied E, Bernstein PS, Li B, Frederick JM, Zhang K, Brantley MA Jr, Lee AY, Zack DJ, Campochiaro B, Campochiaro P, Ripke S, Smith RT, Barile GR, Katsanis N, Allikmets R, Daly MJ, Seddon JM (2010) Genome-wide association study of advanced age-related macular degeneration identifies a role of the hepatic lipase gene (LIPC). Proc Natl Acad Sci U S A 107(16):7395–7400. doi:10.1073/pnas.0912019107
192. McKay GJ, Patterson CC, Chakravarthy U, Dasari S, Klaver CC, Vingerling JR, Ho L, de Jong PT, Fletcher AE, Young IS, Seland JH, Rahu M, Soubrane G, Tomazzoli L, Topouzis F, Vioque J, Hingorani AD, Sofat R, Dean M, Sawitzke J, Seddon JM, Peter I, Webster AR, Moore AT, Yates JR, Cipriani V, Fritsche LG, Weber BH, Keilhauer CN, Lotery AJ, Ennis S, Klein ML, Francis PJ, Stambolian D, Orlin A, Gorin MB, Weeks DE, Kuo CL, Swaroop A, Othman M, Kanda A, Chen W, Abecasis GR, Wright AF, Hayward C, Baird PN, Guymer RH, Attia J, Thakkinstian A, Silvestri G (2011) Evidence of association of APOE with age-related macular degeneration: a pooled analysis of 15 studies. Hum Mutat 32(12):1407–1416. doi:10.1002/humu.21577
193. Chen W, Stambolian D, Edwards AO, Branham KE, Othman M, Jakobsdottir J, Tosakulwong N, Pericak-Vance MA, Campochiaro PA, Klein ML, Tan PL, Conley YP, Kanda A, Kopplin L, Li Y, Augustaitis KJ, Karoukis AJ, Scott WK, Agarwal A, Kovach JL, Schwartz SG, Postel EA, Brooks M, Baratz KH, Brown WL, Complications of Age-Related Macular Degeneration Prevention Trial Research Group, Brucker AJ, Orlin A, Brown G, Ho A, Regillo C, Donoso L, Tian L, Kaderli B, Hadley D, Hagstrom SA, Peachey NS, Klein R, Klein BE, Gotoh N, Yamashiro K, Ferris Iii F, Fagerness JA, Reynolds R, Farrer LA, Kim IK, Miller JW, Corton M, Carracedo A, Sanchez-Salorio M, Pugh EW, Doheny KF, Brion M, Deangelis MM, Weeks DE, Zack DJ, Chew EY, Heckenlively JR, Yoshimura N, Iyengar SK, Francis PJ, Katsanis N, Seddon JM, Haines JL, Gorin MB, Abecasis GR, Swaroop A (2010) Genetic variants near TIMP3 and high-density lipoprotein-associated loci influence susceptibility to age-related macular degeneration. Proc Natl Acad Sci U S A 107(16):7401–7406. doi:10.1073/pnas.0912702107
194. Yu Y, Bhangale TR, Fagerness J, Ripke S, Thorleifsson G, Tan PL, Souied EH, Richardson AJ, Merriam JE, Buitendijk GH, Reynolds R, Raychaudhuri S, Chin KA, Sobrin L, Evangelou E, Lee PH, Lee AY, Leveziel N, Zack DJ, Campochiaro B, Campochiaro P, Smith RT, Barile GR, Guymer RH, Hogg R, Chakravarthy U, Robman LD, Gustafsson O, Sigurdsson H, Ortmann W, Behrens TW, Stefansson K, Uitterlinden AG, van Duijn CM, Vingerling JR, Klaver CC, Allikmets R, Brantley MA Jr, Baird PN, Katsanis N, Thorsteinsdottir U, Ioannidis JP, Daly MJ, Graham RR, Seddon JM (2011) Common variants near FRK/COL10A1 and VEGFA are associated with advanced age-related macular degeneration. Hum Mol Genet 20(18):3699–3709. doi:10.1093/hmg/ddr270
195. Wang G (2014) Chromosome 10q26 locus and age-related macular degeneration: a progress update. Exp Eye Res 119:1–7. doi:10.1016/j.exer.2013.11.009
196. Pahl L, Spangenberg A, Schubert S, Schonmann U, Schmidtke J, Stuhrmann M (2012) Characterization of the 10q26-orthologue in rhesus monkeys corroborates a functional connection between ARMS2 and HTRA1. Exp Eye Res 98:75–78. doi:10.1016/j.exer.2012.03.007
197. Wang G, Spencer KL, Court BL, Olson LM, Scott WK, Haines JL, Pericak-Vance MA (2009) Localization of age-related macular degeneration-associated ARMS2 in cytosol, not mitochondria. Invest Ophthalmol Vis Sci 50(7):3084–3090. doi:10.1167/iovs.08-3240
198. Kanda A, Chen W, Othman M, Branham KE, Brooks M, Khanna R, He S, Lyons R, Abecasis GR, Swaroop A (2007) A variant of mitochondrial protein LOC387715/ARMS2, not

HTRA1, is strongly associated with age-related macular degeneration. Proc Natl Acad Sci U S A 104(41):16227–16232. doi:10.1073/pnas.0703933104

199. Yang Z, Camp NJ, Sun H, Tong Z, Gibbs D, Cameron DJ, Chen H, Zhao Y, Pearson E, Li X, Chien J, Dewan A, Harmon J, Bernstein PS, Shridhar V, Zabriskie NA, Hoh J, Howes K, Zhang K (2006) A variant of the HTRA1 gene increases susceptibility to age-related macular degeneration. Science 314(5801):992–993. doi:10.1126/science.1133811
200. Wang G, Dubovy SR, Kovach JL, Schwartz SG, Agarwal A, Scott WK, Haines JL, Pericak-Vance MA (2013) Variants at chromosome 10q26 locus and the expression of HTRA1 in the retina. Exp Eye Res 112:102–105. doi:10.1016/j.exer.2013.04.019
201. Kwa FA, Thrimawithana TR (2014) Epigenetic modifications as potential therapeutic targets in age-related macular degeneration and diabetic retinopathy. Drug Discov Today 19(9):1387–1393. doi:10.1016/j.drudis.2014.03.026
202. Gemenetzi M, Lotery AJ (2014) The role of epigenetics in age-related macular degeneration. Eye. doi:10.1038/eye.2014.225
203. Hunter A, Spechler PA, Cwanger A, Song Y, Zhang Z, Ying GS, Hunter AK, Dezoeten E, Dunaief JL (2012) DNA methylation is associated with altered gene expression in AMD. Invest Ophthalmol Vis Sci 53(4):2089–2105. doi:10.1167/iovs.11-8449
204. Holz FG, Schmitz-Valckenberg S, Fleckenstein M (2014) Recent developments in the treatment of age-related macular degeneration. J Clin Invest 124(4):1430–1438. doi:10.1172/JCI71029
205. Sieving PA (2012) NEI audacious goals initiative to catalyze innovation. Invest Ophthalmol Vis Sci 53(11):7149–7150. doi:10.1167/iovs.12-11069
206. Salowe RJ, O'Brien JM (2014) NEI's audacious goals initiative. Ophthalmology 121(3):615–616. doi:10.1016/j.ophtha.2013.11.011
207. Zhao C, Yasumura D, Li X, Matthes M, Lloyd M, Nielsen G, Ahern K, Snyder M, Bok D, Dunaief JL, LaVail MM, Vollrath D (2011) mTOR-mediated dedifferentiation of the retinal pigment epithelium initiates photoreceptor degeneration in mice. J Clin Invest 121(1):369–383. doi:10.1172/JCI44303
208. Allison DB, Antoine LH, Ballinger SW, Bamman MM, Biga P, Darley-Usmar VM, Fisher G, Gohlke JM, Halade GV, Hartman JL, Hunter GR, Messina JL, Nagy TR, Plaisance EP, Powell ML, Roth KA, Sandel MW, Schwartz TS, Smith DL, Sweatt JD, Tollefsbol TO, Watts SA, Yang Y, Zhang J, Austad SN (2014) Aging and energetics' 'Top 40' future research opportunities 2010–2013. F1000 Res 3:219. doi:10.12688/f1000research.5212.1
209. Atmaca-Sonmez P, Li Y, Yamauchi Y, Schanie CL, Ildstad ST, Kaplan HJ, Enzmann V (2006) Systemically transferred hematopoietic stem cells home to the subretinal space and express RPE-65 in a mouse model of retinal pigment epithelium damage. Exp Eye Res 83(5):1295–1302. doi:10.1016/j.exer.2006.07.013
210. Sengupta N, Caballero S, Sullivan SM, Chang LJ, Afzal A, Li Calzi S, Kielczewski JL, Prabarakan S, Ellis EA, Moldovan L, Moldovan NI, Boulton ME, Grant MB, Scott EW, Harris JR (2009) Regulation of adult hematopoietic stem cells fate for enhanced tissue-specific repair. Mol Ther 17(9):1594–1604. doi:10.1038/mt.2009.145
211. Fowler BJ, Gelfand BD, Kim Y, Kerur N, Tarallo V, Hirano Y, Amarnath S, Fowler DH, Radwan M, Young MT, Pittman K, Kubes P, Agarwal HK, Parang K, Hinton DR, Bastos-Carvalho A, Li S, Yasuma T, Mizutani T, Yasuma R, Wright C, Ambati J (2014) Nucleoside reverse transcriptase inhibitors possess intrinsic anti-inflammatory activity. Science 346(6212):1000–1003. doi:10.1126/science.1261754
212. Dridi S, Hirano Y, Tarallo V, Kim Y, Fowler BJ, Ambati BK, Bogdanovich S, Chiodo VA, Hauswirth WW, Kugel JF, Goodrich JA, Ponicsan SL, Hinton DR, Kleinman ME, Baffi JZ, Gelfand BD, Ambati J (2012) ERK1/2 activation is a therapeutic target in age-related macular degeneration. Proc Natl Acad Sci U S A 109(34):13781–13786. doi:10.1073/pnas.1206494109

213. Doyle SL, Ozaki E, Brennan K, Humphries MM, Mulfaul K, Keaney J, Kenna PF, Maminishkis A, Kiang AS, Saunders SP, Hams E, Lavelle EC, Gardiner C, Fallon PG, Adamson P, Humphries P, Campbell M (2014) IL-18 attenuates experimental choroidal neovascularization as a potential therapy for wet age-related macular degeneration. Sci Transl Med 6(230):230ra244. doi:10.1126/scitranslmed.3007616
214. Hirano Y, Yasuma T, Mizutani T, Fowler BJ, Tarallo V, Yasuma R, Kim Y, Bastos-Carvalho A, Kerur N, Gelfand BD, Bogdanovich S, He S, Zhang X, Nozaki M, Ijima R, Kaneko H, Ogura Y, Terasaki H, Nagai H, Haro I, Nunez G, Ambati BK, Hinton DR, Ambati J (2014) IL-18 is not therapeutic for neovascular age-related macular degeneration. Nat Med 20(12):1372–1375. doi:10.1038/nm.3671

HIV and Aging: Parallels and Synergistic Mechanisms Leading to Premature Disease and Functional Decline

Anna Hearps, Katherine Schafer, Kevin High, and Alan Landay

Contents

A. Hearps, Ph.D. (✉)
Centre for Biomedical Research, Burnet Institute, Melbourne, Australia
e-mail: annah@burnet.edu.au

K. Schafer, M.D. • K. High, M.D.
Department of Internal Medicine, Section on Infectious Diseases, Wake Forest School of Medicine, Winston-Salem, NC, USA
e-mail: kschafer@wakehealth.edu; khigh@wakehealth.edu

A. Landay, Ph.D.
Department of Immunology/Microbiology, Rush University Medical Center, Chicago, IL, USA
e-mail: Alan_Landay@rush.edu

F. Sierra, R. Kohanski (eds.), *Advances in Geroscience*,
DOI 10.1007/978-3-319-23246-1_17

1 Introduction

Why is there a chapter on HIV in a translational science textbook on aging? Effective antiretroviral therapy (ART) has resulted in many people with HIV infection living far beyond what was thought possible just a few years ago [1, 2]. It is estimated that over half of all U.S. HIV-infected persons will be >50 years by 2015 [3, 4], and the success and availability of ART is even leading to an aging HIV-infected population in developing nations [1, 5, 6], emphasizing the need for aging-related research in those countries where HIV burden of illness is greatest [7].

There is marked debate as to whether HIV accelerates aging itself or is an added risk factor for a number of diseases and conditions that lead to an "aged phenotype." Of course, there is no single pathway that defines "aging" – in fact, two recent, excellent reviews [8, 9] emphasize a number of "hallmarks" of aging – biologic changes that accompany aging, but none is clearly "the" causal pathway. Cardiovascular disease (CVD) and many other diseases increase with age, and advancing age is the leading risk factor for CVD. But we generally don't consider CVD risk factors (e.g. hypercholesterolemia) to be conditions that accelerate aging itself. However aging with HIV is different than aging with hypercholesterolemia; a much broader array of illnesses occurs with greater frequency in people aging with HIV (PAWH). This leads not only to prematurity of a single disease, but multiple diseases, as well as decreased physiologic reserve and increased vulnerability to catastrophic illness, hospitalization and death. Functional decline – physical and/or cognitive – often accompanies multi-morbidity or may occur independently, but in either case functional decline is the strongest risk factor for disability and loss of independence, particularly when social and family support structures are lacking. This state of multi-morbidity, vulnerability, functional decline and loss of independence is what we usually view as "old" – or the aged phenotype – and there is no doubt that this phenotype occurs earlier in PAWH when compared to HIV-uninfected persons [10–14]. In this chapter, we will briefly summarize a few examples of age-related serious non-AIDS events (SNAEs) such as CVD and cancer – and geriatric syndromes (functional decline/frailty and multi-morbidity) to highlight the clinical relevance and translational opportunities to link mechanisms to clinical outcomes in PAWH.

2 Increased Prevalence of Age-Related, Serious Non-AIDS Events (SNAEs) in PAWH

While life expectancy has increased markedly for PAWH, this group experiences a greater frequency of age-associated comorbid conditions, such as CVD, non-AIDS-defining cancers (liver, lung, anal), osteoporosis/osteopenia/bone fractures, metabolic syndrome, and neurocognitive dysfunction. These events are termed SNAEs and increasingly robust data suggest they are very common in PAWH, even those well controlled on ART [15]. CVD and cancer have the most robust database and are therefore examined in greater detail in the following paragraphs.

CVD risk factors and rates of acute coronary syndromes and heart failure are markedly increased in HIV-infected vs. age matched control subjects [16–20], and coronary artery "age" is accelerated on average by about 15 years in treated HIV-infected persons (median duration of ART ~11 years) as assessed by coronary artery calcium (CAC) score comparing PAWH to age-defined norms established in the MultiEthnic Study on Aging (MESA) cohort [21]. Higher levels of C-reactive protein, interleukin-6, and D-dimer have been shown to be significantly associated with an increased risk of all-cause mortality in HIV-infected individuals not on ART, and much of this is cardiovascular mortality [22]. Specific ART drugs also may be causally associated with early heart disease, even after controlling for age and traditional cardiovascular risk factors [23, 24]. Further, lipodystrophy and metabolic syndrome (altered body fat, hyperlipidemia and insulin resistance) are common in HIV-infected patients receiving ART [25, 26]. The redistribution of fat mass and progression to metabolic syndrome (12/100 patient-years) typically occurs within 3 years after the initiation of ART [27], when weight gain is often substantial, thus increasing cardiovascular disease risk. Enhanced cardiovascular "aging" is not limited to coronary artery disease. Left ventricular diastolic dysfunction and increased vascular stiffness [28–30] are more common in HIV-infected subjects versus uninfected, age-matched controls even after controlling for hypertension and other risk factors. Heart failure and atrial fibrillation, typically seen in older adults, is increasingly being reported in younger PAWH [18–20].

As ART use has become widespread, AIDS-defining cancers (Kaposi's Sarcoma, lymphomas) have become less common in this population, but increased survival and perhaps decreased competing causes of AIDS-defining cancer deaths have led to increased numbers of non-AIDS-Defining Cancers (NADC) [31]. A number of NADC occur more frequently in PAWH than age-matched control cohorts [32] and NADC are increasingly a cause of death in PAWH [15]. Initial reports suggested the age of onset of many NADC was much earlier than in those without HIV, but most of this appears to be a cohort effect. PAWH are a younger cohort than the general population [1] so colon, lung or other cancers may appear to only be occurring in younger adults, but there aren't many 70+ year old PAWH so this is often a false impression. As control groups and age-adjustments have been refined, it appears NADC are only minimally "accelerated" with regard to age at diagnosis – perhaps 3–5 years [33] (Table 1). It is important to note that some NADC that are most strongly related to age – breast cancer in women and prostate cancer in men – do not appear to be increased in those with HIV [33, 34], though data are sure to evolve as persons continue to age with HIV infection.

Another way to examine the question of whether HIV directly "ages" individuals or acts in parallel is to assess whether age remains an independent risk factor for SNAEs in PAWH. Within cohorts of PAWH, increased age is an independent predictor of stroke, myocardial infarction, fractures, osteoporosis, diabetes, and non-AIDS associated cancers, while controlling for CD4 count, viral load, intravenous drug use, smoking, and duration of HIV infection [35].

Table 1 Age differences between HIV-infected and HIV-uninfected for select NADC

Select NADC	Median age at diagnosis in AIDS population (years)	Median age at diagnosis in HIV⁻ general population (years)	Apparent difference (years)	Median expected age at diagnosis in HIV⁻ population if cohort limited to the same age distribution as those with AIDS (years)	Real difference (years)
Rectal	46	69	–23	51	–5
Lung	50	70	–20	54	–4
Ovarian	42	63	–21	46	–4
Myeloma	47	70	–23	52	–5

Adapted from [33]

3 Geriatric Syndromes in PAWH

3.1 Multi-morbidity

Despite the success of ART, extensive evidence suggests HIV-infected persons are more likely than their HIV-uninfected counterparts to have multiple comorbidities at a young age. This is perhaps not surprising for illnesses with overlapping risk factors (i.e. hepatitis C, human papillomavirus [HPV]-related cancers), but it is also true across organ systems where intersecting risks are not so clear; early-onset of disease in individual organ systems in PAWH has been observed (e.g. coronary artery disease, arterial stiffness, cerebral blood flow, and bone fractures) [21, 36–39]. Chronic liver and renal diseases are also more common in PAWH compared to HIV uninfected populations [40]. Although behavioral factors such as smoking and illicit drug use are more prevalent in populations of PAWH, controlled studies have shown that these factors do not fully explain the increased risk for age-related conditions such as cardiovascular and liver disease [35, 41, 42]. Where aggressive ART is widely available, 58 % of HIV-infected subjects aged 51–60 have one or more of the following: renal failure, diabetes mellitus, bone fracture, hypertension or overt cardiovascular disease vs. only 35 % of HIV-uninfected controls [10, 35]. The rate of multi-morbidity (> one major chronic illness) at age >50 years is about 2.5 times higher in HIV-infected subjects vs. HIV-uninfected controls [10, 35, 40].

On average, PAWH aged 50 and older have up to three chronic illnesses, in addition to HIV [43] (Fig. 1). Depending on the population, studies have demonstrated increased prevalence of specific comorbidities. The onset of multi-morbidity appears to be accelerated 12–15 years in those with HIV infection [10]. Further, multi-morbidity risk assessments such as the Veterans Aging Cohort Study (VACS) Index derived and validated in HIV-infected subjects correlates with mortality risk and hospitalization [44, 45]. Importantly, the VACS index has now been validated to predict mortality in HIV-uninfected populations [45] demonstrating the generalizability of this integrated measure of cumulative damage to the hematopoietic, immunologic, hepatic and renal systems.

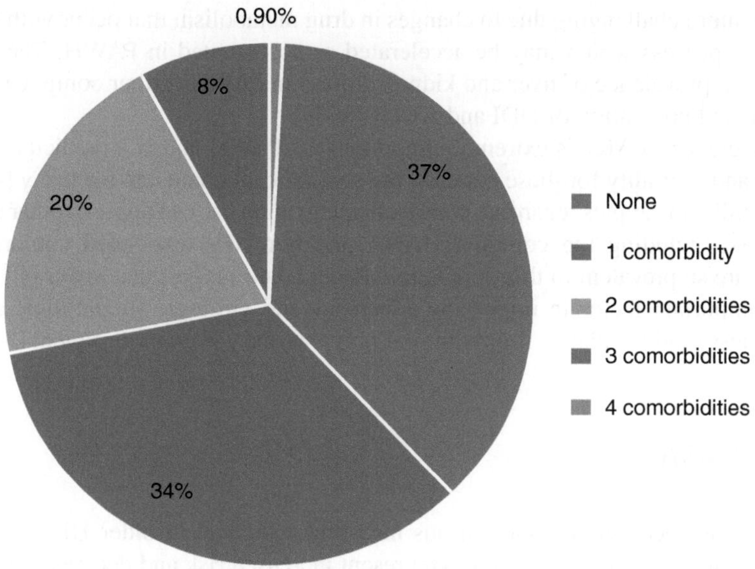

Fig. 1 Prevalence of comorbidity burden, HIV-infected persons age > 50 (Data derived from [10])

3.2 *Polypharmacy*

In the setting of multi-morbidity, PAWH have increased risks of developing both HIV-associated non-AIDS (HANA) and non-HIV related conditions. Consequently, polypharmacy and increased complexity of care are becoming commonplace in the health management of PAWH, noting that the disease courses may be altered depending on the patient's state of virologic suppression [46]. PAWH who are aged 50 and older are more likely to have at least one medication (in addition to ART) compared to PAWH younger than age 50 [47]. Specifically, older PAWH are more likely to take concurrent cardiovascular, gastrointestinal, and hormonal medications than younger patients [47].

The inherent complexity of polypharmacy translates into potential harm for older patients. In older adults without HIV, polypharmacy is a known risk factor for falls, adverse drug events (ADE) including drug-drug interactions (DDI), morbidity, and mortality [48]. These associations remain true for PAWH but may occur at younger ages compared to people without HIV infection [47–49]. At baseline, older patients are at increased risk for ADE, compared to younger patients. In addition to direct toxicity for the patient, ADE and DDI can mean decreased efficacy of therapy, both for HIV and other comorbidities, especially in the case of protease inhibitor (PI)-based ART [49]. The list of potential DDI is extensive and includes virtually every class of therapeutics, including cardiovascular, gastrointestinal, hematologic, anti-neoplastic, antimicrobial, psychiatric, and endocrine (including inhaled steroids which aren't typically considered to have systemic effects) [47–50]. Predicting

DDI is more challenging due to changes in drug metabolism that occur with normal aging, a process which may be accelerated or accentuated in PAWH. The known increased prevalence of liver and kidney disease in PAWH further complicates prediction and prevention of DDI and ADE [48, 49].

Adherence to ART is extremely important for PAWH and is a predictor of morbidity and mortality for these patients, but a significant challenge for many [49, 51–53]. Similar principles regarding consistent medication use can apply to other chronic illnesses, including the common HANA and non-HIV associated comorbidities which are so prevalent in this population. Recent data suggest that lower pill burden is an important factor in improving adherence and virologic suppression, making awareness (and avoidance if possible) of polypharmacy even more salient [54].

3.3 Frailty

Frailty has been defined and various measures validated in older HIV-uninfected adults, but it is generally agreed to represent increased risk and decreased ability to recuperate from illness and injury. Frailty is increased in HIV-infected vs. age-matched HIV-uninfected controls [13, 55–61]. In those PAWH, there is a high correlation between various measures of frailty validated in seniors, though definitions vary from study to study and the reader should be cautious to assess frailty definitions, cohort effects, and control group definitions when comparing individual rates of frailty between studies [62]. Early research measured the prevalence of frailty using the frailty-related phenotype (FRP) in 55 year old men with HIV infection (infected for less than or equal to 4 years) as equivalent to the prevalence of FRP in men 65 years of age or older without HIV [12]. Onen et al. measured a prevalence of 9 % for frailty in an outpatient HIV clinic (mean age of 41.7 years), which was comparable to the prevalence of frailty in Caucasian Europeans aged 65 years and older [13]. In the same study, investigators measured patient-level characteristics; frailty was associated with socioeconomic status, multi-morbidity, lower education level, longer period of HIV infection, history of opportunistic infection, as well as an increased risk of hospitalization, number of hospitalizations, and inpatient length of stay [13]. Much of the early data suggested frailty in PAWH was associated with uncontrolled HIV/weight loss/wasting, but more recent data suggest frailty in HIV has been associated with obesity and intramuscular adiposity, as seen in HIV-uninfected older persons [59, 61, 63].

Frailty is potentially mediated more by inflammation and body composition than by HIV infection itself. This is compounded by the fact that optimal immune function may be hindered by age-related changes that are independent of virologic suppression [46, 64]. In PAWH, frailty is associated with central obesity, sarcopenia, and increased muscle fat density for age [65]. Oursler et al. showed that, despite ART, physical function in PAWH aged 50 years and older was worse compared to HIV-uninfected people [60]. Regardless of age, HIV-infected patients with chronic pulmonary disease had worse physical function compared to HIV-uninfected people, such that a 50 year old person with HIV and chronic obstructive pulmonary disease (COPD) had functional measurements approximating a 68 year old person with COPD, but without HIV [60]. Within populations of PAWH, the prevalence of frailty is increased in people who also use intravenous drugs [43]. Not surprisingly, frail PAWH have a high prevalence of comorbidities, including hypertension, COPD, viral hepatitis, dementia, and cancer; this pattern of multi-morbidity mirrors trends seen in the larger population of PAWH [11].

Beyond the effects that frailty may have on physical health and mental well-being, this phenotype has implications for healthcare delivery and models of care. Guideline-driven care may not be practical or universally applicable to PAWH if their risks of various conditions change at different age breakpoints or based on factors other than what has been measured in the foundational studies. Use of more tailored prediction tools such as the VACS Index may be more applicable due to incorporation of multiple biomarkers [46].

3.4 *Neurocognitive Impairment*

A full examination of the neurologic manifestations of HIV and even discussion limited to cognitive impairment is beyond the scope of this review. Briefly, 50 % of PAWH will develop an HIV-associated neurocognitive disorder (HAND) [43, 66]. HAND is a spectrum of clinical conditions ranging from asymptomatic neurocognitive impairment (ANI – least severe) to HIV-associated dementia (HAD – most severe, previously known as "AIDS Dementia Complex") [67] (Fig. 2). The symptoms can be largely reversed with ART, but the incidence of HAND is associated with worse adherence. The impact of HAND is marked with HAND being associated with decreased ability to complete daily functions, poorer quality of life, and shorter survival. While the incidence and prevalence of HAND are decreasing due to ART, the incidence and prevalence of ANI and mild neurocognitive disorder (MND) are stable to increasing, spurring a recommendation for universal neurocognitive screening of PAWH [67, 68]. Furthermore, HIV itself may alter brain structure, despite ART, thus, the full expression of HIV-related cognitive disorders may require time to become apparent [69].

3.5 *Quality of Life and Mental Health*

Compared to HIV-uninfected people, PAWH (ages $\geq$50 years) are not as happy, optimistic about aging, or resilient [43, 70]. They also experience more perceived stress, anxiety about the future, and lower quality of physical and mental health [70]. Social isolation, a common occurrence in older adults regardless of HIV status, is associated with increased risk for hospitalization and all-cause mortality. The social networks for older PAWH may shrink due to common age-related factors

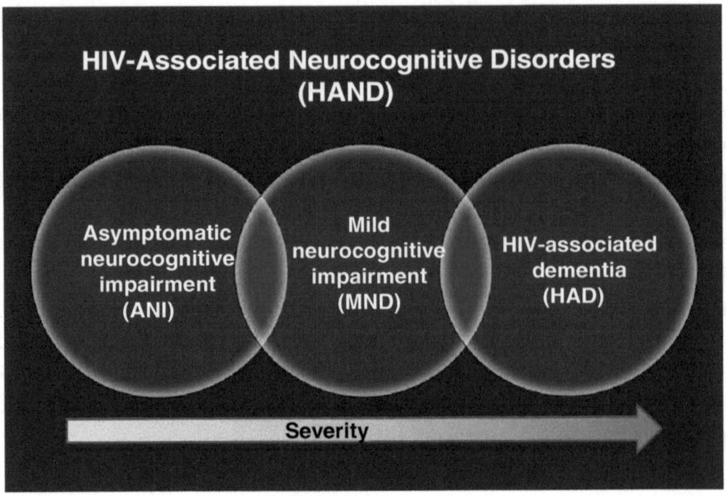

Fig. 2 The spectrum of HIV-associated neurocognitive impairment

(e.g. age-related deaths, limited transportation, geographic isolation) and/or more HIV-specific factors: loss of peers earlier in the HIV/AIDS epidemic, stigma, marginalization in the current make-up of the epidemic [43, 71, 72]. While both HIV-infected and HIV-uninfected older adults may experience social isolation to some degree, HIV infection alone is associated with increased risk and prevalence of social isolation [71].

4 Potential Mechanisms Linking Chronic HIV Infection with Age-Related Conditions

4.1 Immunological Similarities Between HIV Infection and Healthy Aging

The overlapping burden of morbidities and SNAEs in PAWH and aged individuals has led to the hypothesis that similar pathogenic mechanisms are driving the development of these diseases in both populations. Indeed there are many immunological parallels between chronic HIV infection and healthy aging which are summarized in Table 2 and discussed in detail below.

Adaptive Immune Changes

The reduced number of naïve T cells and reduced CD4:CD8 T cell ratio observed in the aged is a hallmark of T cell immunosenescence [73, 74] and also occurs in HIV-infected individuals due in part to thymic involution and reduced regenerative capacity [75, 76]. Low CD4:CD8 T cell ratio is an independent predictor of non-AIDS mortality [77] and cardiovascular disease risk [78, 79] in HIV-infected individuals, suggesting T cell immunosenescence has important clinical implications in the context of PAWH. Importantly, the majority of HIV-infected individuals on long term ART fail to normalize the CD4:CD8 T cell ratio, despite restoration of CD4+ T cell levels [80]. Elderly HIV-uninfected and HIV-infected individuals also exhibit increased T cell activation (as measured by expression of the activation markers HLA-DR and CD38) [76, 81], an increased susceptibility to spontaneous apoptosis [82, 83] and an expansion of 'senescent' memory CD8+ T cells which lack expression of the co-stimulatory molecule CD28, contain shortened telomeres and exhibit a reduced proliferative potential [84–86]. Expansion of this cell population is thought to be largely driven by chronic antigenic stimulation by cytomegalovirus (CMV), with a large proportion of CD8+ T cells in both HIV-infected and aged individuals being specific for CMV epitopes (discussed further below). However, there are phenotypic differences in the T cells expanded due to HIV infection and those observed in CMV+ HIV seronegative individuals, in that the former show an increased number of transitional memory cells and a reduced proportion of CD28-cells expressing CD57 (a marker of reduced proliferative capacity), with low levels of this population being associated with increased risk of mortality [87]. These observations suggest that although there are many phenotypic similarities between HIV infection and aging, the mechanistic drivers, and thus immunological consequences of, senescent T cell expansion in HIV and aging may be subtly different.

While the above mentioned immunological alterations due to HIV are significantly improved by ART, they typically fail to normalize, and defects including reduced naïve T cell proportions, inverted CD4:CD8 T cell ratios and increased T cell activation largely persist in virologically suppressed HIV-infected individuals (reviewed in [88]). Furthermore, aging appears to impact negatively on the immunological benefit of ART and associated reductions in immune activation, with older HIV-infected individuals exhibiting muted naïve T cell regeneration following ART initiation [89], suggesting immunological aging may heighten HIV-related immune dysfunction in older HIV-infected individuals.

During aging, there is a reduction in the number of both total and memory B cells and defects emerge in class switching and antibody production which are thought to contribute to impaired vaccine response in the elderly [81, 90]. Viremic HIV infection is similarly associated with reduced total and memory B cell numbers together with hypergammaglobulinemia, increased cellular activation and increased susceptibility to apoptosis [91]. ART reverses many of these defects, although virologically suppressed HIV-infected individuals continue to show impaired antibody production, reduced vaccine responses and an incomplete restoration of memory B cells [92, 93]. Markers of HIV disease severity including viral load and immune activation are associated with an increased frequency of regulatory B cells (Bregs), which inhibit CD8+ T cell proliferation and function via a mechanism involving IL-10 and PD-1 [94], which may potentiate immune dysfunction in HIV. Bregs from HIV-infected individuals constitutively express higher levels of IL-6, IL-10 and cellular activation markers, suggesting increased Breg activation in vivo [95]. Interestingly, older HIV-infected individuals show an altered pattern of B cell resto-

Table 2 Comparison of immunological changes observed in HIV infection and aging

	HIV Infection	Aging
T cells	Decreased naïve and increased memory T cells Decreased CD4:CD8 T cell ratio Expansion of senescent CD8+ CD28-T cells Expansion of CMV-specific CD8+ T cells	
B cells	Decreased numbers and proportion of mature/memory B cells Impaired class switching and antibody production, leading to impaired vaccine response	
Monocytes/Macrophage	Impaired phagocytosis Increased proportion CD16+ inflammatory monocytes Increased expression of cellular activation markers Increased TLR4-stimulated cytokine/chemokine production Dysfunctional TLR-responses	
Natural Killer (NK) cells	Expansion of CD56dim population (acute HIV infection only) Impaired cytokine production (modest decline in aging)	
	Increased activation Increased spontaneous ADCC Expansion of CD56-population Impaired cytotoxic potential	
Dendritic cells	Reduced numbers in blood (pDC in HIV and mDC in aging)	
	Impaired response to TLR 7/8 stimulation Increased activation in pDC	Impaired chemotaxis and antigen uptake
Neutrophils	Increased basal activation	Decreased chemotaxis
	Impaired phagocytosis, migration, respiratory burst and intracellular killing	
Inflammation	Increased plasma levels of IL-6, TNF, hsCRP	
Soluble markers of immune activation	Increased plasma levels of CXCL10, sCD14, sCD163, neopterin	
Gut integrity	Increased plasma concentrations of LPS Increased levels of gut permeability markers I-FABP and/or zonulin-1	
Oxidative stress	Increased plasma markers of oxidative stress	
Telomere length	Shortened (PBMC, T cells, monocytes)	

Shaded cells indicate immunological changes which occur in both HIV infection and aging

ration after ART initiation, including expansion of the naïve population to levels greater than those in uninfected individuals [96], suggesting that HIV and age may potentiate immune dysfunction in PAWH. These studies collectively indicate that HIV infection induces a phenotype within the adaptive immune system which resembles age-related immunosenescence and immune dysfunction, and viral suppression associated with ART only partially improves these parameters.

4.1.1 Innate Immune Changes

Immunological similarities are also seen between HIV infection and healthy aging within the innate immune system. Monocytes from both HIV-infected individuals and the elderly show impaired phagocytic function, increased TLR4-mediated production of pro-inflammatory cytokines and chemokines and an increase in phenotypic markers of activation, including an expansion of the inflammatory CD16+ monocyte subset [97–101]. Viral suppression associated with ART appears to normalize some of these changes, such as the proportions of CD16+ subsets, whilst other markers of monocyte dysfunction persist [97, 102]. Elevated levels of soluble plasma monocyte/innate immune activation markers including the chemoattractant CXCL10 (released from IFNγ-stimulated monocytes), neopterin and soluble(s) receptors CD14 and CD163 (shed from activated monocytes/macrophages) are elevated in both the elderly and HIV-infected individuals and although ART reduces the levels of these markers in HIV infection, they fail to normalize [97, 98, 103–107].

An increase in total NK cell number, due to expansion of the $CD56^{dim}$ population occurs during aging and in acute HIV infection [74, 108]. Aging is associated with a minimal impairment of NK cell cytotoxic function and cytokine production [109–111], whilst overall cytolytic activity is impaired in HIV infection (most prominently in viremic infection) which may be due in part to expansion of an anergic $CD56^{neg}$ population in HIV-infected individuals [108]. NK cells from both viremic and virologically suppressed HIV-infected individuals show heightened basal activation [112, 113] and spontaneous ADCC activity [112], whilst cytokine production is impaired [114]. The functionality of neutrophils is similarly impaired in HIV infection as in aging, as evidenced by impaired phagocytosis and oxidative burst but a heightened basal level of activation [115]. The impact of age on the number, activation state and function of dendritic cells remains unclear due to conflicting findings (reviewed in [116]), although chemotaxis and antigen uptake are impaired in aged humans [117, 118] whilst HIV infection is associated with impaired ex vivo response of plasmacytoid dendritic cells to TLR7 ligands [119]. Taken together, these data suggest that a signature of increased activation but dysregulated function is a common effect of both HIV infection and aging on innate immune cells, although much work is required to fully define the extent of these effects. It is important to note that many of the above mentioned age-related immunological changes have been observed in cross-sectional studies of HIV-infected individuals with varying degrees of immunosuppression both prior to and following ART initiation. Future longitudinal studies are required in cohorts of individuals who initiate ART early and maintain immunocompetence to adequately determine the impact of virologically suppressed HIV infection on age-related immune changes in PAWH.

4.2 Telomere Shortening

The presence of telomeres at the ends of chromosomes protects the DNA from damage and preserves the replicative potential of the cell. Telomere length progressively decreases with age and triggers replicative senescence, which contributes to immunosenescence and immune aging [120]. Telomere shortening is associated with risk of a range of age-related diseases including malignancies [121], cardiovascular/metabolic disease [122–124] and neurocognitive disease [125, 126] (summarized in Table 3 and reviewed in [195]) and has been linked with premature death in a large prospective study in Denmark [123]. HIV infection is associated with heightened telomere shortening within both T cells [85] and monocytes [97]. However, epidemiological links between shortened telomeres and HIV-related co-morbidities have received little investigation to date.

Telomere length is maintained within cells via the action of telomerase and premature telomere shortening in HIV infection may be due to reduced activity of this critical enzyme. The HIV proteins Vpr [196] and Tat [197] have been shown to inhibit telomerase in vitro whilst HIV-infected individuals appear to have an impaired ability to upregulate telomerase in response to cell stimulation [198]. Antiretroviral therapy may also contribute to premature telomere shortening as the nucleos(t)ide reverse transcriptase inhibitor (NRTI) drugs can inhibit the telomerase reverse transcriptase (TERT) component of human telomerase. In vitro studies have shown that even modern, relatively non-toxic NRTIs such as tenofovir and emtricitabine show inhibitory effects on human TERT [199, 200], and can accelerate telomere loss in cultured cells [199] whilst a small cross-sectional study found telomeres from individuals on NRTI-containing regimens were shorter than HIV negative controls and HIV-infected individuals taking non-NRTI containing regimens [200]. NRTIs remain the backbone of ART regimens throughout the world, but the accumulated consequences of decades of NRTI-treatment on oxidative stress and telomere shortening remain to be defined.

4.3 Oxidative Stress

An imbalance between levels of oxidants and anti-oxidants occurs during aging, resulting in increased plasma markers of oxidative stress in the elderly [201, 202] which contribute to immunosenescence and inflamm-aging (reviewed in [203]). HIV infection is also associated with increased levels of oxidative stress, with decreased plasma levels of anti oxidant factors such as glutathione and increased levels of the oxidative stress marker malondialdehyde found in both viremic and

Table 3 Associations between immunological changes occurring during aging/HIV infection and morbidity/mortality

		HIV-infected	General population
	Marker	Outcome/risk factor	
Inflammation			
Cardiovascular/metabolic disease	IL-6	Cardiovascular events [127, 128], obesity [129]	Sudden cardiac death [130, 131], cardiovascular events [131–134]
	hsCRP	Cardiovascular events [127, 128], progression of cIMT [135], metabolic syndrome [136], diabetes [137]	Cardiovascular events [131, 138], insulin resistance [139]
	sTNFRI/II	Obesity [129], diabetes [137]	Cardiovascular events [138]
	TNF	Coronary artery calcium [140]	
Neurocognitive impairment	IL-6	[141]	[132] Future cognitive decline [142]
	hsCRP		[132, 143]
	sTNFR-I/II	[144]	
	TNF		Alzheimer's disease [145, 146]
Malignancies	IL-6, hsCRP	All cancers [147]	All cancers [148]
	D-Dimer	All cancers [147]	
	TNF		All cancers [148]
Bone disease/osteoporosis	hsCRP		Bone mineral density [149], fracture risk [150], future bone mineral density loss [151]
	IL-6		Future bone mineral density loss [151, 152]
	TNF		Future bone mineral density loss [151]
Frailty/disability	IL-6	[153, 154]	[132, 155]
	TNF	[153]	
	CRP	[153]	[132, 156]
Mortality	sTNFRI, hsCRP	[157]	

(continued)

Table 3 (continued)

		HIV-infected	General population
	Marker	Outcome/risk factor	
Loss of gut integrity/microbial translocation			
Cardiovascular/metabolic disease	LPS	Progression of cIMT [158], endothelial dysfunction [159], hypercholesterolemia [160], insulin resistance [160], hypertension [161]	Metabolic syndrome [162]
	LBP		Metabolic syndrome [163]
Neurocognitive impairment	LPS	[141, 164]	
Malignancies	LPS	Non-Hodgkin's lymphoma [165]	
Mortality	IFAB-1, zonulin	[157]	
Monocyte/innate immune activation			
Cardiovascular/metabolic disease	sCD14	Increased cIMT [166], cIMT progression [158], coronary calcification [167], hypertension [161]	Diabetes [169], hypertension [169]
	sCD163	Arterial inflammation [170], non-calcified coronary artery plaques [171]	Atherosclerosis [172], insulin resistance [139, 173, 174], diabetes [175]
	MCP-1	Coronary artery calcium [140]	
Neurocognitive impairment	sCD14	[141, 144, 176, 177]	
	sCD163	[178]	
	Neopterin		Alzheimer's disease [179]
Malignancies	sCD14	Non-Hodgkin's lymphoma [165]	
Frailty/disability	Neopterin		[180]
Mortality	sCD14	[157]	
T cell activation/senescence			
Cardiovascular/metabolic disease	HLADR+ CD38+ T cells	Carotid artery plaques [181, 182], carotid artery stiffness [183]	
Bone disease/osteoporosis	T cell activation	Bone mineral density [184]	
Mortality	Low CD8+ CD28- CD57+ T cells	[185]	

(continued)

Table 3 (continued)

		HIV-infected	General population
	Marker	Outcome/risk factor	
Telomere shortening			
Cardiovascular/metabolic disease	Leukocyte telomere length		Increased cIMT [122, 185], risk of myocardial infarction and stroke [122–124], Diabetes [122]
	Monocyte telomere length		Type 2 diabetes [187]
Neurocognitive impairment	Leukocyte telomere length		Dementia [125]
	Monocyte telomere length		Alzheimer's disease [126]
Malignancies	Epithelial cell telomere length		Epithelial cancer [121]
	Telomere length in mucosal tissue		Early stages of gastric carcinoma [188]
Early death	Leukocyte telomere length		[123, 125]
Cytomegalovirus (CMV) infection			
Cardiovascular/metabolic disease	CMV seropositivity		Type 2 diabetes [189], mortality in coronary artery disease patients [190]
	CMV-specific T cells responses	cIMT [191]	
	CMV IgG	Carotid artery disease [192]	
Frailty/disability	CMV IgG		[193]
Mortality	CMV IgG CMV seropositivity		All-cause mortality [193], cardiovascular related deaths [194]

virologically suppressed HIV-infected individuals [204, 205]. High intracellular levels of the antioxidant factors N-acetylcysteine and glutathione inhibit HIV replication in infected cells [206] whilst low levels of these factors are associated with increased NF-kB-mediated transcription of HIV and a heightened ability of the pro-inflammatory cytokine TNF to activate HIV transcription [207], suggesting a positive feedback loop between inflammation and HIV replication. The mechanism responsible for decreased anti oxidant levels in HIV may involve the HIV Tat protein, which has been shown in mouse models to decrease production of anti oxidants and induce mitochondrial damage [208]. Certain antiretroviral (ARV) drugs including PIs and NRTIs increase the production of reactive oxygen from cells treated in vitro [209]. Consistent with this, one study reported higher levels of oxidative stress in ART-treated individuals as compared to both untreated HIV-infected and uninfected individuals, however the HIV-infected individuals in this study had significantly higher levels of a number of confounding factors including concurrent hepatitis C infection [210]. More data from virologically suppressed HIV-infected cohorts with adequate control of variables which may potentially influence oxidative stress are required to determine the impact of oxidative stress on immune aging in the modern ART era.

4.4 Chronic Inflammation and Immune Activation

Increased inflammation is one of the cornerstones of immunological aging and geroscience, and appears to be potentiated by HIV infection. Indeed, chronic inflammation and related immune activation likely has the greatest impact on morbidity and mortality in PAWH in the post-ART era. Inflammaging is a well-documented state of chronic, low-grade inflammation occurring progressively with age and is associated with the development of many age-related morbidities and functional decline in the elderly [211]. Markers of inflammation including IL-6, TNFα and high-sensitivity C-reactive protein (hsCRP) are elevated in both HIV-infected individuals and the elderly [212, 213] and are associated with increased risk of SNAEs including CVD, frailty, malignancies, bone disease and neurocognitive decline. Inflammation is intrinsically linked with cellular activation, and biomarkers of immune activation and inflammation are increasingly being recognised as risk predictors of inflammatory diseases in HIV infection, as they are in the aged (see Table 3). Biomarkers of monocyte/macrophage activation including plasma levels of sCD163 and sCD14 are predictive of age-related diseases including neurocognitive impairment/dementia [141, 176–178], malignancies [165] and also mortality [214] in HIV infection (see Table 3). Chronic monocyte/macrophage activation appears to be particularly relevant for the development of CVD in HIV infection; biomarkers of monocyte activation including the proportion of inflammatory CD16+ monocytes, the expression of monocyte activation markers (i.e. CD11b) and the soluble activation markers mentioned above are associated with atherosclerosis and its progression [158, 159, 215], arterial inflammation [170], coronary calcium score [167] and the presence of non-calcified carotid plaques [171] in HIV-infected individuals. Importantly, these associations have been made in cohorts of primarily virologically suppressed individuals, suggesting mechanisms other than overt HIV viremia are involved. Indeed, in the post-ART era, markers of inflammation and/or immune activation are emerging as more relevant predictors of disease outcome and death in virologically suppressed individuals than traditional HIV biomarkers such as viral load and CD4+ T cells count [157, 216]. Recent data reporting an association between sCD163 levels and telomere length [217] provide a direct link between monocyte/macrophage activation and potentiation of immunological aging. Given chronic inflammation/immune activation and resultant disease burden are similar between HIV-infected individuals and the aged, the question arises to what extent the mechanisms driving these phenomena are similar in both populations and what contributing factors may be unique to HIV infection.

5 Factors Potentiating Age-Related Changes and Morbidity in HIV-Infected Individuals

The development of SNAEs in PAWH is multifactorial, and typically results from the combined effects of traditional risk factors, HIV-specific effects, and a potentiation of age-related changes (see Fig. 3).

5.1 Traditional Risk Factors

Traditional risk factors for disease development are highly relevant for the aging HIV-infected population, not only as they are often more readily modifiable but also because they may potentiate HIV-specific factors. Many cohort studies report a higher prevalence of smoking amongst HIV-infected participants [218–220], and whilst illicit drug use is higher within certain high risk HIV-infected populations, this variable is often not adequately assessed or controlled for in HIV cohort studies. Relevant to the development of cardiovascular disease, HIV infection is associated with dyslipidemia and metabolic alterations, which are discussed further below.

5.2 Metabolic Alterations

Hyperglycemia occurs in up to 17 % of HIV-infected individuals receiving ART and diabetes mellitus is more common in HIV infected vs seronegative people [221], with some studies reporting up to a fourfold increased risk due to HIV [222]. Insulin resistance in ART-treated HIV infection is largely associated with the use of protease inhibitor antiretroviral drugs, which act to inhibit the glucose transporter Glut-4 [223], although hepatitis C virus (HCV) co-infection, inflammation and immunodeficiency also contribute to insulin resistance and diabetes in HIV infection [221].

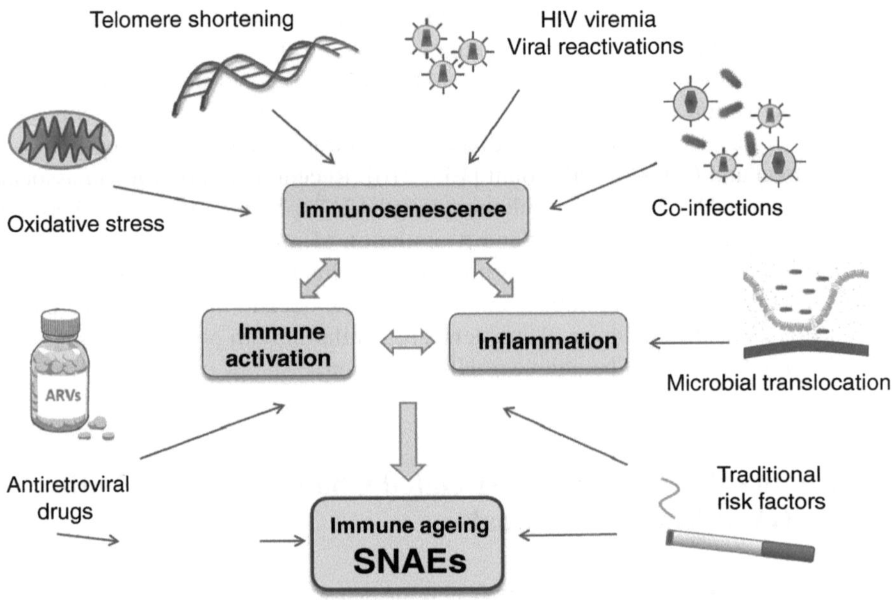

Fig. 3 Mechanism contributing to the pathogenesis of SNAEs in HIV-infected individuals

High glucose levels have been shown to increase the susceptibility of CD4+ T cells in HIV infection in vitro by upregulating the expression of the HIV co-receptor CXCR4 [224], whilst increased expression of Glut-1 on T cells from HIV-infected individuals (irrespective of ART) is associated with T cell activation and immunodeficiency [169]. Taken together, these data suggest that metabolic alterations due to both HIV and its treatment not only increase the risk of co-morbidities such as diabetes, but may also perpetuate HIV replication and immune activation to further drive immune exhaustion and senescence in PAWH.

HIV-related lipodystrophy syndrome is common in HIV infection, and includes lipoatrophy (loss of subcutaneous fat) and dyslipidemia. Lipoatrophy appears to be largely due to PI and NRTI use, particularly the NRTIs stavudine and zidovudine (reviewed in [225]). Whilst HIV infection per se is associated with lipid alterations including high triglyceride and low HDL levels (thought to be due to the effect of inflammation on lipid peroxidation, reactive oxygen species production and vascular changes [226, 227]), the majority of dyslipidemia observed in the post-ART era is due to the specific effects of antiretroviral drugs.

5.3 Antiretroviral Drugs

Although highly effective in inhibiting HIV replication and maintaining immune health, many antiretroviral drugs, particularly the NRTIs, have some degree of toxicity which is at least partially attributable to effects on the mitochondria. The ability of NRTIs to inhibit HIV reverse transcription is due to structural similarities between NRTIs and endogenous nucleos(t)ides, and whilst nuclear DNA polymerases are not significantly affected by NRTIs, the mitochondrial replicase pol γ is inhibited by NRTIs at physiologically relevant levels, resulting in depletion of mitochondrial DNA and increased oxidative stress (reviewed in [208]). Specifically, zidovudine and stavudine have been shown to increase oxidative stress in a number of cell types including adipocytes and macrophages [228]. As discussed above, NRTIs are also able to inhibit the RT component of cellular telomerase and may potentially contribute to premature telomere shortening. Interestingly, certain NRTIs have recently been shown to be able to inhibit NLRP3 inflammasome-mediated activation of caspase-1 and subsequent production of the pro-inflammatory cytokines IL-1β and IL-18 [229], suggesting NRTIs may have an unexpected influence on cytokine production in HIV-infected individuals receiving these drugs.

HIV-infected individuals treated with ART have a relative risk of CVD of 1.52 (95 % CI 1.35–1.70) compared to untreated individuals [230], suggesting ART may contribute to the pathogenesis of CVD. Indeed, recent use of certain PIs and the NRTIs abacavir and didanosine has been associated with increased risk of myocardial infarction [231, 232] although the association with abacavir was not reproduced in a randomised control trial and remains controversial [233]. The increased risk attributable to PIs is largely due to an effect on lipid levels, as 70–80 % of HIV-infected individuals receiving PI-containing ART regimens show elevated lipid levels [226]. Most PIs (with the possible exception of atazanavir) have been shown to induce dyslipidemia involving increased plasma concentrations of triglycerides, total cholesterol and LDL [234], all of which are known risk factors for cardiovascular disease. The mechanism involves a direct effect of PIs on adipocyte differentiation and also an ability of these drugs to inhibit factors involved in lipid transport and metabolism [227].

Untreated HIV infection results in loss of bone mineral density which contributes to increased fracture risk and osteoporosis (as discussed above), but ART-initiation potentiates this effect and results in a further loss of bone mineral density of approximately 2–6 % within the first 2 years of ART initiation. This effect is thought to be due to disruption of the delicate immunological balance in the bone marrow which governs osteogenesis, and specific antiretroviral drugs including the NRTI tenofovir [235] and the protease inhibitor class of drugs have been shown to potentiate bone loss in ART-treated individuals [236].

The relatively recent introduction of ART, combined with the lengthy and multifactorial pathogenesis of many HIV-related co-morbidities, means that significant associations between specific ARVs and disease outcomes are continuing to emerge. HIV-infected individuals initiating therapy in the early days largely did so with low CD4 counts and received ARVs which have since been phased out due to side effects and toxicities. Thus, ongoing and future longitudinal studies will be critical for evaluating the long term effects of current ARVs on immune changes and the development of age-related diseases in HIV-infected individuals who avoid significant immunological damage by initiating ART at higher CD4 T cell counts.

6 Mechanisms That May Contribute to Chronic Inflammation and Immune Activation in HIV

6.1 *Microbial Translocation and Endotoxemia*

HIV infection is associated with increased permeability of the gut to microbial products, which translocate across the gut epithelium and eventually into the bloodstream, resulting in increased plasma levels of the bacterial endotoxin lipopolysaccharide (LPS) and bacterial DNA in HIV-infected individuals [237]. The cause of increased gut barrier permeability in HIV infection is due to immunodeficiency and structural defects within the gut-associated lymphoid tissue (GALT) resulting from HIV-mediated T cell depletion [238]. The majority of lymphocytes in the body are contained in GALT, which is an important site for both pathogenesis and persistence of HIV. CD4+ T cells are rapidly depleted from the GALT during primary HIV infection and remain depleted into chronic infection. Studies in Simian Immunodeficiency Virus (SIV)-infected macaques (a pathogenic animal model of HIV infection) have revealed that peak infection of CD4+ T cells in the lamina propria of the gut occurs within 10 days of infection, at which point 93 % of target CD4+ memory T cells are infected [239]. While effective ART suppresses viral replication and restores peripheral CD4+ T cells, gut-associated CD4+ T cells remain depleted years after ART initiation [240]. Interestingly, a subset of HIV-infected individuals who maintain high CD4+ T cells counts and low/undetectable viral loads in the absence of ART (known as long term non-progressors) maintain normal CD4+ T cell levels in the GALT [241], suggesting the importance of this compartment for disease pathogenesis. The mechanism of increased gut permeability in HIV involves epithelial disruption and decreased production of tight junction proteins in the distal portions of the colon [242] which is consistent with increased levels of intestinal fatty acid binding protein (I-FABP; a marker of enterocyte damage) and zonulin-1 (a regulator of tight junction permeability) in the plasma of HIV-infected individuals [157, 243]. The inability to fully restore GALT structure and function despite effective restoration of peripheral T cells by ART means that chronic endotoxemia (elevated levels of LPS in the blood) persists in virologically suppressed HIV-infected individuals. Lipopolysaccharide (LPS) is a potent immune activator which is recognised by toll-like receptor (TLR)-4 expressing cells such as monocytes/macrophages in an immune complex consisting of LPS-binding protein (LBP), the adaptor protein MD2 and either soluble or cell-bound CD14. LPS signalling stimulates the production of pro-inflammatory cytokines including IL-6, TNF and type I interferons. Microbial translocation is considered a significant driver of both HIV disease and related co-morbidities, with gut translocation markers such as LPS, the LPS binding protein LBP and I-FABP/zonulin-1 associated with immune activation and HIV disease progression [237, 244, 245], cardiovascular and metabolic disease [159, 160], neurocognitive impairment [141] and mortality [157, 216].

In contrast to HIV, relatively little is known regarding the effect of age on the integrity of the gut epithelium in humans [246], however work in *Drosophila* has demonstrated that loss of intestinal barrier integrity occurs with aging and is a better predictor of age-related morbidity and death than chronological age [247]. Increased plasma levels of LPS [98] and LBP [248] in the elderly indicate microbial translocation, may also increase during aging and the inverse association between LBP levels and physical function in the aged [248] suggests it may also contribute to morbidity in this population, although this requires further investigation.

6.1.1 Alterations to the Gut Microbiome

Within the GALT, cytokines including IL-17 and IL-22 play a critical role in maintaining gut integrity and orchestrating the mucosal immune responses to gut pathogens. Depletion of CD4+ T cells from the gut during HIV infection reduces the production of these cytokines and disrupts the delicate mucosal immunological balance. The gut microbiome interacts intimately with mucosal immunity and helps educate and regulate immune cells. Significant alterations are observed in the gut microbiome of HIV-infected individuals, with sequence analysis of bacterial communities from stool/gut mucosa samples revealing an overall increase in genetic diversity, an expansion of *Prevotella* and potentially pathogenic bacteria and a reduced proportion of *Bacteroidia* species [249–252]. Importantly, these changes in microbial communities are associated with inflammation, innate and adaptive immune activation and markers of disease progression in HIV-infected individuals. ART appears to only partially normalize the bacterial composition of the microbiome in a proportion of treated individuals [251]. A higher proportion of bacteria from the order *Lactobacillales* (lactic acid-producing bacteria) in the distal gut of ART naïve individuals has been associated with more favorable immunological parameters including higher pre-ART CD4+ T cells counts and CD4:CD8 T cells ratio but lower viral loads and sCD14 levels [253]. The complex interplay between the gut microbiome, GALT immunity and systemic inflammation/immune activation continues to be elucidated but may reveal an important mechanism of persistent immune dysfunction in HIV which can be targeted therapeutically.

6.1.2 Cytomegalovirus (CMV) and Latent Viral Infections

Accumulative immune stimulation by pathogens and subsequent immune exhaustion is an integral mechanism of immune aging and heightened pathogen burden due to concurrent and reactivated viral infections may hasten this process in PAWH. While CMV-seropositivity rates vary considerably between different countries (ranging from 40 to >90 %), there is a consistent trend of increasing seropositivity with age [254] and CMV is recognized as a significant driver of immunosenescence [255, 256]. CMV infection profoundly shifts the lymphocyte subset proportions towards a differentiated memory T cell phenotype [257, 258]. In aged individuals, the proportion of CD8+ T cells specific for a small number of CMV epitopes can represent up to 27 % of the total CD8+ pool [259], with these cells typically being dysfunctional and exhibiting an immunosenescent phenotype [260]. CMV seropositivity has also been associated with an increased risk of age-related diseases such as cardiovascular disease [261].

CMV disease is a significant cause of morbidity and mortality in HIV-infected individuals with AIDS and/or severe immunodeficiency [262], while asymptomatic CMV infection also appears to potentiate immunosenescence in HIV-infected individuals. CMV infection is almost ubiquitous in the HIV-infected population with seropositivity rates of approximately 95 % [263] and the presence of IgM antibodies suggests viral reactivation/reinfection commonly occurs [264]. Levels of CMV-specific CD8+ T cells are up to twice as high in HIV-infected as in uninfected individuals and persist in ART-treated individuals despite long term virological suppression [265], which is consistent with reactivation and impaired immune control [263]. HIV-infected/CMV seronegative subjects show higher CD4:CD8 T cells ratios and less phenotypic evidence of immunosenescence than HIV/CMV seropositive individuals [266] whilst serum CMV IgG levels, which are increased in HIV-infected individuals, correlate with inflammatory markers [267]. Taken together, these observations suggest that CMV seropositivity may potentiate HIV-related immunosenescence and inflammation and hasten the aging process.

Although ART reduces HIV viral load to near undetectable levels in the plasma, residual HIV replication (up to 20 copies/mL) can be detected in the plasma of the majority of virologically suppressed individuals using ultra-sensitive assays [268]. In addition, ongoing HIV replication may persist at higher levels within anatomical sites such as lymphoid tissue where antiretroviral drugs may fail to penetrate to effective therapeutic concentrations. Reactivation/replication of other latent viruses including Epstein–Barr virus (EBV) and Herpes Simplex Viruses (HSV) also appears to be heightened in HIV-infected individuals, likely due to increased immune activation. HSV-2 reactivation occurs frequently in HIV-infected individuals, is positively associated with HIV viral load [269] and is shed more frequently in HIV-infected vs seronegative individuals [270]. EBV viral loads in HIV-infected individuals are reportedly greater than those in EBV+ HIV-uninfected individuals [271].

Human endogenous retroviruses (HERVs) are a family of replication defective viral elements which comprise up to 8 % of the human genome. Although thought to be largely silent, increased transcription of HML-2 RNA (a member of the HERV-K family) has been demonstrated in PMBCs from HIV-infected individuals [272] and has also been detected at increased levels in plasma in some [273] but not all [272] studies. Increased HERV transcription may be due to heightened immune activation and/or the ability of the HIV Tat protein to activate endogenous retroviral transcriptional elements [274]. Although cause and effect are difficult to delineate, it is clear that heightened inflammation/immune activation and reactivation of latent viral infections may constitute a self-perpetuating cycle contributing to immune exhaustion and immunosenescence in many PAWH.

6.1.3 Concurrent Infections

The development of age-related morbidities in HIV-infected individuals can be influenced by concurrent infection with a range of pathogens. Co-infection with HCV can be up to 90 % in certain high risk HIV+ groups, and is associated with an increased risk of coronary heart disease [275], osteoporotic fracture [276], and neurocognitive impairment [277], suggesting hepatitis C infection may potentiate the pathogenesis of these conditions. The mechanism of this is unclear, although a potentiation of inflammation and immune activation is likely, and increased levels of pro-inflammatory factors such as IL-6 have been demonstrated in HIV/HCV co-infected, as compared to mono or uninfected individuals [278]. Active HCV infection is also associated with shorter leukocyte telomere length in those with HIV [279]. Taken together, these data suggest that HCV co-infection may further heighten inflammation/immune activation and associated immunosenescence in HIV-infected individuals and potentiate the development of age-related diseases.

HIV-infected individuals co-infected with tuberculosis (TB) have significantly increased pro-inflammatory cytokine production [280] and ART initiation in highly immunocompromised HIV+/TB+ individuals often results in TB-associated immune reconstitution inflammatory syndrome, which results in significant pro-inflammatory cytokine production [281]. Heightened CD4+ T cell activation and pro-inflammatory cytokine production also occurs in malaria co-infection [282]. These observations suggest concurrent infections may further potentiate inflammation due to HIV and aging in co-infected individuals, however further studies are required to elucidate the full impact of these effects on age-related disease outcomes.

7 Potential Treatments/Interventions to Alleviate the Effects of HIV on Aging/SNAEs

The immunological similarities between HIV infection and aging (particularly chronic inflammation and its consequences) suggest that addressing mechanism of aging may alleviate premature aging and disease pathogenesis in HIV-infected individuals. A large number of preliminary trials are underway to address immune activation, inflammation, microbial translocation and other mechanisms of enhanced aging in PAWH, but none has yet demonstrated efficacy in definitive clinical trials [283–287]. If this is accomplished in PAWH, it will have vast implications for aging in general and may be applicable to a much broader population.

8 Concluding Remarks

The success of antiretroviral therapy in preventing AIDS and extending the life span of HIV-infected people has revealed unexpected parallels between the impact of HIV infection and aging on immune function. Current research is only beginning to uncover how HIV may be potentiating age-related changes and the consequences of this for premature aging and increased risk of age-related comorbidities in those living and aging with HIV. It is still unclear whether HIV-associated ‘aging’ is the result of chronic infection, or whether those infected with HIV at an older age may experience similar effects. Furthermore, the impact of long-term ARV drug use on age-related process remains to be fully elucidated. HIV infection further complicates the many health challenges experienced by aging individuals including multimorbidity, polypharmacy, impaired physical and mental health and reduced quality of life. Uncovering the critical processes which drive age-related changes and identifying therapeutic strategies to ameliorate the residual effects of HIV will be important for ongoing management of the increasingly aging HIV-infected population.

Acknowledgements Support for KH and AL from NIA 1R24AG044325.

Editor Robin Huebner, National Institute of Allergy and Infectious Diseases (NIAID), NIH.

References

1. Costagliola D (2014) Demographics of HIV and aging. Curr Opin HIV AIDS 9(4):294–301. doi:10.1097/coh.0000000000000076
2. High KP, Brennan-Ing M, Clifford DB, Cohen MH, Currier J, Deeks SG, Deren S, Effros RB, Gebo K, Goronzy JJ, Justice AC, Landay A, Levin J, Miotti PG, Munk RJ, Nass H, Rinaldo CR Jr, Shlipak MG, Tracy R, Valcour V, Vance DE, Walston JD, Volberding P, Aging OARWGoHa (2012) HIV and aging: state of knowledge and areas of critical need for research. A report to the NIH Office of AIDS Research by the HIV and Aging Working Group. J Acquir Immune Defic Syndr 60 Suppl 1:S1–S18. doi:10.1097/QAI.0b013e31825a3668
3. Kirk JB, Goetz MB (2009) Human immunodeficiency virus in an aging population, a complication of success. J Am Geriatr Soc 57(11):2129–2138. doi:10.1111/j.1532-5415.2009.02494.x
4. Luther VP, Wilkin AM (2007) HIV infection in older adults. Clin Geriatr Med 23(3):567–583, vii. doi:10.1016/j.cger.2ar007.02.004
5. Mills EJ, Barnighausen T, Negin J (2012) HIV and aging – preparing for the challenges ahead. N Engl J Med 366(14):1270–1273. doi:10.1056/NEJMp1113643
6. Bloomfield GS, Khazanie P, Morris A, Rabadan-Diehl C, Benjamin LA, Murdoch D, Radcliff VS, Velazquez EJ, Hicks C (2014) HIV and noncommunicable cardiovascular and pulmonary diseases in low- and middle-income countries in the ART era: what we know and best directions for future research. J Acquir Immune Defic Syndr 67 Suppl 1:S40–S53. doi:10.1097/qai.0000000000000257
7. Narayan KM, Miotti PG, Anand NP, Kline LM, Harmston C, Gulakowski R 3rd, Vermund SH (2014) HIV and noncommunicable disease comorbidities in the era of antiretroviral therapy: a vital agenda for research in low- and middle-income country settings. J Acquir Immune Defic Syndr 67 Suppl 1:S2–S7. doi:10.1097/qai.0000000000000267
8. Burch JB, Augustine AD, Frieden LA, Hadley E, Howcroft TK, Johnson R, Khalsa PS, Kohanski RA, Li XL, Macchiarini F, Niederehe G, Oh YS, Pawlyk AC, Rodriguez H, Rowland JH, Shen GL, Sierra F, Wise BC (2014) Advances in geroscience: impact on healthspan and chronic disease. J Gerontol A Biol Sci Med Sci 69(Suppl 1):S1–S3. doi:10.1093/gerona/glu041
9. Lopez-Otin C, Blasco MA, Partridge L, Serrano M, Kroemer G (2013) The hallmarks of aging. Cell 153(6):1194–1217. doi:10.1016/j.cell.2013.05.039
10. Guaraldi G, Orlando G, Zona S, Menozzi M, Carli F, Garlassi E, Berti A, Rossi E, Roverato A, Palella F (2011) Premature age-related comorbidities among HIV-infected persons compared with the general population. Clin Infect Dis 53(11):1120–1126. doi:10.1093/cid/cir627

11. Ruiz M, Cefalu C (2011) Characteristics of frail patients in a geriatric-HIV program: the experience of an Urban Academic Center at one year follow-up. J Int Assoc Phys AIDS Care 10(3):138–143. doi:10.1177/1545109711399658
12. Desquilbet L, Jacobson LP, Fried LP, Phair JP, Jamieson BD, Holloway M, Margolick JB, Study MAC (2007) HIV-1 infection is associated with an earlier occurrence of a phenotype related to frailty. J Gerontol A Biol Sci Med Sci 62(11):1279–1286
13. Onen NF, Agbebi A, Shacham E, Stamm KE, Onen AR, Overton ET (2009) Frailty among HIV-infected persons in an urban outpatient care setting. J Infect 59(5):346–352. doi:10.1016/j.jinf.2009.08.008
14. Erlandson KM, Schrack JA, Jankowski CM, Brown TT, Campbell TB (2014) Functional impairment, disability, and frailty in adults aging with HIV-infection. Curr HIV/AIDS Rep 11(3):279–290. doi:10.1007/s11904-014-0215-y
15. Miller CJ, Baker JV, Bormann AM, Erlandson KM, Huppler Hullsiek K, Justice AC, Neuhaus J, Paredes R, Petoumenos K, Wentworth D, Winston A, Wolfson J, Neaton JD (2014) Adjudicated morbidity and mortality outcomes by age among individuals with HIV infection on suppressive antiretroviral therapy. PLoS One 9(4), e95061. doi:10.1371/journal.pone.0095061
16. Triant VA, Lee H, Hadigan C, Grinspoon SK (2007) Increased acute myocardial infarction rates and cardiovascular risk factors among patients with human immunodeficiency virus disease. J Clin Endocrinol Metab 92(7):2506–2512. doi:10.1210/jc.2006-2190
17. Hall AM, Hendry BM, Nitsch D, Connolly JO (2011) Tenofovir-associated kidney toxicity in HIV-infected patients: a review of the evidence. Am J Kidney Dis 57(5):773–780. doi:10.1053/j.ajkd.2011.01.022
18. Cannillo M, D'Ascenzo F, Grosso Marra W, Cerrato E, Calcagno A, Omede P, Bonora S, Mancone M, Vizza D, DiNicolantonio JJ, Pianelli M, Barbero U, Gili S, Annone U, Raviola A, Salera D, Mistretta E, Vilardi I, Colaci C, Abbate A, Zoccai GB, Moretti C, Gaita F (2014) Heart failure in patients with human immunodeficiency virus: a review of the literature. J Cardiovasc Med. doi:10.2459/jcm.0000000000000168
19. Triant VA (2014) Epidemiology of coronary heart disease in patients with human immunodeficiency virus. Rev Cardiovasc Med 15(Suppl 1):S1–S8
20. Remick J, Georgiopoulou V, Marti C, Ofotokun I, Kalogeropoulos A, Lewis W, Butler J (2014) Heart failure in patients with human immunodeficiency virus infection: epidemiology, pathophysiology, treatment, and future research. Circulation 129(17):1781–1789. doi:10.1161/circulationaha.113.004574
21. Guaraldi G, Zona S, Alexopoulos N, Orlando G, Carli F, Ligabue G, Fiocchi F, Lattanzi A, Rossi R, Modena MG, Esposito R, Palella F, Raggi P (2009) Coronary aging in HIV-infected patients. Clin Infect Dis 49(11):1756–1762. doi:10.1086/648080
22. Kuller LH, Tracy R, Belloso W, De Wit S, Drummond F, Lane HC, Ledergerber B, Lundgren J, Neuhaus J, Nixon D, Paton NI, Neaton JD (2008) Inflammatory and coagulation biomarkers and mortality in patients with HIV infection. PLoS Med 5(10), e203. doi:10.1371/journal.pmed.0050203
23. Deeks SG, Phillips AN (2009) HIV infection, antiretroviral treatment, ageing, and non-AIDS related morbidity. BMJ 338:a3172. doi:10.1136/bmj.a3172
24. Gedela K, Vibhuti M, Pozniak A, Ward B, Boffito M (2014) Pharmacological management of cardiovascular conditions and diabetes in older adults with HIV infection. HIV Med 15(5):257–268. doi:10.1111/hiv.12116
25. Martinez E, Mocroft A, Garcia-Viejo MA, Perez-Cuevas JB, Blanco JL, Mallolas J, Bianchi L, Conget I, Blanch J, Phillips A, Gatell JM (2001) Risk of lipodystrophy in HIV-1-infected patients treated with protease inhibitors: a prospective cohort study. Lancet 357(9256):592–598. doi:10.1016/s0140-6736(00)04056-3
26. Paula AA, Falcao MC, Pacheco AG (2013) Metabolic syndrome in HIV-infected individuals: underlying mechanisms and epidemiological aspects. AIDS Res Ther 10(1):32. doi:10.1186/1742-6405-10-32

27. Wand H, Calmy A, Carey DL, Samaras K, Carr A, Law MG, Cooper DA, Emery S (2007) Metabolic syndrome, cardiovascular disease and type 2 diabetes mellitus after initiation of antiretroviral therapy in HIV infection. AIDS 21(18):2445–2453. doi:10.1097/QAD.0b013e3282efad32
28. Cerrato E, D'Ascenzo F, Biondi-Zoccai G, Calcagno A, Frea S, Grosso Marra W, Castagno D, Omede P, Quadri G, Sciuto F, Presutti D, Frati G, Bonora S, Moretti C, Gaita F (2013) Cardiac dysfunction in pauci symptomatic human immunodeficiency virus patients: a meta-analysis in the highly active antiretroviral therapy era. Eur Heart J 34(19):1432–1436. doi:10.1093/eurheartj/ehs471
29. Papita A, Albu A, Fodor D, Itu C, Carstina D (2011) Arterial stiffness and carotid intima-media thickness in HIV infected patients. Med Ultrason 13(2):127–134
30. Seaberg EC, Benning L, Sharrett AR, Lazar JM, Hodis HN, Mack WJ, Siedner MJ, Phair JP, Kingsley LA, Kaplan RC (2010) Association between human immunodeficiency virus infection and stiffness of the common carotid artery. Stroke 41(10):2163–2170. doi:10.1161/strokeaha.110.583856
31. Cobucci RN, Lima PH, de Souza PC, Costa VV, Cornetta MD, Fernandes JV, Goncalves AK (2014) Assessing the impact of HAART on the incidence of defining and non-defining AIDS cancers among patients with HIV/AIDS: a systematic review. J Infect Public Health. doi:10.1016/j.jiph.2014.08.003
32. Burgi A, Brodine S, Wegner S, Milazzo M, Wallace MR, Spooner K, Blazes DL, Agan BK, Armstrong A, Fraser S, Crum NF (2005) Incidence and risk factors for the occurrence of non-AIDS-defining cancers among human immunodeficiency virus-infected individuals. Cancer 104(7):1505–1511. doi:10.1002/cncr.21334
33. Shiels MS, Pfeiffer RM, Engels EA (2010) Age at cancer diagnosis among persons with AIDS in the United States. Ann Intern Med 153(7):452–460. doi:10.7326/0003-4819-153-7-201010050-00008
34. Shiels MS, Goedert JJ, Moore RD, Platz EA, Engels EA (2010) Reduced risk of prostate cancer in U.S. men with AIDS. Cancer Epidemiol Biomarkers Prev 19(11):2910–2915. doi:10.1158/1055-9965.epi-10-0741
35. Hasse B, Ledergerber B, Furrer H, Battegay M, Hirschel B, Cavassini M, Bertisch B, Bernasconi E, Weber R, Swiss HIVCS (2011) Morbidity and aging in HIV-infected persons: the Swiss HIV cohort study. Clin Infect Dis 53(11):1130–1139. doi:10.1093/cid/cir626
36. Ances BM, Vaida F, Yeh MJ, Liang CL, Buxton RB, Letendre S, McCutchan JA, Ellis RJ (2010) HIV infection and aging independently affect brain function as measured by. J Infect Dis 201(3):336–340. doi:10.1086/649899
37. Womack JA, Goulet JL, Gibert C, Brandt CA, Skanderson M, Gulanski B, Rimland D, Rodriguez-Barradas MC, Tate J, Yin MT, Justice AC (2013) Physiologic frailty and fragility fracture in HIV-infected male veterans. Clin Infect Dis 56(10):1498–1504. doi:10.1093/cid/cit056
38. Womack JA, Goulet JL, Gibert C, Brandt C, Chang CC, Gulanski B, Fraenkel L, Mattocks K, Rimland D, Rodriguez-Barradas MC, Tate J, Yin MT, Justice AC (2011) Increased risk of fragility fractures among HIV infected compared to uninfected male veterans. PLoS One 6(2), e17217. doi:10.1371/journal.pone.0017217
39. Chan W, Dart AM (2011) Vascular stiffness and aging in HIV. Sex Health 8(4):474–484. doi:10.1071/sh10160
40. Goulet JL, Fultz SL, Rimland D, Butt A, Gibert C, Rodriguez-Barradas M, Bryant K, Justice AC (2007) Aging and infectious diseases: do patterns of comorbidity vary by HIV status. Clin Infect Dis 45(12):1593–1601. doi:10.1086/523577
41. El-Sadr WM, Lundgren J, Neaton JD, Gordin F, Abrams D, Arduino RC, Babiker A, Burman W, Clumeck N, Cohen CJ, Cohn D, Cooper D, Darbyshire J, Emery S, Fatkenheuer G, Gazzard B, Grund B, Hoy J, Klingman K, Losso M, Markowitz N, Neuhaus J, Phillips A, Rappoport C (2006) CD4+ count-guided interruption of antiretroviral treatment. N Engl J Med 355(22):2283–2296. doi:10.1056/NEJMoa062360

42. Freiberg MS, Chang CC, Kuller LH, Skanderson M, Lowy E, Kraemer KL, Butt AA, Bidwell Goetz M, Leaf D, Oursler KA, Rimland D, Rodriguez Barradas M, Brown S, Gibert C, McGinnis K, Crothers K, Sico J, Crane H, Warner A, Gottlieb S, Gottdiener J, Tracy RP, Budoff M, Watson C, Armah KA, Doebler D, Bryant K, Justice AC (2013) HIV infection and the risk of acute myocardial infarction. JAMA Intern Med 173(8):614–622. doi:10.1001/jamainternmed.2013.3728
43. Rueda S, Law S, Rourke SB (2014) Psychosocial, mental health, and behavioral issues of aging with HIV. Curr Opin HIV AIDS 9(4):325–331. doi:10.1097/coh.0000000000000071
44. Justice AC, McGinnis KA, Skanderson M, Chang CC, Gibert CL, Goetz MB, Rimland D, Rodriguez-Barradas MC, Oursler KK, Brown ST, Braithwaite RS, May M, Covinsky KE, Roberts MS, Fultz SL, Bryant KJ (2010) Towards a combined prognostic index for survival in HIV infection: the role of "non-HIV" biomarkers. HIV Med 11(2):143–151. doi:10.1111/j.1468-1293.2009.00757.x
45. Akgun KM, Gordon K, Pisani M, Fried T, McGinnis KA, Tate JP, Butt AA, Gibert CL, Huang L, Rodriguez-Barradas MC, Rimland D, Justice AC, Crothers K (2013) Risk factors for hospitalization and medical intensive care unit (MICU) admission among HIV-infected Veterans. J Acquir Immune Defic Syndr 62(1):52–59. doi:10.1097/QAI.0b013e318278f3fa
46. Justice AC, Braithwaite RS (2012) Lessons learned from the first wave of aging with HIV. AIDS 26 Suppl 1:S11–S18. doi:10.1097/QAD.0b013e3283558500
47. Marzolini C, Back D, Weber R, Furrer H, Cavassini M, Calmy A, Vernazza P, Bernasconi E, Khoo S, Battegay M, Elzi L (2011) Ageing with HIV: medication use and risk for potential drug-drug interactions. J Antimicrob Chemother 66(9):2107–2111. doi:10.1093/jac/dkr248
48. Gleason LJ, Luque AE, Shah K (2013) Polypharmacy in the HIV-infected older adult population. Clin Interv Aging 8:749–763. doi:10.2147/cia.s37738
49. Nachega JB, Hsu AJ, Uthman OA, Spinewine A, Pham PA (2012) Antiretroviral therapy adherence and drug-drug interactions in the aging HIV population. AIDS 26 Suppl 1:S39–S53. doi:10.1097/QAD.0b013e32835584ea
50. Frankel JK, Packer CD (2011) Cushing's syndrome due to antiretroviral-budesonide interaction. Ann Pharmacother 45(6):823–824. doi:10.1345/aph.1P731
51. Ingle SM, May MT, Gill MJ, Mugavero MJ, Lewden C, Abgrall S, Fatkenheuer G, Reiss P, Saag MS, Manzardo C, Grabar S, Bruyand M, Moore D, Mocroft A, Sterling TR, D'Arminio Monforte A, Hernando V, Teira R, Guest J, Cavassini M, Crane HM, Sterne JA, Antiretroviral Therapy Cohort C (2014) Impact of risk factors for specific causes of death in the first and subsequent years of antiretroviral therapy among HIV-infected patients. Clin Infect Dis 59(2):287–297. doi:10.1093/cid/ciu261
52. Park WB, Choe PG, Kim SH, Jo JH, Bang JH, Kim HB, Kim NJ, Oh M, Choe KW (2007) One-year adherence to clinic visits after highly active antiretroviral therapy: a predictor of clinical progress in HIV patients. J Intern Med 261(3):268–275. doi:10.1111/j.1365-2796.2006.01762.x
53. Giordano TP, Gifford AL, White AC Jr, Suarez-Almazor ME, Rabeneck L, Hartman C, Backus LI, Mole LA, Morgan RO (2007) Retention in care: a challenge to survival with HIV infection. Clin Infect Dis 44(11):1493–1499. doi:10.1086/516778
54. Nachega JB, Parienti JJ, Uthman OA, Gross R, Dowdy DW, Sax PE, Gallant JE, Mugavero MJ, Mills EJ, Giordano TP (2014) Lower pill burden and once-daily antiretroviral treatment regimens for HIV infection: A meta-analysis of randomized controlled trials. Clin Infect Dis 58(9):1297–1307. doi:10.1093/cid/ciu046
55. Desquilbet L, Jacobson LP, Fried LP, Phair JP, Jamieson BD, Holloway M, Margolick JB (2011) A frailty-related phenotype before HAART initiation as an independent risk factor for AIDS or death after HAART among HIV-infected men. J Gerontol A Biol Sci Med Sci 66(9):1030–1038. doi:10.1093/gerona/glr097
56. Önen NF, Overton ET (2011) A review of premature frailty in HIV-infected persons; another manifestation of HIV-related accelerated aging. Curr Aging Sci 4(1):33–41

57. Terzian AS, Holman S, Nathwani N, Robison E, Weber K, Young M, Greenblatt RM, Gange SJ (2009) Factors associated with preclinical disability and frailty among HIV-infected and HIV-uninfected women in the era of cART. J Womens Health 18(12):1965–1974. doi:10.1089/jwh.2008.1090
58. Desquilbet L, Margolick JB, Fried LP, Phair JP, Jamieson BD, Holloway M, Jacobson LP (2009) Relationship between a frailty-related phenotype and progressive deterioration of the immune system in HIV-infected men. J Acquir Immune Defic Syndr 50(3):299–306
59. Shah K, Hilton TN, Myers L, Pinto JF, Luque AE, Hall WJ (2012) A new frailty syndrome: central obesity and frailty in older adults with the human immunodeficiency virus. J Am Geriatr Soc 60(3):545–549. doi:10.1111/j.1532-5415.2011.03819.x
60. Oursler KK, Goulet JL, Crystal S, Justice AC, Crothers K, Butt AA, Rodriguez-Barradas MC, Favors K, Leaf D, Katzel LI, Sorkin JD (2011) Association of Age and comorbidity with physical function in HIV-infected and uninfected patients: results from the Veterans Aging Cohort Study. AIDS Patient Care STDS 25(1):13–20. doi:10.1089/apc.2010.0242
61. Bauer LO, Wu Z, Wolfson LI (2011) An obese body mass increases the adverse effects of HIV/AIDS on balance and gait. Phys Ther 91(7):1063–1071. doi:10.2522/ptj.20100292
62. Erlandson KM, Allshouse AA, Jankowski CM, Duong S, Mawhinney S, Kohrt WM, Campbell TB (2012) Comparison of functional status instruments in HIV-infected adults on effective antiretroviral therapy. HIV Clin Trials 13(6):324–334. doi:10.1310/hct1306-324
63. Houston DK, Ding J, Nicklas BJ, Harris TB, Lee JS, Nevitt MC, Rubin SM, Tylavsky FA, Kritchevsky SB, Health ABCS (2009) Overweight and obesity over the adult life course and incident mobility limitation in older adults: the health, aging and body composition study. Am J Epidemiol 169(8):927–936. doi:10.1093/aje/kwp007
64. Ruiz M, Cefalu C (2011) Frailty syndrome in patients with HIV infection. Clin Geriatr 19(2):46–49
65. Pathai S, Bajillan H, Landay A, High K (2014) Is HIV a model of accelerated or accentuated aging? J Gerontol A Biol Sci Med Sci 69(7):833–842
66. Valcour V, Paul R, Chiao S, Wendelken LA, Miller B (2011) Screening for cognitive impairment in human immunodeficiency virus. Clin Infect Dis 53(8):836–842. doi:10.1093/cid/cir524
67. Assessment, diagnosis, and treatment of HIV-associated neurocognitive disorder: a consensus report of the mind exchange program (2013) Clin Infect Dis 56(7):1004–1017. doi:10.1093/cid/cis975
68. Aberg JA, Gallant JE, Ghanem KG, Emmanuel P, Zingman BS, Horberg MA, America IDSo (2014) Primary care guidelines for the management of persons infected with HIV: 2013 update by the HIV medicine association of the Infectious Diseases Society of America. Clin Infect Dis 58(1):e1–e34. doi:10.1093/cid/cit665
69. Ances BM, Ortega M, Vaida F, Heaps J, Paul R (2012) Independent effects of HIV, aging, and HAART on brain volumetric measures. J Acquir Immune Defic Syndr 59(5):469–477. doi:10.1097/QAI.0b013e318249db17
70. Moore RC, Moore DJ, Thompson WK, Vahia IV, Grant I, Jeste DV (2013) A case-controlled study of successful aging in older HIV-infected adults. J Clin Psychiatry 74(5):e417–e423. doi:10.4088/JCP.12m08100
71. Greysen SR, Horwitz LI, Covinsky KE, Gordon K, Ohl ME, Justice AC (2013) Does social isolation predict hospitalization and mortality among HIV+ and uninfected older veterans? J Am Geriatr Soc 61(9):1456–1463. doi:10.1111/jgs.12410
72. Shippy RA, Karpiak SE (2005) The aging HIV/AIDS population: fragile social networks. Aging Ment Health 9(3):246–254. doi:10.1080/13607860412331336850
73. Utsuyama M, Hirokawa K, Kurashima C, Fukayama M, Inamatsu T, Suzuki K, Hashimoto W, Sato K (1992) Differential age-change in the numbers of CD4+CD45RA+ and CD4+CD29+ T cell subsets in human peripheral blood. Mech Ageing Dev 63(1):57–68
74. Xu X, Beckman I, Ahern M, Bradley J (1993) A comprehensive analysis of peripheral blood lymphocytes in healthy aged humans by flow cytometry. Immunol Cell Biol 71(Pt 6):549–557. doi:10.1038/icb.1993.61

75. Roederer M, Dubs JG, Anderson MT, Raju PA, Herzenberg LA, Herzenberg LA (1995) CD8 naive T cell counts decrease progressively in HIV-infected adults. J Clin Invest 95(5):2061–2066. doi:10.1172/JCI117892
76. Kalayjian RC, Landay A, Pollard RB, Taub DD, Gross BH, Francis IR, Sevin A, Pu M, Spritzler J, Chernoff M, Namkung A, Fox L, Martinez A, Waterman K, Fiscus SA, Sha B, Johnson D, Slater S, Rousseau F, Lederman MM (2003) Age-related immune dysfunction in health and in human immunodeficiency virus (HIV) disease: association of age and HIV infection with naive CD8+ cell depletion, reduced expression of CD28 on CD8+ cells, and reduced thymic volumes. J Infect Dis 187(12):1924–1933. doi:10.1086/375372
77. Serrano-Villar S, Perez-Elias MJ, Dronda F, Casado JL, Moreno A, Royuela A, Perez-Molina JA, Sainz T, Navas E, Hermida JM, Quereda C, Moreno S (2014) Increased risk of serious non-AIDS-related events in HIV-infected subjects on antiretroviral therapy associated with a low CD4/CD8 ratio. PLoS One 9(1), e85798. doi:10.1371/journal.pone.0085798
78. Menozzi M, Zona S, Santoro A, Carli F, Stentarelli C, Mussini C, Guaraldi G (2014) CD4/CD8 ratio is not predictive of multi-morbidity prevalence in HIV-infected patients but identify patients with higher CVD risk. J Int AIDS Soc 17(4 Suppl 3):19709. doi:10.7448/IAS.17.4.19709
79. Bernal E, Serrano J, Perez A, Valero S, Garcia E, Marin I, Munoz A, Verdu JM, Vera C, Cano A (2014) The CD4:CD8 ratio is associated with IMT progression in HIV-infected patients on antiretroviral treatment. J Int AIDS Soc 17(4 Suppl 3):19723. doi:10.7448/IAS.17.4.19723
80. Saracino A, Bruno G, Scudeller L, Volpe A, Caricato P, Ladisa N, Monno L, Angarano G (2014) Chronic inflammation in a long-term cohort of HIV-infected patients according to the normalization of the CD4:CD8 ratio. AIDS Res Hum Retroviruses 30(12):1178–1184. doi:10.1089/aid.2014.0080
81. Sansoni P, Cossarizza A, Brianti V, Fagnoni F, Snelli G, Monti D, Marcato A, Passeri G, Ortolani C, Forti E et al (1993) Lymphocyte subsets and natural killer cell activity in healthy old people and centenarians. Blood 82(9):2767–2773
82. Phelouzat MA, Arbogast A, Laforge T, Quadri RA, Proust JJ (1996) Excessive apoptosis of mature T lymphocytes is a characteristic feature of human immune senescence. Mech Ageing Dev 88(1–2):25–38
83. de Oliveira Pinto LM, Garcia S, Lecoeur H, Rapp C, Gougeon ML (2002) Increased sensitivity of T lymphocytes to tumor necrosis factor receptor 1 (TNFR1)- and TNFR2-mediated apoptosis in HIV infection: relation to expression of Bcl-2 and active caspase-8 and caspase-3. Blood 99(5):1666–1675
84. Effros RB (1997) Loss of CD28 expression on T lymphocytes: a marker of replicative senescence. Dev Comp Immunol 21(6):471–478
85. Effros RB, Allsopp R, Chiu CP, Hausner MA, Hirji K, Wang L, Harley CB, Villeponteau B, West MD, Giorgi JV (1996) Shortened telomeres in the expanded CD28-CD8+ cell subset in HIV disease implicate replicative senescence in HIV pathogenesis. AIDS (London, England) 10(8):F17–F22
86. Rufer N, Brummendorf TH, Kolvraa S, Bischoff C, Christensen K, Wadsworth L, Schulzer M, Lansdorp PM (1999) Telomere fluorescence measurements in granulocytes and T lymphocyte subsets point to a high turnover of hematopoietic stem cells and memory T cells in early childhood. J Exp Med 190(2):157–167
87. Lee SA, Sinclair E, Hatano H, Hsue PY, Epling L, Hecht FM, Bangsberg DR, Martin JN, McCune JM, Deeks SG, Hunt PW (2014) Impact of HIV on CD8+ T cell CD57 expression is distinct from that of CMV and aging. PLoS One 9(2), e89444. doi:10.1371/journal.pone.0089444
88. Deeks SG (2011) HIV infection, inflammation, immunosenescence, and aging. Annu Rev Med 62:141–155. doi:10.1146/annurev-med-042909-093756
89. Kalayjian RC, Spritzler J, Pu M, Landay A, Pollard RB, Stocker V, Harthi LA, Gross BH, Francis IR, Fiscus SA, Tebas P, Bosch RJ, Valcour V, Lederman MM, Adult ACTG, Study T

(2005) Distinct mechanisms of T cell reconstitution can be identified by estimating thymic volume in adult HIV-1 disease. J Infect Dis 192(9):1577–1587. doi:10.1086/466527
90. Frasca D, Diaz A, Romero M, Landin AM, Blomberg BB (2011) Age effects on B cells and humoral immunity in humans. Ageing Res Rev 10(3):330–335. doi:10.1016/j.arr.2010.08.004
91. Moir S, Fauci AS (2009) B cells in HIV infection and disease. Nat Rev 9(4):235–245. doi:10.1038/nri2524
92. Hart M, Steel A, Clark SA, Moyle G, Nelson M, Henderson DC, Wilson R, Gotch F, Gazzard B, Kelleher P (2007) Loss of discrete memory B cell subsets is associated with impaired immunization responses in HIV-1 infection and may be a risk factor for invasive pneumococcal disease. J Immunol 178(12):8212–8220
93. De Milito A, Nilsson A, Titanji K, Thorstensson R, Reizenstein E, Narita M, Grutzmeier S, Sonnerborg A, Chiodi F (2004) Mechanisms of hypergammaglobulinemia and impaired antigen-specific humoral immunity in HIV-1 infection. Blood 103(6):2180–2186. doi:10.1182/blood-2003-07-2375
94. Siewe B, Stapleton JT, Martinson J, Keshavarzian A, Kazmi N, Demarais PM, French AL, Landay A (2013) Regulatory B cell frequency correlates with markers of HIV disease progression and attenuates anti-HIV CD8(+) T cell function in vitro. J Leukoc Biol 93(5):811–818. doi:10.1189/jlb.0912436
95. Siewe B, Keshavarzian A, French A, Demarais P, Landay A (2013) A role for TLR signaling during B cell activation in antiretroviral-treated HIV individuals. AIDS Res Hum Retroviruses 29(10):1353–1360. doi:10.1089/AID.2013.0115
96. Van Epps P, Matining RM, Tassiopoulos K, Anthony DD, Landay A, Kalayjian RC, Canaday DH (2014) Older age is associated with peripheral blood expansion of naive B cells in HIV-infected subjects on antiretroviral therapy. PLoS One 9(9), e107064. doi:10.1371/journal.pone.0107064
97. Hearps AC, Maisa A, Cheng WJ, Angelovich TA, Lichtfuss GF, Palmer CS, Landay AL, Jaworowski A, Crowe SM (2012) HIV infection induces age-related changes to monocytes and innate immune activation in young men that persist despite combination antiretroviral therapy. AIDS (London, England) 26(7):843–853. doi:10.1097/QAD.0b013e328351f756
98. Hearps AC, Martin GE, Angelovich TA, Cheng WJ, Maisa A, Landay AL, Jaworowski A, Crowe SM (2012) Aging is associated with chronic innate immune activation and dysregulation of monocyte phenotype and function. Aging Cell 11(5):867–875. doi:10.1111/j.1474-9726.2012.00851.x
99. Martin GE, Gouillou M, Hearps AC, Angelovich TA, Cheng AC, Lynch F, Cheng WJ, Paukovics G, Palmer CS, Novak RM, Jaworowski A, Landay AL, Crowe SM (2013) Age-associated changes in monocyte and innate immune activation markers occur more rapidly in HIV infected women. PLoS One 8(1), e55279. doi:10.1371/journal.pone.0055279
100. Clark JA, Peterson TC (1994) Cytokine production and aging: overproduction of IL-8 in elderly males in response to lipopolysaccharide. Mech Ageing Dev 77(2):127–139. doi:0047-6374(94)90020-5 [pii]
101. Delpedro AD, Barjavel MJ, Mamdouh Z, Faure S, Bakouche O (1998) Signal transduction in LPS-activated aged and young monocytes. J Interferon Cytokine Res 18(6):429–437
102. Jaworowski A, Ellery P, Maslin CL, Naim E, Heinlein AC, Ryan CE, Paukovics G, Hocking J, Sonza S, Crowe SM (2006) Normal CD16 expression and phagocytosis of Mycobacterium avium complex by monocytes from a current cohort of HIV-1-infected patients. J Infect Dis 193(5):693–697
103. Spencer ME, Jain A, Matteini A, Beamer BA, Wang NY, Leng SX, Punjabi NM, Walston JD, Fedarko NS (2010) Serum levels of the immune activation marker neopterin change with age and gender and are modified by race, BMI, and percentage of body fat. J Gerontol A Biol Sci Med Sci 65(8):858–865. doi:10.1093/gerona/glq066, glq066 [pii]
104. Antonelli A, Rotondi M, Fallahi P, Ferrari SM, Paolicchi A, Romagnani P, Serio M, Ferrannini E (2006) Increase of CXC chemokine CXCL10 and CC chemokine CCL2 serum levels in normal ageing. Cytokine 34(1–2):32–38. doi:10.1016/j.cyto.2006.03.012, S1043-4666(06)00132-3 [pii]

105. Burdo TH, Lentz MR, Autissier P, Krishnan A, Halpern E, Letendre S, Rosenberg ES, Ellis RJ, Williams KC (2011) Soluble CD163 made by monocyte/macrophages is a novel marker of HIV activity in early and chronic infection prior to and after anti-retroviral therapy. J Infect Dis 204(1):154–163. doi:10.1093/infdis/jir214, jir214 [pii]
106. Kamat A, Misra V, Cassol E, Ancuta P, Yan Z, Li C, Morgello S, Gabuzda D (2012) A plasma biomarker signature of immune activation in HIV patients on antiretroviral therapy. PLoS One 7(2), e30881. doi:10.1371/journal.pone.0030881
107. Méndez-Lagares G, Romero-Sánchez MC, Ruiz-Mateos E, Genebat M, Ferrando-Martínez S, Muñoz-Fernández MÁ, Pacheco YM, Leal M (2013) Long-term suppressive combined antiretroviral treatment does not normalize the serum level of soluble CD14. J Infect Dis 207(8):1221–1225. doi:10.1093/infdis/jit025
108. Alter G, Teigen N, Davis BT, Addo MM, Suscovich TJ, Waring MT, Streeck H, Johnston MN, Staller KD, Zaman MT, Yu XG, Lichterfeld M, Basgoz N, Rosenberg ES, Altfeld M (2005) Sequential deregulation of NK cell subset distribution and function starting in acute HIV-1 infection. Blood 106(10):3366–3369
109. Borrego F, Alonso MC, Galiani MD, Carracedo J, Ramirez R, Ostos B, Pena J, Solana R (1999) NK phenotypic markers and IL2 response in NK cells from elderly people. Exp Gerontol 34(2):253–265. doi:S0531-5565(98)00076-X [pii]
110. Ogata K, Yokose N, Tamura H, An E, Nakamura K, Dan K, Nomura T (1997) Natural killer cells in the late decades of human life. Clin Immunol Immunopathol 84(3):269–275
111. Fernandes G, Gupta S (1981) Natural killing and antibody-dependent cytotoxicity by lymphocyte subpopulations in young and aging humans. J Clin Immunol 1(3):141–148
112. Lichtfuss GF, Cheng WJ, Farsakoglu Y, Paukovics G, Rajasuriar R, Velayudham P, Kramski M, Hearps AC, Cameron PU, Lewin SR, Crowe SM, Jaworowski A (2012) Virologically suppressed HIV patients show activation of NK cells and persistent innate immune activation. J Immunol 189(3):1491–1499. doi:10.4049/jimmunol.1200458
113. Fogli M, Costa P, Murdaca G, Setti M, Mingari MC, Moretta L, Moretta A, De Maria A (2004) Significant NK cell activation associated with decreased cytolytic function in peripheral blood of HIV-1-infected patients. Eur J Immunol 34(8):2313–2321. doi:10.1002/eji.200425251
114. Azzoni L, Papasavvas E, Chehimi J, Kostman JR, Mounzer K, Ondercin J, Perussia B, Montaner LJ (2002) Sustained impairment of IFN-gamma secretion in suppressed HIV-infected patients despite mature NK cell recovery: evidence for a defective reconstitution of innate immunity. J Immunol 168(11):5764–5770
115. Moroni F, Di Paolo ML, Rigo A, Cipriano C, Giacconi R, Recchioni R, Marcheselli F, Malavolta M, Mocchegiani E (2005) Interrelationship among neutrophil efficiency, inflammation, antioxidant activity and zinc pool in very old age. Biogerontology 6(4):271–281. doi:10.1007/s10522-005-2625-0
116. Agrawal A, Agrawal S, Tay J, Gupta S (2008) Biology of dendritic cells in aging. J Clin Immunol 28(1):14–20. doi:10.1007/s10875-007-9127-6
117. Agrawal A, Agrawal S, Cao JN, Su H, Osann K, Gupta S (2007) Altered innate immune functioning of dendritic cells in elderly humans: a role of phosphoinositide 3-kinase-signaling pathway. J Immunol 178(11):6912–6922
118. Plackett TP, Boehmer ED, Faunce DE, Kovacs EJ (2004) Aging and innate immune cells. J Leukoc Biol 76(2):291–299. doi:10.1189/jlb.1103592
119. Martinson JA, Roman-Gonzalez A, Tenorio AR, Montoya CJ, Gichinga CN, Rugeles MT, Tomai M, Krieg AM, Ghanekar S, Baum LL, Landay AL (2007) Dendritic cells from HIV-1 infected individuals are less responsive to toll-like receptor (TLR) ligands. Cell Immunol 250(1–2):75–84
120. Kaszubowska L (2008) Telomere shortening and ageing of the immune system. J Physiol Pharmacol 59(Suppl 9):169–186
121. Meeker AK, Hicks JL, Iacobuzio-Donahue CA, Montgomery EA, Westra WH, Chan TY, Ronnett BM, De Marzo AM (2004) Telomere length abnormalities occur early in the initiation of epithelial carcinogenesis. Clin Cancer Res 10(10):3317–3326. doi:10.1158/1078-0432.CCR-0984-03

122. Fitzpatrick AL, Kronmal RA, Gardner JP, Psaty BM, Jenny NS, Tracy RP, Walston J, Kimura M, Aviv A (2007) Leukocyte telomere length and cardiovascular disease in the cardiovascular health study. Am J Epidemiol 165(1):14–21. doi:10.1093/aje/kwj346
123. Weischer M, Bojesen SE, Cawthon RM, Freiberg JJ, Tybjaerg-Hansen A, Nordestgaard BG (2012) Short telomere length, myocardial infarction, ischemic heart disease, and early death. Arterioscler Thromb Vasc Biol 32(3):822–829. doi:10.1161/ATVBAHA.111.237271
124. Willeit P, Willeit J, Brandstatter A, Ehrlenbach S, Mayr A, Gasperi A, Weger S, Oberhollenzer F, Reindl M, Kronenberg F, Kiechl S (2010) Cellular aging reflected by leukocyte telomere length predicts advanced atherosclerosis and cardiovascular disease risk. Arterioscler Thromb Vasc Biol 30(8):1649–1656. doi:10.1161/ATVBAHA.110.205492
125. Honig LS, Kang MS, Schupf N, Lee JH, Mayeux R (2012) Association of shorter leukocyte telomere repeat length with dementia and mortality. Arch Neurol 69(10):1332–1339. doi:10.1001/archneurol.2012.1541
126. Hochstrasser T, Marksteiner J, Humpel C (2012) Telomere length is age-dependent and reduced in monocytes of Alzheimer patients. Exp Gerontol 47(2):160–163. doi:10.1016/j.exger.2011.11.012
127. Duprez DA, Neuhaus J, Kuller LH, Tracy R, Belloso W, De Wit S, Drummond F, Lane HC, Ledergerber B, Lundgren J, Nixon D, Paton NI, Prineas RJ, Neaton JD (2012) Inflammation, coagulation and cardiovascular disease in HIV-infected individuals. PLoS One 7(9), e44454. doi:10.1371/journal.pone.0044454
128. De Luca A, de Gaetano DK, Colafigli M, Cozzi-Lepri A, De Curtis A, Gori A, Sighinolfi L, Giacometti A, Capobianchi MR, D'Avino A, Iacoviello L, Cauda R, D'Arminio MA (2013) The association of high-sensitivity c-reactive protein and other biomarkers with cardiovascular disease in patients treated for HIV: a nested case--control study. BMC Infect Dis 13(1):414. doi:10.1186/1471-2334-13-414
129. Koethe JR, Dee K, Bian A, Shintani A, Turner M, Bebawy S, Sterling TR, Hulgan T (2013) Circulating interleukin-6, soluble CD14, and other inflammation biomarker levels differ between obese and nonobese HIV-infected adults on antiretroviral therapy. AIDS Res Hum Retroviruses 29(7):1019–1025. doi:10.1089/aid.2013.0016
130. Hussein AA, Gottdiener JS, Bartz TM, Sotoodehnia N, Defilippi C, See V, Deo R, Siscovick D, Stein PK, Lloyd-Jones D (2013) Inflammation and sudden cardiac death in a community-based population of older adults: The Cardiovascular Health Study. Heart Rhythm 10(10):1425–1432. doi:10.1016/j.hrthm.2013.07.004
131. Empana JP, Jouven X, Canoui-Poitrine F, Luc G, Tafflet M, Haas B, Arveiler D, Ferrieres J, Ruidavets JB, Montaye M, Yarnell J, Morange P, Kee F, Evans A, Amouyel P, Ducimetiere P (2010) C-reactive protein, interleukin 6, fibrinogen and risk of sudden death in European middle-aged men: the PRIME study. Arterioscler Thromb Vasc Biol 30(10):2047–2052. doi:10.1161/atvbaha.110.208785
132. Jenny NS, French B, Arnold AM, Strotmeyer ES, Cushman M, Chaves PH, Ding J, Fried LP, Kritchevsky SB, Rifkin DE, Sarnak MJ, Newman AB (2012) Long-term assessment of inflammation and healthy aging in late life: the Cardiovascular Health Study All Stars. J Gerontol A Biol Sci Med Sci 67(9):970–976. doi:10.1093/gerona/glr261
133. Ridker PM, Rifai N, Stampfer MJ, Hennekens CH (2000) Plasma concentration of interleukin-6 and the risk of future myocardial infarction among apparently healthy men. Circulation 101(15):1767–1772
134. Jenny NS, Tracy RP, Ogg MS, le Luong A, Kuller LH, Arnold AM, Sharrett AR, Humphries SE (2002) In the elderly, interleukin-6 plasma levels and the -174G>C polymorphism are associated with the development of cardiovascular disease. Arterioscler Thromb Vasc Biol 22(12):2066–2071
135. Hsue PY, Scherzer R, Hunt PW, Schnell A, Bolger AF, Kalapus SC, Maka K, Martin JN, Ganz P, Deeks SG (2012) Carotid intima-media thickness progression in HIV-infected adults occurs preferentially at the carotid bifurcation and is predicted by inflammation. J Am Heart Assoc 1(2). pii: jah3-e000422. doi:10.1161/jaha.111.000422

136. Biron A, Bobin-Dubigeon C, Volteau C, Piroth L, Perre P, Leport C, Prazuck T, Jovelin T, Billard M, Sebille V, Bard JM, Raffi F, Biron C (2012) Metabolic syndrome in French HIV-infected patients: prevalence and predictive factors after 3 years of antiretroviral therapy. AIDS Res Hum Retroviruses 28(12):1672–1678. doi:10.1089/aid.2012.0048
137. Brown TT, Tassiopoulos K, Bosch RJ, Shikuma C, McComsey GA (2010) Association between systemic inflammation and incident diabetes in HIV-infected patients after initiation of antiretroviral therapy. Diabetes Care 33(10):2244–2249. doi:10.2337/dc10-0633
138. Schnabel RB, Yin X, Larson MG, Yamamoto JF, Fontes JD, Kathiresan S, Rong J, Levy D, Keaney JF Jr, Wang TJ, Murabito JM, Vasan RS, Benjamin EJ (2013) Multiple inflammatory biomarkers in relation to cardiovascular events and mortality in the community. Arterioscler Thromb Vasc Biol 33(7):1728–1733. doi:10.1161/atvbaha.112.301174
139. Parkner T, Sorensen LP, Nielsen AR, Fischer CP, Bibby BM, Nielsen S, Pedersen BK, Moller HJ (2012) Soluble CD163: a biomarker linking macrophages and insulin resistance. Diabetologia 55(6):1856–1862. doi:10.1007/s00125-012-2533-1
140. Shikuma CM, Barbour JD, Ndhlovu LC, Keating SM, Norris PJ, Budoff M, Parikh N, Seto T, Gangcuangco LM, Ogata-Arakaki D, Chow D (2013) Plasma monocyte chemoattractant protein-1 and tumor necrosis factor-alpha levels predict the presence of coronary artery calcium in HIV-infected individuals independent of traditional cardiovascular risk factors. AIDS Res Hum Retroviruses. doi:10.1089/aid.2013.0183
141. Ancuta P, Kamat A, Kunstman KJ, Kim EY, Autissier P, Wurcel A, Zaman T, Stone D, Mefford M, Morgello S, Singer EJ, Wolinsky SM, Gabuzda D (2008) Microbial translocation is associated with increased monocyte activation and dementia in AIDS patients. PLoS One 3(6), e2516. doi:10.1371/journal.pone.0002516
142. Weaver JD, Huang MH, Albert M, Harris T, Rowe JW, Seeman TE (2002) Interleukin-6 and risk of cognitive decline: MacArthur studies of successful aging. Neurology 59(3):371–378
143. Noble JM, Manly JJ, Schupf N, Tang MX, Mayeux R, Luchsinger JA (2010) Association of C-reactive protein with cognitive impairment. Arch Neurol 67(1):87–92. doi:10.1001/archneurol.2009.308
144. Ryan LA, Zheng J, Brester M, Bohac D, Hahn F, Anderson J, Ratanasuwan W, Gendelman HE, Swindells S (2001) Plasma levels of soluble CD14 and tumor necrosis factor-alpha type II receptor correlate with cognitive dysfunction during human immunodeficiency virus type 1 infection. J Infect Dis 184(6):699–706. doi:10.1086/323036
145. Engelhart MJ, Geerlings MI, Meijer J, Kiliaan A, Ruitenberg A, van Swieten JC, Stijnen T, Hofman A, Witteman JC, Breteler MM (2004) Inflammatory proteins in plasma and the risk of dementia: the rotterdam study. Arch Neurol 61(5):668–672. doi:10.1001/archneur.61.5.668
146. Bruunsgaard H, Andersen-Ranberg K, Jeune B, Pedersen AN, Skinhoj P, Pedersen BK (1999) A high plasma concentration of TNF-alpha is associated with dementia in centenarians. J Gerontol A Biol Sci Med Sci 54(7):M357–M364
147. Borges AH, Silverberg MJ, Wentworth D, Grulich AE, Fatkenheuer G, Mitsuyasu R, Tambussi G, Sabin CA, Neaton JD, Lundgren JD (2013) Predicting risk of cancer during HIV infection: the role of inflammatory and coagulation biomarkers. AIDS (London, England) 27(9):1433–1441. doi:10.1097/QAD.0b013e32835f6b0c
148. Il'yasova D, Colbert LH, Harris TB, Newman AB, Bauer DC, Satterfield S, Kritchevsky SB (2005) Circulating levels of inflammatory markers and cancer risk in the health aging and body composition cohort. Cancer Epidemiol Biomarkers Prev 14(10):2413–2418. doi:10.1158/1055-9965.EPI-05-0316
149. de Pablo P, Cooper MS, Buckley CD (2012) Association between bone mineral density and C-reactive protein in a large population-based sample. Arthritis Rheum 64(8):2624–2631. doi:10.1002/art.34474
150. Eriksson AL, Moverare-Skrtic S, Ljunggren O, Karlsson M, Mellstrom D, Ohlsson C (2013) High sensitive CRP is an independent risk factor for all fractures and vertebral fractures in elderly men: The MrOS Sweden study. J Bone Min Res. doi:10.1002/jbmr.2037

151. Ding C, Parameswaran V, Udayan R, Burgess J, Jones G (2008) Circulating levels of inflammatory markers predict change in bone mineral density and resorption in older adults: a longitudinal study. J Clin Endocrinol Metab 93(5):1952–1958. doi:10.1210/jc.2007-2325
152. Scheidt-Nave C, Bismar H, Leidig-Bruckner G, Woitge H, Seibel MJ, Ziegler R, Pfeilschifter J (2001) Serum interleukin 6 is a major predictor of bone loss in women specific to the first decade past menopause. J Clin Endocrinol Metab 86(5):2032–2042
153. Margolick J, Jacobson L, Lopez J (2012) Frailty and circulating concentrations of proinflammatory cytokines and chemokines in HIV-infected and -uninfected men in the Multicenter AIDS Cohort Study (MACS). Paper presented at the 3rd international workshop on HIV and aging, Baltimore, 5–6 Nov 2012
154. Erlandson KM, Allshouse AA, Jankowski CM, Lee EJ, Rufner KM, Palmer BE, Wilson CC, MaWhinney S, Kohrt WM, Campbell TB (2013) Association of functional impairment with inflammation and immune activation in HIV type 1-infected adults receiving effective antiretroviral therapy. J Infect Dis 208(2):249–259. doi:10.1093/infdis/jit147
155. Ferrucci L, Harris TB, Guralnik JM, Tracy RP, Corti MC, Cohen HJ, Penninx B, Pahor M, Wallace R, Havlik RJ (1999) Serum IL-6 level and the development of disability in older persons. J Am Geriatr Soc 47(6):639–646
156. Walston J, McBurnie MA, Newman A, Tracy RP, Kop WJ, Hirsch CH, Gottdiener J, Fried LP (2002) Frailty and activation of the inflammation and coagulation systems with and without clinical comorbidities: results from the Cardiovascular Health Study. Arch Intern Med 162(20):2333–2341
157. Hunt PW, Sinclair E, Rodriguez B, Shive C, Clagett B, Funderburg N, Robinson J, Huang Y, Epling L, Martin JN, Deeks SG, Meinert CL, Van Natta ML, Jabs DA, Lederman MM (2014) Gut epithelial barrier dysfunction and innate immune activation predict mortality in treated HIV infection. J Infect Dis 210(8):1228–1238. doi:10.1093/infdis/jiu238
158. Kelesidis T, Kendall MA, Yang OO, Hodis HN, Currier JS (2012) Biomarkers of microbial translocation and macrophage activation: association with progression of subclinical atherosclerosis in HIV-1 infection. J Infect Dis 206(10):1558–1567
159. Blodget E, Shen C, Aldrovandi G, Rollie A, Gupta SK, Stein JH, Dube MP (2012) Relationship between microbial translocation and endothelial function in HIV infected patients. PLoS One [Electronic Resource] 7(8), e42624
160. Pedersen KK, Pedersen M, Troseid M, Gaardbo JC, Lund TT, Thomsen C, Gerstoft J, Kvale D, Nielsen SD (2013) Microbial translocation in HIV infection is associated with dyslipidemia, insulin resistance, and risk of myocardial infarction. J Acquir Immune Defic Syndr. doi:10.1097/QAI.0b013e31829f919d
161. Manner IW, Baekken M, Kvale D, Oektedalen O, Pedersen M, Nielsen SD, Nowak P, Os I, Trøseid M (2013) Markers of microbial translocation predict hypertension in HIV-infected individuals. HIV Med 14(6):354–361. doi:10.1111/hiv.12015
162. Lassenius MI, Pietilainen KH, Kaartinen K, Pussinen PJ, Syrjanen J, Forsblom C, Porsti I, Rissanen A, Kaprio J, Mustonen J, Groop PH, Lehto M (2011) Bacterial endotoxin activity in human serum is associated with dyslipidemia, insulin resistance, obesity, and chronic inflammation. Diabetes Care 34(8):1809–1815. doi:10.2337/dc10-2197
163. Gonzalez-Quintela A, Alonso M, Campos J, Vizcaino L, Loidi L, Gude F (2013) Determinants of serum concentrations of lipopolysaccharide-binding protein (LBP) in the adult population: the role of obesity. PLoS One 8(1), e54600. doi:10.1371/journal.pone.0054600
164. Vassallo M, Dunais B, Durant J, Carsenti-Dellamonica H, Harvey-Langton A, Cottalorda J, Ticchioni M, Laffon M, Lebrun-Frenay C, Dellamonica P, Pradier C (2013) Relevance of lipopolysaccharide levels in HIV-associated neurocognitive impairment: the Neuradapt study. J Neurovirol 19(4):376–382. doi:10.1007/s13365-013-0181-y
165. Marks MA, Rabkin CS, Engels EA, Busch E, Kopp W, Rager H, Goedert JJ, Chaturvedi AK (2013) Markers of microbial translocation and risk of AIDS-related lymphoma. AIDS (London, England) 27(3):469–474. doi:10.1097/QAD.0b013e32835c1333

166. Merlini E, Luzi K, Suardi E, Barassi A, Cerrone M, Martinez JS, Bai F, D'Eril GV, Monforte AD, Marchetti G (2012) T-cell phenotypes, apoptosis and inflammation in HIV+ patients on virologically effective cART with early atherosclerosis. PLoS One 7(9), e46073. doi:10.1371/journal.pone.0046073
167. Longenecker CT, Jiang Y, Orringer CE, Gilkeson RC, Debanne S, Funderburg NT, Lederman MM, Storer N, Labbato DE, McComsey GA (2014) Soluble CD14 is independently associated with coronary calcification and extent of subclinical vascular disease in treated HIV infection. AIDS (London, England) 28(7):969–977. doi:10.1097/QAD.0000000000000158
168. Palmer CS, Ostrowski M, Gouillou M, Tsai L, Yu D, Zhou J, Henstridge DC, Maisa A, Hearps AC, Lewin SR, Landay A, Jaworowski A, McCune JM, Crowe SM (2013) Increased glucose metabolic activity is associated with CD4+ T-cell activation and depletion during chronic HIV infection. AIDS (London, England). doi:10.1097/QAD.0000000000000128
169. Reiner AP, Lange EM, Jenny NS, Chaves PHM, Ellis J, Li J, Walston J, Lange LA, Cushman M, Tracy RP (2013) Soluble CD14: genomewide association analysis and relationship to cardiovascular risk and mortality in older adults. Arterioscler Thromb Vasc Biol 33(1):158–164. doi:10.1161/atvbaha.112.300421
170. Subramanian S, Tawakol A, Burdo TH, Abbara S, Wei J, Vijayakumar J, Corsini E, Abdelbaky A, Zanni MV, Hoffmann U, Williams KC, Lo J, Grinspoon SK (2012) Arterial inflammation in patients with HIV. JAMA 308(4):379–386. doi:10.1001/jama.2012.6698
171. Burdo TH, Lo J, Abbara S, Wei J, Delelys ME, Preffer F, Rosenberg ES, Williams KC, Grinspoon S (2011) Soluble CD163, a novel marker of activated macrophages, is elevated and associated with noncalcified coronary plaque in HIV-infected patients. J Infect Dis 204(8):1227–1236. doi:10.1093/infdis/jir520, jir520 [pii]
172. Aristoteli LP, Moller HJ, Bailey B, Moestrup SK, Kritharides L (2006) The monocytic lineage specific soluble CD163 is a plasma marker of coronary atherosclerosis. Atherosclerosis 184(2):342–347. doi:10.1016/j.atherosclerosis.2005.05.004, S0021-9150(05)00339-4 [pii]
173. Fjeldborg K, Christiansen T, Bennetzen M, J Møller H, Pedersen SB, Richelsen BR (2013) The macrophage-specific serum marker, soluble CD163, is increased in obesity and reduced after dietary-induced weight loss. Obesity (Silver Spring). doi:10.1002/oby.20376
174. Zanni MV, Burdo TH, Makimura H, Williams KC, Grinspoon SK (2012) Relationship between monocyte/macrophage activation marker soluble CD163 and insulin resistance in obese and normal-weight subjects. Clin Endocrinol (Oxf) 77(3):385–390. doi:10.1111/j.1365-2265.2011.04284.x
175. Moller HJ, Frikke-Schmidt R, Moestrup SK, Nordestgaard BG, Tybjaerg-Hansen A (2011) Serum soluble CD163 predicts risk of type 2 diabetes in the general population. Clin Chem 57(2):291–297. doi:10.1373/clinchem.2010.154724, clinchem.2010.154724 [pii]
176. Lyons JL, Uno H, Ancuta P, Kamat A, Moore DJ, Singer EJ, Morgello S, Gabuzda D (2011) Plasma sCD14 is a biomarker associated with impaired neurocognitive test performance in attention and learning domains in HIV infection. J Acquir Immune Defic Syndr 57(5):371–379. doi:10.1097/QAI.1090b1013e3182237e3182254
177. Kamat A, Lyons JL, Misra V, Uno H, Morgello S, Singer EJ, Gabuzda D (2012) Monocyte activation markers in cerebrospinal fluid associated with impaired neurocognitive testing in advanced HIV infection. J Acquir Immune Defic Syndr 60(3):234–243. doi:10.1097/QAI.1090b1013e318256f318253bc
178. Burdo TH, Weiffenbach A, Woods SP, Letendre S, Ellis RJ, Williams KC (2013) Elevated sCD163 in plasma but not cerebrospinal fluid is a marker of neurocognitive impairment in HIV infection. AIDS (London, England) 27(9):1387–1395. doi:10.1097/QAD.0b013e32836010bd
179. Blasko I, Knaus G, Weiss E, Kemmler G, Winkler C, Falkensammer G, Griesmacher A, Wurzner R, Marksteiner J, Fuchs D (2007) Cognitive deterioration in Alzheimer's disease is accompanied by increase of plasma neopterin. J Psychiatr Res 41(8):694–701. doi:10.1016/j.jpsychires.2006.02.001

180. Leng SX, Tian X, Matteini A, Li H, Hughes J, Jain A, Walston JD, Fedarko NS (2011) IL-6-independent association of elevated serum neopterin levels with prevalent frailty in community-dwelling older adults. Age Ageing 40(4):475–481. doi:10.1093/ageing/afr047, afr047 [pii]
181. Kaplan RC, Sinclair E, Landay AL, Lurain N, Sharrett AR, Gange SJ, Xue X, Hunt P, Karim R, Kern DM, Hodis HN, Deeks SG (2011) T cell activation and senescence predict subclinical carotid artery disease in HIV-infected women. J Infect Dis 203(4):452–463
182. Longenecker C, Funderburg N, Jiang Y, Debanne S, Storer N, Labbato D, Lederman M, McComsey G (2013) Markers of inflammation and CD8 T-cell activation, but not monocyte activation, are associated with subclinical carotid artery disease in HIV-infected individuals. HIV Med. doi:10.1111/hiv.12013
183. Kaplan RC, Sinclair E, Landay AL, Lurain N, Sharrett AR, Gange SJ, Xue X, Parrinello CM, Hunt P, Deeks SG, Hodis HN (2011) T cell activation predicts carotid artery stiffness among HIV-infected women. Atherosclerosis 217(1):207–213
184. Gazzola L, Bellistri GM, Tincati C, Ierardi V, Savoldi A, del Dole A, Tagliabue L, d'Arminio Monforte A, Marchetti G (2013) Association between peripheral T-Lymphocyte activation and impaired bone mineral density in HIV-infected patients. J Transl Med 11:51. doi:10.1186/1479-5876-11-51
185. Lee SA, Sinclair E, Jain V, Huang Y, Epling L, Van Natta M, Meinert CL, Martin JN, McCune JM, Deeks SG, Lederman MM, Hecht FM, Hunt PW (2014) Low proportions of CD28–CD8+ T cells expressing CD57 can be reversed by early ART initiation and predict mortality in treated HIV infection. J Infect Dis 210(3):374–382. doi:10.1093/infdis/jiu109
186. Sanders JL, Fitzpatrick AL, Boudreau RM, Arnold AM, Aviv A, Kimura M, Fried LF, Harris TB, Newman AB (2012) Leukocyte telomere length is associated with noninvasively measured age-related disease: The Cardiovascular Health Study. J Gerontol A Biol Sci Med Sci 67(4):409–416. doi:10.1093/gerona/glr173
187. Sampson MJ, Winterbone MS, Hughes JC, Dozio N, Hughes DA (2006) Monocyte telomere shortening and oxidative DNA damage in type 2 diabetes. Diabetes Care 29(2):283–289
188. Mu Y, Zhang Q, Mei L, Liu X, Yang W, Yu J (2012) Telomere shortening occurs early during gastrocarcinogenesis. Med Oncol 29(2):893–898. doi:10.1007/s12032-011-9866-3
189. Chen S, de Craen AJ, Raz Y, Derhovanessian E, Vossen AC, Westendorp RG, Pawelec G, Maier AB (2012) Cytomegalovirus seropositivity is associated with glucose regulation in the oldest old. Results from the Leiden 85-plus Study. Immun Ageing 9(1):18. doi:10.1186/1742-4933-9-18
190. Muhlestein JB, Horne BD, Carlquist JF, Madsen TE, Bair TL, Pearson RR, Anderson JL (2000) Cytomegalovirus seropositivity and C-reactive protein have independent and combined predictive value for mortality in patients with angiographically demonstrated coronary artery disease. Circulation 102(16):1917–1923
191. Hsue PY, Hunt PW, Sinclair E, Bredt B, Franklin A, Killian M, Hoh R, Martin JN, McCune JM, Waters DD, Deeks SG (2006) Increased carotid intima-media thickness in HIV patients is associated with increased cytomegalovirus-specific T-cell responses. AIDS (London, England) 20(18):2275–2283. doi:10.1097/QAD.0b013e3280108704
192. Parrinello CM, Sinclair E, Landay AL, Lurain N, Sharrett AR, Gange SJ, Xue X, Hunt PW, Deeks SG, Hodis HN, Kaplan RC (2012) Cytomegalovirus immunoglobulin G antibody is associated with subclinical carotid artery disease among HIV-infected women. J Infect Dis 205(12):1788–1796. doi:10.1093/infdis/jis276
193. Wang GC, Kao WH, Murakami P, Xue QL, Chiou RB, Detrick B, McDyer JF, Semba RD, Casolaro V, Walston JD, Fried LP (2010) Cytomegalovirus infection and the risk of mortality and frailty in older women: a prospective observational cohort study. Am J Epidemiol 171(10):1144–1152. doi:10.1093/aje/kwq062
194. Savva GM, Pachnio A, Kaul B, Morgan K, Huppert FA, Brayne C, Moss PA (2013) Cytomegalovirus infection is associated with increased mortality in the older population. Aging Cell 12(3):381–387. doi:10.1111/acel.12059

195. Kong CM, Lee XW, Wang X (2013) Telomere shortening in human diseases. FEBS J 280(14):3180–3193. doi:10.1111/febs.12326
196. Wang X, Singh S, Jung HY, Yang G, Jun S, Sastry KJ, Park JI (2013) HIV-1 Vpr protein inhibits telomerase activity via the EDD-DDB1-VPRBP E3 ligase complex. J Biol Chem 288(22):15474–15480. doi:10.1074/jbc.M112.416735
197. Comandini A, Naro C, Adamo R, Akbar AN, Lanna A, Bonmassar E, Franzese O (2013) Molecular mechanisms involved in HIV-1-Tat mediated inhibition of telomerase activity in human CD4(+) T lymphocytes. Mol Immunol 54(2):181–192. doi:10.1016/j.molimm.2012.12.003
198. Reynoso R, Laufer N, Bolcic F, Quarleri J (2010) Telomerase activity in peripheral blood mononuclear cells from HIV and HIV-HCV coinfected patients. Virus Res 147(2):284–287. doi:10.1016/j.virusres.2009.11.006
199. Hukezalie KR, Thumati NR, Cote HC, Wong JM (2012) In vitro and ex vivo inhibition of human telomerase by anti-HIV nucleoside reverse transcriptase inhibitors (NRTIs) but not by non-NRTIs. PLoS One 7(11), e47505. doi:10.1371/journal.pone.0047505
200. Leeansyah E, Cameron PU, Solomon A, Tennakoon S, Velayudham P, Gouillou M, Spelman T, Hearps A, Fairley C, de Smit V, Pierce AB, Armishaw J, Crowe SM, Cooper DA, Koelsch KK, Liu JP, Chuah J, Lewin SR (2013) Inhibition of telomerase activity by human immunodeficiency virus (HIV) nucleos(t)ide reverse transcriptase inhibitors: a potential factor contributing to HIV-associated accelerated aging. J Infect Dis 207(7):1157–1165. doi:10.1093/infdis/jit006
201. Mutlu-Turkoglu U, Ilhan E, Oztezcan S, Kuru A, Aykac-Toker G, Uysal M (2003) Age-related increases in plasma malondialdehyde and protein carbonyl levels and lymphocyte DNA damage in elderly subjects. Clin Biochem 36(5):397–400
202. Gil L, Siems W, Mazurek B, Gross J, Schroeder P, Voss P, Grune T (2006) Age-associated analysis of oxidative stress parameters in human plasma and erythrocytes. Free Radic Res 40(5):495–505. doi:10.1080/10715760600592962
203. Cannizzo ES, Clement CC, Sahu R, Follo C, Santambrogio L (2011) Oxidative stress, inflamm-aging and immunosenescence. J Proteomics 74(11):2313–2323. doi:10.1016/j.jprot.2011.06.005
204. Gil L, Martinez G, Gonzalez I, Tarinas A, Alvarez A, Giuliani A, Molina R, Tapanes R, Perez J, Leon OS (2003) Contribution to characterization of oxidative stress in HIV/AIDS patients. Pharmacol Res 47(3):217–224. doi:S1043661802003201 [pii]
205. Kalinowska M, Bazdar DA, Lederman MM, Funderburg N, Sieg SF (2013) Decreased IL-7 responsiveness is related to oxidative stress in HIV disease. PLoS One 8(3), e58764. doi:10.1371/journal.pone.0058764
206. Mihm S, Ennen J, Pessara U, Kurth R, Droge W (1991) Inhibition of HIV-1 replication and NF-kappa B activity by cysteine and cysteine derivatives. AIDS (London, England) 5(5):497–503
207. Staal FJ, Roederer M, Herzenberg LA, Herzenberg LA (1990) Intracellular thiols regulate activation of nuclear factor kappa B and transcription of human immunodeficiency virus. Proc Natl Acad Sci U S A 87(24):9943–9947
208. Torres RA, Lewis W (2014) Aging and HIV/AIDS: pathogenetic role of therapeutic side effects. Lab Invest 94(2):120–128. doi:10.1038/labinvest.2013.142
209. Lagathu C, Eustace B, Prot M, Frantz D, Gu Y, Bastard JP, Maachi M, Azoulay S, Briggs M, Caron M, Capeau J (2007) Some HIV antiretrovirals increase oxidative stress and alter chemokine, cytokine or adiponectin production in human adipocytes and macrophages. Antivir Ther 12(4):489–500
210. Mandas A, Iorio EL, Congiu MG, Balestrieri C, Mereu A, Cau D, Dessi S, Curreli N (2009) Oxidative imbalance in HIV-1 infected patients treated with antiretroviral therapy. J Biomed Biotechnol 2009:749575. doi:10.1155/2009/749575
211. Franceschi C, Capri M, Monti D, Giunta S, Olivieri F, Sevini F, Panourgia MP, Invidia L, Celani L, Scurti M, Cevenini E, Castellani GC, Salvioli S (2007) Inflammaging and anti-inflammaging: a systemic perspective on aging and longevity emerged from studies in humans. Mech Ageing Dev 128(1):92–105

212. Deeks SG, Tracy R, Douek DC (2013) Systemic effects of inflammation on health during chronic HIV infection. Immunity 39(4):633–645. doi:10.1016/j.immuni.2013.10.001
213. Franceschi C, Bonafe M, Valensin S, Olivieri F, De Luca M, Ottaviani E, De Benedictis G (2000) Inflamm-aging. An evolutionary perspective on immunosenescence. Ann N Y Acad Sci 908:244–254
214. Sandler NG, Wand H, Roque A, Law M, Nason MC, Nixon DE, Pedersen C, Ruxrungtham K, Lewin SR, Emery S, Neaton JD, Brenchley JM, Deeks SG, Sereti I, Douek DC (2011) Plasma levels of soluble CD14 independently predict mortality in HIV infection. J Infect Dis 203(3):780–790. doi:10.1093/infdis/jiq118, jiq118 [pii]
215. Westhorpe CL, Maisa A, Spelman T, Hoy JF, Dewar EM, Karapanagiotidis S, Hearps AC, Cheng W, Trevillyan J, Lewin SR, Sviridov D, Elliott JH, Jaworowski A, Dart AM, Crowe SM (2013) Associations between blood monocyte markers and carotid atherosclerosis in HIV-positive patients. Immunol Cell Biol 92(2):133–138. doi:10.1038/icb.2013.84
216. Tenorio AR, Zheng Y, Bosch RJ, Krishnan S, Rodriguez B, Hunt PW, Plants J, Seth A, Wilson CC, Deeks SG, Lederman MM, Landay AL (2014) Soluble markers of inflammation and coagulation but not T-cell activation predict non-AIDS-defining morbid events during suppressive antiretroviral treatment. J Infect Dis 210(8):1248–1259. doi:10.1093/infdis/jiu254
217. Srinivasa S, Fitch KV, Petrow E, Burdo TH, Williams KC, Lo J, Cote HC, Grinspoon SK (2014) Soluble CD163 is associated with shortened telomere length in HIV-infected patients. J Acquir Immune Defic Syndr 67(4):414–418. doi:10.1097/QAI.0000000000000329
218. Chew D, Steinberg MB, Thomas P, Swaminathan S, Hodder SL (2014) Evaluation of a smoking cessation program for HIV infected individuals in an urban HIV clinic: challenges and lessons learned. AIDS Res Treat 2014:237834. doi:10.1155/2014/237834
219. Tron L, Lert F, Spire B, Dray-Spira R, Group AN-Vs (2014) Tobacco smoking in HIV-infected versus general population in France: heterogeneity across the various groups of people living with HIV. PLoS One 9(9), e107451. doi:10.1371/journal.pone.0107451
220. Browning KK, Wewers ME, Ferketich AK, Diaz P (2013) Tobacco use and cessation in HIV-infected individuals. Clin Chest Med 34(2):181–190. doi:10.1016/j.ccm.2013.01.005
221. Kalra S, Agrawal N (2013) Diabetes and HIV: current understanding and future perspectives. Curr Diab Rep 13(3):419–427. doi:10.1007/s11892-013-0369-9
222. Brown TT, Cole SR, Li X, Kingsley LA, Palella FJ, Riddler SA, Visscher BR, Margolick JB, Dobs AS (2005) Antiretroviral therapy and the prevalence and incidence of diabetes mellitus in the multicenter AIDS cohort study. Arch Intern Med 165(10):1179–1184. doi:10.1001/archinte.165.10.1179
223. Dube MP (2000) Disorders of glucose metabolism in patients infected with human immunodeficiency virus. Clin Infect Dis 31(6):1467–1475. doi:10.1086/317491
224. Lan X, Cheng K, Chandel N, Lederman R, Jhaveri A, Husain M, Malhotra A, Singhal PC (2013) High glucose enhances HIV entry into T cells through upregulation of CXCR4. J Leukoc Biol 94(4):769–777. doi:10.1189/jlb.0313142
225. de Waal R, Cohen K, Maartens G (2013) Systematic review of antiretroviral-associated lipodystrophy: lipoatrophy, but not central fat gain, is an antiretroviral adverse drug reaction. PLoS One 8(5), e63623. doi:10.1371/journal.pone.0063623
226. Souza SJ, Luzia LA, Santos SS, Rondo PH (2013) Lipid profile of HIV-infected patients in relation to antiretroviral therapy: a review. Rev Assoc Med Bras 59(2):186–198. doi:10.1016/j.ramb.2012.11.003
227. Galescu O, Bhangoo A, Ten S (2013) Insulin resistance, lipodystrophy and cardiometabolic syndrome in HIV/AIDS. Rev Endocr Metab Disord 14(2):133–140. doi:10.1007/s11154-013-9247-7
228. Caron M, Auclairt M, Vissian A, Vigouroux C, Capeau J (2008) Contribution of mitochondrial dysfunction and oxidative stress to cellular premature senescence induced by antiretroviral thymidine analogues. Antivir Ther 13(1):27–38

229. Fowler BJ, Gelfand BD, Kim Y, Kerur N, Tarallo V, Hirano Y, Amarnath S, Fowler DH, Radwan M, Young MT, Pittman K, Kubes P, Agarwal HK, Parang K, Hinton DR, Bastos-Carvalho A, Li S, Yasuma T, Mizutani T, Yasuma R, Wright C, Ambati J (2014) Nucleoside reverse transcriptase inhibitors possess intrinsic anti-inflammatory activity. Science 346(6212):1000–1003. doi:10.1126/science.1261754
230. Islam FM, Wu J, Jansson J, Wilson DP (2012) Relative risk of cardiovascular disease among people living with HIV: a systematic review and meta-analysis. HIV Med 13(8):453–468. doi:10.1111/j.1468-1293.2012.00996.x
231. Sabin CA, Worm SW, Weber R, Reiss P, El-Sadr W, Dabis F, De Wit S, Law M, D'Arminio Monforte A, Friis-Moller N, Kirk O, Pradier C, Weller I, Phillips AN, Lundgren JD (2008) Use of nucleoside reverse transcriptase inhibitors and risk of myocardial infarction in HIV-infected patients enrolled in the D:A:D study: a multi-cohort collaboration. Lancet 371(9622):1417–1426. doi:10.1016/S0140-6736(08)60423-7
232. Worm SW, Sabin C, Weber R, Reiss P, El-Sadr W, Dabis F, De Wit S, Law M, Monforte AD, Friis-Moller N, Kirk O, Fontas E, Weller I, Phillips A, Lundgren J (2010) Risk of myocardial infarction in patients with HIV infection exposed to specific individual antiretroviral drugs from the 3 major drug classes: the data collection on adverse events of anti-HIV drugs (D:A:D) study. J Infect Dis 201(3):318–330. doi:10.1086/649897
233. Bavinger C, Bendavid E, Niehaus K, Olshen RA, Olkin I, Sundaram V, Wein N, Holodniy M, Hou N, Owens DK, Desai M (2013) Risk of cardiovascular disease from antiretroviral therapy for HIV: a systematic review. PLoS One 8(3), e59551. doi:10.1371/journal.pone.0059551
234. Wohl DA, McComsey G, Tebas P, Brown TT, Glesby MJ, Reeds D, Shikuma C, Mulligan K, Dube M, Wininger D, Huang J, Revuelta M, Currier J, Swindells S, Fichtenbaum C, Basar M, Tungsiripat M, Meyer W, Weihe J, Wanke C (2006) Current concepts in the diagnosis and management of metabolic complications of HIV infection and its therapy. Clin Infect Dis 43(5):645–653. doi:10.1086/507333
235. Yin MT, Zhang CA, McMahon DJ, Ferris DC, Irani D, Colon I, Cremers S, Shane E (2012) Higher rates of bone loss in postmenopausal HIV-infected women: a longitudinal study. J Clin Endocrinol Metab 97(2):554–562. doi:10.1210/jc.2011-2197
236. Hoy J (2011) Bone, fracture and frailty. Curr Opin HIV AIDS 6(4):309–314. doi:10.1097/COH.0b013e3283478741
237. Brenchley JM, Price DA, Schacker TW, Asher TE, Silvestri G, Rao S, Kazzaz Z, Bornstein E, Lambotte O, Altmann D, Blazar BR, Rodriguez B, Teixeira-Johnson L, Landay A, Martin JN, Hecht FM, Picker LJ, Lederman MM, Deeks SG, Douek DC (2006) Microbial translocation is a cause of systemic immune activation in chronic HIV infection. Nat Med 12(12):1365–1371
238. Douek D (2007) HIV disease progression: immune activation, microbes, and a leaky gut. Top HIV Med 15(4):114–117
239. Li Q, Duan L, Estes JD, Ma ZM, Rourke T, Wang Y, Reilly C, Carlis J, Miller CJ, Haase AT (2005) Peak SIV replication in resting memory CD4+ T cells depletes gut lamina propria CD4+ T cells. Nature 434(7037):1148–1152. doi:10.1038/nature03513
240. Guadalupe M, Reay E, Sankaran S, Prindiville T, Flamm J, McNeil A, Dandekar S (2003) Severe CD4+ T-cell depletion in gut lymphoid tissue during primary human immunodeficiency virus type 1 infection and substantial delay in restoration following highly active antiretroviral therapy. J Virol 77(21):11708–11717
241. Ciccone EJ, Greenwald JH, Lee PI, Biancotto A, Read SW, Yao MA, Hodge JN, Thompson WL, Kovacs SB, Chairez CL, Migueles SA, Kovacs JA, Margolis LB, Sereti I (2011) CD4+ T cells, including Th17 and cycling subsets, are intact in the gut mucosa of HIV-1-infected long-term nonprogressors. J Virol 85(12):5880–5888. doi:10.1128/JVI.02643-10
242. Chung CY, Alden SL, Funderburg NT, Fu P, Levine AD (2014) Progressive proximal-to-distal reduction in expression of the tight junction complex in colonic epithelium of virally-suppressed HIV+ individuals. PLoS Pathog 10(6), e1004198. doi:10.1371/journal.ppat.1004198

243. Vesterbacka J, Nowak P, Barqasho B, Abdurahman S, Nystrom J, Nilsson S, Funaoka H, Kanda T, Andersson LM, Gisslen M, Sonnerborg A (2013) Kinetics of microbial translocation markers in patients on efavirenz or lopinavir/r based antiretroviral therapy. PLoS One 8(1), e55038. doi:10.1371/journal.pone.0055038
244. Marchetti G, Cozzi-Lepri A, Merlini E, Bellistri GM, Castagna A, Galli M, Verucchi G, Antinori A, Costantini A, Giacometti A, di Caro A, D'Arminio Monforte A (2011) Microbial translocation predicts disease progression of HIV-infected antiretroviral-naive patients with high CD4+ cell count. AIDS 25(11):1385–1394. doi:10.1097/QAD.0b013e3283471d10
245. Nowroozalizadeh S, Mansson F, da Silva Z, Repits J, Dabo B, Pereira C, Biague A, Albert J, Nielsen J, Aaby P, Fenyo EM, Norrgren H, Holmgren B, Jansson M (2010) Microbial translocation correlates with the severity of both HIV-1 and HIV-2 infections. J Infect Dis 201(8):1150–1154. doi:10.1086/651430
246. Man AL, Gicheva N, Nicoletti C (2014) The impact of ageing on the intestinal epithelial barrier and immune system. Cell Immunol 289(1–2):112–118. doi:10.1016/j.cellimm.2014.04.001
247. Rera M, Clark RI, Walker DW (2012) Intestinal barrier dysfunction links metabolic and inflammatory markers of aging to death in Drosophila. Proc Natl Acad Sci U S A 109(52):21528–21533. doi:10.1073/pnas.1215849110
248. Stehle JR Jr, Leng X, Kitzman DW, Nicklas BJ, Kritchevsky SB, High KP (2012) Lipopolysaccharide-binding protein, a surrogate marker of microbial translocation, is associated with physical function in healthy older adults. J Gerontol A Biol Sci Med Sci 67(11):1212–1218. doi:10.1093/gerona/gls178
249. Mutlu EA, Keshavarzian A, Losurdo J, Swanson G, Siewe B, Forsyth C, French A, Demarais P, Sun Y, Koenig L, Cox S, Engen P, Chakradeo P, Abbasi R, Gorenz A, Burns C, Landay A (2014) A compositional look at the human gastrointestinal microbiome and immune activation parameters in HIV infected subjects. PLoS Pathog 10(2), e1003829. doi:10.1371/journal.ppat.1003829
250. Dillon SM, Lee EJ, Kotter CV, Austin GL, Dong Z, Hecht DK, Gianella S, Siewe B, Smith DM, Landay AL, Robertson CE, Frank DN, Wilson CC (2014) An altered intestinal mucosal microbiome in HIV-1 infection is associated with mucosal and systemic immune activation and endotoxemia. Mucosal Immunol 7(4):983–994. doi:10.1038/mi.2013.116
251. Vujkovic-Cvijin I, Dunham RM, Iwai S, Maher MC, Albright RG, Broadhurst MJ, Hernandez RD, Lederman MM, Huang Y, Somsouk M, Deeks SG, Hunt PW, Lynch SV, McCune JM (2013) Dysbiosis of the gut microbiota is associated with HIV disease progression and tryptophan catabolism. Sci Transl Med 5(193):193ra191. doi:10.1126/scitranslmed.3006438
252. Lozupone CA, Li M, Campbell TB, Flores SC, Linderman D, Gebert MJ, Knight R, Fontenot AP, Palmer BE (2013) Alterations in the gut microbiota associated with HIV-1 infection. Cell Host Microbe 14(3):329–339. doi:10.1016/j.chom.2013.08.006
253. Perez-Santiago J, Gianella S, Massanella M, Spina CA, Karris MY, Var SR, Patel D, Jordan PS, Young JA, Little SJ, Richman DD, Smith DM (2013) Gut Lactobacillales are associated with higher CD4 and less microbial translocation during HIV infection. AIDS (London, England) 27(12):1921–1931
254. Cannon MJ, Schmid DS, Hyde TB (2010) Review of cytomegalovirus seroprevalence and demographic characteristics associated with infection. Rev Med Virol 20(4):202–213. doi:10.1002/rmv.655
255. Pawelec G, Derhovanessian E, Larbi A, Strindhall J, Wikby A (2009) Cytomegalovirus and human immunosenescence. Rev Med Virol 19(1):47–56
256. Derhovanessian E, Larbi A, Pawelec G (2009) Biomarkers of human immunosenescence: impact of Cytomegalovirus infection. Curr Opin Immunol 21(4):440–445
257. Chidrawar S, Khan N, Wei W, McLarnon A, Smith N, Nayak L, Moss P (2009) Cytomegalovirus-seropositivity has a profound influence on the magnitude of major lymphoid subsets within healthy individuals. Clin Exp Immunol 155(3):423–432. doi:10.1111/j.1365-2249.2008.03785.x

258. Derhovanessian E, Maier AB, Hahnel K, Beck R, de Craen AJ, Slagboom EP, Westendorp RG, Pawelec G (2011) Infection with cytomegalovirus but not herpes simplex virus induces the accumulation of late-differentiated CD4+ and CD8+ T-cells in humans. J Gen Virol 92(Pt 12):2746–2756. doi:10.1099/vir.0.036004-0
259. Ouyang Q, Wagner WM, Wikby A, Walter S, Aubert G, Dodi AI, Travers P, Pawelec G (2003) Large numbers of dysfunctional CD8+ T lymphocytes bearing receptors for a single dominant CMV epitope in the very old. J Clin Immunol 23(4):247–257
260. Ouyang Q, Wagner WM, Zheng W, Wikby A, Remarque EJ, Pawelec G (2004) Dysfunctional CMV-specific CD8(+) T cells accumulate in the elderly. Exp Gerontol 39(4):607–613. doi:10.1016/j.exger.2003.11.016
261. Simanek AM, Dowd JB, Pawelec G, Melzer D, Dutta A, Aiello AE (2011) Seropositivity to cytomegalovirus, inflammation, all-cause and cardiovascular disease-related mortality in the United States. PLoS One 6(2), e16103. doi:10.1371/journal.pone.0016103
262. Arama V, Mihailescu R, Radulescu M, Arama SS, Streinu-Cercel A, Youle M, Group C-HS (2014) Clinical relevance of the plasma load of cytomegalovirus in patients infected with HIV-A survival analysis. J Med Virol 86(11):1821–1827. doi:10.1002/jmv.24027
263. Stone SF, Price P, French MA (2006) Cytomegalovirus (CMV)-specific CD8+ T cells in individuals with HIV infection: correlation with protection from CMV disease. J Antimicrob Chemother 57(4):585–588. doi:10.1093/jac/dkl049
264. Anuradha B, Mane Pratibha M, Vijayadurga S (2011) The reactivation of the cytomegalovirus (CMV) infection in HIV infected patients. J Clin Diagn Res 5(4):749–751
265. Naeger DM, Martin JN, Sinclair E, Hunt PW, Bangsberg DR, Hecht F, Hsue P, McCune JM, Deeks SG (2010) Cytomegalovirus-specific T cells persist at very high levels during long-term antiretroviral treatment of HIV disease. PLoS One 5(1), e8886. doi:10.1371/journal.pone.0008886
266. Barrett L, Stapleton SN, Fudge NJ, Grant MD (2014) Immune resilience in HIV-infected individuals seronegative for cytomegalovirus. AIDS (London, England) 28(14):2045–2049. doi:10.1097/QAD.0000000000000405
267. Solana R, Tarazona R, Aiello AE, Akbar AN, Appay V, Beswick M, Bosch JA, Campos C, Cantisan S, Cicin-Sain L, Derhovanessian E, Ferrando-Martinez S, Frasca D, Fulop T, Govind S, Grubeck-Loebenstein B, Hill A, Hurme M, Kern F, Larbi A, Lopez-Botet M, Maier AB, McElhaney JE, Moss P, Naumova E, Nikolich-Zugich J, Pera A, Rector JL, Riddell N, Sanchez-Correa B, Sansoni P, Sauce D, van Lier R, Wang GC, Wills MR, Zielinski M, Pawelec G (2012) CMV and immunosenescence: from basics to clinics. Immun Ageing 9(1):23. doi:10.1186/1742-4933-9-23
268. Palmer S, Maldarelli F, Wiegand A, Bernstein B, Hanna GJ, Brun SC, Kempf DJ, Mellors JW, Coffin JM, King MS (2008) Low-level viremia persists for at least 7 years in patients on suppressive antiretroviral therapy. Proc Natl Acad Sci U S A 105(10):3879–3884
269. Schacker T, Zeh J, Hu HL, Hill E, Corey L (1998) Frequency of symptomatic and asymptomatic herpes simplex virus type 2 reactivations among human immunodeficiency virus-infected men. J Infect Dis 178(6):1616–1622
270. Kim HN, Meier A, Huang ML, Kuntz S, Selke S, Celum C, Corey L, Wald A (2006) Oral herpes simplex virus type 2 reactivation in HIV-positive and -negative men. J Infect Dis 194(4):420–427. doi:10.1086/505879
271. Dehee A, Asselot C, Piolot T, Jacomet C, Rozenbaum W, Vidaud M, Garbarg-Chenon A, Nicolas JC (2001) Quantification of Epstein-Barr virus load in peripheral blood of human immunodeficiency virus-infected patients using real-time PCR. J Med Virol 65(3):543–552
272. Bhardwaj N, Maldarelli F, Mellors J, Coffin JM (2014) HIV-1 infection leads to increased transcription of human endogenous retrovirus HERV-K (HML-2) proviruses in vivo but not to increased virion production. J Virol 88(19):11108–11120. doi:10.1128/JVI.01623-14
273. Contreras-Galindo R, Kaplan MH, Markovitz DM, Lorenzo E, Yamamura Y (2006) Detection of HERV-K(HML-2) viral RNA in plasma of HIV type 1-infected individuals. AIDS Res Hum Retroviruses 22(10):979–984. doi:10.1089/aid.2006.22.979

274. Gonzalez-Hernandez MJ, Swanson MD, Contreras-Galindo R, Cookinham S, King SR, Noel RJ Jr, Kaplan MH, Markovitz DM (2012) Expression of human endogenous retrovirus type K (HML-2) is activated by the Tat protein of HIV-1. J Virol 86(15):7790–7805. doi:10.1128/JVI.07215-11
275. Freiberg MS, Chang CC, Skanderson M, McGinnis K, Kuller LH, Kraemer KL, Rimland D, Goetz MB, Butt AA, Rodriguez Barradas MC, Gibert C, Leaf D, Brown ST, Samet J, Kazis L, Bryant K, Justice AC, Veterans Aging Cohort S (2011) The risk of incident coronary heart disease among veterans with and without HIV and hepatitis C. Circ Cardiovasc Qual Outcomes 4(4):425–432. doi:10.1161/CIRCOUTCOMES.110.957415
276. Maalouf NM, Zhang S, Drechsler H, Brown GR, Tebas P, Bedimo R (2013) Hepatitis C co-infection and severity of liver disease as risk factors for osteoporotic fractures among HIV-infected patients. J Bone Min Res 28(12):2577–2583. doi:10.1002/jbmr.1988
277. Richardson JL, Nowicki M, Danley K, Martin EM, Cohen MH, Gonzalez R, Vassileva J, Levine AM (2005) Neuropsychological functioning in a cohort of HIV- and hepatitis C virus-infected women. AIDS (London, England) 19(15):1659–1667
278. Salter ML, Lau B, Mehta SH, Go VF, Leng S, Kirk GD (2013) Correlates of elevated interleukin-6 and C-reactive protein in persons with or at high risk for HCV and HIV infections. J Acquir Immune Defic Syndr 64(5):488–495. doi:10.1097/QAI.0b013e3182a7ee2e
279. Zanet DL, Thorne A, Singer J, Maan EJ, Sattha B, Le Campion A, Soudeyns H, Pick N, Murray M, Money DM, Cote HC, Therapy CETGoH, Aging C (2014) Association between short leukocyte telomere length and HIV infection in a cohort study: No evidence of a relationship with antiretroviral therapy. Clin Infect Dis 58(9):1322–1332. doi:10.1093/cid/ciu051
280. Bentwich Z, Maartens G, Torten D, Lal AA, Lal RB (2000) Concurrent infections and HIV pathogenesis. AIDS (London, England) 14(14):2071–2081
281. Chang CC, Crane M, Zhou J, Mina M, Post JJ, Cameron BA, Lloyd AR, Jaworowski A, French MA, Lewin SR (2013) HIV and co-infections. Immunol Rev 254(1):114–142. doi:10.1111/imr.12063
282. Alemu A, Shiferaw Y, Addis Z, Mathewos B, Birhan W (2013) Effect of malaria on HIV/AIDS transmission and progression. Parasites & Vectors 6:18. doi:10.1186/1756-3305-6-18
283. Calza L, Trapani F, Bartoletti M, Manfredi R, Colangeli V, Borderi M, Grossi G, Motta R, Viale P (2012) Statin therapy decreases serum levels of high-sensitivity C-reactive protein and tumor necrosis factor-α in HIV-infected patients treated with ritonavir-boosted protease inhibitors. HIV Clin Trials 13(3):153–161. doi:10.1310/hct1303-153
284. Tenorio AR, Chan ES, Bosch RJ, Macatangay BJ, Read SW, Yesmin S, Taiwo B, Margolis DM, Jacobson JM, Landay AL, Wilson CC, for the AT (2014) Rifaximin has a marginal impact on microbial translocation, T-cell activation and inflammation in HIV-positive immune Non-responders to antiretroviral therapy – ACTG A5286. J Infect Dis. doi:10.1093/infdis/jiu515
285. Paton Ni GRLDDT et al (2012) Effects of Hydroxychloroquine on immune activation and disease progression among hiv-infected patients not receiving antiretroviral therapy: a randomized controlled trial. JAMA 308(4):353–361. doi:10.1001/jama.2012.6936
286. Baker JV, Huppler Hullsiek K, Prosser R, Duprez D, Grimm R, Tracy RP, Rhame F, Henry K, Neaton JD (2012) Angiotensin converting enzyme inhibitor and HMG-CoA reductase inhibitor as adjunct treatment for persons with HIV infection: *a feasibility randomized trial*. PLoS One 7(10), e46894. doi:10.1371/journal.pone.0046894
287. Sandler NG, Zhang X, Bosch RJ, Funderburg NT, Choi AI, Robinson JK, Fine DM, Coombs RW, Jacobson JM, Landay AL, Douek DC, Tressler R, Read SW, Wilson CC, Deeks SG, Lederman MM, Gandhi RT, Team ACTGA (2014) Sevelamer does not decrease lipopolysaccharide or soluble CD14 levels but decreases soluble tissue factor, low-density lipoprotein (LDL) cholesterol, and oxidized LDL cholesterol levels in individuals with untreated HIV infection. J Infect Dis 210(10):1549–1554. doi:10.1093/infdis/jiu305

Pain in the Elderly

Roger B. Fillingim, Dennis C. Turk, and Robert P. Yezierski

Contents

R.B. Fillingim, Ph.D. (✉) • R.P. Yezierski, Ph.D.
Pain Research and Intervention Center of Excellence, University of Florida, Gainesville, FL, USA
e-mail: rfilling@ufl.edu; ryezierski@dental.ufl.edu

D.C. Turk, Ph.D.
Department of Anesthesiology & Pain Medicine, University of Washington, Seattle, WA, USA
e-mail: turkdc@uw.edu

F. Sierra, R. Kohanski (eds.), *Advances in Geroscience*,
DOI 10.1007/978-3-319-23246-1_18

1 Introduction

Pain, especially chronic pain, arguably represents the most prevalent and costly public health condition in the United States. The Institute of Medicine (IOM) noted that chronic pain affects 100 million U.S. adults with costs exceeding $600 billion annually [1, 2]. While pain affects individuals throughout the lifespan, older adults are at increased risk for chronic pain and pain-related disability [3]. Despite the greater prevalence and adverse impact of pain among older adults, the relationship between pain and aging remains a surprisingly underexplored area of inquiry. This chapter provides a broad overview of past and current research regarding chronic pain in older adults, including a discussion of biopsychosocial mechanisms contributing to age-related influences on pain. First, we will describe findings from epidemiologic and clinical studies examining age differences in pain prevalence and impact. Then, after briefly reviewing the impact of aging on biological processes that contribute to pain we will discuss the biopsychosocial model of pain. Next, we discuss human laboratory studies examining age-related changes in pain processing, followed by consideration of psychosocial factors that contribute to pain perception among older adults. Finally, we review the impact of aging on medical and non-medical therapies. We conclude the chapter with a discussion of future directions for pain and aging research.

2 Epidemiological and Clinical Aspects of Pain and Aging

Epidemiological studies emphasize pain prevalence in the population, which is subject to all of the methodological vagaries inherent in such research (e.g., case definition, sampling strategies). However, prevalence alone inadequately captures the overall burden of pain; and for this reason it is critical to characterize the impact of pain through measures of pain severity, impact on physical function and disability, quality of life, and psychological morbidity. Below we address the influence of aging on these issues.

Several studies have investigated the prevalence of chronic pain across the lifespan. For example, Blyth and colleagues [4] surveyed more than 17,000 Australians and found that chronic pain (i.e., pain experienced daily for three of the previous 6 months) frequency increased with age, though sex differences emerged in the pattern of pain prevalence (see Fig. 18.1). Other studies show a similar pattern of increases in chronic pain prevalence until approximately age 70, at which point pain prevalence plateaus or even declines slightly [5–7].

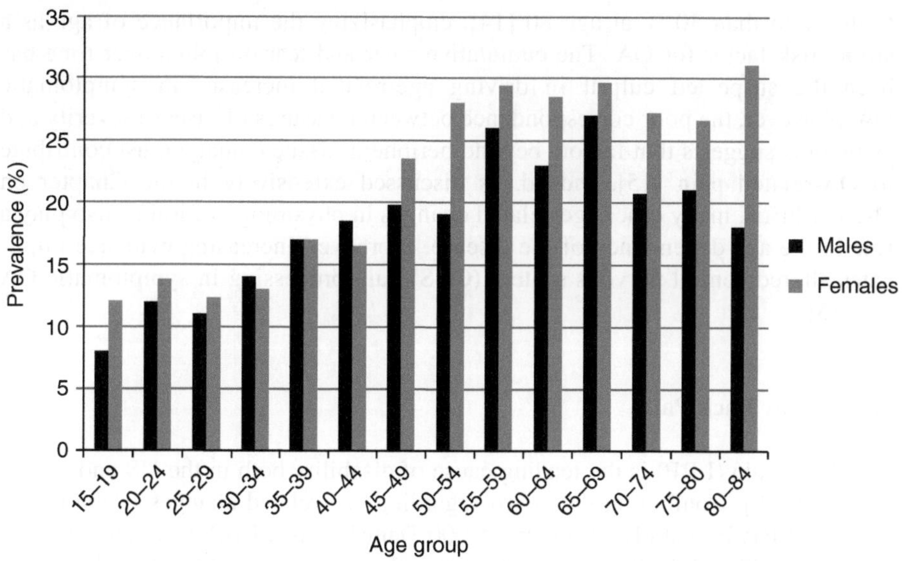

Fig. 18.1 Pain prevalence across the lifespan. Data are from interviews of 17,543 Australians. Chronic pain was defined as pain experienced every day for 3 months in the 6 months prior to interview (Adapted from Blyth and colleagues [4])

2.1 *Specific Pain Conditions Associated with Aging*

The findings summarized above relate to chronic pain in general; however, age-related influences on chronic pain prevalence vary across different pain conditions. Specifically, compared to younger people, older adults are more likely to experience musculoskeletal pain, including pain of the joints, lower extremities and back [7]. A recent systematic review based on data from more than 116 thousand elderly Brazilians reported that lower limb and spine pain were the most common pain conditions, reported by over half of the sample [8]. Similarly, the incidence of osteoarthritis (OA) of the knee, hip, and hand all increase through seventh decade of life and then decrease above age 80 [9]. In fact, a recent analysis from Sweden estimates that the odds of developing musculoskeletal pain become one and a half times greater for each decade of increased age [10]. In contrast, the prevalence of abdominal pain, migraine headache, and pain due to temporomandibular disorders peak in the third to fifth decade and decrease thereafter [11, 12]. Below we discuss the impact of age-related influences on pain in several specific clinical conditions.

2.1.1 Osteoarthritis Pain

OA represents one of the most common sources of pain and disability in the elderly [13]. The knee is the most frequently affected joint, with the lifetime risk of symptomatic knee OA estimated to be 40 % for men and 47 % for women [9]. In one study, the prevalence of symptomatic knee OA increased from 16 % at age

55 to more than 30 % at age 80 [14], emphasizing the importance of age as a strong risk factor for OA. The cumulative wear and tear on joints over time has been the suspected culprit in driving age-related increases in symptomatic OA. However, the poor correspondence between measures of disease severity and symptoms suggests that factors beyond peripheral tissue damage must contribute to OA-related pain [15]. Indeed, as discussed extensively in the Chapter on Osteoarthritis, many other age-related changes in physiology of joints also play a role in the age dependency of the disease. Moreover, increasing evidence implicates altered central nervous system (CNS) pain processing in symptomatic OA [16, 15].

2.1.2 Low Back Pain

Low back pain (LBP) is the leading cause of disability both in the US and worldwide [17, 18]. Conflicting data exist regarding age-related changes in the prevalence of LBP. In a study of nearly 35,000 Danish twins, LBP prevalence peaked around age 45 and declined thereafter for both women and men [19]. Others have reported that prevalence increases with age until age 60–65, gradually declining in subsequent years [20, 21]. A systematic review found that the prevalence of *severe*, but not "benign or mixed," back pain increases with age [22]. Thus, while overall back pain prevalence may decline slightly in older age groups, more severe pain increases in frequency, suggesting a greater burden of back pain among older adults.

2.1.3 Neuropathic Pain

Neuropathic pain refers to "pain arising as direct consequence of a lesion or disease of the somatosensory nervous system [23]." Common neuropathic pain conditions include diabetic neuropathy, post-herpetic neuralgia, and central post-stroke pain. Because the medical conditions producing these neuropathies are more common in older adults, the prevalence of these neuropathic pain conditions increases with age [24]. However, age can increase the risk for neuropathic pain independent of its effects on the parent medical condition. For example, among patients with acute herpes zoster, age represents a risk factor for progression to post-herpetic neuralgia [25]. Moreover, risk of diabetic neuropathy increases with age, thereby increasing the likelihood of painful diabetic neuropathy in older adults [26]. Less commonly studied neuropathic pain conditions that show increased prevalence with advancing age include trigeminal neuralgia and glossopharyngeal neuralgia, both of which show peak incidence in the seventh decade [27]. Among patients with multiple sclerosis, central neuropathic pain is also more prevalent with age, peaking around age 60 [28]. Also, older age increases risk for HIV-associated sensory neuropathy [29]. Thus, older adults are at substantially greater risk for many forms of neuropathic pain.

2.1.4 Visceral Pain

Pain emanating from internal organs (e.g., gut, heart, bladder) represents a considerable problem. Evidence suggests, however, that older adults experience reduced visceral pain relative to their younger counterparts [30]. For example, prevalence of irritable bowel syndrome declines after the fifth decade [31]. Also, older adults are more likely to experience silent myocardial ischemia and painless myocardial infarction [32]. Similarly, visceral conditions, such as peptic ulcer disease, pneumothorax, and intestinal blockage are characterized by reduced or absent pain in older compared to younger adults [33]. Although the reduction of visceral pain might be considered a positive aspect of aging, pain from internal structures often signals the presence of a potentially life threatening condition. Therefore, the decreased ability to detect visceral pain may increase older adults' risk for morbidity and mortality.

2.1.5 Pain in the Cognitively Impaired

Cognitive impairment represents an enormous public health issue in the elderly. Importantly, pain and cognitive impairment may be reciprocally related. Several studies have demonstrated reduced cognitive performance in chronic pain populations [34, 35]. There is conflicting evidence regarding whether dementia is associated with altered prevalence and severity of pain [36, 37], because the ability to self-report pain is often compromised [37]. In fact, from a clinical perspective, pain among older adults with dementia presents management challenges due to the difficulty of assessing pain and determining treatment effectiveness. Among non-demented individuals, cognitive performance is inversely correlated with pain severity and mediates the influence of pain on physical performance [38]. Thus, cognitive impairment may be a risk factor for increased pain and pain-related physical dysfunction.

2.2 *Symptoms and Impact of Chronic Pain in Older Adults*

The burden of chronic pain on older adults extends well beyond age differences in pain prevalence. Several aspects of the severity of pain differ across the lifespan, and multiple clinical features that are comorbid with chronic pain can disproportionately impact older adults.

2.2.1 Pain Characteristics

Chronic pain is characterized not only by the intensity of the pain, but also by other pain characteristics (e.g., pain quality, temporal pattern, locations) and interference with daily living. Conflicting evidence exists regarding age-related influences on the intensity of pain. As noted above, studies of severe LBP (e.g., LBP interfering with daily activities) reported a higher prevalence in older versus younger adults, while studies of back pain in general (e.g., LBP lasting 1 day or longer in the past 4 weeks) reported decreased LBP frequency after middle age [22]. In contrast, in a sample of mixed chronic pain conditions, no age differences in numerical pain ratings emerged; however, older adults reported lower sensory and affective (i.e., the emotional component of pain, which includes how unpleasant or bothersome the pain is) pain scores on the McGill Pain Questionnaire (MPQ) [39]. Also, older adults with metastatic cancer reported lower pain intensity ratings than their younger counterparts [40], although a more recent study showed no age-related differences in MPQ scores among patients with cancer pain [41]. Regarding pain duration, some evidence links increased age with persistence of pain [42]; however, age appears to be protective against development of chronic postoperative pain [43, 44]. In terms of body dispersion of pain, both regional and widespread pain conditions increase from early adulthood to middle age, and prevalence seems to peak in the fifth or sixth decade, declining slightly thereafter [45, 46]. Thus, there are inconsistent associations of age with various pain characteristics, and additional research is needed to clarify these findings and elucidate their mechanisms.

2.2.2 Pain, Quality of Life, and Function

Chronic pain is commonly associated with activity limitations and symptoms that negatively impact quality of life, including fatigue and sleep disturbance, and these issues may represent particular problems in older adults. For example, pain is associated with mobility impairment among older adults [6, 47], and the association of musculoskeletal pain with activity interference and disability increases with age [48, 49]. As described below, falls are a major clinical concern among the elderly, and pain significantly increases fall risk among older adults [50].

Fatigue is associated with aging, is an important symptom of frailty, and predicts functional limitations, disability, and mortality in older adults [51]. Because pain and fatigue often coexist, fatigue is a particularly pertinent clinical issue among older adults with chronic pain. Indeed, independent of pain, fatigue is a significant negative predictor of physical activity in older adults with OA [52]. Further, sleep disturbance increases significantly with age [53], and sleep disturbance confers increased susceptibility to chronic pain, and vice versa [54]. Thus, sleep disturbance may be an age-related risk factor for development and/or exacerbation of pain. These findings suggest that while chronic pain severity is not consistently greater among older adults, pain nevertheless predicts substantially diminished quality of life and mobility and increased fall risk in the elderly.

3 Aging and Pain Biology

Aging is a complex, multifactorial process occurring across many different levels of functional organization – psychological, physiological, and molecular events that occur with the passage of time [55]. Similarly, chronic pain is recognized as a multifactorial experience driven by complex interactions among multiple biopsychosocial processes [56]. Important characteristics of aging include a declining ability to respond to stress, increasing homeostatic imbalance, and an increase in the risk for onset of pathological changes, which parallels descriptions of chronic pain. The overlapping mechanisms contributing to aging and chronic pain provide a broad foundation for future study, as demonstrated by the examples presented below.

Multiple biological systems, including the nervous, endocrine, and immune systems, are responsible for the maintenance of homeostasis and for the activation of defense responses. Environmental exposures can produce a dysregulation of these systems, resulting in maladaptive changes that contribute to both aging and chronic pain. For example, leukocyte telomere length is a marker of cellular aging that is also associated with the intensity and persistence of stress. Shorter leukocyte telomeres have been associated with age-related diseases, as well as chronic mental and physical health conditions, while longer telomeres are associated with a healthy life [57, 58]. Recent reports have linked chronic pain with shorter telomeres, particularly among individuals with chronic pain and high stress, supporting a potential link between cellular aging and chronic pain [59, 60]. Historically, this relationship has been overlooked; recently however, it is attracting attention and providing a new perspective on the overlapping mechanism(s) of aging and chronic pain [61].

Another factor common to the biology of aging and chronic pain is the role of free radicals. Free radicals have been suggested to be involved in the development and maintenance of capsaicin-induced hyperalgesia through central sensitization and the elevation of spinal reactive oxygen species due to the increased production of mitochondrial superoxides in dorsal horn neurons [62, 63]. It may be argued that oxidative damage is linked to aging in a universal manner and that changes in endocrine function are secondary to changes in oxidant production in endocrine cells [64, 65]. Thus, similar cellular mechanisms (free radical production) may broadly affect aging and simultaneously contribute to the development of pain.

Another example of the overlap in the biology of chronic pain, and aging is in the case of central pain resulting from spinal cord injury. Specifically, neuronal injury and changes in neuronal excitability have been shown to involve the extracellular signal-regulated kinases (ERK) cellular signaling pathway [66, 67]. Similar changes in neuronal function are commonly observed in uninjured older animals, suggesting an overlap in the cellular mechanisms responsible for neuronal changes in excitability resulting from injury and aging. Similarities in cellular mechanism(s) underlying pain and aging can also be found with the results of the drug rapamycin, initially developed as an antifungal agent and found to have immunosuppressive properties [68]. Rapamycin has been shown to extend the lifespan of mice by affecting a large number of processes regulated by the Target of Rapamycin (TOR) signaling pathway [69]. A rapamycin sensitive signaling pathway was also shown to be essential for the expression of persistent pain states [70], and rapamycin reduces clinical signs of neuropathic pain in a model of experimental autoimmune encephalomyelitis [71]. These examples of overlapping mechanisms reveal the need for collaborative efforts between the fields of pain and aging research to more fully explore the biological processes that contribute to both senescence and chronic pain.

3.1 Age-Related Changes in the Somatosensory System

Understanding the impact of advancing age on pain-related biological systems represents a major challenge in the field of pain research [72, 73]. Age-related anatomical and functional changes have been described in both human and non-human studies of the somatosensory system [74]. For example, peripheral nerves show an age-related reduction of both myelinated and unmyelinated fibers [75, 76] as well as signs of Wallerian degeneration [77, 78]. The number and size of sensory neurons in dorsal root ganglia (DRG) also decreases with age [79, 80]. Reduction in the number of peripheral afferents, the presence of demyelination, together with increasing inflammation are similar to the cascade of pathological events associated with nerve and tissue injury-induced pain in younger animals, thereby suggesting a common pathway between the events associated with neuropathic pain and those common to age-related changes in nociception [81, 82].

In addition to peripheral changes, central changes in the expression of neurotransmitters and receptors may contribute to age-related alterations in somatosensation. The decreased presence of calcitonin gene-related peptide (CGRP), substance P, nitric oxide, and somatostatin in the dorsal horn of aged rats has been reported [83–87], and loss of serotonergic and noradrenergic terminals in the dorsal horn also suggests the potential for age-related changes in descending pain modulatory pathways [81, 86, 88]. A decrease in the number of opiate receptors and decreased efficacy of opiate-mediated antinociception may also contribute to age-related changes in the processing of nociceptive information [89–91]. Finally, age-related altered gene expression for trophic factors, neuropeptides, cell adhesion molecules, ion channels, and genes related to mitochondrial function and calcium handling, as well as changes in the functional state of spinal and supraspinal glial cells may drive enhanced pain sensitivity with aging [92].

3.2 Effects of Age on Autonomic Function: Impact on Pain Sensitivity

Age-related changes in autonomic function may contribute to the changing pain sensitivity in older adults [93]. Experimental evidence has shown that tonic whole-body sympathetic nervous system (SNS) activity increases with age [94]. Further, autonomic dysfunction can be instrumental in the generation and maintenance of chronic pain. Two reviews summarize an extensive literature supporting the following relationships: (1) psychological stress activates limbic structures projecting to the hypothalamo-pituitary axis (HPA), resulting in an increase in sympathetic tone; (2) activation of stress circuitry increases pain sensitivity by central actions leading to stress-induced hyperalgesia; (3) chronic sympathetic activation and associated peripheral vasoconstriction produces muscular ischemia and a microenvironment conducive to myofascial pain; and (4) nociceptors in deep tissues are particularly sensitive to sympathetic induced ischemia and are potent generators of central sensitization when tonically active [95, 96]. Thus, older adults may experience maladaptive physiological responses to stressors, including psychosocial and environmental events, as well as inflammation and pain, which can potentiate hyperalgesia thereby exacerbating pain.

Aging is associated with changes in the sympathetic nerve supply to a number of targets, and quantitative changes in sympathetic nerve fibers result in changes in transmitter expression [97]. Although the exact relationship between pain and the sympathetic nervous system remains unclear, the sympathetic nervous system is known to be involved in maintaining protective body reactions associated with pain. Moreover, in certain pathological conditions, the SNS may be involved in the generation of pain. For example, SNS activity can elicit spontaneous pain and enhance pain evoked by mechanical and cold stimulation [98, 99]. Clinical and preclinical studies have documented that physiological activation of sympathetic neurons can enhance pain and blockade of sympathetic activity can relieve pain. The SNS also contributes to the generation of pain during inflammation [100]. Given age-related changes in SNS structure and function and that the SNS influences the generation and maintenance of chronic pain, sympathetic function represents an important potential mediator of age-related changes in pain.

3.3 Aging, Pain and Immune Function

Changes in immune function, including microglial responses, may contribute importantly to age-related increases in chronic pain. The glial response to injury contributes to neuronal hypersensitivity leading to the production of inflammatory mediators such as cytokines and chemokines. This "glia cascade" has been related to the regulation of synaptic strength and plasticity and the generation of central sensitization [101, 102]. However, the contribution of glia to the induction or maintenance of chronic pain in aged rats is unknown. Stuesse and colleagues [103] found that ligation-induced hyperreflexia was correlated with increased staining for activated microglia regardless of age. Activated microglia have also been implicated in the initiation of chronic pain via the local release of neuroactive substances, including cytokines, ATP, substance P, reactive oxygen species, nitric oxide, arachidonic acid, fractalkine, and nerve growth factors [104, 105]. Selective inhibition of activated microglia can alleviate acute and chronic pain behaviors [105], though clinical evidence of a beneficial effect of microglia inhibition in persistent pain conditions is lacking [106, 107].

The microglia-to-neuron signaling link has also been shown to involve a molecular pathway in the spinal cord that includes Toll-like receptors, phosphorylated mitogen-activated protein kinase and purinergic P2X4 receptors on microglia [108, 109]. Interestingly, the microglia-to-neuron signaling pathway involving prostaglandin E2 (PGE2) is involved in producing excitability changes underlying chronic pain following spinal cord injury [110]. This same pathway could contribute to the emergence of age-related chronic pain conditions. Thus, substantial evidence exists that immune responses elicit a well-orchestrated temporal pattern of activation of different immune cells, including microglia and astrocytes, which may contribute to chronic pain development. At present the involvement of glia in the induction or maintenance of chronic pain in aged rats is an evolving story. However, age-related morphological changes in microglia may reflect an important mechanism mediating age-dependent increases in pain sensitivity.

3.4 *Effects of Age on Pain Sensitivity: Pre-clinical Studies*

Pre-clinical studies examining the effects of age on pain sensitivity have resulted in conflicting observations that include increases, decreases, or no change in nociceptive responses with advancing age [88]. It is important to point out that the majority of these studies employed reflex-based behavioral measures to determine changes in thermal and/or mechanical sensitivity. The execution of these reflex-based measures do not require cerebral processing for the conscious perception of sensory events and are subsequently thought to be less relevant to clinical pain [111]. In order to address these deficiencies an operant escape task was developed to evaluate thermal nociceptive sensitivity in awake, unrestrained rats [112]. This test overcomes the limitations inherent with reflexive responses by providing a measure of *pain sensitivity* and *affective response* to nociceptive stimuli. Use of operant (learned) tests provides a measure of pain involving neuronal pathways extending throughout the neuraxis. Importantly, reflex-based and operant assays often yield substantively different results [113–115], and the findings from operant assays are typically more consistent with predictions from available human reports than are the results from reflex-based tests [111].

3.4.1 Findings from Reflex-Based Assays

A review of studies using reflex-based strategies to examine age-related changes in pain sensitivity reveals inconsistent and confusing results. Using paw lick and tail flick latencies in young (2–3 months), adult (6–12 months), and aged (24 months) rats, Hess et al. [90] described a decrease in sensitivity to thermal and electric shock with increasing age. These results correlated with a decrease in the number of opiate receptors in the frontal poles, striatum, and hippocampus. Another evaluation of thermal response latencies showed that young mice (6–8 weeks) had significantly shorter latencies than animals 24 months of age [116]. The decreased sensitivity in older animals was greater for females and correlated with a decrease in the expression of Nav1.8 sodium channels. Similarly, Wang and Albers [92] showed that aged male and female mice had decreased sensitivity to thermal stimuli, accompanied by a decrease in receptor expression for the growth factor artemin and the ion channel TRPV1. In contrast, Chan and Lai [117] showed decreased hot plate response latencies (i.e., increased sensitivity) for 11.2 versus 1.6 month old rats.

In contrast to the above findings, age-dependent increases in sensitivity to mechanical [118] and thermal stimuli [119] in the rat have been reported. These latter results paralleled the response profiles of wide dynamic range and nociceptive specific neurons recorded in the spinal cord of adult versus aged animals [119]. Significantly lower mean background activity and after-discharge responses were recorded in adult animals compared to those recorded in aged animals. Similar increases in neuronal excitability and receptive field sizes for neurons recorded in the dorsal column nuclei in aged versus adult animals have also been reported [120]. In summary, the results of 25 studies addressing age-related changes in pain sensitivity using reflex-based behavioral responses showed decreased sensitivity (9/25), increased sensitivity (12/25), or no changes (4/25) with advancing age. In addition to the behavioral assessment strategies used to evaluate responses to different stimulus conditions, there are many possible factors contributing to the variability of results, including the role of different sexes, species, and ages of animals when tested [121].

3.4.2 Findings from Operant Assays

In a cross-sectional study with rats ranging in age from 8 to 32 months, both operant escape and reflex testing methods were used to evaluate the effects of age on thermal sensitivity [122]. Operant measures of pain assessment revealed an increase in thermal sensitivity at older ages. By contrast, reflex responding did not show any age-related differences in sensitivity to 44.5 °C. In the case of cold sensitivity, operant escape revealed increased sensitivity from 8 to 32 months. Decreased latencies for licking/guarding responses to 1.5 °C were observed for animals ranging in age from 11 to 27 months of age. Interestingly, there was an increase in lick/guard latencies (i.e., decreased sensitivity) when animals were tested at 35 months. Thus, when comparing the results obtained from different age groups using operant escape and a reflex-based assessment task, consistent changes in thermal sensitivity were not observed. These results provide additional evidence that there are significant differences when comparing results of reflex versus cortically dependent outcome measures [111].

3.4.3 Inflammatory Pain

An important factor leading to central sensitization is the change in excitability of peripheral nociceptors by chronic inflammation [123]. The influence of injury- or age-induced inflammation on pain sensitivity was evaluated by Zhang et al. [123] in a study evaluating hindpaw withdrawal latencies in adult and aged animals following injection of complete Freund's adjuvant (CFA). Aged animals had a significant increase in nociceptive sensitivity after CFA compared to adult animals. Kitagawa et al. [120] showed greater excitability of dorsal horn nociceptive neurons with advancing age, but the excitability could not be further increased by inflammation. Using a different inflammatory agent, Gagliese and Melzack [124] showed that formalin injections elicited similar nociceptive responses in animals 3 and 24 months, which were significantly lower than animals 18 months of age, suggesting that sensitivity to inflammatory pain may peak at mid-life. Formalin injections showed a larger number of c-fos (a marker of neuronal activation) positive cells in the medullary dorsal horn of older rats compared to their younger counterparts [125], which correlated with increased nociceptive sensitivity in an older cohort of animals.

The effects of formalin injection on thermal sensitivity were also evaluated using an operant escape task over 5 weeks of testing [122]. A significant formalin-induced increase in thermal sensitivity was obtained for cold and heat stimulation in animals 16 and 24 months old, but not in 8-month old animals. Similarly, paw injections of complete Freund's adjuvant increased thermal sensitivity and expression of the peptide dynorphin (DYN), a pronociceptive peptide [126], in the spinal cords of 18 month old rats, compared to 3 month old rats [123].

3.4.4 Neuropathic Pain

Considerable evidence suggests that aging and nerve injury may elicit similar anatomical, physiological, and behavioral changes. Age-dependent changes in pain sensitivity following nerve injury were evaluated following sciatic nerve ligation in young (4–6 months), mature (14–16 months) and aged (24–26 months) rats. This study observed prolonged increases in thermal sensitivity at 3 and 21 days following injury. The effects were most pronounced in the oldest animals, lasting a period of 35 days [127]. In a comparison of chronic constriction injury (CCI) and partial sciatic nerve ligation (PSNL) Crisp et al. [128] showed aged (24–26 months) rats undergoing PSNL, but not CCI, developed a more vigorous and longer duration thermal hyperalgesia and tactile allodynia compared to their younger (4–6 months) counterparts. Chung et al. [129], using a model of L5/L6 spinal nerve ligation, showed a decrease in mechanical sensitivity in middle-aged (15 months) versus young (40 days) and mature (4 months) animals using reflex based behavioral testing. A decrease in sensitivity to neuropathic pain for senescent (37–39 months) animals compared to old (20–22 months) and young (4–6 months) animals was observed by Pickering et al. [118]. Others [130] found no differences in responses to thermal stimuli for animals 7–8 weeks versus 18 months of age following partial denervation of the tail, while older animals showed increased mechanical allodynia [130]. The variable results across studies may be due to the use of reflex-based assays, different ages of animals, and differences in neuropathic pain models.

4 The Biopsychosocial Model Applied to Pain and Aging

The biopsychosocial model has become a guiding framework for conceptualizing the experience of disease and illness, including chronic pain [131]. The biopsychosocial model recognizes that while primary biological disease processes are important in human health, our understanding of illness is enhanced by incorporating the additional contributions of psychological and social factors. Importantly, these three sets of factors interact to influence the development, manifestations, and

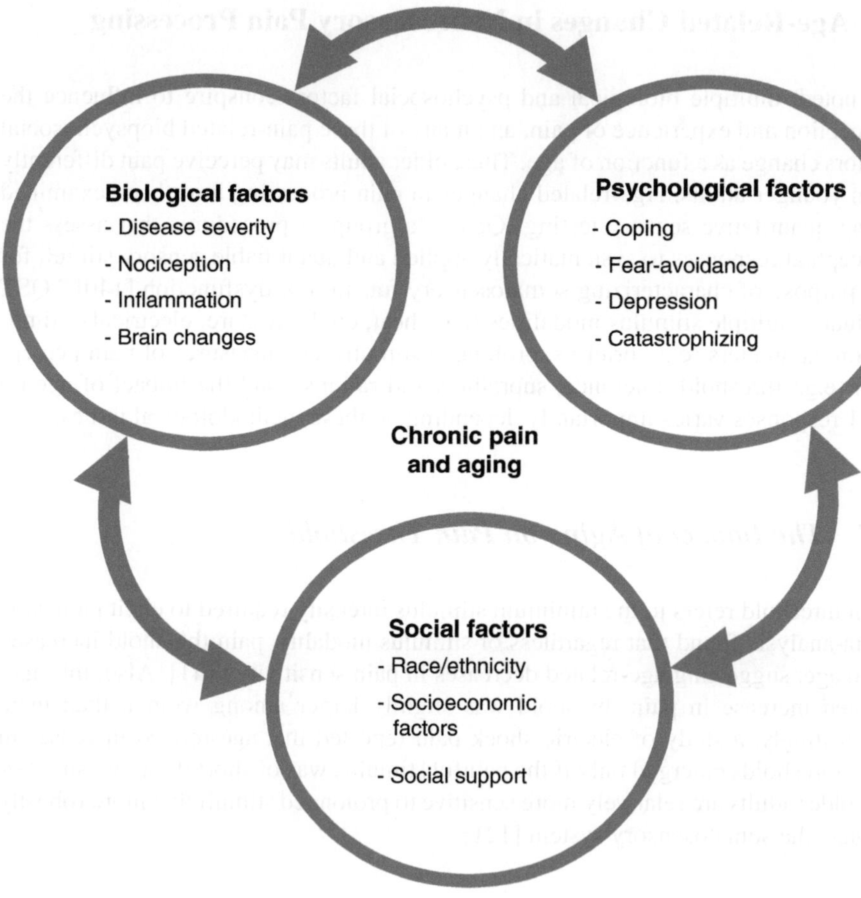

Fig. 18.2 The biopsychosocial model of pain and aging

course of clinical conditions such as chronic pain. Moreover, the biopsychosocial model represents a more comprehensive approach for understanding health and disease in aging [132, 133]. Figure 18.2 depicts the biopsychosocial model as applied to pain and aging. By way of example, while disease processes in OA, including reductions in cartilage volume, bony changes (e.g., osteophytes, subchondral bone cysts), and local inflammation (e.g., synovitis) are clearly important, measures of joint damage are at best only modest predictors of OA-related pain and disability [15, 134]. Interestingly, psychosocial factors contribute significantly to symptoms of OA. Indeed, recent systematic reviews found moderate to strong evidence that psychological factors, including depression [135] pain coping, self-efficacy, somatic symptoms, and catastrophizing, contribute to knee OA pain [136]. Further, social factors, including race and socioeconomic status, are also associated with OA-related pain and disability [137–139]. Hence, OA provides a prototypical example of the application of the biopsychosocial model's relevance to an age-related pain condition.

5 Age-Related Changes in Neurosensory Pain Processing

As noted, multiple biological and psychosocial factors conspire to influence the perception and experience of pain, and many of these pain-related biopsychosocial factors change as a function of age. Thus, older adults may perceive pain differently than younger adults. Age-related changes in pain processing have been examined using quantitative sensory testing (QST), "a group of procedures that assess the perceptual responses to systematically applied and quantifiable sensory stimuli for the purpose of characterizing somatosensory function or dysfunction [140]." QST includes multiple stimulus modalities (e.g., heat, cold, pressure, electrical), stimulation parameters (e.g., brief vs. prolonged stimuli), and measures of pain perception (e.g., threshold, tolerance, suprathreshold ratings), and the impact of age on QST responses varies importantly depending on these methodological issues.

5.1 The Impact of Aging on Pain Threshold

Pain threshold refers to the minimum stimulus intensity required to elicit pain. One meta-analysis found that regardless of stimulus modality, pain threshold increased with age, suggesting age-related decreases in pain sensitivity [141]. Also, this age-related increase in pain threshold was slightly larger among women than men. Interestingly, a study of electric shock pain reported that age-related increases in pain thresholds emerged only if the painful stimulus was of short duration, suggesting older adults are relatively more sensitive to prolonged stimuli that more robustly engage the somatosensory system [121].

5.2 The Impact of Aging on Pain Tolerance

Pain tolerance refers to the maximum stimulus intensity that an individual is willing or able to withstand. Studies of pain tolerance have revealed no age differences in response to thermal and electrical stimuli, but decreased pressure pain tolerance with age [141]. In addition, although not included in the meta-analysis, one study demonstrated dramatically lower tolerance for ischemic pain among older adults [142]. This finding is notable, because this experimental stimulus produces a sustained, deep muscle pain that is qualitatively similar to some forms of clinical musculoskeletal pain.

5.3 The Impact of Aging on Pain Facilitation

Temporal summation of pain refers to the increase in pain evoked by repeated rapid stimulation at a fixed stimulus intensity, which reflects a transient form of central sensitization and is therefore considered a measure of pain facilitation. In general, temporal summation of heat pain is greater in older versus younger adults [141], suggesting an age-related increase in pain facilitation. Another form of pain facilitation occurs in response to capsaicin, a chemical that selectively activates peripheral nociceptive neurons (e.g., c-fibers). After the primary pain from capsaicin subsides, there remains an increased pain response to mechanical stimuli (i.e., mechanical hyperalgesia), which reflects a reversible central sensitization. Zheng and colleagues [143] reported that capsaicin-induced mechanical hyperalgesia lasted substantially longer in older versus younger adults, which may indicate prolonged pain facilitation.

5.4 The Impact of Aging on Pain Inhibition

Several QST methods have been used to investigate pain inhibitory function, the most common of which is conditioned pain modulation (CPM). CPM refers to the decrease in pain evoked by one stimulus (the test stimulus) produced by contemporaneous application of a second pain stimulus at a different body site (the conditioning stimulus). Consistent age differences have been observed in CPM, with older adults showing reduced pain inhibition [141]. In fact, some studies have found that older adults report pain facilitation in response to the conditioning stimulus. Importantly, a recent study found the age-related reduction in CPM was independent of expectations and depression [144] suggesting that other central mechanisms are involved and need to be investigated.

Another measure of pain inhibition is offset analgesia. This occurs when a prolonged heat stimulus is delivered, in which the stimulus is slightly increased in intensity and then returned to the original temperature. This slight reduction in heat evokes a disproportionate decrease in pain [145]. A recent study reported that older adults showed reduced offset analgesia compared to younger adults [146]. Riley and colleagues [147] also reported that older adults showed a reduced decay of pain following offset of a prolonged heat pain stimulus, which may reflect impaired pain inhibition.

6 Cognitive Function and Pain Perception in Older Adults

Several QST studies have examined whether age-related declines in cognitive function influence pain processing. Benedetti and colleagues [148] reported that, compared to cognitively intact older adults, those with Alzheimer's disease showed similar pain thresholds but higher tolerance for electrical and ischemic pain, and pain tolerance was positively correlated with scores on the mini-mental status exam (MMSE). They reported similar findings for patients with frontotemporal dementia [149]. Also, autonomic responses to painful stimuli were reduced in patients with Alzheimer's as well as those with mild cognitive impairment [150, 151]. In contrast, Gibson and colleagues reported that patients with Alzheimer's disease showed similar pain thresholds and evoked potential amplitudes compared to cognitively intact older adults in response to laser-induced painful stimulation [152]. Marouf and colleagues [153] recently demonstrated that reduced CPM demonstrated by pain-free older adults was associated with poorer performance on a cognitive inhibition task (i.e., the Stroop task), suggesting that diminished cognitive function is associated with reduced pain inhibitory capacity.

The findings from QST studies suggest that aging is associated with decreased sensitivity to mild painful stimuli (e.g., pain threshold) but increased pain in response to higher intensity pain stimuli. Moreover, dynamic QST measures reveal increased pain facilitation and reduced pain inhibition among older adults. While dementia appears to be associated with attenuated pain responses, cognitive performance in cognitively intact older adults may positively predict pain inhibitory function. Yarnitsky and colleagues [154] recently suggested the concept of a pain modulation profile, which reflects an individual's balance of pain inhibition versus pain facilitation. This profile can be ascertained through QST measures such as CPM and temporal summation, and those who show a pro-nociceptive imbalance (i.e., high pain facilitation, low pain inhibition) are at increased risk for adverse pain outcomes. Based on the studies described above, aging is characterized by a pro-nociceptive pain modulation profile that may contribute to the increased risk of certain clinical pain conditions or of more severe or widespread pain in older adults. The factors driving these age-related changes in pain modulation are largely unknown, but are likely to include multiple biological (e.g., inflammation, endogenous opioid function, changes in brain structure and function) and psychosocial (e.g., mood, cognition, coping) processes and their interactions.

7 Neurobiological Factors Contributing to Age-Related Changes in Pain Processing

7.1 Brain Structure and Function

Aging is associated with global reductions in grey matter and white matter volume, though some brain regions are more affected than others [155, 156]. While these brain changes have been linked to declines in cognitive function, it seems plausible that aging effects on the brain could also impact pain processing. Indeed, chronic pain has been associated with decreases in grey and white matter volume [157–159]. Also, some evidence links structural brain changes to QST measures of pain perception. For example, total grey matter volume in fibromyalgia patients was negatively correlated with sensitivity to digital palpation [160]. Also, in pain-free adults, grey matter volume in several brain regions has been inversely associated with visceral sensitivity [161] and heat pain sensitivity [162]. While no study has yet linked changes in brain structure with age differences in pain processing, a plausible hypothesis is that age-related changes in brain morphology contribute to the enhanced pain facilitation and/or reduced pain inhibition observed in older adults.

Painful stimuli elicit patterns of neural activity in a variety of brain regions, and age differences in this pain-evoked cerebral activation could help explain age-related changes in pain perception. Two studies using heat pain showed reduced pain-evoked brain activity among older compared to younger adults. In response to painful heat older adults showed lower activation in several cortical regions, including somatosensory cortex, anterior insula, and supplementary motor area [163]. More recently, age was inversely associated with pain-related activation in somatosensory, insular, and premotor cortices [164] and grey matter volumes in the anterior and mid-cingulate cortex were positively correlated with pain ratings. Both of these studies showed decreased pain-related cerebral activation evoked by mild to moderate heat pain, which older adults typically report to be less painful. In contrast, Cole and colleagues [165] examined brain responses to pressure pain, to which older adults were more sensitive compared to their younger counterparts. While no age differences emerged in response to a mild pressure pain stimulus, younger adults showed greater activation in the contralateral putamen and caudate nucleus in response to moderate pressure pain. The authors suggested that these age differences may reflect an impairment of endogenous pain modulation among older adults.

7.2 Inflammation and Age Differences in Pain Processing

Aging is characterized by increases in systemic inflammation (i.e., "inflammaging"), which could contribute to increases in pain sensitivity. Systemic inflammation could influence pain processing via both peripheral and CNS mechanisms, as systemic inflammation can sensitize nociceptors [166, 167], and increasing evidence demonstrates that systemic inflammation can alter brain structure and function [168, 169]. Indeed, induction of systemic inflammation in humans reliably increases functional activation in several brain regions implicated in pain processing, including the brainstem, amygdala, anterior and posterior cingulate, and the insula [170]. Recent studies have reported that inducing systemic inflammation via endotoxemia in humans significantly increases pain sensitivity in response to visceral and somatic stimuli [171, 172]. Thus, age-related increases in systemic inflammation could contribute to the imbalanced pain modulatory profile that has been observed in older adults.

8 Psychosocial Factors Contributing to Pain Perception and Response in Older Adults

8.1 Beliefs/Attitudes About Pain

Regardless of age, people's beliefs and appraisal processes contribute to the exacerbation, attenuation, or maintenance of chronic pain, and to behavioral responses including affective distress, adjustment, health care seeking, response to treatment, and disability (i.e., [173, 174]). Furthermore, people's expectations about the impact of their pain and the likelihood of recovery following a painful injury have been shown to be more predictive of long-term disability than objective levels of physical pathology [175, 176]. Consequently, pain that persists over time should not be viewed as solely physical or solely psychological. Rather, as described earlier, persistent pain is a complex, biopsychosocial phenomenon that results from the reciprocal interactions between psychological, social, as well as physiological components.

Individuals' perceptions and responses have important implications for the persistence of pain and associated disability across the life cycle. However, some beliefs, expectations, and responses are particularly prevalent and important for older adults experiencing pain. For example, community-dwelling older adults often restrict their activity in response to pain [177, 178]. People may reduce their activities as a direct attempt to diminish the pain but also may reduce activity because they believe that pain is a signal of harm and wish to prevent further tissue damage or exacerbation of their pain. For example, a study of older adults reported that *all* study participants changed their activity in response to the experience of persistent pain by deliberately substituting activities perceived as physically demanding with more passive ones or stopping certain activities altogether [178]. Activity restriction was viewed as a way of safeguarding function and avoiding conventional treatments, such as medications and surgery [178]. Paradoxically, although activity restriction was approached strategically to preserve function and avoid medical interventions, the associated physical constraints, and loss of social contact were emotionally distressing [178]. Moreover, reduction in activity may reduce muscle strength, endurance, and flexibility, thus contributing to deconditioning, increased activity-related pain, and ultimately further disability. These findings highlight the trade-off that many people with chronic pain, older persons included, face between wanting to participate in valued activities and safeguarding function through reduced activity.

8.2 Emotional Distress – Anxiety/Fear, Depression

The sequelae of chronic pain include depression, anxiety and impaired cognitive function, as well as reduced socialization, compromised sleep and ambulation [35, 179]. These consequences are particularly important for older adults as they may contribute to falls, hospitalization, and increased dependence. Emotional distress may be a precipitant of symptoms, be a modulating factor amplifying or inhibiting the severity of pain, be a consequence of persistent pain, or a perpetuating factor. Moreover, these potential roles are not mutually exclusive and any number of them may be involved in a particular circumstance interacting with cognitive appraisals. For example, the literature is replete with studies demonstrating that current mood state modulates tolerance for acute pain (e.g., [180]). Levels of pre-surgery anxiety have been shown to influence not only pain severity, but also complications and length of stay following surgery [181, 182].

Fear of movement and fear of (re)injury are better predictors of functional limitations than biomedical parameters or even pain severity and duration [183, 184]. For example, Crombez et al. [185] reported that approximately two-thirds of individuals with chronic LBP (CLBP) avoid back-straining activities because of fear of (re) injury. Further, pain-related fear was the best predictor of behavioral performance in trunk-extension, flexion, and weight lifting tasks, even after separating out the effects of pain intensity. Moreover, fear of movement/(re)injury was the best predictor of self-reported disability among CLBP patients, and sensory perception of pain and biomedical findings did not add any predictive value [186]. Interestingly, reduction in pain-related anxiety predicted improvements in functioning, affective distress, pain, and pain-related interference with activity [187]. Pain-related fear of movement can be an important issue among older adults and may be further complicated by fear of falling. Clearly, fear, pain-related anxiety, and concerns about harm-avoidance all play important roles in age-related chronic pain and need to be assessed and addressed in treatment.

Falls are a common, costly, and often devastating problem among older people, causing a significant amount of morbidity, use of health care services, and increased risk of loss of independent living status, premature nursing home admissions, and mortality [188–190]. Analysis of a large national sample of Medicare beneficiaries found the prevalence of falls and the fear of falls that limits activity are three times higher in older adults with pain than in those without pain [191], and longitudinal studies of older adults show that chronic pain is associated with decreased mobility function and increased falls over time [192]. Importantly, *concerns* about falls [193] may be a crucial determinant of activity limitations, regardless of the objective fall risk [194, 195]. Indeed, fear of falling itself may become a risk factor, as it contributes to deconditioning, loss of flexibility, weakness, and abnormal gait, and in the long run may increase risk of falls [196, 197].

Depression represents another pain-related concern among older adults. A large survey of community-dwelling older adults found that baseline depression symptoms increased the odds of disabling LBP after two years, independent of sociodemographic characteristics and functional status. Conversely, disabling LBP at baseline also increased the odds of depressive symptoms after 2 years to a similar degree [198]. Depressed mood is quite common in OA, and is associated with increased pain sensitivity and disability [199]. Williamson and Schulz [200] found that activity restriction mediated the relationship between pain and symptoms of depression, and accounted for differences in pain intensity between non-depressed people and those at risk for developing depression. More recently a longitudinal study in older adults with OA found that pain interference in activity was a risk factor for developing depressive symptomatology [201].

8.3 Family and Other Social Supports

Consistent with the biopsychosocial model, persistent pain occurs in a social context. Social support, relationships with others, and resources can be defined as the availability of tangible (e.g., help with meals, transportation, assistance with personal care), affectionate (expressing love and affection), emotional (empathic understanding), informational (e.g., advice), and positive social interaction support (sharing of activities, companionship) [202]. Older adults with chronic health conditions often have difficulty participating in everyday activities [203, 204], thus affecting their quality of life and ability to participate in their communities.

Social isolation has an especially important impact on pain and disability in older adults. The elderly often face significant social losses (e.g., death of loved ones, reduction in social status, loss of independence) and difficulty maintaining social relationships, which can contribute to the exacerbation of persistent pain conditions and can negatively impact adherence and response to treatments [205]. In turn, persistent pain contributes to increased social isolation, as older adults with chronic pain spend less time in previous social roles and experience greater restrictions in social and leisure activities [206, 207].

Variations in the family, community, home, and healthcare environments can play important roles in how older adults adjust to pain. Significant others may express sympathy and excuse the individual from responsibilities, and encourage passivity, thereby fostering further functional impairment. Nursing homes are often perceived as coercive settings, promoting non-autonomous orientation that restricts activities. When events are objectively coercive, people may perceive a lack of autonomy and hence be at greater risk of depression. What may really be important to emotional well-being is not so much pain itself, but the way in which pain alters older people's lives.

9 Treatment of Pain in Older Adults

A substantial armamentarium is available to treat patients with chronic pain. These can be grouped into several general classes: (1) educational, (2) pharmacological, (3) activation (physical exercises), (4) psychological (e.g., cognitive-behavior therapy (CBT), (5) surgical, (6) neuroaugmentative, and (7) a variety of complementary health approaches and modalities. This chapter will focus on some of the most commonly studied and prescribed – medication, physical activation, complementary medicine, and CBT. Although we discuss these separately, they are often used in combinations (e.g., multidisciplinary programs).

9.1 Pharmacological Treatments

The first-line treatment for pain, regardless of age, is analgesic medication either over-the-counter or prescribed. The most commonly used and prescribed analgesic medications include nonsteroidal anti-inflammatory agents (NSAIDs) such as acetaminophen, opioids, antidepressants (e.g., amitriptyline, duloxetine, milnacipran) and anticonvulsants (calcium channel alpha2-delta ligands; gabapentin, pregabalin).

There are also a plethora of newer agents prescribed (sodium channel modulators [Lidocaine patch 5 %; mexiletine]; alpha$_2$-receptor agonists [clonidine, tizanidine], NMDA receptor antagonists [ketamine, memantine], cannabinoid receptor agonists, and vanilloid receptor ligands [topical capsaicin]). Each of these classes of medications may provide some level of pain relief, but often there are significant limitations and some adverse effects associated with each of them, particularly when used in older adults. For example, NSAIDS have known gastrointestinal and nephrotoxicities [208, 209]; NSAIDS and opioids have cardiotoxic and problematic hormonal effects [210, 211]; opioids, antidepressants, and antiepileptic drugs are all associated with increased susceptibility to falls and fractures [212, 213]; and opioids have potential for abuse (see Table 18.1 for a summary of the advantages and limitations of the most commonly prescribed medications).

Medical comorbidities are an important consideration in treating pain in older persons. Older adults often have several medical conditions in addition to the particular pain diagnosis (e.g., cardiac diseases, diabetes, cancer, dementia, osteoporosis). Aging itself and associated diseases affect gastrointestinal and physiological processes including motility, secretions, blood flow, and absorptive surface [214], and these changes can affect drug absorption, bioavailability, and transit time, as can reductions in plasma albumin, increased fat to lean mass ratios, and decreased total body water [214, 215]. In addition, liver mass, liver blood flow, and the glomerular filtration rate of kidneys decrease with age. Of particular clinical importance, reduced renal clearance leads to a decline in the excretion of water-soluble drugs [215]. Lowered activities of most of the cytochrome P450 enzymes reduce the drug-elimination clearance rate of the liver, especially in the presence of chronic disease [216]. The potential for drug accumulation and increased CNS sensitivity increases the risks for cognitive impairment and respiratory depression in conjunction with concomitant CNS medication or with underlying pulmonary conditions [217].

In a United States study published over a decade ago, 50 % of patients aged 65 or older consumed five or more prescription drugs and 10 % were using ten or more medications [218]. Polypharmacy can be a confounding risk factor when prescribing pain medications as there are both known and unknown drug-drug interactions that need to be considered. For example, individuals aged 60 or older are provided with an average of 40.8 prescriptions per year according to a study conducted in the United Kingdom [215]. With polypharmacy, dose-limiting adverse effects of pain-relieving medications may limit the potential achievable efficacy. Because of the increased likelihood of drug-drug and drug-disease interactions, as well as the pharmacokinetic and pharmacodynamic challenges associated with polypharmacy in older adults, frequent monitoring is critical when analgesic medications are prescribed.

Age-related changes in body composition and organ function can also alter metabolic and pharmacokinetic responses to medications. These changes along with medical and psychiatric comorbidities and concomitant polypharmacy (see also [219, 220]), suggest that conventional pharmacological therapies may not always be appropriate for older adults and should be used with caution [221–224].

Table 18.1 Common treatment options: advantages and limitations with elderly populations

Treatment	Advantages	Limitations
Pharmacologic treatments		
Analgesics in general	Pain relief	Age-related physical changes Interactions with comorbidities/frailty Drug-Drug interactions Adverse events (AEs)
NSAIDS/acetaminophen	Pain relief Reduced inflammation	AEs (e.g., Gastrointestinal, cardiac and nephrotoxicities, immune function, increased bleeding)
Opioids	Pain relief	Societal stigma Acceptance AEs (e.g., constipation, sedation, hormonal balance, dizziness, increased fall/fracture risk, cognitive impairment, cardiotoxicity) Interaction with alcohol Misuse potential
Anticonvulsants	Pain relief Improved sleep	Avoid in renally impaired AEs (e.g., sedation, cognitive impairment, weight gain, increased fall/fracture risk)
Antidepressants	Pain relief Improved mood	AEs (e.g., impact on blood pressure, dizziness, increased fall/fracture risk, nausea, sleep disruption)
Nonpharmacologic treatments		
Information/education	Prevention of falls	Cognitive limitations Sensory limitations Time intensive
Physical activation in general	Improved function	Acceptance Generalizability Adherence/maintenance Time/effort Motivation
Supervised exercise (e.g., EnhancedFitness)	Improved function Possible pain relief	Possible increased fall risk
Home exercise	Improved function Enhanced adherence	Possible increased fall risk
Tai Chi	Improved function Improved balance Reduced fall risk	Adherence/maintenance Time/effort
Psychological treatments in general	Improved mood Enhanced coping	Cognitive requirements Acceptance Motivation Time/effort
Cognitive Behavioral Therapy (CBT)	Improved mood Reduced fear of falls	Access/availability

Despite these cautions, older individuals are among the highest consumers of analgesics [225] with up to 20 % of the elderly taking analgesics for more than 6 months [226, 227].

Analgesics are often inappropriately prescribed for elderly patients, failing to follow clinical practice guidelines [228, 229]. For example, one study of community-dwelling men over age 70 noted that 8.2 % reported regular NSAID use compared to 2.9 % who reported as-needed use. Furthermore, the mean treatment duration for regular NSAID use was 4.9 years, suggesting long-term rather than short-term use as is recommended by guidelines [219]. Opioids are poorly tolerated by elderly patients [228, 230], and antidepressants and anticonvulsants are limited due to their effects on hepatic and renal function that may already be compromised because of the aging process. As noted, opioids, anticonvulsants, and antidepressants all increase the risk of falls and are major contributors to morbidity, increased hospitalization, and mortality in the elderly.

In sum, although conventional pharmacological treatments for pain can provide some relief for symptoms, they have significant hazards in older adults that need to be balanced in treatment decisions. Further, the long-term effectiveness of analgesic treatments for this population is unclear as few randomized controlled trials (RCTs) involve older adults with multiple morbidities [231, 232]. Moreover, in general, pharmacologic treatments provide only modest reductions in pain (30 % in fewer than 50 % of treated patients) and little impact on improving function [233]. There have been few studies that specifically address the issue of treatment effectiveness with older adults. In view of the limited evidentiary base and well-established adverse effects of current analgesic medications, there is an urgent need to develop both safe and effective pharmacological and non-pharmacological therapies for the rapidly growing older population. Greater emphasis on non-pharmacological approaches, alone or in combination with lower doses of pharmacological agents, may be particularly important for older adults with chronic pain.

9.2 Non-pharmacological Treatments

Clinical practice guidelines frequently recommend non-pharmacological interventions for pain in older adults. For example, in the 2013 guidance on the management of hip and knee OA, the European League Against Rheumatism (EULAR; [234]) recommended that all patients should have an individualized management plan that includes information and education on all aspects of OA management, advice on how to maintain activity, an exercise program consisting of aerobic and muscle-strengthening exercises, advice on weight loss as necessary and advice on home adaptations, among others.

Exercise is widely recognized as an approach for reducing pain and improving physical function in patients with chronic pain regardless of age [235, 236]. Despite recommendations for exercise, several studies have shown that objectively measured levels of physical activity are significantly lower in older chronic pain populations [6]. Indeed, as noted, activity restriction is a commonly reported strategy older adults use to reduce pain [177]. An important target for physical activation in older adults is improved balance [237–239]. Two examples of activity-based programs that have been specifically developed or adapted for use with older adults include – EnhanceFitness™ and Tai Chi (TC). The former is a more traditional approach to physical activity whereas TC makes use of complementary and alternative medicine modalities of relaxation and structured movements. Both approaches may be particularly useful when targeted to improve balance.

EnhanceFitness (EF) is a community-based multicomponent, supervised group physical exercise program for older adults, involving balance, strength, and endurance training that is widely disseminated throughout the United States [240]. Each EF class uses a standardized format that includes several phases: a warm-up, moderate-intensity aerobic conditioning/walking phase, progressive strength training, and a cool down. EF focuses on flexibility and static and dynamic balance exercises that are known to reduce falls risk. Two RCTs [241, 242] have shown that EF modestly improves health-related quality of life, objective measures of balance, functional and gait performance, and reduce health care costs. However, in a large RCT, adherence to EF was variable with a median of 58 % (interquartile range: 15–75 %) of exercise classes attended over 12 months [241]. Unpublished exit interview data identified pain as a leading cause of non-adherence. Greater adherence may lead to better outcomes but as noted below, adherence with any self-management regimen is a significant concern.

Tai Chi (TC) is a traditional Chinese martial art that incorporates aerobic activity, diaphragmatic breathing, relaxation, and meditation with postures that are designed to flow imperceptibly and smoothly through slow, gentle movements (low impact, low velocity) that are particularly appropriate for elderly adults. The slow speed and constant weight shifting associated with TC increases the load on the lower limbs in a gradual fashion and may have a direct effect on improving balance. TC enhances self-awareness of balance and thereby may contribute to the amelioration of fear of falling [243, 244]. It has emerged as a viable exercise intervention, and it is recommended for older populations by the American Geriatrics Society [221]. TC improved function in people with knee OA [245], CLBP [246, 247], and improved balance and reduced risk of falls in older adults [238, 248]. It has been shown to be more effective than other exercises for improving mobility and reducing fear of falling in older adults [246, 249].

In general, increasing exercise is a key challenge to address in the geriatric population as relatively few older adults use exercise and other behavioral strategies to cope with pain [177, 250]. Instead, passive strategies and avoidant behaviors are more common and associated with increased disability [250]. Thus, efforts to increase physical activity in the geriatric pain population will require restructuring maladaptive beliefs about exercise and fears of falling, which seems to occur with EF and TC, as well as addressing common comorbid symptoms that are prevalent and often contribute to activity limitation, such as difficulty sleeping and depressed mood.

9.3 Psychological Treatments

Although there are a range of psychological approaches to managing pain, Cognitive-behavior therapy (CBT) is perhaps the most commonly prescribed and thoroughly investigated [251]. CBT is a generic term; there are many components (e.g., exposure, addressing maladaptive beliefs, skills training [252]). The primary assumptions shared by all CBT interventions are that: (1) people are active processors of information; (2) people are capable of gaining control over their thoughts, feelings, behaviors, and to some extent physiological processes; and (3) there are interrelationships among thoughts, feelings, behaviors, and physiological processes; changes in one or more of these factors may result in changes in the others.

Meta-analyses and systematic reviews have shown beneficial effects of CBT for a number of pain conditions that often affect older adults (e.g., OA pain, cancer pain, CLBP) [253–255]. In general they demonstrate that CBT produce significant decreases in pain (typically small to medium effect sizes) and significant improvements in indices of adjustment to pain (e.g., activity, depression, anxiety, self-confidence, maladaptive beliefs). However the results are not consistent across studies, which may relate to the specific content, mode of delivery, duration of treatment, and extent of therapist training. In designing CBT for older patients, clinicians should consider providing rationales, guided practice, experiential learning (more active, less didactic), home practice assignments in written and verbal forms, using audio recording to guide home practice to eliminate the need for memorization, and possibly involving significant others [256]. Of particular relevance, CBT has demonstrated significant yet modest efficacy in the treatment of fear of falling in older adults [257]. The results are comparable to those reported for exercise in community-dwelling older adults [249, 258]. CBT has been well received in small samples of older adults with chronic pain [259, 260]. Tailored CBT for older populations has included a set of components including:

- Education about the long-term benefits of exercise for chronic pain
- Restructuring beliefs about activity and pain
- Teach coping skills to address pain and related symptoms including problem-solving skills to overcome barriers to exercise
- Foster activity pacing
- Set realistic goals for increasing activity incorporating any limitations in physical capacity to ensure safety and promote the adoption of a physically active lifestyle
- Expose participants to feared and avoided activities that are not restricted by physical limitations
- Address fear of falling by restructuring misconceptions to promote a view of falls risk as controllable
- Teach coping skills to address pain and related symptoms, including regulating daily activities and sleep schedule, and modifying cognitive and emotional responses to pain
- In general, emphasize self-control and self-efficacy

- Set realistic goals for activity and promoting the adoption of a physically active lifestyle
- Advise about the environment to reduce fall hazards

9.4 Mechanisms of Psychological and CAM Interventions

With the advent of sophisticated imaging procedures such as fMRI, it has become possible to identify regions of the brain associated with self-reports of pain as described above. Recent studies using fMRI have been able to document unique physiological responses in the brain that are associated with prolonged pain, emotional responses accompanying pain, attentional foci, and processing of noxious stimulation (e.g., [261, 262]).

Of particular interest is the use of neuroimaging technologies to identify changes in brain function that accompany alterations in pain perception and responses following pain treatments. Jensen [263] has hypothesized that different psychological pain treatments and changes in the psychological factors targeted by these treatments (e.g., maladaptive cognitions, reassuring thoughts) could potentially have different effects on the activity of different brain structures and processes. CBT has been shown to produce structural and functional changes in pain related regions of the brain in several populations [264–267]. Non-pharmacological therapies, including CBT, evoke increased activation of brain regions involved in executive cognitive function, while decreasing activity in regions associated with pain transmission [268]. In addition, neuroimaging studies suggest that non-pharmacological approaches may produce positive structural brain changes in areas that often decline with aging (i.e., cortical thickness in the prefrontal and right anterior insula and occipito-temoral region) [209]. Seminowicz and colleagues [269] reported increased prefrontal cortex gray matter volume following CBT for chronic pain, and greater reductions in pain catastrophizing after CBT were associated with greater increases in gray matter volume in several brain regions. The impact of CBT on brain structure and function in older adults with chronic pain represents an important yet unexplored area.

9.5 Adherence to Prescribed Treatments

The non-pharmacological treatments described above require patient understanding, motivation, and adherence. These may be of particular concern for older adults who have cognitive and sensory limitations. The impact of pain on cognitive functioning (e.g., memory) might impede following prescribed regimens. Treatments that were originally developed for younger individuals need to have appropriate adaptations and adjustments in content and format when prescribed for the elderly to accommodate any age-related limitations. Another impediment to adherence is cost, which impacts adherence to prescription medications among older adults, especially among individuals who have poor health, multiple morbidities, and limited drug coverage through insurance [270]. Successful treatment of older people with chronic pain will require that problems associated with treatment adherence be addressed regardless of the intervention – pharmacological or non-pharmacological.

10 Conclusions and Future Directions

Age-related changes in pain are complex and remain poorly understood. Epidemiological and clinical studies demonstrate that pain prevalence and impact change with age, although patterns vary for different types of pain; some pain conditions increase while others decrease in prevalence with age. Preclinical models reveal conflicting findings regarding age-related changes in nociceptive sensitivity, likely due to methodological variations. In humans, QST studies reveal an age-related pain modulatory imbalance, as older adults tend to show increased pain facilitation and diminished pain inhibition. Additional research is needed to more clearly define the biopsychosocial factors that contribute to age-related changes in pain processing. Potential biological mechanisms include anatomical, physiological, immune, neuroendocrine, inflammatory, and autonomic changes. Likewise, multiple psychosocial factors influence pain experiences among older adults, including beliefs and perceptions, negative mood (e.g., fear, depression), and social changes and support (e.g., isolation, reduced social activity). However, limited information exists regarding the extent to which each of these factors individually contributes to age-related influences on pain, let alone their interactions. Likewise, aging complicates the treatment of pain. Pharmacologic therapies offer limited clinical efficacy and produce increased adverse effects in older adults, and non-pharmacologic treatments, while effective, are often underprescribed in elderly patients.

Based on the current state of the evidence, we recommend the following lines of investigation to move the field of pain and aging forward.

- Investigators should take a more systematic and standardized approach to characterizing chronic pain in studies of pain and aging, including: increased consistency in case-definitions for pain conditions (e.g., time-frame – pain lasting 3 months, last 30 days), assessment of pain impact, pain severity, temporal patterns, etc.
- Studies are needed to evaluate the interaction of how systems known to change as a function of age (immune, HPA-axis, autonomic nervous system) impact age-related changes in pain sensitivity and the prevalence of chronic pain conditions.
- Advancing age needs to be a relevant variable in preclinical models of pain. Specifically, additional cross-sectional and longitudinal studies comparing vari-

ous pain models and assays (e.g., operant vs. reflex-based, neuropathic vs. visceral) could help elucidate mechanisms underlying differential age-related patterns across different clinical pain conditions.

- Preclinical studies are needed in older animals to determine the efficacy and drug-drug interactions of pain medications and the potential side effect profile of drugs compared to those in younger animals.
- Studies are needed investigating how age may be protective against development or severity of certain pain conditions.
- There is a need for increasing the representation of older adults in clinical trials of pain therapies and for secondary analysis of age as a potential moderating factor for efficacy in existing clinical trials.
- Given the modest outcomes for existing treatments, and currently limited treatment armamentarium, evaluation of treatment combinations for pain in older adults requires investigation.
- Given the importance of self-management strategies in managing pain in older adults, research examining adherence and maintenance enhancement strategies is essential.
- Evaluation of changes in brain structure and function that may predict positive outcomes from non-pharmacological interventions in older adults is warranted.
- Investigators should explore adaptations to pain management approaches required to compensate for cognitive and sensory limitations associated with aging.

Addressing these issues and adopting these methodological enhancements should help reduce inconsistencies in the literature, thereby substantially improving our understanding of age-related influences on pain.

Acknowledgments The preparation of this chapter was supported in part by NIH grants K07AG04637 and R37AG033906 (RBF). Preclinical studies related to the effects of age on thermal sensitivity were supported by RAG031821 (RPY).

Editors: Wen G. Chen, National Center for Complementary and Integrative Health (NCCIH/NIH); Joseph Frascella, National Institute on Drug Abuse (NIDA/NIH) and Partap Khalsa (National Center for Complementary and Integrative Health (NCCIH/NIH).

References

1. Gaskin DJ, Richard P (2012) The economic costs of pain in the United States. J Pain 13(8):715–724
2. Institute of Medicine (US) Committee on Advancing Pain Research, Care, and Education (2011) Relieving pain in America: a blueprint for transforming prevention, care, education, and research. National Academies Press, Washington, DC
3. Verhaak PF, Kerssens JJ, Dekker J, Sorbi MJ, Bensing JM (1998) Prevalence of chronic benign pain disorder among adults: a review of the literature. Pain 77(3):231–239
4. Blyth FM, March LM, Brnabic AJ, Jorm LR, Williamson M, Cousins MJ (2001) Chronic pain in Australia: a prevalence study. Pain 89(2–3):127–134

5. Tsang A, Von KM, Lee S, Alonso J, Karam E, Angermeyer MC, Borges GL, Bromet EJ, Demytteneare K, de Girolamo G, de Graaf R, Gureje O, Lepine JP, Haro JM, Levinson D, Oakley Browne MA, Posada-Villa J, Seedat S, Watanabe M (2008) Common chronic pain conditions in developed and developing countries: gender and age differences and comorbidity with depression-anxiety disorders. J Pain 9(10):883–891
6. Patel KV, Guralnik JM, Dansie EJ, Turk DC (2013) Prevalence and impact of pain among older adults in the United States: findings from the 2011 National Health and Aging Trends Study. Pain 154(12):2649–2657. doi:10.1016/j.pain.2013.07.029
7. Gibson SJ, Lussier D (2012) Prevalence and relevance of pain in older persons. Pain Med (Malden, MA) 13(Suppl 2):S23–S26. doi:10.1111/j.1526-4637.2012.01349.x
8. Miranda VS, Decarvalho VB, Machado LA, Dias JM (2012) Prevalence of chronic musculoskeletal disorders in elderly Brazilians: a systematic review of the literature. BMC Musculoskelet Disord 13:82. doi:10.1186/1471-2474-13-82
9. Neogi T, Zhang Y (2013) Epidemiology of osteoarthritis. Rheum Dis Clin North Am 39(1):1–19
10. Ahacic K, Kareholt I (2010) Prevalence of musculoskeletal pain in the general Swedish population from 1968 to 2002: age, period, and cohort patterns. Pain 151(1):206–214. doi:10.1016/j.pain.2010.07.011
11. LeResche L (2000) Epidemiologic perspectives on sex differences in pain. In: Fillingim RB (ed) Sex, gender, and pain. IASP Press, Seattle, pp 233–249
12. Smitherman TA, Burch R, Sheikh H, Loder E (2013) The prevalence, impact, and treatment of migraine and severe headaches in the United States: a review of statistics from national surveillance studies. Headache 53(3):427–436. doi:10.1111/head.12074
13. Anderson AS, Loeser RF (2010) Why is osteoarthritis an age-related disease? Best Pract Res Clin Rheumatol 24(1):15–26. doi:10.1016/J.Berh.2009.08.006
14. Jordan JM, Helmick CG, Renner JB, Luta G, Dragomir AD, Woodard J, Fang F, Schwartz TA, Abbate LM, Callahan LF, Kalsbeek WD, Hochberg MC (2007) Prevalence of knee symptoms and radiographic and symptomatic knee osteoarthritis in African Americans and Caucasians: the Johnston County Osteoarthritis Project. J Rheumatol 34(1):172–180
15. Phillips K, Clauw DJ (2011) Central pain mechanisms in chronic pain states–maybe it is all in their head. Best Pract Res Clin Rheumatol 25(2):141–154
16. King CD, Sibille KT, Goodin BR, Cruz-Almeida Y, Glover TL, Bartley E, Riley JL, Herbert MS, Sotolongo A, Schmidt J, Fessler BJ, Redden DT, Staud R, Bradley LA, Fillingim RB (2013) Experimental pain sensitivity differs as a function of clinical pain severity in symptomatic knee osteoarthritis. Osteoarthritis Cartilage 21(9):1243–1252
17. Hoy D, March L, Brooks P, Blyth F, Woolf A, Bain C, Williams G, Smith E, Vos T, Barendregt J, Murray C, Burstein R, Buchbinder R (2014) The global burden of low back pain: estimates from the Global Burden of Disease 2010 study. Ann Rheum Dis 73:968–974. doi:10.1136/annrheumdis-2013-204428
18. Collaborators USBoD (2013) The state of US health, 1990–2010: burden of diseases, injuries, and risk factors. JAMA 310(6):591–608. doi:10.1001/jama.2013.13805
19. Leboeuf-Yde C, Nielsen J, Kyvik KO, Fejer R, Hartvigsen J (2009) Pain in the lumbar, thoracic or cervical regions: do age and gender matter? A population-based study of 34,902 Danish twins 20–71 years of age. BMC Musculoskelet Disord 10:39. doi:10.1186/1471-2474-10-39
20. Lawrence RC, Helmick CG, Arnett FC, Deyo RA, Felson DT, Giannini EH, Heyse SP, Hirsch R, Hochberg MC, Hunder GG, Liang MH, Pillemer SR, Steen VD, Wolfe F (1998) Estimates of the prevalence of arthritis and selected musculoskeletal disorders in the United States. Arthritis Rheum 41(5):778–799
21. Loney PL, Stratford PW (1999) The prevalence of low back pain in adults: a methodological review of the literature. Phys Ther 79(4):384–396
22. Dionne CE, Dunn KM, Croft PR (2006) Does back pain prevalence really decrease with increasing age? A systematic review. Age Ageing 35(3):229–234. doi:10.1093/ageing/afj055

23. Treede RD, Jensen TS, Campbell JN, Cruccu G, Dostrovsky JO, Griffin JW, Hansson P, Hughes R, Nurmikko T, Serra J (2008) Neuropathic pain: redefinition and a grading system for clinical and research purposes. Neurology 70(18):1630–1635. doi:10.1212/01.wnl.0000282763.29778.59
24. Schmader KE, Baron R, Haanpaa ML, Mayer J, O'Connor AB, Rice AS, Stacey B (2010) Treatment considerations for elderly and frail patients with neuropathic pain. Mayo Clin Proc 85(3 Suppl):S26–S32. doi:10.4065/mcp.2009.0646
25. Jung BF, Johnson RW, Griffin DR, Dworkin RH (2004) Risk factors for postherpetic neuralgia in patients with herpes zoster. Neurology 62(9):1545–1551
26. Sima AA, Greene DA (1995) Diabetic neuropathy in the elderly. Drugs Aging 6(2):125–135
27. Sadosky A, McDermott AM, Brandenburg NA, Strauss M (2008) A review of the epidemiology of painful diabetic peripheral neuropathy, postherpetic neuralgia, and less commonly studied neuropathic pain conditions. Pain Pract 8(1):45–56
28. Osterberg A, Boivie J, Thuomas KA (2005) Central pain in multiple sclerosis–prevalence and clinical characteristics. Eur J Pain (London, England) 9(5):531–542. doi:10.1016/j.ejpain.2004.11.005
29. Kamerman PR, Wadley AL, Cherry CL (2012) HIV-associated sensory neuropathy: risk factors and genetics. Curr Pain Headache Rep 16(3):226–236. doi:10.1007/s11916-012-0257-z
30. Giamberardino MA (2005) Visceral pain. IASP Pain Clinical Updates 8:6
31. Jones R, Lydeard S (1992) Irritable bowel syndrome in the general population. BMJ 304(6819):87–90
32. Tresch DD (1998) Management of the older patient with acute myocardial infarction: difference in clinical presentations between older and younger patients. J Am Geriatr Soc 46(9):1157–1162
33. Moore AR, Clinch D (2004) Underlying mechanisms of impaired visceral pain perception in older people. J Am Geriatr Soc 52(1):132–136
34. Berryman C, Stanton TR, Bowering KJ, Tabor A, McFarlane A, Moseley GL (2014) Do people with chronic pain have impaired executive function? A meta-analytical review. Clin Psychol Rev 34(7):563–579. doi:10.1016/j.cpr.2014.08.003
35. Moriarty O, McGuire BE, Finn DP (2011) The effect of pain on cognitive function: a review of clinical and preclinical research. Prog Neurobiol 93(3):385–404
36. Monroe TB, Gore JC, Chen LM, Mion LC, Cowan RL (2012) Pain in people with Alzheimer disease: potential applications for psychophysical and neurophysiological research. J Geriatr Psychiatry Neurol 25(4):240–255. doi:10.1177/0891988712466457
37. Hadjistavropoulos T, Herr K, Prkachin KM, Craig KD, Gibson SJ, Lukas A, Smith JH (2014) Pain assessment in elderly adults with dementia. Lancet Neurol 13(12):1216–1227. doi:10.1016/S1474-4422(14)70103-6
38. Weiner DK, Rudy TE, Morrow L, Slaboda J, Lieber S (2006) The relationship between pain, neuropsychological performance, and physical function in community-dwelling older adults with chronic low back pain. Pain Med 7(1):60–70
39. Gagliese L, Melzack R (2003) Age-related differences in the qualities but not the intensity of chronic pain. Pain 104(3):597–608
40. Cheung WY, Le LW, Gagliese L, Zimmermann C (2011) Age and gender differences in symptom intensity and symptom clusters among patients with metastatic cancer. Support Care Cancer (Official Journal of the Multinational Association of Supportive Care in Cancer) 19(3):417–423. doi:10.1007/s00520-010-0865-2
41. Gauthier LR, Young A, Dworkin RH, Rodin G, Zimmermann C, Warr D, Librach SL, Moore M, Shepherd FA, Pillai Riddell R, Macpherson A, Melzack R, Gagliese L (2014) Validation of the short-form McGill pain questionnaire-2 in younger and older people with cancer pain. J Pain (Official Journal of the American Pain Society) 15(7):756–770. doi:10.1016/j.jpain.2014.04.004

42. van Oostrom SH, Monique Verschuren WM, de Vet HC, Picavet HS (2011) Ten year course of low back pain in an adult population-based cohort–the Doetinchem cohort study. Eur J Pain (London, England) 15(9):993–998. doi:10.1016/j.ejpain.2011.02.007
43. Kehlet H, Jensen TS, Woolf CJ (2006) Persistent postsurgical pain: risk factors and prevention. Lancet 367(9522):1618–1625
44. Wang HT, Liu W, Luo AL, Ma C, Huang YG (2012) Prevalence and risk factors of chronic post-thoracotomy pain in Chinese patients from Peking Union Medical College Hospital. Chin Med J (Engl) 125(17):3033–3038
45. Hardt J, Jacobsen C, Goldberg J, Nickel R, Buchwald D (2008) Prevalence of chronic pain in a representative sample in the United States. Pain Med 9(7):803–812
46. McBeth J, Jones K (2007) Epidemiology of chronic musculoskeletal pain. Best Pract Res Clin Rheumatol 21(3):403–425
47. Covinsky KE, Lindquist K, Dunlop DD, Yelin E (2009) Pain, functional limitations, and aging. J Am Geriatr Soc 57(9):1556–1561. doi:10.1111/j.1532-5415.2009.02388.x
48. Thomas E, Peat G, Harris L, Wilkie R, Croft PR (2004) The prevalence of pain and pain interference in a general population of older adults: cross-sectional findings from the North Staffordshire Osteoarthritis Project (NorStOP). Pain 110(1–2):361–368
49. Wijnhoven HA, de Vet HC, Picavet HS (2007) Sex differences in consequences of musculoskeletal pain. Spine 32(12):1360–1367
50. Stubbs B, Schofield P, Binnekade T, Patchay S, Sepehry A, Eggermont L (2014) Pain is associated with recurrent falls in community-dwelling older adults: evidence from a systematic review and meta-analysis. Pain Med (Malden, MA) 15(7):1115–1128. doi:10.1111/pme.12462
51. Harrington ME (2012) Neurobiological studies of fatigue. Prog Neurobiol 99(2):93–105. doi:10.1016/j.pneurobio.2012.07.004
52. Murphy SL, Smith DM, Clauw DJ, Alexander NB (2008) The impact of momentary pain and fatigue on physical activity in women with osteoarthritis. Arthritis Rheum 59(6):849–856. doi:10.1002/art.23710
53. Neikrug AB, Ancoli-Israel S (2010) Sleep disorders in the older adult – a mini-review. Gerontology 56(2):181–189. doi:10.1159/000236900
54. Finan PH, Goodin BR, Smith MT (2013) The association of sleep and pain: an update and a path forward. J Pain (Official Journal of the American Pain Society) 14(12):1539–1552. doi:10.1016/j.jpain.2013.08.007
55. Troen BR (2003) The biology of aging. Mt Sinai J Med (New York) 70(1):3–22
56. Gatchel RJ, Peng YB, Peters ML, Fuchs PN, Turk DC (2007) The biopsychosocial approach to chronic pain: scientific advances and future directions. Psychol Bull 133(4):581–624
57. Olovnikov AM (1996) Telomeres, telomerase, and aging: origin of the theory. Exp Gerontol 31(4):443–448
58. Shalev I, Entringer S, Wadhwa PD, Wolkowitz OM, Puterman E, Lin J, Epel ES (2013) Stress and telomere biology: a lifespan perspective. Psychoneuroendocrinology 38(9):1835–1842. doi:10.1016/j.psyneuen.2013.03.010
59. Hassett AL, Epel E, Clauw DJ, Harris RE, Harte SE, Kairys A, Buyske S, Williams DA (2012) Pain is associated with short leukocyte telomere length in women with fibromyalgia. J Pain (Official Journal of the American Pain Society) 13(10):959–969. doi:10.1016/j.jpain.2012.07.003
60. Sibille KT, Langaee T, Burkley B, Gong Y, Glover TL, King C, Riley JL III, Leeuwenburgh C, Staud R, Bradley LA, Fillingim RB (2012) Chronic pain, perceived stress, and cellular aging: an exploratory study. Mol Pain 8:12. doi:10.1186/1744-8069-8-12.:12[–]18
61. Sibille KT, Witek-Janusek L, Mathews HL, Fillingim RB (2012) Telomeres and epigenetics: potential relevance to chronic pain. Pain 153(9):1789–1793
62. Chung JM (2004) The role of reactive oxygen species (ROS) in persistent pain. Mol Interv 4(5):248–250. doi:10.1124/mi.4.5.3

63. Lee DZ, Chung JM, Chung K, Kang MG (2012) Reactive oxygen species (ROS) modulate AMPA receptor phosphorylation and cell-surface localization in concert with pain-related behavior. Pain 153(9):1905–1915. doi:10.1016/j.pain.2012.06.001
64. Knight JA (2000) The biochemistry of aging. Adv Clin Chem 35:1–62
65. Pollack M, Leeuwenburgh C (2000) Molecular mechanisms of oxidative stress in aging: free radicals, aging, antioxidents and disease. In: Sen CK, Packer LH, Hannien O (eds) Handbook of oxidants and antioxidants in exercise. Elsevier, Amsterdam, pp 881–923
66. Yu CG, Yezierski RP (2005) Activation of the ERK1/2 signaling cascade by excitotoxic spinal cord injury. Brain Res Mol Brain Res 138(2):244–255. doi:10.1016/j.molbrainres.2005.04.013
67. Yu CG, Yezierski RP, Joshi A, Raza K, Li Y, Geddes JW (2010) Involvement of ERK2 in traumatic spinal cord injury. J Neurochem 113(1):131–142. doi:10.1111/j.1471-4159.2010.06579.x
68. Richardson A, Galvan V, Lin AL, Oddo S (2014) How longevity research can lead to therapies for Alzheimer's disease: the rapamycin story. Exp Gerontol 68:51–58. doi:10.1016/j.exger.2014.12.002
69. Harrison DE, Strong R, Sharp ZD, Nelson JF, Astle CM, Flurkey K, Nadon NL, Wilkinson JE, Frenkel K, Carter CS, Pahor M, Javors MA, Fernandez E, Miller RA (2009) Rapamycin fed late in life extends lifespan in genetically heterogeneous mice. Nature 460(7253):392–395. doi:10.1038/nature08221
70. Geranton SM, Jimenez-Diaz L, Torsney C, Tochiki KK, Stuart SA, Leith JL, Lumb BM, Hunt SP (2009) A rapamycin-sensitive signaling pathway is essential for the full expression of persistent pain states. J Neurosci (The Official Journal of the Society for Neuroscience) 29(47):15017–15027. doi:10.1523/jneurosci.3451-09.2009
71. Lisi L, Navarra P, Cirocchi R, Sharp A, Stigliano E, Feinstein DL, Dello Russo C (2012) Rapamycin reduces clinical signs and neuropathic pain in a chronic model of experimental autoimmune encephalomyelitis. J Neuroimmunol 243(1–2):43–51. doi:10.1016/j.jneuroim.2011.12.018
72. Gagliese L (2009) Pain and aging: the emergence of a new subfield of pain research. J Pain 10(4):343–353
73. Gibson SJ, Weiner DK (2005) Pain in older persons. IASP Press, Seattle
74. Gagliese L, Farrell MJ (2005) The neurobiology of aging, nociception and pain: an integration of animal and human experimental evidence. In: Gibson SJ, Weiner DK (eds) Pain in older persons. IASP Press, Seattle, pp 25–44
75. Ceballos D, Cuadras J, Verdu E, Navarro X (1999) Morphometric and ultrastructural changes with ageing in mouse peripheral nerve. J Anat 195(Pt 4):563–576
76. Ochoa J, Mair WG (1969) The normal sural nerve in man. II. Changes in the axons and Schwann cells due to ageing. Acta Neuropathol 13(3):217–239
77. Drac H, Babiuch M, Wisniewska W (1991) Morphological and biochemical changes in peripheral nerves with aging. Neuropatol Pol 29(1–2):49–67
78. O'Sullivan DJ, Swallow M (1968) The fibre size and content of the radial and sural nerves. J Neurol Neurosurg Psychiatry 31(5):464–470
79. Cecchini T, Cuppini R, Ciaroni S, Barili P, De Matteis R, Del Grande P (1995) Changes in the number of primary sensory neurons in normal and vitamin-E-deficient rats during aging. Somatosens Mot Res 12(3–4):317–327
80. Devor M (1991) Chronic pain in the aged: possible relation between neurogenesis, involution and pathophysiology in adult sensory ganglia. J Basic Clin Physiol Pharmacol 2(1–2):1–15
81. Iwata K, Tsuboi Y, Shima A, Harada T, Ren K, Kanda K, Kitagawa J (2004) Central neuronal changes after nerve injury: neuroplastic influences of injury and aging. J Orofac Pain 18(4):293–298
82. Zimmermann M (2001) Pathobiology of neuropathic pain. Eur J Pharmacol 429(1–3):23–37
83. Amenta F, Zaccheo D, Collier WL (1991) Neurotransmitters, neuroreceptors and aging. Mech Ageing Dev 61(3):249–273

84. Chung YH, Kim D, Lee KJ, Kim SS, Kim KY, Cho DY, Sohn DS, Lee WB (2005) Immunohistochemical study on the distribution of neuronal nitric oxide synthase-immunoreactive neurons in the spinal cord of aged rat. J Mol Histol 36(5):325–329. doi:10.1007/s10735-005-9001-x
85. Hoskins DL, Gordon TL, Crisp T (1998) The effects of aging on mu and delta opioid receptors in the spinal cord of Fischer-344 rats. Brain Res 791(1–2):299–302
86. Ko ML, King MA, Gordon TL, Crisp T (1997) The effects of aging on spinal neurochemistry in the rat. Brain Res Bull 42(2):95–98
87. Ma W, Bisby MA (1998) Increase of calcitonin gene-related peptide immunoreactivity in the axonal fibers of the gracile nuclei of adult and aged rats after complete and partial sciatic nerve injuries. Exp Neurol 152(1):137–149. doi:10.1006/exnr.1998.6827
88. Gagliese L, Melzack R (2000) Age differences in nociception and pain behaviours in the rat. Neurosci Biobehav Rev 24(8):843–854
89. Crisp T, Stafinsky JL, Hoskins DL, Dayal B, Chinrock KM, Uram M (1994) Effects of aging on spinal opioid-induced antinociception. Neurobiol Aging 15(2):169–174
90. Hess GD, Joseph JA, Roth GS (1981) Effect of age on sensitivity to pain and brain opiate receptors. Neurobiol Aging 2:49–55
91. Jourdan D, Pickering G, Marchand F, Gaulier JM, Alliot J, Eschalier A (2002) Impact of ageing on the antinociceptive effect of reference analgesics in the Lou/c rat. Br J Pharmacol 137(6):813–820. doi:10.1038/sj.bjp.0704944
92. Wang S, Albers KM (2009) Behavioral and cellular level changes in the aging somatosensory system. Ann N Y Acad Sci 1170:745–749. doi:10.1111/j.1749-6632.2009.04011.x
93. Heiden M, Barnekow-Bergkvist M, Nakata M, Lyskov E (2005) Autonomic activity, pain, and perceived health in patients on sick leave due to stress-related illnesses. Integr Physiol Behav Sci (The Official Journal of the Pavlovian Society) 40(1):3–16
94. Seals DR, Esler MD (2000) Human ageing and the sympathoadrenal system. J Physiol 528(Pt 3):407–417
95. Elenkov IJ, Wilder RL, Chrousos GP, Vizi ES (2000) The sympathetic nerve–an integrative interface between two supersystems: the brain and the immune system. Pharmacol Rev 52(4):595–638
96. Vierck CJ Jr (2006) Mechanisms underlying development of spatially distributed chronic pain (fibromyalgia). Pain 124(3):242–263. doi:10.1016/j.pain.2006.06.001
97. Kuchel GA, Cowen T (2001) The aged sympathetic nervous system. In: Hof P, Mobbs C (eds) Fundamental neurobiology of aging. Academic, New York, pp 929–939
98. Baron R, Blumberg H, Janig W (1996) Clinical characteristics of patients with complex regional pain syndrome in Germany with special emphasis on vasomotor function. In: Stanton-Hicks M, Janig W (eds) Progress in pain research and therapy, vol 6, Reflex sympathetic dystrophy: a reappraisal. IASP Press, Seattle, pp 25–48
99. Harden RN, Baron R, Janig W (2001) Complex regional pain syndrome. IASP Press, Seattle
100. Khasar SG, Miao JP, Janig W, Levine JD (1998) Modulation of bradykinin-induced mechanical hyperalgesia in the rat by activity in abdominal vagal afferents. Eur J Neurosci 10(2):435–444
101. De Leo JA, Tawfik VL, LaCroix-Fralish ML (2006) The tetrapartite synapse: path to CNS sensitization and chronic pain. Pain 122(1–2):17–21. doi:10.1016/j.pain.2006.02.034
102. Watkins LR, Milligan ED, Maier SF (2001) Spinal cord glia: new players in pain. Pain 93(3):201–205
103. Stuesse SL, Cruce WL, Lovell JA, McBurney DL, Crisp T (2000) Microglial proliferation in the spinal cord of aged rats with a sciatic nerve injury. Neurosci Lett 287(2):121–124
104. McMahon SB, Cafferty WB, Marchand F (2005) Immune and glial cell factors as pain mediators and modulators. Exp Neurol 192(2):444–462. doi:10.1016/j.expneurol.2004.11.001
105. Watkins LR, Maier SF (2002) Beyond neurons: evidence that immune and glial cells contribute to pathological pain states. Physiol Rev 82(4):981–1011. doi:10.1152/physrev.00011.2002

106. Martinez V, Szekely B, Lemarie J, Martin F, Gentili M, Ben Ammar S, Lepeintre JF, Garreau de Loubresse C, Chauvin M, Bouhassira D, Fletcher D (2013) The efficacy of a glial inhibitor, minocycline, for preventing persistent pain after lumbar discectomy: a randomized, double-blind, controlled study. Pain 154(8):1197–1203. doi:10.1016/j.pain.2013.03.028
107. Vanelderen P, Van Zundert J, Kozicz T, Puylaert M, De Vooght P, Mestrum R, Heylen R, Roubos E, Vissers K (2014) Effect of minocycline on lumbar radicular neuropathic pain: a randomized, placebo-controlled. Double-blind clinical trial with amitriptyline as a comparator. Anesthesiology 122:399–406. doi:10.1097/aln.0000000000000508
108. Horvath RJ, DeLeo JA (2009) Morphine enhances microglial migration through modulation of P2X4 receptor signaling. J Neurosci (The Official Journal of the Society for Neuroscience) 29(4):998–1005. doi:10.1523/jneurosci.4595-08.2009
109. Tanga FY, Nutile-McMenemy N, DeLeo JA (2005) The CNS role of Toll-like receptor 4 in innate neuroimmunity and painful neuropathy. Proc Natl Acad Sci U S A 102(16):5856–5861. doi:10.1073/pnas.0501634102
110. Zhao P, Waxman SG, Hains BC (2007) Extracellular signal-regulated kinase-regulated microglia-neuron signaling by prostaglandin E2 contributes to pain after spinal cord injury. J Neurosci (The Official Journal of the Society for Neuroscience) 27(9):2357–2368. doi:10.1523/jneurosci.0138-07.2007
111. Vierck CJ, Yezierski RP (2015) Comparison of operant escape and reflex tests of nociceptive sensitivity. Neurosci Biobehav Rev 51:223–242
112. Mauderli AP, Acosta-Rua A, Vierck CJ (2000) An operant assay of thermal pain in conscious, unrestrained rats. J Neurosci Methods 97(1):19–29
113. Vierck CJ, Acosta-Rua A, Nelligan R, Tester N, Mauderli A (2002) Low dose systemic morphine attenuates operant escape but facilitates innate reflex responses to thermal stimulation. J Pain (Official Journal of the American Pain Society) 3(4):309–319
114. King CD, Devine DP, Vierck CJ, Mauderli A, Yezierski RP (2007) Opioid modulation of reflex versus operant responses following stress in the rat. Neuroscience 147(1):174–182. doi:10.1016/j.neuroscience.2007.04.012
115. King CD, Devine DP, Vierck CJ, Rodgers J, Yezierski RP (2003) Differential effects of stress on escape and reflex responses to nociceptive thermal stimuli in the rat. Brain Res 987(2):214–222
116. Wang S, Davis BM, Zwick M, Waxman SG, Albers KM (2006) Reduced thermal sensitivity and Nav1.8 and TRPV1 channel expression in sensory neurons of aged mice. Neurobiol Aging 27(6):895–903. doi:10.1016/j.neurobiolaging.2005.04.009
117. Chan SH, Lai YY (1982) Effects of aging on pain responses and analgesic efficacy of morphine and clonidine in rats. Exp Neurol 75:112–119
118. Pickering G, Jourdan D, Millecamps M, Chapuy E, Alliot J, Eschalier A (2006) Age-related impact of neuropathic pain on animal behaviour. Eur J Pain 10(8):749–755
119. Iwata K, Fukuoka T, Kondo E, Tsuboi Y, Tashiro A, Noguchi K, Masuda Y, Morimoto T, Kanda K (2002) Plastic changes in nociceptive transmission of the rat spinal cord with advancing age. J Neurophysiol 87(2):1086–1093
120. Kitagawa J, Tsuboi Y, Ogawa A, Ren K, Hitomi S, Saitoh K, Takahashi O, Masuda Y, Harada T, Hanzawa N, Kanda K, Iwata K (2005) Involvement of dorsal column nucleus neurons in nociceptive transmission in aged rats. J Neurophysiol 94(6):4178–4187
121. Helme RD, Meliala A, Gibson SJ (2004) Methodologic factors which contribute to variations in experimental pain threshold reported for older people. Neurosci Lett 361(1–3):144–146. doi:10.1016/j.neulet.2003.12.014
122. Yezierski RP, King CD, Morgan D, Carter CS, Vierck CJ (2010) Effects of age on thermal sensitivity in the rat. J Gerontol A Biol Sci Med Sci 65(4):353–362
123. Zhang RX, Lao L, Qiao JT, Ruda MA (2004) Effects of aging on hyperalgesia and spinal dynorphin expression in rats with peripheral inflammation. Brain Res 999(1):135–141
124. Gagliese L, Melzack R (1999) Age differences in the response to the formalin test in rats. Neurobiol Aging 20(6):699–707

125. Iwata K, Kanda K, Tsuboi Y, Kitajima K, Sumino R (1995) Fos induction in the medullary dorsal horn and C1 segment of the spinal cord by acute inflammation in aged rats. Brain Res 678(1–2):127–139
126. Wang Z, Gardell LR, Ossipov MH, Vanderah TW, Brennan MB, Hochgeschwender U, Hruby VJ, Malan TP Jr, Lai J, Porreca F (2001) Pronociceptive actions of dynorphin maintain chronic neuropathic pain. J Neurosci (The Official Journal of the Society for Neuroscience) 21(5):1779–1786
127. Novak JC, Lovell JA, Stuesse SL, Cruce WL, McBurney DL, Crisp T (1999) Aging and neuropathic pain. Brain Res 833(2):308–310
128. Crisp T, Giles JR, Cruce WL, McBurney DL, Stuesse SL (2003) The effects of aging on thermal hyperalgesia and tactile-evoked allodynia using two models of peripheral mononeuropathy in the rat. Neurosci Lett 339(2):103–106
129. Chung JM, Choi Y, Yoon YW, Na HS (1995) Effects of age on behavioral signs of neuropathic pain in an experimental rat model. Neurosci Lett 183(1–2):54–57
130. Kim YI, Na HS, Yoon YW, Nahm SH, Ko KH, Hong SK (1995) Mechanical allodynia is more strongly manifested in older rats in an experimental model of peripheral neuropathy. Neurosci Lett 199(2):158–160
131. Gatchel RJ, Peng YB, Peters ML, Fuchs PN, Turk DC (2007) The biopsychosocial approach to chronic pain: scientific advances and future directions. Psychol Bull 133(4):581–624. doi:10.1037/0033-2909.133.4.581
132. Inui TS (2003) The need for an integrated biopsychosocial approach to research on successful aging. Ann Intern Med 139(5 Pt 2):391–394
133. Alkema GE, Alley DE (2006) Gerontology's future: an integrative model for disciplinary advancement. Gerontologist 46(5):574–582
134. Hunter DJ, Guermazi A, Roemer F, Zhang Y, Neogi T (2013) Structural correlates of pain in joints with osteoarthritis. Osteoarthritis Cartilage (OARS, Osteoarthritis Research Society) 21(9):1170–1178. doi:10.1016/j.joca.2013.05.017
135. Phyomaung PP, Dubowitz J, Cicuttini FM, Fernando S, Wluka AE, Raaijmaakers P, Wang Y, Urquhart DM (2014) Are depression, anxiety and poor mental health risk factors for knee pain? A systematic review. BMC Musculoskelet Disord 15:10. doi:10.1186/1471-2474-15-10
136. Urquhart DM, Phyomaung PP, Dubowitz J, Fernando S, Wluka AE, Raajmaakers P, Wang Y, Cicuttini FM (2014) Are cognitive and behavioural factors associated with knee pain? A systematic review. Semin Arthritis Rheum 44:445–455. doi:10.1016/j.semarthrit.2014.07.005
137. Cleveland RJ, Luong ML, Knight JB, Schoster B, Renner JB, Jordan JM, Callahan LF (2013) Independent associations of socioeconomic factors with disability and pain in adults with knee osteoarthritis. BMC Musculoskelet Disord 14:297. doi:10.1186/1471-2474-14-297
138. Allen KD (2010) Racial and ethnic disparities in osteoarthritis phenotypes. Curr Opin Rheumatol 22(5):528–532
139. Cruz-Almeida Y, Sibille KT, Goodin BR, Petrov ME, Bartley EJ, Riley JL 3rd, King CD, Glover TL, Sotolongo A, Herbert MS, Schmidt JK, Fessler BJ, Staud R, Redden D, Bradley LA, Fillingim RB (2014) Racial and ethnic differences in older adults with knee osteoarthritis. Arthritis Rheumatol (Hoboken, NJ) 66(7):1800–1810. doi:10.1002/art.38620
140. Cruz-Almeida Y, Fillingim RB (2014) Can quantitative sensory testing move us closer to mechanism-based pain management? Pain Med (Malden, MA) 15(1):61–72. doi:10.1111/pme.12230
141. Lautenbacher S (2012) Experimental approaches in the study of pain in the elderly. Pain Med (Malden, MA) 13(Suppl 2):S44–S50. doi:10.1111/j.1526-4637.2012.01326.x
142. Edwards RR, Fillingim RB (2001) Age-associated differences in responses to noxious stimuli. J Gerontol A Biol Sci Med Sci 56(3):M180–M185
143. Zheng Z, Gibson SJ, Khalil Z, Helme RD, McMeeken JM (2000) Age-related differences in the time course of capsaicin-induced hyperalgesia. Pain 85(1–2):51–58

144. Grashorn W, Sprenger C, Forkmann K, Wrobel N, Bingel U (2013) Age-dependent decline of endogenous pain control: exploring the effect of expectation and depression. PLoS One 8(9):e75629. doi:10.1371/journal.pone.0075629
145. Grill JD, Coghill RC (2002) Transient analgesia evoked by noxious stimulus offset. J Neurophysiol 87(4):2205–2208
146. Naugle KM, Cruz-Almeida Y, Fillingim RB, Riley JL III (2013) Offset analgesia is reduced in older adults. Pain 154(11):2381–2387
147. Riley JL III, King CD, Wong F, Fillingim RB, Mauderli AP (2010) Lack of endogenous modulation and reduced decay of prolonged heat pain in older adults. Pain 150(1):153–160
148. Benedetti F, Vighetti S, Ricco C, Lagna E, Bergamasco B, Pinessi L, Rainero I (1999) Pain threshold and tolerance in Alzheimer's disease. Pain 80(1–2):377–382
149. Carlino E, Benedetti F, Rainero I, Asteggiano G, Cappa G, Tarenzi L, Vighetti S, Pollo A (2010) Pain perception and tolerance in patients with frontotemporal dementia. Pain 151(3):783–789. doi:10.1016/j.pain.2010.09.013
150. Benedetti F, Arduino C, Vighetti S, Asteggiano G, Tarenzi L, Rainero I (2004) Pain reactivity in Alzheimer patients with different degrees of cognitive impairment and brain electrical activity deterioration. Pain 111(1–2):22–29. doi:10.1016/j.pain.2004.05.015
151. Kunz M, Mylius V, Schepelmann K, Lautenbacher S (2009) Effects of age and mild cognitive impairment on the pain response system. Gerontology 55(6):674–682. doi:10.1159/000235719
152. Gibson SJ, Voukelatos X, Ames D, Flicker L, Helme RD (2001) An examination of pain perception and cerebral event-related potentials following carbon dioxide laser stimulation in patients with Alzheimer's disease and age-matched control volunteers. Pain Res Manag (The Journal of the Canadian Pain Society = Journal de la societe canadienne pour le traitement de la douleur) 6(3):126–132
153. Marouf R, Caron S, Lussier M, Bherer L, Piche M, Rainville P (2014) Reduced pain inhibition is associated with reduced cognitive inhibition in healthy aging. Pain 155(3):494–502. doi:10.1016/j.pain.2013.11.011
154. Yarnitsky D, Granot M, Granovsky Y (2014) Pain modulation profile and pain therapy: between pro- and antinociception. Pain 155(4):663–665. doi:10.1016/j.pain.2013.11.005
155. Fjell AM, McEvoy L, Holland D, Dale AM, Walhovd KB, Alzheimer's Disease Neuroimaging Initiative (2014) What is normal in normal aging? Effects of aging, amyloid and Alzheimer's disease on the cerebral cortex and the hippocampus. Prog Neurobiol 117:20–40. doi:10.1016/j.pneurobio.2014.02.004
156. Yap QJ, Teh I, Fusar-Poli P, Sum MY, Kuswanto C, Sim K (2013) Tracking cerebral white matter changes across the lifespan: insights from diffusion tensor imaging studies. J Neural Transm 120(9):1369–1395. doi:10.1007/s00702-013-0971-7
157. Borsook D, Erpelding N, Becerra L (2013) Losses and gains: chronic pain and altered brain morphology. Expert Rev Neurother 13(11):1221–1234. doi:10.1586/14737175.2013.846218
158. Yoon EJ, Kim YK, Shin HI, Lee Y, Kim SE (2013) Cortical and white matter alterations in patients with neuropathic pain after spinal cord injury. Brain Res 1540:64–73. doi:10.1016/j.brainres.2013.10.007
159. Smallwood RF, Laird AR, Ramage AE, Parkinson AL, Lewis J, Clauw DJ, Williams DA, Schmidt-Wilcke T, Farrell MJ, Eickhoff SB, Robin DA (2013) Structural brain anomalies and chronic pain: a quantitative meta-analysis of gray matter volume. J Pain 14(7):663–675
160. Fallon N, Alghamdi J, Chiu Y, Sluming V, Nurmikko T, Stancak A (2013) Structural alterations in brainstem of fibromyalgia syndrome patients correlate with sensitivity to mechanical pressure. Neuroimage Clin 3:163–170. doi:10.1016/j.nicl.2013.07.011
161. Elsenbruch S, Schmid J, Kullmann JS, Kattoor J, Theysohn N, Forsting M, Kotsis V (2014) Visceral sensitivity correlates with decreased regional gray matter volume in healthy volunteers: a voxel-based morphometry study. Pain 155(2):244–249. doi:10.1016/j.pain.2013.09.027
162. Emerson NM, Zeidan F, Lobanov OV, Hadsel MS, Martucci KT, Quevedo AS, Starr CJ, Nahman-Averbuch H, Weissman-Fogel I, Granovsky Y, Yarnitsky D, Coghill RC (2014) Pain

sensitivity is inversely related to regional grey matter density in the brain. Pain 155(3):566–573. doi:10.1016/j.pain.2013.12.004
163. Quiton RL, Roys SR, Zhuo J, Keaser ML, Gullapalli RP, Greenspan JD (2007) Age-related changes in nociceptive processing in the human brain. Ann NY Acad Sci 1097:175–178
164. Tseng MT, Chiang MC, Yazhuo K, Chao CC, Tseng WY, Hsieh ST (2013) Effect of aging on the cerebral processing of thermal pain in the human brain. Pain 154(10):2120–2129. doi:10.1016/j.pain.2013.06.041
165. Cole LJ, Farrell MJ, Gibson SJ, Egan GF (2010) Age-related differences in pain sensitivity and regional brain activity evoked by noxious pressure. Neurobiol Aging 31(3):494–503
166. Villarreal CF, Funez MI, Cunha FQ, Parada CA, Ferreira SH (2013) The long-lasting sensitization of primary afferent nociceptors induced by inflammation involves prostanoid and dopaminergic systems in mice. Pharmacol Biochem Behav 103(3):678–683
167. Xiao WH, Bennett GJ (2008) C-fiber spontaneous discharge evoked by chronic inflammation is suppressed by a long-term infusion of lidocaine yielding nanogram per milliliter plasma levels. Pain 137(1):218–228
168. Gianaros PJ, Marsland AL, Sheu LK, Erickson KI, Verstynen TD (2013) Inflammatory pathways link socioeconomic inequalities to white matter architecture. Cereb Cortex 23(9):2058–2071
169. Rosano C, Marsland AL, Gianaros PJ (2012) Maintaining brain health by monitoring inflammatory processes: a mechanism to promote successful aging. Aging Dis 3(1):16–33
170. Linnman C, Becerra L, Borsook D (2013) Inflaming the brain: CRPS a model disease to understand neuroimmune interactions in chronic pain. J Neuroimmune Pharmacol 8(3):547–563
171. de Goeij M, van Eijk LT, Vanelderen P, Wilder-Smith OH, Vissers KC, van der Hoeven JG, Kox M, Scheffer GJ, Pickkers P (2013) Systemic inflammation decreases pain threshold in humans in vivo. PLoS One 8(12):e84159. doi:10.1371/journal.pone.0084159
172. Benson S, Kattoor J, Wegner A, Hammes F, Reidick D, Grigoleit JS, Engler H, Oberbeck R, Schedlowski M, Elsenbruch S (2012) Acute experimental endotoxemia induces visceral hypersensitivity and altered pain evaluation in healthy humans. Pain 153(4):794–799. doi:10.1016/j.pain.2011.12.001
173. Burns JW, Kubilus A, Bruehl S, Harden RN, Lofland K (2003) Do changes in cognitive factors influence outcome following multidisciplinary treatment for chronic pain? A cross-lagged panel analysis. J Consult Clin Psychol 71(1):81–91
174. Turner JA, Jensen MP, Romano JM (2000) Do beliefs, coping, and catastrophizing independently predict functioning in patients with chronic pain? Pain 85(1–2):115–125
175. Carragee EJ, Alamin TF, Miller JL, Carragee JM (2005) Discographic, MRI and psychosocial determinants of low back pain disability and remission: a prospective study in subjects with benign persistent back pain. Spine J 5(1):24–35
176. Jarvik JG, Hollingworth W, Heagerty PJ, Haynor DR, Boyko EJ, Deyo RA (2005) Three-year incidence of low back pain in an initially asymptomatic cohort: clinical and imaging risk factors. Spine 30(13):1541–1548
177. Barry LC, Gill TM, Kerns RD, Reid MC (2005) Identification of pain-reduction strategies used by community-dwelling older persons. J Gerontol A Biol Sci Med Sci 60(12):1569–1575
178. Mackichan F, Adamson J, Gooberman-Hill R (2013) 'Living within your limits': activity restriction in older people experiencing chronic pain. Age Ageing 42(6):702–708. doi:10.1093/ageing/aft119
179. Gerrits MM, Vogelzangs N, van Oppen P, van Marwijk HW, van der Horst H, Penninx BW (2012) Impact of pain on the course of depressive and anxiety disorders. Pain 153(2):429–436. doi:10.1016/j.pain.2011.11.001
180. Fernandez E, Turk DC (1992) Sensory and affective components of pain: separation and synthesis. Psychol Bull 112(2):205–217

181. de Groot KI, Boeke S, van den Berge HJ, Duivenvoorden HJ, Bonke B, Passchier J (1997) The influence of psychological variables on postoperative anxiety and physical complaints in patients undergoing lumbar surgery. Pain 69(1–2):19–25
182. Pavlin DJ, Rapp SE, Polissar NL, Malmgren JA, Koerschgen M, Keyes H (1998) Factors affecting discharge time in adult outpatients. Anesth Analg 87(4):816–826
183. Crombez G, Eccleston C, Baeyens F, van Houdenhove B, van den Broeck A (1999) Attention to chronic pain is dependent upon pain-related fear. J Psychosom Res 47(5):403–410
184. Vlaeyen JW, Kole-Snijders AM, Boeren RG, van Eek H (1995) Fear of movement/(re)injury in chronic low back pain and its relation to behavioral performance. Pain 62(3):363–372
185. Crombez G, Vlaeyen JW, Heuts PH, Lysens R (1999) Pain-related fear is more disabling than pain itself: evidence on the role of pain-related fear in chronic back pain disability. Pain 80(1–2):329–339
186. Vlaeyen JW, Kole-Snijders AM, Rotteveel AM, Ruesink R, Heuts PH (1995) The role of fear of movement/(re)injury in pain disability. J Occup Rehabil 5(4):235–252. doi:10.1007/bf02109988
187. McCracken LM, Gross RT, Eccleston C (2002) Multimethod assessment of treatment process in chronic low back pain: comparison of reported pain-related anxiety with directly measured physical capacity. Behav Res Therapy 40(5):585–594
188. Blyth FM, Cumming R, Mitchell P, Wang JJ (2007) Pain and falls in older people. Eur J Pain (London, England) 11(5):564–571. doi:10.1016/j.ejpain.2006.08.001
189. Bohl AA, Fishman PA, Ciol MA, Williams B, Logerfo J, Phelan EA (2010) A longitudinal analysis of total 3-year healthcare costs for older adults who experience a fall requiring medical care. J Am Geriatr Soc 58(5):853–860. doi:10.1111/j.1532-5415.2010.02816.x
190. Gibson SJ (2007) IASP global year against pain in older persons: highlighting the current status and future perspectives in geriatric pain. Expert Rev Neurother 7(6):627–635
191. Patel KV, Phelan EA, Leveille SG, Lamb SE, Missikpode C, Wallace RB, Guralnik JM, Turk DC (2014) High prevalence of falls, fear of falling, and impaired balance in older adults with pain in the United States: findings from the 2011 national health and aging trends study. J Am Geriatr Soc 62(10):1844–1852. doi:10.1111/jgs.13072
192. Leveille SG, Bean J, Ngo L, McMullen W, Guralnik JM (2007) The pathway from musculoskeletal pain to mobility difficulty in older disabled women. Pain 128(1–2):69–77. doi:10.1016/j.pain.2006.08.031
193. della Volpe R, Popa T, Ginanneschi F, Spidalieri R, Mazzocchio R, Rossi A (2006) Changes in coordination of postural control during dynamic stance in chronic low back pain patients. Gait Posture 24(3):349–355. doi:10.1016/j.gaitpost.2005.10.009
194. (1987) The prevention of falls in later life. A report of the Kellogg International Work Group on the Prevention of Falls by the Elderly. Dan Med Bull 34(Suppl 4):1–24
195. Howland J, Peterson EW, Levin WC, Fried L, Pordon D, Bak S (1993) Fear of falling among the community-dwelling elderly. J Aging Health 5(2):229–243
196. Campbell AJ, Borrie MJ, Spears GF (1989) Risk factors for falls in a community-based prospective study of people 70 years and older. J Gerontol 44(4):M112–M117
197. Hindmarsh JJ, Estes EH Jr (1989) Falls in older persons. Causes and interventions. Arch Intern Med 149(10):2217–2222
198. Meyer T, Cooper J, Raspe H (2007) Disabling low back pain and depressive symptoms in the community-dwelling elderly: a prospective study. Spine (Phila Pa 1976) 32(21):2380–2386. doi:10.1097/BRS.0b013e3181557955
199. Edwards RR, Cahalan C, Mensing G, Smith M, Haythornthwaite JA (2011) Pain, catastrophizing, and depression in the rheumatic diseases. Nat Rev Rheumatol 7(4):216–224. doi:10.1038/nrrheum.2011.2
200. Williamson GM, Schulz R (1992) Pain, activity restriction, and symptoms of depression among community-residing elderly adults. J Gerontol 47(6):P367–P372
201. Arola HM, Nicholls E, Mallen C, Thomas E (2010) Self-reported pain interference and symptoms of anxiety and depression in community-dwelling older adults: can a temporal

relationship be determined? Eur J Pain (London, England) 14(9):966–971. doi:10.1016/j.ejpain.2010.02.012
202. Sherbourne CD, Stewart AL (1991) The MOS social support survey. Soc Sci Med 32(6):705–714
203. Cardol M, de Jong BA, van den Bos GA, Beelem A, de Groot IJ, de Haan RJ (2002) Beyond disability: perceived participation in people with a chronic disabling condition. Clin Rehabil 16(1):27–35
204. Perruccio AV, Power JD, Badley EM (2007) The relative impact of 13 chronic conditions across three different outcomes. J Epidemiol Community Health 61(12):1056–1061
205. Karp JF, Shega JW, Morone NE, Weiner DK (2008) Advances in understanding the mechanisms and management of persistent pain in older adults. Br J Anaesth 101(1):111–120. doi:10.1093/bja/aen090
206. Gignac MA, Backman CL, Davis AM, Lacaille D, Mattison CA, Montie P, Badley EM (2008) Understanding social role participation: what matters to people with arthritis? J Rheumatol 35(8):1655–1663
207. Machado GP, Gignac MA, Badley EM (2008) Participation restrictions among older adults with osteoarthritis: a mediated model of physical symptoms, activity limitations, and depression. Arthritis Rheum 59(1):129–135. doi:10.1002/art.23259
208. Lanas A, Fernandez A (2007) Inappropriate prevention of NSAID-induced gastrointestinal events among long-term users in the elderly. Drugs Aging 24:121–131
209. Lazar SW, Kerr CE, Wasserman RH, Gray JR, Greve DN, Treadway MT, Fischl B (2005) Meditation experience is associated with increased cortical thickness. Neuroreport 16(17):1893–1897
210. Caughey GE, Roughead EE, Pratt N, Killer G, Gilbert AL (2011) Stroke risk and NSAIDs: an Australian population-based study. Med J Aust 195(9):525–529
211. Solomon DH, Rassen JA, Glynn RJ, Garneau K, Levin R, Lee J, Schneeweiss S (2010) The comparative safety of opioids for nonmalignant pain in older adults. Arch Intern Med 170(22):1979–1986. doi:10.1001/archinternmed.2010.450
212. Huang AR, Mallet L, Rochefort CM, Eguale T, Buckeridge DL, Tamblyn R (2012) Medication-related falls in the elderly: causative factors and preventive strategies. Drugs Aging 29(5):359–376. doi:10.2165/11599460-000000000-00000
213. Shorr RI, Griffin MR, Daugherty JR, Ray WA (1992) Opioid analgesics and the risk of hip fracture in the elderly: codeine and propoxyphene. J Gerontol 47(4):M111–M115
214. Italiano D, Perucca E (2013) Clinical pharmacokinetics of new-generation antiepileptic drugs at the extremes of age: an update. Clin Pharmacokinet 52(8):627–645. doi:10.1007/s40262-013-0067-4
215. Milton JC, Hill-Smith I, Jackson SH (2008) Prescribing for older people. BMJ 336(7644):606–609. doi:10.1136/bmj.39503.424653.80
216. Benoist M (2003) Natural history of the aging spine. Eur Spine J (Official Publication of the European Spine Society, the European Spinal Deformity Society, and the European Section of the Cervical Spine Research Society) 12(Suppl 2):S86–S89. doi:10.1007/s00586-003-0593-0
217. Pergolizzi J, Boger RH, Budd K, Dahan A, Erdine S, Hans G, Kress HG, Langford R, Likar R, Raffa RB, Sacerdote P (2008) Opioids and the management of chronic severe pain in the elderly: consensus statement of an International Expert Panel with focus on the six clinically most often used World Health Organization Step III opioids (buprenorphine, fentanyl, hydromorphone, methadone, morphine, oxycodone). Pain Prac (The Official Journal of World Institute of Pain) 8(4):287–313. doi:10.1111/j.1533-2500.2008.00204.x
218. Kaufman DW, Kelly JP, Rosenberg L, Anderson TE, Mitchell AA (2002) Recent patterns of medication use in the ambulatory adult population of the United States: the Slone survey. JAMA 287(3):337–344

219. Gnjidic D, Blyth FM, Le Couteur DG, Cumming RG, McLachlan AJ, Handelsman DJ, Seibel M, Waite L, Naganathan V (2014) Nonsteroidal anti-inflammatory drugs (NSAIDs) in older people: prescribing patterns according to pain prevalence and adherence to clinical guidelines. Pain 155(9):1814–1820. doi:10.1016/j.pain.2014.06.009
220. Jyrkka J, Enlund H, Lavikainen P, Sulkava R, Hartikainen S (2011) Association of polypharmacy with nutritional status, functional ability and cognitive capacity over a three-year period in an elderly population. Pharmacoepidemiol Drug Saf 20(5):514–522. doi:10.1002/pds.2116
221. American Geriatrics Society Panel on Pharmacological Management of Persistent Pain in Older Persons (2009) Pharmacological management of persistent pain in older persons. J Am Geriatr Soc 57(8):1331–1346. doi:10.1111/j.1532-5415.2009.02376.x
222. Buckeridge D, Huang A, Hanley J, Kelome A, Reidel K, Verma A, Winslade N, Tamblyn R (2010) Risk of injury associated with opioid use in older adults. J Am Geriatr Soc 58(9):1664–1670. doi:10.1111/j.1532-5415.2010.03015.x
223. Day RO, Graham GG (2013) Non-steroidal anti-inflammatory drugs (NSAIDs). BMJ 346:f3195. doi:10.1136/bmj.f3195
224. McLachlan AJ, Bath S, Naganathan V, Hilmer SN, Le Couteur DG, Gibson SJ, Blyth FM (2011) Clinical pharmacology of analgesic medicines in older people: impact of frailty and cognitive impairment. Br J Clin Pharmacol 71(3):351–364. doi:10.1111/j.1365-2125.2010.03847.x
225. Arneric SP, Laird JM, Chappell AS, Kennedy JD (2014) Tailoring chronic pain treatments for the elderly: are we prepared for the challenge? Drug Discov Today 19(1):8–17. doi:10.1016/j.drudis.2013.08.017
226. Gibson SJ (2003) Pain and ageing: a comparison of the pain experience over the adult life span. Prog Pain Res Manage 24:767–790
227. Ljung R, Lu Y, Lagergren J (2011) High concomitant use of interacting drugs and low use of gastroprotective drugs among NSAID users in an unselected elderly population: a nationwide register-based study. Drugs Aging 28(6):469–476. doi:10.2165/11589070-000000000-00000
228. National Institute for Health and Clinical Excellence (2008) Clinical guideline for care and management of osteoarthritis in adults. http://www.nice.org.uk/guidance/CG59, Accessed, 2013.
229. Abdulla A, Adams N, Bone M, Elliott AM, Gaffin J, Jones D, Knaggs R, Martin D, Sampson L, Schofield P (2013) Guidance on the management of pain in older people. Age Ageing 42(Suppl 1):i1–i57. doi:10.1093/ageing/afs200
230. Zhang W, Nuki G, Moskowitz RW, Abramson S, Altman RD, Arden NK, Bierma-Zeinstra S, Brandt KD, Croft P, Doherty M, Dougados M, Hochberg M, Hunter DJ, Kwoh K, Lohmander LS, Tugwell P (2009) OARSI recommendations for the management of hip and knee osteoarthritis: part III: Changes in evidence following systematic cumulative update of research published through January 2009. Osteoarthritis Cartilage (OARS, Osteoarthritis Research Society) 18(4):476–499. doi:10.1016/j.joca.2010.01.013
231. Bayer A, Tadd W (2000) Unjustified exclusion of elderly people from studies submitted to research ethics committee for approval: descriptive study. BMJ 321(7267):992–993
232. Rochon PA, Fortin PR, Dear KB, Minaker KL, Chalmers TC (1993) Reporting of age data in clinical trials of arthritis. Deficiencies and solutions. Arch Intern Med 153(2):243–248
233. Turk DC, Wilson HD, Cahana A (2011) Treatment of chronic non-cancer pain. Lancet 377(9784):2226–2235. doi:10.1016/s0140-6736(11)60402-9
234. Fernandes L, Hagen KB, Bijlsma JW, Andreassen O, Christensen P, Conaghan PG, Doherty M, Geenen R, Hammond A, Kjeken I, Lohmander LS, Lund H, Mallen CD, Nava T, Oliver S, Pavelka K, Pitsillidou I, da Silva JA, de la Torre J, Zanoli G, Vliet Vlieland TP, European League Against Rheumatism (2013) EULAR recommendations for the non-pharmacological core management of hip and knee osteoarthritis. Ann Rheum Dis 72(7):1125–1135. doi:10.1136/annrheumdis-2012-202745
235. van Middelkoop M, Rubinstein SM, Verhagen AP, Ostelo RW, Koes BW, van Tulder MW (2010) Exercise therapy for chronic nonspecific low-back pain. Best Pract Res Clin Rheumatol 24(2):193–204. doi:10.1016/j.berh.2010.01.002
236. Zhang W, Moskowitz RW, Nuki G, Abramson S, Altman RD, Arden N, Bierma-Zeinstra S, Brandt KD, Croft P, Doherty M, Dougados M, Hochberg M, Hunter DJ, Kwoh K, Lohmander

LS, Tugwell P (2008) OARSI recommendations for the management of hip and knee osteoarthritis, Part II: OARSI evidence-based, expert consensus guidelines. Osteoarthritis Cartilage (OARS, Osteoarthritis Research Society) 16(2):137–162. doi:10.1016/j.joca.2007.12.013
237. Muir SW, Berg K, Chesworth B, Klar N, Speechley M (2010) Quantifying the magnitude of risk for balance impairment on falls in community-dwelling older adults: a systematic review and meta-analysis. J Clin Epidemiol 63(4):389–406. doi:10.1016/j.jclinepi.2009.06.010
238. Logghe IH, Verhagen AP, Rademaker AC, Bierma-Zeinstra SM, van Rossum E, Faber MJ, Koes BW (2010) The effects of Tai Chi on fall prevention, fear of falling and balance in older people: a meta-analysis. Prev Med 51(3–4):222–227. doi:10.1016/j.ypmed.2010.06.003
239. Ruhe A, Fejer R, Walker B (2011) Center of pressure excursion as a measure of balance performance in patients with non-specific low back pain compared to healthy controls: a systematic review of the literature. Eur Spine J (Official Publication of the European Spine Society, the European Spinal Deformity Society, and the European Section of the Cervical Spine Research Society) 20(3):358–368. doi:10.1007/s00586-010-1543-2
240. Belza B, Snyder S, Thompson M, Logefro J (2010) From research to practice: EnhanceFitness, an innovative community-based senior exercise program. Top Geriatr Rehabil 26:299–309
241. Shumway-Cook A, Silver IF, LeMier M, York S, Cummings P, Koepsell TD (2007) Effectiveness of a community-based multifactorial intervention on falls and fall risk factors in community-living older adults: a randomized, controlled trial. J Gerontol A Biol Sci Med Sci 62(12):1420–1427
242. Wallace JI, Buchner DM, Grothaus L, Leveille S, Tyll L, LaCroix AZ, Wagner EH (1998) Implementation and effectiveness of a community-based health promotion program for older adults. J Gerontol A Biol Sci Med Sci 53(4):M301–M306
243. Howe TE, Rochester L, Neil F, Skelton DA, Ballinger C (2011) Exercise for improving balance in older people. Cochrane Database Syst Rev 11:CD004963. doi:10.1002/14651858.CD004963.pub3
244. Zijlstra GA, van Haastregt JC, Ambergen T, van Rossum E, van Eijk JT, Tennstedt SL, Kempen GI (2009) Effects of a multicomponent cognitive behavioral group intervention on fear of falling and activity avoidance in community-dwelling older adults: results of a randomized controlled trial. J Am Geriatr Soc 57(11):2020–2028. doi:10.1111/j.1532-5415.2009.02489.x
245. Tsai PF, Chang JY, Beck C, Kuo YF, Keefe FJ (2013) A pilot cluster-randomized trial of a 20-week Tai Chi program in elders with cognitive impairment and osteoarthritic knee: effects on pain and other health outcomes. J Pain Symptom Manage 45(4):660–669. doi:10.1016/j.jpainsymman.2012.04.009
246. Hall AM, Maher CG, Lam P, Ferreira M, Latimer J (2011) Tai chi exercise for treatment of pain and disability in people with persistent low back pain: a randomized controlled trial. Arthritis Care Res (Hoboken) 63(11):1576–1583. doi:10.1002/acr.20594
247. Peng PW (2012) Tai chi and chronic pain. Reg Anesth Pain Med 37(4):372–382. doi:10.1097/AAP.0b013e31824f6629
248. Low S, Ang LW, Goh KS, Chew SK (2009) A systematic review of the effectiveness of Tai Chi on fall reduction among the elderly. Arch Gerontol Geriatr 48(3):325–331. doi:10.1016/j.archger.2008.02.018
249. Huang TT, Yang LH, Liu CY (2011) Reducing the fear of falling among community-dwelling elderly adults through cognitive-behavioural strategies and intense Tai Chi exercise: a randomized controlled trial. J Adv Nurs 67(5):961–971. doi:10.1111/j.1365-2648.2010.05553.x
250. Blyth FM, March LM, Nicholas MK, Cousins MJ (2005) Self-management of chronic pain: a population-based study. Pain 113(3):285–292. doi:10.1016/j.pain.2004.12.004
251. Williams AC, Eccleston C, Morley S (2012) Psychological therapies for the management of chronic pain (excluding headache) in adults. Cochrane Database Syst Rev 11:CD007407. doi:10.1002/14651858.CD007407.pub3
252. Skinner MS, Wilson HD, Turk DC (2012) Cognitive-behavioral perspective and cognitive-behavioral therapy for people with chronic pain – distinctions, outcomes, & innovations. J Cogn Psychother 26:93–113

253. Abernethy AP, Currow D (2006) A framework for managing comorbid conditions in palliative care. J Palliat Med 9:620–622
254. Dixon KE, Keefe FJ, Scipio CD, Perri LM, Abernethy AP (2007) Psychological interventions for arthritis pain management in adults: a meta-analysis. Health Psychol 26(3):241–250
255. Hoffman BM, Papas RK, Chatkoff DK, Kerns RD (2007) Meta-analysis of psychological interventions for chronic low back pain. Health Psychol 26(1):1–9
256. Keefe FJ, Porter L, Somers T, Shelby R, Wren AV (2013) Psychosocial interventions for managing pain in older adults: outcomes and clinical implications. Br J Anaesth 111(1):89–94. doi:10.1093/bja/aet129
257. Delbaere K, Crombez G, Vanderstraeten G, Willems T, Cambier D (2004) Fear-related avoidance of activities, falls and physical frailty. A prospective community-based cohort study. Age Ageing 33(4):368–373. doi:10.1093/ageing/afh106
258. Dorresteijn TA, Zijlstra GA, Delbaere K, van Rossum E, Vlaeyen JW, Kempen GI (2011) Evaluating an in-home multicomponent cognitive behavioural programme to manage concerns about falls and associated activity avoidance in frail community-dwelling older people: design of a randomised control trial [NCT01358032]. BMC Health Serv Res 11:228. doi:10.1186/1472-6963-11-228
259. Kerns RD, Otis JD, Marcus KS (2001) Cognitive-behavioral therapy for chronic pain in the elderly. Clin Geriatr Med 17(3):503–522
260. Reid MC, Otis J, Barry LC, Kerns RD (2003) Cognitive-behavioral therapy for chronic low back pain in older persons: a preliminary study. Pain Med (Malden, MA) 4(3):223–230
261. Apkarian AV (2011) The brain in chronic pain: clinical implications. Pain Manag 1(6):577–586. doi:10.2217/pmt.11.53
262. Seminowicz DA, Davis KD (2006) Cortical responses to pain in healthy individuals depends on pain catastrophizing. Pain 120(3):297–306. doi:10.1016/j.pain.2005.11.008
263. Jensen MP (2010) A neuropsychological model of pain: research and clinical implications. J Pain (Official Journal of the American Pain Society) 11(1):2–12. doi:10.1016/j.jpain.2009.05.001
264. Beauregard M (2007) Mind does really matter: evidence from neuroimaging studies of emotional self-regulation, psychotherapy, and placebo effect. Prog Neurobiol 81(4):218–236. doi:10.1016/j.pneurobio.2007.01.005
265. Bryant RA, Felmingham K, Kemp A, Das P, Hughes G, Peduto A, Williams L (2008) Amygdala and ventral anterior cingulate activation predicts treatment response to cognitive behaviour therapy for post-traumatic stress disorder. Psychol Med 38(4):555–561. doi:10.1017/s0033291707002231
266. Felmingham K, Kemp A, Williams L, Das P, Hughes G, Peduto A, Bryant R (2007) Changes in anterior cingulate and amygdala after cognitive behavior therapy of posttraumatic stress disorder. Psychol Sci 18(2):127–129. doi:10.1111/j.1467-9280.2007.01860.x
267. Frewen PA, Dozois DJ, Lanius RA (2008) Neuroimaging studies of psychological interventions for mood and anxiety disorders: empirical and methodological review. Clin Psychol Rev 28(2):228–246. doi:10.1016/j.cpr.2007.05.002
268. Jensen KB, Berna C, Loggia ML, Wasan AD, Edwards RR, Gollub RL (2012) The use of functional neuroimaging to evaluate psychological and other non-pharmacological treatments for clinical pain. Neurosci Lett 520(2):156–164. doi:10.1016/j.neulet.2012.03.010
269. Seminowicz DA, Shpaner M, Keaser ML, Krauthamer GM, Mantegna J, Dumas JA, Newhouse PA, Filippi CG, Keefe FJ, Naylor MR (2013) Cognitive-behavioral therapy increases prefrontal cortex gray matter in patients with chronic pain. J Pain (Official Journal of the American Pain Society) 14(12):1573–1584. doi:10.1016/j.jpain.2013.07.020
270. Soumerai SB, Pierre-Jacques M, Zhang F, Ross-Degnan D, Adams AS, Gurwitz J, Adler G, Safran DG (2006) Cost-related medication nonadherence among elderly and disabled medicare beneficiaries: a national survey 1 year before the medicare drug benefit. Arch Intern Med 166(17):1829–1835. doi:10.1001/archinte.166.17.1829

The Way Forward: Translation

James L. Kirkland and Tamar Tchkonia

Contents

J.L. Kirkland, M.D., Ph.D. (✉)
Aging Research, Robert and Arlene Kogod Center on Aging, Mayo Clinic, 200 First Street S.W., Rochester, MN 55905, USA
e-mail: Kirkland.James@Mayo.edu

T. Tchkonia, Ph.D.
Department of Medicine, Robert and Arlene Kogod Center on Aging, Mayo Clinic, 200 First Street S.W., Rochester, MN 55905, USA
e-mail: Tchkonia.Tamar@mayo.edu

F. Sierra, R. Kohanski (eds.), *Advances in Geroscience*,
DOI 10.1007/978-3-319-23246-1_19

1 Introduction

Major age-related chronic diseases are the leading drivers of morbidity, loss of independence, hospitalization, mortality, and health costs throughout most of the world. These include atherosclerosis, most cancers, mild cognitive impairment, dementias, Parkinson's and other neurodegenerative diseases, type 2 diabetes, renal dysfunction, arthritis, blindness, frailty, chronic obstructive pulmonary disease, sarcopenia, and many others [1–4], as discussed throughout this book. For each of these conditions, chronological aging is a major risk factor and for most, aging leads all other known predictors combined. Numbers of chronic disorders *per individual* increase with aging, associated with loss of independence, frailty, and increased risk of death. Although more epidemiological research is needed to be certain about this, it seems that while many elderly individuals are healthy, those who are not have multiple comorbidities that often begin around the same time. If true, this suggests there are shared pathogenic mechanisms. Also as discussed in the preceding chapters, the major age-related disorders often share the disturbances in tissue, cellular, and molecular function that occur with chronological aging. These include chronic "sterile" inflammation, cellular senescence, macromolecular changes (DNA, proteins, carbohydrates, and lipids), and stem and progenitor cell dysfunction.

Based on these points, the "geroscience hypothesis" has been proposed: by targeting fundamental aging processes, it may be possible to alleviate the major age-related chronic disorders as a group, instead of one at a time. Therefore, as summated in the previous chapters of this book, targeting fundamental aging mechanisms may be a way to delay, prevent, alleviate, or treat chronic disorders as a group, instead of one at a time. Even if a single major chronic disease such as atherosclerosis were eradicated, as transformative as such an advance would be, it would only add 2–4 years to life expectancy, only to be followed by another potentially fatal age-related chronic disease [5, 6]. However, targeting the intersection between fundamental aging mechanisms and processes that lead to chronic diseases could alleviate multiple age-related disorders and extend healthspan.

2 Is Aging a Modifiable Risk Factor?

In the first chapter of this book, Austad argues that nature has achieved changes in longevity multiple times and seemingly by independent, distinct mechanisms. Nevertheless, and although aging is the leading predictor for chronic diseases and disabilities, it has only recently become viewed as a potentially modifiable risk factor. Supporting the contention that aging can be modified in several species in the laboratory (*i.e.*, independently from evolution) are the findings that: (1) maximum lifespan is extended and age-related diseases can be delayed in experimental animals by a number of single gene mutations [7], suggesting that pathways impacted by these mutations could be therapeutic targets. (2) Humans who live beyond age 100 can have delayed onset of clinically overt age-related diseases, such as Alzheimer's [8], contributing to compression of morbidity and enhancing healthspan. (3) As discussed below, a number of agents, including rapamycin, metformin, acarbose, 17α-estradiol, angiotensin converting enzyme inhibitors (ACEi), senolytic agents (a new class of drugs that selectively eliminate senescent cells [9]), flavonoids, and others increase healthspan and/or lifespan in experimental animals [3, 10–13]. For example, rapamycin appears to delay age-related cognitive decline and cancers [14]. A pipeline is developing of yet more agents that show promise for enhancing lifespan and perhaps healthspan in experimental animals that have not yet been published. In general, these agents alleviate inflammation, cellular senescence, metabolic dysfunction (in some cases by targeting pathways related to caloric restriction), macromolecular damage or processing, or stem cell and progenitor dysfunction. (4) Caloric restriction, which can increase maximum lifespan in some experimental animals, is associated with delayed onset of multiple chronic diseases [15]. (5) Factors in the circulation of young animals alleviate muscle, cardiac, brain, and potentially other forms of dysfunction in older animals [16, 17]. (6) Senescent cell accumulation is associated with many age-related chronic diseases and frailty [9, 18], and genetic or pharmacological senescent cell elimination enhances healthspan and delays age-related dysfunction, at least in mice [19, 13].

Since interventions that increase lifespan and healthspan in mammals exist, it might be possible to circumvent an issue that has made studying the pathogenesis of many of these diseases in humans difficult: many of these chronic diseases, such as Alzheimer's or atherosclerosis, occur only in humans or a very limited number of species. Furthermore, many of them become manifest clinically once the disease is advanced at the molecular and cellular levels. These issues make delineation of initiating mechanisms difficult because of the impracticality of obtaining tissue samples for analysis sufficiently early during disease development in humans. By targeting upstream, fundamental aging processes that predispose to these human diseases, these difficulties could be circumvented.

3 Do We Have Interventions That May Work?

Recent, important advances have been made in our understanding of the basic biology of aging. The field has moved from an era of description to hypothesis-driven research with a focus on elucidating mechanisms and, most recently, into developing interventions that target fundamental aging processes. Modulators and interventions that delay age-related changes in experimental animals include caloric restriction, several hundred single gene mutations across species, and, most recently, drugs. Among the alterations that extend lifespan or healthspan are mutations in the growth hormone (GH)/ insulin-like growth factor-1 (IGF-1)/ insulin signaling pathway and pathways related to protein, carbohydrate, or lipid metabolism, caloric restriction, inflammation, and the renin-angiotensin system. The NIA-funded

Interventions Testing Program (ITP) has been particularly effective in identifying drugs that extend lifespan in mice. Several are beginning to show promise in extending healthspan and delaying age-related chronic diseases as well. In general, these genetic and pharmacologic interventions are related to inflammation, cell-survival, cellular senescence, macromolecular processing, fuel sensing and processing, stem and progenitor cell function, and integrity of the stem cell niche. Agents resulting from research conducted through the ITP that might be among the first to be translated into humans are considered below.

3.1 *Rapamycin and Rapalogs*

Rapamycin is an immunosuppressant approved by the FDA-for solid organ transplantation, while derivative "rapalogs" are approved for treating particular cancers [20]. Rapamycin inhibits the eponymous mechanistic target of rapamycin (mTOR) kinase, which comprises two distinct protein complexes. The mTOR Complex 1 (mTORC1) is an integrative node for cellular energy signaling, activated by glucose and amino acids, IGF-1, insulin, and other growth signals. Accordingly, mTOR inhibition recapitulates some of the effects of caloric restriction. Rapamycin extends lifespan in yeast, flies, and worms, as well as in C57BL/6, 129/Sv, and genetically heterogeneous mice, even when started after mid-life, as found by the ITP [10, 21]. A net effect of mTOR inhibition by rapamycin is an increase in protein quality. Rapamycin has a number of effects in mice that might explain how it extends longevity, including modulating stem cell function and inflammation, promoting autophagy and alleviating the pro-inflammatory senescence-associated secretory phenotype (SASP), thus resulting in improvements in cognitive decline, heart failure, and neurodegeneration [22]. It has side effects, including increased rates of kyphosis and cataracts in rodents and metabolic dysfunction, impaired wound healing, and aphthous ulcers in humans [22, 23]. Despite these challenges, rapamycin and rapalogs are currently being considered as an adjuvant to cardiac rehabilitation, to reduce cognitive impairment in Alzheimer's disease [24], and to enhance influenza vaccine responses in the elderly [25].

3.2 *Metformin*

Metformin is an approved first-line drug for type 2 diabetes mellitus (T2DM). It has been used for over 60 years with an excellent safety record. Metformin can prevent the progression of impaired glucose tolerance to overt diabetes in overweight subjects [26], including overweight subjects older than 60 [27]. Metformin has been shown to be associated with increased longevity in rodents [28–31] as well as nematodes [32], suggesting evolutionarily conserved mechanisms.

Metformin reduces oxidative stress and inflammation, with prolongation of both lifespan and healthspan in mice [11]. The mechanism of action in treating T2DM includes decreased hepatic glucose production through several mechanisms [33, 34] and improved insulin action, leading to lower glucose and insulin levels. Metformin also decreases IGF-1 signaling and inhibits the pro-inflammatory SASP [35]. Although the precise molecular target of metformin is not known with certainty, it inhibits mitochondrial complex 1, which is associated with activation of AMP-activated protein kinase (AMPK) and inhibition of mTOR. In the United Kingdom Prospective Diabetes Study and other studies, metformin was associated with a decreased risk of cardiovascular disease in human subjects compared to other anti-diabetes drugs [36–41]. In vitro studies have indicated that metformin attenuates tumorigenesis [42–48], and epidemiologic studies have suggested an association between metformin use and decreased risk of cancer and cancer mortality [49–53]. The potential protective effect of metformin against cancer is being studied intensively, with over 100 studies registered at the Clinical Trials.gov website. Data about effects of metformin on dementia are emerging, but controversial [54, 55]. Importantly, a recent observational study indicated that metformin treatment of diabetics was associated with a 15 % increase in overall survival of subjects in their 70's compared with matched control subjects without diabetes [56].

3.3 Acarbose

Acarbose has been used for treating T2DM for over 15 years. It slows processing of starch into disaccharides by inhibiting intestinal α-glucosidases in the intestine, thus reducing peaks in glucose absorption [57–59]. Acarbose has an excellent safety record, although it frequently leads to minor gastrointestinal side effects, especially on North American diets. It was tested in the ITP based on the notion that post-prandial glucose spikes might contribute to aging [60]. Acarbose increased median lifespan by 22 % in male mice, but by only 5 % in females [12]. Conversely, acarbose increased maximum lifespan by around 10 % in both sexes. Despite increased food intake, acarbose led to decreased body weight and glucose, insulin, and IGF-1 levels [12, 61]. In humans, acarbose can prevent progression of impaired glucose tolerance to T2DM [62]. It was associated with a significant decline in the risk for cardiovascular events, including myocardial infarction [63]. Acarbose was also associated with reduced incidence of hypertension and silent myocardial infarction [64]. Glucose-lowering efforts using classes of drugs (other than acarbose or metformin) in attempts to prevent cardiovascular effects associated with T2DM have failed. Thus, acarbose may impact aging processes through mechanisms other than reduced postprandial glucose spikes, perhaps including effects on the intestinal microbiome, by inducing intestinal cells to release protective peptides, or systemic absorption, perhaps leading to direct effects of acarbose on cellular function.

3.4 17-α Estradiol

17-α-estradiol is a non-feminizing endogenous estrogen that has lower affinity for estrogen receptors than its feminizing and more widely studied enantiomer, 17-β-estradiol. For many years, it has been given to post-menopausal women as a component of Premarin [65].17-α-estradiol has neuroprotective properties in animal models of Parkinson's disease through its anti-oxidative effects [66]. These effects do not seem to be mediated through estrogen receptors [67, 68]. 17-α-estradiol administration to mice in the ITP caused significant extension (12 %) in median lifespan in male mice, but not in females, with wide variations in findings among the three ITP test sites [12]. 17-α-estradiol appears to be safe in humans [66], and it is approved for topical use in Europe for treating alopecia, with few reports of side effects [69].

3.5 Growth and Differentiation Factor Analogs

Stem cell and progenitor pools can become depleted or lose functionality with aging. Cell autonomous and non-autonomous changes can occur during aging that restrict cellular replicative potential, interfering with repair or regeneration following injury or disease [70]. Non-cell autonomous changes in the stem cell niche or microenvironment can contribute to declines in adult stem cell recruitment. The chronic, low grade, sterile (non-microbial) inflammation associated with aging may result in a toxic microenvironment that leads to stem cell or progenitor dysfunction, as can dysregulated crosstalk among organ systems, for example between adipose tissue and bone [71]. Findings from parabiosis experiments, in which old and young mice are joined surgically so that they share circulations for several weeks or months, implicate age-related changes of the progenitor cell microenvironment in the observed age-related decrease of tissue repair capacity [16]. For example, following skeletal muscle injury in the older animal, circulating factors from the young animal in the parabiotic pair lead to faster repair by the older animal's progenitors than occurs in parabiotic pairs of old-with-old animals. Conversely, factors from old animals induce dysfunction and impede neurogenesis in young animals, compared to parabiotic pairs of young animals cross-circulated with young animals. Thus, tissue environmental influences, potentially including inflammation, SASP products, or circulating or paracrine factors, appear to contribute to age-related adult stem cell dysfunction, suggesting that aging progenitor cells may have at least a degree of preserved inherent function, which is suppressed by the aging environment. Provision of circulating factors from young animals to old animals, including growth and differentiation factor (GDF)-11 and oxytocin, enhances progenitor potential and repair capacity of the brain, heart, and muscle of old mice [72, 73], suggesting that pharmaceutical agents based on GDF-related factors or oxytocin might enhance regeneration in elderly humans.

3.6 Senolytics

Cellular senescence refers to the essentially irreversible cell cycle arrest induced by oncogenic and metabolic insults that appears to have evolved as a defense against tumor formation or to facilitate wound healing [18]. Senescent cells can develop a senescence-associated secretory phenotype (SASP), with release of pro-inflammatory cytokines, chemokines, thrombotic factors, extracellular matrix proteases and proteins, and growth factors [74, 75]. The pathogenic mechanisms promoted by senescence at the tissue level include inflammation, loss of functional progenitor cells, clotting, extracellular matrix dysfunction, and altered tissue architecture. Senescent cells accumulate in multiple tissues with aging [9, 18, 76–78].

Senescent cell burden is associated with lifespan. At 18 months of age, there are fewer senescent cells in fat tissue of long-lived Ames dwarf, Snell dwarf, and growth hormone receptor knockout mice than age-matched control animals, while short-lived growth hormone over-expressing mice have increased senescent cell burden [79]. Caloric restriction sufficient to increase lifespan is associated with decreased expression of p16^{INK4A}, a cell senescence marker, in multiple tissues of mice compared to ad libitum-fed controls [80]. Conversely, senescent cells accumulate in fat and other tissues in obesity, particularly when associated with diabetes [81, 82]. Senescent cell burden is also increased in several types of progeroid mice [13, 18, 83–85]. In longer- compared with shorter-lived mouse cohorts, senescent cell burden predicts mean and maximum lifespan [86].

Cellular senescence contributes to age-related dysfunction and is frequently evident at the sites of pathology that underlie chronic, age-related diseases, including atherosclerosis, hypertension, dementias, other neurodegenerative diseases, cancers, arthritis, osteoporosis, chronic obstructive pulmonary disease, renal dysfunction, adverse effects of chemotherapy and radiation, diabetes, and many others [4, 9, 18, 78]. The senescent cells at sites of pathology in these conditions might have systemic effects through the SASP that predispose to other pathologies. Genetically eliminating senescent cells by activating a drug-inducible "suicide" gene only in senescent cells enhances healthspan, at least in progeroid mice [19]. Genetically targeting senescent cells led to partial reversal of age-related lipodystrophy and delayed progression of frailty, sarcopenia, and cataracts [18, 19]. These findings support a link between senescent cell burden and age-related dysfunction, raising interest in developing drugs that eliminate senescent cells – *senolytics* [9]. These drugs selectively eliminate senescent cells without clearing normal cells [13]. They do so without interfering with the mechanisms that permit generation of new senescent cells when they are needed, for example as a defense against cancer or for wound healing [87]. The first senolytic agents discovered act by interfering with the pro-survival pathways that confer resistance to apoptosis to senescent cells [13]. These agents enhance cardiac and carotid vascular function in old mice, reverse gait impairment due to radiation of a leg in younger mice, and delay development of frailty, neurological dysfunction, and osteoporosis in progeroid mice. Alleviation of senescence-related dysfunction by senolytic agents is sustained for many months even after a single oral dose. Another strategy for reducing adverse effects of senescent cells would be to target components of the SASP – *SASP protectors* [9].

For senolytic agents, SASP protectors, or other drugs that target fundamental aging mechanisms to be translated into application in humans: (1) Better animal models of human age-related diseases need to be developed. (2) Models have to be developed in which it can be tested whether any beneficial effects of candidate senolytic drugs are caused directly by clearing senescent cells or targeting the SASP, rather than "off-target" effects of the drugs on non-senescent cells. One such model has been described: gait is improved for 7 months after a single dose of senolytic drugs in mice. These agents have a half-life of <12 h. The classes of drugs that have sustained effects after a single dose modify cellular or extracellular matrix composition of tissues, and include antibiotics, chemotherapy agents, or teratogens. Few, if any, other examples exist of drugs that have sustained effects after the drug is cleared from the system. Thus, the sustained effects of senolytic drugs are consistent with senescent cell ablation. (3) Models are needed to test for possible side effects of senolytic agents. Even though continuous clearance of senescent cells from genetically-modified mice did not lead to any overt side effects in over a year of observation [19], it has been shown that cellular senescence has beneficial effects under some circumstances. Indeed, cellular senescence protects against cancer development, helps to resolve tissue fibrosis during healing, is involved in immune responses, promotes skin wound resolution, and can contribute to tissue remodeling [18, 87–89].

There are potential important advantages of senolytics over other treatment approaches. It may be feasible to administer these agents intermittently, perhaps for a day or two every few months or once a year during periods of good health. Furthermore, unlike microbes or cancer cells, senescent cells do not divide, suggesting that acquired drug resistance to senolytics is unlikely to occur. Finally, it has been shown that removing only a fraction of senescent cells can have beneficial effects. Indeed, while only 30 % of senescent cells were removed from *INK-ATTAC;BubR1*$^{H/H}$ mice treated with AP20187 to activate the "suicide" gene expressed in their senescent cells, healthspan was enhanced [19]. Similar dramatic improvements in age- and senescence-related dysfunction were found in mice treated with senolytic agents that remove from 20 to 70 %, but not all senescent cells from tissues [13]. Thus, senolytic agents hold promise for treating age-related diseases and dysfunction.

3.7 Angiotensin Converting Enzyme Inhibitors and Angiotensin II Receptor Blockers

Angiotensin converting enzyme inhibitors (ACEi) and angiotensin II receptor blockers (ARB's) are widely used antihypertensive drugs. Independently from their antihypertensive effects, ACEi and ARB's reduce mortality due to heart disease and protect the kidneys in subjects with diabetes [90]. ARB's appear to be associated with reduced cancer risk and mortality [91, 92] and decrease incidence and progression of dementia [93–95]. Genetic disruption of angiotensin II receptor expression prolongs mouse lifespan and variations in the angiotensin II receptor gene are associated with exceptional human longevity [96, 97]. ACEi increase lifespan in healthy as well as hypertensive rodents [98–101]. ACEi and ARB's increase lifespan and improve cardiovascular function in rats [102], as well as delaying age-related decreases in renal and cognitive function [103]. Thus, ACEi and ARB's alleviate a range of age-related disorders and phenotypes in humans and rodents, as well as increasing lifespan in rodents. How they impact molecular pathways tied to fundamental aging mechanisms remains to be elucidated.

3.8 Sirtuin Activators and Flavonoids

Sirtuins are mammalian NAD-dependent deacetylases related to Sir2, one of the first longevity genes discovered in yeast. Genetically enhancing the activity of Sir2 or its homologs increases yeast and possibly fruit fly and round worm lifespan [104–107]. There are seven mammalian Sir2 homologs. Overexpression of SIRT6 increases longevity in male mice and protects against the deleterious effects of high-fat feeding, possibly due to repression of IGF signaling [108]. Overexpressing other mammalian sirtuins has not been found to increase lifespan, but does have beneficial effects in physiology and a number of age-related diseases [109]. Mice overexpressing SIRT1 have reduced incidence of cancer and osteoporosis and enhanced wound healing and glucose tolerance [110]. Possibly, sirtuin activating compounds may phenocopy some of these effects. Resveratrol, a plant flavonoid that leads to activation of sirtuin-related pathways, extends yeast and fly but not mouse lifespan [111–113]. Newer sirtuin activators, including SRT1720 and SRT2104, extend healthspan and lifespan in non-obese mice [114, 115]. Sirtuin activators alleviate the effects of high-calorie feeding in mice, rhesus monkeys on a high calorie diet, and possibly humans [116–120]. However, it is unclear if sirtuin activators improve glucose tolerance in humans [121, 122]. Flavonoids related to resveratrol may eventually prove to be effective in targeting certain age-related disorders.

Flavonoids and pathways impacted by them are involved in cellular senescence. Interfering with sirtuin signaling by targeting the deleted breast cancer-1 gene product protects against generation of senescent cells in obesity [123]. Quercetin, a flavonoid, is senolytic, particularly in combination with dasatinib, a tyrosine kinase inhibitor [13].

3.9 Aspirin and Salicylic Acid

Aspirin is an analgesic, anti-inflammatory agent and an anti-platelet thrombosis inhibitor. Aspirin appears to reduce cancer risk [124–126], and it was found in the Intervention Testing Program to increase median but not maximum lifespan in mice

[127]. Many of its clinical effects are due to irreversible acetylation of serine residues in the active site of cyclooxygenases. Salicylic acid, which is a metabolite of aspirin as well as the related anti-inflammatory drug, salsalate, is an allosteric activator of AMPK [128]. Salicylic acid increases lipid utilization in mice through AMPK activation and also enhances glucose homeostasis through an independent, as yet uncharacterized mechanism [128]. Salsalate improves glucose control in diabetics [129]. Aspirin activates AMPK in human colorectal cancer cells [130]. It enhances lifespan though activation of AMPK and Daf-16/Foxo3 in worms [131]. These results from human and experimental animal studies indicate that aspirin and related drugs may affect fundamental aging processes and age-related diseases through several mechanisms, including inflammation-related and nutrient processing-related pathways.

3.10 Others

At least a dozen more agents that target basic aging mechanisms are being developed but are as yet unpublished. New rapalogs (analogs of rapamycin) are being developed that have lower side effect profiles than rapamycin, including less glucose intolerance and gastrointestinal irritation [132, 133]. Drugs affecting mitochondrial function, drugs that target protein synthesis or enhance autophagy, and caloric restriction mimetics are among compounds that are currently under development. No doubt, many other interventions are being developed or will be devised in the near future that will increase life- and/or healthspan in experimental animals and potentially humans. Some of the agents under development are already used clinically for other conditions. Others are new chemical entities and new classes of compounds. Pre-clinical and proof of concept clinical trials paradigms need to be developed to identify those compounds emerging from the expanding basic biology of aging pipeline that merit continued translational effort.

3.11 Combinations

Agents that target diverse aging pathways are being combined to test if their effects are synergistic in experimental animals. New combinations will probably be devised in the near future. Using drug combinations follows the path used for developing clinical interventions in other emerging fields, such as for treatment-resistant cancers or viral diseases. In these cases, combinations are often used initially in order to enhance treatment effectiveness, reduce doses of each individual drug, and thereby minimize side effect severity. As more becomes known about interventions that target fundamental aging processes, next generation single drugs that are more precisely targeted could supplant combinations of earlier generation drugs. Alternatively, combination therapy involving multiple drugs with non-overlapping mechanisms of action has remained a mainstay for treating many common diseases, including coronary artery disease, chronic obstructive pulmonary disease, tuberculosis, and most cancers, and could remain the strategy of choice for targeting fundamental aging processes and age-related chronic diseases.

4 Translation

The promise of interventions that target fundamental aging mechanisms is that they could have a huge impact on important outcomes: independence, function, quality of life, and freedom from pain and disability through alleviating many different age-related diseases, conditions, and syndromes as a group. Data from preclinical experimental animal studies and limited human studies of the interventions described above suggest they might be effective in slowing age-related decline in multiple organs, tissues, and functions, potentially shifting the curve of aging phenotypes and chronic disease predisposition to the right. If clinical trials of these interventions broadly affect multiple age-related disorders, the implications for health care and society would be enormous [2, 9].

The first step in this path will be one or several focused, small, short term clinical trials with several objectives. Ideally, each of these trials would test proof of a principle in a specific, but generalizable group of subjects. They would provide information needed to design and scale-up for larger trials. They would also provide biological data that could spur reverse translational studies in preclinical animal models to dissect mechanisms and determine next steps in translating each aging intervention into humans. We consider here the potential proof of concept clinical trial strategies for studies of interventions that target basic aging mechanisms.

Successfully translating interventions that are effective in experimental animals into clinical application is difficult, lengthy, and expensive. It can take over 17 years to complete translation into clinical practice, even in areas with an established translational tradition, such as infectious diseases or oncology. The needed steps include pre-clinical studies to test efficacy, safety, and pharmacokinetics in mammals, usually in at least two species and following good laboratory practices, as stipulated by regulatory agencies [134]. For pharmaceutical interventions that target basic aging mechanisms and that are intended to treat age-related chronic diseases, it would make sense to conduct these safety studies in old animals, rather than the young mice that are currently commonly used.

Before preclinical animal studies begin, the nature and goals of clinical studies need to be planned so that the preclinical test program can be designed to yield data useful for designing and refining the clinical trials. Even for preclinical studies in animals, it is important to select outcome measures that have been validated, are reproducible, can be measured in a short time-frame, are as non-invasive as possible, are accepted by regulators, and are related to the outcomes to be tested in subsequent clinical trials. Iterative bench-to-bedside coupled to reverse translational bedside-to-bench developmental phases are generally necessary, particularly for investigational new drugs (IND's). This requires a close partnership between basic biologists and clinicians with a strong basic biology background. While it is feasible to repurpose existing agents not covered by patent protection by conducting publicly funded trials in academic settings, without early attention to protecting intellectual property, the chances of getting a treatment to patients through commercial partnerships are reduced. On the other hand, commercial entities might find attractive the possibility that seeking FDA approval for adding a label indication to an existing drug might require less effort and take fewer years than a de novo drug approval. Early attention to marketing and the potential interest of payers for interventions can motivate commercialization and moving treatments into the clinic.

4.1 Potential Clinical Trial Scenarios

Elderly subjects have generally been excluded from clinical trials, especially those involving new formulations. Since several interventions that target basic aging mechanisms and extend lifespan appear to be effective in mice, there is an opportunity to select those that are more readily translatable. Some interventions, such as lifestyle modifications, are particularly challenging (e.g., caloric restriction in the face of an obesity epidemic). Desirable characteristics of interventions suitable for translation include: (1) low toxicity and few side effects, (2) effectiveness of oral as opposed to parenteral administration, (3) low dosing frequency (i.e., relatively long half-life), (4) stability, (5) scalability and low manufacturing cost, (6) detectability in blood, and (7) importantly, effectiveness if initiated in later life or once symptoms have started to develop. Interventions that need to be applied in earlier life while subjects are still asymptomatic with an expectation of affecting health much later in life would be difficult to translate into humans, particularly if they exhibit virtually any side effects.

It will of course be impractical to study the success of experimental strategies to extend lifespan in humans within a reasonable time frame. Therefore, feasible and clinically relevant trials paradigms must be devised to test if agents that target fundamental aging processes can be translated into clinical use by developing more rapidly detectable outcomes. Short-term clinical trials in symptomatic individuals are feasible, and some early clinical trials have already commenced using some of the drugs discussed above that enhance life- or healthspan in rodents. For example, trials of rapamycin in Alzheimer's disease are currently underway or about to begin and a trial of rapalogs on enhancing immune responses of elderly subjects to vaccination has already been reported [25]. Resveratrol congeners are being developed to treat type 2 diabetes [135]. Collection of ancillary biological data would help inform later studies by providing additional ways to predict risk, response to the intervention, and data examining effects on basic aging mechanisms.

At least six potential drug development scenarios can be envisaged in which agents that target basic aging processes might first be tested. These include treatment of: (1) multiple co-morbidities, (2) otherwise fatal conditions, (3) frailty and geriatric syndromes, (4) resilience enhancement, (5) localized diseases related to fundamental aging processes, and (6) accelerated aging-like states.

4.1.1 Multiple Co-morbidities

Most clinical trials have been focused on younger subjects with a single target condition and have excluded older subjects with co-morbidities. However, multiple age-related chronic diseases often occur within the same older individuals. Therefore, agents that target basic aging processes might simultaneously alleviate several age-related chronic diseases within the same older subjects or delay the appearance of new chronic diseases in subjects who have already developed their first age-related chronic disease. Such scenarios will require novel clinical study designs. A potential scenario for initial small-scale proof-of-principle trials of candidate drugs would be to study their effect in elderly subjects with combinations of two or more of: atherosclerosis, hypertension, memory impairment, diabetes, chronic obstructive pulmonary disease, renal dysfunction, or other age- or senescence-related conditions. Outcomes could be surrogate endpoints already recognized by regulatory agencies, such as blood pressure, psychometric indices of cognitive function, fasting glucose or HbA_1C, circulating lipids, left ventricular function or hypertrophy, pulmonary function tests, etc. The endpoints could be combined into a composite score, although this carries the risk that an effective drug may appear less than effective if one of the composite endpoint components is affected in a direction opposite to that expected. For example, rapamycin may lead to improvements in several age-related measures of function, but also causes decreased glucose tolerance [136].

4.1.2 Otherwise Fatal Conditions

Another scenario for initial proof-of-concept trials of agents that target basic aging processes may be to test their impact on otherwise fatal conditions for which either very invasive or no effective treatments are available. In the case of senolytics, these include certain cancers, cancer predisposition syndromes, idiopathic pulmonary fibrosis, chronic obstructive pulmonary disease, and primary biliary cirrhosis, among others [3, 9, 137]. In some cancers, caloric restriction, senolytics, or other approaches that target fundamental aging mechanisms might facilitate use of higher chemotherapy or radiation doses or enhance effectiveness of these treatments.

4.1.3 Frailty and Geriatric Syndromes

Frailty is an age-related syndrome that involves weakness, loss of function, and decreased resilience [138–147]. It can be diagnosed through clinical scales that are moderately sensitive and specific, involving assessments of a combination of weight loss, decreased activity, weakness, fatigue, and burden of chronic disease and disabilities [144, 146, 148, 149]. The prevalence of frailty increases with aging [138–149], and it is associated with chronic diseases, loss of independence, high mortality, and the "geriatric syndromes" of sarcopenia, immobility, falling, cachexia, depression, and confusion, as well as chronic inflammation.

Subjects with mild or moderate degrees of frailty or sarcopenia would likely be better candidates for initial trials of agents that modulate fundamental aging processes, especially sterile inflammation and perhaps cellular senescence, than subjects with advanced, potentially irreversible frailty. Effects of agents that target fundamental aging processes could be studied on outcomes indices that are related to frailty, predictive of mortality, and are recognized or close to being recognized by drug regulators [18, 150, 151]. These include timed walking distances, strength measurements, inflammation-associated and SASP markers such as circulating IL-6, MCP-1, or PAI-1, pulmonary function (e.g., VO_2), renal function tests, among others.

4.1.4 Resilience Enhancement

Resilience in this context refers to the ability to recover from perturbations such as surgery, anesthesia, chemotherapy, radiation, a heart attack, fracture, or stroke. Resilience declines with aging and is associated with concurrent or subsequent overt frailty. Studies involving a stressor and relevant outcome measures could be a means for testing proof of principle for interventions targeting aging processes to determine if they enhance resilience and impact multiple clinically relevant outcomes in relatively short clinical trials. The stressor tested could be major, such as hip fracture after a fall, or minor, such as routine vaccination. The stressor could be elective and planned, such as elective surgery, chemotherapy, therapeutic radiation, or immunization, or unplanned, for example an acute illness, such as pneumonia, a heart attack, or a stroke. In the case of elective stressors, the drug could be administered before the stress occurs and in the case of either elective or unplanned stressors, during recovery after the stress-inducing event.

Some of these types of clinical resilience trials have already been reported. For example, the response to influenza immunization is improved a few weeks following a brief course with low doses of a rapalog [25]. Responses to chemotherapy are enhanced and its side effects reduced if the chemotherapy is preceded by a brief caloric restriction [152]. The effects of candidate agents that target basic aging mechanisms could be tested in multiple clinical trial scenarios, for example: (1) strength, endurance, nausea, and appetite after chemotherapy; (2) time needed for return of function, wound healing, incidence of delirium, or discharge disposition after elective surgery; or (3) recovery after myocardial infarction in elderly subjects. Secondary outcomes of these types of studies could include measures of function, comorbidity, and blood or tissue biomarkers of basic aging mechanisms.

4.1.5 Localized Diseases Related to Fundamental Aging Processes

Certain localized disorders involve operation of pathogenic processes shared with those that are associated with chronological aging. These conditions might be alleviated by local administration of agents that target those basic aging processes, such as aerosols, local injections, or topical skin solutions. For example, idiopathic pulmonary fibrosis is associated with accumulation of senescent cells in the lung [153, 154]. Subjects with this potentially fatal condition that lacks effective treatment options might benefit from aerosol delivery of senolytic or SASP-protective agents. This would need to be tested first in animals, for example in mice with pulmonary fibrosis induced by aerosolized bleomycin [155] crossed with animals from which senescent cells can be cleared genetically or treated with senolytic drugs. A similar strategy could be used for chronic obstructive pulmonary disease, which is associated with senescent cell accumulation in the lungs, following cigarette smoke exposure. Another example is osteoarthritis, an inflammatory condition that can affect multiple joints and becomes increasingly common in old age, and which is at least partially associated with accumulation of senescent cells and SASP factors (inflammatory cytokines, matrix metalloproteinases) in joints [156]. Currently, this disorder is treated with oral analgesics, anti-inflammatories, and intra-articular steroid injections. The oral agents have to be administered frequently, often at least daily. Steroid injections have effects that last for weeks, but often have to be administered repeatedly. Unfortunately, repeated steroid injections eventually contribute to worsening joint damage. Potentially, injected or systemic agents that target basic aging processes would have a more sustained effect and fewer adverse effects than currently used treatments.

4.1.6 Accelerated Aging-Like States

Several conditions have features resembling an accelerated aging-like state, including obesity and diabetes, long term effects of chemotherapy or radiation, HIV or side effects of the drugs used for treating it, and progeroid syndromes [3, 9, 18]. These conditions represent a potential scenario for proof-of-principle studies of candidate agents that target basic aging processes. Obesity and diabetes are associated with accelerated onset of other age-associated conditions, including atherosclerosis, vascular dysfunction, sarcopenia, early menopause, cancers, cognitive impairment and dementia [81, 157, 158]. Survivors of cancers who were treated during childhood with chemotherapy or radiation can develop frailty and sarcopenia, diabetes, cardiac disease, cognitive impairment ("chemo-brain"), and second, unrelated cancers by mid-adulthood [159, 160]. Progeroid syndromes with phenotypes resembling an accelerated aging-like state have been associated with increased senescent cell burden and accentuation of other fundamental processes that are also associated with chronological aging [161]. Short term effects of candidate agents on muscle strength, metabolic, cardiovascular, cognitive, or other functional measures in these subjects could be tested in initial proof-of-principle trials. For example, trials of rapalogs for Hutchinson-Guilford progeria are currently being planned.

4.2 Preclinical Studies

Animal models reflective of the potential indications for agents that target aging mechanisms in humans are needed to facilitate the preclinical studies required before proceeding to proof-of-principle human trials. An advantage to the aging biology field is that because aging is not a disease, unlike disease-focused studies, it is not necessary to 'model' aging: natural aging occurs in virtually all species as a natural model that does not need to be modelled. Mortality can be followed in animals that are otherwise 'normal', not genetically or otherwise modified. For studying some age-related diseases, relevant genetically modified mice are available or disease-inducing pharmacological or dietary manipulations have been devised. However, in many cases there are either no models of age-related chronic diseases or only models that are imperfect. Some mouse models are available that are reasonably close approximations to human progerias that result from single gene mutations [18, 83–85]. Drug candidates need to be tested in such mice before human subjects with these diseases. In addition to helping in devising interventions for these diseases, such studies would indicate if the drug candidates can resolve aging-like or healthspan phenotypes, such as impaired glucose homeostasis, grip strength, exercise endurance, activity, cardiac, or neurological or cognitive dysfunction. Senolytic agents have been reported to alleviate frailty, neurologic dysfunction, and osteoporosis in progeroid mice [13].

For multifactorial polygenic human diseases that become clinically manifest in later life, animals with single gene mutations that develop superficially similar syndromes in early life have drawbacks for drug development. For example, single gene mutations that lead to phenotypes resembling Alzheimer's disease in young mice do not fully phenocopy human Alzheimer's disease. Animals with dysfunction-provoking mutations that are inducible in later life may be better for testing agents that target basic aging mechanisms. An aging tissue microenvironment would be recapitulated in such mice. Furthermore, a range of mammalian species beyond mice is needed to test generalizability and meet regulatory requirements, especially for investigational new drugs.

Manipulations can be used in experimental animals to model human age-associated disorders or clinical stresses, including high fat feeding, localized or systemic radiation, pharmacological interventions (*e.g.*, chemotherapy, streptozotocin, Parkinson's-inducing agents), localized pharmacological interventions (*e.g.*, inhaled bleomycin or cigarette smoke), cancer xenografts, skin wounding, or surgically-induced arthritis. Effects of candidate agents on a panel of such models could be helpful in selecting potential clinical applications for each new drug, as long as the test animals are of the appropriate ages. In some cases, such as Alzheimer's disease, screening for agents might be more practical in human cell culture systems that mimic disease pathology more faithfully than currently available animal models [162]. Also, for investigational new drugs, systems for medicinal chemical optimization and testing toxicology and pharmacokinetics (absorption, distribution, metabolism, and excretion) need to be developed using aging cell culture and animal models, rather than young cultures or animals.

4.3 Clinical Trials

Clinical trials are prospective human studies used to determine whether new biomedical or behavioral interventions are safe, efficacious, and effective. Clinical trials are generally categorized in phases. Phase 0 studies are used to determine if investigational new drugs act in humans as expected from preclinical animal studies, to acquire preliminary data about their pharmacokinetics or pharmacodynamics, to select the most promising lead candidates, or to determine biodistribution characteristics. Phase 1 trials are used to provide information about the metabolism and pharmacologic actions of the candidate drug in humans, side effects associated with escalating doses, and early evidence for effectiveness. These trials may include healthy participants or patients with the disease under study. Phase 2 studies are controlled clinical trials that evaluate the effectiveness of the candidate drug for a particular indication in subjects with the disease or condition under study and to identify the most common short-term side effects and risks. Some trials are combinations of Phases I and II, investigating both efficacy and toxicity. Phase 3 studies are expanded controlled or uncontrolled trials that follow the acquisition of preliminary evidence about the candidate agent in Phases 0–2, and are designed to provide additional information about overall risk-benefit relationships as well as for drug labeling. Generally, these trials compare new candidate drugs to drugs already in use for that condition. Phase 4 studies are conducted after clinical use at the population level has begun. They provide additional information about the risks, benefits, comparative effectiveness, and optimal use of the drug. These studies monitor effectiveness of the approved drug in the general population and provide data about any adverse effects that become apparent with widespread use.

Information collected during proof of concept and later phase trials can be used for following subjects to determine long-term clinical outcomes. Long-term outcomes can provide useful information about whether the agent also delays or prevents other chronic diseases, functional pre-frailty or frailty, loss of resilience, or loss of independence.

5 Biomarkers

There are three types of biomarkers that can catalyze clinical trials: surrogate endpoint biomarkers, drug activity biomarkers, and biomarkers related to mechanisms. Surrogate endpoint biomarkers are those that can be substituted for a clinical event endpoint as the outcome of a clinical trial. An example is fasting blood sugar as a surrogate outcome for a drug treating diabetes, as opposed to hard clinical endpoints such as weight loss, polyuria, or diabetic crises. Another surrogate biomarker is hypertension as a predictor for risk of stroke. These types of surrogate endpoints take years or decades to achieve acceptance by the medical community and regulators, for usage in place of the hard clinical event endpoints that the drug is expected to affect. There is a lack of surrogate biomarkers that: (1) predict lifespan or healthspan in humans, (2) have been demonstrated to vary in the same direction as lifespan or healthspan in response to interventions, and (3) are generally accepted as potential endpoints for clinical studies. It could take decades to do the studies that would convince the medical community and regulators that a surrogate biomarker reliably predicts and tracks longevity in humans and would be an acceptable clinical trial primary outcome endpoint. Some work has begun to validate surrogate outcomes for studies of frailty or age-related disability, such as clinical frailty scales involving physical function tests or assays of circulating cytokines. Although frailty, healthspan, and resilience biomarkers are some way from gaining regulatory acceptance and need further refinement in experimental animals and human studies, they are more within reach for use in clinical trials than biomarkers predictive of human longevity.

The second types of biomarker are those that reflect drug delivery, activity, or efficacy. As an example, in the case of rapamycin or related agents, assays of blood levels of the drug would be useful to ensure compliance. Assays of cellular S6 kinase activity in blood cells or tissue biopsies could be used to follow drug delivery and mTOR pharmacodynamic activity and would be useful in adjusting dose intensity of the intervention.

The third type of biomarkers includes those that test the mechanism of action in humans. In studies of candidate agents that are intended to target fundamental aging mechanisms, it would be desirable to measure a range of parameters suspected of being associated with aging processes to test if this is really how the drug could work. These biomarkers can be helpful, but it should be noted that precise information about the mechanism of action is not always needed for drug development. Indeed, for many drugs currently in widespread use in humans, the molecular mechanism of action was not determined before clinical use. Metformin is an example of this: its molecular mechanism of action is still not precisely understood. On the other hand, knowledge of the mechanism of action might help to predict side effects, although for many side effects this knowledge has not helped. Nevertheless, markers related to molecular aspects of aging might help to define the mechanism of action of the intervention in humans. This may stimulate reverse translational laboratory studies that could inform future clinical trials, spur discovery of new drug targets and development of new agents, and lead to optimization and refinement of preclinical animal models.

6 Personnel Needed

There is a divide between biologists studying aging processes and clinicians treating the elderly. Geriatricians rarely have basic biology training. Few geriatricians have investigational new drug (IND) experience. In the US, there are 7,000 Board-certified geriatricians. Only around a dozen have R01 grants from the Division of Aging Biology at the National Institute on Aging. Worldwide, there are few geriatricians who are also basic aging researchers. This is unlike most other biomedical fields, such as endocrinology or hematology, in which academic clinicians frequently conduct laboratory research. Few geriatricians attend meetings in the basic biology of aging. Conversely, basic biologists only rarely attend clinical geriatrics meetings. Unlike in other areas of medicine, such as infectious diseases or oncology, few geriatricians have experience in translating IND's into clinical application. Thus, there is a shortage of investigators with the combination of basic biological, geriatric medical care, and human trials skills needed to design and conduct the pre-clinical and clinical studies and navigate the regulatory framework necessary to translate recent advances from the basic biology of aging into clinical practice. A new group of investigators needs to be trained in the basic biology of aging who have a thorough grasp of translational strategies and clinical geriatrics, as well as a group of geriatricians with sufficient understanding of the fundamental biology of aging and clinical trials methodology, to spearhead the process of taking IND's through pre-clinical studies, clinical trials, and regulatory approval. This could take well over 5 years, even if we begin training these investigators right away. In the meantime, we need to encourage collaborative strategies that bring teams of basic biologists, geriatricians, and clinical trials investigators together to translate agents that target fundamental aging mechanisms into the clinic. We also need to develop clinical trials networks, perhaps emulating successful approaches taken by the networks in the cancer field.

7 Conclusions

We are on the verge of a new era in the basic biology of aging. There is tremendous potential for drugs that target aging fundamental mechanisms to prevent or treat age-related disorders as a whole, rather than one at a time. We are at the point where it seems increasingly likely that interventions targeting aging mechanisms could begin to be tested to delay, prevent, alleviate, or reverse multiple age-related chronic diseases and disabilities that afflict the elderly. If true, and if the interventions effective in targeting fundamental aging mechanisms in mice can be translated into humans, geriatrics practice and all of medicine as we know it would be transformed. Despite this, financial, infrastructure, and personnel resources are insufficient to expand or even sustain the important discovery, mechanistic, and interventional basic biology of aging pipeline, let alone to fuel translation of promising drug candidates into clinical practice. Creation of new experimental paradigms, development and validation of relevant, measureable outcomes, sufficient funding, and training of investigators with new skills are necessary. A strategy to increase resources and personnel and to optimize existing resources is needed to accelerate progress and avoid duplication without depriving the basic biology pipeline.

While it seems possible that drugs acting on basic aging processes will alleviate many age-related conditions, it is unlikely that it will be feasible to reverse every age-related change in the foreseeable future, if ever. Some age-related changes involve progenitor cell depletion, some inflammation, others extracellular matrix and structural changes (*e.g.*, cataracts, osteoporosis), and yet others involve macromolecular dysfunction. Different fundamental age-related disruptions are implicated to different extents in different organs and cell types. This suggests that several strategies will need to be combined to be maximally effective: no one strategy is very likely to be a panacea. A process involving at least two overall steps may ultimately be needed, based on the longstanding principle of first removing damaged tissues and then replacing them with good tissue. The first step may involve eliminating damaged or senescent cells, blunting the senescence-associated secretory phenotype, reducing inflammation, and removing cytotoxic lipids and lipofuscin, protein aggregates, damaged macromolecules, and advanced glycation endproducts. The second step may involve transplanting stem cells, differentiated cells, tissues, or organs or restoring endogenous stem cell or progenitor function to repopulate damaged organs. Advances are being made in each of these areas, and effective combined approaches may one day be feasible. If any or all of this comes to fruition, if we are able to push back age-related diseases as a group and extend healthspan, and if this can be translated into clinical treatments, health care as we know it would be transformed with myriad economic and social consequences and benefits.

Acknowledgements The author is grateful for the advice and ideas shared by members of the Geroscience Network supported by NIH grant R24AG044396.

Editor: Francesca Macchiarini, National Institute of Allergy and Infectious Diseases (NIAID), NIH.

References

1. Research AfA (2012) The Silver Book: chronic disease and medical innovation in an aging nation. Washington, DC
2. Goldman DP, Cutler D, Rowe JW, Michaud PC, Sullivan J, Peneva D, Olshansky SJ (2013) Substantial health and economic returns from delayed aging may warrant a new focus for medical research. Health Aff (Millwood) 32(10):1698–1705. doi:10.1377/hlthaff.2013.0052
3. Kirkland JL (2013) Translating advances from the basic biology of aging into clinical application. Exp Gerontol 48(1):1–5
4. Kirkland JL (2013) Inflammation and cellular senescence: potential contribution to chronic diseases and disabilities with aging. Public Policy Aging Rep 23:12–15
5. Olshansky SJ, Carnes BA, Cassel C (1990) In search of Methuselah: estimating the upper limits to human longevity. Science 250(4981):634–640
6. Fried LP, Xue QL, Cappola AR, Ferrucci L, Chaves P, Varadhan R, Guralnik JM, Leng SX, Semba RD, Walston JD, Blaum CS, Bandeen-Roche K (2009) Nonlinear multisystem physiological dysregulation associated with frailty in older women: implications for etiology and treatment. J Gerontol A Biol Sci Med Sci 64(10):1049–1057
7. Bartke A (2011) Single-gene mutations and healthy ageing in mammals. Philos Trans R Soc Lond B Biol Sci 366(1561):28–34
8. Lipton RB, Hirsch J, Katz MJ, Wang C, Sanders AE, Verghese J, Barzilai N, Derby CA (2010) Exceptional parental longevity associated with lower risk of Alzheimer's disease and memory decline. J Am Geriatr Soc 58(6):1043–1049

9. Kirkland JL, Tchkonia T (2014) Clinical strategies and animal models for developing senolytic agents. Exp Gerontol. doi:10.1016/j.exger.2014.10.012
10. Harrison DE, Strong R, Sharp ZD, Nelson JF, Astle CM, Flurkey K, Nadon NL, Wilkinson JE, Frenkel K, Carter CS, Pahor M, Javors MA, Fernandez E, Miller RA (2009) Rapamycin fed late in life extends lifespan in genetically heterogeneous mice. Nature 460(7253): 392–395
11. Martin-Montalvo A, Mercken EM, Mitchell SJ, Palacios HH, Mote PL, Scheibye-Knudsen M, Gomes AP, Ward TM, Minor RK, Blouin MJ, Schwab M, Pollak M, Zhang Y, Yu Y, Becker KG, Bohr VA, Ingram DK, Sinclair DA, Wolf NS, Spindler SR, Bernier M, de Cabo R (2013) Metformin improves healthspan and lifespan in mice. Nat Commun 4:2192. doi:10.1038/ncomms3192
12. Harrison DE, Strong R, Allison DB, Ames BN, Astle CM, Atamna H, Fernandez E, Flurkey K, Javors MA, Nadon NL, Nelson JF, Pletcher S, Simpkins JW, Smith D, Wilkinson JE, Miller RA (2014) Acarbose, 17-alpha-estradiol, and nordihydroguaiaretic acid extend mouse lifespan preferentially in males. Aging Cell 13(2):273–282. doi:10.1111/acel.12170
13. Zhu Y, Tchkonia T, Pirtskhalava T, Gower A, Ding H, Giorgadze N, Palmer AK, Ikeno Y, Borden G, Lenburg M, O'Hara SP, LaRusso NF, Miller JD, Roos CM, Verzosa GC, LeBrasseur NK, Wren JD, Farr JN, Khosla S, Stout MB, McGowan SJ, Fuhrmann-Stroissnigg H, Gurkar AU, Zhao J, Colangelo D, Dorronsoro A, Ling YY, Barghouthy AS, Navarro DC, Sano T, Robbins PD, Niedernhofer LJ, Kirkland JL (2015) The Achilles' heel of senescent cells: from transcriptome to senolytic drugs. Aging Cell. doi:10.1111/acel.12344
14. Majumder S, Caccamo A, Medina DX, Benavides AD, Javors MA, Kraig E, Strong R, Richardson A, Oddo S (2012) Lifelong rapamycin administration ameliorates age-dependent cognitive deficits by reducing IL-1beta and enhancing NMDA signaling. Aging Cell 11(2): 326–335
15. Anderson RM, Weindruch R (2012) The caloric restriction paradigm: implications for healthy human aging. Am J Hum Biol 24(2):101–106
16. Conboy IM, Conboy MJ, Wagers AJ, Girma ER, Weissman IL, Rando TA (2005) Rejuvenation of aged progenitor cells by exposure to a young systemic environment. Nature 433:760–764
17. Lavasani M, Robinson AR, Lu A, Song M, Feduska JM, Ahani B, Tilstra JS, Feldman CH, Robbins PD, Niedernhofer LJ, Huard J (2012) Muscle-derived stem/progenitor cell dysfunction limits healthspan and lifespan in a murine progeria model. Nat Commun 3:608
18. Tchkonia T, Zhu Y, van Deursen J, Campisi J, Kirkland JL (2013) Cellular senescence and the senescent secretory phenotype: therapeutic opportunities. J Clin Invest 123(3):966–972
19. Baker DJ, Wijshake T, Tchkonia T, LeBrasseur NK, Childs BG, van de Sluis B, Kirkland JL, van Deursen JM (2011) Clearance of p16Ink4a-positive senescent cells delays ageing-associated disorders. Nature 479(7372):232–236
20. Li J, Kim SG, Blenis J (2014) Rapamycin: one drug, many effects. Cell Metab 19(3): 373–379. doi:10.1016/j.cmet.2014.01.001
21. Lamming DW, Ye L, Sabatini DM, Baur JA (2013) Rapalogs and mTOR inhibitors as anti-aging therapeutics. J Clin Invest 123(3):980–989. doi:10.1172/JCI64099
22. Kennedy BK, Pennypacker JK (2014) Drugs that modulate aging: the promising yet difficult path ahead. Transl Res 163(5):456–465. doi:10.1016/j.trsl.2013.11.007
23. Wilkinson JE, Burmeister L, Brooks SV, Chan CC, Friedline S, Harrison DE, Hejtmancik JF, Nadon N, Strong R, Wood LK, Woodward MA, Miller RA (2012) Rapamycin slows aging in mice. Aging Cell 11(4):675–682. doi:10.1111/j.1474-9726.2012.00832.x
24. Richardson A, Galvan V, Lin AL, Oddo S (2014) How longevity research can lead to therapies for Alzheimer's disease: the rapamycin story. Exp Gerontol. doi:10.1016/j.exger.2014.12.002
25. Mannick JB, Del Giudice G, Lattanzi M, Valiante NM, Praestgaard J, Huang B, Lonetto MA, Maecker HT, Kovarik J, Carson S, Glass DJ, Klickstein LB (2014) mTOR inhibition improves immune function in the elderly. Sci Transl Med 6(268):268ra179. doi:10.1126/scitranslmed.3009892

26. Knowler WC, Barrett-Connor E, Fowler SE, Hamman RF, Lachin JM, Walker EA, Nathan DM (2002) Reduction in the incidence of type 2 diabetes with lifestyle intervention or metformin. N Engl J Med 346(6):393–403. doi:10.1056/NEJMoa012512
27. Crandall J, Schade D, Ma Y, Fujimoto WY, Barrett-Connor E, Fowler S, Dagogo-Jack S, Andres R (2006) The influence of age on the effects of lifestyle modification and metformin in prevention of diabetes. J Gerontol A Biol Sci Med Sci 61(10):1075–1081
28. Anisimov VN, Berstein LM, Egormin PA, Piskunova TS, Popovich IG, Zabezhinski MA, Tyndyk ML, Yurova MV, Kovalenko IG, Poroshina TE, Semenchenko AV (2008) Metformin slows down aging and extends life span of female SHR mice. Cell Cycle 7(17):2769–2773
29. Anisimov VN, Berstein LM, Popovich IG, Zabezhinski MA, Egormin PA, Piskunova TS, Semenchenko AV, Tyndyk ML, Yurova MN, Kovalenko IG, Poroshina TE (2011) If started early in life, metformin treatment increases life span and postpones tumors in female SHR mice. Aging (Milano) 3(2):148–157
30. Anisimov VN, Egormin PA, Piskunova TS, Popovich IG, Tyndyk ML, Yurova MN, Zabezhinski MA, Anikin IV, Karkach AS, Romanyukha AA (2010) Metformin extends life span of HER-2/neu transgenic mice and in combination with melatonin inhibits growth of transplantable tumors in vivo. Cell Cycle 9(1):188–197
31. Smith DL Jr, Elam CF Jr, Mattison JA, Lane MA, Roth GS, Ingram DK, Allison DB (2010) Metformin supplementation and life span in Fischer-344 rats. J Gerontol A Biol Sci Med Sci 65(5):468–474. doi:10.1093/gerona/glq033
32. Cabreiro F, Au C, Leung KY, Vergara-Irigaray N, Cocheme HM, Noori T, Weinkove D, Schuster E, Greene ND, Gems D (2013) Metformin retards aging in C. elegans by altering microbial folate and methionine metabolism. Cell 153(1):228–239. doi:10.1016/j.cell.2013.02.035
33. Baur JA, Birnbaum MJ (2014) Control of gluconeogenesis by metformin: does redox trump energy charge? Cell Metab 20(2):197–199. doi:10.1016/j.cmet.2014.07.013
34. Madiraju AK, Erion DM, Rahimi Y, Zhang XM, Braddock DT, Albright RA, Prigaro BJ, Wood JL, Bhanot S, MacDonald MJ, Jurczak MJ, Camporez JP, Lee HY, Cline GW, Samuel VT, Kibbey RG, Shulman GI (2014) Metformin suppresses gluconeogenesis by inhibiting mitochondrial glycerophosphate dehydrogenase. Nature 510(7506):542–546. doi:10.1038/nature13270
35. Moiseeva O, Deschenes-Simard X, St-Germain E, Igelmann S, Huot G, Cadar AE, Bourdeau V, Pollak MN, Ferbeyre G (2013) Metformin inhibits the senescence-associated secretory phenotype by interfering with IKK/NF-kappaB activation. Aging Cell 12(3):489–498. doi:10.1111/acel.12075
36. Anfossi G, Russo I, Bonomo K, Trovati M (2010) The cardiovascular effects of metformin: further reasons to consider an old drug as a cornerstone in the therapy of type 2 diabetes mellitus. Curr Vasc Pharmacol 8(3):327–337
37. Hong J, Zhang Y, Lai S, Lv A, Su Q, Dong Y, Zhou Z, Tang W, Zhao J, Cui L, Zou D, Wang D, Li H, Liu C, Wu G, Shen J, Zhu D, Wang W, Shen W, Ning G (2013) Effects of metformin versus glipizide on cardiovascular outcomes in patients with type 2 diabetes and coronary artery disease. Diabetes Care 36(5):1304–1311. doi:10.2337/dc12-0719
38. Johnson JA, Simpson SH, Toth EL, Majumdar SR (2005) Reduced cardiovascular morbidity and mortality associated with metformin use in subjects with Type 2 diabetes. Diabet Med 22(4):497–502. doi:10.1111/j.1464-5491.2005.01448.x
39. Lamanna C, Monami M, Marchionni N, Mannucci E (2011) Effect of metformin on cardiovascular events and mortality: a meta-analysis of randomized clinical trials. Diabetes Obes Metab 13(3):221–228. doi:10.1111/j.1463-1326.2010.01349.x
40. Roumie CL, Hung AM, Greevy RA, Grijalva CG, Liu X, Murff HJ, Elasy TA, Griffin MR (2012) Comparative effectiveness of sulfonylurea and metformin monotherapy on cardiovascular events in type 2 diabetes mellitus: a cohort study. Ann Intern Med 157(9):601–610. doi:10.7326/0003-4819-157-9-201211060-00003
41. Whittington HJ, Hall AR, McLaughlin CP, Hausenloy DJ, Yellon DM, Mocanu MM (2013) Chronic metformin associated cardioprotection against infarction: not just a glucose lowering phenomenon. Cardiovasc Drugs Ther 27(1):5–16. doi:10.1007/s10557-012-6425-x

42. Anisimov VN, Bartke A (2013) The key role of growth hormone-insulin-IGF-1 signaling in aging and cancer. Crit Rev Oncol Hematol 87(3):201–223. doi:10.1016/j.critrevonc.2013.01.005
43. Karnevi E, Said K, Andersson R, Rosendahl AH (2013) Metformin-mediated growth inhibition involves suppression of the IGF-I receptor signalling pathway in human pancreatic cancer cells. BMC Cancer 13:235. doi:10.1186/1471-2407-13-235
44. Liu B, Fan Z, Edgerton SM, Yang X, Lind SE, Thor AD (2011) Potent anti-proliferative effects of metformin on trastuzumab-resistant breast cancer cells via inhibition of erbB2/IGF-1 receptor interactions. Cell Cycle 10(17):2959–2966
45. Quinn BJ, Dallos M, Kitagawa H, Kunnumakkara AB, Memmott RM, Hollander MC, Gills JJ, Dennis PA (2013) Inhibition of lung tumorigenesis by metformin is associated with decreased plasma IGF-I and diminished receptor tyrosine kinase signaling. Cancer Prev Res (Phila) 6(8):801–810. doi:10.1158/1940-6207.CAPR-13-0058-T
46. Salani B, Maffioli S, Hamoudane M, Parodi A, Ravera S, Passalacqua M, Alama A, Nhiri M, Cordera R, Maggi D (2012) Caveolin-1 is essential for metformin inhibitory effect on IGF1 action in non-small-cell lung cancer cells. FASEB J 26(2):788–798. doi:10.1096/fj.11-192088
47. Seibel SA, Chou KH, Capp E, Spritzer PM, von Eye CH (2008) Effect of metformin on IGF-1 and IGFBP-1 levels in obese patients with polycystic ovary syndrome. Eur J Obstet Gynecol Reprod Biol 138(1):122–124. doi:10.1016/j.ejogrb.2007.02.001
48. Tosca L, Rame C, Chabrolle C, Tesseraud S, Dupont J (2010) Metformin decreases IGF1-induced cell proliferation and protein synthesis through AMP-activated protein kinase in cultured bovine granulosa cells. Reproduction 139(2):409–418. doi:10.1530/REP-09-0351
49. Landman GW, Kleefstra N, van Hateren KJ, Groenier KH, Gans RO, Bilo HJ (2010) Metformin associated with lower cancer mortality in type 2 diabetes: ZODIAC-16. Diabetes Care 33(2):322–326. doi:10.2337/dc09-1380
50. Lee MS, Hsu CC, Wahlqvist ML, Tsai HN, Chang YH, Huang YC (2011) Type 2 diabetes increases and metformin reduces total, colorectal, liver and pancreatic cancer incidences in Taiwanese: a representative population prospective cohort study of 800,000 individuals. BMC Cancer 11:20. doi:10.1186/1471-2407-11-20
51. Libby G, Donnelly LA, Donnan PT, Alessi DR, Morris AD, Evans JM (2009) New users of metformin are at low risk of incident cancer: a cohort study among people with type 2 diabetes. Diabetes Care 32(9):1620–1625. doi:10.2337/dc08-2175
52. Monami M, Colombi C, Balzi D, Dicembrini I, Giannini S, Melani C, Vitale V, Romano D, Barchielli A, Marchionni N, Rotella CM, Mannucci E (2011) Metformin and cancer occurrence in insulin-treated type 2 diabetic patients. Diabetes Care 34(1):129–131. doi:10.2337/dc10-1287
53. Tseng CH (2012) Diabetes, metformin use, and colon cancer: a population-based cohort study in Taiwan. Eur J Endocrinol 167(3):409–416. doi:10.1530/EJE-12-0369
54. Moore EM, Mander AG, Ames D, Kotowicz MA, Carne RP, Brodaty H, Woodward M, Boundy K, Ellis KA, Bush AI, Faux NG, Martins R, Szoeke C, Rowe C, Watters DA (2013) Increased risk of cognitive impairment in patients with diabetes is associated with metformin. Diabetes Care 36(10):2981–2987. doi:10.2337/dc13-0229
55. Ng TP, Feng L, Yap KB, Lee TS, Tan CH, Winblad B (2014) Long-term metformin usage and cognitive function among older adults with diabetes. J Alzheimers Dis 41(1):61–68. doi:10.3233/JAD-131901
56. Bannister CA, Holden SE, Jenkins-Jones S, Morgan CL, Halcox JP, Schernthaner G, Mukherjee J, Currie CJ (2014) Can people with type 2 diabetes live longer than those without? A comparison of mortality in people initiated with metformin or sulphonylurea monotherapy and matched, non-diabetic controls. Diabetes Obes Metab 16(11):1165–1173. doi:10.1111/dom.12354
57. Frantz S, Calvillo L, Tillmanns J, Elbing I, Dienesch C, Bischoff H, Ertl G, Bauersachs J (2005) Repetitive postprandial hyperglycemia increases cardiac ischemia/reperfusion injury: prevention by the alpha-glucosidase inhibitor acarbose. FASEB J 19(6):591–593. doi:10.1096/fj.04-2459fje

58. Miyamura M, Schnell O, Yamashita C, Yoshioka T, Matsumoto C, Mori T, Ukimura A, Kitaura Y, Matsumura Y, Ishizaka N, Hayashi T (2010) Effects of acarbose on the acceleration of postprandial hyperglycemia-induced pathological changes induced by intermittent hypoxia in lean mice. J Pharmacol Sci 114(1):32–40
59. Balfour JA, McTavish D (1993) Acarbose. An update of its pharmacology and therapeutic use in diabetes mellitus. Drugs 46(6):1025–1054
60. Archer VE (2003) Does dietary sugar and fat influence longevity? Med Hypotheses 60(6):924–929
61. Yamamoto M, Otsuki M (2006) Effect of inhibition of alpha-glucosidase on age-related glucose intolerance and pancreatic atrophy in rats. Metabolism 55(4):533–540. doi:10.1016/j.metabol.2005.11.007
62. Chiasson JL, Josse RG, Gomis R, Hanefeld M, Karasik A, Laakso M (2002) Acarbose for prevention of type 2 diabetes mellitus: the STOP-NIDDM randomised trial. Lancet 359(9323):2072–2077. doi:10.1016/S0140-6736(02)08905-5
63. Chiasson JL, Josse RG, Gomis R, Hanefeld M, Karasik A, Laakso M (2003) Acarbose treatment and the risk of cardiovascular disease and hypertension in patients with impaired glucose tolerance: the STOP-NIDDM trial. J Am Med Assoc 290(4):486–494. doi:10.1001/jama.290.4.486
64. Zeymer U, Schwarzmaier-D'assie A, Petzinna D, Chiasson JL (2004) Effect of acarbose treatment on the risk of silent myocardial infarctions in patients with impaired glucose tolerance: results of the randomised STOP-NIDDM trial electrocardiography substudy. Eur J Cardiovasc Prev Rehabil 11(5):412–415
65. Dey M, Lyttle CR, Pickar JH (2000) Recent insights into the varying activity of estrogens. Maturitas 34(Suppl 2):S25–S33
66. Dykens JA, Moos WH, Howell N (2005) Development of 17alpha-estradiol as a neuroprotective therapeutic agent: rationale and results from a phase I clinical study. Ann N Y Acad Sci 1052:116–135. doi:10.1196/annals.1347.008
67. Littlefield BA, Gurpide E, Markiewicz L, McKinley B, Hochberg RB (1990) A simple and sensitive microtiter plate estrogen bioassay based on stimulation of alkaline phosphatase in Ishikawa cells: estrogenic action of delta 5 adrenal steroids. Endocrinology 127(6):2757–2762. doi:10.1210/endo-127-6-2757
68. Toran-Allerand CD (2004) Minireview: A plethora of estrogen receptors in the brain: where will it end? Endocrinology 145(3):1069–1074. doi:10.1210/en.2003-1462
69. Orfanos CE, Vogels L (1980) Local therapy of androgenetic alopecia with 17 alpha-estradiol. A controlled, randomized double-blind study (author's transl). Dermatologica 161(2):124–132
70. Jones DL, Rando TA (2011) Emerging models and paradigms for stem cell ageing. Nat Cell Biol 13(5):506–512. doi:10.1038/ncb0511-506
71. Gimble JM, Nuttall ME (2012) The relationship between adipose tissue and bone metabolism. Clin Biochem 45(12):874–879. doi:10.1016/j.clinbiochem.2012.03.006
72. Sinha M, Jang YC, Oh J, Khong D, Wu EY, Manohar R, Miller C, Regalado SG, Loffredo FS, Pancoast JR, Hirshman MF, Lebowitz J, Shadrach JL, Cerletti M, Kim MJ, Serwold T, Goodyear LJ, Rosner B, Lee RT, Wagers AJ (2014) Restoring systemic GDF11 levels reverses age-related dysfunction in mouse skeletal muscle. Science 344(6184):649–652. doi:10.1126/science.1251152
73. Katsimpardi L, Litterman NK, Schein PA, Miller CM, Loffredo FS, Wojtkiewicz GR, Chen JW, Lee RT, Wagers AJ, Rubin LL (2014) Vascular and neurogenic rejuvenation of the aging mouse brain by young systemic factors. Science 344(6184):630–634. doi:10.1126/science.1251141
74. Coppé JP, Patil C, Rodier F, Sun Y, Muñoz DP, Goldstein J, Nelson PS, Desprez PY, Campisi J (2008) Senescence-associated secretory phenotypes reveal cell-nonautonomous functions of oncogenic RAS and the p53 tumor suppressor. PLoS Biol 6:2853–2868
75. Kuilman T, Peeper DS (2009) Senescence-messaging secretome: SMS-ing cellular stress. Nat Rev Cancer 9:81–94
76. Waaijer ME, Parish WE, Strongitharm BH, van Heemst D, Slagboom PE, de Craen AJ, Sedivy JM, Westendorp RG, Gunn DA, Maier AB (2012) The number of p16INK4a positive cells in human skin reflects biological age. Aging Cell 11:722–725

77. Dimri GP, Lee X, Basile G, Acosta M, Scott G, Roskelley C, Medrano EE, Linskens M, Rubelj I, Pereira-Smith O, Peacocke M, Campisi J (1995) A biomarker that identifies senescent human cells in culture and in aging skin in vivo. Proc Natl Acad Sci U S A 92:9363–9367
78. Zhu Y, Armstrong JL, Tchkonia T, Kirkland JL (2014) Cellular senescence and the senescent secretory phenotype in age-related chronic diseases. Curr Opin Clin Nutr Metab Care 17(4):324–328. doi:10.1097/MCO.0000000000000065
79. Stout MB, Tchkonia T, Pirtskhalava T, Palmer AK, List EO, Berryman DE, Lubbers ER, Escande C, Spong A, Masternak MM, Oberg AL, LeBrasseur NK, Miller RA, Kopchick JJ, Bartke A, Kirkland JL (2014) Growth hormone action predicts age-related white adipose tissue dysfunction and senescent cell burden in mice. Aging (Milano)
80. Krishnamurthy J, Torrice C, Ramsey MR, Kovalev GI, Al-Regaiey K, Su L, Sharpless NE (2004) Ink4a/Arf expression is a biomarker of aging. J Clin Invest 114(9):1299–1307
81. Tchkonia T, Morbeck DE, von Zglinicki T, van Deursen J, Lustgarten J, Scrable H, Khosla S, Jensen MD, Kirkland JL (2010) Fat tissue, aging, and cellular senescence. Aging Cell 9:667–684
82. Minamino T, Orimo M, Shimizu I, Kunieda T, Yokoyama M, Ito T, Nojima A, Nabetani A, Oike Y, Matsubara H, Ishikawa F, Komuro I (2009) A crucial role for adipose tissue p53 in the regulation of insulin resistance. Nat Med 15(9):1082–1087
83. Eren M, Boe AE, Murphy SB, Place AT, Nagpal V, Morales-Nebreda L, Urich D, Quaggin SE, Budinger GR, Mutlu GM, Miyata T, Vaughan DE (2014) PAI-1-regulated extracellular proteolysis governs senescence and survival in Klotho mice. Proc Natl Acad Sci U S A 111(19):7090–7095. doi:10.1073/pnas.1321942111
84. Chen Q, Liu K, Robinson AR, Clauson CL, Blair HC, Robbins PD, Niedernhofer LJ, Ouyang H (2013) DNA damage drives accelerated bone aging via an NF-kappaB-dependent mechanism. J Bone Miner Res 28(5):1214–1228
85. Baker DJ, Perez-Terzic C, Jin F, Pitel KS, Niederlander NJ, Jeganathan K, Yamada S, Reyes S, Rowe L, Hiddinga HJ, Eberhardt NL, Terzic A, van Deursen JM (2008) Opposing roles for p16Ink4a and p19Arf in senescence and ageing caused by BubR1 insufficiency. Nat Cell Biol 10(7):825–836. doi:10.1038/ncb1744
86. Jurk D, Wilson C, Passos JF, Oakley F, Correia-Melo C, Greaves L, Saretzki G, Fox C, Lawless C, Anderson R, Hewitt G, Pender SL, Fullard N, Nelson G, Mann J, van de Sluis B, Mann DA, von Zglinicki T (2014) Chronic inflammation induces telomere dysfunction and accelerates ageing in mice. Nat Commun 2:4172. doi:10.1038/ncomms5172
87. Demaria M, Ohtani N, Youssef SA, Rodier F, Toussaint W, Mitchell JR, Laberge RM, Vijg J, Van Steeg H, Dolle ME, Hoeijmakers JH, de Bruin A, Hara E, Campisi J (2014) An essential role for senescent cells in optimal wound healing through secretion of PDGF-AA. Dev Cell 31(6):722–733. doi:10.1016/j.devcel.2014.11.012
88. Xue W, Zender L, Miething C, Dickins RA, Hernando E, Krizhanovsky V, Cordon-Cardo C, Lowe SW (2007) Senescence and tumour clearance is triggered by p53 restoration in murine liver carcinomas. Nature 445(7128):656–660
89. Krizhanovsky V, Yon M, Dickins RA, Hearn S, Simon J, Miething C, Yee H, Zender L, Lowe SW (2008) Senescence of activated stellate cells limits liver fibrosis. Cell 134:657–667
90. Min TZ, Stephens MW, Kumar P, Chudleigh RA (2012) Renal complications of diabetes. Br Med Bull 104:113–127. doi:10.1093/bmb/lds030
91. Pasternak B, Svanstrom H, Callreus T, Melbye M, Hviid A (2011) Use of angiotensin receptor blockers and the risk of cancer. Circulation 123(16):1729–1736. doi:10.1161/CIRCULATIONAHA.110.007336
92. Huang CC, Chan WL, Chen YC, Chen TJ, Lin SJ, Chen JW, Leu HB (2011) Angiotensin II receptor blockers and risk of cancer in patients with systemic hypertension. Am J Cardiol 107(7):1028–1033. doi:10.1016/j.amjcard.2010.11.026
93. Hanon O, Berrou JP, Negre-Pages L, Goch JH, Nadhazi Z, Petrella R, Sedefdjian A, Sevenier F, Shlyakhto EV, Pathak A (2008) Effects of hypertension therapy based on eprosartan on systolic arterial blood pressure and cognitive function: primary results of the Observational Study on Cognitive function and Systolic Blood Pressure Reduction open-label study. J Hypertens 26(8):1642–1650. doi:10.1097/HJH.0b013e328301a280

94. Davies NM, Kehoe PG, Ben-Shlomo Y, Martin RM (2011) Associations of anti-hypertensive treatments with Alzheimer's disease, vascular dementia, and other dementias. J Alzheimers Dis 26(4):699–708. doi:10.3233/JAD-2011-110347
95. Li NC, Lee A, Whitmer RA, Kivipelto M, Lawler E, Kazis LE, Wolozin B (2010) Use of angiotensin receptor blockers and risk of dementia in a predominantly male population: prospective cohort analysis. BMJ 340:b5465. doi:10.1136/bmj.b5465
96. Benigni A, Orisio S, Noris M, Iatropoulos P, Castaldi D, Kamide K, Rakugi H, Arai Y, Todeschini M, Ogliari G, Imai E, Gondo Y, Hirose N, Mari D, Remuzzi G (2013) Variations of the angiotensin II type 1 receptor gene are associated with extreme human longevity. Age 35(3):993–1005. doi:10.1007/s11357-012-9408-8
97. Benigni A, Corna D, Zoja C, Sonzogni A, Latini R, Salio M, Conti S, Rottoli D, Longaretti L, Cassis P, Morigi M, Coffman TM, Remuzzi G (2009) Disruption of the Ang II type 1 receptor promotes longevity in mice. J Clin Invest 119(3):524–530. doi:10.1172/JCI36703
98. Linz W, Heitsch H, Scholkens BA, Wiemer G (2000) Long-term angiotensin II type 1 receptor blockade with fonsartan doubles lifespan of hypertensive rats. Hypertension 35(4):908–913
99. Linz W, Jessen T, Becker RH, Scholkens BA, Wiemer G (1997) Long-term ACE inhibition doubles lifespan of hypertensive rats. Circulation 96(9):3164–3172
100. Santos EL, de Picoli SK, da Silva ED, Batista EC, Martins PJ, D'Almeida V, Pesquero JB (2009) Long term treatment with ACE inhibitor enalapril decreases body weight gain and increases life span in rats. Biochem Pharmacol 78(8):951–958. doi:10.1016/j.bcp.2009.06.018
101. Ferder L, Inserra F, Romano L, Ercole L, Pszenny V (1993) Effects of angiotensin-converting enzyme inhibition on mitochondrial number in the aging mouse. Am J Physiol 265(1 Pt 1): C15–C18
102. Basso N, Cini R, Pietrelli A, Ferder L, Terragno NA, Inserra F (2007) Protective effect of long-term angiotensin II inhibition. Am J Physiol Heart Circ Physiol 293(3):H1351–H1358. doi:10.1152/ajpheart.00393.2007
103. Basso N, Paglia N, Stella I, de Cavanagh EM, Ferder L, del Rosario Lores Arnaiz M, Inserra F (2005) Protective effect of the inhibition of the renin-angiotensin system on aging. Regul Pept 128(3):247–252. doi:10.1016/j.regpep.2004.12.027
104. Burnett C, Valentini S, Cabreiro F, Goss M, Somogyvari M, Piper MD, Hoddinott M, Sutphin GL, Leko V, McElwee JJ, Vazquez-Manrique RP, Orfila AM, Ackerman D, Au C, Vinti G, Riesen M, Howard K, Neri C, Bedalov A, Kaeberlein M, Soti C, Partridge L, Gems D (2011) Absence of effects of Sir2 overexpression on lifespan in C. elegans and Drosophila. Nature 477(7365):482–485. doi:10.1038/nature10296
105. Kaeberlein M, McVey M, Guarente L (1999) The SIR2/3/4 complex and SIR2 alone promote longevity in Saccharomyces cerevisiae by two different mechanisms. Genes Dev 13(19): 2570–2580
106. Rogina B, Helfand SL (2004) Sir2 mediates longevity in the fly through a pathway related to calorie restriction. Proc Natl Acad Sci U S A 101(45):15998–16003. doi:10.1073/pnas.0404184101
107. Viswanathan M, Guarente L (2011) Regulation of Caenorhabditis elegans lifespan by sir-2.1 transgenes. Nature 477(7365):E1–E2. doi:10.1038/nature10440
108. Kanfi Y, Naiman S, Amir G, Peshti V, Zinman G, Nahum L, Bar-Joseph Z, Cohen HY (2012) The sirtuin SIRT6 regulates lifespan in male mice. Nature 483(7388):218–221. doi:10.1038/nature10815
109. Guarente L (2011) Franklin H. Epstein Lecture: Sirtuins, aging, and medicine. N Engl J Med 364(23):2235–2244. doi:10.1056/NEJMra1100831
110. Herranz D, Munoz-Martin M, Canamero M, Mulero F, Martinez-Pastor B, Fernandez-Capetillo O, Serrano M (2010) Sirt1 improves healthy ageing and protects from metabolic syndrome-associated cancer. Nat Commun 1:3. doi:10.1038/ncomms1001
111. Howitz KT, Bitterman KJ, Cohen HY, Lamming DW, Lavu S, Wood JG, Zipkin RE, Chung P, Kisielewski A, Zhang LL, Scherer B, Sinclair DA (2003) Small molecule activators of

sirtuins extend Saccharomyces cerevisiae lifespan. Nature 425(6954):191–196. doi:10.1038/nature01960
112. Wood JG, Rogina B, Lavu S, Howitz K, Helfand SL, Tatar M, Sinclair D (2004) Sirtuin activators mimic caloric restriction and delay ageing in metazoans. Nature 430(7000):686–689. doi:10.1038/nature02789
113. Strong R, Miller RA, Astle CM, Baur JA, de Cabo R, Fernandez E, Guo W, Javors M, Kirkland JL, Nelson JF, Sinclair DA, Teter B, Williams D, Zaveri N, Nadon NL, Harrison DE (2013) Evaluation of resveratrol, green tea extract, curcumin, oxaloacetic acid, and medium-chain triglyceride oil on life span of genetically heterogeneous mice. J Gerontol A Biol Sci Med Sci 68(1):6–16. doi:10.1093/gerona/gls070
114. Mitchell SJ, Martin-Montalvo A, Mercken EM, Palacios HH, Ward TM, Abulwerdi G, Minor RK, Vlasuk GP, Ellis JL, Sinclair DA, Dawson J, Allison DB, Zhang Y, Becker KG, Bernier M, de Cabo R (2014) The SIRT1 activator SRT1720 extends lifespan and improves health of mice fed a standard diet. Cell Rep 6(5):836–843. doi:10.1016/j.celrep.2014.01.031
115. Mercken EM, Mitchell SJ, Martin-Montalvo A, Minor RK, Almeida M, Gomes AP, Scheibye-Knudsen M, Palacios HH, Licata JJ, Zhang Y, Becker KG, Khraiwesh H, Gonzalez-Reyes JA, Villalba JM, Baur JA, Elliott P, Westphal C, Vlasuk GP, Ellis JL, Sinclair DA, Bernier M, de Cabo R (2014) SRT2104 extends survival of male mice on a standard diet and preserves bone and muscle mass. Aging Cell 13(5):787–796. doi:10.1111/acel.12220
116. Baur JA, Pearson KJ, Price NL, Jamieson HA, Lerin C, Kalra A, Prabhu VV, Allard JS, Lopez-Lluch G, Lewis K, Pistell PJ, Poosala S, Becker KG, Boss O, Gwinn D, Wang M, Ramaswamy S, Fishbein KW, Spencer RG, Lakatta EG, Le Couteur D, Shaw RJ, Navas P, Puigserver P, Ingram DK, de Cabo R, Sinclair DA (2006) Resveratrol improves health and survival of mice on a high-calorie diet. Nature 444(7117):337–342. doi:10.1038/nature05354
117. Minor RK, Baur JA, Gomes AP, Ward TM, Csiszar A, Mercken EM, Abdelmohsen K, Shin YK, Canto C, Scheibye-Knudsen M, Krawczyk M, Irusta PM, Martin-Montalvo A, Hubbard BP, Zhang Y, Lehrmann E, White AA, Price NL, Swindell WR, Pearson KJ, Becker KG, Bohr VA, Gorospe M, Egan JM, Talan MI, Auwerx J, Westphal CH, Ellis JL, Ungvari Z, Vlasuk GP, Elliott PJ, Sinclair DA, de Cabo R (2011) SRT1720 improves survival and healthspan of obese mice. Sci Rep 1:70. doi:10.1038/srep00070
118. Jimenez-Gomez Y, Mattison JA, Pearson KJ, Martin-Montalvo A, Palacios HH, Sossong AM, Ward TM, Younts CM, Lewis K, Allard JS, Longo DL, Belman JP, Malagon MM, Navas P, Sanghvi M, Moaddel R, Tilmont EM, Herbert RL, Morrell CH, Egan JM, Baur JA, Ferrucci L, Bogan JS, Bernier M, de Cabo R (2013) Resveratrol improves adipose insulin signaling and reduces the inflammatory response in adipose tissue of rhesus monkeys on high-fat, high-sugar diet. Cell Metab 18(4):533–545. doi:10.1016/j.cmet.2013.09.004
119. Timmers S, Konings E, Bilet L, Houtkooper RH, van de Weijer T, Goossens GH, Hoeks J, van der Krieken S, Ryu D, Kersten S, Moonen-Kornips E, Hesselink MK, Kunz I, Schrauwen-Hinderling VB, Blaak EE, Auwerx J, Schrauwen P (2011) Calorie restriction-like effects of 30 days of resveratrol supplementation on energy metabolism and metabolic profile in obese humans. Cell Metab 14(5):612–622. doi:10.1016/j.cmet.2011.10.002
120. Hubbard BP, Sinclair DA (2014) Small molecule SIRT1 activators for the treatment of aging and age-related diseases. Trends Pharmacol Sci 35(3):146–154. doi:10.1016/j.tips.2013.12.004
121. Hausenblas HA, Schoulda JA, Smoliga JM (2014) Resveratrol treatment as an adjunct to pharmacological management in type 2 diabetes mellitus-systematic review and meta-analysis. Mol Nutr Food Res. doi:10.1002/mnfr.201400173
122. de Ligt M, Timmers S, Schrauwen P (2014) Resveratrol and obesity: can resveratrol relieve metabolic disturbances? Biochim Biophys Acta. doi:10.1016/j.bbadis.2014.11.012
123. Escande C, Nin V, Pirtskhalava T, Chini CC, Thereza Barbosa M, Mathison A, Urrutia R, Tchkonia T, Kirkland JL, Chini EN (2014) Deleted in breast cancer 1 regulates cellular senescence during obesity. Aging Cell. doi:10.1111/acel.12235
124. Burn J, Gerdes AM, Macrae F, Mecklin JP, Moeslein G, Olschwang S, Eccles D, Evans DG, Maher ER, Bertario L, Bisgaard ML, Dunlop MG, Ho JW, Hodgson SV, Lindblom A,

Lubinski J, Morrison PJ, Murday V, Ramesar R, Side L, Scott RJ, Thomas HJ, Vasen HF, Barker G, Crawford G, Elliott F, Movahedi M, Pylvanainen K, Wijnen JT, Fodde R, Lynch HT, Mathers JC, Bishop DT (2011) Long-term effect of aspirin on cancer risk in carriers of hereditary colorectal cancer: an analysis from the CAPP2 randomised controlled trial. Lancet 378(9809):2081–2087. doi:10.1016/S0140-6736(11)61049-0
125. Chang ET, Froslev T, Sorensen HT, Pedersen L (2011) A nationwide study of aspirin, other non-steroidal anti-inflammatory drugs, and Hodgkin lymphoma risk in Denmark. Br J Cancer 105(11):1776–1782. doi:10.1038/bjc.2011.443
126. Flossmann E, Rothwell PM (2007) Effect of aspirin on long-term risk of colorectal cancer: consistent evidence from randomised and observational studies. Lancet 369(9573):1603–1613. doi:10.1016/S0140-6736(07)60747-8
127. Strong R, Miller RA, Astle CM, Floyd RA, Flurkey K, Hensley KL, Javors MA, Leeuwenburgh C, Nelson JF, Ongini E (2008) Nordihydroguaiaretic acid and aspirin increase lifespan of genetically heterogeneous male mice. Aging Cell 7:641–650
128. Hawley SA, Fullerton MD, Ross FA, Schertzer JD, Chevtzoff C, Walker KJ, Peggie MW, Zibrova D, Green KA, Mustard KJ, Kemp BE, Sakamoto K, Steinberg GR, Hardie DG (2012) The ancient drug salicylate directly activates AMP-activated protein kinase. Science 336(6083):918–922. doi:10.1126/science.1215327
129. Goldfine AB, Fonseca V, Jablonski KA, Chen YD, Tipton L, Staten MA, Shoelson SE (2013) Salicylate (salsalate) in patients with type 2 diabetes: a randomized trial. Ann Intern Med 159(1):1–12. doi:10.7326/0003-4819-159-1-201307020-00003
130. Din FV, Valanciute A, Houde VP, Zibrova D, Green KA, Sakamoto K, Alessi DR, Dunlop MG (2012) Aspirin inhibits mTOR signaling, activates AMP-activated protein kinase, and induces autophagy in colorectal cancer cells. Gastroenterology 142(7):1504–1515 e1503. doi:10.1053/j.gastro.2012.02.050
131. Wan QL, Zheng SQ, Wu GS, Luo HR (2013) Aspirin extends the lifespan of Caenorhabditis elegans via AMPK and DAF-16/FOXO in dietary restriction pathway. Exp Gerontol 48(5):499–506. doi:10.1016/j.exger.2013.02.020
132. Pallet N, Legendre C (2013) Adverse events associated with mTOR inhibitors. Expert Opin Drug Saf 12(2):177–186. doi:10.1517/14740338.2013.752814
133. Verges B, Walter T, Cariou B (2014) Endocrine side effects of anti-cancer drugs: effects of anti-cancer targeted therapies on lipid and glucose metabolism. Eur J Endocrinol 170(2):R43–R55. doi:10.1530/EJE-13-0586
134. Steinmetz KL, Spack EG (2009) The basics of preclinical drug development for neurodegenerative disease indications. BMC Neurol 9(Suppl 1):S2
135. Baur JA, Ungvari Z, Minor RK, Le Couteur DG, de Cabo R (2012) Are sirtuins viable targets for improving healthspan and lifespan? Nat Rev Drug Discov 11(6):443–461
136. Lamming DW, Ye L, Katajisto P, Goncalves MD, Saitoh M, Stevens DM, Davis JG, Salmon AB, Richardson A, Ahima RS, Guertin DA, Sabatini DM, Baur JA (2012) Rapamycin-induced insulin resistance is mediated by mTORC2 loss and uncoupled from longevity. Science 335(6076):1638–1643
137. Calhoun C, Shivshankar P, Saker M, Sloane LB, Livi CB, Sharp ZD, Orihuela CJ, Adnot S, White ES, Richardson A, Jourdan Le Saux C (2015) Senescent cells contribute to the physiological remodeling of aged lungs. J Gerontol A Biol Sci Med Sci. doi:10.1093/gerona/glu241
138. Kanapuru B, Ershler WB (2009) Inflammation, coagulation, and the pathway to frailty. Am J Med 122(7):605–613
139. Bandeen-Roche K, Walston JD, Huang Y, Semba RD, Ferrucci L (2009) Measuring systemic inflammatory regulation in older adults: evidence and utility. Rejuvenation Res 12(6):403–410
140. Qu T, Walston JD, Yang H, Fedarko NS, Xue QL, Beamer BA, Ferrucci L, Rose NR, Leng SX (2009) Upregulated ex vivo expression of stress-responsive inflammatory pathway genes by LPS-challenged CD14(+) monocytes in frail older adults. Mech Ageing Dev 130(3):161–166

141. Walston JD, Matteini AM, Nievergelt C, Lange LA, Fallin DM, Barzilai N, Ziv E, Pawlikowska L, Kwok P, Cummings SR, Kooperberg C, LaCroix A, Tracy RP, Atzmon G, Lange EM, Reiner AP (2009) Inflammation and stress-related candidate genes, plasma interleukin-6 levels, and longevity in older adults. Exp Gerontol 44(5):350–355
142. Leng SX, Xue QL, Tian J, Walston JD, Fried LP (2007) Inflammation and frailty in older women. J Am Geriatr Soc 55(6):864–871
143. Walston J, Hadley E, Ferrucci L, Guralnick JM, Newman AB, Studenski SA, Ershler WB, Harris T, Fried LP (2006) Research agenda for frailty in older adults: toward a better understanding of physiology and etiology: summary from the American Geriatrics Society/National Institute on Aging research conference on frailty in older adults. J Am Geriatr Soc 54:991–1001
144. Fried LP, Tangen CM, Walston J, Newman AB, Hirsch C, Gottdiener J, Seeman T, Tracy R, Kop WJ, Burke G, McBurnie MA (2001) Frailty in older adults: evidence for a phenotype. J Gerontol A Biol Sci Med Sci 56(3):M146–M156
145. Walston J, McBurnie MA, Newman A, Tracy RP, Kop WJ, Hirsch CH, Gottdiener J, Fried LP (2002) Frailty and activation of the inflammation and coagulation systems with and without clinical comorbidities: results from the Cardiovascular Health Study. Arch Intern Med 162(20):2333–2341
146. Bandeen-Roche K, Xue QL, Ferrucci L, Walston J, Guralnik JM, Chaves P, Zeger SL, Fried LP (2006) Phenotype of frailty: characterization in the women's health and aging studies. J Gerontol A Biol Sci Med Sci 61(3):262–266
147. Rockwood K, Mitnitski A, Song X, Steen B, Skoog I (2006) Long-term risks of death and institutionalization of elderly people in relation to deficit accumulation at age 70. J Am Geriatr Soc 54(6):975–979
148. Rockwood K, Mitnitski A (2011) Frailty defined by deficit accumulation and geriatric medicine defined by frailty. Clin Geriatr Med 27(1):17–26
149. Lucicesare A, Hubbard RE, Searle SD, Rockwood K (2010) An index of self-rated health deficits in relation to frailty and adverse outcomes in older adults. Aging Clin Exp Res 22(3):255–260
150. Villareal DT, Chode S, Parimi N, Sinacore DR, Hilton T, Armamento-Villareal R, Napoli N, Qualls C, Shah K (2011) Weight loss, exercise, or both and physical function in obese older adults. N Engl J Med 364(13):1218–1229. doi:10.1056/NEJMoa1008234
151. McLean RR, Shardell MD, Alley DE, Cawthon PM, Fragala MS, Harris TB, Kenny AM, Peters KW, Ferrucci L, Guralnik JM, Kritchevsky SB, Kiel DP, Vassileva MT, Xue QL, Perera S, Studenski SA, Dam TT (2014) Criteria for clinically relevant weakness and low lean mass and their longitudinal association with incident mobility impairment and mortality: the foundation for the National Institutes of Health (FNIH) sarcopenia project. J Gerontol A Biol Sci Med Sci 69(5):576–583. doi:10.1093/gerona/glu012
152. Lee C, Raffaghello L, Brandhorst S, Safdie FM, Bianchi G, Martin-Montalvo A, Pistoia V, Wei M, Hwang S, Merlino A, Emionite L, de Cabo R, Longo VD (2012) Fasting cycles retard growth of tumors and sensitize a range of cancer cell types to chemotherapy. Sci Transl Med 4(124):124ra127
153. Takasaka N, Araya J, Hara H, Ito S, Kobayashi K, Kurita Y, Wakui H, Yoshii Y, Yumino Y, Fujii S, Minagawa S, Tsurushige C, Kojima J, Numata T, Shimizu K, Kawaishi M, Kaneko Y, Kamiya N, Hirano J, Odaka M, Morikawa T, Nishimura SL, Nakayama K, Kuwano K (2014) Autophagy induction by SIRT6 through attenuation of insulin-like growth factor signaling is involved in the regulation of human bronchial epithelial cell senescence. J Immunol 192(3):958–968. doi:10.4049/jimmunol.1302341
154. Minagawa S, Araya J, Numata T, Nojiri S, Hara H, Yumino Y, Kawaishi M, Odaka M, Morikawa T, Nishimura SL, Nakayama K, Kuwano K (2011) Accelerated epithelial cell senescence in IPF and the inhibitory role of SIRT6 in TGF-beta-induced senescence of human bronchial epithelial cells. Am J Physiol Lung Cell Mol Physiol 300(3):L391–L401. doi:10.1152/ajplung.00097.2010

155. Moore BB, Hogaboam CM (2008) Murine models of pulmonary fibrosis. Am J Physiol Lung Cell Mol Physiol 294(2):L152–L160. doi:10.1152/ajplung.00313.2007
156. Ryu JH, Moua T, Daniels CE, Hartman TE, Yi ES, Utz JP, Limper AH (2014) Idiopathic pulmonary fibrosis: evolving concepts. Mayo Clin Proc 89(8):1130–1142. doi:10.1016/j.mayocp.2014.03.016
157. De Felice FG, Ferreira ST (2014) Inflammation, defective insulin signaling, and mitochondrial dysfunction as common molecular denominators connecting type 2 diabetes to Alzheimer disease. Diabetes 63(7):2262–2272. doi:10.2337/db13-1954
158. Joost HG (2014) Diabetes and cancer: epidemiology and potential mechanisms. Diab Vasc Dis Res. doi:10.1177/1479164114550813
159. Hudson MM, Ness KK, Gurney JG, Mulrooney DA, Chemaitilly W, Krull KR, Green DM, Armstrong GT, Nottage KA, Jones KE, Sklar CA, Srivastava DK, Robison LL (2013) Clinical ascertainment of health outcomes among adults treated for childhood cancer. J Am Med Assoc 309(22):2371–2381. doi:10.1001/jama.2013.6296
160. Ness KK, Armstrong GT, Kundu M, Wilson CL, Tchkonia T, Kirkland JL (2014) Frailty in childhood cancer survivors. Cancer. doi:10.1002/cncr.29211
161. Benson EK, Lee SW, Aaronson SA (2010) Role of progerin-induced telomere dysfunction in HGPS premature cellular senescence. J Cell Sci 123(Pt 15):2605–2612
162. Choi SH, Kim YH, Hebisch M, Sliwinski C, Lee S, D'Avanzo C, Chen H, Hooli B, Asselin C, Muffat J, Klee JB, Zhang C, Wainger BJ, Peitz M, Kovacs DM, Woolf CJ, Wagner SL, Tanzi RE, Kim DY (2014) A three-dimensional human neural cell culture model of Alzheimer's disease. Nature. doi:10.1038/nature13800

MIX
Papier aus verantwortungsvollen Quellen
Paper from responsible sources
FSC® C105338

If you have any concerns about our products,
you can contact us on
ProductSafety@springernature.com

In case Publisher is established outside the EU,
the EU authorized representative is:
Springer Nature Customer Service Center GmbH
Europaplatz 3, 69115 Heidelberg, Germany

Printed by Libri Plureos GmbH
in Hamburg, Germany